Gas-Phase Combustion Chemistry

Springer
New York
Berlin
Heidelberg
Barcelona
Hong Kong
London
Milan
Paris
Singapore
Tokyo

W.C. Gardiner, Jr.
Editor

Gas-Phase Combustion Chemistry

With 174 Figures

Springer

W.C. Gardiner, Jr.
Department of Chemistry and Biochemistry
University of Texas at Austin
Austin, TX 78712
USA

Library of Congress Cataloging-in-Publication Data
Gas-Phase Combustion Chemistry / edited by W. C. Gardiner, Jr.
p. cm.
Includes bibliographical references and index.
ISBN 0-387-98861-0
1. Combustion. I. Gardiner, William C. (William Cecil), 1933–.
QD516.C6147 1999
541.3′61—dc21 99-15020

Printed on acid-free paper.

Production managed by Allan Abrams; manufacturing supervised by Jacqui Ashri.
Photocomposed copy prepared using the editor's PostScript files.
Printed and bound by Sheridan Books, Inc., Ann Arbor, MI.
Printed in the United States of America.

9 8 7 6 5 4 3 2 1

ISBN 0-387-98861-0 Springer-Verlag New York Berlin Heidelberg SPIN 10728498

Preface

This book differs from its out of print 1984 predecessor[1] primarily by lacking theoretical chapters on combustion modeling and elementary reaction rate coefficients. While noteworthy advances in these subjects have been made since 1984, it was decided to mention theory in this book only where appropriate in chapters describing combustion chemistry itself. Otherwise, space limitation would have forced us to discuss only new developments in theoretical areas, thereby abandoning our goal of keeping this book readable by newcomers to the field of combustion modeling. Contemporary modeling and rate coefficient theory as applied to combustion deserve a book of their own.

A second omission is a chapter devoted to reviewing the elementary reactions that contribute to the combustion chemistry of hydrogen, carbon monoxide, and hydrocarbon or alternate fuels. Readers looking for guidance to the current knowledge we have in this area will find a broad outline and extensive references to the review and archival literature in Chapter 1, where the essential features of combustion chemistry modeling are surveyed.

The heart of this book is its chapters on the combustion chemistry of nitrogen, sulfur, and chlorine. Nitrogen and sulfur draw interest mostly because their oxides are the primary pollutants formed in combustion, chlorine because it is the prototype flame inhibitor and because incineration of toxic waste faces the challenge of reducing the concentrations of chlorine-containing organic compounds in waste streams to extremely low levels. It will be clear to all readers that while many molecular-level details have been discovered about the high-temperature chemistry of these elements, our ability to describe in combustion simulations what has been measured in combustion experiments is still limited by the incompleteness of our chemical understanding.

Like most basic combustion research, this book deals with what happens in the gas phase. Condensed phase chemistry relevant to explosives and propellants is not specifically addressed, although many elementary reactions relevant to that chemistry also play roles in gas-phase combustion. Two-phase combustion reactions, including formation and oxidation of soot, combustion of coal char and formation of inorganic ash, and the influence of chamber walls on nearby flames are likewise not described. These are certainly important chemical processes, but, aside from soot particle nucleation, modeling of two-phase combustion has so far been mostly limited to empirical descriptions that do not attempt to capture chemical detail in a fundamental way. On the other hand, there are some high temperature gas-phase reactions—such as chemiluminescence, ion formation, and reactions of metals—for which we do have molecular-level knowledge but which have remained at the periphery of combustion science. Readers interested in topics like these can readily gain entrance to the relatively limited literature[1] on them;

---

[1] *Combustion Chemistry*, W.C. Gardiner, Ed., Springer-Verlag, New York 1984.

[1] Beginning with the biennial International Combustion Symposium volumes published by the Combustion Institute in Pittsburgh.

thoughtful reading of Chapter 1 will provide all the background needed to place such special interests in context with the mainstream of combustion chemistry modeling.

Much of the content found here can be supplemented with resources on the Internet and the World Wide Web. We have tried to include URLs for all of the relevant sites that we know about, but some have surely been overlooked and many new ones will be established before this book becomes obsolete. Despite the limited lifetimes and uncertain reliability of Internet resources, creators and users of combustion chemistry knowledge generally have such Internet-friendly personalities that we expect to see essentially all of the combustion chemistry database and most of the computational resources one needs to utilize it on-line before this book is out of print.

Austin, Texas William C. Gardiner, Jr.

Contents

Contributors

Joseph W. Bozzelli, Department of Chemical Engineering and Chemistry, New Jersey Institute of Technology, Newark, New Jersey 07102, U.S.A.

Alexander Burcat, Department of Aeronautical Engineering, Technion—Israel Institute of Technology, Haifa 31000, Israel

Anthony M. Dean, Corporate Research Laboratory, Exxon Research and Engineering Company, Annandale, New Jersey 08801, U.S.A.

William C. Gardiner, Jr., Department of Chemistry and Biochemistry, The University of Texas at Austin, Austin, Texas 78712, U.S.A.

Anthony J. Hynes, Division of Marine and Atmospheric Chemistry, Rosenstiel School of Marine and Atmospheric Sciences, University of Miami, Miami, Florida 33149, U.S.A.

Vitali V. Lissianski, GE Energy and Environmental Research Corporation, 18 Mason, Irvine, California 92618, U.S.A.

Selim M. Senkan, Department of Chemical Engineering, University of California, Los Angeles, California 90024, U.S.A.

Paul H. Wine, School of Chemistry and Biochemistry, School of Earth and Atmospheric Sciences, and Georgia Tech Research Institute, Georgia Institute of Technology, Atlanta, Georgia 30332, U.S.A.

Vladimir M. Zamansky, GE Energy and Environmental Research Corporation, 18 Mason, Irvine, California 92618, U.S.A.

1

Combustion Chemistry Modeling

Vitali V. Lissianski [1]

Vladimir M. Zamansky [1]

William C. Gardiner, Jr. [2]

1.1 INTRODUCTION

Combustion is a fast chemical reaction that liberates heat.

For many purposes the chemistry can be ignored, and combustion scientists and engineers can focus their attention on other aspects of the combustion process like mixing, heat transfer, and flame stabilization. For other purposes, such as pollution control and safety analysis, the chemical details may be all important.

The chemical details can be captured in computer models of combustion reactions. To do this, one needs molecular-level knowledge about how combustion occurs and computer programs which express computed chemistry in ways that can be compared to experiments. In this chapter, we introduce the chemical thinking necessary to devise and use such programs and present an overview of what has been learned by comparing computational predictions with experiments.[3]

---

[1] GE Energy and Environmental Research Corporation, 18 Mason, Irvine, California 92618, U.S.A.

[2] Department of Chemistry and Biochemistry, The University of Texas at Austin, Austin, Texas 78712, U.S.A.

[3] The literature on combustion chemistry modeling has grown so large that mentioning all important contributions to it in a review has not been possible for some time. The authors apologize to the many significant contributors to the field whose work is not discussed or adequately cited in this chapter.

1.1.1 Terms used in dynamic modeling of chemical reaction

The language of computer modeling of combustion chemistry is the language of chemical kinetics, the study of how chemical composition and physical conditions combine to determine the evolution of both.

First the basic terms. A *chemical reaction* is the conversion of one kind of matter into another chemically different form; a *combustion reaction* is a fast chemical reaction in which a *fuel* combines with an *oxidizer* (usually oxygen from air) to form *combustion products.* Combustion reactions are described by writing the chemical symbols for the starting materials—the *reactants*—and the final *products* separated by an arrow[1] that means "react with one another to form". The combustion of heptane with oxygen as the oxidizer is expressed as

$$C_7H_{16} + 11\,O_2 \longrightarrow 7\,CO_2 + 8\,H_2O\,,$$

where the *stoichiometric coefficients* 11, 7, and 8 denote the relative numbers of molecules of oxygen, carbon dioxide. and water participating in the combustion process per molecule of heptane.[2]

The terms *chemical equation* and *chemical reaction* are used interchangeably for such symbolic descriptions. The stoichiometric coefficients have the property that the chemical equation can also be interpreted as a special kind of algebraic one, in which the sum of the number of atoms of each element represented on the left of the arrow must be the same as the corresponding sum on the right-hand side of the arrow. If the stoichiometric coefficients are present, the chemical equation is said to be *balanced.* There is no inflexible requirement to write balanced equations, however, and there is also no assertion that stoichiometric amounts of reactants must be present in order for reaction to occur. Chemical equations like biomass + air $\longrightarrow$ energy + CO_2 appear now and then to draw attention to the occurrence of a chemical reaction that one would be reluctant to try to describe in stoichiometric terms.

If the molar ratio of heptane to oxygen is 1:11, then just enough fuel is present to burn out all of the oxidizer present; the *equivalence ratio* of a combustion reaction, usually denoted by ϕ, is the actual starting fuel:oxidizer ratio divided by the ratio required for complete combustion to fully burnt products. That is, for heptane burning in air the equivalence ratio is $\phi = (n_{C_7H_{16}}/n_{O_2})/(1/11)$, where n_i represents the number of moles of i present. (Or, in a flow situation, moles introduced into a combustor per unit time.)

---

[1] In the combustion modeling literature, an equals sign is sometimes used in place of an arrow, particularly if the reaction may occur in both directions.

[2] There is no occult significance to the term "stoichiometry"; it is only a reminder that chemical conversions involve consumption and production of substances in exact relative amounts, the amount of matter being defined in terms of the *mole*, 6.022×10^{23} molecules. The derived adjective *molar* has two meanings: In one sense it means "on the basis of moles", rather than, say, on the basis of kilograms; in the other it is a unit of concentration, the number of moles of some substance in 1 liter of solution.

Chemical equations describe only what the transformation is, not how it happens at the molecular scale. It is not the case that heptane burns as one molecule of it after another is intercepted by 11 oxygen molecules with a consequent reshuffling of all 45 atoms such that seven molecules of carbon dioxide and eight molecules of water fly away from the scene. Instead, heptane combustion is the composite result of a multitude of different kinds of molecular-level events, the end result of which is that the heptane and oxygen molecules have been replaced by carbon dioxide and water molecules. Each of the molecular events, which are called *elementary reactions,* or *elementary steps* or simply *steps,* can also be described in the form of a chemical equation. It turns out that the pace of many combustion reactions is set by the elementary reaction between H atoms and O_2 molecules

$$H + O_2 \longrightarrow OH + O. \tag{1}$$

Whether a chemical or an elementary reaction is being described has to be inferred from the context in which the reaction appears. It is important to distinguish between the two uses of the same abstract form.

Elementary reactions are called *unimolecular, bimolecular* or *termolecular* depending on whether there are 1, 2, or 3 reactant molecules. Each type behaves in a characteristically different way, as discussed later in this section.

1.1.2 Chain reactions

The set of all elementary reactions that one thinks are responsible for a chemical transformation is called its *reaction mechanism.*[1] Combustion mechanisms prove to have a special kind of mechanism; they are all *chain reactions.*

The basic idea of a chain mechanism is that the product(s) of one step may serve as reactant(s) of one or more subsequent steps. Often the reactive *chain centers,* or *active species,* are single atoms or molecular fragments called *free radicals.*[2] The free radical OH from reaction (1) may react with a molecule of ethane

$$OH + C_2H_6 \longrightarrow H_2O + C_2H_5 \tag{2}$$

[1] The word mechanism is also used in chemical kinetics in a second way, a more mechanical one, whereby one is describing the motions of atoms and electrons that occur during the course of a single elementary reaction. Many mechanisms in this sense are discussed in Chapter 2. In this chapter the term is used only in the sense of a set of elementary reactions.

[2] A common feature of free radicals that is responsible for their reactivity is having an odd number of electrons; thus, fragmenting the stable molecule CH_4 gives an H atom with one electron and a CH_3 radical with nine electrons. This rule is not universal, however; reactions of CH_4, for example, can also lead to an eight-electron CH_2 fragment that is more reactive than H or CH_3. None of the radicals discussed here are "ions", that is, they do not have net electrical charge; the OH hydroxyl radical and the H atom are not the OH^- hydroxyl ion and H^+ ion to which liquid water dissociates. While ions do appear in flames, their concentrations are very small compared to those of neutral atoms and radicals.

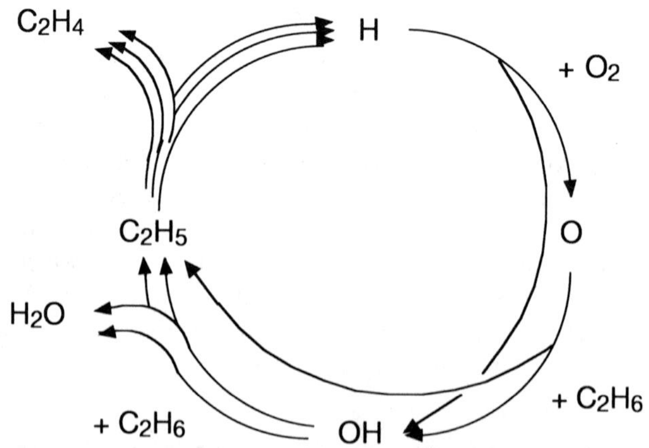

FIGURE 1.1. Schematic diagram of the branching chain sequence illustrated by reactions (1–4). Starting with a single H atom reacting with O_2, the effect of the sequence is producing 3 H atoms in the course of converting 3 molecules of C_2H_6 to C_2H_4. Exponential growth of all reaction products until the starting materials start to be exhausted is implied by this set of reactions. Oxidation of C_2H_6 actually involves over a hundred significant elementary reactions, only a few of which contribute to chain branching.

to form the stable product water and the unstable ethyl radical, which at combustion temperatures rapidly decomposes to ethylene and a hydrogen atom

$$C_2H_5 \longrightarrow C_2H_4 + H \tag{3}$$

so as to continue the reaction chain. Similarly, the O atom produced in the $H + O_2$ reaction (1) could also react with ethane to produce an ethyl radical

$$O + C_2H_6 \longrightarrow C_2H_5 + OH\,. \tag{4}$$

The combined effect of the four elementary reactions (1–4) can be thought of as starting and ending with an H atom—the combined chemical transformation

$$H + O_2 + 3\,C_2H_6 \longrightarrow 3\,C_2H_4 + 2\,H_2O + 3\,H \tag{5}$$

is derived by adding together reaction (1), twice reaction (2), three times reaction (3), and reaction (4), as illustrated in Figure 1.1. Because the number of chain centers increases from one to three, the combination (5) is called a *branching chain* sequence. Elementary steps in which the number of chain centers increases are *chain branching* reactions; steps in which the number of chain centers stay the same are *chain propagation* steps. The exponential growth nature of the process is illustrated in Figure 1.1.

Other important types of elementary reactions besides branching and propagation steps occur. If chain centers are produced from stable molecules, as in

$$C_7H_{16} \longrightarrow C_3H_7 + C_4H_9\,, \tag{6}$$

one speaks of a *chain initiation step*, while loss of chain centers, as in

$$H + CH_3 \longrightarrow CH_4\,, \tag{7}$$

happens in *chain termination* steps, also called *recombination* reactions.

1.1.3 Reaction rates, rate laws, and rate coefficients

Chemical reactions can be slow or fast, depending on many things. Iron left outdoors may rust away over years, concrete hardens in hours, gasoline in the cylinder of an internal combustion engine burns up in a millisecond. It is conventional to express *reaction rates* in terms of the time rate of change of concentration rather than in terms of amount reacted. In this book, concentrations are expressed in mol/cm^3 units and time in seconds. Two conventions are used to denote concentrations. The more common is to enclose a chemical "species" name in brackets, for example, a concentration of heptane $[C_7H_{16}]$ might be 2.0×10^{-6} mol/cm^3 in some experiment and its reaction rate r might be -3.1×10^4 mol cm^{-3}s^{-1}, the minus sign indicating that the heptane concentration is decreasing. In mathematical contexts, it is more convenient to denote concentration by a subscripted variable c_i, the i indexing chemical species. Thus the time rate of change of concentration would be expressed as $\mathrm{d}c_i/\mathrm{d}t$.

There is no general way to predict in advance how chemical reaction rates depend on concentrations, because that depends on the details of the reaction mechanism involved. For elementary reactions, however, the concentration dependence is more straightforward, being governed by what are called *mass action rate laws.* (But see the important exception at the end of this section.)

While elementary reactions are commonly written down with an arrow going in one direction only—"the forward reaction"—it is common in combustion chemistry to find that elementary reactions occur in both directions, denoted in this chapter by replacing the forward reaction arrow with the symbol $\rightleftharpoons$. For this reason we call r the *net rate* of reaction and write

$$r = r_{\text{forward}} - r_{\text{reverse}} \,. \tag{8}$$

The rate in each direction is proportional to the concentration of each species participating, the proportionality coefficient being called a *rate coefficient* or a *rate constant.*[1] Thus for elementary reaction (1), bimolecular in both directions, we have

$$r_1 = k_{\text{f}1}[\text{H}][\text{O}_2] \; - \; k_{\text{r}1}[\text{OH}][\text{O}] \,, \tag{9}$$

where $k_{\text{f}1}$ and $k_{\text{r}1}$ are the forward and reverse rate coefficients, respectively; for elementary reaction (6) the forward reaction is unimolecular while the reverse one is bimolecular, so

$$r_6 = k_{\text{f}6}[\text{C}_7\text{H}_{16}] \; - \; k_{\text{r}6}[\text{C}_3\text{H}_7][\text{C}_4\text{H}_9] \,, \tag{10}$$

---

[1] In combustion reactions the temperature, upon which rate coefficients may depend strongly, changes over the course of reaction. For this reason we avoid the possibly misleading term "rate constant". Equilibrium "constants" may also change by many orders of magnitude over the combustion temperature range. Common ways to express the temperature dependence of rate coefficients are discussed in Section 1.9.1.

and for the termolecular reaction[1]

$$H + H + N_2 \longrightarrow H_2 + N_2 \tag{11}$$

one has

$$r_{11} = k_{f11}[H]^2[N_2] - k_{r11}[H_2][N_2]\,. \tag{12}$$

These examples illustrate the basic chemical kinetics ideas underlying elementary reaction rates. The structure of the mathematical problem to be solved in dynamic modeling is that all of the elementary reaction rates, each of which depends on part of the current composition of the reacting mixture (which in typical combustion reactions may contain tens or even hundreds of different chemical species in significant concentrations), have to be computed; once these rates are known, their effects upon all the concentrations can be evaluated from the form of the elementary reactions.

Exception to the mass action forms: The mass action rate laws of equations (9, 10, and 12) have to be replaced by more complicated expressions for accurate calculation of unimolecular and "chemically activated" bimolecular reactions. These complications are explained in Section 2.3.1. For the remainder of this chapter it is only important to understand that net rates of elementary reactions can be computed by equation (8) and that the forward and reverse reaction rates can be computed from the current composition and temperature. Formally, this is the assertion that we will be able to compute, for all elementary reactions j,

$$r_j = F_{j,\text{forward}}(T, \{c_i\}) - F_{j,\text{reverse}}(T, \{c_i\}), \tag{13}$$

where the forward and reverse rate functions F depend in known ways on the temperature T and the set of concentrations $\{c_i\}$.

1.1.4 Model constraints

In non-combustion computer modeling of chemical reaction, the physical constraints under which reaction occurs can usually be ignored, because the temperature and specific volume (the inverse of the density of the reacting system) remain

[1] In most termolecular reactions of combustion, one of the chemical species appears on both sides of the reaction. The reason for this is that such species are required for the formation of a stable chemical bond even though they are unchanged by the reaction. In the reaction (11) example, the two H atoms would fly apart immediately after encountering one another if it were not for the nitrogen molecule being there at the same time to absorb part of the energy released in forming the chemical bond between the hydrogen atoms. Different such "collision partners", "chaperones", or "bath gas molecules" prove to have different efficiencies in stabilizing newly bonded molecules. Essentially, larger bath gas molecules have larger rate coefficients. It is common to denote such non-reacting molecules generically as M and to rate their efficiencies in a consistent way in combustion chemistry models.

constant during reaction. In combustion, however, large and rapid changes of temperature, pressure, density, and flow speeds go hand in hand with the progress of chemical reaction. Combustion modeling therefore has to be structured so that the changing physical conditions are taken into account together with the reaction rates. The chemistry affects the physics and vice versa.

There are practical limitations on the degree to which computer models of combustion chemistry can approach the actual physical conditions under which laboratory experiments and real-world combustion happen.

The simplest model of chemical reaction is to assume that the reaction takes place at constant volume and temperature, formally called *isometric* and *isothermal* constraints. While unrealistic for combustion itself, it is common when setting up laboratory experiments to take measures to insure that heat liberated by reaction is lost to the surroundings (to a laboratory thermostat, for example) faster than the reaction proceeds.

An alternative is to assume that the pressure and temperature remain constant. Formally, reaction takes place under *isobaric, isothermal* constraints. These would pertain for example to gas flowing through a tube furnace, providing again that heat liberated by chemical reaction is compensated by heat exchange with a fixed-temperature flow tube and that the kinetic energy associated with the flow is so much smaller than the chemical and thermal energies that it can be neglected.[1]

More common in combustion modeling, however, are *adiabatic* constraints, whereby one assumes that none of the heat liberated by reaction is lost to the surroundings. In this case the temperature will change during combustion, and with it the rate coefficients of most of the elementary reactions. One may have *isometric, adiabatic* or *isobaric, adiabatic* constraints depending on whether one assumes that the reacting gas is free to expand as reaction proceeds.

The cases described so far assume stationary gas, or that the flow energy is small enough to neglect without loss of accuracy. In combustion, however, one may be interested in how chemical reaction interacts with gas flow. The simplest case would be *steady flow*, or *steady duct flow*, one-dimensional gas motion such that the flow speed and gas composition at any point in space stay constant. Duct flow is often assumed to take place at constant duct area, but this is not required by the form of the steady flow equations, and assuming variable-area duct flow is a simple way to make corrections for boundary layer effects. (Or to model reaction that actually takes place in a variable-area duct.) In setting up the mathematics of a steady flow model, it is convenient to solve the problem using distance rather than time as independent variable. Constant-area duct flow is often called *plug flow*, and it may be assumed to be either adiabatic or isothermal. In either case, pressure and specific volume change along the flow coordinate as required by steady flow mechanics.

---

[1] In the mathematical language introduced in the following sections, this means that the actual energy per unit mass $h + u^2/2$, where h is the specific enthalpy and u the flow velocity, is taken to be h alone.

All of the above possibilities presume that the modeler does not wish to intrude into the mathematics of the assumed physical model by forcing any of the physical (or, for that matter, chemical) variables to evolve to match an experimental profile. This is not required. One can, for example, set up one's model to include a measured temperature or pressure profile as a constraint.

A final model commonly used in combustion chemistry is the *continuously stirred tank reactor* model. One supposes that one or more reactant streams flow into a reactor, of fixed volume and temperature, from which an outflow removes the same mass per unit time as is supplied by the inflow streams, and further that the mixing of the gas inside the reactor is very fast compared to chemical reaction. The outflow stream has the same composition as the interior of the reactor, and reaction proceeds at that constant composition, the outflow stream removing matter as fast as the inflow stream(s) supply it and chemical reaction transforms it. Adiabatic and isothermal variants are possible, as described in the next section.

Nearly all combustion chemistry modeling is done assuming that the gas obeys the ideal gas law and that no reactions with the vessel walls need be taken into account. It is possible, however, to include real gas effects or wall reactions in any of the models mentioned in this section if one thinks that these may be important.

1.1.5 Differential equations of chemical reaction without transport

Each of the elementary reactions j composing a mechanism has a net rate r_j defined by equation (13) at all times. For each reaction a positive or negative integer β_{ij} can be defined to give the change in number of moles of species i implied by the stoichiometric coefficients of reaction j. Defining the differential operator $\Delta = \mathrm{d}/\mathrm{d}t$ we can write the time derivative of the concentration of chemical species i as

$$\Delta c_i = \sum \beta_{ij} r_j \,, \tag{14}$$

where the sum is over all $\mathcal{R}$ reactions in the mechanism. For a mechanism with $\mathcal{S}$ species there are $\mathcal{S}$ ordinary differential equations like (14) that describe the chemical part of the model. The right hand sides of the $\mathcal{S}$ equations (14) depend upon concentrations c_i (explicit variables in the r_j) and implicitly upon the temperature T through the forward and reverse rate coefficients appearing in the F_j functions.

In combustion modeling reactions the forward and reverse rate coefficients of elementary reactions are usually assumed to be related to one another through

$$\frac{k_\mathrm{f}}{k_\mathrm{r}} = K_c \,, \tag{15}$$

where the equilibrium constant K_c is a function of temperature only that can be computed from the thermochemical properties of the chemical species participating in the reaction, as described in Section 5.9 of Chapter 5. This *Principle of Detailed Balance* or *Microscopic Reversibility* has to be valid for true elementary chemical reactions, but one may wish to take liberties with this for special purposes (see Section 1.8).

The chemical part of the ordinary differential equation set needed for combustion modeling is defined by equations (13–15) once a set of species and elementary reactions has been decided. Before one can integrate these differential equations for a set of starting concentration and temperature values, additional differential equations that express the physical constraints for the model have to be added.

Isothermal reaction at constant volume

The only physical variable that can change due to chemical reaction in a system where the density and temperature are held constant is the pressure. One way to keep track of the pressure as the set of concentrations evolves is to solve the ideal gas law equation

$$P = (n/V)\mathrm{R}T = \mathrm{R}T\textstyle\sum c_i \,, \tag{16}$$

A second way is to take the derivative of equation (16)

$$\Delta P = \mathrm{R}T\textstyle\sum \Delta c_i \tag{17}$$

and add equation (17) to the $\mathcal{S}$ equations (14) to get $\mathcal{S}+1$ equations to be integrated.

Adiabatic reaction at constant volume

Heat release increases the temperature of an adiabatic system. To compute the physical and chemical evolution of the system we proceed as follows.

The total derivative of the ideal gas law divided by the gas law itself is

$$\frac{\Delta P}{P} - \frac{\sum \Delta c_i}{\sum c_i} - \frac{\Delta T}{T} = 0 \,. \tag{18}$$

Because no expansion work can be done, the thermodynamic energy equation for an adiabatic constant volume system equates to zero

$$\Delta U = q + w = 0 \,. \tag{19}$$

It is more convenient to follow changes in enthalpy than changes in internal energy, so in place of equation (19) we use

$$\Delta U = \Delta(h - Pv) = 0 \,, \tag{20}$$

where h is the enthalpy of the system per unit mass (the *specific enthalpy*) and $v = 1/\rho$ is the *specific volume*. The specific enthalpy can be found from the molar enthalpies H_i of the species present by

$$h = (1/\rho)\textstyle\sum c_i H_i \,. \tag{21}$$

Substituting equation (21) for h and the inverse density for v in equation (28) gives

$$\frac{\Delta\sum c_i H_i}{\rho} - \frac{\Delta P}{\rho} = 0 \quad \text{or}$$
$$\Delta\sum c_i H_i - \Delta P = 0\,. \tag{22}$$

The derivative of the product $c_i H_i$ is

$$\Delta(c_i H_i) = c_i \Delta H_i + H_i \Delta c_i\,. \tag{23}$$

Because the molar enthalpy of an ideal gas depends on temperature only, that is,

$$\Delta H_i = C_{Pi} \Delta T\,, \tag{24}$$

equations (22) and (23) allow the energy equation to be recast into

$$\sum H_i \Delta c_i + \sum c_i C_{Pi}\, \Delta T - \Delta P = 0\,. \tag{25}$$

Eliminating ΔP between equations (18) and (25) and factoring gives

$$\Delta T = \sum (H_i - P/\sum c_i) \Delta c_i / (P/T - \sum c_i C_{Pi})\,. \tag{26}$$

In equation (26) we see that once the concentration changes Δc_i have been calculated from equation (14), the temperature change can be found from equation (26). Then the pressure change can be found by rearranging equation (18) to

$$\Delta P = P\left(\frac{\Delta T}{T} + \frac{\sum \Delta c_i}{\sum c_i}\right). \tag{27}$$

Equations (14), (26), and (27) then provide $\mathcal{S} + 2$ equations for the $\mathcal{S} + 2$ derivatives to be computed. An additional complication in the adiabatic case over the isothermal one is that the F terms needed in equation (13) to compute the reaction rates r_j depend on temperature; one has to recompute these anew at each step of the solution before evaluating equation (14).

Adiabatic reaction at constant pressure

Combustion gas is normally free to expand as reaction proceeds. For this reason the constraints considered in the previous sections may be inappropriate for realistic modeling of combustion, although they may well represent useful approximations in some laboratory situations. Constant pressure reaction may be the closest approximation to real combustion conditions that one can use without taking flow or transport into account. We will again require $\mathcal{S} + 2$ equations to complete the model, this time for the density ρ, the temperature T, and the $\mathcal{S} + 1$ species concentrations.

With the density of the system free to change, the change in volume will mean that PV work must be included in the energy equation for the system. In place of equations (18) and (19), we have for adiabatic reaction at constant pressure

$$\Delta U = \Delta(h - PV) = \mathrm{d}w = -P\mathrm{d}V\,. \tag{28}$$

Expanding the derivative of PV converts equation (28) into

$$\Delta h = V\mathrm{d}P = 0 \tag{29}$$

at constant pressure. Taking the derivative of $h = \left(\sum c_i H_i\right)/\rho$ and multiplying by ρ converts equation (29) into

$$\sum c_i \Delta H_i + \sum H_i \Delta c_i - \sum H_i c_i \Delta\rho/\rho = 0. \tag{30}$$

For ideal gases, H_i is a function of temperature only, so $\Delta H_i = C_{Pi}\Delta T$ and the energy equation for adiabatic reaction at constant pressure is

$$\sum c_i C_{Pi} \Delta T + \sum H_i \Delta c_i - \sum H_i c_i \Delta\rho/\rho = 0. \tag{31}$$

The second governing equation is the derivative of the ideal gas law for $\Delta P = 0$,

$$\frac{\sum \Delta c_i}{\sum c_i} + \frac{\Delta T}{T} = 0\,. \tag{32}$$

The concentration changes have to be found as total derivatives, that is, concentrations change both by chemical reaction and by density changes

$$\Delta c_i = \frac{\partial c_i}{\partial t} + \frac{\partial c_i}{\partial \rho}\Delta\rho\,. \tag{33}$$

The partial derivative with respect to density can be found from the concentrations and the average molecular weight $\langle W \rangle = \sum W_i c_i / \sum c_i$, where W_i is the molecular weight of species i, through

$$\sum c_i = \frac{\rho}{\langle W \rangle}\,. \tag{34}$$

The partial derivative of equation (34) is

$$\frac{\partial \sum c_i}{\partial \rho} = \frac{1}{\langle W \rangle} = \frac{\sum c_i}{\rho}\,. \tag{35}$$

Because the c_i are independent variables, their partial derivatives must each be given by

$$\frac{\partial c_i}{\partial \rho} = \frac{c_i}{\rho}\,. \tag{36}$$

The first partial derivative in equation (33) is the sum of the net rates at which reaction j produces species i, while the second one is given by equation (36). The total derivatives of the concentrations are therefore

$$\Delta c_i = \sum_j \beta_{ij} r_j + c_i \Delta\rho/\rho\,, \tag{37}$$

where β_{ij} is the number of moles of species i produced in reaction j. Eliminating ΔT between equations (31) and (32) gives

$$\sum H_i \Delta c_i - \frac{T \sum \Delta c_i \sum c_i C_{Pi}}{\sum c_i} = \frac{\sum c_i H_i \Delta\rho}{\rho}\,, \tag{38}$$

which contains only $\Delta\rho$ and the Δc_i. Substituting for the Δc_i from equation (37) leads after some algebra to

$$\frac{\Delta\rho}{\rho} = \frac{\sum_i \sum_j H_i \beta_{ij} r_j}{T \sum C_i C_{Pi}} - \frac{\sum_i \sum_j \beta_{ij} r_j}{\sum c_i}\,, \tag{39}$$

which has only known quantities on the right hand side.

An equation for the temperature derivative in terms of the concentration derivatives can be found from equation (32)

$$\Delta T = \frac{-T \sum \Delta c_i}{\sum c_i}\,. \tag{40}$$

In order to evaluate the $S + 2$ derivatives of ρ, T, and the S species concentrations one has to (1) evaluate the sums on the right-hand side of equation (39) to find $\Delta\rho$, (2) evaluate the right-hand sides of the S equations (37) to find the Δc_i values, And (3) evaluate the right-hand side of equation (40) to find ΔT. Numerical integration of these $S + 1$ equations provides the solution for adiabatic reaction at constant pressure.

Steady reactive flow

When exothermic or endothermic chemical reaction takes place in moving gas, energy released or absorbed in the course of chemical reaction does not act only in the form of heat, raising or lowering the gas temperature; it also manifests itself by changes in the flow velocity. In order to cast the flow equations into a form useful for modeling combustion chemistry, an additional restriction has to be made beyond what we have encountered in the static cases discussed so far. Namely, the flow has to be assumed to be unaffected by frictional forces, or *inviscid,* and to be constant in time, or *steady.* Both isothermal and adiabatic flows can be modeled; for combustion purposes, the adiabatic choice is appropriate. Normally one uses a *constant-area* flow model; the variable-area computation is done with essentially identical equations, and we include the area function in the equations that follow.

There are four constraints to embody in the governing equations: the ideal gas law, conservation of matter as a *continuity equation*, conservation of momentum in *Euler's equation*, and conservation of total (heat plus flow) energy.

Conservation of matter is expressed by stating that the total mass flowing through any cross section along a flow must be a constant m called the *mass flow*. Using x for the space coordinate along the flow and $A(x)$ for the cross-sectional area, the continuity equation is

$$A(x)\rho u = m\,, \tag{41}$$

where u is the flow velocity. Euler's equation is

$$\frac{\mathrm{d}P}{\mathrm{d}x} + \rho u \frac{\mathrm{d}u}{\mathrm{d}x} = 0\,, \tag{42}$$

and conservation of energy per unit mass requires

$$h + \frac{u^2}{2} = h^*\,, \tag{43}$$

where h is the enthalpy per unit mass and h^* is a constant called the *stagnation enthalpy*. Equations (41–43) and the ideal gas law express the physical constraints of steady flow. By a series of algebraic manipulations one can avoid using matrix manipulations to solve for the $\mathcal{S}$ + 4 derivatives of P, T, ρ, u, and the $\mathcal{S}$ species concentrations (Gardiner et al. 1981). With the definitions

$$\begin{aligned} &\Sigma c = \sum c_i \quad && \Sigma C_P = \sum c_i C_{Pi} \quad && D_i = \sum \beta_{ij} r_j / u \\ &\Sigma D = \sum D_i \quad && \Sigma HD = \sum H_i D_i \quad && \Omega = \frac{\mathrm{d}\,(A(0)/A(x))}{\mathrm{d}x} \end{aligned} \tag{44}$$

one derives the following equations for the derivatives, now with respect to distance x rather than with respect to time as we did for the static case. The distance derivative of the first concentration is

$$\begin{aligned} \Delta_x c_1 = {} & \frac{\rho u^2 (T\Sigma C_P/P - 1)D_1 - c_1 \Sigma HD + T\Sigma C_P(c_1 \Sigma D/\Sigma c - D_1)}{\rho u^2(T\Sigma C_P/P - 1) - T\Sigma C_P} \\ & - \frac{c_1 u (1 - T\Sigma C_P/P)\rho(0)u(0)\Omega}{\rho u^2(T\Sigma C_P/P - 1) - T\Sigma C_P}\,. \end{aligned} \tag{45}$$

The remaining distance derivatives are

$$\Delta_x c_i = D_i + \frac{c_i(\Delta_x c_1 - D_1)}{c_1} \tag{46}$$

$$\Delta_x \rho = \frac{\rho(\Delta_x c_1 - D_1)}{c_1} \tag{47}$$

$$\Delta_x u = \frac{-u(\Delta_x c_1 - D_1)}{c_1} + \frac{\rho(0)u(0)\Omega}{\rho} \tag{48}$$

$$\Delta_x P = -\rho u \Delta_x u \tag{49}$$

$$\Delta_x T = T\left(\frac{\Delta_x P}{P} - \frac{\sum \Delta_x c_i}{\Sigma c}\right)\,. \tag{50}$$

For constant area flow, $\Omega = 0$ and the second terms drop out of equations (45) and (48). Species 1 should be one whose concentration never approaches zero, as it appears in the denominators of equations (45), (46) and (47). To obtain the elapsed reaction time, one can expand the set of integrated equations to $S + 5$ and integrate $1/u$ along with the other variables. While there is more computing to do, the adiabatic steady flow problem is not intrinsically more complex than the other cases described.

The Continuously Stirred Tank Reactor

The last model we describe is the continuously stirred tank reactor or CSTR. This is sometimes described as a "perfectly stirred reactor" or PSR on the basis of the underlying computational assumption, and sometimes as a "jet stirred reactor" or JSR to indicate the nature of the stirring, One imagines a reactor into which a steady mass flow m continuously supplies new reactant species at known concentrations, while the same mass flow is taken out of the reactor with the set of concentrations that prevails, thanks to the assumed vigorous mixing within the reactor. After initial transients have decayed, the internal and output concentrations reach steady values. Solving the CSTR problem requires one to find the values of these concentrations.

From a practical point of view, a CSTR experiment offers the advantage of extending the accessible range of reaction time well beyond what can be done with an ordinary flow reactor, the characteristic "residence" time being essentially the ratio of the mass of gas contained in the reactor to the mass flow. It is argued that the error caused by ignoring the insufficiencies of the mixing process is no larger than that caused by ignoring wall interactions and viscosity losses in ordinary flow reactors. From a theoretical point of view, the CSTR model is a reasonable first approximation to model gas-phase reaction at essentially constant pressure in a gas flow field where intense mixing is expected.

As in the models discussed earlier, one can consider the energy released during reaction as being compensated at steady state by heat exchange with the environment so as to produce an enforced temperature, or one can assume that the temperature of the reactor, and accordingly of the output stream, is determined by a balance between heat release in reaction and heat loss to the output stream. The temperature may therefore be a stated one or a variable whose steady value is to be computed. For combustion modeling the latter case is more common; it has occasionally been termed "adiabatic" even though reaction heat is constantly lost to the output stream. In the equations derived here, we include both the heat loss to the output flow and an arbitrary additional heat exchange $\dot{q}$, which could be positive or negative; $\dot{q} = 0$ applies to a perfectly insulated reactor.

One can make different assumptions about the temperature of the input stream. For combustion applications, this has to be taken as an independent input variable. All combustion applications consider the pressure and volume of the reactor to be fixed. Once a converged solution is obtained, the errors introduced by ignoring pressure drop and flow velocity changes can be computed from the characteristics of the particular reactor being modeled.

There is a serious computational problem with the CSTR that is absent in the models considered so far, namely, that the steady state may not be uniquely defined by the mass flow and the input stream composition.[1] In the chemical industry this can pose serious safety and economic problems, and for this reason CSTR models have been thoroughly studied computationally and tested at many scales. For normal combustion modeling, the multiple steady-state problem is more a computational nuisance than a safety concern. It can be approached by testing CSTR models for varied assumed sets of input concentrations. For example, an "ignited" steady state can be computed by insuring that the assumed input stream composition is rich enough in reactive species that the steady state reached is one in which the composition is determined by competition of the equilibration reactions with loss of products and heat to the output stream.

The CSTR equations can be derived by adding flow and heat exchange terms to the equations introduced earlier for adiabatic constant pressure reaction. Because mass loss to the output flow must balance mass gained from the input flow, mass fractions computed from mole fractions $X_i = c_i/\sum c_i$ and the average molecular weight $\langle W\rangle = \sum W_i c_i/\sum c_i$, where W_i is the molecular weight of species i, are also needed.[2]

The number of moles of species i added per second to the reactor is its mole fraction in the input stream X_i^{inflow} times the total molar inflow $m/\langle W\rangle_{\text{inflow}}$ or $m\,X_i^{\text{inflow}}/\langle W\rangle_{\text{inflow}}$, which adds the term $m\,X_i^{\text{inflow}}/(\langle W\rangle_{\text{inflow}} V)$, where V is the volume of the CSTR, to the total concentration derivative for constant pressure reaction given in equation (37). The corresponding loss to the output stream is $-m\,X_i^{\text{reactor}}/(\langle W\rangle_{\text{reactor}} V)$, which can be expressed in terms of the species concentrations in the reactor as $mc_i RT/(\langle W\rangle_{\text{reactor}} PV)$. Adding these terms to equation (37) gives the continuity equation for the CSTR

$$\begin{aligned}\Delta c_i &= \sum_j \beta_{ij} r_j + c_i \Delta\rho/\rho + \frac{m}{V}\left(\frac{X_i^{\text{inflow}}}{\langle W\rangle_{\text{inflow}}} - \frac{c_i RT}{\langle W\rangle_{\text{reactor}} P}\right)\\ &= \sum_j \beta_{ij} r_j + c_i \Delta\rho/\rho + A_i\,. \end{aligned} \tag{51}$$

We also have to add the flow contributions to the energy equation for constant pressure reaction given by equation (31). The inflow stream adds

$$\frac{m \sum X_i^{\text{inflow}} H_i^{\text{inflow T}}}{\langle W\rangle_{\text{inflow}} V}\,. \tag{52}$$

where we have divided by the reactor volume to get the energy added per unit time and volume, to be consistent with equation (31), and where $H_i^{\text{inflow T}}$ is the molar

[1] For a combustion example, see Park and Vlachos (1997).

[2] For programming the CSTR equations in SI units, molecular weights must be converted to kg/mol, for example, the molecular weight of O_2 is 0.032 kg/mol.

enthalpy of species i at the temperature of the input stream. The outflow stream removes energy, again per unit volume of the reactor, at the rate

$$\frac{m \sum c_i H_i RT}{\langle W \rangle_{\text{reactor}} PV}, \tag{53}$$

and $\dot{q}$ represents the net gain or loss of energy by any other processes one may assume to be present, such as radiation or heat conduction through the reactor walls. The CSTR energy equation is then the sum of equations (31), (52), and (53) plus, if desired, $\dot{q}/V$:

$$\sum c_i C_{Pi} \Delta T + \sum H_i \Delta c_i - \sum H_i c_i \Delta\rho/\rho + \frac{m \sum X_i^{\text{inflow}} H_i^{\text{inflow T}}}{\langle W \rangle_{\text{inflow}} V} - \frac{m \sum c_i H_i RT}{\langle W \rangle_{\text{reactor}} PV} + \frac{\dot{q}}{V} = 0,$$

or

$$\sum c_i C_{Pi} \Delta T + \sum H_i \Delta c_i - \sum H_i c_i \Delta\rho/\rho + B = 0. \tag{54}$$

Proceeding as in the adiabatic constant pressure case, ΔT and Δc_i can be replaced in equation (54) by solving equation (32) for ΔT and using equation (51) for Δc_i, which leaves an equation for $\Delta\rho/\rho$ in terms of known quantities:

$$\frac{\Delta\rho}{\rho} = \frac{\sum H_i \left(A_i + \sum \beta_{ij} r_i\right) + B}{T \sum c_i C_{Pi}} - \frac{\sum \left(\sum \beta_{ij} + A_i\right)}{\sum c_i} \tag{55}$$

where A_i and B are the flow terms in equations (51) and (54), respectively. Comparison with the corresponding equation (39) for adiabatic constant pressure reaction without flow shows that the CSTR solution reflects the structure of the problem itself, that is, the flow terms appear as supplements to the reactive changes in composition and energy.

Solution for the derivatives can be done in a systematic way by finding $\Delta\rho$ from equation (55), the Δc_i from equation (51), and ΔT from equation (32). Numerical integration starting from a set of assumed starting concentrations, density, and temperature will eventually lead to the CSTR steady-state condition implied by the initial assumptions.

In contrast to the cases previously discussed, however, numerical integration may not be the most efficient way to find the steady-state solution, and for near-critical combinations of flow and initial conditions convergence may be very slow. For non-sensitive reactor conditions, one may be able to find solutions more quickly by converting the CSTR equations into a system of non-linear algebraic equations and using simultaneous equation solvers rather than numerical integration of the ordinary differential equation set. Unless one is particularly interested in studying stability boundaries, however, CSTR solutions can readily be found by testing different starting compositions—for example, trial compositions identified by starting the integration from static near-equilibrium conditions, which can be found by setting the mass flow to zero—and integrating with any of the efficient algorithms discussed in the next section.

1.1.6 Methods of numerical integration

The reaction mechanisms encountered in combustion chemistry lead to differential equations with rates that span many orders of magnitude for typical combustion conditions. For this reason, the common textbook methods for integrating systems of ordinary differential equations are inefficient, as the time steps of the numerical procedures are forced by the fast rates to be small, while the time required to complete the numerical integration is governed by the slow rates. This problem, which came to be called "stiffness" (from the prototype differential equation describing vibrations of a long stiff board), was solved by the development of numerical procedures able to maintain accuracy while using time steps large enough to complete the integration within an acceptable computer time (Gear 1971). A number of different algorithms have evolved and been tested on stiff chemical kinetics problems. (For examples, see Young and Boris 1977; Bader and Deuflhard 1983; Pratt 1985; Deuflhard and Nowak 1986; and Radhakrishnan 1993a.) Appropriate well-tested computer codes are available at moderate (Press et al. 1992) or no cost.[1] A comparison of some of them for typical combustion problems was undertaken by Radhakrishnan (1993a), who found that the LSODE program (Brown et al. 1989; Radhakrishnan and Hindmarsh 1993) was generally the most accurate and efficient one. Byrne and Dean (1993) found that the VODE program outperformed LSODE.

Aside from the intrinsic speed and accuracy of the subroutine used to integrate the set of differential equations, significant acceleration of chemical kinetics simulations can be obtained by rationalizing the program itself, by observing the common efficient computing rules (minimal indexing and recomputing, etc.), and by applying chemical common sense (by (1) not reevaluating parameters such as rate coefficients or thermochemical quantities until temperature changes make it necessary to do so, and (2) not requesting unneeded accuracy from the integration). Even extensive sensitivity scans (next section) for large mechanisms can be accomplished in short times with contemporary workstations if the simulation software is intelligently written.

1.1.7 Sensitivity and flux analysis of reaction profiles

Numerical simulation of dynamic systems has interpretive and predictive value only when the constraints, the system of differential equations, and the parameters

---

[1] A robust implementation of the Gear backward differentiation method with useful interfacing additions can be obtained by anonymous ftp to `netlib.lucent.com`. The integration subroutine is in the directory `/netlib/ode` as the compressed FORTRAN source file `vode.f.gz`; the required subroutines `daxpy`, `dcopy`, `ddot`, `dgbfa`, `dgbsl`, `dgefa`, `dgesl`, `dscal`, and `idamax` are to be found in the `/netlib/linpack` and `/netlib/blas` directories. The earlier subroutine `lsode` can be obtained from the `/netlib/odepack` directory. Comments within the code explain the usage.

of the model are complete enough and accurate enough for the purpose at hand. For physical models, however, including detailed combustion modeling, the parameter set is often much larger, by any measure, than the set of model outputs that can be compared to experimental data. Moreover, in combustion modeling the parameters describing the thermochemical and transport properties of the chemical species and the rates of the elementary reactions may include many values that are not based on measurements at all, but are just estimates based on analogies or theory. Indeed, it has been said that the only solidly established parameters in combustion modeling are the molecular weights.

Two factors mitigate the situation. First, there is an abundant and well-understood theoretical background constraining the values that the parameters used in combustion modeling may reasonably have. (Cf. Section 1.9.1 and Chapters 2 and 4 for rate parameter theory and estimation methods, Chapter 5 for thermochemical parameter estimation methods, and Reid et al. (1987) for transport parameter values and estimation methods.) Second, it is a common characteristic of dynamic models that the relative importance of their parameters spans a very wide range, which means that model predictions usually are affected by the values of only a small fraction of the model parameters. In other words, only a few parameters really matter; no harm will be done to model predictions as long as the rest are "more or less correct".

The study of relationships between parameter values and model predictions is called *sensitivity analysis.* Parameters which prove to strongly influence model predictions are *sensitive parameters,* those with little influence are *insensitive* ones; the set of sensitivity values for the whole parameter set is its *sensitivity spectrum.* There is an extensive literature on the subject (see, for example, Frank 1978 and Frenklach 1984).

The basic definition for applications of sensitivity analysis in combustion modeling is that of the *normalized* or *logarithmic response sensitivity*

$$S_{ij} = \frac{\theta_j}{\eta_i}\frac{\partial \eta_i}{\partial \theta_j} \tag{56}$$

where η_i is a *response* of the model—for example, the concentration of species i at a given time—and θ_j is one of the parameters of the model—for example, the A-factor (see Section 1.9.1) of the rate coefficient expression of reaction j. The partial derivative is understood as being evaluated for a small change in θ_j at fixed values of the other parameters. While a number of advanced mathematical methods for computing response sensitivities have been developed, the most common procedure is to estimate each one by integrating the model twice with different values for a parameter, θ_j and $\theta_{j'}$, and computing

$$S_{ij} = \frac{\theta_j}{\eta_i}\frac{(\eta_{i'} - \eta_i)}{(\theta_{j'} - \theta_j)} \approx \frac{\ln(\eta_{i'}/\eta_i)}{\ln(\theta_{j'}/\theta_j)}\,. \tag{57}$$

Equation (56) suggests that normalized response sensitivity has the character of the slope of a logarithmic graph of a response versus a parameter value, determined at a particular pair of points. Sensitivities are thus "local" in two senses: they refer to just one segment of the range of the parameter under study, θ_j, and to the particular set of values of all other parameters that were held constant for the evaluation of S_{ij}. In order to derive a meaningful sensitivity spectrum for a dynamic model, one must therefore derive it using a parameter set that is already close to the correct one, or at least close to an optimal approximate one. An example sensitivity spectrum is shown in Figure 1.2.

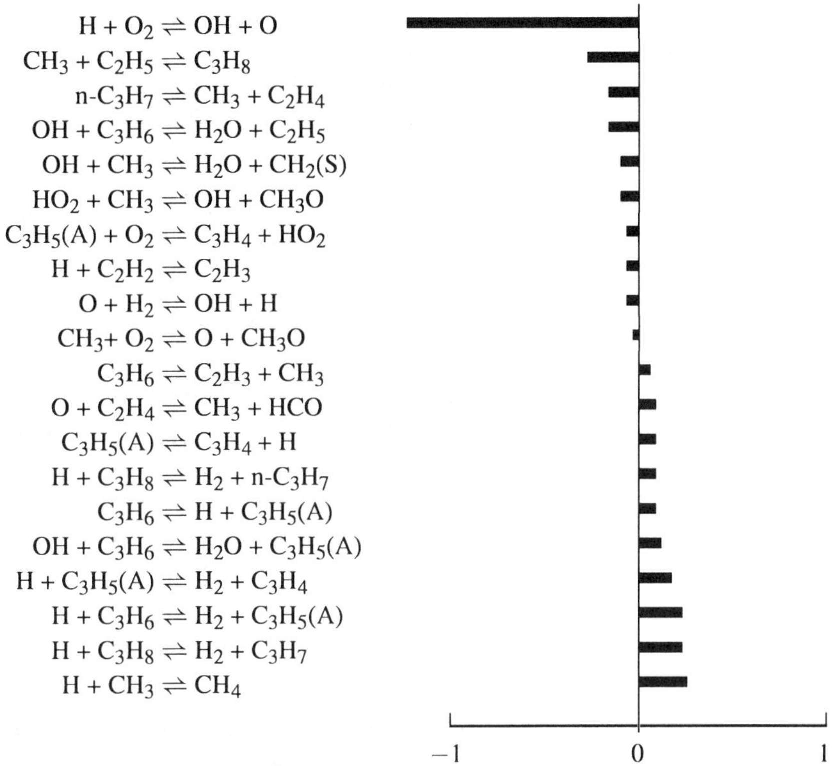

FIGURE 1.2 Sensitivity spectrum for ignition delay τ (defined by the time required to reach the maximum slope of OH radical concentration profile) in $\phi = 1$ propane combustion (of 0.2% C_3H_8 and 1.0% O_2 in Ar) starting at 1450 K and 1 atm. The pressure dependence of unimolecular reaction rates is not noted in the chemical equations. $CH_2(S)$ denotes an "electronically excited" methylene radical, $C_3H_5(A)$ denotes the H_2CCHCH_2 isomer of C_3H_5. The computational resolution used for sensitivity was 0.03, corresponding to 10 μs at an ignition delay of 470 μs for the mechanism under study. Multiplication by two was used to evaluate equation (57); because the modeling software computed reverse reaction rate coefficient values from the forward reaction rate coefficient using Equation (15), this multiplication increased the forward and reverse reaction rates by the same factor. In this example, more rapid ignition corresponds to negative sensitivity (Qin 1998).

According to equations (56) and (57), a normalized response sensitivity value of 1 corresponds to direct proportionality between η_i and θ_j, while a value of -1 would indicate inverse proportionality. In combustion modeling, all but the very most sensitive parameters are an order of magnitude or more smaller than this, reflecting that combustion phenomena are the composite effects of large numbers of elementary chemical processes. The sensitivity spectrum of Figure 1.2 is typical in this respect. A sensitivity spectrum is nonetheless very informative in identifying the elementary reaction rate coefficients, or other parameters,[1] that are important in controlling the course of combustion. Many references to sensitivity studies are made in later chapters.

A second general-purpose tool for interpreting profiles computed by dynamic modeling is *flux analysis.* As the name implies, one seeks to discover the "channels" in which chemical reaction "flows" by extracting information about rates of elementary reactions. In the simplest form, one would just scan all the net rates r (equation 8) at interesting times during the reaction to see which are the fastest, which contribute nothing, which reactions are so fast in both directions that they are essentially equilibrated, and so on. While a comprehensive scan like this can be quite informative, still more information can be derived by associating elementary reaction rates, preferably in both forward and reverse directions, with species of interest, that is, by examining the individual terms that contribute to equation (14). For example, one could identify the important reactions of CH_3 radicals at a given time by rank-ordering all of the reactions producing and all of the reactions consuming CH_3.

Sensitivity and flux analysis as described in the foregoing paragraphs have the limitation of enabling one to identify important elementary reactions only one at a time. In real combustion mechanisms, the reaction networks are so complex[2] that studying the sensitivity or flux spectra for a single response or species may be misleading, because numerical experiments in which part of the mechanism is examined will not reveal the many-species or many-reaction correlations that characterize combustion reactions. Although a number of graphical and numerical measures have been proposed to enable broader understanding of complex reaction networks, none of them are able to replace the dedicated human intellect for interpretation of dynamic modeling results. They are only useful supplements.

---

[1] Sensitivity studies are primarily focused on the rates of elementary reactions, and response sensitivities are normally computed by varying the corresponding A-factors (cf. Section 1.9.1.). It is not a straightforward matter to compare sensitivities for rate and other model parameters. For example, suppose that one evaluates the normalized sensitivity for the standard enthalpy of formation of some species by adding 5 kJ/mol to the reference value. Reverse reaction rate coefficients for reactions of that species would change by factors of 9.4 at 300 K, 1.8 at 1000 K, and 1.35 at 2000 K.

[2] See, for example, the nitrogen chemistry networks of Figures 2.62 and 2.80 in Chapter 2.

1.2 OXIDATION OF HYDROGEN AND CARBON MONOXIDE

1.2.1 Hydrogen oxidation at high temperatures

Hydrogen oxidation is the classic example of branching chain reaction. It is also one of the simplest combustion reactions, as water is the only reaction product. The reaction mechanism of hydrogen oxidation is not simple, however, because it includes a number of elementary steps of the four active species, O, H, OH, and HO_2, and under some conditions hydrogen peroxide can also be present in detectable amounts.

Quantitative studies of hydrogen oxidation have a rich history, starting in the 1920s when the theory of chain branching reactions was developed (Semenov 1934). It has been studied over wide ranges of temperature, pressure, and composition. There are three "ignition limits"—boundaries on a pressure–temperature diagram that divide regions of different ignition behavior.

Our primary interest for combustion modeling is the high-temperature reaction. It has been shown not to depend on processes taking place on the walls of the vessel in which the reaction occurs, and abundant data have been obtained for it in high-temperature shock tube and flame studies as well as in low-temperature studies of individual elementary reactions. While all the important elementary reactions of high-temperature H_2 oxidation are thought to be known, the publication flow on aspects of H_2 oxidation continues. Recent reports on H_2–O_2 studies discuss rate coefficient refinements, report more precise measurements of ignition and flame phenomena, explain influences of additives, and serve for developing new modeling and theoretical approaches.

Many versions of the high-temperature H_2 oxidation mechanism can be found in the literature, differing from one another in the expressions used to compute rate coefficients for the elementary reactions and by the inclusion of various reactions of secondary importance to combustion. Some differences between reported mechanisms are due not only to different choices of rate coefficient expressions, but also to the fact that the developers may have intended to describe regimes where different elementary reactions are important.

Comprehensive combustion chemistry models always include a complete H_2 oxidation mechanism as a component, because in the early stages of hydrocarbon combustion, molecular hydrogen is formed rapidly. The mechanism of hydrocarbon oxidation discussed later in this chapter includes 19 elementary reactions describing H_2 oxidation. High-temperature H_2 oxidation mechanisms have been described by Dixon-Lewis and Williams (1977), Dixon-Lewis (1979), Glarborg et al. (1986), Tsang and Hampson (1986), and many others; rate coefficient compilations and critical reviews have been presented by Westley et al. (1994), Baulch et al. (1994), Atkinson et al. (1992), and Ryu et al. (1995). Aside from many minor differences among the proposed H_2–O_2 reaction mechanisms, all of them include the main initiation, propagation, termination, and chain branching steps that we now discuss.

A chain initiation process is needed for any H_2 oxidation mechanism. For high temperature modeling, one or the other of the reactions

$$H_2 + O_2 \longrightarrow OH + OH \text{ (a)} \qquad \text{or} \qquad H_2 + O_2 \longrightarrow H + HO_2 \text{ (b)} \tag{58}$$

is usually included; the computed results have been shown to be affected by the initiation rate but not by the choice of products.[1] Because the initiation reaction(s) provide(s) only a very low initial level of active species, that is, most of the chain centers are formed later in the chain branching reactions, the rate coefficient(s) of the initiation step(s) has/have not been defined experimentally. Fortunately, only the correct initiation rate is important for modeling.

The most important elementary steps of hydrogen oxidation prove to be the ones that provide chain branching and propagation:

$$\begin{aligned} H + O_2 &\longrightarrow OH + O && (1)\\ O + H_2 &\longrightarrow OH + H && (59)\\ OH + H_2 &\longrightarrow H + H_2O. && (60) \end{aligned}$$

The reaction of hydrogen atoms with O_2 is also found to be the most important chain branching step for other combustion systems: As a rule, the higher the temperature, the larger its role in governing the rate of chain reactions.

The main chain termination reactions include recombinations of H and O atoms and OH radicals, formation of HO_2 radicals (which are less reactive than other chain centers), and, at lower temperatures, formation of H_2O_2:[2]

$$\begin{aligned} H + H + M &\longrightarrow H_2 + M && (61)\\ H + OH + M &\longrightarrow H_2O + M && (62)\\ H + O_2 + M &\longrightarrow HO_2 + M && (63)\\ OH + OH + M &\longrightarrow H_2O_2 + M\,. && (64) \end{aligned}$$

---

[1] The reverse of reaction (58a) has not been observed in the laboratory, while the reverse of reaction (58b) has. Using the measured rate coefficient for reaction (58b) in a program that automatically includes both forward and reverse directions of reaction provides a satisfactory chain initiation rate at high temperatures. At temperatures below 1000 K one has to include chain initiation by reactions on a vessel wall, or chain centers introduced by diffusion, as in H_2–O_2 flames. At extremely high temperatures (over around 2800 K), thermal dissociation of H_2 also becomes a significant chain initiation reaction.

[2] See footnote on Page 6 for an explanation of the M symbol.

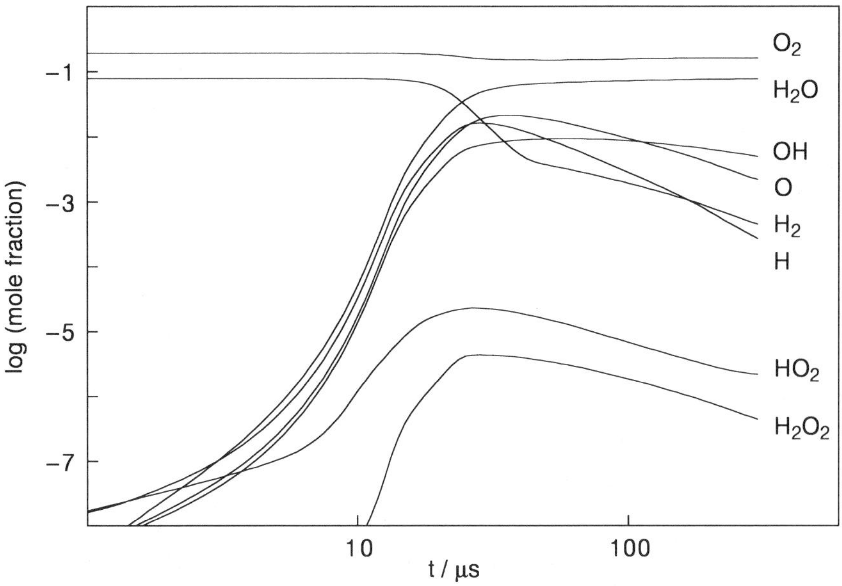

FIGURE 1.3. Log–log graph of species profiles for H_2 oxidation in an H_2–air mixture with $\phi = 0.2$ at 1400 K and 1 atm, calculated using the mechanism described in Section 1.9.4. Adiabatic constant volume reaction was assumed.

The relative importance of these four paths changes with equivalence ratio. The final concentrations of active species are very low at temperatures below 1000 K, but with increasing temperature their concentrations steadily increase. Above 2500 K they begin to exceed the water concentration.

A variety of computed profiles are shown in Figures 1.3–1.9 for a hydrogen–air mixture igniting with equivalence ratio 0.2 at 1400 K. The modeling was done with two different mechanisms, one assembled for the illustrations of this chapter as described in Section 1.9.4 and one reported by Miller and Bowman (1989, 1991), using an adiabatic constant-volume constraint. One sees that the results calculated by the two mechanisms are similar for all species. The features of the reaction include the early chain initiation period, the exponential growth of radical concentrations, temporary steady-state chain center concentrations, ignition delay, and an approach to equilibrium that is much slower than the ignition part of the reaction. The different forms of graphical presentation shown are all common in combustion chemistry modeling.

In the double logarithmic presentations of Figures 1.3–1.6, one can see that the slightly different early chemistry of the two models results in only a slightly different ignition delay. Once the exponential increase of chain center concentrations dominates, the remaining chemistry runs the same course for both models shown, as it would for any other H_2–O_2 combustion model in current use.

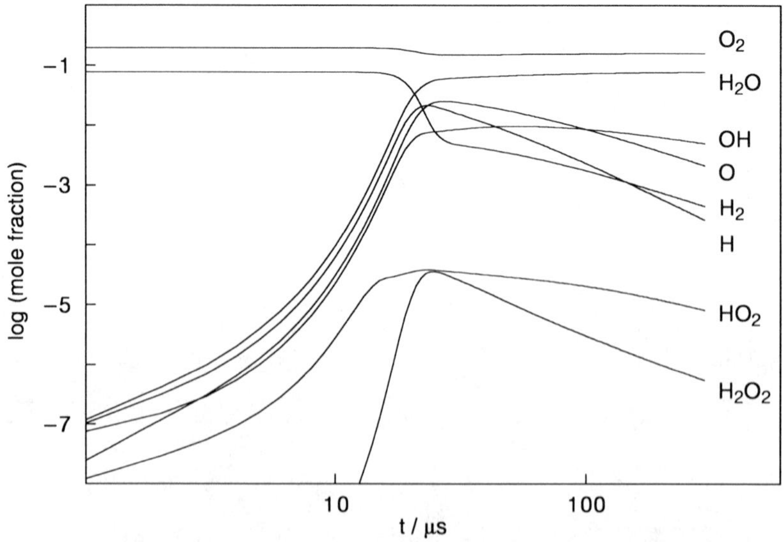

FIGURE 1.4. Species profiles for H_2 oxidation in an H_2–air mixture with $\phi = 0.2$ at 1400 K and 1 atm, calculated using the Miller and Bowman (1989, 1991) mechanism. Adiabatic reaction at constant volume was assumed.

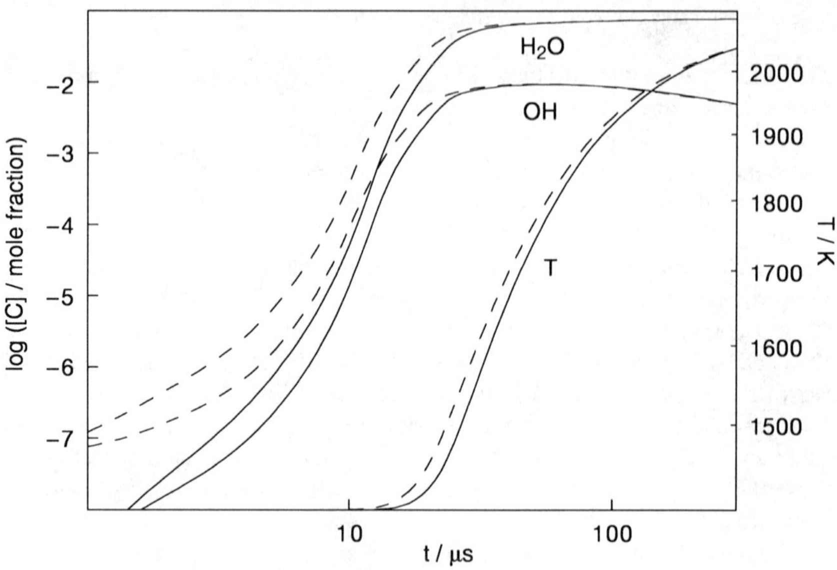

FIGURE 1.5. Comparison of temperature, OH, and H_2O profiles for H_2 oxidation in an H_2–air mixture with $\phi = 0.2$ at 1400 K and 1 atm, calculated using the mechanism described in Section 1.9.4 (solid lines) and the one described by Miller and Bowman (1989, 1991) (dashed lines). Adiabatic reaction at constant volume was assumed.

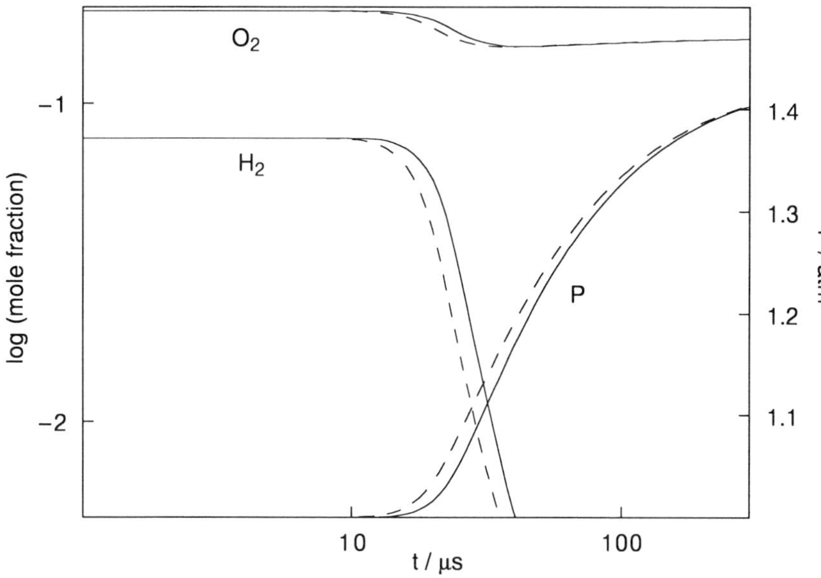

FIGURE 1.6. Comparison of pressure, H_2, and O_2 profiles for H_2 oxidation in an H_2–air mixture with ϕ = 0.2 at 1400 K and 1 atm, calculated using the mechanism described in Section 1.9.4 (solid lines) and the one described by Miller and Bowman (1989, 1991) (dashed lines). Adiabatic reaction at constant volume was assumed.

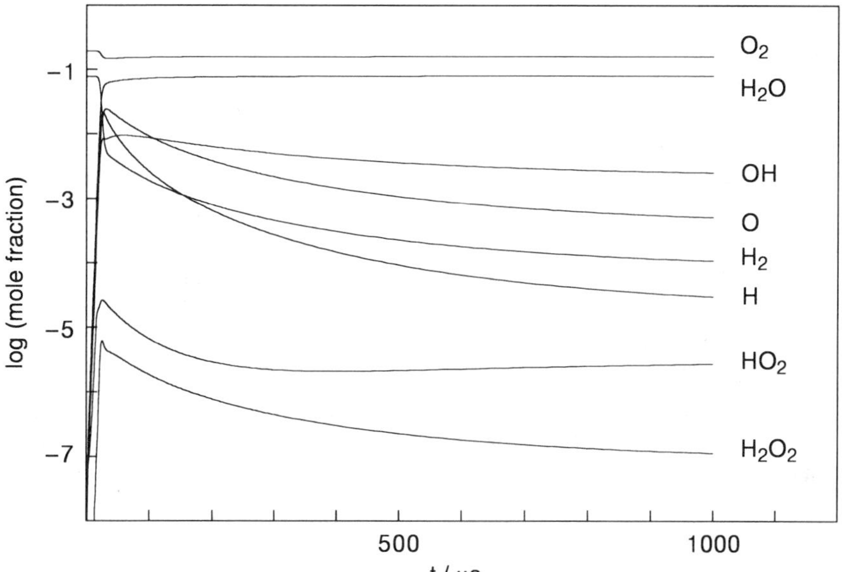

FIGURE 1.7. Approach to equilibrium in H_2 oxidation in an H_2–air mixture with ϕ = 0.2 at 1400 K and 1 atm, calculated using the mechanism described in Section 1.9.4.

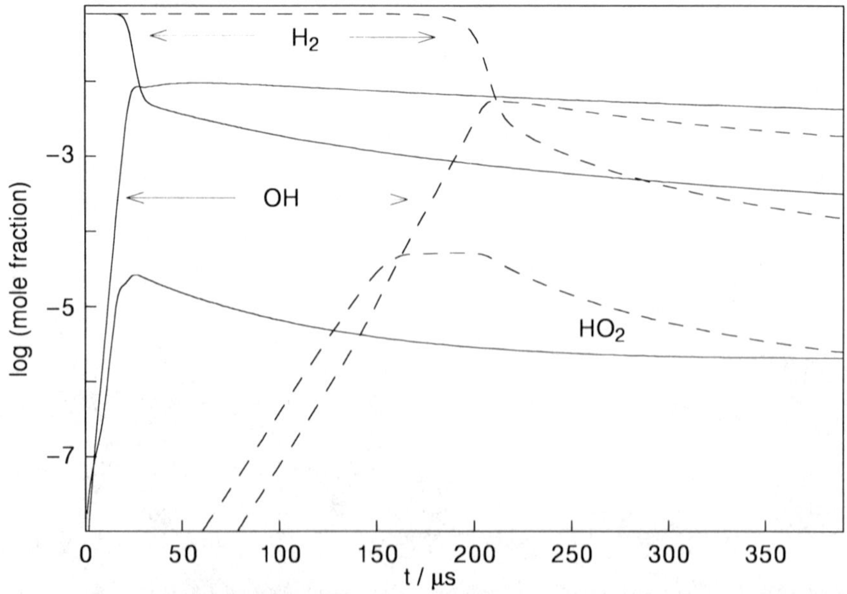

FIGURE 1.8. H_2, OH and HO_2 profiles for a $\phi = 0.2$ mixture of H_2 in air at 1000 (dashed lines) and 1400 K (solid lines). In semilogarithmic presentation the exponential concentration growth of intermediate species appears as essentially parallel straight lines. It is seen that the main effect of temperature change is to accelerate the induction zone chemistry.

In Figure 1.7, one sees that the approach to final equilibrium is much slower than the induction chemistry for H_2 oxidation. This behavior, caused by the different stoichiometries of the branched chain and recombination reactions, is not the same chemical effect as one has in hydrocarbon oxidation, although the post-induction zone chemistry is also slower for hydrocarbons (Schott 1960; Schott and Getzinger 1973).

In Figures 1.8 and 1.9, the effects of temperature and pressure changes are shown. The semilogarithmic presentations allow the course of the branched chain reaction of the induction zone to be distinguished clearly from the later reaction. One sees that only the rate of reaction is affected, not the character of the explosion.

In every combustion chemistry modeling study, one asks which of the many reactions are "important" and for what particular reasons. One wonders which reactions are responsible for most of the chemical conversion, that is, one does the rank-ordered complete flux analysis introduced in Section 1.1.7. The fastest elementary reactions for the hydrogen–oxygen ignition example are shown in Figure 1.10.

Similarly, one asks which reactions affect the concentration of a particular species. The rates of the reactions affecting O atoms for the same conditions are shown in Figure 1.11.

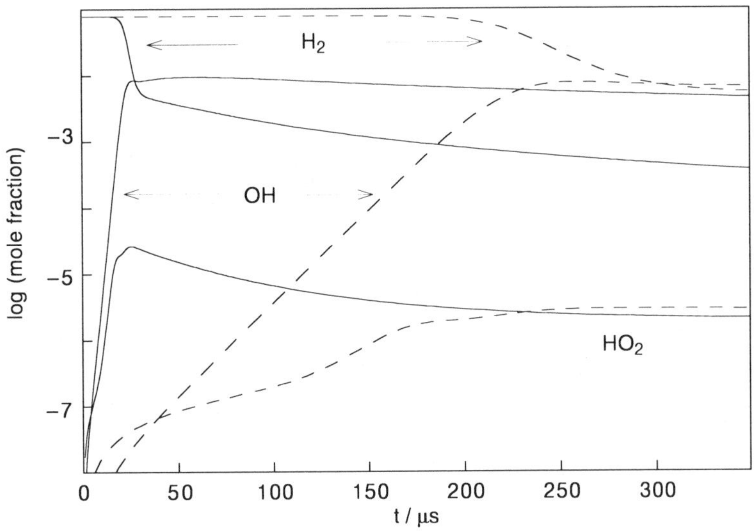

FIGURE 1.9. Induction zone profiles for oxidation of a $\phi = 0.2$ mixture of H_2 in air at 1 atm (solid lines) and 0.1 atm (dashed lines) pressure. Except for the HO_2 profile, depressed by the decrease in pressure because of the slower $H + O_2 + M$ reaction, the main effect of pressure decrease is seen to be slower induction zone chemistry.

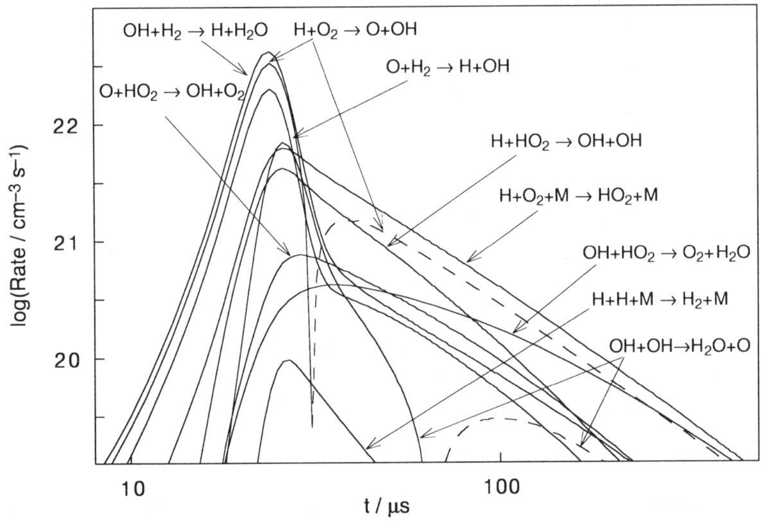

FIGURE 1.10. Reaction rates in oxidation of a $\phi = 0.2$ mixture of H_2 in air at 1 atm and a starting temperature of 1400 K. Dashed lines indicate reactions proceeding in the reverse direction late in the reaction. The rate units used are molecules per cubic centimeter per second, i.e., the normal units multiplied by 6×10^{23}, in order to avoid negative logarithms. Reactions prominent in the induction zone recede quickly at the end of the ignition delay, replaced by reactions leading to final equilibrium.

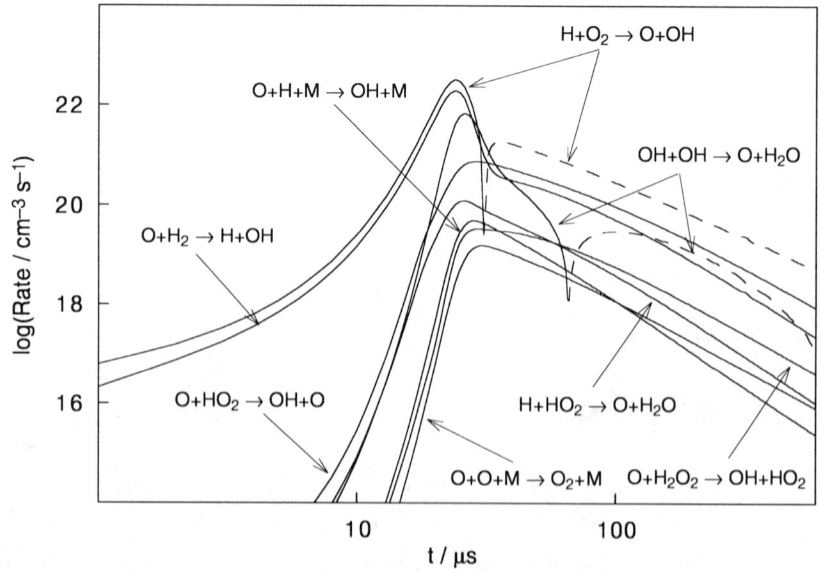

FIGURE 1.11. Rates of the elementary reactions that produce or consume O atoms for the same initial conditions as in the preceding figures.

1.2.2 Role of peroxides at low temperatures

The H_2–O_2 reaction at pressures near atmospheric and temperatures over 1000 K is the classic example of a gas-phase branched chain explosion. Clear distinctions are seen between the processes of chain initiation, exponential growth of chain center concentrations, chain center concentrations above equilibrium levels, slow approach to final equilibrium through termolecular reactions, and effects of stoichiometry changes that we have not illustrated (Schott and Getzinger 1973; Dixon-Lewis and Williams 1977). At temperatures below 1000 K, the H_2–O_2 reaction becomes more complicated for two reasons. First, at lower temperatures HO_2 and H_2O_2 are stable enough to participate as chain centers, and some of the reactions they undergo may have decisive influence upon the reaction kinetics observed. One such reaction is that of two HO_2 radicals to form H_2O_2

$$HO_2 + HO_2 \longrightarrow H_2O_2 + O_2 \tag{65}$$

which can then decompose in the reverse of reaction (64) to two OH radicals, which are able to attack H_2 to regenerate H atoms. That is, the sum of twice reactions (60) and (63), (–64) and (65) is

$$2\,H + O_2 + 2\,H_2 \longrightarrow 2\,H + 2\,H_2O. \tag{66}$$

Reaction (66) represents an exothermic straight chain reaction. Its essential consequence is thus not exponential growth of chain centers, but steady release of heat and resulting temperature increase.

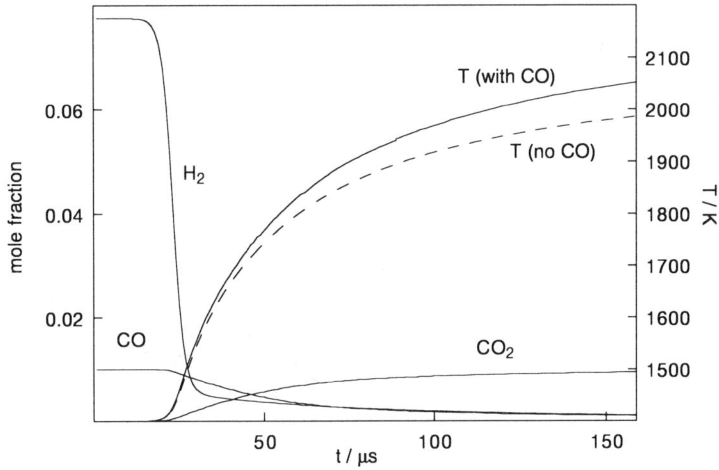

FIGURE 1.12. Profiles for 7.75% H_2 and 15% O_2 in N_2, and with 0.1% CO added. While H_2 consumption in the induction zone is not perturbed significantly by CO addition, the rate of heat release after ignition is increased by CO_2 formation.

The second difference is that when the reactions are slow compared to heat and matter transport by thermal conduction, convection, and diffusion, the nature of the reaction is influenced by the size of the reaction vessel and the nature of its walls. The low-temperature reaction, despite being much more complicated than the high-temperature reaction, has, because of sustained experimental and computational study, been well characterized (Dixon-Lewis and Williams 1977).

Reactions of HO_2 are also found to be significant in hydrocarbon oxidation kinetics, particularly at above-atmospheric pressure.

1.2.3 Carbon monoxide oxidation

Carbon monoxide itself can serve as a fuel component in "water gas" combustion; indeed, mixtures of H_2 and CO were widely used as fuels before introduction and expansion of modern methods of natural gas production and distribution displaced water gas from the marketplace. "Synthesis gas" combustion is still a common industrial energy source. A more prominent role of CO, however, is as an intermediate in hydrocarbon combustion. The main elementary reaction oxidizing CO to CO_2 is

$$OH + CO \longrightarrow CO_2 + H. \tag{67}$$

Reaction (67) is slow compared to other radical reactions, which leads to significant amounts of CO remaining in automobile exhausts and smokestack plumes. A comparison of the CO oxidation process to that of H_2 is shown in Figure 1.12. Similar behavior is also found in hydrocarbon flames.

1.2.4 Rate coefficients of the rate-limiting steps of H_2 and CO oxidation

The participation of H_2 and CO oxidation in all combustion processes has led to sustained attention being paid to determining the rate coefficients of the rate-limiting steps in their oxidation mechanisms. As a result, the rate coefficients of reactions (1) and (67) in particular are now thought to be known with enough accuracy that modeling the high-temperature chemistry of H_2 and CO, in which these two elementary reactions control the reaction rate, is more accurate than most experimental measurements made under reaction conditions where H_2–CO chemistry dominates. Arrhenius graphs[1] for these prominent elementary reactions are shown in Figures 1.13 and 1.14. Extensive discussion of the data and of the rate coefficient expressions assigned to these reactions can be found in the cited publications.

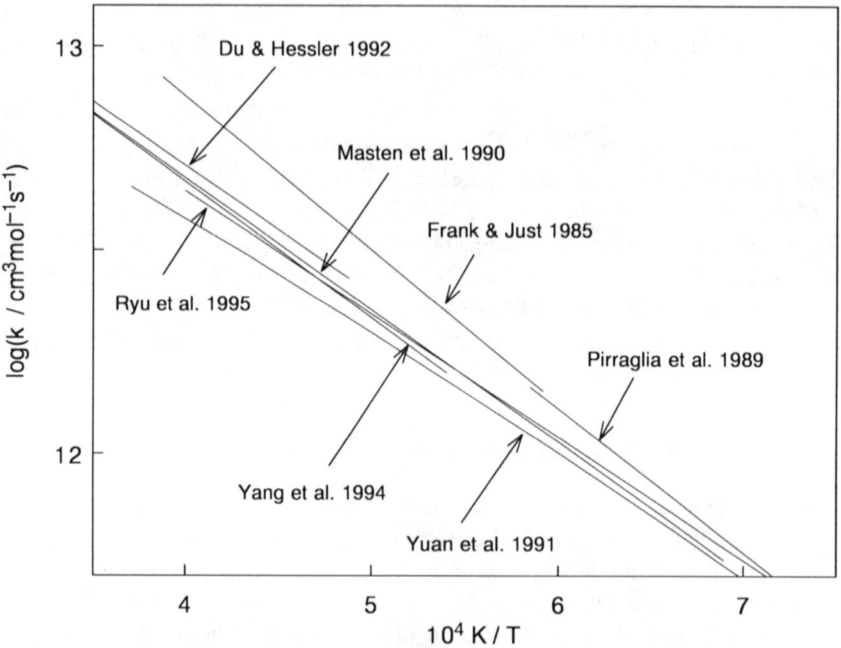

FIGURE 1.13. Rate coefficient expressions for the $H + O_2 \longrightarrow OH + O$ reaction reported for experiments done since the review of Warnatz (1984). Not shown are the expressions of Yu et al. (1994), who reported a combined analysis of the Masten et al. (1990) and Yuan et al. (1991) data, and Shin and Michael (1991). These studies gave expressions nearly identical to those reported by Ryu et al. (1995), who presented a critical evaluation of all experimental data and theoretical analyses for this reaction.

[1] See Section 1.9.1.

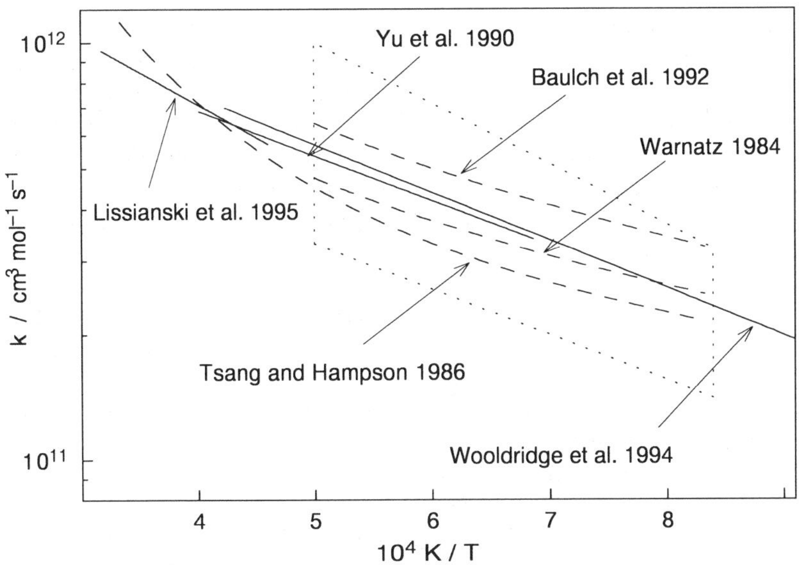

FIGURE 1.14. Overview of rate coefficient expressions for the reaction OH + CO $\longrightarrow$ CO_2 + H reported for experiments done since the review of Warnatz (1984). The box indicates an area bounding 13 earlier proposals for the rate coefficient. While the data obtained in individual studies are well represented in Arrhenius form (see Section 1.9.1), there are convincing reasons to believe that the rate coefficient of this reaction, in contrast to the rate coefficient of the H + O_2 $\longrightarrow$ OH + O reaction, does not follow an Arrhenius temperature dependence. The expected one is indicated by the Tsang and Hampson (1986) curve shown here.

1.3 HYDROCARBON COMBUSTION CHEMISTRY

A wide variety of hydrocarbons, mostly in the form of complex mixtures, are used as practical fuels. The basic chemical features of their combustion reactions vary little from one hydrocarbon to another. The main differences are found to arise from the relative amounts of aliphatic (with single bonds connecting the carbon atoms), olefinic (such as ethylene, C_2H_4), and aromatic (such as benzene, C_6H_6) hydrocarbons. Methane, CH_4, is an exception for reasons that are discussed below. There are important differences, however, in the chemistry of hydrocarbon oxidation at low (hundreds of K) and high (over 1000 K) temperatures. In this section, we present outlines of the observations that characterize hydrocarbon oxidation chemistry and discuss a few of the elementary reactions that play important roles in it.

1.3.1 General features of hydrocarbon oxidation

Hydrocarbon oxidation shows self-acceleration and induction periods that are chemically quite different from those discussed in the previous section. While induction periods alone suggest that some form of branched chain reaction is involved, simple branched chain mechanisms can not explain the induction behavior observed in hydrocarbon-oxygen mixtures at low temperatures: (a) The self-acceleration rates are much lower than in explosive chain branching reactions, and (b) very large fractions of the fuel are converted to stable partially oxidized intermediate species long before onset of significant energy release. Explaining these observations in chemical detail became one of the main goals of basic combustion chemistry research as soon as quantitative data were available.

Semenov (1958) suggested that during low-temperature hydrocarbon oxidation, the chain branching process requires "secondary" reactions of the partially oxidized intermediate products that are formed early in the reaction in the course of an essentially unbranched chain mechanism. Such intermediates would be relatively stable products able to undergo further transformations to give the active centers that eventually lead to "degenerate" chain branching. Gradual accumulation of intermediate products in significant amounts eventually leads to exponential growth of radical concentrations and fuel oxidation rates. At the end of the induction period, the reaction is also self-accelerated by the increasing temperature.

Roles like these for intermediate products are indeed a general feature of hydrocarbon oxidation, and they are observed over wide ranges of reaction conditions. While Semenov's basic insight has been confirmed, it has also been learned that the oxidation of hydrocarbon fuels is also characterized by essentially complete fragmentation of large aliphatic fuel molecules to small olefinic ones; it is mixtures of the smaller fuel fragments that eventually react to form the final products of combustion and release the reaction heat.

Significant differences in the identity of the intermediate species—and their roles in degenerate chain branching—are found to prevail at high and low temperatures. In general, Semenov's definition of degenerate chain reaction, as one where chain branching occurs in reactions of intermediate products, is found to be more appropriate for low- and intermediate-temperature ranges of hydrocarbon oxidation. At high temperatures, intermediate products do play important roles in chain branching but the mechanism of branching is different.

1.3.2 Low- and intermediate-temperature oxidation

Detailed analysis (Shtern 1960; Minkoff and Tipper 1962; Benson 1981, 1982) of the low temperature reactions of hydrocarbons shows that at temperatures around 700 K, where laboratory investigations are conveniently carried out, the core mechanism of hydrocarbon oxidation involves the formation and subsequent reactions of peroxide free radicals produced by association of O_2 with alkyl radicals

$$\mathrm{R} + \mathrm{O_2} \longrightarrow \mathrm{ROO}\,. \tag{68}$$

With large hydrocarbons, isomerization to form the hydroperoxy radical can occur

$$\mathrm{ROO} \longrightarrow \mathrm{R'OOH}\,. \tag{69}$$

R′OOH then decomposes to give chain branching by fragmenting the O–O bond

$$R'OOH \longrightarrow R'O + OH\,. \tag{70}$$

The degenerate branching species ROO accumulate to significant concentrations before chain branching becomes important. Only after this accumulation has taken place is self-acceleration observed.

At higher reaction temperatures, ROO decomposition by the reverse of reaction (68) becomes faster and the rate of degenerate branching slows down. The existence of the upper temperature limit of "cool" hydrocarbon flames (Minkoff and Tipper 1962) is due mainly to the instability of ROO radicals at temperatures above 700 K.

H_2O_2 production is more significant at higher reaction temperatures, and HO_2 and OH become the dominant chain centers. Functionally, HO_2 radicals replace ROO, while H_2O_2 replaces R′OOH. These changes characterize the transition between the low- and intermediate-temperature mechanisms of hydrocarbon oxidation.

1.3.3 High-temperature oxidation

While the low- and intermediate-temperature chemistry of hydrocarbon oxidation has drawn attention for decades, in part because of the prospects for making industrial use of partial oxidation processes, for understanding the chemistry of flames it is more important to focus attention on what happens at temperatures above 1000 K. We begin with CH_4, the smallest hydrocarbon fuel.

For CH_4, the general sequence of oxidation processes at high temperatures is

$$CH_4 \longrightarrow CH_3 \longrightarrow CH_2O \longrightarrow CO \longrightarrow CO_2\,. \tag{71}$$

Thus four interlinked chains must be considered: oxidation of CH_4, H_2, CH_2O, and CO. The four chains interact with one another such that, for example, radicals produced during CH_2O oxidation accelerate CH_4 oxidation. The connection between these chains is hierarchical, that is, the CO–H_2–O_2 reaction mechanism is part of that for CH_2O oxidation, which in turn is part of the CH_4 oxidation mechanism. The mechanisms of ethane, C_2H_6, and ethylene, C_2H_4, oxidation are also necessarily parts of the CH_4 oxidation mechanism, because these species are formed in the course of CH_4 oxidation as a result of

$$CH_3 + CH_3 \longrightarrow C_2H_6\,. \tag{72}$$

The reaction mechanism of hydrocarbon oxidation at high temperature has been discussed extensively (by Lifshitz et al. 1971, Crossley et al. 1972, Olson and Gardiner 1977, Westbrook 1979, Borisov et al. 1983a, Miller and Bowman 1989, Sloane 1989, Chevalier et al. 1992, Corre et al. 1992, Frenklach et al.

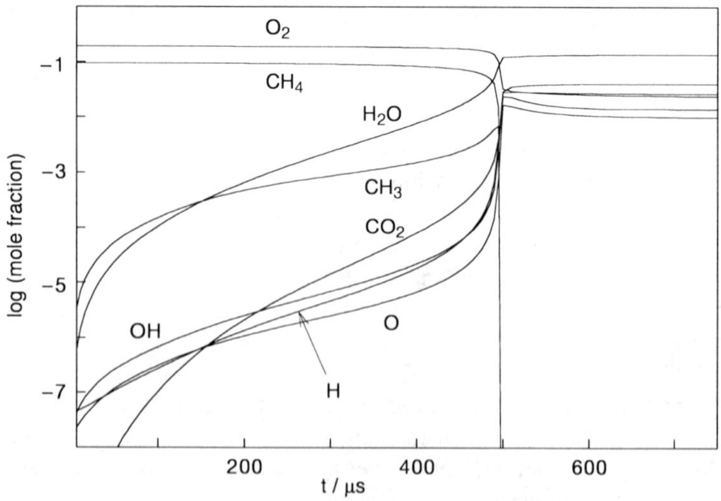

FIGURE 1.15. Major species and radical concentrations computed for a mixture of 9.5% CH_4 and 19% O_2 in N_2 starting at 1600 K and 1 atm. An adiabatic constant volume constraint was used in the simulation. After an initial jump, the CH_3 concentration rises only slowly until the burst of radical concentrations at the very end of the induction zone. Because of the high temperature at the end of reaction (see Figure 1.16), the major radical concentrations persist after ignition.

1992 and 1995, Konnov 1996, Pilling et al. 1996 and many others). There are abundant data for high temperature hydrocarbon oxidation, information for CH_4 being overwhelming. Reaction mechanisms proposed for CH_4 oxidation (for example, Frenklach et al. 1992; Miller and Bowman 1989; Sloane 1989) differ in the numbers of species and elementary reactions, and in numerous details. Even though there are still many quantitatively uncertain chemical aspects—examples would be the ways CH_3 and C_2H_3 interact with O_2 during CH_4 oxidation—its general features are believed to be established. We will illustrate them using the mechanism described in Section 1.9.4. Other contemporary mechanisms predict somewhat different concentration profiles, but the course of the reaction remains the same.

Conversion of CH_4 and O_2 to the final products CO_2 and H_2O at constant volume (such as in experiments in shock waves and in static systems) is illustrated in Figures 1.15 and 1.16. At first, fuel decomposition supplies most of the radicals. This process is slow, and radical concentrations at the beginning of reaction remain small (Figure 1.15). Once chain centers have been created, the important step becomes attack on the fuel by O, OH, H, and HO_2. These reactions effectively convert CH_4 to CH_3, so that the CH_3 concentration is the highest among radicals during ignition delay. This part of the reaction dominates most of the induction zone. It is characterized by relatively constant reaction rates and accumulation of intermediate species—mostly CH_2O, H_2, C_2H_6, and C_2H_4—as a result of CH_3 reactions with O_2 or with other radicals (Figure 1.16).

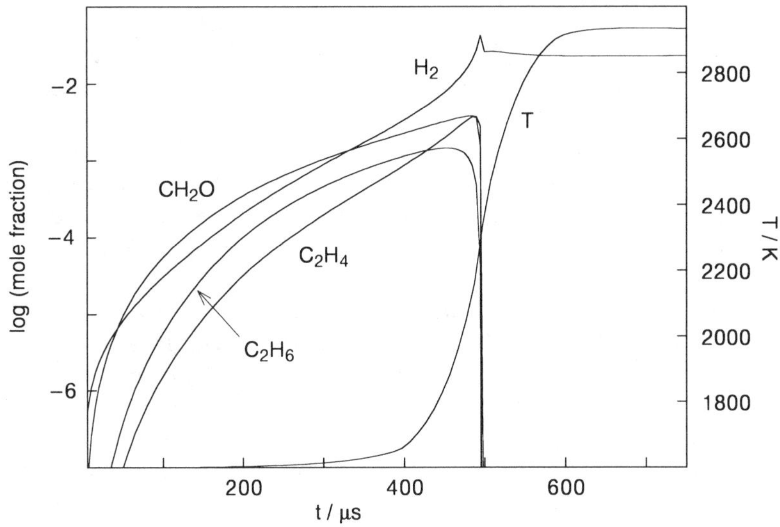

FIGURE 1.16. Stable intermediates produced during the induction zone of CH_4 ignition, computed for the same conditions as the preceding figure. All of the carbon-containing intermediates vanish at the end of the induction zone.

(Because the concentrations of radicals and intermediate species change slowly, a "steady-state approximation" is often used to simplify analysis of the reaction at this stage. According to this approximation, the concentrations of some radicals and other intermediate species are more or less constant during ignition delay, so that the mathematical assertion $d[C]/dt = 0$ may be used to derive "reduced" combustion mechanisms from the detailed chemistry of hydrocarbon oxidation. Cf. Section 1.8 and Peters and Rogg (1993).)

The induction zone reactions lead to production of large amounts of formaldehyde, CH_2O. While CH_2O reactions themselves do not give chain branching (CH_2O decomposition is relatively slow), they are actually responsible for it because CH_2O reactions produce H atoms, which do lead to chain branching through the same reaction that determines the rate of chain branching in the H_2–O_2 system

$$H + O_2 \longrightarrow OH + O\,. \qquad (1)$$

Because of chain branching through H atoms, CH_2O accumulation in significant amounts eventually leads to the sharp increase in radical concentrations, and as a consequence to rapid fuel consumption, intensive product formation, and heat release. Thus, the time required for CH_2O production in significant amounts can be considered to define the length of the induction zone of CH_4 oxidation.

Oxidation and decomposition of CH_2O lead to formation of CO, which at high temperatures is subsequently oxidized to produce CO_2. This process is relatively slow and mainly occurs later in the reaction, after the fuel has essentially vanished. For this reason the oxidation rate of CO has little effect on ignition delays but is very important for determining the rate of flame propagation.

The foregoing picture makes it clear that high-temperature oxidation of hydrocarbons depends very much on the extent to which H atoms are present in the system, and this in turn depends on the stoichiometric ratio of the reactants. In the oxygen-rich case, alkyl radicals are rapidly oxidized to CH_2O. In a fuel-rich situation, attack by H atoms becomes important and the radicals formed have a greater opportunity to react among themselves to give small quantities of larger hydrocarbons. Figure 1.16 shows that even for a stoichiometric CH_4–O_2 mixture, significant amounts of C_2H_6 and C_2H_4 accumulate during the induction zone reactions.

In general, the characteristics of the oxidation of C_2H_6 and larger aliphatics are similar to CH_4 oxidation, but CH_4 does exhibit oxidation characteristics different from all other hydrocarbons. The dissociation energy of the C–H bond in CH_4(435 kJ/mol; cf. Chupka 1968, Dobis and Benson 1987, Bauschlicher and Partridge 1993, Martin 199a, and Peterson and Dunning, 1997) is much higher than that of the C–C bonds in heavier hydrocarbons (370 kJ/mol in C_2H_6), and also greater than the C–H bond dissociation energy in long-chain hydrocarbons (410 kJ/mol in C_2H_6).

For these reasons, it is more difficult to ignite CH_4 than other hydrocarbons. Usually CH_3 radicals do not significantly decompose to give other hydrocarbon radicals (CH_2 or CH) even at high temperatures. Instead, their reactions, together with those of CH_2O, define the kinetics of CH_4 oxidation.

Because larger hydrocarbon radicals are unstable, the radicals initially resulting from removal of one of the H atoms from the fuel usually fragment to olefins and smaller radicals. (In the case of C_2H_5, the olefin is C_2H_4, and an H atom becomes the chain center.) The result is that total radical concentrations are higher, reactions are faster, and ignition occurs more quickly. Because of the lower thermal stability of the fuel molecules, the initial fuel is largely consumed early in the induction zone, as illustrated for propane (C_3H_8) in Figure 1.17. This situation is therefore quite different from the CH_4 case, where significant consumption of fuel occurs only at the end of the induction zone (Figure 1.15). Figure 1.17 shows that C_3H_8 decomposes during the initial phase of oxidation, generating C_3H_6, CH_3, CH_4, C_2H_4, C_2H_6, CH_2O, and CO; oxidation of these species defines the main part of the induction zone reaction. The concentration of CH_3 is still the highest among the radicals (Figure 1.18) because of fast conversion of larger hydrocarbon radicals to CH_3. Figures 1.17 and 1.18 show the same three stages of oxidation occurring in C_3H_8 oxidation as in the case of CH_4: an initial stage of fuel decomposition, an intermediate stage with essentially steady concentrations of radicals and intermediate products, and a final self-acceleration stage. The similarities in combustion behavior of different hydrocarbons are mainly due to the fact that larger hydrocarbon molecules and radicals quickly decompose to smaller molecules and radicals.

The main processes that define the chemistry of hydrocarbon oxidation at high temperatures are thus (a) reactions of fuel decomposition, (b) reactions of radicals with initial molecules, (c) chain branching reactions, and (d) recombination reactions.

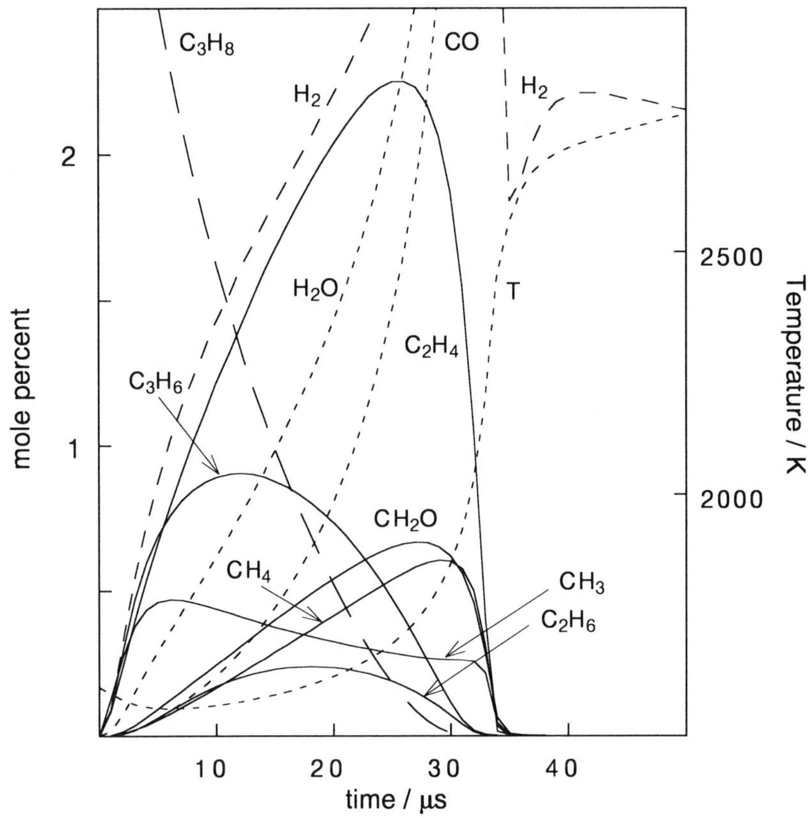

FIGURE 1.17. Major species and temperature profiles during ignition of 4% propane in air (ϕ = 1.0) starting at 1600 K and 1 atm. Initially, the products of fuel decomposition are CH_3, C_2H_4, and C_3H_6 in equivalent amounts; later, the most prominent hydrocarbon species left is C_2H_4. Ultimately, only CO and H_2 remain to be oxidized.

Let us return to CH_4 oxidation to consider these reactions in more detail. In methane ignition at high temperatures, CH_4 thermal decomposition,

$$CH_4 + M \longrightarrow CH_3 + H + M\,, \tag{73}$$

is the chain initiation step. For larger hydrocarbons, the chain initiation step is one in which a C–C bond is broken to form hydrocarbon radicals, for example

$$C_3H_8 \longrightarrow CH_3 + C_2H_5\,. \tag{74}$$

This step dominates over H-atom loss because the C–C bond is substantially weaker than any of the C–H bonds. At combustion temperatures, however, some C–H bonds are also broken and H atoms also appear in initiation steps, such as

$$C_2H_6 \longrightarrow C_2H_5 + H\,. \tag{75}$$

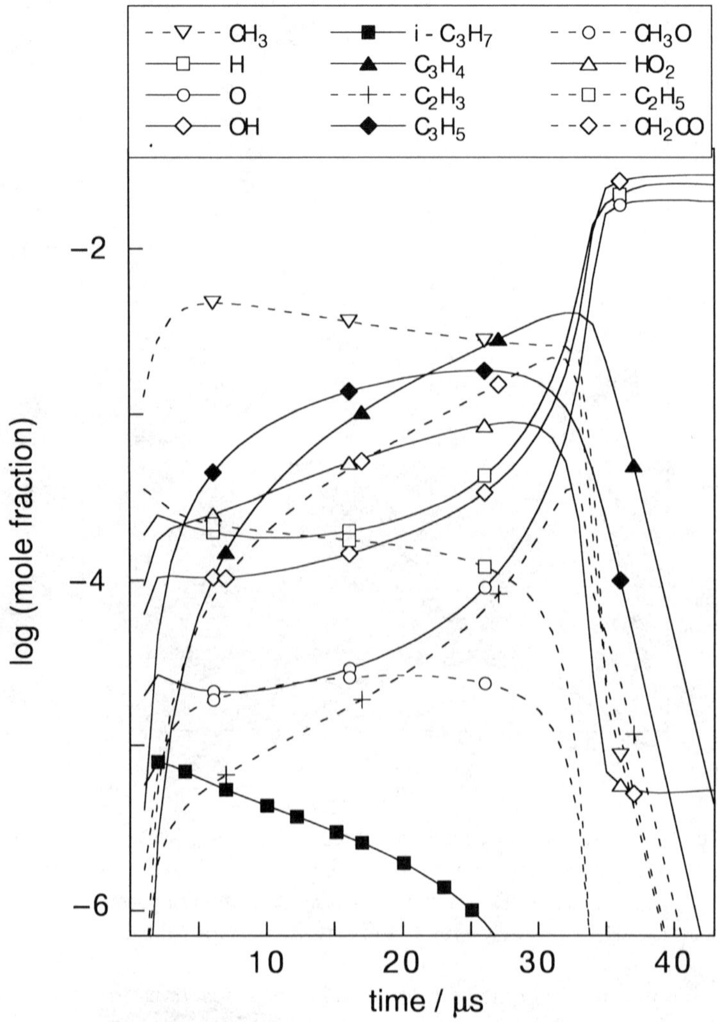

FIGURE 1.18. The intermediate radical and stable species profiles computed for oxidation of 4% C_3H_8 in air starting at 1600 K and 1 atm. Comparison with the previous figure shows that the intermediate products persist until onset of significant temperature rise, after which they all quickly disappear.

At low temperatures, direct reaction of fuel with oxygen is the most important chain initiation step

$$CH_4 + O_2 \longrightarrow CH_3 + HO_2 . \tag{76}$$

Because of its high endothermicity, the chain initiating reaction is not a significant route to radical formation once even a relatively small concentration of radicals is present. The important steps very quickly become radical attack on fuel molecules. In the case of CH_4 combustion, the main ones are

$$OH + CH_4 \longrightarrow CH_3 + H_2O \tag{77}$$
$$O + CH_4 \longrightarrow CH_3 + OH \tag{78}$$
$$H + CH_4 \longrightarrow CH_3 + H_2 \tag{79}$$
$$HO_2 + CH_4 \longrightarrow CH_3 + H_2O_2 . \tag{80}$$

These reactions, together with

$$CH_3 + O_2 \longrightarrow CH_2O + OH \tag{81}$$
$$\longrightarrow CH_3O + O \tag{82}$$
$$CH_3 + O \longrightarrow CH_2O + H \tag{83}$$

generate "radical pools" that control fuel disappearance and CH_2O generation. Reactions (81) and (82) are important under radical-poor conditions, most importantly in ignition, while reaction (83) dominates in radical-rich conditions, that is, in flame fronts. Reaction (82) is also an important chain branching reaction, because methoxy radicals are unstable at flame temperatures and quickly decompose

$$CH_3O + M \longrightarrow CH_2O + H + M . \tag{84}$$

There is a long-standing controversy about the channel distribution of the reaction of CH_3 (methyl) radicals with oxygen (reactions (81) and (82)) (Brabbs and Brokaw, 1975; Baldwin and Golden, 1978; Bhaskaran et al. 1980; Saito et al. 1986; Ewig et al. 1987; Zellner and Ewig 1988; Wu et al. 1990; Braun-Unkhoff et al. 1994; Yu et al. 1995; Tischer et al. 1996). While the rate coefficient of reaction (82) can be considered well-established (Bhaskaran et al. 1980; Brabbs and Brokaw, 1975; Saito et al. 1986; Wu et al. 1990), rate coefficients proposed for reaction (81) scatter by almost one order of magnitude, so as to render it somewhere from negligible (Baldwin and Golden 1978; Bhaskaran et al. 1980; Brabbs and Brokaw 1975; Wu et al. 1990) to dominant (Borisov et al. 1983a; Saito et al. 1986; Ewig et al. 1987; Zellner and Ewig, 1988) at combustion temperatures.

Recombination reactions are very important for high-temperature oxidation of hydrocarbons. At temperatures between 800 and 1200 K, chain termination mainly occurs in the reverse of reaction (72), but at higher temperatures, the role of this reaction decreases because of C_2H_6 decomposition. Reaction (72) is especially important in fuel-rich mixtures, and is the principal path by which CH_3 radicals disappear. In such mixtures there are essentially two parallel fuel oxidation paths: one via the oxidation of CH_3 radicals and the other via oxidation of C_2H_6. Reactions of C_2H_6 lead to the formation of ethyl radical (C_2H_5), ethylene (C_2H_4), and acetylene (C_2H_2), so that the oxidation of the C_2 species must also be considered in CH_4 oxidation. Among others, the important termination reactions are the reverse of reaction (73) and

$$H + O_2 + M \longrightarrow HO_2 + M . \tag{63}$$

The sensitivity of the rate of oxidation to particular elementary reactions depends on the conditions under which oxidation occurs. An example for C_3H_8 ignition is shown in Figure 1.2.

Flame propagation and ignition have similarities and essential differences. One of the main differences is that in self-propagating flames, diffusive transport of radicals before the main reaction zone makes chain initiation irrelevant. As a result of the high temperature in flame fronts, all reactions are faster, and because of the higher radical concentrations, radical-radical reactions are more important than during ignition. Another difference between flames and ignition is the presencse of pre-heat zones in flames but the absense of them ignition. For stable fuels like CH_4, little or no pyrolysis occurs in this zone, but for other hydrocarbons considerable fuel degradation occurs; the fuel fragments surviving this zone comprise mainly smaller hydrocarbons, alkenes, and hydrogen. A major consequence of this is that the reaction zone composition of flames burning larger hydrocarbons varies little with the fuel. For this reason, flame temperatures and burning velocities vary little for such fuels. Sensitivity calculations show that in the main reaction zone the dominant reactions are

$$H + O_2 \longrightarrow OH + O \tag{1}$$

$$OH + CO \longrightarrow CO_2 + H. \tag{67}$$

Reactions (1) and (66) drive a chain center pool for effecting the key flame process of oxidizing CO to CO_2.

Reactions in the post-flame zone, where recombination reactions are important, determine the final product distribution. It is often useful to consider the final composition of the burnt gas as largely determined by reactions equilibrating the water-gas shift reaction

$$CO + H_2O \rightleftharpoons H_2 + CO_2. \tag{85}$$

1.3.4 Combustion of larger hydrocarbons

In nearly every respect, the oxidation of propane (illustrated in Figures 1.17 and 1.18) is a qualitatively correct paradigm for the combustion of all large aliphatic hydrocarbons, including the primary reference fuels for internal combustion engines, n-heptane (C_7H_{16}) and iso-octane (C_8H_{16}) for spark-ignited engines, cetane (n-$C_{16}H_{34}$) for diesel engines, and jet aircraft fuels. The course of reaction begins with thermal decomposition and abstraction of H atoms from the fuel molecules, forming radicals that fragment and react in various ways to give smaller olefins and partially oxidized carbon-containing species similar to CH_2O. A mixture mostly comprising C_2H_4, CH_4, CH_2O, H_2, and CO is the fuel that "actually burns" as the initial fragmentation process finishes converting the original fuel into small stable intermediates and low concentrations of free radicals.

There are, nonetheless, some characteristic differences in the combustion chemistry of hydrocarbon fuels that are practically important, particularly as regards ignition properties and varying tendencies to form partially oxidized products, hydrogen-poor aromatic hydrocarbons, and soot. The reasons why high-octane fuels offer better performance in internal combustion engines, and why unsaturated hydrocarbons like butadiene and benzene produce copious amounts of soot, are rooted in the details of their respective combustion chemistries.

Studying the combustion chemistry of large fuel molecules by means of dynamic modeling is a more ambitious undertaking than studying propane combustion chemistry. This is because larger fuel molecules generate so many different stable and radical intermediates that it is hopeless to attempt to write differential equations for all of their concentrations. While modeling propane combustion adds only about a dozen new composition variables, and fewer than 100 elementary reactions to a model describing ethane combustion (Sloane 1989; Tan et al. 1994; Qin 1998), a comparably complete model for the combustion chemistry of a liquid fuel would require myriads of composition variables and vastly more elementary reactions, none of which would have experimentally determined rate coefficients.

A detailed model need not be complete in the sense of accounting for all isomers and all plausible elementary reactions, however, if its purpose is the more limited one of describing the chemistry faithfully in terms of *representative* pathways of *representative* species in a manner that correctly incorporates the characteristics of real elementary reactions. There are models for liquid fuels that meet this criterion and are both computationally feasible and scientifically productive.

A few examples will convey the scope of such mechanisms. Curran et al. (1996, 1998) modeled the autoignition chemistry of n-heptane and iso-octane using a model with 3445 elementary reactions of 805 species. Their model was an extension of earlier work on n-heptane and iso-octane ignition in which a model with 765 reactions of 212 species was developed (Westbrook et al. 1988). Roberts et al. (1996) did the same using a model with 131 reactions of 103 species, achieving the reduction in size by eliminating a number of pathways not supported by experimental studies of exhaust gas composition and by revising the degenerate branching chain reactions.

While such numbers of reactions can be composed manually, possible reaction lists can also be generated automatically by rule-based examination of the reaction possibilities. Such a procedure was developed inter alia by Chevalier et al. (1992) for implementation not just at the n-heptane level (2400 reactions of 620 species), but also at the cetane(n-$C_{16}H_{34}$) level with 7000 reactions of 1200 species.

Such studies demonstrate not only that detailed dynamic modeling is possible for the combustion chemistry of liquid fuels, but also that it provides insight into the underlying chemical process of ignition. It is not a barrier to the investigation of liquid fuel combustion that species and reactions are conceived, and rate coefficient expressions are parameterized, by following rules based upon experience with reactions of smaller species.

Extensive references to the literature on combustion modeling with very large reaction mechanisms are given in Section 1.9.2.

1.4 NITROGEN, SULFUR, AND HALOGENS IN FLAMES

1.4.1 Oxidation of ammonia and hydrogen cyanide

Chemical reactions involving the element nitrogen have been one of the main focal points of combustion chemistry research, and numerous reaction mechanisms have been developed to model high-temperature processes in which nitrogen compounds participate. These include the combustion of ammonia and its organic derivatives; combustion of coals and oils that contain significant amounts of chemically bound nitrogen; ignition of nitrates, nitrites, and nitrogen-containing organic compounds (as explosives or as additives to promote ignition of other fuels); and the formation and destruction of nitrogen oxides derived from atmospheric nitrogen in the course of combustion. The reactive species and elementary reactions involved in these processes are described in detail in Chapter 2.

Although the chemical basis for most of the nitrogen oxide formation in flames was discovered long ago, by Zeldovich in 1946, predictive high-temperature nitrogen chemistry models have emerged only in the last 20 years. Nowadays, it is possible to model the essential chemistry of all of the above-listed processes, at least qualitatively. This modeling capability opens a door to better understanding of nitrogen chemistry, to more effective combustion of nitrogen compounds, and to development of more effective methods for controlling emissions of nitrogen oxides. A variety of comprehensive models of nitrogen combustion chemistry have been proposed (Hanson and Salimian 1984; Glarborg et al. 1986; Miller and Bowman 1989, 1991; Kilpinen et al. 1992; Glarborg et al. 1998; and Dean and Bozzelli, Chapter 2). In this section, the simulations done to illustrate nitrogen chemistry used the mechanism described in Section 1.9.4.

Ammonia oxidation

Ammonia pyrolysis and oxidation have been studied extensively by modern experimental techniques. Profiles of H and N atoms, NH, OH, and NH_2 radicals, and of nitrogen oxides have been measured by various diagnostic methods. The results of these investigations give a clear picture of the NH_3 oxidation process and serve as a good starting point for describing nitrogen chemistry in flames.

The most important initiation step in NH_3 pyrolysis is found to be

$$NH_3 + M \longrightarrow NH_2 + H + M \,. \tag{86}$$

Secondary pyrolytic reactions include reactions of H and NH with NH_3, recombination reactions of H, NH, and NH_2, and thermal decomposition of NH_2 and N_2H_3. For near-stoichiometric reaction, the main abstraction reaction is

$$NH_3 + OH \longrightarrow NH_2 + H_2O \,, \tag{87}$$

while under fuel-rich and fuel-lean conditions, H-atom abstraction by H and O atoms becomes important:

$$NH_3 + H \longrightarrow NH_2 + H_2 \tag{88}$$
$$NH_3 + O \longrightarrow NH_2 + OH\,. \tag{89}$$

NH radicals and N atoms are formed by corresponding hydrogen abstraction reactions of NH_2 and NH with H and OH. Nitrogen-containing active species can either be oxidized to NO (nitric oxide) by O and O_2, or react with NO to form N_2. These reactions are summarized in the following diagram. Whether the NH_i radicals formed react with NO to form N_2, or are oxidized through several intermediate steps to form NO, depends on the reaction conditions.

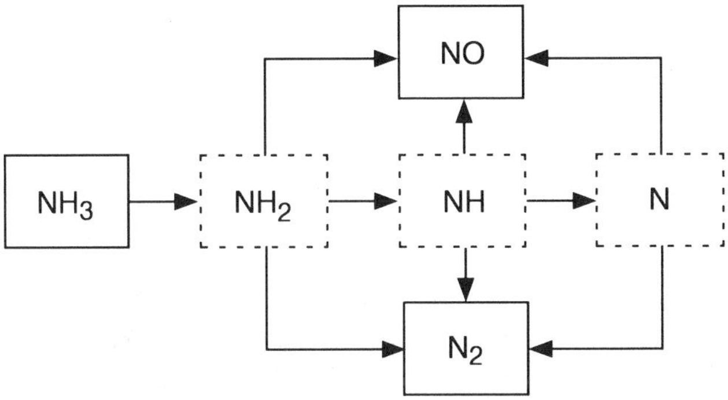

Major species profiles in NH_3 oxidation at 1450 K are shown in Figure 1.19. NH_i radical concentrations under these conditions remain below 1 ppm. The reaction products are entirely N_2 and NO; nitrous oxide (N_2O) formed early in the reaction disappears rapidly at this high temperature, while the concentration of HO_2 is too low to generate significant amounts of nitrogen dioxide (NO_2) by the main reaction known to produce it

$$HO_2 + NO \longrightarrow NO_2 + OH\,. \tag{90}$$

Hydrogen cyanide oxidation

Hydrogen cyanide appears early in the thermal decomposition and oxidation reactions of nitrogen-containing fuel molecules. As described in later sections, it also appears during "Prompt NO" formation and in several emission control technologies. Its most important initial reactions are with atomic oxygen

$$HCN + O \longrightarrow NCO + H \tag{91}$$
$$HCN + O \longrightarrow NH + CO\,. \tag{92}$$

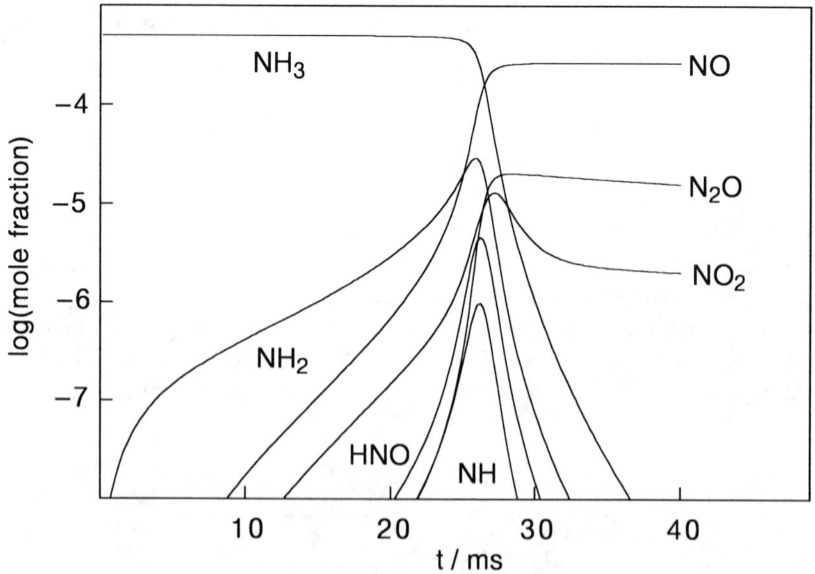

FIGURE 1.19. Concentration profiles in oxidation of 500 ppm NH_3 and 3% O_2 (balance N_2) for adiabatic constant volume reaction starting at 1450 K and 1 atm. About half of the ammonia nitrogen is converted to NO under these conditions.

The NCO and NH formed react mostly with H atoms to produce N atoms that lead either to NO or to N_2, mostly in the reactions

$$N + NO \longrightarrow N_2 + O \tag{93}$$

$$N + OH \longrightarrow NO + H \tag{94}$$

$$N + O_2 \longrightarrow NO + O\,. \tag{95}$$

The interactions between the basic elementary steps of HCN oxidation are shown in the following diagram.

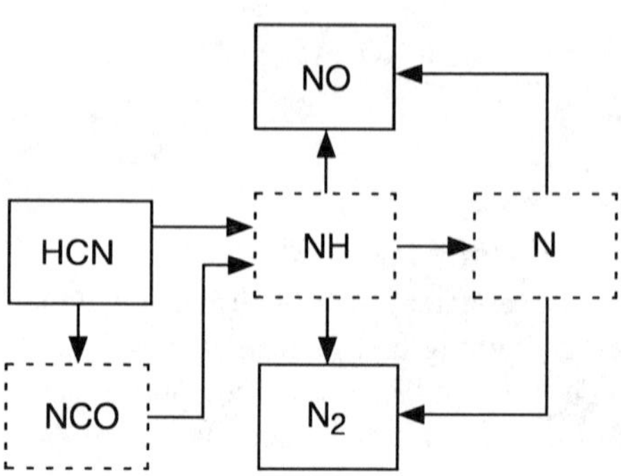

Species profiles illustrating HCN oxidation are shown in Figures 1.20 and 1.21. In contrast to the induction zone characterizing the branched chain explosion

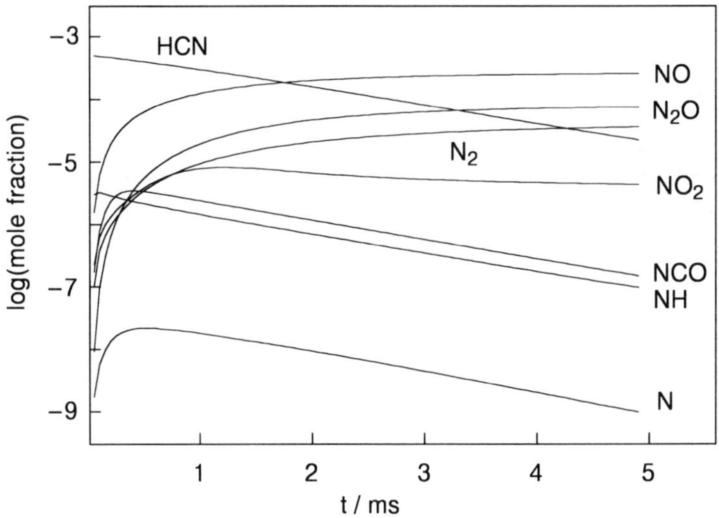

FIGURE 1.20. Initial course of reaction for adiabatic oxidation of 500 ppm HCN by 3% O_2 in Ar at 1 atm constant pressure as computed using the mechanism described in Section 1.9.4. The initial temperature was 1300 K.

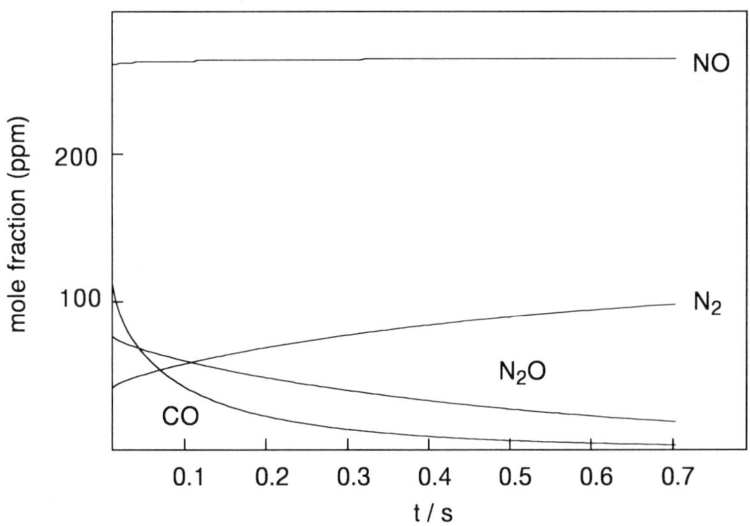

FIGURE 1.21. Later course of reaction for adiabatic oxidation of 500 ppm HCN by 3% O_2 in Ar at 1 atm constant pressure. The initial temperature was 1300 K.

nature of NH_3 oxidation, HCN equilibrates with the active species in a few hundred microseconds and then disappears in a first-order manner on the millisecond time scale to form N_2, NO, N_2O, H_2O, and CO. At 1300 K, N_2O and CO are oxidized within one second, while the NO concentration remains constant in the 200 ppm concentration range. At lower temperatures the N_2O and CO oxidation processes are slower, and these species survive as final products.

1.4.2 Formation and destruction of nitrogen oxides in flames

Nitrogen oxides present in urban atmospheres induce photochemical smog formation. Tropospheric chemical processes, thought to occur by heterogeneous catalysis on aerosol particle surfaces, convert them to the nitric acid component of acid rain. (The N–OH bond of nitric acid (HNO_3) is so weak that direct HNO_3 emission from flames is not an environmental concern.)

The oxides of interest are NO, N_2O, and NO_2. The more common NO and NO_2 are collectively referred to as NO_X. The more stable NO almost always predominates over the others in combustion flue gas. Nitrogen oxides are formed in combustion processes both from nitrogen in the combustion air and from nitrogen bound in organic compounds present in fossil fuels, mainly coals and heavy oils. Were NO_X emissions governed by thermodynamic equilibrium constraints, the concentrations of nitrogen oxides at exhaust temperatures would be less than 1 ppm. As the temperature of combustion products decreases, NO_X concentrations do start to decrease, but at flame temperatures the time available for NO_X destruction in ordinary combustion equipment is too short to reach equilibrium, and the nitrogen oxides are quenched at concentrations from tens to thousands of ppm depending on the combustion conditions. Thus, the chemical processes of NO_X formation and destruction are controlled by chemical kinetics, not by thermodynamics.

NO and NO_2 concentrations are coupled to one another by fast radical reactions: NO_2 reacts with O, H, and OH to form NO, and NO reacts with HO_2 to form NO_2. As discussed in detail in Chapter 2, while the reactions of nitrogen oxides with inorganic species like these are fairly well understood, there are significant gaps in our knowledge of the reactions of NO with carbon-containing species, an area of intensive research interest. In the following paragraphs, the main chemical processes of NO_X formation under combustion conditions and several high-temperature methods for NO_X control in combustion and post-combustion zones are briefly reviewed. The details are discussed in Chapter 2.

There are several NO formation routes in flames. The most important ones are Thermal NO, Prompt NO, Fuel NO, and NO formed from N_2O.

Thermal NO

Thermal NO comes from fixation of molecular nitrogen in the combustion air. The concentration of O atoms in flames grows with temperature due to dissociation processes, and if the temperature is high enough, O atoms attack N–N chemical bonds in the direct high activation energy process

$$O + N_2 \longrightarrow NO + N \,. \qquad (-93)$$

The N atoms formed react quickly with O_2 to regenerate O atoms by

$$N + O_2 \longrightarrow NO + O \,. \qquad (95)$$

This pair of reactions is known as the Zeldovich mechanism (Zeldovich 1946). The concentration of NO is further increased by a radical-radical reaction

$$N + OH \longrightarrow NO + H \tag{94}$$

followed by fast return of OH

$$H + O_2 \longrightarrow OH + O\,. \tag{1}$$

The rate of NO formation in these reactions is very sensitive to temperature and radical concentrations. Thus, Thermal NO is formed mostly in the flame regions with the highest temperatures. Measured Thermal NO levels can be higher than calculated ones if radical concentrations exceed their equilibrium values, which often happens in flame fronts. Because oxygen participates in Thermal NO formation, more Thermal NO is formed under fuel-lean conditions. When the equivalence ratio is not close to stoichiometric, the maximum flame temperature drops and Thermal NO formation is suppressed.

The main strategies to avoid high Thermal NO levels are reduction of the peak combustion temperature, reduction of the residence time, and reduction of O_2 concentrations at the peak combustion temperature.

Figure 1.22 shows species profiles that illustrate Thermal NO chemistry under conditions where significant Thermal NO formation occurs, a temperature of 2400 K in a stoichiometric CH_4-air mixture. It can be seen that at the ignition time the NO concentration is small, between 1 and 10 ppm, but it subsequently grows by nearly 3 orders of magnitude.

Prompt NO

NO formation in fuel-rich, lower temperature hydrocarbon flames is greater than one would expect from Thermal NO production. It is attributed to reaction of one or more hydrocarbon radicals with N_2. Because it appears early in the flame, this NO was called "Prompt NO" by Fenimore (1971). It is generally thought that the most important reaction is that of CH radicals with N_2

$$CH + N_2 \longrightarrow HCN + N\,. \tag{96}$$

The CH radicals themselves are formed in fuel-rich zones by serial degradation of other carbon-containing radicals, that is, $CH_3 \rightarrow CH_2 \rightarrow CH$, primarily by reactions with H and OH

$$H + CH_3 \longrightarrow H_2 + CH_2 \tag{97}$$
$$OH + CH_3 \longrightarrow H_2O + CH_2 \tag{98}$$
$$H + CH_2 \longrightarrow H_2 + CH \tag{99}$$
$$OH + CH_2 \longrightarrow H_2O + CH\,. \tag{100}$$

If sufficient O_2 is present, the N atoms formed in reaction (96) react with it to form NO in reaction (95). The HCN also formed in the CH + N_2 reaction can form still more NO, as described in Section 1.4.1.

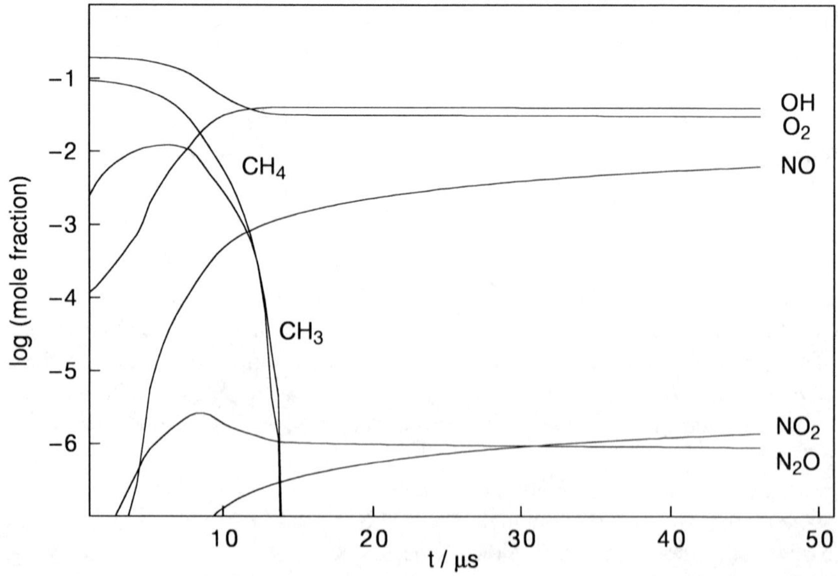

FIGURE 1.22. Typical species profiles for conditions favoring formation of high levels of Thermal NO_X. Adiabatic reaction at constant volume in a stoichiometric CH_4–air mixture initially at 2400 K was modelled using the mechanism described in Section 1.9.4. In a flame, expansion of the hot gas suppresses NO formation below the levels shown here. In agreement with experimental observation, only very low levels of NO_2 and N_2O formation are computed.

Suppression of Prompt NO formation can best be accomplished by ensuring rapid fuel–air mixing.

Major species concentration profiles during ignition of a $\phi = 1.2$ CH_4–air mixture at 1800 K and 1 atm are shown in Figure 1.23. While Prompt NO formation is significant under these conditions, NO generated by this mechanism remains at lower concentrations than NO formed at higher temperatures by the Zeldovich reactions.

Fuel NO

Coals and heavy oils contain from less than 0.1% to about 2.5% nitrogen bound in organic molecules. They include NH_3 derivatives, where one or more H atoms are replaced by organic radicals, and compounds with one or more N atoms in one or more aromatic rings, for instance pyridine (C_5H_5N).

The rate of Fuel NO formation depends strongly on the rate of fuel-air mixing. Usually Fuel NO is only a small fraction of the available fuel nitrogen. Fuel NO levels increase when the nitrogen content in the fuel grows, but this dependence is not proportional and is strongly affected by combustion conditions, that is, by local stoichiometry and temperature. The main reaction path for destruction of fuel nitrogen and for formation of Fuel NO is shown in the following diagram.

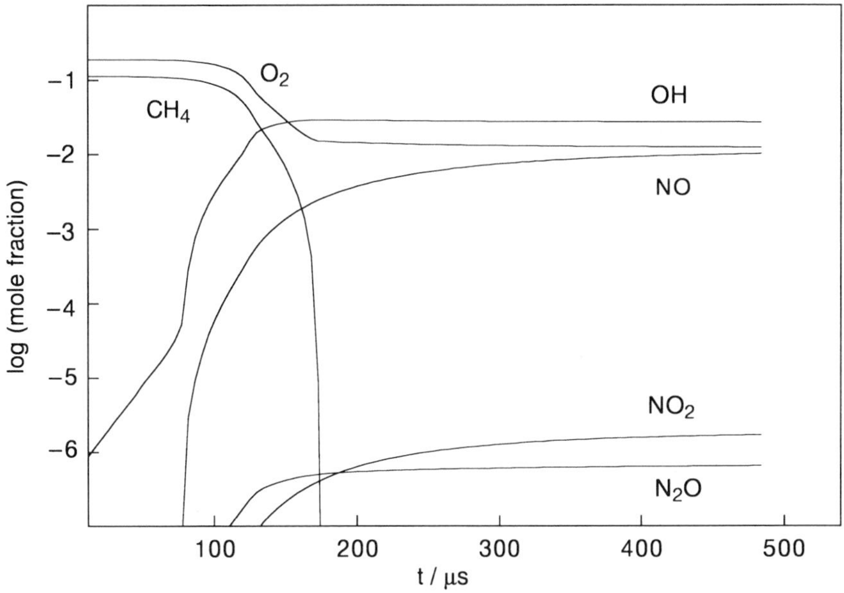

FIGURE 1.23. Species profiles for conditions with significant Prompt NO_X formation. The computation was done for adiabatic reaction at 1 atm in a $\phi = 1.2$ mixture of CH_4 in air using the reaction mechanism described in Section 1.9.4. The starting temperature was 1800 K. The main reason for the chemistry being slower here than in the previous figure is the 600 K lower starting temperature; the additional CH_4 in the mixture also has a small retarding effect.

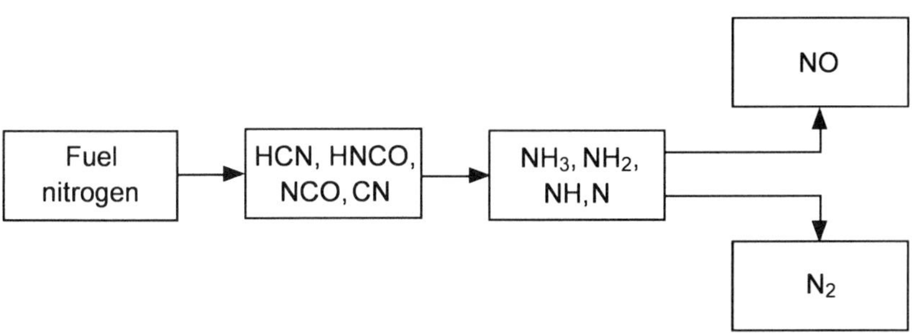

The amount of Fuel NO formed is essentially independent of the chemical nature of the fuel nitrogen source, because all nitrogen compounds are rapidly converted to HCN, NH_3, and their fragments in flames. Oxidation of these species to NO is the rate-limiting factor.

In order to reduce the fraction of the fuel nitrogen converted to NO, one can use the strategy of staged combustion, where the fuel is primarily oxidized under fuel-rich conditions. Fuel nitrogen released in a reducing atmosphere tends to be converted to molecular nitrogen rather than to be oxidized.

NO formation from N_2O intermediate

Nitrous oxide formed in a combustion zone (see below) is partially converted to NO. Two reactions contribute, those of O and H atoms with N_2O,

$$O + N_2O \longrightarrow NO + NO \tag{101}$$

and

$$H + N_2O \longrightarrow NH + NO\,. \tag{102}$$

The first reaction has a second channel with about the same rate coefficient, forming N_2 and O_2. The reaction of H atoms with N_2O also has a second channel, into N_2 and OH, which is about two orders of magnitude faster than NO formation. (See Sections 2.5.4 and 2.6.3–2.6.6 of Chapter 2.)

NO formed from N_2O can be a significant component of the NO_X under fuel-lean conditions at low flame temperatures.

Nitrogen dioxide emissions

Concentrations of NO_2 in combustion flue gas are typically less than 5% of total NO_X levels. Nevertheless, it is important to understand the chemical reactions of NO_2 formation and destruction because these reactions can define NO_X control strategies. There are several reactions that convert NO_2 to NO, and under some conditions a significant amount of NO can be converted to NO_2. The termolecular reaction

$$2\,NO + O_2 \longrightarrow 2\,NO_2 \tag{103}$$

is too slow to be significant at typical flame NO concentrations within the residence times typical of industrial combustion and post-combustion equipment. The fastest flame reaction forming NO_2 is

$$HO_2 + NO \longrightarrow NO_2 + OH\,. \tag{90}$$

This reaction is important whenever the HO_2 concentration is significant, typically in the temperature range from 600 to 1000 K under fuel-lean conditions. At higher temperatures HO_2 dissociates quickly into H atoms and O_2, and the higher prevailing concentrations of H, O, and OH lead to more rapid NO_2 loss by

$$O + NO_2 \longrightarrow NO + O_2 \tag{104}$$

$$H + NO_2 \longrightarrow NO + OH \tag{105}$$

$$OH + NO_2 \longrightarrow NO + HO_2\,. \tag{−90}$$

Nitrous oxide emissions

At low combustion temperatures and high pressures, nitrous oxide formed by the termolecular reaction

$$O + N_2 + M \longrightarrow N_2O + M \tag{106}$$

can appear in emissions. Nitrous oxide is of importance for combustion chemistry not only as an air pollutant that can be formed during combustion, but also because N_2O is a monofuel, that is, it can be ignited without additives. The initiation reaction of the ignition is the reverse reaction of the N_2O formation process. The oxygen atoms produced thereby react exothermically with N_2O

$$O + N_2O \longrightarrow NO + NO \tag{101}$$

$$O + N_2O \longrightarrow N_2 + O_2, \tag{107}$$

releasing more energy than is lost in the N_2O decomposition step.

In the presence of hydrocarbon fuel molecules, the reactions of N_2O with O atoms are not important. This is because O atoms attack the fuel molecules more rapidly to form other active species which react with N_2O, the reaction with H atoms

$$H + N_2O \longrightarrow N_2 + OH \tag{108}$$

being the fastest. The reactions of N_2O with CH_3 (and other hydrocarbon radicals) and with OH

$$N_2O + OH \longrightarrow N_2 + HO_2 \tag{109}$$

are found to be much slower.

Combustion of nitrogen-containing fuels has also been shown to lead to measurable levels of N_2O in flue gas. Various nitrogen-containing radicals contribute to this process, for example

$$NH + NO \longrightarrow N_2O + H \tag{-102}$$

$$NCO + NO \longrightarrow N_2O + CO. \tag{110}$$

1.4.3 Chemistry of NOx control methods

Combustion modification

For natural gas and other low-nitrogen fuels, the Thermal NO mechanism is the main contributor to NOx formation. Where this is the case, temperature control is the most important factor in combustion modification methods for when the total excess combustion air is close to the value that is optimal for engineering considerations, typically near 10%. For fuels with high nitrogen content, Fuel NO can be comparably important, in which case NOx emissions are most effectively minimized by reducing the O_2 concentration in the main combustion zone.

Because the rate of NOx formation from Thermal NO and Fuel NO depends strongly on operating conditions, controlling them appropriately has significant impact on NOx emissions. Combustion modification methods for NOx control are cost effective and can reduce NOx levels typically by 40%, or in some circumstances even to 60–80%. Thus, in most situations the main operating strategies for reducing NOx emissions are minimizing the excess air supply, reducing the maximum combustion temperature, and staging the combustion process.

Numerous technologies have been put into practice for reducing available air in high-temperature zones, lowering peak flame temperatures, reducing residence times in high-temperature zones, providing flue gas recirculation, injecting water or steam, fuel staging, air staging, reducing air preheat, and automating temperature and excess air control. While NO_X reduction by means of combustion modification is an engineering task, it has to be accomplished by applying chemical understanding of the mechanism of NO_X formation and destruction.

Combustion modification technologies for NO_X control are continually being refined as experience with them is gained. It has become clear, however, that combustion modifications alone can not achieve the levels of NO_X suppression that are needed to meet current societal demands for cleaner air in urban areas. For this reason, post-combustion technologies for treatment of combustion products ("flue" or "exhaust" gas) have been put into practice and new options are under active development. As will be seen in the following paragraphs, these technologies are even more strongly dependent on detailed understanding of high-temperature nitrogen chemistry.

Reburning

Reburning, a variant of the fuel staging method for NO_X control suggested by Wendt et al. in 1973, exploits the sequence of combustion stages shown in the following diagram.

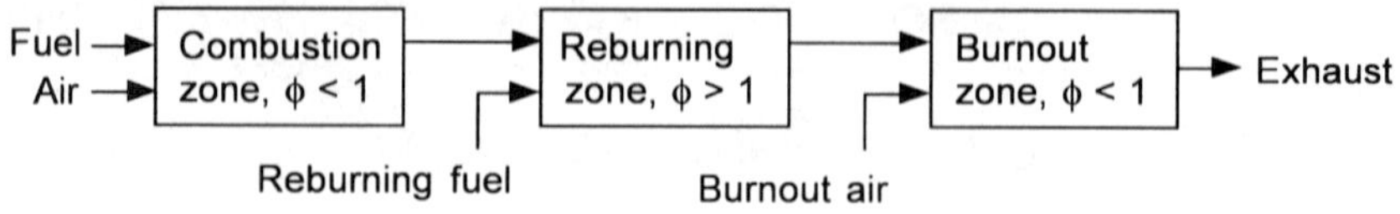

In the main combustion zone, 80–90% of the fuel is burned, normally with about 10% excess air, forming the corresponding amount of NO_X. In the second stage, more fuel (the "reburning fuel") is injected into a secondary combustion zone, the so-called reburning zone, at temperatures from 1400 to 1700 K, to maintain a fuel-rich environment. In this reducing atmosphere, both NO_X forming and NO_X removing reactions occur. The experimental results indicate that over a specific range of reburning conditions (equivalence ratio, temperature, and residence time in the reburning zone) NO_X concentrations can be reduced typically by 50–70%, or even up to 90% if reburning is used together with injection of other additives (Seeker 1990; Zamansky et al. 1996). In the third stage, "burnout air" is injected at a lower temperature to complete the combustion. Any type of fuel can be used in the first and second combustion stages, but natural gas is found to be the most effective reburning fuel.

Studies of the chemistry of fuel-rich combustion as it relates to reburning have been reported in many publications, starting from that of Meyerson et al. in 1957 to the comprehensive study by Kilpinen et al. in 1992. While many questions of detail persist (Dean and Bozzelli, Chapter 2), there is basic agreement about the main features. The gas mixture leaving the main combustion zone contains N_2, the main combustion products CO_2 and H_2O, excess O_2, the NO formed, and low concentrations of active species not far from equilibrium levels. Addition of reburning fuel leads to rapid oxidation of some of it to CO and H_2. The reburning fuel provides a reducing environment with effective concentrations of carbon-containing radicals, including CH_3, CH_2, CH, C, HCCO, and others in smaller amounts. These active species can participate both in formation of NO precursors, by reactions with N_2, and in reactions with NO. There are so many elementary steps which can share responsibility for NO reduction that there is so far no commonly accepted opinion about their relative importance. For example, five competing reactions of CH + NO are included in the mechanism presented in Table 2.19 of Chapter 2, the fastest of them being

$$\mathrm{CH} + \mathrm{NO} \longrightarrow \mathrm{H} + \mathrm{NCO} \qquad (111)$$

$$\longrightarrow \mathrm{N} + \mathrm{HCO}\,. \qquad (112)$$

Miller and Bowman (1989) suggest otherwise, that the main reaction paths converting NO to N_2 include NO reactions with C and CH,

$$\mathrm{C} + \mathrm{NO} \longrightarrow \mathrm{CN} + \mathrm{O} \qquad (113)$$

$$\mathrm{CH} + \mathrm{NO} \longrightarrow \mathrm{HCN} + \mathrm{O}\,, \qquad (114)$$

the CN and HCN products subsequently being oxidized to NCO, while Kilpinen et al. (1992) suggest that the most important reactions are those of HCCO, singlet CH_2, and CH_3 with NO. Stapf and Leuckel (1996) and Staph (1998) interpreted their experimental results from a laboratory-scale reburning experiment in terms of HCCO attack. As discussed in Chapter 2, two reaction channels are important:

$$\mathrm{HCCO} + \mathrm{NO} \longrightarrow \mathrm{HCNO} + \mathrm{CO} \qquad (115)$$

$$\longrightarrow \mathrm{HCN} + \mathrm{CO_2}\,. \qquad (116)$$

A schematic diagram of the main reburning zone reactions is shown in the following diagram.

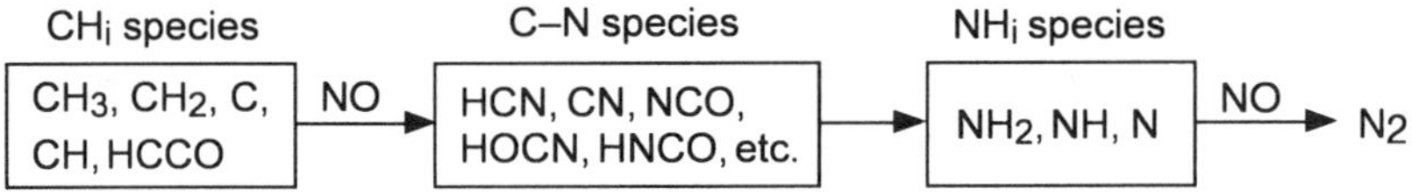

Carbon-containing radicals (CH_i) formed in the reburning zone reduce the NO concentration by converting it to various intermediate species with C–N bonds. These species are reduced in reactions with various radicals to NH_i species that react with NO to form N_2. There are thus two types of chemical reactions of NO conversion, those with CH_i and those with NH_i radicals.

If a fuel is burned in the primary combustion zone at $\phi = 0.9$, the combustion products entering the reburning zone contain no fuel and about 2% O_2. Addition

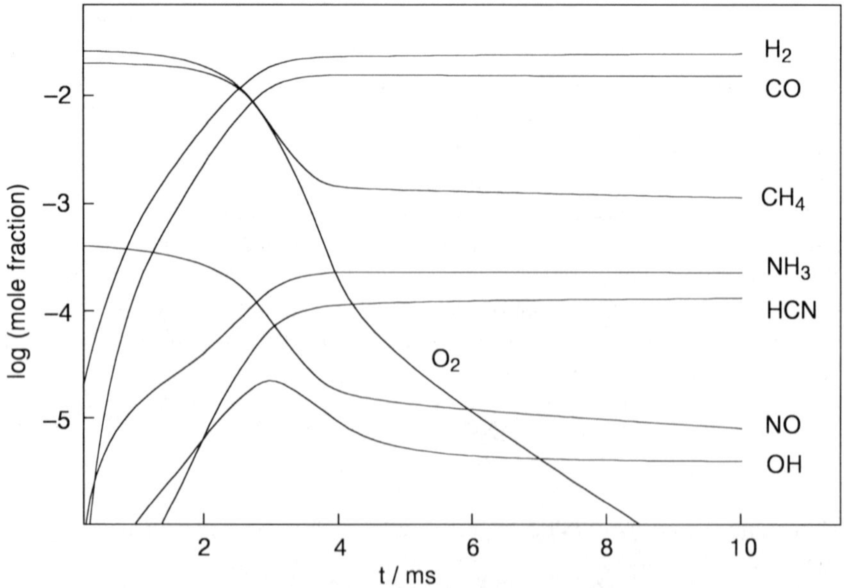

FIGURE 1.24. Simulation of major species profiles for conditions typical in a reburning zone. The simulation was done using the mechanism described in Section 1.9.4 for adiabatic reaction at 1 atm ensuing upon addition of 2.6% CH_4 to a mixture of 2% O_2, 8% CO_2, 16% H_2O, and 400 ppm NO initially at 1650 K.

of about 2.6% CH_4 into the reburning zone will establish fuel-rich conditions. Figure 1.24 shows major species profiles under such typical reburning zone conditions. Here 2.6% CH_4 is mixed with 2% O_2 in the presence of 400 ppm NO, 8% CO_2, and 16% H_2O, amounts typical of practical flue gas environments. The CH_4–O_2 reactions form CO and H_2 and some HCN and NH_3. At the same time, the NO concentration decreases.

Combination of reburning with injection of nitrogen-containing compounds, as described in the following paragraphs, affords NO_X control strategies that achieve up to 90% NO_X reduction (Chen et al. 1991; Pont et al. 1993; Zamansky et al. 1996, 1998). These advanced reburning methods include injection of N-agents (ammonia or urea) into the reburning zone or with the overfire air. For effective reduction of NO with N-agents, it is necessary to provide near-stoichiometric conditions in the zone, which requires a lower amount of the reburning fuel, typically about 10%. At 10% reburning fuel the reburning agents can increase the NO reduction efficiency from about 50% to 70–80%. Further addition of alkali promoters (cf. Section 1.7) increases NO reduction efficiency to 80–90%.

Post-combustion NO_X control by selective non-catalytic reduction methods

In situations where combustion modification does not reduce NO_X emissions to an acceptable level, post-combustion "selective reduction" methods can be used. There are two groups of such methods, called the selective catalytic (SCR) and selective non-catalytic (SNCR) reduction methods. In the context of NO_X control, the SCR acronym is understood to refer to the use of catalysis on various solid surfaces, while SNCR refers to gas-phase reactions. They achieve NO_X reduction by injection of "SNCR agents", nitrogen compounds that generate active species capable of reacting with NO in flue gas. The most effective SNCR agents are NH_3, urea (H_2NCOH_2), and cyanuric acid, $(HOCN)_3$. The corresponding control technologies are called Thermal DeNO_X, NO_XOUT, and RapreNO_X. Typical NO_X reduction by SNCR agents ranges from 60–80% in small-scale experiments and from 30–50% in industrial applications. Further reduction by means of alkali promoters is discussed in Section 1.7.

Thermal DeNOx

The Thermal DeNO_X process, invented by Lyon (1975), has been described in detail by Lyon and Hardy (1986). Injection of ammonia into combustion flue gas containing NO and oxygen at temperatures between 1100 and 1400 K, which has become known as the "temperature window" for the process, leads to chemical reactions that promote conversion of NO to N_2. The reaction starts with formation of NH_2, mostly in reaction with OH

$$NH_3 + OH \longrightarrow NH_2 + H_2O \tag{87}$$

and to a lesser extent in reactions with H and O

$$NH_3 + H \longrightarrow NH_2 + H_2 \tag{88}$$

$$NH_3 + O \longrightarrow NH_2 + OH\,. \tag{89}$$

The main elementary reaction causing NO to N_2 conversion is

$$NH_2 + NO \longrightarrow N_2 + H_2O\,. \tag{117}$$

There is extensive discussion in the literature concerning the rate of the second pathway for the NH_2 + NO reaction, which forms NNH and OH; see Section 2.6.14 of Chapter 2 for a detailed review.

In the presence of additives that accelerate radical formation, such as H_2, CO, hydrocarbons, or H_2O_2, the window is shifted to lower temperatures. Figures 2.69 and 2.70 show calculated and experimental NO removal windows for injection of NH_3 and NH_3 + H_2O mixtures. The temperature windows are controlled by the availability of active species (OH, H, and O) to form NH_2 in reactions with NH_3. At low temperatures, the concentrations of active species are low and NH_2 is not formed. In the presence of H_2 or other fuels, more active species are formed through the H_2–O_2 reactions at lower temperatures and the window is shifted. At high temperatures, the NH_2 radicals are oxidized by OH, O, and O_2 into HNO radicals, which can either dissociate or react with other radicals to form NO. Thus, the main reaction path of the Thermal DeNO_X process is formation of NH_2 from NH_3; it can be converted either to N_2 in reaction with NO or into NO by high-temperature oxidation.

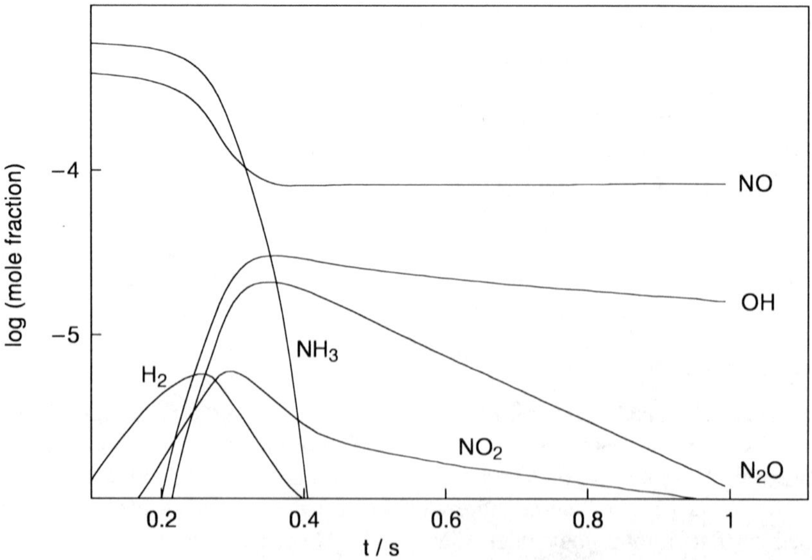

FIGURE 1.25. Species profiles for typical Thermal DeNOx conditions. The simulation was done for adiabatic constant pressure reaction at 1 atm and initial conditions of 3% O_2, 8% CO_2, 15% H_2O, 400 ppm NO, and 600 ppm NH_3 in N_2, at an initial temperature of 1300 K using the mechanism described in Section 1.9.4.

Figure 1.25 shows major species profiles for typical conditions of the Thermal DeNOx process. For the conditions shown, the added ammonia is seen to be completely consumed in the chain reactions that generate the NO-reducing species. While, under well-controlled conditions, NO_X reduction by NH_3 injection may range from 60–70%, the finite times required to mix ammonia with flue gas and the inevitable temperature gradients present in industrial boiler applications limit the practical applicability of the process. Thus the shortcomings of the Thermal DeNOx process, which are shared by the other SNCR methods, are the characteristic narrow temperature window and, especially when mixing is incomplete, possible NH_3 breakthrough.

The RapreNOx process

Another effective SNCR agent is cyanuric acid, $(HOCN)_3$, suggested by Perry in 1988. The experimental observations (Caton and Siebers, 1988) have been subjected to modeling study by Lyon and Cole (1990) and by Miller and Bowman (1991). Upon injection into a hot flue gas stream, cyanuric acid decomposes on surfaces or in the gas phase into isocyanic acid

$$(HOCN)_3 \longrightarrow 3\,HNCO \tag{118}$$

and decomposition and radical reactions of HNCO lead to NH_2 and NCO radicals capable of reacting with NO. Along with NO removal, another nitrogen oxide, N_2O, is formed. The temperature window of the RapreNOx process extends from 1000 to 1350 K. At temperatures below 1000 K, NO concentrations are also reduced by cyanuric acid injection due to heterogeneous catalytic reactions.

The most important homogeneous HNCO reactions with radicals are

$$\mathrm{HNCO + H \longrightarrow NH_2 + CO} \tag{119}$$

and

$$\mathrm{HNCO + OH \longrightarrow NCO + H_2O}\,. \tag{120}$$

As in the Thermal DeNOx process, NH_2 can either remove NO

$$\mathrm{NH_2 + NO \longrightarrow N_2 + H_2O} \tag{117}$$

or form it via HNO. NCO can react with NO to form N_2O

$$\mathrm{NCO + NO \longrightarrow N_2O + CO} \tag{121}$$

and CO and N_2O are oxidized by OH and H

$$\mathrm{OH + CO \longrightarrow CO_2 + H} \tag{67}$$

$$\mathrm{H + N_2O \longrightarrow N_2 + OH}\,. \tag{108}$$

Species profiles describing NO reduction by HNCO injection into flue gas at 1200 K are presented in Figure 1.26. 1800 ppm HNCO corresponds to 600 ppm cyanuric acid, $(HOCN)_3$. It is of interest that HNCO does not react with NO in absence of CO, as is also the case for simulations with the Section 1.9.4 mechanism. Oxidation of CO in the presence of water provides the active species (H and OH) which drive the reactions with HNCO. Numerous differences in predicted species concentrations appear when RapreNOx profiles are computed with different nitrogen chemistry mechanisms.

The NOxOUT process

A final SNCR agent is urea, $(NH_2)_2CO$, suggested by Arand et al. in 1982. It has become known as the NOxOUT process. The chemistry that ensues upon urea injection follows the main features of the NH_3 and HNCO reactions with NO, because urea is rapidly converted to NH_3 and HNCO at high temperatures (Caton and Siebers, 1988)

$$\mathrm{(NH_2)_2CO \longrightarrow NH_3 + HNCO}\,. \tag{122}$$

The NOxOUT process has a temperature window comparable to the other SNCR processes, and like cyanuric acid injection can be complicated by N_2O formation. For comparison with NH_3 and HNCO injection, Figure 1.27 shows profiles of NH_3, HNCO, H_2, OH, NO, N_2O, and NO_2 for a similar mixture composition.

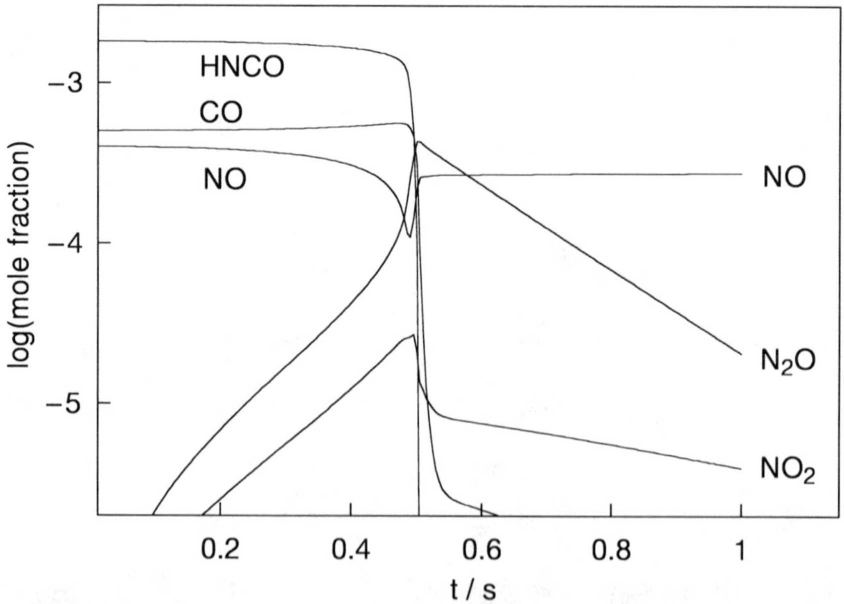

FIGURE 1.26. Simulation for conditions typical of the RapreNOx process for NOx reduction. Adiabatic reaction at 1 atm pressure was modeled using the mechanism described in Section 1.9.4 for a mixture assumed to contain 3% O_2, 8% CO_2, 15% H_2O, 500 ppm CO, 400 ppm NO, and 1800 ppm HNCO at the start of reaction. At the high starting temperature assumed for the simulation, 1200 K, cyanuric acid dissociates very rapidly to form HNCO. In this simulation, higher levels of N_2O are predicted to be formed than in the NH_3 example shown in Figure 1.25.

Oxidation of NO to NO_2

The above-described SNCR agents convert NO to N_2. In another NOx control strategy, concurrent NO-to-NO_2 and SO_3-to-SO_2 conversions are followed by simultaneous scrubbing of NO_2 and SO_2 from the flue gas. Because flue gas desulfurization systems are required for SO_2 removal after combustion of sulfur-containing fuels anyway, and since it has been shown that NO_2 can be removed efficiently by SO_2 scrubbers (Zamansky et al. 1993), conversion of the relatively inert compound NO into the much more reactive one NO_2, and the very corrosive SO_3 into SO_2, becomes an indirect method for combined NOx and SOx control.

An economical method for NO-to-NO_2 and SO_3-to-SO_2 conversion by chain reactions induced by injection of methanol was suggested by Jones in 1988 and Lyon in 1989. In Lyon's experiments, about 90% NO-to-NO_2 and SO_3-to-SO_2 conversion was achieved by injection of methanol (CH_3OH) into simulated flue gas. Unfortunately, there is a serious problem in using CH_3OH, because significant levels of CO by-product are formed. Essentially every molecule of NO or SO_3 converted to NO_2 or SO_2 produces one molecule of CO, which is not oxidized sufficiently rapidly to CO_2 at the relatively low CH_3OH injection temperatures that have to be used (800 to 1100 K).

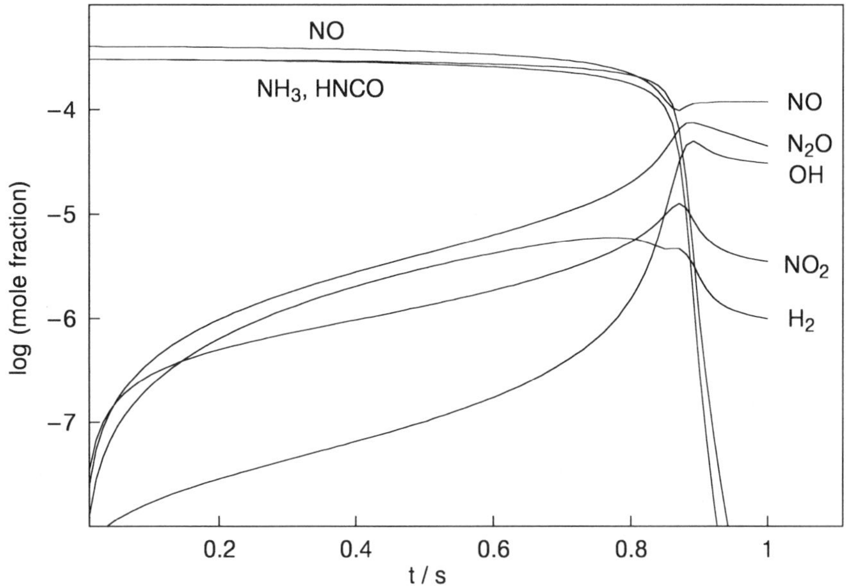

FIGURE 1.27. Simulation for conditions used in the NOxOUT process for reduction of NOx emissions. Adiabatic reaction at 1 atm pressure was modeled using the mechanism described in Section 1.9.4 for a mixture containing 3% O_2, 8% CO_2, 15% H_2O, 400 ppm NO, 300 ppm HNCO, and 300 ppm NH_3 in N_2.

Methanol reacts with OH to form CH_2OH, which is readily oxidized

$$CH_3OH + OH \longrightarrow CH_2OH + H_2O \tag{123}$$

$$CH_2OH + O_2 \longrightarrow CH_2O + HO_2 . \tag{124}$$

At these temperatures, CH_2O is rapidly converted to HCO,

$$CH_2O + OH(H, O) \longrightarrow HCO + H_2O\ (H_2, OH) , \tag{125}$$

which in turn rapidly dissociates to CO and H. The active species that converts NO to NO_2 and SO_3 to SO_2 in the methanol injection process is HO_2:

$$NO + HO_2 \longrightarrow NO_2 + OH \tag{90}$$

$$SO_3 + HO_2 \longrightarrow HSO_3 + O_2 \tag{126}$$

$$HSO_3 + M \longrightarrow SO_2 + OH + M . \tag{127}$$

The OH radicals formed can react with CH_3OH in reaction (98) to continue the chain. Figure 1.28 illustrates the process.

Zamansky et al. (1995) found that H_2O_2 and H_2O_2–CH_3OH mixtures can also oxidize NO to NO_2 and reduce SO_3 to SO_2 via similar mechanisms. Hydrogen peroxide is an effective source of HO_2 radicals through the sequence

$$H_2O_2 + M \longrightarrow OH + OH + M \tag{128}$$

$$OH + H_2O_2 \longrightarrow H_2O + HO_2 . \tag{129}$$

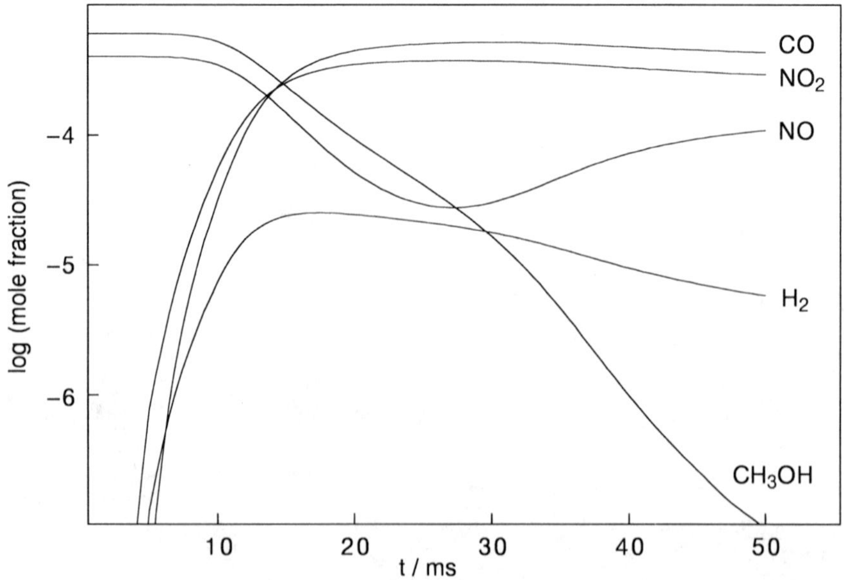

FIGURE 1.28. Species profiles for conditions illustrating the methanol injection process for NO_X control. Adiabatic reaction at 1 atm pressure was modeled using the mechanism described in Section 1.9.4 for a mixture containing 3% O_2, 8% CO_2, 15% H_2O, 400 ppm NO, and 600 ppm CH_3OH in N_2 starting at 950 K.

The HO_2 radical reacts quickly with NO in reaction (90) to give a chain process with the stoichiometry

$$H_2O_2 + NO \longrightarrow H_2O + NO_2\,. \tag{130}$$

As with CH_3OH injection, reactions (126) and (127) convert SO_3 to SO_2. Because H_2O_2 supplies HO_2 radical at lower temperatures than CH_3OH does, the temperature at which the maximum conversion of NO to NO_2 and SO_3 to SO_2 occurs is lowered to about 800 K. The main advantage of using H_2O_2 rather than CH_3OH for this combined process is that no CO is formed. Moreover, the increased concentrations of OH provided by H_2O_2 injection promote removal of CO and CH_4, as well as any other hydrocarbons present, by reactions (67) and (77). Zamansky et al. found that H_2O_2–CH_3OH mixtures are effective around 800 K and are more effective than either one alone.

Injecting other hydrocarbons to form HO_2 was found to be less efficient than addition of CH_3OH and H_2O_2: much higher hydrocarbon concentrations are required, and significant amounts of CO are produced.

Another low-temperature method for NO oxidation, developed by Chang and Lee (1992), involves addition of yellow phosphorus into wet flue gas scrubbing systems. Ozone is formed, which attacks NO in

$$O_3 + NO \longrightarrow O_2 + NO_2\,. \tag{131}$$

1.4.4 Sulfur

Sulfur compounds introduced into the environment from fossil fuel combustion have long been recognized as major pollutants. They lead to health risks from pollution of urban atmospheres, erosion of structural and ornamental stonework, and corrosion of industrial equipment. In recent times, the enforced use of coals and oils with low sulfur content has diminished the negative environmental influences of fuel sulfur. Because of the decreasing availability of such fuels and the increasing stringency of air quality standards, however, there is now renewed concern about sulfur emissions. Another aspect of fuel sulfur is that some sulfur compounds in petroleum-derived fuels promote engine knock (Cullis et al. 1986). Still another is that sulfur compounds deactivate the noble metal catalysts used for control of automobile exhaust emissions. Dealing with these problems in the future will require a deeper understanding than we now have of the combustion chemistry of sulfur-containing compounds (described in detail in Chapter 3).

The main sulfur-containing pollutants emitted from combustion sytems are SO_2 (up to several thousand ppm) and much smaller amounts of SO_3 in fuel-lean systems; the two compounds together are called the SO_X emission. SO_2 is always the major product; under fuel-lean conditions SO_3 accounts for at most 2–3% of the total sulfur. Over wide ranges of initial temperatures, the concentration of SO_2 in combustion products is close to the equilibrium value, while the SO_3 concentration, because of fast reactions in the main reaction zone, can be higher than the equilibrium value. Under fuel-rich conditions in hydrocarbon flames, the situation becomes more complicated and H_2S, COS, CS_2, and C_4H_4S (thiophene) can account for significant fractions of the total sulfur. Combination of SO_3 and water leads to sulfuric acid (H_2SO_4), the main source of the erosive and corrosive effects of atmospheric sulfur, and acid rain. As noted above for nitric acid, combustion itself does not form appreciable amounts of any oxyacids of sulfur; these emerge as products of subsequent tropospheric chemistry of SO_2, probably catalyzed on dust or aerosol particles.

Studies of the combustion chemistry of sulfur have developed in two directions, the combustion of sulfur-containing species themselves and the effect of sulfur-containing species on the combustion chemistry of common fuels (Zachariah and Smith 1987; Kramlich et al. 1981). The H_2S–O_2 system is the most thoroughly investigated example in the first category. It resembles the H_2–O_2 system in several ways, for example in having three limits of explosion. Because the combustion mechanism of H_2 came to be understood earlier than that of the simple hydrocarbons, studies of the influence of sulfur chemistry on combustion of other fuels started with hydrogen flames.

The presence of sulfur can profoundly affect basic flame characteristics, mainly through the direct coupling of sulfur chemistry to radical reactions. For example, starting from early researches it was noted that addition of small amounts of sulfur species resulted in substantial net decreases in postflame radical concentrations. This appears to happen in the following way. Sulfur oxidation occurs as a progressive oxidation sequence, that is, fuel-S $\rightarrow$ SO $\rightarrow$ SO_2 $\rightarrow$ SO_3. SO_2 formation in flames is rapid, essentially because only bimolecular reactions are required. SO_3 is mostly formed in slower termolecular reactions, notably

$$SO_2 + O + M \longrightarrow SO_3 + M\,. \tag{132}$$

Coupling of SO_2 to HSO_2 is generally believed to be responsible for the catalytic removal of radicals, especially H atoms. The main elementary reactions are

$$H + SO_2 + M \longrightarrow HSO_2 + M \tag{133}$$
$$H + HSO_2 \longrightarrow H_2 + SO_2 \tag{134}$$
$$OH + HSO_2 \longrightarrow H_2O + SO_2\,. \tag{135}$$

Most of the HSO_2 is formed early in the flame, where low temperatures favor the termolecular recombination reaction.

Interactions of sulfur and nitrogen chemistry are discussed in Chapter 3.

1.4.5 Halogens

World wide use of incineration for disposal of industrial and municipal wastes has led to increasing interest in the high-temperature chemistry of halogens. Chlorine has drawn the most attention, since chlorine-containing compounds are prominent in consumer and industrial wastes (Senkan, Chapter 4; Seeker and Koshland, 1990; Seeker 1990; Lighty and Veranth 1998). A major problem arising in the incineration of halogen-containing organic compounds is that their presence slows down the combustion process and enhances the output of products of incomplete combustion, some of them partially halogenated and toxic. On the other hand, the flame-retarding property of halogen-containing compounds is also exploited in safety engineering, that is, for suppression of hydrocarbon fuel oxidation in flames and detonation—quite small additions of halogen-containing inhibitors reduce flame speeds and thus impede the combustion process (Biordi et al. 1974; Westbrook 1983; Wilson et al. 1969).

There are therefore two major aspects to halogen combustion chemistry: halogens in organic molecules to be destroyed, and hydrocarbon oxidation in the presence of inorganic (X_2 or HX, where X is a halogen atom) and organic (RX) halogen-containing compounds. The chemical basis for the flame-inhibiting effect is better understood than the combustion mechanisms of halogen-containing organic compounds, mainly because larger molecules are then involved. As described in Chapter 4, the combustion of chlorinated organic molecules involves both pyrolytic and oxidative reactions, and the specific nature of the compound being burned is important in determining the by-products that are formed. The important elementary reactions of the combustion of chlorine-containing organic compounds are discussed in Chapter 4 and in the review of Tsang (1990b). Modeling studies of the combustion chemistry of a number of chlorinated organics have been reported (Chang and Senkan 1989; Senkan et al. 1992; Ho and Bozzelli 1992; Ho et al. 1992a, 1992b, 1995).

The slowing effect of halogen-containing species on the combustion of hydrocarbons is caused mainly by reactions of these species with hydrogen atoms; such reactions compete against the H + O_2 reaction (Westbrook 1983). H atoms react more rapidly with halogen-containing species (HX or X_2) than with O_2, especially at low temperatures. For example, the rate coefficient of

$$H + HCl \longrightarrow H_2 + Cl \tag{136}$$

is about 50 times larger than the rate coefficient of the H + O_2 reaction at 1000 K. The reason for this difference is that the activation energies of the halogen reactions (16 kJ/mol for the H + HCl reaction) are lower than the 68 kJ/mol value for the H + O_2 reaction. This allows small amounts of halogen species to compete for H atoms with large amounts of O_2. In flames, the important reactions of halogen species occur early in the primary reaction zone, where the temperature is still relatively low. Mixtures in which the H atom concentration ahead of the flame is higher (H_2–O_2 mixtures) are less affected by the inhibiting reactions than mixtures which have low levels of H atoms (hydrocarbon–O_2 mixtures) (Westbrook 1983). At higher temperatures, the H + O_2 reaction becomes faster than the inhibiting reactions, which makes the inhibiting influence of halogen compounds on ignition and detonation processes less prominent than it is on flames.

Our present understanding of hydrocarbon oxidation and halogen chemistry suffices to undertake detailed modeling studies of the influence of halogen-containing species on hydrocarbon oxidation. This approach has been used for studying inhibition of CO, H_2, CH_4, CH_3OH, and C_2H_4 oxidation (Brouwer et al. 1992; Chang and Senkan 1985; Dixon-Lewis 1979; Westbrook 1982, 1983) by a variety of halogen-containing compounds.

Different classes of compounds (such as Br_2, HBr, CH_3Br, or CF_3Br) have slightly different inhibiting effectiveness, but the basic mechanism of inhibition is the same. To illustrate the basic chemistry involved, we consider the influence of a small addition of HBr to a near-stoichiometric mixture of C_2H_6 reacting at an initial temperature of 1400 K.

In Figures 1.29–1.31 the influence of HBr addition on major species and radical concentrations is shown. The modeling was done using the mechanism described in Section 9.4 for a constant pressure of 1 atm. In presence of HBr, the ignition delay of the C_2H_6–O_2 mixture increases from 150 to 180 μs, as the rates of C_2H_6 oxidation and CO formation are slower (Figure 1.29). This is a result of smaller radical concentrations in the induction zone (Figure 1.30). The inhibition is caused by removal of hydrogen atoms, the most active species in oxidation of hydrocarbon-oxygen mixtures, mostly in reactions with HBr and Br_2. Inhibiting species are then regenerated by reactions with hydrocarbon molecules and by recombination reactions:

$$H + HBr \longrightarrow H_2 + Br \tag{137}$$

$$H + Br_2 \longrightarrow HBr + Br \tag{138}$$

$$Br + Br + M \longrightarrow Br_2 + M\,. \tag{139}$$

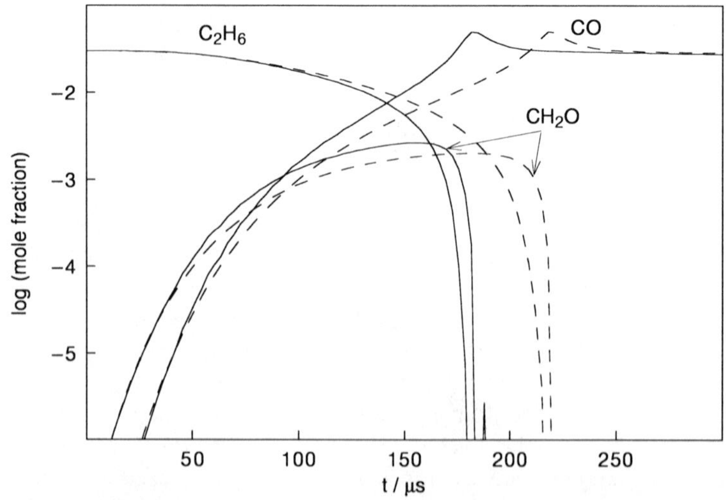

FIGURE 1.29. Species concentrations profiles during ignition of a 3% C_2H_6–10% O_2–N_2 mixture (solid lines) compared to the behavior of the same mixture with 0.3% HBr added (dashed lines). The initial conditions are T = 1400 K, P = 1 atm.

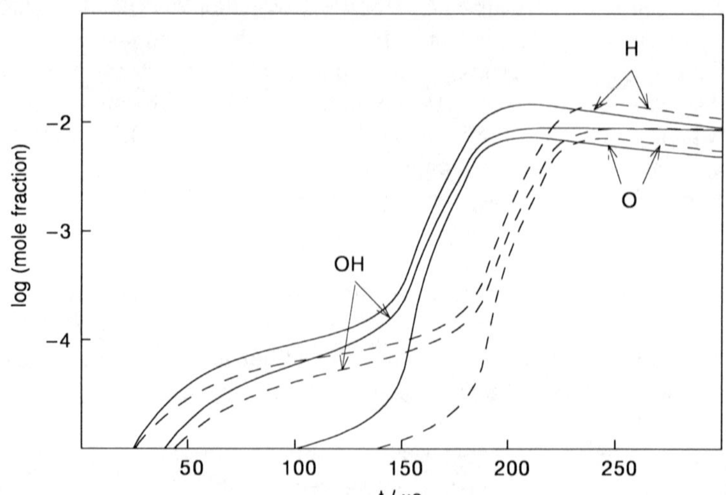

FIGURE 1.30. Radical concentration profiles during ignition of a 3% C_2H_6–10% O_2–N_2 mixture (solid lines) compared to the behavior of the same mixture with 0.3% HBr added (dashed lines). The initial conditions are T = 1400 K, P = 1 atm.

These reactions show that there are two paths for removal of H atoms—through HBr and through Br_2, the two being coupled by the H + Br_2 reaction producing HBr. These paths provide a catalytic mechanism for removal of H atoms. Suppression of the ignition rate is due not only to decreased concentration of H atoms, however, but also due to an indirect effect on the concentrations of

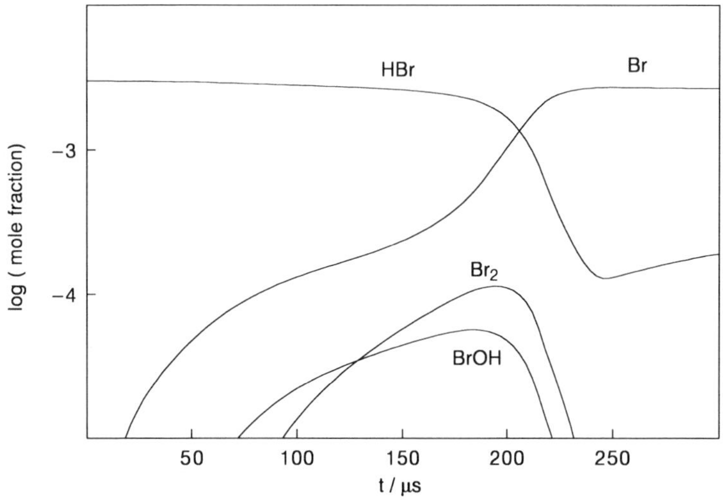

FIGURE 1.31. Profiles of bromine-containing species during ignition of a 0.3% HBr–3% C_2H_6–10% O_2–N_2 mixture. The initial conditions are T = 1400 K, P = 1 atm.

other active species. The decreased CH_2O concentration in presence of HBr (Figure 1.29) means that part of the CH_2O is replaced by CH_3Br, which leads to decreased concentrations of other radicals normally available for chain branching reactions. These radicals also participate in bromine chemistry. In absence of bromine, the consumption of OH radicals is almost exclusively through reactions with C_2H_6, H_2, and CO. With bromine in the system, however, new channels for OH consumption appear:

$$OH + HBr \longrightarrow H_2O + Br \tag{140}$$
$$OH + Br_2 \longrightarrow HOBr + Br\,. \tag{141}$$

The concentration of O atoms is similarly affected:

$$O + HBr \longrightarrow OH + Br \tag{142}$$
$$O + Br_2 \longrightarrow BrO + Br\,. \tag{143}$$

As a result of the lower O and OH concentrations in the presence of bromine, the CO/CO_2 ratio is higher and the concentration of products of incomplete oxidation increases.

Another aspect of the influence of halogens on hydrocarbon oxidation is that OH radicals and H and O atoms are actually replaced by halogen atoms. At high halogen levels, oxidative consumption of fuel and initial intermediates is supplanted by reactions with halogens, such that reactions of halogen species with unburned hydrocarbons become an important source of undesirable halogenated products. Thus, addition of halogens both dramatically changes the structure of hydrocarbon oxidation (degenerate chain branching being slower than simple chain branching) and leads to a different set of products.

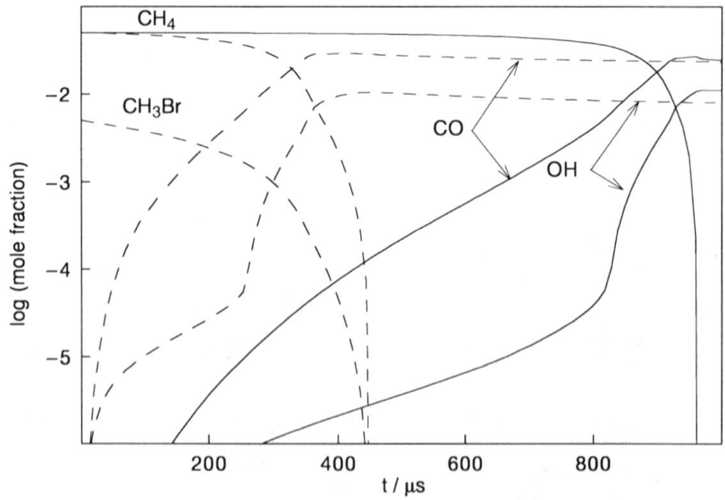

FIGURE 1.32. Species profiles during ignition of a 5% CH_4–10% O_2–N_2 mixture (solid lines) compared to those computed for the same mixture with 0.5% CH_3Br added (dashed lines). The initial conditions are T = 1600 K, P = 1 atm.

As temperature increases, the H + O_2 reaction becomes very fast and the stabilities of halogen species such as HBr and Br_2 diminish, such that they no longer divert H atoms and other chain centers from the main oxidation process. These effects lead to a markedly decreased inhibition effect at high temperatures and even to promotion of ignition, as found for methane oxidation in presence of CH_3Br by Suzuki et al. (1991). Because of the weak C–Br bond (350 kJ/mol), CH_3Br decomposes more rapidly than CH_4 (C–H bond dissociation energy of 435 kJ/mol) and generates active species by dissociation

$$CH_3Br \longrightarrow CH_3 + Br, \tag{144}$$

which is faster than normal chain initiation in methane-oxygen ignition. Figure 1.32 shows that the mechanism described in Section 9.4 also predicts that, at high temperatures, CH_3Br addition promotes the oxidation of methane, in other words, that the ignition delay in presence of CH_3Br is shorter. The concentration of CH_3 increases, leading to a higher rate of converting CH_4 to CO.

The influence of other halogens on hydrocarbon oxidation is expected to have the same basic mechanism, that is, capture of H atoms. It is found experimentally (Garner et al. 1957; Homann and Poss 1972; Rosser et al. 1958) that iodine and bromine atoms are the most effective inhibitors, whilw chlorine and fluorine atoms are successively less effective. This ranking can be understood from the C–X bond strengths in halogenated hydrocarbons, ordered as F > Cl > Br > I. Thus, because the CH_3–F bond strength is 452 kJ/mol, 18 kJ/mol greater than the CH_3–I bond strength of 234 kJ/mol (Benson 1968), I atoms are more readily supplied to the system during iodide decomposition.

The use of halogen-containing compounds in fire suppression is discussed in Section 1.6.

1.5 COMBUSTION OF ALTERNATIVE FUELS

Methanol and ethanol are thought to be the most favorable candidates for economical alternatives to hydrocarbon fuels, particularly for motor vehicles, in view of their low pollutant emission levels and high octane numbers (Gordon and Austin 1992). Their other advantages are convenient storage in liquid form under normal pressure and temperature and the fact that there are petroleum-independent ways to produce them from coal and renewable biomass. Alcohols are already used as blends with gasoline to increase the octane number and to increase fuel oxygen content, which leads to more complete combustion. The development of practical applications of neat alcohols as alternative fuels will require chemical understanding of their pyrolysis and oxidation reactions.

Most studies of the combustion chemistry of alcohols have been carried out during the past 15 to 20 years. The presence of the C–O bond in alcohols has been found to define their combustion properties and kinetic behavior at high temperatures. A chemical kinetics database for methanol-related reactions, including a critical evaluation of the available data, was developed by Tsang (1987), and a review of methanol and ethanol combustion chemistry was reported by Zamansky and Borisov (1989). Detailed reaction mechanisms were tested by Norton and Dryer (1992) and Held and Dryer (1998). Here we describe the qualitative features of alcohol combustion chemistry.

1.5.1 Methanol

The primary reaction of methanol pyrolysis is dissociation into CH_3 and OH

$$CH_3OH + M \longrightarrow CH_3 + OH + M\,. \tag{145}$$

It has been studied in shock tubes, static reactors and flow systems (Aronowitz et al. 1977; Spindler and Wagner 1983; Cribb et al. 1984; Borisov et al. 1988; Norton and Dryer 1990). The dissociation of C–H and O–H bonds is less important. The active species formed abstract H atoms from CH_3OH molecules:

$$H + CH_3OH \longrightarrow CH_2OH + H_2 \tag{146}$$

$$CH_3 + CH_3OH \longrightarrow CH_2OH + CH_4 \tag{147}$$

$$OH + CH_3OH \longrightarrow CH_2OH + H_2O \tag{123}$$

and CH_2OH radicals dissociate:

$$CH_2OH + M \longrightarrow CH_2O + H + M\,. \tag{148}$$

(Whether the radical initially formed has the CH_3O or the CH_2OH structure is thought to be unimportant, as isomerization and formaldehyde (CH_2O) formation are generally faster than the abstraction reactions.)

The mechanism of high-temperature methanol oxidation has been studied by Bowman (1975), Aronowitz et al. (1979), Westbrook and Dryer (1979), Natarajan and Bhaskaran (1981), Tsuboi and Hashimoto (1981), Bhaskaran et al. (1981), Borisov et al. (1989), and Egolfopoulos et al. (1992). A variety of reaction mechanisms for methanol combustion are suggested by these authors. Methanol–air flames have been modeled by Chen and Echekki (1996) and Li and Williams (1996).

The main features of the oxidation process are characterized by the following elementary reactions:

$$O + CH_3OH \longrightarrow CH_2OH + OH \tag{149}$$
$$HO_2 + CH_3OH \longrightarrow CH_2OH + H_2O_2 \tag{150}$$
$$O_2 + CH_2OH \longrightarrow CH_2O + HO_2 . \tag{124}$$

The main reaction path of methanol oxidation therefore includes formation of both intermediate radicals (CH_2OH and HCO) and stable species (CH_2O and CO), as illustrated in the following diagram.

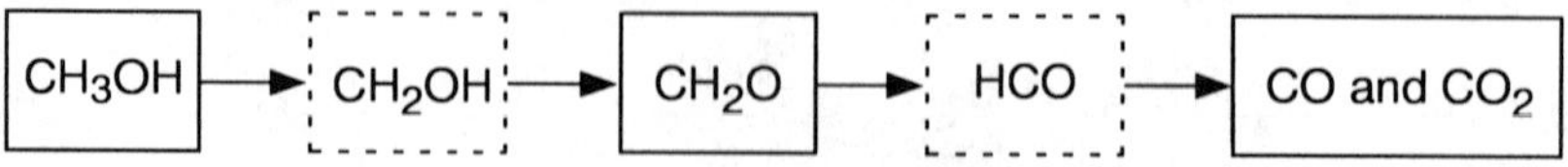

Reactions in which C_2 species participate are less important than they are in hydrocarbon oxidation because the concentrations of CH_3 radicals are substantially lower in the CH_3OH–O_2 system.

Figure 1.33 presents computed species profiles for methanol ignition in a fuel-lean mixture, 3% CH_3OH in air at 1 atm and 1300 K. It can be seen that at reaction times close to the induction period, almost all of the methanol has been converted to CH_2O, CO, and H_2. It is therefore these compounds that actually ignite even though the ignition delay time is defined by methanol reactions. Several ppm of CH_2OH are also formed during the induction period. At ignition, the concentrations of all active species suddenly increase and then immediately decrease. Under these conditions, CO_2 and H_2O are the major products.

Detailed mechanisms of methanol combustion have been used not only for modeling ignition and low-concentration interactions, but also for comparison with experimental profiles in flames (Olson et al. 1986, 1987).

Recent developments in the study of methanol high-temperature chemistry include low-temperature (750–1100 K) and high-pressure (20–30 atm) experiments in a flow system (Held and Dryer 1993) and in a rapid compression machine (Lee et al. 1993). These data have the potential to improve the current models in the direction of gaining greater understanding of methanol chemistry in internal combustion engines.

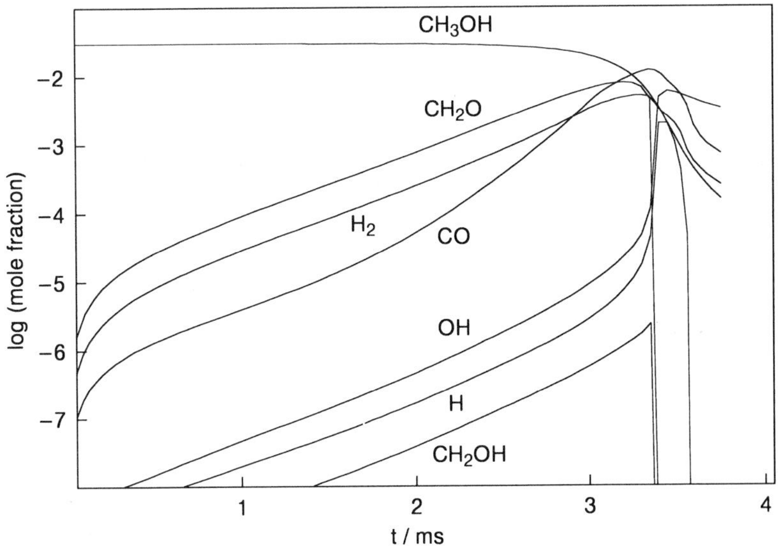

FIGURE 1.33. Species profiles for methanol ignition for a fuel-lean ($\phi = 0.22$) mixture of 3% CH_3OH in air at 1300 K and 1 atm, computed for a constant volume constraint using the mechanism of Section 1.9.4.

1.5.2 Ethanol

Ethanol is a renewable fuel readily obtained from cellulosic materials such as corn, wood, grass, and biomass wastes (Lynd 1996). The physical properties of ethanol are similar to those of methanol, and the differences between the alcohols and gasoline are much greater than the differences between alcohols. At high temperatures, the chemical reactions of ethanol are, as would be expected, significantly more complex than those of methanol.

The pyrolysis and oxidation reactions of ethanol have been studied by Tsang (1976), Bhaskaran et al. (1981), Borisov et al. (1985, 1989a, 1989b), Norton and Dryer (1992), Dagaut et al. (1992), and Curran et al. (1993). The main features are shown in the following diagram.

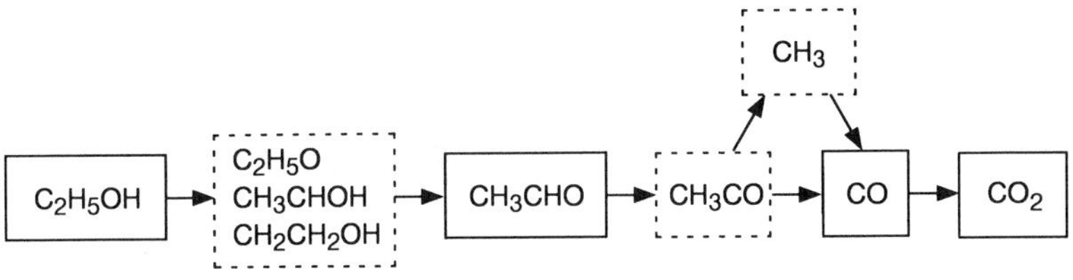

The fastest initiation reaction is

$$C_2H_5OH \longrightarrow CH_3 + CH_2OH \tag{151}$$

although other dissociation channels may also be important. The difficulties of modeling the ethanol reactions are defined by the presence of three fundamentally different possibilities for H atom abstraction,

$$\begin{aligned} C_2H_5OH + R &\longrightarrow RH + CH_2CH_2OH &(152)\\ &\longrightarrow RH + CH_3CHOH &(153)\\ &\longrightarrow RH + C_2H_5O\,. &(154)\end{aligned}$$

There is unfortunately only limited information in the literature on the rate coefficients of these processes, and the predicted product distributions under different conditions depend on the rate coefficient ratios of these abstraction channels. They are therefore of special importance for ethanol combustion chemistry, because a substantial part of the fuel is converted to products at the beginning of the process where the abstraction reactions take place.

Other special features of ethanol combustion include the secondary reactions of the C_2 radicals with active species and with oxygen. The main product of these reactions is acetaldehyde, CH_3CHO, and the mechanisms of CH_3CHO pyrolysis and oxidation must be included as part of the ethanol combustion mechanism. Acetaldehyde molecules mainly dissociate or transfer an H atom to molecular oxygen:

$$\begin{aligned} CH_3CHO + M &\longrightarrow CH_3 + HCO + M &(155)\\ CH_3CHO + O_2 &\longrightarrow CH_3CO + HO_2\,. &(156)\end{aligned}$$

Acetyl (CH_3CO) radicals are the main intermediates in acetaldehyde oxidation. They are formed readily in losing the aldehydic H atom of CH_3CHO in reactions with active species. Their main loss route is rapid dissociation into CH_3 and CO:

$$CH_3CO \longrightarrow CH_3 + CO\,. \tag{157}$$

Thus, the main reaction path of ethanol oxidation includes formation of C_2 radicals, their conversion to the intermediate product acetaldehyde, and further oxidation of CH_3CHO to CO and then CO_2.

Figure 1.34 shows species profiles during the ignition of a mixture of 3% C_2H_5OH in air at 1 atm and 1300 K. As in methanol ignition, high concentrations of the intermediate products CH_2O, CH_3CHO, CO, and H_2 are formed during the induction period. The β-radicals CH_2CH_2OH are computed in the mechanism used to have the highest concentration among the C_2 radicals. After ignition, all intermediate species are rapidly converted to CO_2 and H_2O.

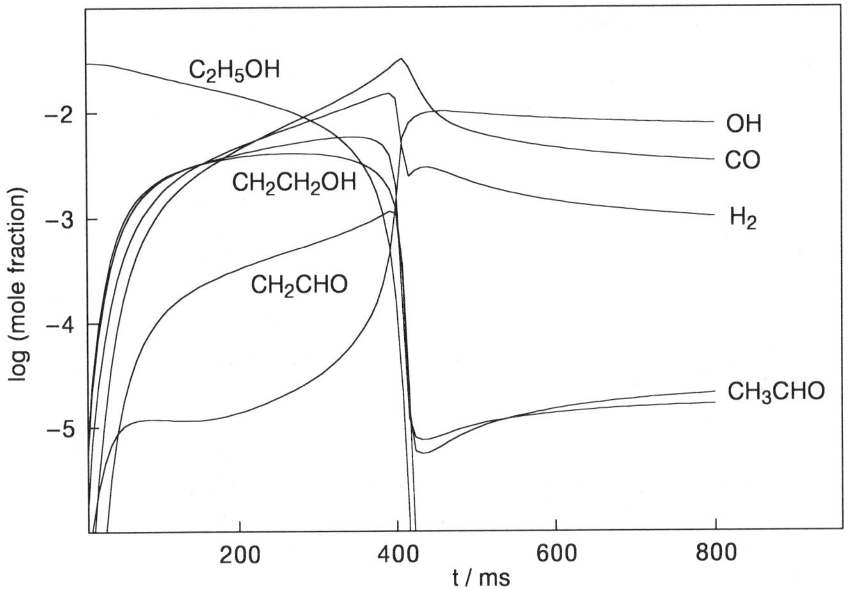

FIGURE 1.34. Species profiles for ethanol ignition for a fuel-lean ($\phi = 0.44$) mixture of 3% C_2H_5OH in air at 1300 K and 1 atm, computed for constant volume constraints using the mechanism of Section 1.9.4.

1.5.3 Higher alcohols and MTBE

Propanol pyrolysis has been studied in a static system (Trenwith 1975), a flow reactor (Norton and Dryer 1990b), and in a flame (Smith and Gordon 1956). A variety of hydrocarbon intermediates were observed, including ethylene, methane, propene, acetylene, ethane, acetaldehyde, propionaldehyde, and formaldehyde. Propionaldehyde is produced by H atom abstraction at the α-carbon of n-propanol and subsequent dissociation of the radical produced:

$$R + C_2H_5CH_2OH \longrightarrow RH + C_2H_5CHOH \longrightarrow RH + H + C_2H_5CHO\,. \quad (158)$$

Propionaldehyde is very reactive and was found only in small quantities during the early stages of n-propanol decomposition. The oxidation mechanism of n-propanol was found to be similar to that of ethanol (Norton and Dryer 1990b). The authors demonstrated that oxidation of iso-propanol is slower, due to the lower reactivity of its major intermediate species—for example acetone—compared to propionaldehyde in n-propanol oxidation. Iso-propanol produces more C_3H_6 and less acetaldehyde than does n-propanol.

It has been reported in a number of studies (Daly and Wentrup 1968; Choo et al. 1974; Brocard et al. 1983; Brocard and Baronnet 1987; Norton and Dryer 1990b) that the pyrolysis reactions of tert-butylalcohol (TBA) and methyl-tert-butylether (MTBE) appear to proceed unimolecularly, through "four-center activated complexes" to form iso-butene and water or methanol:

$$(CH_3)_3COH \longrightarrow (CH_3)_2C{=}CH_2 + H_2O \tag{159}$$
$$(CH_3)_3COCH_3 \longrightarrow (CH_3)_2C{=}CH_2 + CH_3OH\,. \tag{160}$$

Both MTBE and the iso-butene product of its decomposition have been found to be effective anti-knock fuel components for internal combustion engines. They are being increasingly substituted for toxic anti-knock substances, such as tetraethyl lead and aromatic additives. Although MTBE has heretofore been used in gasoline formulations mainly to increase the octane number, in some areas it is now required as an additive by reason of its ability to reduce vehicle emissions. MTBE is currently the second largest end use of methanol, consuming about one-fourth of the global methanol production.

The high-temperature oxidation of MTBE and iso-butene was studied by Brezinsky and Dryer (1986), Nikitin et al. (1987), Norton and Dryer (1990b), Dunphy and Simmie (1991), Curran et al. (1992), and Gray and Westbrook (1994). The mechanism of MTBE oxidation includes rapid unimolecular decomposition of MTBE, as shown above, followed by H atom abstraction reactions and β-scission of the radicals formed:

$$\text{t-}C_4H_9OCH_3 + R \longrightarrow \text{t-}C_4H_9OCH_2 + RH \tag{161}$$
$$\text{t-}C_4H_9OCH_3 + R \longrightarrow \text{t-}C_4H_8OCH_3 + RH \tag{162}$$
$$\text{i-}C_4H_8 + R \longrightarrow \text{i-}C_4H_7 + RH \tag{163}$$
$$\text{t-}C_4H_9OCH_2 \longrightarrow \text{t-}C_4H_9 + CH_2O \tag{164}$$
$$\text{t-}C_4H_8OCH_3 \longrightarrow \text{i-}C_4H_8 + CH_3O \tag{165}$$
$$\text{t-}C_4H_9 \longrightarrow \text{i-}C_4H_8 + H \tag{166}$$
$$CH_3O + M \longrightarrow CH_2O + H + M\,. \tag{167}$$

The 2-methyl-allyl radicals i-C_4H_7 then decompose to form allene and methyl radicals

$$\text{i-}C_4H_7 \longrightarrow C_3H_4 + CH_3\,, \tag{168}$$

which are relatively non-reactive and return active species to the radical pool only slowly. At lower temperatures, reactions of hydrocarbon radicals with molecular oxygen have to be taken into account, as discussed by Brocard et al. (1983). Gray and Westbrook (1994) noted the likelihood of isomerizations such as

$$\text{t-}C_4H_9OCH_2 \longrightarrow \text{t-}C_4H_8OCH_3 \tag{169}$$

at higher temperatures. Whether or not they dominate, the reaction mechanisms and intermediate species are very similar for MTBE and iso-butene oxidation, and so, accordingly, are their inhibiting effects. Both compounds remove active species from the radical pool and lead to replacement of more reactive species by less reactive ones. Net removal of radicals results in the overall decrease in oxidation rate observed experimentally.

1.6 COMBUSTION INHIBITORS

Combustion is usually regulated by physical means, such as by changing the equivalence ratio, the combustion pressure, or the extent and/or location of heat exchange between the combustion zone and the surroundings. There are chemical ways to affect combustion as well. Combustion can be accelerated by adding small amounts of promoting additives (Section 1.7) or slowed by adding inhibitors. As noted in Section 1.4.5, halogen-containing compounds have been widely used for flame inhibition. The chemical basis of their influence is removal of H atoms from the reaction zone. For inhibition by HBr, the elementary reactions primarily responsible for suppressing flame propagation are

$$\mathrm{H + HBr \longrightarrow H_2 + Br} \tag{137}$$

$$\mathrm{H + Br_2 \longrightarrow HBr + Br} \tag{138}$$

$$\mathrm{Br + Br} + M \longrightarrow \mathrm{Br_2} + M\,. \tag{139}$$

Removal of H atoms occurs as a result of these reactions, for their sum is just H + H $\longrightarrow$ H_2; thus, the rate of chain branching by H + O_2 $\longrightarrow$ OH + O ahead of the flame front decreases.

The inhibiting effect of halogen-containing species on ignition is more complicated. During the ignition process, because the concentrations of intermediate species are very low, decomposition and other reactions of inhibitor molecules with the initial and intermediate species are able to suppress the inhibiting effect (Takahashi et al. 1989). It is even possible that hydrocarbon ignition can be promoted by addition of bromine-containing compounds, as was found by Inomata et al. (1985) to be the case for methane ignition in the presence of CH_3Br.

Chemical reaction during ignition can be separated into two phases: the induction zone, which is generally long and determines the total duration of the process, and ignition itself. Babushok et al. (1996) found in a modeling study that halogen-containing additives reduce ignition delays of hydrocarbon fuels because the additives are usually less stable than the main fuel and therefore decompose more rapidly early in the induction zone. Since the ignition delay time is sensitive to the rate of chain initiation, it becomes shorter. The chemistry of the second part of the ignition process—ignition itself—is similar in nature to the corresponding flame chemistry of the same fuel and is characterized by high concentrations of radicals and intermediate species. This part of the ignition chemistry is inhibited by halogen-containing additives through mechanisms similar to the HBr reactions.

In general, inhibition results from a complex chemical and physical interaction of the additive with the flame (Babushok et al. 1997a; Noto et al. 1998). When an inhibitor concentration is relatively small, the inhibiting effect is mostly due to radical scavenging by a mechanism similar to the HBr case. However, as concentrations of additives increase, their chemical influence saturates and the contributions of physical factors (for example, the flame temperature decrease due to heat absorption by the additive) become increasingly important. Some additives (CO_2, H_2O, etc.) affect combustion mostly in a physical way even at

TABLE 1.1. Effectiveness of flame inhibitors.[1]

Additive	Inhibitor requirement[2]	Relative efficiency[3]
CO_2	10	0.12
CH_3Cl	4.9	0.24
Cl_2	3.8	0.32
CHF_3	3	0.4
HBr	1.8	0.67
CH_3I	1.7	0.71
CH_3Br	1.6	0.75
CF_3Br	1.2	1
Br_2	0.83	1.4
$(CH_3)_3PO_4$	0.3	4
$TiCl_3$	0.22	5.5
PCl_3	0.175	6.9
$NaHCO_3$	0.1	12
$Pb(C_2H_5)_4$	0.022	55
$Fe(CO)_5$	0.02	60

1. After Babushok et al. 1998. 2. Molecules of inhibitor, per 100 molecules of CH_4, required to reduce the laminar flame speed of a stoichiometric methane–air mixture by 10%. 3. Relative to CF_3Br (Halon 1301).

small concentrations. They are less effective than "chemical inhibitors" and must be present in large amounts to have noticeable effects. Roughly speaking, halogen-containing compounds are about ten times more efficient fire suppressants than inert additives.

The inhibiting efficiencies of different chemical inhibitors for hydrocarbon flames were found by Babushok et al. (1998) to differ by almost two orders of magnitude. Table 1.1 shows the efficiencies of various flame suppression agents relative to CF_3Br (Halon 1301) as a standard. It can be seen that the inert inhibitor CO_2 is 500 times less effective than $Fe(CO)_5$, the most effective chemical inhibitor. The large range of effectiveness suggests that the chemical pathways by which these agents affect combustion may be completely different.

While the details of these pathways are specific to each inhibitor, general trends in behavior do permit some classification. There are threE Basic ways that reactive species can be removed from the reaction zone by an inhibitor: (1) by removing active species in direct chemical reactions, followed by regeneration of the inhibitor in a cycle similar to that of HBr; (2) by removing active species indirectly through reactions with intermediate species present in the combustion zone, again followed by regeneration of the inhibitor; and (3) by removing active species without regeneration of the inhibitor. Other mechanisms are possible in principle, but they prove to be essentially variations of these three. The first pathway was found by Babushok et al. (1998) to be the most efficient. The "ideal" inhibitor acting that way is able to decrease the flame velocity of an atmospheric methane-air flame by about 30% when present at a level of 100 ppm. Such an ideal inhibitor sets the upper limit of chemical inhibition of flames. Babushok et al. found that bromine-containing compounds are typically about 50 times less efficient than an ideal inhibitor.

Bromine-containing compounds are nonetheless commonly considered to be the most effective combustion inhibitors. Halon 1301 (CF_3Br), in particular, has long been the fire-fighting agent of choice in many situations. It is able to significantly inhibit combustion at low cost and it has favorable physical characteristics—a low boiling point, low electrical conductivity, low toxicity, high liquid density, and good chemical stability. Its mechanism of flame inhibition (Biordi et al. 1974; Westbrook 1982; Noto et al. 1998; Babushok et al. 1996, 1997b) is similar to that of other halogen-containing compounds: the CF_3Br disappears early in the flame to form HBr, primarily in the reaction with H atoms

$$H + CF_3Br \longrightarrow HBr + CF_3 , \qquad (170)$$

and the HBr formed catalyzes H atom removal through the cycle discussed above.

Due to its high ozone depletion potential, caused by the presence of the Br atoms that convey the effectiveness in fire suppression, Halon 1301 was banned from production in 1994. Possibilities for effective and environmentally safe alternatives have been reviewed by Miziolek and Tsang (1995). Hydrofluorocarbons such as HFC-227ea (C_3HF_7), HFC-125 (CHF_2CF_3), CF_3H, and CF_3I are widely thought to be the most promising prospective substitutes for Halon 1301, and extensive study of the thermochemical and kinetic properties of such compounds has been undertaken (Burgess et al. 1995, 1996, 1997). The mechanism of their influence on combustion is similar to that for CH_3Br: fast decomposition of the inhibitor and subsequent reactions of HF (HI) and F_2 (I_2) with radicals in the combustion zone. However, the prospects for using HFCs for fire suppression are limited, because they do not contain the Br atoms which are more effective than other halogen atoms in removal of active species from the flame zone, and they are thus generally less effective fire suppression agents than Halon 1301 (Figure 1.35). Studies by numerical simulations of the influence of different halogen-containing hydrocarbons on propagation of laminar flames ranked the inhibition efficiencies of the compounds tested as CF_3Br, $CF_3I > C_2F_6 > C_2HF_5 > CHF_3 > CF_4$ (Noto et al. 1996; Linteris et al. 1998; see Figure 1.36). The bromine- and iodine-containing compounds were found, as expected, to be the most effective flame inhibitors.

Some compounds that do not contain halogens are effective combustion inhibitors. Many metal-containing compounds (for example, inorganic salts such as Na_2CO_3 and $NaHCO_3$) have long been known to be effective inhibitors of flame propagation (Jensen and Jones, 1982; Mitani and Nioka, 1984). Babushok et al. (1998) reviewed the literature on flame inhibitors and found that there are metal compounds that can serve as fire extinguishing agents with higher efficiency than even Halon 1301. They identified several potassium and sodium compounds with efficiencies 12 to 17 times greater than Halon 1301. Iron pentacarbonyl, $Fe(CO)_5$,

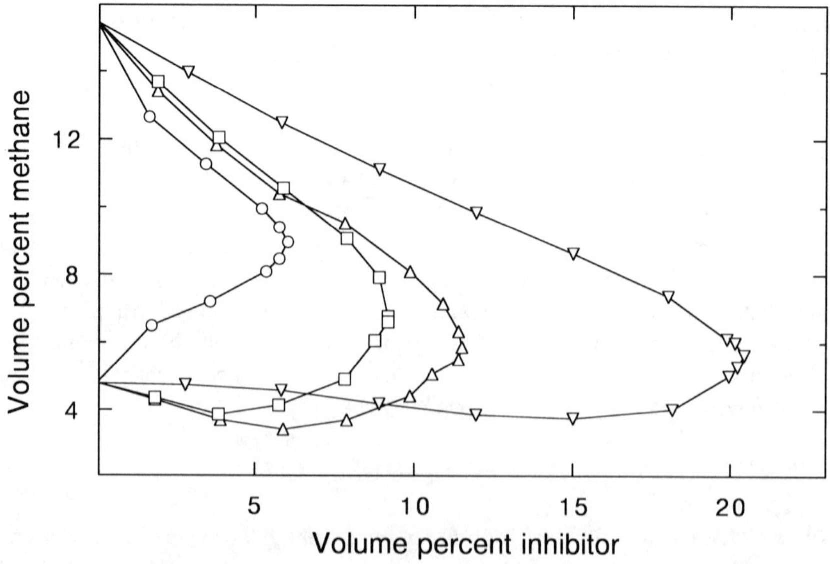

FIGURE 1.35. Flammability limits for addition of fluorinated hydrocarbons to methane-air mixtures (Liao et al. 1996) ○ CF_3Br (Halon 1301) □ C_4F_{10} (FC-3-1-10) △ C_3HF_7 (HFC227ea) ▽ CHF_3 (HFC-23).

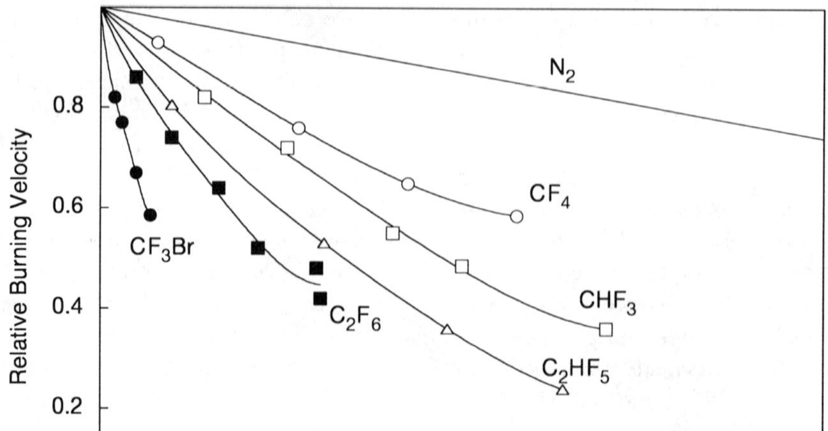

FIGURE 1.36. Dependence of burning velocity of an atmospheric pressure methane–air flame on the concentration of various inhibitors (Noto et al. 1996). The N_2 line illustrates the extent of inhibition caused by dilution alone.

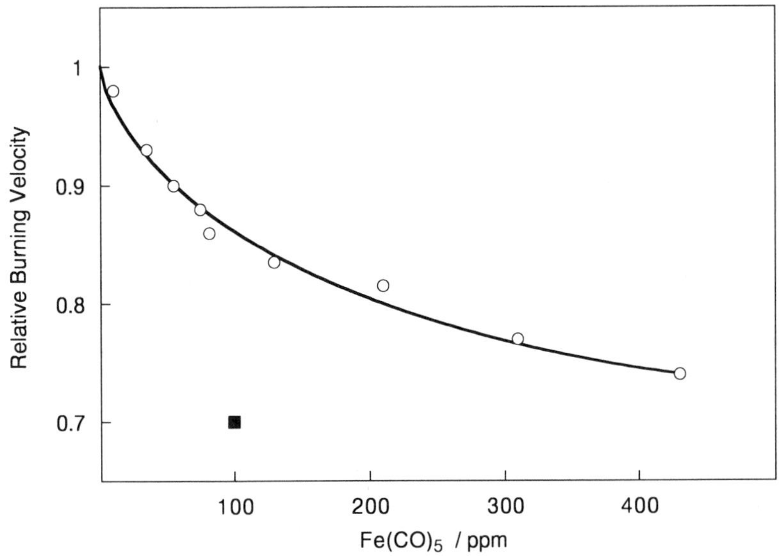

FIGURE 1.37. Dependence of the burning velocity of an atmospheric pressure methane–air flame on the concentration of added $Fe(CO)_5$. A "perfect inhibitor" would give the result indicated by ■ , i.e., reduce the burning velocity by 30% when present at a concentration of 100 ppm (Babushok et al. 1998).

was found to have the greatest efficiency among the compounds tested, being 60 times more efficient than Halon 1301. The high sensitivity of the burning velocity of a premixed methane–air flame to trace additions of iron pentacarbonyl is shown in Figure 1.37.

The mechanism of flame inhibition by metal-containing compounds is not well understood. There are suggestions (Lask and Wagner 1962; Jensen and Webb 1976) that both homogeneous and heterogeneous reactions are involved. Rumminger et al. (1998) suggested a gas-phase inhibition mechanism that involves catalytic removal of H atoms by iron-containing species in a way similar to that for halogen-containing compounds. Metal-containing compounds are most effective at small concentrations (100 ppm and less) and their relative efficiencies decrease as the additive concentration increases (Figure 1.37). This effect was interpreted by Rumminger et al. as condensation of metal-containing compounds in the reaction zone. Condensation simply removes metals from the gas phase and reduces the concentration of gas-phase metal atoms available for reaction in the flame. Nonetheless, metal-containing compounds are more efficient flame inhibitors than halogen-containing compounds. A comparison of the performance of $Fe(CO)_5$ with that of an ideal inhibitor (Reinelt and Linteris 1996; Babushok et al. 1998) showed that at low concentrations it is only four times less efficient and thus approaches the chemical limit of inhibiting efficiency.

The toxicity and other unfavorable characteristics of metal-containing compounds, however, have to be taken into account before these compounds can be considered for fire extinguishing. Iron pentacarbonyl, for example, is extremely toxic and autoignites spontaneously at 50 °C, while tetraethyl lead, $Pb(C_2H_5)_4$, has been banned in many countries as a knock-suppressing additive to gasoline because of its toxicity. It is therefore no surprise that benign alkali salt powders like carbonates and bicarbonates are thought to be more attractive Halon replacements (Miziolek and Tsang 1995).

1.7 COMBUSTION PROMOTERS

Over the years, number of fuel additives have been investigated for their ability to improve upon the combustion performance of the common hydrocarbon fuels derived from petroleum. The main goal has been to provide more rapid or more reliable initiation of combustion, particularly under unfavorable environmental conditions or combustor geometry. A related target of promoter research is enhanced removal of atmospheric pollutants from combustion products. Because both types of promotion are primarily chemical in nature, significant efforts have been made to understand the important elementary reactions that promoters introduce.

Organic nitrates, nitrites, some metal-organic compounds, and peroxides have proved to be effective as fuel additives, for example to increase the cetane number of diesel fuels. The majority of these compounds decompose quickly to form active radicals that accelerate the chain reactions of hydrocarbon oxidation. Fast decomposition of promoters into radicals, however, is readily shown not to be the most efficient way to increase the overall reaction rate, and so other promoting mechanisms have been studied experimentally and by kinetic modeling.

Kinetic analyses of promotion mechanisms, undertaken by Borisov and coworkers between 1988 and 1990, have been reviewed by Zamansky and Borisov (1989, 1992). These analyses were done by deriving analytic solutions for the kinetic equations of a variety of prototype reaction mechanisms. About a dozen different accelerating mechanisms, for both non-branched and branched chain reaction mechanisms, were tested against experimental data on reactions of various promoters and on their capacity to accelerate high-temperature oxidation of various fuels, including hydrogen, carbon monoxide, methane, higher hydrocarbons, methanol, ethanol, ammonia, and nitrous oxide.

The most efficient promotion mechanisms are different for non-branched and branched chain reactions. For non-branched reactions with long chains, the maximum promoting effect is achieved with additives having a relatively fast decomposition rate and forming active species which support the main chain reaction. If the recombination rate of the radicals is high, there is an "ideal promoter" with an optimum decomposition rate coefficient for specific reaction conditions. Model simulations of these effects are shown in Figure 1.38. A hypothetical promoter B was added to supply, at an adjustable rate, H atoms and an inert product:

$$B \longrightarrow H + \text{Products}\,. \qquad (171)$$

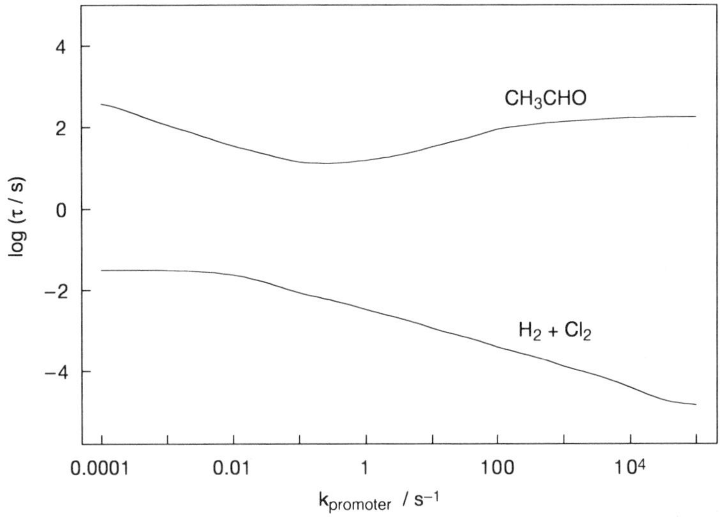

FIGURE 1.38. Effect of promoter decomposition rate coefficient on the rates of CH_3CHO pyrolysis and the H_2–Cl_2 reaction. For the CH_3CHO pyrolysis case, 0.1% promoter was added to 10% CH_3CHO in N_2 at 800 K and 1 atm; for the H_2–Cl_2 case, 0.1% promoter was added to 10% H_2 and 10% Cl_2 in N_2 at 750 K and 1 atm. Adiabatic reaction at constant pressure was assumed. For the CH_3CHO case, the reaction time was assigned to 50% consumption of CH_3CHO; for the H_2–Cl_2 case, the sharp temperature rise at ignition was used to define the reaction time.

Two cases were studied, the H_2–Cl_2 reaction at 750 K and CH_3CHO pyrolysis at 800 K. In the H_2–Cl_2 reaction, where the loss of active species occurs through slow termolecular atom recombination reactions, the ignition delay is seen to drop by about two orders of magnitude and then level off as the rate coefficient of the promoting reaction is increased. In CH_3CHO pyrolysis, fast bimolecular recombination of active species occurs, mostly by reaction (72). In this case, there is an optimum rate coefficient for reaction (171) where the greatest promotion effect is observed. This "ideal promotion rate" can be readily understood: A promoter with a "too slow" decomposition rate has only a small effect because it forms radicals too slowly to make a difference, while a promoter with a "too fast" decomposition rate has only a small effect because the quick burst of radicals it provides disappears rapidly by bimolecular recombination and no more can be supplied because the promoter is gone.

Effective promoting mechanisms for branched chain reactions are quite different. Addition of chain initiators is inefficient because the chain branching mechanism itself initiates many active species. The most effective promoter for these reactions is instead a homogeneous catalyst that supplements the elementary reactions of chain propagation and is recovered in them, as was shown in the Section 1.4 example of promoting NO-to-NO_2 conversion. Such promoters may introduce an effective reaction, that is, one that summarizes a sequence

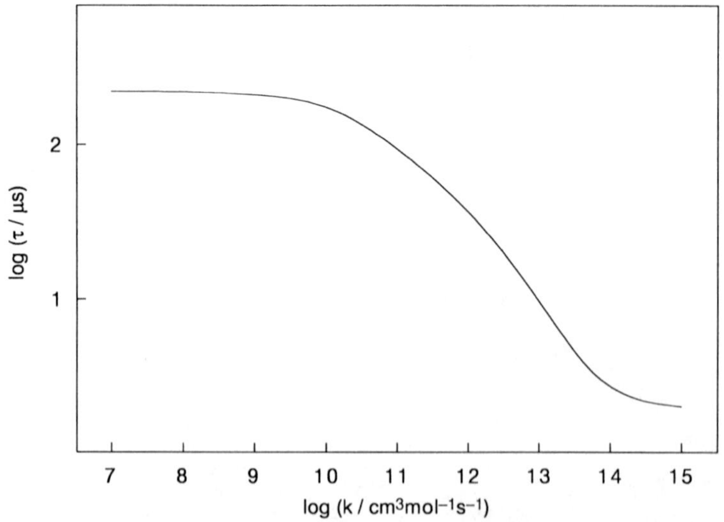

FIGURE 1.39. Promotion of ignition. The conditions modeled were adiabatic constant pressure reaction with 950 ppm promoter, decomposing in the manner of reaction (173), 9.5% CH_4 and 19% O_2 in N_2 at 1700 K and 1 atm.

of elementary reactions that regenerates the promoter, such as

$$C + A \longrightarrow C + R + \text{Products}. \tag{172}$$

Figure 1.39 shows that the effect of adding the reaction

$$\text{Pr} + O_2 \longrightarrow \text{Pr} + O + O \tag{173}$$

upon the ignition delay of methane at 1700 K is up to two orders of magnitude.

Numerical analysis of promoting effects leads to a number of general conclusions (Zamansky and Borisov 1992):

1. The commonly adopted assumption that rapidly dissociating species are good promoters is not justified in many cases, since the promoting effect can decrease as the decomposition rate increases. Additives that provide radicals may reduce ignition delays by orders of magnitude when the underlying ignition mechanism is a non-branching chain, but have only weak effects on a branched chain reaction.
2. "Ideal promoters" do exist for several promotion mechanisms; their characteristic feature is having a promoter decay time close to the time scale of the promoted reaction itself.
3. The amount of promoter needed for obtaining a strong promoting effect—several orders of magnitude—in a chain reaction is typically 0.5% to 5%.
4. The strongest promoting effect is achieved with promoters that are regenerated during the reaction.
5. The promoting effect generally decreases with increasing temperature.

TABLE 1.2. Computed and observed acceleration of methane ignition at 1000 K.[1]

Promoter	k / $cm^3mol^{-1}s^{-1}$	$\tau_\circ/\tau$	
		Computed	Experimental
H_2	10^{-6}	2.0	1.2
C_2H_6	0.01	2.0	1.2
C_2H_5OH	0.02	1.1	1.3
CH_3OH	0.02	1.1	1.2
Cl_2	0.1	1.3	1.2
CH_3CHO	0.4	1.5	2.0
C_4H_{10}	0.6	1.6	1.5–2.0
CH_3I	50	1.4	1.2
$(CH_3)_2N_2$	10^5	2.4	2.5–3
N_2F_4	10^{13}	2–3	3.0

[1] The reference composition was 6% CH_4 and 12% O_2 in Ar, to which 1% of each promoter was added. $\tau_\circ$ and τ are the unpromoted and promoted ignition delays, respectively. The data are from Borisov et al. (1983b) and Lissianski (1982).

Table 1.2 shows results computed for a number of promoters, again for methane ignition.

Several examples of well-known high-temperature promotion mechanisms are presented below. Additional ones can be found in the review of Zamansky and Borisov (1992).

The classic example of combustion promotion is the oxidation of carbon monoxide. Very pure CO reacts with oxygen very slowly, in only two elementary reactions

$$CO + O_2 \longrightarrow CO_2 + O \qquad (174)$$

$$CO + O + M \longrightarrow CO_2 + M\,. \qquad (175)$$

Trace amounts of hydrogen, water, or hydrocarbons, even at sub-ppm levels, can accelerate CO oxidation by generating OH radicals, which support the chain process

$$OH + CO \longrightarrow CO_2 + H \qquad (67)$$

$$H + O_2 \longrightarrow OH + O\,. \qquad (1)$$

Another example of effective promotion is oxidation of N_2O in the presence of additives and radical-providing impurities. At temperatures between 1300 and 1400 K, the bimolecular exchange reaction

$$CO + N_2O \longrightarrow CO_2 + N_2 \qquad (176)$$

is relatively slow because of its over 200 kJ/mol activation energy. Low impurity levels of water or hydrogen-containing fuels generate active species, H and OH, which in turn support a rapid chain process with the same $CO_2 + N_2$ products

$$H + N_2O \longrightarrow N_2 + OH \tag{108}$$

$$OH + CO \longrightarrow CO_2 + H\,. \tag{67}$$

The pronounced promoting effect is predicted and observed at temperatures from 1300 to 1400 K and promoter concentrations near 10 ppm (Borisov et al. 1979).

Slack and Grillo (1981) studied the effect of NO_2 and NO on ignition of CH_4–O_2 mixtures in shock waves. They found a significant promotion effect that could be explained by a chain reaction in which non-reactive HO_2 radicals are replaced by more reactive OH radicals

$$NO + HO_2 \longrightarrow NO_2 + OH \tag{90}$$

$$CH_3 + NO_2 \longrightarrow CH_3O + NO\,. \tag{177}$$

Similar conclusions were drawn from NO and NO_2 promotion of CH_4 oxidation at lower temperatures by Bromly et al. (1996) and Bendtsen et al. (1998).

There is a practical application of promotion suited to enhancing the interaction of NO and ammonia in combustion flue gas (Ho et al. 1993; Lissianski et al. 1999). It was found that small additives of sodium compounds, such as sodium carbonate or sodium hydroxide, significantly increase the rate of NO removal in the SNCR process by means of a chain reaction that forms additional active radicals (Lissianski et al. 1999). Addition of just 5 to 15 ppm Na_2CO_3 to combustion flue gas resulted in significantly more efficient NO removal. Modeling (Figure 1.40) confirmed that the sodium promoter effect can be explained as a chain reaction of sodium atoms and oxides according to

$$NaOH \longrightarrow NaO_2 \longrightarrow Na \longrightarrow NaO \longrightarrow NaOH\,. \tag{178}$$

The net effect of these reactions,

$$H_2O + HO_2 \longrightarrow 3\,OH\,, \tag{179}$$

is formation of OH radicals from water and non-reactive HO_2 radicals. The OH radicals promote NO reduction by enhancing the formation of NH_2 radicals in

$$NH_3 + OH \longrightarrow NH_2 + H_2O\,, \tag{87}$$

which is followed by their interaction with NO,

$$NH_2 + NO \longrightarrow N_2 + H_2O \tag{117}$$

$$\longrightarrow NNH + OH\,. \tag{180}$$

As these examples demonstrate, small amounts of additives are able to introduce supplemental chain processes that influence the dominant chemistry in two different ways—by forming new active species or by enhancing the concentrations of reactive species already present.

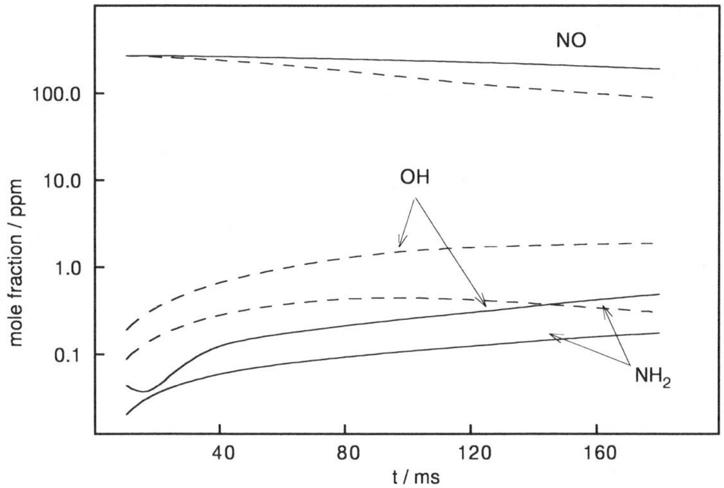

FIGURE 1.40. Promotion of NO reduction in the reburning process by addition of sodium carbonate. The modeling was done for a simulated reburning environment having 500 ppm NO, 750 ppm NH_3, 3.8% O_2, 8% CO_2, and 15% H_2O in N_2 at 1205 K and 1 atm. The solid lines show unpromoted NO and radical profiles; the dashed lines show the profiles computed for addition of 20 ppm NaOH, corresponding to addition of 10 ppm $NaCO_3$.

1.8 REDUCED CHEMISTRY MODELS OF COMBUSTION

At the beginning of this chapter it was emphasized that combustion, even of small fuel molecules, invariably proceeds in a complex way and involves many elementary chemical reactions. There are, nonetheless, valid reasons for seeking means to describe at least parts of the chemistry in simple ways. First, there is computational convenience, sometimes asserted to be computational necessity, when other aspects of combustion, such as heat transfer or flame structure, occupy the attention of a modeler. Second, one may wish to describe a body of combustion data with a simple formula, say for the purpose of establishing correlations among experimental composition variables. Finally, for oxidation conditions restricted to limited ranges, locally valid dependences of combustion features upon composition and temperature, expressed in simple chemical language, may be useful summaries. These motivations may or may not overlap one another.

Because of the diversity of approaches to composing abbreviated chemical descriptions of combustion, no consensus has emerged about how it can best be done. Discussions of the issues involved have been given by Frenklach (1990), Turányi (1990b), Dryer (1991), and Tomlin et al. (1997); a variety of applications have been described by Peters and Rogg (1993), Seshadri and Williams (1994), and Griffiths (1995). These sources provide access to the extensive literature on reduced chemistry models, which is cited only by examples in this section.

1.8.1 One-step chemistry

Two essential requirements have to be fulfilled in order for a simple combustion model to make chemical sense.

The first is reasonably faithful representation of the chemical reaction itself. For some purposes, this could mean adherence to a balanced chemical equation like $C_3H_8 + 5\,O_2 \longrightarrow 3\,CO_2 + 4\,H_2O$. As we have seen, however, combustion reactions do not yield final products like CO_2 and H_2O until the gas has cooled, and then only if excess air has been used so as to overcome the slowness of CO oxidation. This makes it likely that a chemical equation somehow expressing limited extent of reaction may be more useful. The second requirement is that heat release due to reaction has to be accounted for in a way that either asserts the full reaction exothermicity or deliberately underasserts it as a way of simulating limited extent of reaction.

Aside from the obligation to consider mass and energy conservation carefully, a model designer not interested in the chemical detail is free to compose whatever expressions of the dynamic aspects of combustion chemistry that may be suitable to provide reaction and heat release rates that lead to agreement between computed and observed combustion phenomena of interest. If this can be done with a single formal reaction step, the purpose is served. For propane flames, for example, Westbrook and Dryer (1981) developed a model in which the rate of fuel consumption was expressed as

$$\frac{-\mathrm{d}[C_3H_8]}{\mathrm{d}t} = 8.6 \times 10^{11} \exp(-15000\,\mathrm{K}/T)[C_3H_8]^{0.1}[O_2]^{1.65} \ \mathrm{mol\,cm^{-3}s^{-1}}$$

that allowed them to account for laminar premixed propane–air flame speeds over a wide range of equivalence ratio. Expressions of the same form but with different parameters were found to give similar results for a large number of other fuels.

One-step chemistry can readily be amended to account for incomplete reaction (Dryer 1991).

1.8.2 The steady-state approximation and global reaction models

The traditional approach in chemical kinetics to reduce the number of rate equations to be solved, and the number of chemical species appearing in them, is the "steady-state approximation", based on the idea that for given reaction conditions the rates of formation and destruction of one or more of the reactive intermediates may be approximately equal. If there is good reason to believe (for example, on the basis of examining the results of simulations done with a detailed chemical model) that this approximation is appropriate for a combustion system of interest, then setting equal to one another the rate equations for formation and destruction of each intermediate asserted to be in such a steady state can lead to equations for steady-state concentrations of the form

$$\text{Sum of formation rates} = \text{Sum of destruction rates} \tag{181}$$

in which each term in the destruction rate sum is proportional to the intermediate concentration. Whenever such equations are good approximations, they allow the intermediate concentrations to be eliminated from the rate laws for the rest of the species in the model. (Equation (181) can be solved directly for the steady-state intermediate concentration if all of the destruction reactions are first order in the concentration of the steady-state species; when this is not the case, complex quadratic solutions may still be useful.)

As an example of this procedure, Müller and Peters (1993) studied premixed CH_3OH–air flames first with an 18-species, 43-reaction detailed mechanism, then with the 9-species, 5-reaction "global" mechanism

$$CH_3OH + 2\,H \longrightarrow 2\,H_2 + CH_2O \tag{182}$$

$$CH_2O \longrightarrow CO + H_2 \tag{183}$$

$$CO + H_2O \longrightarrow H_2 + CO_2 \tag{85}$$

$$H + H + M \longrightarrow H_2 + M \tag{61}$$

$$3\,H_2 + O_2 \longrightarrow 2\,H + 2\,H_2O \tag{184}$$

that they derived from the detailed mechanism by applying steady-state approximations to the species O, OH, HO_2, H_2O_2, CH, CH_2, CH_3, HCO, and CH_2OH.

Because the global reactions were derived as linear combinations among the 43 elementary reactions, their rates are related to the rate coefficients of the elementary reactions and the steady-state concentrations of the eliminated intermediates. Deriving expressions for the rates of the global reactions (the task was mostly in discovering appropriate approximations to the formulas for the steady-state concentrations) was tedious, and the algebraic formulas that eventually emerged are both approximate and complex. Nonetheless, the end result was found to be that the 5-reaction global mechanism generated concentration profiles of its 9 species that mirrored well, for most of the conditions studied, the corresponding profiles computed using the 43-reaction detailed mechanism from which it was derived. The authors do not comment on the relative computational demand of the premixed flame calculations that were done using the two mechanisms.

Examples of reduced mechanisms derived in similar ways are given by Paczko et al. (1986), Peters and Kee (1987), Bilger et al. (1990), and Glarborg et al. (1992). They share the common limitation that procedures which assume steady-state concentrations may be expected to give results valid only for conditions where steady states in the intermediate concentrations really exist. Accordingly, global mechanisms derived to account for flame properties should not be expected to describe ignition phenomena well. It is not possible to judge the reduction in computational burden gained in using this type of reduced mechanism by comparing it to the detailed one from which it was derived; reducing the number of differential equations to be integrated is offset by the complicated formulas for the rates and steady-state concentrations that have to be evaluated at every integration step.

1.8.3 Empirically derived global mechanisms

The success of one-step chemistry models in reproducing observed combustion phenomena leads naturally to the concept that adding more global steps, preferably guided by appropriate chemical ideas, is a way to widen the scope of the model, with minor additional computational burden, while avoiding all consideration of the complexity of the real chemistry. This concept proves to be valid.

For example, Hautman et al. (1981) reasoned that the essential dominant chemistry of large hydrocarbon combustion is thermal decomposition to olefins and H_2, oxidation of these to CO and H_2O, and finally, oxidation of CO to CO_2. On this basis they proposed the four-step global mechanism

$$C_nH_{2n+2} \longrightarrow \tfrac{n}{2}C_2H_4 + H_2 \quad (185)$$

$$C_2H_4 + O_2 \longrightarrow 2\,CO + 2\,H_2 \quad (186)$$

$$CO + \tfrac{1}{2}O_2 \longrightarrow CO_2 \quad (187)$$

$$H_2 + \tfrac{1}{2}O_2 \longrightarrow H_2O\,. \quad (188)$$

Rate laws for each step were invented, without regard for stoichiometry, and assigned rate coefficients with Arrhenius[1] temperature dependence. After parameter optimization to fit flow reactor and shock tube ignition data, it was found that the model indeed gave satisfactory results.

Replacing one-step chemistry with a model that reflects at least part of the observed sequence of chemical events is clearly a step toward more realistic accounting for the combustion process, and so it is no surprise that such a model can be parameterized more successfully to account in a reasonable way for wider ranges of combustion conditions. As is the case for one-step chemistry, empirically parameterized multi-step global models are and will continue to be useful.

On the other hand, there have to be inherent limitations to any combustion model that captures part of the underlying chemistry while omitting essential features, in this case the facts that combustion chemistry is driven by reactions of active radicals and that additional intermediate species are present. These limitations are addressed in the generalized chemistry models described in Section 1.8.5.

1.8.4 Automated mechanism reduction by sensitivity analysis

An alternative to searching for steadyr-state situations in the predictions of a detailed mechanism is to do automated sensitivity testing of detailed mechanisms so as to identify species and elementary reactions that do not significantly affect the flow of reaction under any of the conditions of interest, then successively eliminate such species and reactions until further truncation introduces deviations larger than one wants to tolerate. In addition to sparing human labor, automated sensitivity analysis is capable of uncovering insensitive species and reactions that one would not find by a steady-state analysis.

---

[1] See Section 1.9.1

This approach, applied to small hydrocarbon combustion, typically leads to mechanisms with 20 to 25 species and 30 to 40 elementary reactions, much smaller than their starting mechanisms, which usually have well over 100 elementary reactions. Because the sensitivity testing can sample any combustion conditions for which the detailed mechanism itself is valid, the resulting reduced mechanism can be developed to retain the full range of validity of the detailed mechanism from which it was derived.

Examples of this approach are provided by Nicol and Malte (1994), Yoshihara et al. (1994), Kazakov and Frenklach (1995), and Sung et al. (1998). The merits of focusing the sensitivity analysis on reaction rates rather than species concentrations are discussed by Turányi (1990c).

1.8.5 Generalized mechanisms: Combustion chemistry in outline form

A completely empirical approach to generating useful models of combustion chemistry is to work not in the "bottom up" direction, starting from a detailed chemical description, but rather "top down", beginning with a combination of generalized reactions that reflects all of the dominant stoichiometric, thermal, and kinetic features of hydrocarbon oxidation. It is not necessary in this approach that any real chemical species appear in the chemical equations, although there may be advantages to doing this. The rate parameters of such a mechanism—like those of the global mechanisms described in Section 1.8.3—have to be calibrated against experimental results, such as ignition delay times or flame speeds, after which the mechanism can be tested for its range of validity by exploring its predictions for other conditions.

This idea was introduced in the "Shell" model of Halstead et al. (1977) and has been extended by a number of authors including Bond et al. (1982), Cox and Cole (1985), Keck and Hu (1986), Cowert et al. (1990), and Griffiths et al. (1994). While purely empirical parameterization, that is, against laboratory data only, is possible in principle for generalized chemistry mechanisms, it does prove to be effective to optimize the model parameters against synthetic data derived from appropriate detailed modeling. One can expect that mechanisms developed this way can provide accurate interpolations within the ranges of conditions where the parameters were established. Their method of creation implies, however, that extrapolations to other conditions may not be reliable. For this reason, test and calibration ranges spanning the entire target set of conditions for which the resulting model is needed are used in model development.

An application of this approach that succeeded in spanning a very wide range of combustion conditions was reported by Schreiber et al. (1994), who were interested in the autoignition properties, in particular the ignition delays, of IC engine fuels of varying octane number. They developed the model shown in Table 1.3 for the primary reference fuels n-heptane and i-octane and their mixtures. The fitting ranges were chosen to represent IC engine ignition conditions—temperatures from 600 to 1500 K, pressures from 3 to 42 bar, and equivalence ratios from 0.8 to 1.6.

Both experimental data and detailed modeling results (Griffiths et al. 1994) were used for calibration and validation.

Another approach to generalized chemistry modeling involves selecting a minimal set of "generalized intermediates" (GI) such that the characteristic dynamic properties which real intermediates have in combustion reactions are all included and can be expressed in mass action rate laws with conventional expressions for the temperature (and in the case of unimolecular reactions, pressure) dependence of rate coefficients. Species profiles generated by detailed combustion chemistry can be matched quite closely with GI mechanisms. The set of reactions used has to take a form suitable for describing the fundamental combustion processes of radical generation, true and degenerate chain branching, and chain termination (Zamansky and Borisov 1989, 1992; Gardiner et al. 1993; Lissianski et al. 1994, 1995a, 1996). GI mechanisms account for the effects of changing fuel composition, equivalence ratio, pressure, and temperature over wide ranges with about half the number of dependent variables reported for lumped species mechanisms derived by the methods described in Section 1.8.3. The essential reason for this is that the structure of the rate equations used in a GI mechanism inherently mimics the real chain nature of combustion reactions. (Global rate equations like those given in Section 1.8.1 do not have this character.)

A GI mechanism for methane oxidation, optionally with added ethane, and concurrent NO_x chemistry is shown in Table 1.4. Its parameters were derived from simulations done using the detailed chemistry of the GRI-Mech 1.2 and 2.11 mechanisms (Frenklach et al. 1995, cf. Section 1.9.4; Bowman et al. 1997). In the Table 1.4 reactions, R and U collectively represent two classes of free radicals (which may include atoms) that have significant concentrations during methane oxidation. R represents the more reactive ones; their concentration during ignition delay is accordingly smaller than that of the less reactive radicals U. In methane ignition, U behaves similarly to the CH_3 radical, while the behavior of R would resemble the behavior of such radicals as OH, H, and O atoms. The stable intermediate species I would include CH_2O, H_2, and C_2H_4, depending on reaction conditions. The product molecules P_1, P_2, and P_3 are surrogates for H_2O, CO, and CO_2, while P_4 mimics in a summary way the chemical behavior of reduced nitrogen-containing species like HCN, HNO, HNCO, and NH_3. Discussions of the relationships between the Table 1.4 reactions and the real elementary reactions participating in combustion of CH_4 and other fuels are given by Gardiner et al. (1993) and Lissianski et al. (1994, 1995a, 1996), as are comparisons between the predictions of the GI and the detailed mechanisms. Examples of the ability of the generalized species to mimic concentration profiles of the real ones are shown in Figures 1.41 and 1.42.

Other generalized chemistry mechanisms with similar scope and performance have been reported by Kiehne et al. (1987) for propane combustion, Revel et al. (1994) for methane combustion and NO_x chemistry, and Jiang et al. (1995) for methane ignition.

TABLE 1.3. Generalized chemistry mechanism for n-heptane and i-octane ignition. Adapted from Schreiber et al. (1994).

Reaction	$\Delta_r H^\circ_{300}$/kJ mol^{-1}	Rate/mol m^{-3} s^{-1}
$F \rightarrow X$	709.9	$5.0 \times 10^8\,[F](p/p_0)^{0.5} e^{-18{,}050/T}$
$X + (11 + 1.5f)\,O_2 \rightarrow P$	−4709.9	$7.0 \times 10^6\,[X][O_2][M]\,e^{-7200/T}$
$F + 2\,O_2 \rightarrow I$	−53.9	$(38.5 - 35f)(11 - 10f) \times 10^9\,[F][O_2][M](p/p_0)^{-2.2}\,e^{-19{,}500/T}$
$I \rightarrow F + 2\,O_2$	53.9	$6.0 \times 10^{27}\,[I](p/p_0)^{-3.5}\,e^{-37{,}500/T}$
$I \rightarrow 2\,Y$	−60.0	$(20 - 14f)(11 - 10f)^{1/2} \times 10^7 [I]\,e^{-5000/T}$
$Y + 0.5\,F + (10 + 1.5f)\,O_2 \rightarrow P$	−3913.1	$1.0 \times 10^9\,[O_2][Y]\,e^{-16{,}500/T}$

Notes: Concentration units are mol/m^3. M represents the set of all species present, such that $[M] = p/RT$.
The generalized chemistry symbols have the following meanings. F represents fuel, either n-heptane or i-octane or a mixture of both, with f representing the fraction of i-octane in the mixture. Since these two hydrocarbons are the primary reference fuels used to define octane number (ON), the value of f is also ON/100. X represents intermediate species prevailing at high combustion temperatures; its concentration can be thought of as $[X] = 3\,[C_2H_4] + [CH_2] + [CH_3] + [H]$. Y represents chain propagating species, which can be thought of as OH or other partially oxygenated active species, that are active at both low and high combustion temperatures. I represents the radicals that dominate at low combustion temperatures; its concentration can be thought of as $[I] = [OC_8H_{15}O_2H]$, which would be the radical product formed from i-octane upon addition of two molecules of oxygen and loss of one molecule of water. P represents final products, in the case of i-octane $8\,CO_2 + 9\,H_2O$.
The influence of fuel composition on the reaction rates enters, through the factor f, in a manner that accounts for the higher rate of chain branching in n-heptane compared with i-octane. The effect of pressure is expressed in the p/p_0 factors, where the reference pressure p_0 was taken to be 10 bar.
Detailed descriptions of the way the parameters were obtained and comparisons with data are given by Schreiber et al. (1994).

TABLE 1.4. Reactions and rate coefficients of a GI mechanism for CH_4 oxidation and NO formation and reduction. (Lissianski et al. 1995a)

	Reaction	A	m	E_a
1	$R + U \rightleftharpoons CH_4$	3.5×10^{33}	-4.88	9.3
	Falloff parameters: 5.0×10^{16} -0.82 1317;		0.354 1370 880	
2	$U + U \rightleftharpoons C_2H_6$	1.26×10^{50}	-9.67	26
	Falloff parameters: 1.51×10^{16} -0.97 2593;		0.231 848 454	
3	$R + CH_4 \rightarrow U + P_1$	7.0×10^{14}		63
4	$R + C_2H_6 \rightarrow 2U + P_1$	5.4×10^{2}	3.5	22
5	$U + O_2 \rightarrow I + R$	1.6×10^{12}		56
6	$U + O_2 \rightarrow I + 2R$	1.2×10^{14}		125
7	$R + I \rightarrow 2R + P_2$	4.0×10^{13}		42
8	$P_2 + O_2 \rightarrow P_3 + R$	1.0×10^{11}		21
9	$R + R + M \rightarrow P_1 + M$	1.0×10^{18}	-0.6	0
	Reactions of NO			
10	$R + N_2 \rightarrow NO + R$	4.8×10^{18}		170
11	$U + N_2 \rightleftharpoons P_4 + NO$	2.5×10^{11}		310
12	$R + P_4 \rightarrow P_1 + NO$	5.0×10^{15}	-0.6	64

Notes: Rate coefficients are given in the form $k = AT^m \exp(-E_a/RT)$. M represents the set of all species present. Units are cm, mol, s, kJ. For reactions (1) and (2), the tabulated parameters refer to the low pressure limit rate coefficients; the first three of the falloff parameters listed for these reactions are A, m, and E_A for their high pressure limit rate coefficients, and the remaining falloff parameters are the a, b, and c values that define the temperature dependence of the broadening factor (Stewart et al. 1989). The $\rightarrow$ notation means that the reaction is taken as irreversible, the $\rightleftharpoons$ notation that forward and reverse reactions are both included. For modeling applications, the thermochemical properties of R, U, I, P_1, P_2, P_3, and P_4 were taken to be those of H, CH_3, CH_2O, H_2O, CO, CO_2, and NH_3, respectively.

1.8.6 Local linearization and eigenvalue analysis

While combustion chemistry is inherently governed by sets of non-linear differential equations, it does not follow that analyzing the behavior of these equations by linearization methods is inappropriate. Indeed, it has been shown that large gains in computational efficiency can be realized by examining the eigenvalues and eigenvectors of locally linearized rate equations for detailed mechanisms and using only the dominant ones to compute the evolution of the system. The fundamental finding is that the apparent complexity expressed in the chemical and rate equations does not lead to a comparably complex eigenvalue spectrum of the linearized equations; the reacting system evolves toward equilibrium in a simpler way than one would assume from looking at the rate equations. Moreover, taking advantage of this simplification is not burdened by the limitation of the steady-state approach that steady-state concentrations must exist. The sacrifice one must make in following this path to computational efficiency, as in the global and steady-state

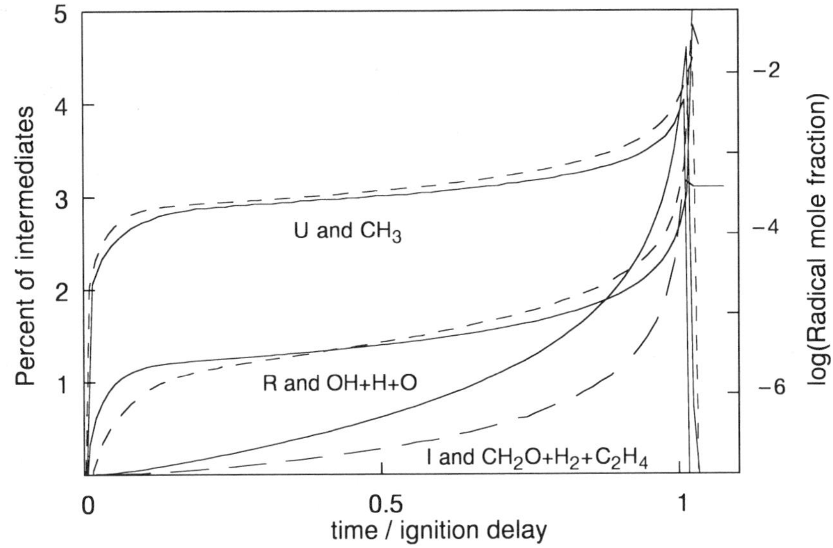

FIGURE 1.41. Comparison of GI (solid lines) with GRI-Mech 2.11 (dashed lines) radical and intermediate profiles for adiabatic constant volume reaction in 9.5% CH_4, 0.6% C_2H_6, and 19.1% O_2 in Ar starting at 1450 K and 3.6 atm.

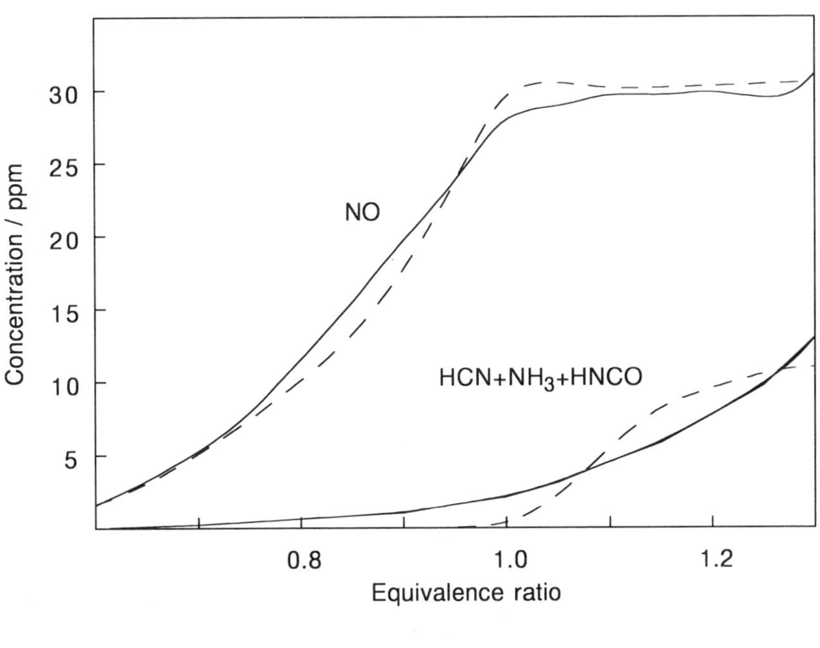

FIGURE 1.42. Comparison of GI (solid lines) and GRI-Mech 2.11 (dashed lines) predictions for postflame NO and reduced fixed nitrogen mole fractions in a freely propagating CH_4–air flame at 1 atm.

cases, is obscuring the chemical flavor and detail, which may or may not make a difference in the modeling effort undertaken.

One can use a linearized equation analysis to assess the importance of species and elementary reactions in a detailed mechanism, and by means of this analysis generate a reduced mechanism. This has been reported for CH_3OH combustion by Goussis and Lam (1992) and for CH_4 combustion by Trevino and Mendez (1992). Such mechanisms necessarily share the limitations of any global mechanism, as described in Section 1.8.2. Alternatively, linearization analysis can be used directly to compute the trajectory in concentration space (the many-dimensional hyperspace in which the coordinates represent the concentrations of chemical species) followed by a combustion reaction. Application of this idea, which was named an "intrinsic low-dimensional manifold" (ILDM) method, to oxidation of a CO–H_2–air mixture in a static system and a flow reactor was described by Maas and Pope (1992a, 1992b). The method was expanded in computational organization to achieve "in situ adaptive tabulation in principal directions" (ISATPD) and applied to methane combustion by Yang and Pope (1998).

1.8.6 Algebraic representation of databases generated from detailed models: Repro-models

Yet another way to compute the evolution of a combustion system efficiently, not necessarily looking away from the chemical detail and its requirement to integrate a large number of differential equations, is to create a large database of information about the hyperspace describing the composition and physical variables of a combustion system by computing many profiles with a detailed model, then represent the evolution so described by algebraic fitting formulas (Tomlin et al. 1997). For use in computational fluid dynamics simulations, the idea is that one can alternately compute the chemical and the physical changes at each step of the solution, the former being derived by algebraic interpolations within the chemical database. The size of the required database is much smaller than one might think from the number of variables, as most of the hyperspace is never traversed by the simulations done with the detailed model. Expressed chemically, if one radical concentration is high, so are the other radical concentrations and so is the temperature.

It is not obvious how the inherently differential nature of a system of rate equations can be transformed at all into an algebraic form. The basic idea is that for a fixed time step, any set of initial values for the chemical and physical variables can be associated, by integrating the system of rate equations, with its corresponding set of values at the end of the time step. It was recognized by Dunker (1986) and Marsden et al. (1987) that one can systematically derive polynomial expressions that describe the chemical transformations between initial and later values even for the highly nonlinear differential equations of combustion reactions. In a combustion model, one can then alternate between solving the equations of fluid dynamics without chemical reaction and solving the algebraic equations describing chemical reaction without fluid dynamics, an approximation called "operator splitting".

It is also not obvious that the "data" generated by numerical integrations that span the concentration hyperspace of interest in a combustion problem can be trusted to be represented correctly and usefully in algebraic form. This question has been studied in some depth, and it has been shown that useful algebraic forms can indeed be found and systematically parameterized for typical combustion applications (Meisel and Collins 1973; Spivakovsky et al. 1990; Turányi 1994a, 1994b).

An example of this "repro-modeling" approach to the constant pressure ignition of a stoichiometric mixture of 20 CO + H_2 in air initially at 1000 K was reported by Turányi (1994b). The procedure was to compute the parameters of an algebraic representation from data at 100 time steps of 10 ms generated by 300 numerical integrations (giving 30,000 data sets) of the ignition of the 20:1 CO-H_2 mixture over the ranges $0.5 \leq \phi \leq 1.5$ and $990 \leq T/\mathrm{K} \leq 1010$ using the 13-species, 67-reaction mechanism constructed by Maas and Warnatz (1989). The data chosen for representation were the values of [CO], [O_2], and the temperature before and after each step; excluding the values at temperatures below 1100 K reduced the number of data sets to about 20,000. The [CO], [O_2], and temperature profiles computed by the repro-model were nearly identical to those computed by numerical integration, but took a factor of 12,000 less computer time. Detailed comparisons of repro-modeling with the steady-state approximation and the Maas and Pope (1992a, 1992b) ILDM method are given both in the original report and by Tomlin et al. (1997). A combination of repro-modeling with the ILDM method was presented by Turányi et al. (1998).

1.8.7 Chemical lumping methods

Combustion scientists accept that chemical representations of liquid and solid fossil fuel combustion are inevitably simplifications, because there is neither a possibility nor a purpose to create a model taking notice of all of the species contained in or derived from burning these fuels. The best that one can ever do is develop models of the combustion of surrogate molecules, or their mixtures, as in the practice of representing spark-ignited IC engine fuel as a mixture of n-heptane and i-octane. Extension of this idea can be done in ways that go beyond the global and generalized chemistry representations discussed so far. In particular, one can use chemical knowledge of the structure of reaction mechanisms to create combustion models that self-consistently express the basic chemical ideas to be tested in a manner that can be used in normal combustion modeling.

An example of this approach in connection with modeling the formation of soot was described by Frenklach (1990). Growth of polyaromatic hydrocarbon (PAH) concentrations in the early stages of soot formation in flames was formulated as a set of chemical reactions of aromatic molecules with i fused rings, successively larger i-values being created in radical reactions with acetylene:

$$A_i + \mathrm{H} \rightleftharpoons A_i^* + \mathrm{H_2} \tag{189}$$
$$A_i^* + \mathrm{C_2H_2} \rightleftharpoons A_i\mathrm{CHCH}^* \tag{190}$$
$$A_i\mathrm{CHCH}^* + \mathrm{C_2H_2} \longrightarrow A_{i+1} + \mathrm{H} \tag{191}$$
$$A_{i+1} + \mathrm{H} \rightleftharpoons A_{i+1}^* + \mathrm{H_2} \tag{192}$$
$$A_{i+1}^* + \mathrm{C_2H_2} \rightleftharpoons A_{i+1}\mathrm{CHCH}^* \tag{193}$$
$$A_{i+1}\mathrm{CHCH}^* + \mathrm{C_2H_2} \longrightarrow A_{i+2} + \mathrm{H} \tag{194}$$

and so on. Reactions (189–191) and (192–194) are chemically similar to one another, that is, they describe formation of an aromatic radical by abstraction of an H atom, followed by successive addition of two acetylene molecules, ring closure, and loss of an H atom to create another fused ring system with one more ring than at the start. The sets may be expected to have rate coefficients that are equal or related to one another in a simple way. By assuming that this sequence continues indefinitely, a set of differential equations for the "lumped" species concentrations $c_1 = \sum[A_i]$, $c_2 = \sum[A_i^*]$, and $c_3 = \sum[A_i\mathrm{CHCH}^*]$ could be derived and solved numerically. In combination with methods to describe the more complex process that develops when the PAH concentration becomes large, causing bimolecular reactions of two PAH molecules to become significant, it was found that the entire course of soot formation could be modeled efficiently.

1.9 RESOURCES FOR COMBUSTION CHEMISTRY MODELING

Physical models can only be as reliable as the parameters chosen for them. For combustion modeling one has three essentially different parameter classes, for thermochemistry, transport properties, and reaction rates. Thermochemical parameters for combustion modeling are discussed in Chapter 5. The discussion of transport property parameters given by Dixon-Lewis (1984) and the computer coding suggested by Kee et al. (1983) remain basically valid, with more realistic accounting for strong intermolecular forces being the main concern for future improvements (Paul and Warnatz 1998).[1]

What remains to be resolved are the problems one encounters in selecting a set of elementary reactions and assigning values for their rate parameters. The selection process is an evolutionary one; it is commonly based on review of the antecedent literature and modifying earlier mechanisms to make them better suited to the task at hand. An overview of the literature on combustion mechanisms is provided in Section 1.9.2. First we address the question of assigning rate parameters.

[1] The estimation procedures described by Reid et al. (1987) can be used to compute transport coefficients for uncommon species.

1.9.1 Elementary reaction rate coefficient data

In combustion modeling, the temperature dependence of rate coefficients for bimolecular and termolecular elementary reactions[1] is nearly always represented by equations of the form

$$k = A\,T^m\,\mathrm{e}^{E_\mathrm{A}/\mathrm{R}T} \tag{195}$$

where R is the gas constant (8.314 kJ/molK in the SI units used in this book) and the three "rate parameters" are called the "Arrhenius A-factor", which in combustion modeling is generally expressed in the units s^{-1} for unimolecular reactions, $\mathrm{cm^3mol^{-1}s^{-1}}$ for bimolecular reactions and $\mathrm{cm^6mol^{-2}s^{-1}}$ for termolecular reactions, "temperature exponent", and "activation energy". Sometimes experimental data or theoretically derived temperature dependences may be well represented with m or E_A values of zero, which implies two-parameter formulas. (Occasionally, authors choose to report E_A/R rather than E_A values in order to avoid questions of energy units.) In constructing a database for elementary reaction rate coefficients one is therefore confronted with the need to find, for each reaction included in the mechanism, the best values one can of its two or three rate parameters.

Fortunately, there is a substantial review literature available to help in the search for these parameters. The resources listed in this section provide access to nearly all of the experimentally derived values ever reported for gas-phase elementary reactions. Some additional possibilities may be found in the citations of archival literature provided by the authors of comprehensive reaction mechanisms mentioned in the following section.

Less fortunately, nearly all of the experimental measurements of elementary reaction rate coefficients at combustion temperatures are for reactions of small molecules, that is, with fewer than four atoms other than hydrogen. A consequence of this is that many of the rate parameters found in combustion models are perforce estimated by various methods. Some estimation methods are based on theories of elementary reaction rates, described for example by Zellner (1984), Laidler (1987), and Westmoreland (1992); others are based on rules developed to account for trends in experimental rate parameters for small-molecule reactions, for example, the procedures described in Section 2.4 of Chapter 2, Section 4.6 of Chapter 4, and by Ranzi et al. (1992); and some are based entirely on theoretically calculated molecular properties and reaction rate theory. References to descriptions of the various estimation methods that have been proposed can be found in the validated reaction mechanism literature discussed in the following section.

In this section, we provide a guide to the review literature on elementary reactions of C/H/O species complementary to the reviews of combustion-relevant elementary reactions of N-, S-, and Cl-containing species in Chapters 2, 3, and 4.

Critical evaluations of the rate parameters for about 200 elementary combustion reactions were carried out by Baulch et al. (1992, 1994), expanding on a series of

[1] The more complex problem of representing the pressure-dependent rate coefficients needed for unimolecular reactions is discussed in Section 2.3.1 of Chapter 2.

earlier efforts at Leeds University extending over more than two decades. Their comprehensive reports have served as the basis for a number of mechanisms used in combustion modeling. An attempt was made to include all relevant reactions of C_1 and C_2 species.

A number of critical reviews of elementary reaction rate coefficients have been prepared at the U.S. National Institute of Science and Technology. These include the following. Tsang and Hampson (1986) compiled an evaluation for elementary reactions involved in the combustion of methane and related compounds. Tsang (1987) evaluated kinetic data for elementary reactions involved in the combustion of methanol. Tsang (1988) evaluated kinetic data for elementary reactions involved in the combustion of propane. Tsang (1990a) evaluated kinetic data for elementary reactions involved in the combustion of isobutane. Reactions of atomic oxygen with organic compounds in the gas phase were reviewed by Herron (1988). Reactions of propene (C_3H_6) were reviewed by Tsang (1991). Rate parameters for reactions of fluorinated hydrocarbons are available at the URL `http://www.nist.gov/cstl/div836/ckmech.html`.

Other critical reviews of elementary reactions relevant to combustion modeling have been prepared by Kondratiev (1972, with a comprehensive review of the older literature and including many reactions not to be found in other evaluations), Warnatz (1984, including reactions up to C_4 species), and Cohen and Westberg (1991). Atkinson (1997) reviewed a large number of elementary reactions of alkanes and alkenes, including the reactions with O and OH. Elementary gas reactions related to atmospheric modeling have been reviewed on a regular basis by expert committees working under the auspices of the International Union of Pure and Applied Chemistry and U.S. National Aeronautics and Space Administration (Baulch et al. 1980, 1982, 1984; Atkinson et al. 1989, 1992, 1997; Carver and Cox 1999). While the primary focus of these reviews has been on elementary reactions at low (300 K and below) temperatures, the recommendations given in these reports are nonetheless relevant to combustion modeling, particularly for photochemical and chemiluminescent reactions, and reactions of electronically excited species. Information about the current NASA recommendations can be found at the URL `http://remus.jpl.nasa.gov/jpl97`;the current IUPAC recommendations can be found at `http://www.iupac-kinetic.ch.cam.ac.uk/`.

The most comprehensive reference to the archival and review literature of gas-phase elementary reactions is that of Mallard et al. (1998), who offer a large collection of data and literature citations but make no rate parameter recommendations. The literature through early 1998 was reviewed; over 15,000 gas-phase elementary reactions are included, and tools for searching, plotting, and saving data are provided. Information can be found at `http://www.nist.com/srd/nist17.htm`.

1.9.2 Validated reaction mechanisms

Combustion chemistry models have evolved in a complicated manner, characterized not only by inheritance of elementary reaction sets and rate parameters as reaction mechanisms pass through the hands of successive collaborators, but also by frequent cross-fertilization between mechanisms that do not have common ancestors. Following this evolution in the literature is difficult, because different lead authors appear in publications of what are actually successive generations of the same mechanism, principal authors may move from one institution to another as their mechanisms evolve, the names of some authors appear in more than one genealogical branch, and there are historically important mechanisms that are not comprehensively documented in the open literature at all or are described in fragmented form over several publications, some with limited distribution.

A consequence of the evolutionary process is that earlier versions of most combustion mechanisms should be regarded as obsolete. The following summaries therefore mainly offer comments on and access to only the current versions of hydrocarbon combustion mechanisms. Their history can usually be discovered by working backwards through the cited literature.

An apology is offered to the many authors whose mechanism development efforts are not reflected in the citations given below.[1]

Long-term mechanism development projects

The previous edition of this book included the hydrocarbon combustion mechanism of Warnatz (1984), which has since been extended to describe liquid fuels, been revised several times, been subjected to extensive validation studies, and served as the base mechanism for a number of mechanism reduction projects. Descriptions can be found in Chevalier et al. (1990a, 1990b, 1992), Chevalier (1992), Warnatz et al. (1996), and Nehse et al. (1996).

Combustion mechanisms developed earlier as individual efforts at Louisiana State University, Pennsylvania State University, SRI International, Stanford University, and the University of Texas at Austin were merged into a collaborative effort, sponsored by the Gas Research Institute, that developed the GRI-Mech model. The version current at the time of this writing includes nitrogen chemistry and describes combustion of gaseous fuels up to propane. Information can be obtained at `http://www.me.berkeley.edu/gri_mech/`.

A long-term combustion mechanism development project has been sustained at Lawrence Livermore National Laboratory. Some of the later results have been described by Castaldi et al. (1996), Curran et al. (1995, 1996a, 1996b, 1998), Pitz et al. (1984), Westbrook and Pitz (1984), and Westbrook et al. (1988, 1991). These publications describe tests of mechanisms developed to describe a very

---

[1] Neither Table 1.5 nor any of the discussion in this Chapter address modeling of soot formation and oxidation chemistry. This very important modeling research area has not reached a level of maturity that would allow outsiders to access and use soot chemistry models independently, as is possible with modeling the combustion of the common fuels. An overview of the chemical and physical issues involved can be gained from the review of Wagner (1979), the reports on soot modeling in Bockhorn (1994), and the many articles on soot formation appearing in the proceedings of the biennial Combustion Symposia.

wide variety of combustion conditions and fuels, including the primary reference fuels used as surrogates for common liquid hydrocarbon fuels.

An often-cited mechanism that combines a small hydrocarbon oxidation mechanism with nitrogen chemistry is that of Miller and Bowman (1989, 1991), which has as a partial descendant the Glarborg et al. (1998) mechanism.

Another long-term study of detailed combustion mechanisms for hydrocarbon fuels has been carried out at the Centre National de la Recherche Scientifique in Orléans. Validations have been reported for fuels ranging from methane to kerosene over the full range of conditions for which there are laboratory data or practical interest. Some recent reports are those of Chakir et al. (1992), Corre et al. (1992), Dagaut et al. (1991, 1994, 1995) and Tan et al. (1994a, 1994b, 1994c).

Mechanism development projects at the École Nationale Supérieure des Industries Chimiques de Nancy and the Université de Nancy have been described by Côme et al. (1996), Barbé et al. (1995), Baugé et al. (1997), Glaude et al. (1997) and Warth et al. (1998). A URL describing the Nancy mechanisms and programs for generating rate and thermochemical parameters is `http://www.ensic.u-nancy.fr/ENSIC/DCPR/chiminfo/mecanismes`.

Yet another long-term effort has been sustained at the Politecnico di Milano. Some recent mechanism studies are reported by Callahan et al. (1996), Faravelli et al. (1998), Goldaniga et al. (1998), and by Ranzi et al. (1994a, 1994b, 1995a, 1995b, 1997a, 1997b).

Mechanism development projects for small hydrocarbon combustion were undertaken at the Institute of Chemical Physics in Moscow in the 1980s and early 1990s. Konnov (1996) built upon experience gained in those studies and composed a mechanism of over 1000 elementary reactions of 121 species describing combustion of hydrocarbons and their derivatives up to C_3-species. It can be found at the URL `http://homepages.vub.ac.be/~akonnov`.

A mechanism development effort addressing a wide range of practical combustion problems, including proposals for the detailed chemistry of reactions leading to formation of heavy hydrocarbons and soot, was undertaken at Imperial College of Science, London. Results have been described by Lindstedt and Maurice (1995), Lindstedt and Skevis (1996), and Lindstedt (1998).

A cooperative mechanism development project that involves the School of Chemistry and the Fuel and Energy Department at Leeds University, and the Central Research Institute for Chemistry of the Hungarian Academy of Sciences in Budapest, led to a mechanism for methane combustion, with extensions to include nitrogen and sulfur combustion chemistry. They are described, together with a variety of programs developed to analyze reaction mechanisms, at the URL `http://chem.leeds.ac.uk/Combustion/Combustion.html`.

Mechanism development projects at Princeton University, in part as collaborations with the Lawrence Livermore Laboratory effort noted above, have included applications of sensitivity analysis in mechanism development and detailed validations against flow reactor data covering the intermediate temperature range between experiments in static reactors and experiments in flames and shock tubes. Recent reports include those of Curran et al. (1998) and Held and Dryer (1998).

Readily accessed combustion mechanisms

The sets of elementary reactions used in the projects mentioned above, and the corresponding thermochemical and kinetic parameters, are in many cases difficult to extract from the literature. In Table 1.5, an overview is given of well-studied mechanisms that are also well documented in the archival literature and/or available on the Internet.

TABLE 1.5. Sources for detailed hydrocarbon combustion mechanisms.[1]

Mechanisms for modeling H_2, CO, and CH_2O oxidation.	
Frenklach et al. (1995)	The GRI-Mech H_2–CO–CH_2O chemistry; 45 elementary reactions of 12 species.
Konnov (1996)	An H_2 oxidation submechanism with 23 reactions of 8 species.
`http://chem.leeds.ac.uk/Combustion/Combustion.html`	The Leeds H_2 oxidation submechanism has 22 reactions of 8 species; forward and reverse reactions are entered separately.
Yetter et al. (1991) Kim et al. (1994) Held and Dryer (1998) Lee and Hochgreb (1998)	The Princeton H_2–CO–CH_2O submechanism has 41 reactions of 12 species.
Dagaut et al. (1991)	The CNRS model has 36 reactions describing the oxidation of H_2 and CO.
Warnatz (1992) Warnatz et al. (1996) Chevalier (1992)	Update of the Warnatz (1984) chemistry; 19 reactions of 8 species for H_2–O_2.
Miller and Bowman (1989, 1991)	The Miller–Bowman H_2–CO oxidation submechanism; 30 reactions of 12 species.
Mechanisms for modeling CH_4, C_2H_6, C_2H_4, and C_2H_2 combustion.	
Frenklach et al. (1995) `http://www.me.berkeley.edu/gri_mech`	The GRI-Mech Version 1.2 small hydrocarbon chemistry adds 132 elementary reactions of 18 species to its H_2–CO–CH_2O chemistry.
`http://chem.leeds.ac.uk/Combustion/Combustion.html`	The Leeds methane oxidation mechanism adds 152 reactions of 27 additional species to its H_2 oxidation mechanism. Forward and reverse reactions are entered separately.
Battin-Lerclerc and Barbé (1997) `http://www.ensic.u-nancy.fr/ENSIC/DCPR/chiminfo/mecanismes/COC2ReactionsDataBase`	The Nancy mechanism for oxidation of C_1 and C_2 hydrocarbons has 426 reactions of 52 species.[2]

Konnov (1996)	A mechanism with 1027 reactions of 121 species. Nitrogen chemistry is included. Extensive validations are presented.
Ranzi et al. (1994a)	A mechanism with 1600 reactions of 70 species; includes CH_3CHO chemistry.
Warnatz et al. (1996) Chevalier (1992)	Contains revisions and extensions of the Warnatz (1984) mechanism.
Sloane (1989)	An independently assembled, comprehensive mechanism for CH_4 oxidation.
Dagaut et al. (1991)	A 200-reaction small hydrocarbon oxidation mechanism.
Kiefer et al. (1992)	A thoroughly tested 86-reaction mechanism for C_2H_2 pyrolysis.

Mechanisms for modeling C_3H_8 combustion.

Qin (1998)	A supplement to the GRI-Mech chemistry; 76 reactions of 10 C_3H_n species, $3 \le n \le 8$, describing C_3H_8 oxidation.
Dagaut et al. (1992b)	391 reactions of 57 species validated against a variety of C_3H_8 oxidation data.
Sloane (1992)	An extension of the Sloane (1989) methane oxidation mechanism to describe C_3H_8 oxidation.
Konnov (1996)	1027 reactions of 121 species.
Leung and Lindstedt (1995)	Oxidation of C_1 through C_3 hydrocarbons. Many of the 451 reactions concern larger species involved in PAH and soot formation.

Mechanisms for modeling combustion of alcohols.[3]

Held and Dryer (1998)	Methanol oxidation, described by 89 elementary reactions of 22 species including Ar and N_2. It includes the Princeton H_2–CO–CH_2O chemistry; extensive validations and discussions of earlier mechanisms are presented.
Norton and Dryer (1992)	Ethanol oxidation, described by 142 elementary reactions. It is based on the Princeton H_2–CO–CH_2O chemistry; extensive validation is presented.
Egolfopoulos et al. (1992)	A 171-reaction mechanism for methanol oxidation.

Mechanisms for modeling liquid hydrocarbon combustion.[4]

Côme et al. (1996) Glaude et al. (1997, 1998) Baugé et al. (1998) Warth et al. (1998) `http://www.ensic.u-nancy.fr/ENSIC/DCPR/chiminfo/textes/chiminfo.htm`	The Nancy butane oxidation mechanism has 352 reactions of 112 species added to the Nancy base C_0–C_2 oxidation mechanism. Links on this URL also provide mechanism files for oxidation of n-heptane, n-octane, iso-octane, and n-decane.
Ranzi et al. (1994a, 1994b, 1995b) Callahan et al. (1996) Gaffuri (1995) Gaffuri et al. (1996)	Mechanisms describing oxidation of n-heptane and iso-octane.
Nehse et al. (1996)	N-heptane and decane oxidation; the method of mechanism formulation is described, but the actual reaction set, generated automatically by a computer program, is not listed.
Lindstedt and Maurice (1995)	N-heptane oxidation.
Roberts et al. (1996)	An iso-octane oxidation mechanism containing 131 reactions of 103 species. Validated for exhaust gas composition under autoignition conditions.
Curran et al. (1998a) `curran6@llnl.gov`	Mechanism with 4060 reactions of 990 species describing oxidation of n-heptane and iso-octane and their mixtures.

Mechanisms for modeling combustion of aromatic compounds.

Tan and Frank (1996)	A benzene oxidation mechanism; 75 reactions of C_5 and C_6 species were added to the Tan et al. (1994a) reactions to give a mechanism with 338 reactions of 64 species.
Zhang and McKinnon (1995)	Benzene combustion, described by 514 reactions.

Note 1. The purpose of this table is to provide reasonably direct access to a variety of hydrocarbon combustion mechanisms in current use, not to give a systematic or exhaustive survey of the recent literature of detailed combustion modeling. Citations of relatively difficult-to-locate sources and obsolete or incompletely documented mechanisms have been minimized. Careful reading of the papers and searches through several citations are nonetheless necessary in order to reconstruct some of the mechanisms listed. Contacting the authors may enable easier acquisition of mechanism files than transcription from print versions. Note 2. The file name contains the capital letter "O", not the numeral "0". Note 3. Methanol, and sometimes also ethanol, chemistry is usually included in

1.9.3 Combustion modeling software

In many combustion modeling applications, the computer code needed for solving the chemical part of the modeling is embedded in large special-purpose programs developed for investigating a particular class of reactive fluid dynamics problems. Even in pursuing such applications, however, it is informative to survey the relevant chemistry using general-purpose chemical kinetics modeling programs.

The most popular general-purpose program for combustion modeling is the Chemkin program developed at Sandia National Laboratory (Kee et al. 1980, 1983, 1985, 1986, 1989; Kee and Miller 1986; Glarborg et al. 1986; Lutz et al. 1988). It includes thermochemical and transport property databases and Fortran subroutines to model CSTR reactions and premixed flames as well as carry out sensitivity analysis. Information about the current version, Chemkin-III, can be obtained at the URL `http://www.ReactionDesign.com`; a PostScript version of the user's manual for the previous version, Chemkin-II, can be found at the URL `http://stokes.lance.colostate.edu/chemkinmanualslist.html`. Auxiliary programs for Chemkin have been described. These include programs for automatic modification of Chemkin input files and analysis of Chemkin output files, MECHMOD and KINALC (Turányi 1990a, 1990b), available at `http://chem.leeds.ac.uk/combustion/combustion.html`. For graphical display of Chemkin output profiles and related information there is XSenkPlot, a freeware UNIX and XWindows/OSF/Motif program, available through the URL `http://fluid.nist.gov/836.03/xsenkplot/index.html`.

A program that accepts Chemkin-style input and performs basic combustion simulations is described at the URL `http://members.aol.Jelezniak/`.

A program similar to Chemkin is HCT, developed at Lawrence Livermore National Laboratory (Lund 1978; Lund and Chase 1995). LSENS, a set of general-purpose programs offering somewhat different options, was developed at NASA Lewis Research Center (Radhakrishnan 1993b, 1994; Radhakrishnan and Bittker 1994; Bittker and Radhakrishnan 1994). Earlier programs of similar nature were described by Turner et al. (1970), Walker (1970), Bittker and Scullin (1972), and Deuflhard et al. (1981).

It is sometimes useful to have a small program able to integrate the simplest case of many-reaction kinetics, isothermal constant volume reaction without automatic inclusion of reverse reactions. Early programs of this kind were described and made available for public use by Stabler and Chesick (1978), Brown (1981) and Braun et al. (1988). A Wintel application able to integrate, using the

---

modern small hydrocarbon mechanisms, including those of Chevalier (1992), Dagaut et al. (1992b), Ranzi et al. (1994a), Frenklach et al. (1995), Konnov (1996), and others. The extent of validation against alcohol combustion data varies.

Note 4. Recognizing that the full isomeric complexity of all pyrolytic and oxidative pathways cannot be included in liquid fuel combustion models, most authors describe these as "semi-detailed" mechanisms. The rate and thermochemical parameters used are nonetheless derived as for ordinary elementary reactions.

Gear method, up to 100 reversible reactions of up to 50 species, with up to five reactants and five products in each reaction, is offered by ARSoftware at the URL `http://www.ari.net/ars/`. This company also offers Wintel and Macintosh applications for computing thermochemical properties, including equilibrium composition of multiphase systems.

Modeling services are also being provided over the World Wide Web, albeit without, at the time of this writing, user control of the reaction mechanism. URLs of interest for combustion modeling are `http://adam.caltech.edu/tcc/` and `http://diesel.berkeley.edu/~gri_mech/cal22/`. A variety of combustion thermodynamics calculations (and some related ones) can be done at the URL `http://kahuna.sdsu.edu/testcenter`. Commercial programs capable of accepting small chemistry models in fluid dynamics calculations that include turbulence and heat transfer are available for modeling geometrically complex flows. The URL for information about one family of them, the FLUENT and GAMBIT programs, is `http://www.fluent.com/software/software.htm`.

1.9.4 Notes on the mechanism used in this chapter

The hydrogen–oxygen, hydrogen–oxygen–carbon monoxide and hydrocarbon oxidation modeling reported in this chapter was done with the GRI-Mech 1.2 model (Frenklach et al. 1995), which comprises 177 elementary reactions of 32 chemical species, two of which (Ar and N_2) are inert. The GRI-Mech 1.2 thermochemistry was used as well.

For modeling ethanol oxidation, 37 reactions of C_2H_5OH, C_2H_5O, CH_3CHOH, CH_2CHO, and CH_3CHO were added to the GRI-Mech 1.2 mechanism from the mechanism of Chevalier (1992). For modeling acetaldehyde oxidation, a 28-reaction submechanism for the reactions of CH_3CHO, CH_3CO, and CH_2CHO was added. Two of the 28 were thermal decompositions, of CH_3CHO and CH_3CO; 24 of the remainder were abstractions of each methyl and aldehydic hydrogen atom by attack of H, O, OH, HO_2, CH_3, and O_2; the final two were triplet CH_2 abstraction of the aldehydic hydrogens of CH_3CHO and CH_2CHO. The rate parameters for all abstractions of aldehydic hydrogens were those of Baulch et al. (1992) for abstraction from CH_3CHO itself, except for the CH_2 reactions, for which the parameters were taken from Böhland et al. (1985). Rate coefficient expressions for abstraction of methyl and methylene hydrogens were taken as one-half of the corresponding expressions of Baulch et al. (1992) for abstraction of a hydrogen atom from C_2H_6. The overall result of this submechanism was conversion of CH_3CHO into ketene (CH_2CO), which is already described by the GRI-Mech 1.2 chemistry, and other products.

The nitrogen chemistry in Table 2.19 of Chapter 2 was added for the modeling described in Section 1.4. Bromine chemistry modeling was done with the addition of the Babushok (1998) bromofluorocarbon chemistry, 67 reactions of 13 bromine-containing species.

The additional thermochemistry needed for the ethanol and acetaldehyde modeling was taken from Burcat and McBride (1997); the additional thermochemistry needed for nitrogen- and bromine-containing species was that described in Chapter 2 and at the Babushok et al. (1998) URL, respectively.

The effect of sodium carbonate addition on the SNCR process was modeled by adding to the C–H–O–N combination a set of 23 reactions of the four sodium-containing species NaOH, NaO, NaO_2, and Na (Lissianski et al. 1999). Because the evaporation and decomposition of liquid sodium carbonate is very fast at SNCR temperatures, NaOH, the most stable sodium-containing species in the gas phase, was assumed to be the only sodium-containing species initially present. The sodium thermochemistry was that of McBride et al. (1993) except for the thermochemistry of NaO_2, for which the standard enthalpy of formation was taken as the recommendation of Partridge et al. (1992).

The composition of "air" used in our calculations was taken to be that of sea level dry air set as the 1962 U.S. Standard Atmosphere (NASA 1962).[1] It contains 78.084% N_2, 20.948% O_2, 0.934% Ar, 0.031% CO_2, 0.002% Ne, 0.001% He, and several additional components at ppm levels and below, on a molar basis. For the calculations in this chapter, the Ne and He content were merged into the Ar content, while the CO_2 content was retained.

1.10 REFERENCES

Arand, J.K., Muzio, L.J., & Sotter, J.G. (1982). *Urea Reduction of NOx in Combustion Effluents,* U.S. Patent 4,208,386.

Aronowitz, D., Naegeli, D.W., & Glassman, I. (1979). 17th Symposium (International) on Combustion, p. 633.

Aronowitz, D., Santoro, R.J., Dryer, F.L., & Glassman, I. (1977). J. Chem. Phys., **81,** 2555.

Atkinson, R. (1997). J. Phys. Chem. Ref. Data, **26,** 215.

Atkinson, R., Baulch, D.L., Cox, R.A., Hampson, R.F., Jr., Kerr, J.A., & Troe, J. (1989). J. Phys. Chem. Ref. Data, **18,** 881.

Atkinson, R., Baulch, D.L., Cox, R.A., Hampson, R.F., Jr., Kerr, J.A., & Troe, J. (1992). J. Phys. Chem. Ref. Data, **21,** 1125.

Atkinson, R., Baulch, D.L., Cox, R.A., Hampson, R.F., Jr., Kerr, J.A., Rossi, M.J., & Troe, J. (1997). J. Phys. Chem. Ref. Data, **26,** 521, 1329.

[1] The composition of real air varies by location as well as in moisture and dust content. Of possible interest in combustion research are the CH_4 content, variously quoted in the 1 to 2 ppm range, the 0.5 ppm H_2 content, and the 0.5 ppm N_2O content.

Babushok, V. (1998). NIST CKMech bromofluorocarbon mechanism; at the URL `http://fluid.nist.gov/836.03/ckmech/Mechanism.html`.

Babushok, V., Noto, T., Burgess, D.R.F., Hamins, A., & Tsang, W. (1996). Combust. Flame, **107,** 351.

Babushok, V., Noto, T., Burgess, D.R.F., Hamins, A., & Tsang, W. (1997a). Combust. Flame, **108,** 61.

Babushok, V., Tsang, W., Linteris, G.T., & Reinelt, D. (1998). Combust. Flame, **115,** 551.

Babushok, V., Tsang, W., Linteris, G.T., & Reinelt, D. (1997b). Proc. Halon Options Tech. Working Conf., Center for Global Environmental Technologies, New Mexico Engineering Research Institute, p. 55..

Bader, G., & Deuflhard, P. (1983). Numerische Mathematik, **41,** 399.

Baer, T., & Hase, W.S. (1996). *Unimolecular Reaction Dynamics: Theory and Experiments,* Oxford University Press, Oxford.

Barnard, J.A. (1977). In *Comprehensive Chemical Kinetics,* Vol. 17, C.H. Bamford and C.F.H. Tipper, Eds., Elsevier, Amsterdam, p. 441.

Barbé, P., Battin-Leclerc, F., & Côme, G.M. (1995). J. Chim. Phys., **92,** 1666.

Battin-Leclerc, F., & Barbé, P. (1997). `http://www.ensic.u-nancy.fr/ENSIC/DCPR/chiminfo/mecanismes/COC2ReactionsDataBase`.

Baugé, J.C., Battin-Leclerc, F., & Baronnet, F. (1998). Int. J. Chem. Kinet., **30,** 629.

Baugé, J.C., Glaude, P.A., Pommier, P., Battin-Leclerc, F., Scacchi, G., Côme, G.M., & Paillard, C. (1997). J. Chim. Phys., **94,** 460.

Baulch, D.L., Cox, R.A., Hampson, R.F., Jr., Kerr, J.A., Troe, J., & Watson, R.T. (1980). J. Phys. Chem. Ref. Data, **9,** 295.

Baulch, D.L., Cox, R.A., Crutzen, P.J., Hampson, R.F., Jr., Kerr, J.A., Troe, J., & Watson, R.T. (1982). J. Phys. Chem. Ref. Data, **11,** 327.

Baulch, D.L., Cox, R.A., Hampson, R.F., Jr., Kerr, J.A., Troe, J., & Watson, R.T. (1984). J. Phys. Chem. Ref. Data, **13,** 1259.

Baulch, D.L., Cobos, C.J., Cox, R.A., Esser, C., Frank, P., Just, Th., Kerr, J.A., Pilling, M.J., Troe, J., Walker, R.W., & Warnatz, J. (1992). J. Phys. Chem. Ref. Data, **21,** 411.

Baulch, D.L., Cobos, C.J., Cox, R.A., Frank, P., Hayman, G., Just, Th., Kerr, J.A., Murrells, T., Pilling, M.J., Troe, J., Walker, R.W., & Warnatz, J. (1994). Combust. Flame, **98,** 59.

Bauschlicher, C.W., & Partridge, H. (1993). Chem. Phys. Lett., **215,** 451.

Bendtsen, A.B., Glarborg, P., Dam-Johansen, K., Kristensen, P.G., & Karll, B. (1998). Poster W2A05, 27th Symposium (International) on Combustion, Boulder, to be published.

Benson, S.W. (1981). Prog. Energy Combust. Sci., **7,** 125.

Benson, S.W. (1982). Oxid. Commun., **2,** 169.

Bhaskaran, K.A., Frank, P., & Just, Th. (1980). 13th International Symposium on Shock Tubes and Waves, p. 503.

Bhaskaran, K.A., Ravikumar, R., Karuppannan, K.M., & Natarajan, K. (1981). 1st Specialists Meeting (International) of the Combustion Institute, France, p. 278.

Bilger, R., Stâmer, S.H., & Kee, R.J. (1990). Combust. Flame, **80,** 135.

Biordi, J.C., Lazzara, C.P., & Papp, J.F. (1974a). 15th Symposium (International) on Combustion, p. 367.

Biordi, J.C., Lazzara, C.P., & Papp, J.P. (1974b). 15th Symposium (International) on Combustion, p. 917.

Bittker, D.A., & Radhakrishnan, K. (1994). "LSENS, A general chemical kinetics and sensitivity analysis code for homogeneous gas-phase reactions III. Illustrative test problems," *NASA Reference Publication 1329.*

Bittker, D.A., & Scullin, V.J. (1972). "General chemical kinetics computer program for static or flow reactors, with application to combustion and shock tube kinetics," *NASA Technical Note D-6586.*

Bockhorn, H. (1994). *Soot Formation in Combustion: Mechanisms and Models,* Springer-Verlag, Heidelberg.

Böhland, T., Dobe, S., Temps, F., & Wagner, H.Gg. (1985). Ber. Bunsen-Ges. Phys. Chem., **89,** 1110.

Bond, J.R., Gray, P., Griffiths, J.F., & Scott, S.K. (1982). Proc. Roy. Soc. (London), **A381,** 293.

Borisov, A.A., Dragalova, E.V., Lissianski, V.V., Skachkov, G.I., & Zamansky, V.M. (1983a). Oxid. Commun., **4,** 45.

Borisov, A.A., Egorkin, S.V., Zamansky, V.M., & Skachkov, G.I. (1979). Kinetika i Kataliz, **20,** 1395.

Borisov, A.A., Gel'fand, B.E., Dragalova, E.V., Zamansky, V.M., Lissianski, V.V., Skatchkov, G.I., & Tsyganov, S.A (1983b). Khim. Fiz., **2,** 838.

Borisov, A.A., Lissianski, V.V., Skachkov, G.I., Zamansky, V.M., & Troshin, K.Y. (1989). 22nd Symposium (International) on Combustion, p. 903.

Borisov A.A., Zamansky, V.M., Lissianski, V.V., Skachkov, G.I., & Troshin, K.Y. (1990a). Sov. J. Chem. Phys., **6,** 472.

Borisov A.A., Zamansky, V.M., Lissianski, V.V., Skachkov, G.I., & Troshin, K.Y. (1990b). Khim. Fiz., **9,** 472.

Borisov, A.A., Zamansky, V.M., Konnov, A.A., Lissianski, V.V., Rusakov, S.A., & Skachkov, G.I. (1985). Khim. Fiz., **4,** 236.

Borisov, A.A., Zamansky, V.M., Konnov, A.A., Lissianski, V.V., Rusakov, S.A., & Skachkov, G.I. (1989a). Khim. Fiz., **8,** 89.

Borisov, A.A., Zamansky, V.M., Konnov, A.A., Lissianski, V.V., Rusakov, S.A., & Skachkov, G.I. (1989b). Khim. Fiz., **8,** 1532.

Borisov, A.A., Zamansky, V.M., Lissianski, V.V., & Rusakov, S.A. (1988a). Khim. Fiz., **7,** 773.

Borisov, A.A., Zamansky, V.M., Lissianski, V.V., & Rusakov, S.A. (1989c). Khim. Fiz., **8,** 195.

Borisov, A.A., Zamansky, V.M., Lissianski, V.V., Skachkov, G.I., & Troshin, K.Y. (1988a). 22nd Symposium (International) on Combustion, p. 903.

Bowman, C.T. (1975). Combust. Flame, **25,** 343.

Bowman, C.T., Hanson, R.K., Gardiner, W.C., Jr., Lissianski, V.V., Frenklach, M., Goldenberg, M., & Smith, G.P. (1997). *GRI-Mech 2.11—An Optimized Detailed Chemical Reaction Mechanism for Methane Combustion and NO Formation and Reburning*, Gas Research Institute Topical Report GRI-97/0020; `http://euler.me.berkeley.edu/gri_mech/`.

Brabbs, T.A., & Brokaw, R.S. (1975). 15th Symposium (International) on Combustion, p. 893.

Braun, W., Herron, J.T., & Kahaner, D.K. (1988). Int. J. Chem. Kinet., **20,** 51.

Braun-Unkhoff, M., Horn, Ch., Naumann, C., & Frank, P. (1994). 25th Symposium (International) on Combustion, Abstracts of work-in-progress poster session presentations, p. 262.

Brezinsky, K., & Dryer, F.L. (1986). Combust. Sci. Technol., **45,** 225.

Brocard, J.C., Baronnet, F., & O'Neal, H.E. (1983). Combust. Flame, **52,** 25.

Brocard, J.C., & Baronnet, F. (1987). J. Chim. Phys., **84,** 19.

Bromly, J.H., Barnes, F.J., Muris, S., You, X., & Haynes, B.S. (1996). Combust. Sci. Technol., **115,** 259.

Brouwer, J., Longwell, J.P., Sarofim, A.F., Barat, R.B., & Bozzelli, J.W. (1992). Combust. Sci. Technol., **85,** 87.

Brown, P.N., Byrne, G.D., & Hindmarsh, A.C. (1989). SIAM J. Sci. Stat. Comput., **10,** 1038.

Brown, R.L. (1981). *HICHM: A Fortran code for homogeneous, isothermal systems*, National Bureau of Standards Report NBSIR81-2281, Gaithersburg.

Burcat, A., & McBride, B. (1997). *Ideal Gas Thermodynamic Data for Combustion and Air Pollution Use.* TAE Report No. 804, Techion, Haifa.

Burgess, D.R.F., Jr., Zachariah, M.R., Tsang, W., & Westmoreland, P.R. (1995). NIST Tech. Note 1412.

Burgess, D.R.F., Jr., Zachariah, M.R., Tsang, W., & Westmoreland, P.R. (1996). Prog. Energy Combust. Sci., **21,** 453.

Burgess, D.R.F., Jr., Zachariah, M.R., Tsang, W., & Westmoreland, P.R. (1997). *WWW CKMech: Thermochemical and Chemical Kinetic Data for Fluorinated Hydrocarbons,* at the URL `http://www.nist.gov/cstl/div836/836.03/papers/NistTNIntro.html`.

Byrne, G.D., & Dean, A.M. (1993). Comput. Chem., **17,** 297.

Callahan, C.V., Held, T.J., Dryer, F.L., Minetti, R., Ribaucour, M., Sochet, L.R., Faravelli, T., Gaffuri, P., & Ranzi, E. (1996). 26th Symposium (International) on Combustion, p. 739.

Carver, G.D., & Cox, R.A. (1999). *IUPAC Subcommittee for Gas Kinetic Data Evaluation, Evaluated Kinetic Data Web Edition,* `http://www.iupac-kinetic.ch.cam.ac.uk/`.

Castaldi, M.J., Marinov, N.M., Melius, C.F., Huang, J., Senkan, S.M., Pitz, W.J., & Westbrook, C.K. (1996). 26th Symposium (International) on Combustion, p. 693.

Caton, J.A., & Siebers, D.L. (1988). *Comparison of Nitric Oxide Removal by Cyanuric Acid and by Ammonia,* Fall Meeting of the Western States Section of the Combustion Institute, Dana Point, CA.

Chakir, A., Belliman, M., Boettner, J.C., & Cathonnet, M. (1992). Int. J. Chem. Kinet., **24,** 385.

Chang, S.G., & Lee, S.G. (1992). Environ. Prog., **11,** 66.

Chang, W.D., & Senkan, S.M. (1985). Combust. Sci. Technol., **43,** 49.

Chang, W.D., & Senkan, S.M. (1989). Environ. Sci. Technol., **23,** 442.

Chen, J.H., & Echekki, T. (1996). *Direct numerical simulation of a laminar methanol triple flame,* presented at the 49th annual meeting of the Division of Fluid Dynamics of the American Physical Society, Syracuse, November 1996.

Chen, S.L., Lyon, R.K., & Seeker, W.R. (1991). Environ. Prog., **10,** 182.

Chevalier, C. (1992). *Entwicklung eines detaillierten Reaktionsmechanismus zur Modellierung der Verbrennungsprozesse von Kohlenwasserstoffen bei Hoch- und Niedertemperaturebedingungen,* Dissertation, Universität Stuttgart.

Chevalier, C., Louessard, P., Müller, U.C., & Warnatz, J. (1990a). *Second International Symposium on Diagnostics and Modeling of Combustion in Reciprocating Engines,* Japanese Society of Mechanical Engineers, Tokyo.

Chevalier, C., Pitz W.J., Warnatz, J., Westbrook, C.K., & Melenk, H. (1992). 24th Symposium (International) on Combustion, p. 93.

Chevalier, C., Warnatz, J., & Melenk, H. (1990b). Ber. Bunsen-Ges. Phys. Chem., **94,** 1362.

Choo, K.Y., Golden, D.M., & Benson, S.W. (1974). Int. J. Chem. Kinet., **6,** 631.

Chupka, W.A. (1968). J. Chem. Phys., **48,** 2337.

Cohen, N., & Westberg, K.R. (1991). J. Phys. Chem. Ref. Data, **20,** 1211.

Côme, G.M., Warth, V., Glaude, P.A., Fournet, R., Battin-Leclerc, F., & Scacchi, G. (1996). 26th Symposium (International) on Combustion, p. 755.

Corre, C., Dryer, F.L., Pitz, W.J., & Westbrook, C.K. (1992). 24th Symposium (International) on Combustion, p. 843.

Coward, H.F., & Jones, G.W. (1952). U.S. Bureau of Mines Bulletin 503.

Cowert, J.S., Keck, J.C., Heywood, J.B., Westbrook, C.K., & Pitz, W.J. (1990). 23rd Symposium (International) on Combustion, p. 1055.

Cox, R.A., & Cole, J.A. (1985). Combust. Flame, **60,** 109.

Cribb, P.H., Dove, J.E., & Yamazaki, S. (1984). 20th Symposium (International) on Combustion, p. 779.

Crossley, R.W., Dorko, E.A., Scheller, K., & Burcat, A. (1972). Combust. Flame, **19,** 373.

Cullis, C.F., Hirschler, M.M., & Wall, S.W. (1986). 21st Symposium (International) on Combustion, p. 1223.

Curran, H.J., Dunphy, M.P., Simmie, J.M., Westbrook, C.K., & Pitz, W.J. (1993). 24th Symposium (International) on Combustion, p. 769.

Curran, H.J., Gaffuri, P., Pitz, W.J., & Westbrook, C.K. (1998a). Combust. Flame, **114,** 149.

Curran, H.J., Gaffuri, P., Pitz, W.J., Westbrook, C.K., & Leppard, W.R. (1995). Publication SAE-952406, Society of Automotive Engineers.

Curran, H.J., Gaffuri, P., Pitz, W.J., Westbrook, C.K., & Leppard, W.R. (1996a). 26th Symposium (International) on Combustion, p. 2669.

Curran, H.J., Pitz, W.J., Westbrook, C.K., Callahan, C.V., & Dryer, F.L. (1998b). 27th Symposium (International) on Combustion, p. 379.

Curran, H.J., Pitz, W.J., Westbrook, C.K., Hisham, M.W.M., & Walker, R.W. (1996b). 26th Symposium (International) on Combustion, p. 641.

Dagaut, P., Boettner, J., & Cathonnet, M. (1991). Combust. Sci. Technol., **77,** 127.

Dagaut, P., Boettner, J.C., & Cathonnet, M. (1992a). J. Chim. Phys., **89,** 867.

Dagaut, P., Cathonnet, M., & Boettner, J.-C. (1991). Int. J. Chem. Kinet., **23,** 437.

Dagaut, P., Cathonnet, M., & Boettner, J.-C. (1992b). Int. J. Chem. Kinet., **24,** 813.

Dagaut, P., Reuillon, M., & Cathonnet, M. (1994). Combust. Sci. Technol., **95,** 233.

Dagaut, P., Reuillon, M., & Cathonnet, M. (1995). Combust. Sci. Technol., **101,** 132.

Daly, N.J., & Wentrup, C. (1968). Aust. J. Chem., **21,** 2711.

Dempsey, C.R., & Oppelt, E.T. (1993). Air and Waste Management Association, **43,** 25.

Deuflhard, P., Bader, G., & Nowak, U. (1981). In *Modeling of Chemical Reaction Systems,* K.H. Ebert, P. Deuflhard, and W. Jaeger, Eds., Springer-Verlag Ser. Chem. Phys. 18, pp. 38–55.

Deuflhard, P., & Nowak, U. (1986). "Efficient numerical simulation and identification of large chemical reaction systems," *Konrad Zuse-Zentrum fur Informationstechnik Preprint SC-86-1.*

Dixon-Lewis, G. (1979a). Combust. Flame, **36,** 1.

Dixon-Lewis, G. (1979b). Philos. Trans. R. Soc. London, **A292,** 45 .

Dixon-Lewis, G. (1984). In *Combustion Chemistry,* W.C. Gardiner, Jr., Ed., Springer-Verlag, New York.

Dixon-Lewis, G., & Williams, D.J. (1977). In *Comprehensive Chemical Kinetics,* Vol. 17, C.H. Bamford and C.F.H. Tipper, Eds., Elsevier, Amsterdam, p. 1.

Dobis, O., & Benson, S.W. (1987). Int. J. Chem. Kinet., **19,** 691.

Dryer, F.L. (1991). In *Fossil Fuel Combustion,* W. Bartok and A.F. Sarofim, Eds., Wiley, New York, p. 121.

Du, H., & Hessler, J.P. (1992). J. Chem. Phys., **96,** 1077.

Dunker, A.M. (1986). Atmos. Environ., **20,** 479.

Dunphy, M.P., & Simmie, J.M. (1991). Combust. Flame, **85,** 498.

Egolfopoulos, F.N., Cho, P., & Law, C.K. (1989). Combust. Flame, **76,** 375.

Egolfopoulos, F.N., Du, D.X., & Law, C.K. (1992). Combust. Sci. Technol., **83,** 33.

Emdee, J.L., Brezinsky, K., & Glassman, I. (1992). J. Phys. Chem., **96,** 2151.

Ewig, F., Rhäsa, D., & Zellner, R. (1987). Ber. Bunsen-Ges. Phys. Chem., **91,** 708.

Faravelli, T., Goldaniga, A., & Ranzi, E. (1998). 27th Symposium (International) on Combustion, p. 1489.

Fenimore, C.P. (1971). 13th Symposium (International) on Combustion, p. 373.

Frank, P., & Just, Th. (1985). Ber. Bunsen-Ges. Phys. Chem., **89,** 181.

Frank, P.M. (1978). *Introduction to System Sensitivity Theory,* Academic Press, New York.

Frenklach, M. (1990). In *Numerical Approaches to Combustion Modeling,* Progress in Astronautics and Aeronautics Vol. 135, AIAA, New York.

Frenklach, M. (1984). In *Combustion Chemistry,* W.C. Gardiner, Jr., Ed., Springer-Verlag, New York.

Frenklach, M., Wang, H., & Rabinovich, M.J. (1992). Prog. Energy Combust. Sci., **18,** 47.

Frenklach, M., Wang, H., Goldenberg, M., Smith, G.P., Golden, D.M., Bowman, C.T., Hanson, R.K., Gardiner, W.C., Jr., & Lissianski, V.V. (1995). *GRI-Mech—An Optimized Detailed Chemical Reaction Mechanism for Methane Combustion,* Gas Research Institute Topical Report GRI-95/0058; `http://www.me.berkeley.edu/gri_mech/`.

Gaffuri, P. (1995). Ph.D. Thesis, Dipartimento di Ingegneria Chimica, Politecnico di Milano.

Gaffuri, P., Faravelli, T., Ranzi, E., Callahan, C.V., & Dryer, F.L. (1996). MAE Report No. 2070, Princeton University.

Gardiner, W.C., Jr., Walker, B.F., & Wakefield, C.B. (1981). In *Shock Waves in Chemistry,* A. Lifshitz, Ed., Dekker, New York, p. 319.

Gardiner, W.C., Jr., Lissianski, V.V., & Zamansky, V.M. (1993). *Shock Waves @ Marseille,* Vol. II, Springer-Verlag, Berlin, p. 155.

Garner, F.H., Long, R., Graham, A.J., & Badakhashan, A. (1959). 6th Symposium (International) on Combustion, Reinhold, New York, p. 802.

Gear, C.W. (1971). *Numerical Initial Value Problems in Ordinary Differential Equations,* Prentice-Hall, Englewood Cliffs.

Gilbert, R.G., & Smith, S.C. (1990). *Theory of Unimolecular and Recombination Reactions,* Blackwell Scientific Publishers, Oxford.

Glarborg, P., Alzueta, M.U., Dam-Johansen, K., & Miller, J.A. (1998). Combust. Flame, **115,** 1.

Glarborg, P., Kee, R.J., Grcar, J.F., & Miller, J.A. (1986). *PSR: A Fortran Program for Modeling Well-Stirred Reactors,* Sandia National Laboratory Report 86-8209.

Glarborg, P., Lilleheie, N.I., Byggstöl, S., Magnussen, B.J., Kilpinen, P., & Hupa, M. (1992). 24th Symposium (International) on Combustion, p. 889.

Glarborg, P., Miller, J.A., & Kee, R.J. (1986). Combust. Flame, **65,** 177.

Glaude, P.A., Warth, V., Fournet, R., Battin-Leclerc, F., Côme, G.M., & Scacchi, F. (1997). Bull. Soc. Chim. Belg., **106,** 343.

Glaude, P.A., Warth, V., Fournet, R., Battin-Leclerc, F., Scacchi, F., & Côme, G.M. (1998). Int. J. Chem. Kinet., **30,** 949.

Goldaniga, A., Faravelli, T., Ranzi, E., Dagaut, P., & Cathonnet, M. (1998). 27th Symposium (International) on Combustion, p. 353.

Gordon, A.S., & Austin, T.C. (1992). Prog. Energy Combust. Sci., **18,** 493.

Goussis, D.A., & Lam, S.H. (1992). 24th Symposium (International) on Combustion, p. 113.

Gray, J.A., & Westbrook, C.K. (1994). Int. J. Chem. Kinet., **26,** 757.

Griffiths, J.F. (1995). Prog. Energy Combust. Sci., **21,** 25.

Griffiths, J.F., Hughes, K.J., Schreiber, M., & Poppe, C. (1994). Combust. Flame, **99,** 533.

Griffiths, J.F., & Mohamed, C. (1997). In *Low Temperature Combustion and Autoignition*, M.J. Pilling, Ed., Elsevier, Amsterdam, p. 545.

Halstead, M.P., Kirsch, L.J., & Quinn, C.P. (1977). Combust. Flame, **30,** 45.

Hanson, R.K., & Salimian, S. (1984). In *Combustion Chemistry*, W.C. Gardiner, Jr., Ed., Springer-Verlag, New York.

Hautman, D.J., Dryer, F.L., Schug, K.P., & Glassman, I. (1981). Combust. Sci. Technol., **25,** 219.

Held,T.J., & Dryer, F.L. (1994). 25th Symposium (International) on Combustion, p. 901.

Held,T.J., & Dryer, F.L. (1998). Int. J. Chem. Kinet., **30,** 805.

Herron, J.T. (1988). J. Phys. Chem. Ref. Data, **17,** 967.

Ho, L., Chen, S.L., Seeker, W.R., & Maly, P.M. (1993). U.S. Patent 5,270,025.

Ho, W., & Bozzelli, J.W. (1992). 24th Symposium (International) on Combustion, p. 743.

Ho, W., Barat, R.B., & Bozzelli, J.W. (1992a). Combust. Flame, **88,** 265.

Ho, W., Booty, M.R., Magee, R.S., & Bozzelli, J.W. (1995). Ind. Eng. Chem. Res., **34,** 4185.

Ho, W., Yu, C., & Bozzelli, J.W. (1992b). Combust. Sci. Technol., **85,** 23.

Holbrook, K.A., Pilling, M.J., & Robertson, S.H. (1996). *Unimolecular Reactions*, John Wiley & Sons, New York.

Homann, K.H., & Poss, R. (1972). Combust. Flame, **18,** 300.

Inomata, T., Moriwaki, T., & Okazaki, S. (1985). Combust. Flame, **62,** 183.

Jensen, D.E., & Webb, B.C. (1976). AIAA Journal, **14,** 947.

Jensen, D.E., & Jones, G.A. (1982). J. Chem. Soc. Faraday Trans. 1, **78,** 2843.

Jiang, B., Ingram, D., Causon, D., & Sanders, R. (1995). Shock Waves, **5,** 81.

Jones, D.G. (1988). *Process and Apparatus for Removing Oxides of Nitrogen and Sulfur from Combustion Gases,* U.S. Patent No. 4,783,325.

Kazakov, A., & Frenklach, M. (1995). Reduced mechanism at URL `http://www.me.berkeley.edu/drm/`.

Keck, J.C. & Hu, H. (1986). 21st Symposium (International) on Combustion, p. 521.

Kee, R.J., Dixon-Lewis, G., Warnatz, J., Coltrin, M.E., & Miller, J.A. (1986). *A Fortran Computer Code Package for the Evaluation of Gas-Phase Multicomponent Transport Properties,* Sandia National Laboratory Report 86-8246.

Kee, R.J., Grcar, J.F., Smooke, M.D., & Miller, J.A. (1985). *A Fortran program for modeling steady laminar one-dimensional premixed flames,* Sandia National Laboratory report SAND85-8240.

Kee, R.J., & Miller, J.A. (1986). In *Complex Chemical Reaction Systems,* J. Warnatz and W. Jäger, Eds., Springer-Verlag, Berlin, pp. 196–221.

Kee, R.J., Miller, J.A., & Jefferson, T.H. (1980). *Chemkin: A general-purpose, problem-independent, transportable, Fortran chemical kinetics code package,* Sandia National Laboratory Report SANBO-8003.

Kee, R.J., Rupley, F.M., & Miller, J.A. (1989). *Chemkin-II: A Fortran chemical kinetics package for the analysis of gas-phase chemical kinetics,* Sandia National Laboratory Report SAND89-8009.

Kee, R.J., Warnatz, J., & Miller, J.A. (1983). *A Fortran computer code package for the evaluation of gas-phase viscosities, conductivities, and diffusion coefficients,* Sandia National Laboratory Report SAND83-8209.

Kiefer, J.H., Sidhu, S.S., Kern, R.D., Xie, K., Chen, H., & Harding, L.B. (1992). Combust. Sci. Technol., **82,** 101.

Kiehne, T.M., Matthews, R.D., & Wilson, D.E. (1987). Combust. Sci. Technol., **54,** 1.

Kilpinen, P., Glarborg, P., & Hupa, P. (1992). Ind. Eng. Chem. Res., **31,** 1477.

Kim, T.J., Yetter, R.A., & Dryer, F.L. (1994). 25th Symposium (International) on Combustion, p. 759.

Kondratiev, V.N. (1972). *Rate Constants of Gas Phase Reactions,* NSRDS Report COM-72-10014, Office of Standard Reference Data, National Bureau of Standards, Washington.

Konnov, A.A. (1996). *Detailed reaction mechanism for C1–C2 hydrocarbon combustion,* 26th Symposium (International) on Combustion, Abstracts of work-in-progress posters, p. 286; `http://homepages.vub.ac.be/~akonnov/`.

Kramlich, J.C., Malte, P.C., & Grosshandler, W.L. (1981). 18th Symposium (International) on Combustion, p. 151.

Laidler, K.J. (1987). *Chemical Kinetics,* Third Edition, Harper & Row, New York.

Lask, G., & Wagner, H.G. (1963). 8th Symposium (International) on Combustion, Williams & Wilkins, Baltimore, p. 432.

Lee, D., & Hochgreb, S. (1998). Int. J. Chem. Kinet., **30,** 385.

Lee, D., Hochgreb, S., & Keck, J.C. (1993). Joint Technical Meeting of Central and Eastern States Sections of the Combustion Institute, New Orleans, p. 478.

Leung, K.M. & Lindstedt, R.P. (1995). Combust. Flame, **102,** 130.

Li, S.C., & Williams, F.A. (1996). 26th Symposium (International) on Combustion, p. 1017.

Liao, C., Saito, N., Saso, Y., & Ogawa, Y. (1996). Fire Safety J., **27,** 49.

Lifshitz, A., Bidani, M., Carroll, H.F., Hwang, S.M., Fu, P.-Y., Shin, K.S., & Gardiner, W.C., Jr. (1991). Combust. Flame, **86,** 229.

Lifshitz, A., Scheller, K., Burcat, A., & Skinner, G.B. (1971). Combust. Flame, **16,** 311.

Lighty, J.S., & Veranth, J.M. (1998). 27th Symposium (International) on Combustion, Topical Review 4C03/04, to be published..

Lindstedt, R.P. (1998). 27th Symposium (International) on Combustion, to be published.

Lindstedt, R.P., & Maurice, L.Q. (1995). Combust. Sci. Technol., **107,** 317.

Lindstedt, R.P., & Skevis, G. (1996). 26th Symposium (International) on Combustion, p. 703.

Linteris, G.T., Burgess, D.R., Jr., Babushok, V., Zachariah, M., & Tsang, W. (1998). Combust. Flame, **113,** 164.

Lissianski, V.V. (1982). Dissertation, Institute of Chemical Physics, Moscow.

Lissianski, V.V., Gardiner, W.C., Jr., & Zamansky, V.M. (1994). 25th Symposium (International) on Combustion, Work-in-progress poster session, Irvine.

Lissianski, V.V., Gardiner, W.C., Jr., & Zamansky, V.M. (1995a). 3rd Workshop on Modeling of Chemical Reaction Systems, Heidelberg.

Lissianski, V.V., Gardiner, W.C., Jr., & Zamansky, V.M. (1996). Combust. Sci. Technol., **110,** 169.

Lissianski, V., Yang, H., Qin, Z., Mueller, M.R., Shin, K.S., & Gardiner, W.C., Jr. (1995b). Chem. Phys. Lett., **240,** 57.

Lissianski, V.V., Zamansky, V.M., Maly, P.M., Ho, L., Rusli, D., & Gardiner, W.C., Jr. (1999). Combust. Flame, in press.

Lund, C.M. (1978). *HCT—A general computer program for calculating time-dependent phenomena involving one-dimensional hydrodynamics, transport, and detailed chemical kinetics,* Lawrence Livermore National Laboratory Report UCRL-52504.

Lund, C.M., & Chase, L. (1995). *HCT—A general computer program for calculating time-dependent phenomena involving one-dimensional hydrodynamics, transport, and detailed chemical kinetics,* Lawrence Livermore National Laboratory Report UCRL-52504, revised.

Lutz, A.E., Kee, R.J., & Miller, J.A. (1988). *Senkin: A Fortran Program for Predicting Homogeneous Gas Phase Chemical Kinetics with Sensitivity Analysis,* Sandia National Laboratory Report SAND87-8248.

Lynd, L.R. (1996). Annu. Rev. Energy Environ., **21,** 403.

Lyon, R.K. (1975). *Method for the Reduction of the Concentration of NO in Combustion Effluents Using Ammonia,* U.S. Patent No. 3,900,554.

Lyon, R.K. (1989). EPRI/EPA 1989 Joint Symposium on Stationary Combustion NOx Control, San Francisco, Vol. 2, pp. 7B-29.

Lyon, R.K., & Cole, J.A. (1990). Combust. Flame, **82,** 435.

Lyon, R.K., & Hardy, J.E. (1986). Ind. Eng. Chem. Fundam., **25,** 19.

Maas, U., & Pope, S.B. (1992a). Combust. Flame, **88,** 239.

Maas, U., & Pope, S.B. (1992b). 24th Symposium (International) on Combustion, p. 103.

Maas, U., & Warnatz, J. (1989). 22nd Symposium (International) on Combustion, p. 1695.

Mallard, W.G., Westley, F., Herron, J.T., Hampson, R.F., & Frizzell, D.H. (1998). *NIST Chemical Kinetics Database,* NIST Standard Reference Database 17, Version 2Q98; at the URL `http://www.nist.gov/srd`.

Marsden, A.R., Frenklach, M., & Reible, D.D. (1987). J. Air Pollut. Control Assoc., **37,** 370. Martin, J.M.L. Chem. Phys. Lett. 273 98 1997

Masten, D.A., Hanson, R.K., & Bowman, C.T. (1990). J. Phys. Chem., **94,** 7119.

McBride, B.J., Gordon, S., & Reno, M.J. (1993). *Coefficients for Calculating Thermodynamic and Transport Properties of Individual Species,* NASA Technical Memorandum 4513.

Miller, J.A., & Bowman, C.T. (1989). Prog. Energy Combust. Sci., **15,** 287.

Miller, J.A., & Bowman, C.T. (1991). Int. J. Chem. Kinet., **23,** 289.

Minkoff, G.J., & Tipper, C.F.H. (1962). *Chemistry of Combustion Reactions,* Butterworths, London.

Mitani, T., & Nioka, T. (1984). Combust. Flame, **55,** 13.

Miziolek, A.W., & Tsang, W., Eds. (1995). *Halon Replacements: Technology and Science,* American Chemical Society, Washington.

Müller, C.M., & Peters, N. (1993). In *Reduced Kinetic Mechanisms for Applications in Combustion Systems,* N. Peters and B. Rogg, Eds., Springer-Verlag, p. 142.

Muller, C.H., Schofield, K., Steinberg, M., & Broida, H.P. (1978). 17th Symposium (International) on Combustion, p. 867.

Myerson, A.L., Taylor, F.R., & Faunce, B.G. (1957). 6th Symposium (International) on Combustion, Reinhold, New York, p. 154.

NASA (1962). U.S. National Aeronautics and Space Administration, U.S. Air Force, and U.S. Weather Bureau, *U.S. Standard Atmosphere, 1962,* Government Printing Office, Washington, 1962.

Natarajan, K., & Bhaskaran, K.A. (1981). Combust. Flame, **43,** 35.

Nehse, M., Warnatz, J., & Chevalier, C. (1996). 26th Symposium (International) on Combustion, p. 773.

Nicol, D.G., & Malte, P.C. (1994). 25th Symposium (International) on Combustion, Work-in-progress poster session, Irvine.

Nikitin, I.M., Prostov, V.N., & Gladkikh, V.A. (1987). Khim. Fiz., **6,** 390.

Norton, T.S., & Dryer, F.L. (1990a). Int. J. Chem. Kinet., **22,** 219.

Norton, T.S., & Dryer, F.L. (1990b). 23rd Symposium (International) on Combustion, p. 179.

Norton, T.S., & Dryer, F.L. (1992). Int. J. Chem. Kinet., **24,** 319.

Noto, T., Babushok, V., Burgess, D.R., Jr., Hamins, A., & Tsang, W. (1996). 26th Symposium (International) on Combustion, p. 1377.

Noto, T., Babushok, V., Hamins, A., & Tsang, W. (1998). Combust. Flame, **112,** 147.

Olson, D.B., & Gardiner, W.C., Jr. (1977). J. Phys. Chem., **81,** 2514.

Olson, J.O., Karlsson, L.S., & Andersson, L.L. (1986). J. Phys. Chem., **90,** 1458.

Olson, J.O., Olsson, I.B.M., & Andersson, L.L., (1987). J. Phys. Chem., **91,** 4160.

Paczko, G., Lefdal, P.M., & Peters, N. (1986). 21st Symposium (International) on Combustion, p. 739.

Park, Y.K., & Vlachos, D.G. (1997). AIChE J., **43,** 2083.

Partridge, H., Bauschlicher, C.W., Jr., Sodupe, M., & Langhoff, S.R. (1992). Chem. Phys. Lett., **195,** 200.

Paul, P. & Warnatz, J. (1998). A Re-Evaluation of the Means Used to Calculate Transport Properties of Reacting Flows, Poster P020, 27th Symposium (International) on Combustion, Boulder, to be published.

Perry, R.A. (1988). *NO Reduction Using Sublimation of Cyanuric Acid,* U.S. Patent 4,731,231.

Peters, N. (1985). In *Numerical Simulation of Combustion Phenomena,* M.D. Smooke, Ed., Springer-Verlag, New York, p. 48.

Peters, N., & Kee, R.J. (1987). Combust. Flame, **68,** 17.

Peters, N., & Rogg, B., Eds. (1993). *Reduced Kinetic Mechanisms for Applications in Combustion Systems,* Springer-Verlag, New York.

Peterson, K.A., & Dunning, T.H., Jr. (1997). J. Chem. Phys., **106,** 4119.

Pilling, M.J., Ed. (1997). *Low-temperature Combustion and Autoignition,* Vol. 35 of Comprehensive Chemical Kinetics, Elsevier, Amsterdam.

Pilling, M.J., & Seakins, P.W. (1995). *Reaction Kinetics,* Oxford University Press, Oxford.

Pilling, M.J., & Smith, I.W.M., Eds. (1987). *Modern Gas Kinetics: Theory, Experiment and Application,* Blackwell, Oxford.

Pilling, M.J., Turányi, T., Hughes, K.J., & Clague, A.R. (1996). `http://chem.leeds.ac.uk/Combustion/Combustion.html`.

Pirraglia, A.N., Michael, J.V., Sutherland, J.W., & Klemm, R.B. (1989). J. Phys. Chem., **93,** 282.

Pitz, W.J., Westbrook, C.K., Proscia, W.M., & Dryer, F.L. (1984). 12th Symposium (International) on Combustion, p. 831.

Pollard, R.T. (1977). In *Comprehensive Chemical Kinetics,* Vol. 17, C.H. Bamford and C.F.H. Tipper, Eds., Elsevier, Amsterdam, p. 249.

Pont, J.N., Evans, A.B., England, G.C., Lyon, R.K., & Seeker, W.R. (1993). Environ. Prog., **12,** 140.

Pratt, D.T. (1985). Fast Algorithms for Combustion Kinetics Calculations. In *Stiff Computation,* R.C. Aiken, Ed., Oxford University Press, New York, p. 217.

Press, W.H., Teukolsky, S.A., Vetterling, W.T., & Flannery, B.P. (1992). *Numerical Recipes in FORTRAN, The Art of Scientific Computing*, 2nd Ed., Cambridge University Press, Ch. 16.

Qin, Z. (1998). Dissertation, University of Texas at Austin.

Radhakrishnan, K. (1993a). "A critical analysis of the accuracy of several numerical techniques for combustion kinetic rate equations," *NASA Technical Paper 3315.*

Radhakrishnan, K. (1993b). "LSENS, A general chemical kinetics and sensitivity analysis code for gas-phase reactions: User's Guide," *NASA Technical Memorandum 105851.*

Radhakrishnan, K. (1994). "LSENS, A general chemical kinetics and sensitivity analysis code for gas-phase reactions I. Theory and numerical solution procedures," *NASA Reference Publication 1328.*

Radhakrishnan, K., & Bittker, D.A. (1994). "LSENS, A general chemical kinetics and sensitivity analysis code for homogeneous gas-phase reactions II. Code description and usage," *NASA Reference Publication 1329.*

Radhakrishnan, K., & Hindmarsh, A.C. (1993). "Description and use of LSODE, the Livermore solver for ordinary differential equations," *NASA Reference Publication 1327.*

Ranzi, E., Dente, M., Faravelli, T., & Pennati, G. (1992). Hydrogen abstraction reactions: Analogy rules for the prediction of kinetic parameters. In *From Molecular Dynamics to Combustion Chemistry*, S. Carrà and N. Rahman, Eds., World Scientific Press, Singapore.

Ranzi, E., Faravelli, T., Gaffuri, P., Sogaro, A., D'Anna, A., & Ciajola, A. (1995a). Combust. Flame, **102,** 179.

Ranzi, E., Faravelli, T., Gaffuri, P., Garavaglia, E., & Goldaniga, A. (1997a). Ind. Eng. Chem. Res., **36,** 3336.

Ranzi, E., Faravelli, T., Gaffuri, P., Sogaro, A., D'Anna, A., & Ciajola, A. (1997b). Combust. Flame, **108,** 24.

Ranzi, E., Gaffuri, P., Faravelli, T., & Dagaut, P. (1995b). Combust. Flame, **103,** 91.

Ranzi, E., Sogaro, A., Gaffuri, P., Pennati, G., & Faravelli, T. (1994a). Combust. Sci. Technol., **96,** 279.

Ranzi, E., Sogaro, A., Gaffuri, P., Pennati, G., Westbrook, C.K., & Pitz, W.J. (1994b). Combust. Flame, **99,** 201.

Reid, R.C., Prausnitz, J.M., & Poling, B.E. (1987). *The Properties of Gases and Liquids,* McGraw-Hill, New York.

Reinelt, D., & Linteris, G.T. (1996). 26th Symposium (International) on Combustion, p. 1421.

Revel, J., Boettner, J.C., Cathonnet, M., & Bachman, J.S. (1994). J. Chim. Phys., **91,** 365.

Roberts, C.E., Matthews, R.D., & Leppard, W.R. (1996). *Development of a semi-detailed kinetics mechanism for the autoignition of iso-octane,* SAE Technical Paper Series 962107.

Rosser, W.A., Wise, H., & Miller, J. (1959). 7th Symposium (International) on Combustion, Butterworths, London, p. 175.

Rumminger, M.D., Reinelt, D., Babushok, V., & Linteris, G.T. (1998). Combust. Flame, **116,** 207.

Ryu, S.-O., Hwang, S.M., & Rabinowitz, M.J. (1995). J. Phys. Chem., **99,** 13984.

Saito, K., Ito, R., Kakumoto, T., & Imamura, A. (1986). J. Phys. Chem., **90,** 1422.

Schott, G.L. (1960). J. Chem. Phys., **32,** 710.

Schott, G.L., & Getzinger, R.W. (1973). In *Physical Chemistry of Fast Reactions,* Vol. 1, B.P. Levitt, Ed., Plenum Press, London.

Schreiber, M., Sadat Sakak, A., Lingens, A., & Griffiths, J.F. (1994). 25th Symposium (International) on Combustion, p. 933.

Seeker, W. R. (1990). 23rd Symposium (International) on Combustion, p. 867.

Seeker, W.R. (1990). Proceedings of the Reburning Workshop, Orenas Slott, Sweden, Nov. 26-27.

Seeker, W.R., & Koshland, C.P. (1990). Combust. Sci. Technol., **74,** *i-viii.*

Semenov, N.N. (1934). *Chain Reactions,* Goschimtechizdat, Leningrad.

Semenov, N.N. (1958). *Some Problems in Chemical Kinetics and Reactivity,* Princeton University Press, Princeton.

Senkan, S.M., Yildirim, R., & Gutman, D. (1992). 24th Symposium (International) on Combustion, p. 749.

Seshadri, K., & Williams, F.A. (1994). In *Turbulent Reacting Flows,* P.A. Libby and F.A. Williams, Eds., Academic Press, New York, p. 153.

Shin, K.S., & Michael, J.V. (1991). J. Chem. Phys., **95,** 262.

Shtern, V.Ya. (1960). *The Gas Phase Oxidation of Hydrocarbons,* Pergamon, London.

Slack, M.W., & Grillo, A.R. (1981). Combust. Flame, **40,** 155.

Sloane, T.M. (1989). Combust. Sci. Technol., **63,** 287.

Sloane, T.M. (1992). Combust. Sci. Technol., **83,** 77.

Smith, S.R., & Gordon, A.S. (1956). J. Phys. Chem., **60,** 1059.

Spindler, K., & Wagner, H.Gg. (1983). Ber. Bunsen-Ges. Phys. Chem., **86,** 1.

Stabler, R.N., & Chesick, J.P. (1978). Int. J. Chem. Kinet., **10,** 461.

Staph, D., & Leuckel, W. (1996). 26th Symposium (International) on Combustion, p. 2083.

Staph, D. (1998). *Experimentell basierte Weiterentwicklung von Berechnungsmodellen der NO_X-Emission technischer Verbrennungssysteme,* Fortschritt-Berichte VDI Reihe 6 Nr. 395, VDI Verlag, Düsseldorf.

Stewart, P.H., Larson, C.W., & Golden, D.M. (1989). Combust. Flame, **75,** 25.

Sung, C.J., Law, C.K., & Chen, J.-Y. (1998). An Augmented Reduced Mechanism for Methane Oxidation with Comprehensive Global Parametric Validation, Poster P043, 27th Symposium (International) on Combustion, Boulder, to be published.

Suzuki, A., Inomata, T., Jinno, H., & Moriwaki, T. (1991). Bull. Chem. Soc. Jpn., **64,** 3345.

Takahashi, K., Inomata, T., Moriwaki, T., & Okazaki, S. (1989). Bull. Chem. Soc. Jpn., **62,** 2138.

Tan, Y. (1994). *Cinétique de combustion du gaz naturel: Etude expérimentale et modélisation.* Ph.D. Thesis, Université de P. & M. Curie, Paris.

Tan, Y., Dagaut, P., Cathonnet, M., & Boettner, J.C. (1994a). Combust. Sci. Technol., **102,** 21.

Tan, Y., Dagaut, P., Cathonnet, M., & Boettner, J.C. (1994b). Combust. Sci. Technol., **103,** 131.

Tan, Y., Dagaut, P., Cathonnet, M., & Boettner, J.C. (1995). J. Chim. Phys., **92,** 726.

Tan, Y., Dagaut, P., Cathonnet, M., Boettner, J.C., Bachman, J.S., & Carlier, P. (1994c). 25th Symposium (International) on Combustion, p. 1563.

Tan, Y., & Frank, P. (1996). 26th Symposium (International) on Combustion, p. 677.

Tischer, A.L., IV, Hwang, S.M., & De Witt, K.J. (1996). Western States Section of the Combustion Institute Fall Meeting, Paper 96F-124.

Tomlin, A.S., Turányi, T., & Pilling, M.J. (1997). Mathematical tools for the construction, investigation and reduction of combustion mechanisms, Chapter 4 of *Low-temperature Combustion and Autoignition,* M.J. Pilling, Ed., Elsevier.

Trenwith,, A.B. (1975). J. Chem. Soc. Faraday Trans. 1., **71,** 2405.

Trevino, C., & Mendez, F. (1992). 24th Symposium (International) on Combustion, p. 121.

Tsang, W. (1976). Int. J. Chem. Kinet., **8,** 173.

Tsang, W. (1987). J. Phys. Chem. Ref. Data, **16,** 471.

Tsang, W. (1988). J. Phys. Chem. Ref. Data, **17,** 887.

Tsang, W. (1990a). J. Phys. Chem. Ref. Data, **19,** 1.

Tsang, W. (1990b). Combust. Sci. Technol., **74,** 99.

Tsang, W. (1991). J. Phys. Chem. Ref. Data, **20,** 221.

Tsang, W., & Hampson, R.F. (1986). J. Phys. Chem. Ref. Data, **15,** 1087.

Tsang, W., & Hampson, R.F. (1991). J. Phys. Chem. Ref. Data, **20,** 609.

Tsang, W., & Herron, J.T. (1991). J. Phys. Chem. Ref. Data, **20,** 609.

Tsuboi, T., & Hashimoto, K. (1981). Combust. Flame, **42,** 61.

Turányi, T. (1990a). Comput. Chem., **14,** 253.

Turányi, T. (1990b). J. Math. Chem., **5,** 203.

Turányi, T. (1990c). New J. Chem., **14,** 795.

Turányi, T. (1994a). Comput. Chem., **18,** 45.

Turányi, T. (1994b). 25th Symposium (International) on Combustion, p. 949.

Turányi, T., Büki, A., & Maas, U. (1998). Poster W5B08, 27th Symposium (International) on Combustion, Boulder, to be published.

Turner, E.B., Emanuel, G., & Wilkins, R.L. (1970). *The NEST chemistry computer program, Vol. 1,* Aerospace Corp. Report TR-0059(6420-20)-1.

Wagner, H.Gg. (1979). 17th Symposium (International) on Combustion, p. 3.

Walker, B.F. (1970). Dissertation, University of Texas at Austin.

Warnatz, J. (1984). In *Combustion Chemistry,* W.C. Gardiner, Jr., Ed., Springer-Verlag, New York.

Warnatz, J. (1992). 24th Symposium (International) on Combustion, p. 553.

Warnatz, J., Maas, U., & Dibble, R.W. (1996). *Combustion,* Springer-Verlag, Heidelberg.

Warth, V., Stef, N., Glaude, P.A., Battin-Leclerc, F., Scacchi, G., & Côme, G.M. (1998a). Combust. Flame, **114,** 81.

Warth, V., Stef, N., Glaude, P.A., Battin-Leclerc, F., Scacchi, G., & Côme, G.M. (1998b). URL at `www.ensic.u-nancy.fr/ENSIC/DCPR/chiminfo/textes/chiminfo.htm`.

Wendt, J.O.L. (1996). 25th Symposium (International) on Combustion, p. 277.

Wendt, J.O.L., Sterling, C.V., & Matovich, M.A. (1973). 14th Symposium (International) on Combustion, p. 897.

Westbrook, C.T. (1979). Combust. Sci. Technol., **20,** 5.

Westbrook, C.K. (1982). 19th Symposium (International) on Combustion, p. 127.

Westbrook, C.K. (1983). Combust. Sci. Technol., **34,** 201.

Westbrook, C.K., & Dryer, F.L. (1979). Combust. Sci. Technol., **20,** 125.

Westbrook, C.K., & Dryer, F.L. (1981). Combust. Sci. Technol., **27,** 31.

Westbrook, C.K., & Pitz, W.J. (1984). Combust. Sci. Technol., **37,** 117.

Westbrook, C.K., Pitz, W.J., & Leppard, W.R. (1991). Publication SAE-912314, Society of Automotive Engineers.

Westbrook, C.K., Warnatz, J., & Pitz, W.J. (1988). 22nd Symposium (International) on Combustion, p. 893.

Westmoreland, P.R. (1992). Combust. Sci. Technol., **82,** 151.

Wilson, W.E., Jr., O'Donovan, J.T., & Fristrom, R.M. (1969). 12th Symposium (International) on Combustion, p. 929.

Wooldridge, M.S., Hanson, R.K., & Bowman, C.T. (1994). 25th Symposium (International) on Combustion, p. 741.

Wu, C.H., Lin, C.-Y., Wang, H.-T., & Lin, M.C. (1990). 17th International Symposium on Shock Tubes and Waves, p. 450.

Yang, B., & Pope, S.B. (1998). Combust. Flame, **112,** 85.

Yang, H., Gardiner, W.C., Jr., Shin, K.S., & Fujii, N. (1994). Chem. Phys. Lett., **231,** 449.

Yetter, R.A., Dryer, F.L., & Rabitz, H. (1991). Combust. Sci. Technol., **79,** 97.

Yoshihara, Y., Kazakov, A., Wang, A., & Frenklach, M. (1994). 25th Symposium (International) on Combustion, p. 941.

Yu, C.-L., Frenklach, M., Masten, D.A., Hanson, R.K., & Bowman, C.T. (1994). J. Phys. Chem., **98,** 4770.

Yu, C.-L., Wang, C., & Frenklach, M. (1990). Meeting of the Eastern States Section of the Combustion Institute.

Yu, C.-L., Wang, C., & Frenklach, M. (1995). J. Phys. Chem., **99,** 14377.

Yuan, T. Wang, C., Yu, C.-L., Frenklach, M., & Rabinowitz, M.J. (1991). J. Phys. Chem., **95,** 1258.

Zabetakis, M.G. (1965). U.S. Bureau of Mines Bulletin 627.

Zachariah, M.R., & Smith, O.I. (1987). Combust. Flame, **69,** 125.,

Zamansky, V.M., & Borisov, A.A. (1989). *Mechanizm i Promotirovanie Samovocplameneniya Perspektivnyx Topliv, Itogi Nauki i Techniki, Seria Kinetika, Kataliz, Tom 19,* Bcecoyuznii Institut Nautchnoi i Technitcheskoi Informatsii, Moscow.

Zamansky, V.M., & Borisov, A.A. (1992). Prog. Energy Combust. Sci., **18,** 297.

Zamansky, V.M., Ho, L., Maly, P., & Seeker, W.R. (1995). *Third International Conference on Combustion Technologies for a Clean Environment,* Lisbon, Vol. II, Paper 26.4.

Zamansky, V.M., Ho, L., Maly, P., & Seeker, W.R. (1996). 26th Symposium (International) on Combustion, p. 2075.

Zamansky, V.M., Lyon, R.K., Evans, A.B., Pont, J.N., Seeker, W.R., & Schmidt, C.E. (1993). SO_2 Control Symposium, EPRI/EPA/DOE, Boston, Vol. 3, Session 7.

Zamansky, V.M., Maly, P.M., & Seeker, W.R. (1998). *Improved Advanced Reburning Methods for High Efficiency NO_x Control,* U.S. Patent 5,756,059.

Zeldovich, Ya. B. (1946). Dokl. Akad. Nauk, **51,** 3, 213.

Zellner, R. (1984). In *Combustion Chemistry,* W.C. Gardiner, Jr., Ed., Springer-Verlag, New York.

Zellner, R., & Ewig, F. (1988). J. Phys. Chem., **92,** 2971.

Zhang, H.-Y., & McKinnon, J.T. (1995). Combust. Sci. Technol., **107,** 261.

2

Combustion Chemistry of Nitrogen

Anthony M. Dean [1]

Joseph W. Bozzelli [1,2]

2.1 INTRODUCTION

The purpose of this chapter is to examine reactions of nitrogen-containing species that are important in high-temperature gas-phase systems so as to provide the best set of rate coefficients presently available for use in combustion chemistry modeling. Since the 1984 review of N/H/O rate coefficients by Hanson and Salimian 1984 there has been a major review of nitrogen chemistry under combustion conditions by Miller and Bowman (1989). Several compilations of evaluated rate coefficients have also appeared (Cohen and Westberg 1991; Tsang and Herron 1991; Baulch et al. 1992, 1994; Tsang 1992). We update these discussions where appropriate and then analyze a number of chemically activated reactions that are relevant to understanding nitrogen chemistry.

After an introductory overview of nitrogen chemistry and an outline of our approach to analyzing chemically activated reactions and hydrogen abstraction reactions, we examine closely those elementary reactions considered by Hanson and Salimian for which later measurements have made it possible to define rate coefficient expressions more accurately. We then focus upon a number of chemically activated reactions for which recent research has improved our knowledge of reaction rates and product distributions. Among these reactions are those between hydrocarbon radicals and nitrogen species, such as the ones involved in "prompt NO" formation. We demonstrate that for many of these reactions one can understand otherwise apparently conflicting information by

[1] Corporate Research Laboratory, Exxon Research and Engineering Company, P. O. Box 998, Annandale, NJ 08801

[2] Department of Chemical Engineering and Chemistry, New Jersey Institute of Technology, Newark, NJ 07102

recognizing that chemically activated adducts formed by making new chemical bonds in addition, radical recombination, and insertion reactions often have multiple high-temperature reaction pathways. We then consider reactions for which the database is sparse but which must nonetheless be considered in modeling nitrogen chemistry; our approach is to assign rate coefficients that are consistent with the scant data and thermochemical and reaction kinetics principles. Brief discussion is provided of interactions between nitrogen and chlorine chemistry that may arise in toxic waste incineration. Finally, we present a few modeling examples to illustrate predictions with our rate coefficient set and to point out areas that require additional study.

2.2 OVERVIEW OF NITROGEN CHEMISTRY

Interest in the combustion chemistry of nitrogen compounds derives from the role of nitrogen oxides, collectively known as NO_X, in our environment. Increasingly stringent environmental regulations require increasingly sophisticated approaches for the control of NO_X formation during combustion (Bowman 1992). Detailed chemical modeling plays a major role in the implementation of such approaches. We set the stage for consideration of specific elementary reactions by discussing the mechanisms for production and removal of NO, the primary NO_X species produced in combustion.

NO is known to be formed in a variety of ways: (1) "Thermal NO" (Zeldovich 1946) is primarily a consequence of high flame temperatures; (2) "Prompt NO" (Fenimore 1976) is generated in fuel-rich parts of flames; (3) the "N_2O mechanism" (Wolfrum 1972; Malte and Pratt 1974) can be important in high-pressure flames; (4) "Fuel nitrogen" NO (Fenimore 1976) results from converting nitrogen-containing compounds in the fuel into NO; and (5) the NNH mechanism (Bozzelli and Dean 1995) is active in flame fronts where high atom concentrations appear. All but (4) convert atmospheric nitrogen into NO and relate to all fuels, while pathway (4) pertains to flames of petroleum, coal, and biomass.

2.2.1 Thermal, or Zeldovich, NO

The principal elementary reactions[1] here are:

$$O + N_2 \rightleftharpoons NO + N \quad (1)$$

$$N + O_2 \rightleftharpoons NO + O \quad (26a)$$

$$N + OH \rightleftharpoons NO + H\,. \quad (26c)$$

The first two[2] reactions compose a chain sequence, so a small amount of atomic

[1] Reaction numbers in this section correspond to those introduced serially in later sections of this chapter.

[2] Sometimes the term "Zeldovich mechanism" is used to refer only to reactions (1) and (26a); inclusion of reaction (26c) is then the "extended Zeldovich mechanism".

oxygen can produce large amounts of NO. The first one is rate limiting because of its high activation energy, about 320 kJ/mol. The Zeldovich mechanism is very sensitive to temperature not only because of the high activation energy of reaction (1), but also because the concentration of oxygen atoms in flames increases rapidly with increasing temperature. Since the rate coefficient for reaction (1) is well-established, computation of the Zeldovich NO production rate is straightforward if the oxygen atom concentration is known. Away from the flame front region, one can indeed compute Zeldovich NO production with little concern for details of the chemical kinetics by assuming equilibrium O-atom concentrations. The situation is more complicated in the flame front region, where super-equilibrium concentrations of radicals are present. This is not because of any particular flame front reactions but because flame-front chemistry is driven by fast bimolecular reactions while post-flame chemistry includes slow termolecular recombination reactions. One must describe the kinetics of flame front chemistry properly in order to predict Zeldovich NO production accurately.

2.2.2 Prompt, or Fenimore, NO

This is aptly called Prompt NO, because it is manifested by prompt, that is, rapid, production of NO in a flame front. The experimental observation is that an extrapolation of NO concentration to zero time in a plot of NO against residence time, or distance from the flame front, gives a positive intercept. (It is invariably followed by a slower rate of NO production attributable to Thermal NO.) In a flame front, the concentration of hydrocarbon radicals is large, and NO production is initiated by their reaction with N_2 to break the N–N bond. As discussed in detail later, the most important such reaction is thought to be

$$CH + N_2 \rightleftharpoons HCN + N\,. \tag{24c}$$

The product N atom can form an NO molecule via reactions (26a) or (26c), while the HCN can lead to a second NO molecule through a series of reactions discussed by Miller et al. (1984). A description of NO production in this case requires, in addition to accurate values of rate coefficients for reactions such as (24c), a reliable mechanism for fuel combustion, since one needs to know the concentrations of the hydrocarbon radicals, including CH. Although there has been appreciable progress in recent years in developing such mechanisms, there remains much doubt about hydrocarbon radical profiles. This is still an active research area.

2.2.3 The N_2O pathway

Reaction (1) above is one pathway for reactions of oxygen atoms with molecular nitrogen. Another is intermediate formation of N_2O by a "recombination reaction"

$$O + N_2 + M \rightleftharpoons N_2O + M\,, \tag{-3}$$

where the "collision partner" M collectively represents all molecules present[1] (cf. Section 2.3).

The N_2O formed in reaction (–3) can then react to form NO by

$$O + N_2O \rightleftharpoons 2\,NO \tag{4b}$$

and

$$H + N_2O \rightleftharpoons NO + NH\,. \tag{11c}$$

The termolecular character of reaction (–3) implies that it becomes more important at higher pressures. NO predictions for this pathway are somewhat more complicated than for Thermal NO, but not as difficult as for Prompt NO. A major issue is that there are other pathways besides (4b) and (11c) for the reactions of O and H with N_2O, and one must properly account for the branching ratios in these reactions. Moreover, as with Thermal NO, super-equilibrium concentrations of oxygen atoms in the flame front region must be taken into account.

2.2.4 Fuel nitrogen

It is frequently assumed that most fuel nitrogen is quickly converted to HCN in flames (Morley 1980). The subsequent kinetics of HCN are generally similar to those that apply to HCN in forming Prompt NO. We note, however, there is evidence suggesting that the assumption of rapid conversion of fuel nitrogen to HCN is not always valid (Mackie et al. 1990). It may eventually turn out that correct accounting for NO formation from fuel nitrogen also requires including details of other N-containing species besides HCN.

2.2.5 The NNH mechanism

Under combustion conditions where the concentrations of atoms are high, that is, in flame fronts, the reaction

$$O + NNH \longrightarrow NO + NH \tag{$-$11g}$$

contributes significantly to NO production. This mechanism requires participation of H atoms to form NNH from N_2 as well as O atoms to react with NNH.

---

[1] Reaction numbers preceded by a minus sign refer to the "reverse reaction", that is, the reaction with that number understood to be occurring in the direction opposite to the one indicated by the arrow of the "forward reaction" denoted by the number without the minus sign.

2.2.6 Effects of temperature and pressure

The relative importance of the first three channels for NO production in methane–air flames has been addressed in a series of model calculations reported by Drake et al. (1990; 1991a; 1991b), and these can also be taken as representative of the situation for other fuels. For a 0.1 atm stoichiometric laminar premixed flame, their calculations indicated that the Fenimore mechanism was dominant. At these relatively low pressures, much of the Zeldovich NO production could be traced to super-equilibrium oxygen atom concentrations. In a 20 atm stoichiometric methane–air flame, the major contributor is equilibrium Zeldovich NO, but the higher pressure has increased the contribution of the N_2O pathway to be comparable to that of Fenimore NO. In laminar diffusion flames at 1 atm, these workers concluded that Fenimore NO was the dominant pathway, contributing more than two-thirds of the total NO, with the Zeldovich and N_2O paths being comparably minor components.

These results illustrate the complexity of NO production chemistry. Given the very high sensitivity to temperature of the Zeldovich route, use of different fuels (with different adiabatic flame temperatures) or considering a flame configuration with a different amount of heat loss dramatically affects the predicted amount of Zeldovich NO. Similarly, different fuels change the amount of CH produced in the flame front, thus affecting the amount of Fenimore NO formed.

2.2.7 NO reduction

There are several reactions by which NO can be converted to N_2 or to species that may subsequently be converted to N_2 during the combustion or post-combustion process. As discussed in detail later, these include

$$\begin{aligned} \mathrm{NO} + \mathrm{CH_3} &\rightleftharpoons \mathrm{H_2CN} + \mathrm{OH} && \text{(19d)} \\ + \mathrm{NH_2} &\rightleftharpoons \mathrm{N_2} + \mathrm{H_2O} && \text{(18a)} \\ + \mathrm{NH_2} &\rightleftharpoons \mathrm{NNH} + \mathrm{OH} && \text{(18b)} \\ + \mathrm{NH} &\rightleftharpoons \mathrm{N_2} + \mathrm{OH} && \text{(11e)} \\ + \mathrm{NH} &\rightleftharpoons \mathrm{N_2O} + \mathrm{H}. && \text{(−11c)} \end{aligned}$$

The first can be important in "reburning", where additional fuel is added after the primary combustion stage (Bowman 1992). The second and third play central roles in post-combustion NO removal—Selective Non-Catalytic Reduction (SNCR)—processes such as Thermal DeNOx, wherein ammonia is added in the post-combustion zone to reduce NO_X emission levels in the exhaust gas (Lyon 1975). The last two illustrate the complexity of the chemistry again: N_2 may be a direct product (11e), or the initial product N_2O can subsequently react either to reform NO or to be converted to N_2.

2.3 UNIMOLECULAR AND CHEMICALLY ACTIVATED BIMOLECULAR REACTIONS

2.3.1 Unimolecular reactions

Combustion scientists specializing in chemical kinetics have known for some time that it is necessary to consider the pressure dependence of the rate coefficients of dissociation and recombination reactions. The pressure effect can be qualitatively understood just by recognizing that these reactions are really not one-step processes. Considering the rate coefficient k_{rec} for a recombination reaction, for example,

$$\mathrm{A + B \longrightarrow AB},$$

one has to take into account that formation of the new chemical bond initially produces an energy-rich adduct that can either redissociate to A and B or be stabilized in an energy-transferring collision with a "bath gas" molecule M to form products:

$$\mathrm{A + B \rightleftharpoons AB^*} \tag{a}$$

$$\mathrm{AB^* + M \longrightarrow AB + M}. \tag{b}$$

Assuming that the overall process can be described by assigning rate coefficients k_a and k_b to these two processes and using the steady-state approximation—that the time derivative of the concentration of the energy-rich adduct AB* is small compared to the rate of product formation—one readily derives

$$\frac{\mathrm{d[AB]}}{\mathrm{d}t} = k_{rec}[\mathrm{A}][\mathrm{B}] = \frac{k_a k_b[\mathrm{M}]}{k_{-a} + k_b[\mathrm{M}]}[\mathrm{A}][\mathrm{B}]$$

from which it is seen that k_{rec} is actually a function of [M]

$$k_{rec} = \frac{k_a k_b[\mathrm{M}]}{k_{-a} + k_b[\mathrm{M}]}.$$

Inspection of these equations shows that the reaction rate is predicted to scale linearly with pressure at sufficiently low pressure, that is, sufficiently low [M]. When $k_{-a} >> k_b[\mathrm{M}]$, the "low pressure rate coefficient" becomes

$$k_{rec}(P\rightarrow 0) = \frac{k_a}{k_{-a}} k_b[\mathrm{M}] = K_a k_b[\mathrm{M}] = k_\circ\,[M].$$

The rate coefficient is independent of pressure at high pressures, where $k_{-a} << k_b[\mathrm{M}]$, for which we derive the "high pressure rate coefficient" for recombination to be

$$k_{rec}(P\rightarrow\infty) = \frac{k_a k_b[\mathrm{M}]}{k_b[\mathrm{M}]} = k_a = k_\infty.$$

The rate coefficient of the A + B $\longrightarrow$ AB recombination reaction can be expressed as a function of the high- and low-pressure limiting values k_∞ and $k_\circ[\mathrm{M}]$ in the form

$$k_{rec} = k_\infty \left(\frac{P_r}{1 + P_r}\right)$$

where the dimensionless reduced pressure P_r is given by

$$P_r = k_\circ[\mathrm{M}]/k_\infty \,.$$

Such scaling with [M], that is, with pressure, has been observed experimentally for many reactions of this kind.

For the reverse reaction of dissociation

$$\mathrm{AB} \longrightarrow \mathrm{A} + \mathrm{B}\,,$$

one can follow an analogous line of reasoning to find the pressure dependence of the rate coefficient for dissociation k_{diss} in the first-order decay rate law

$$\frac{\mathrm{d}[\mathrm{AB}]}{\mathrm{d}t} = -\,k_{diss}[\mathrm{AB}]\,.$$

Because the recombination and dissociation reactions represent the forward and reverse directions of the same chemical process, their rate coefficients are related to one another by the equilibrium constant K_{eq} for the reaction A + B $\rightleftharpoons$ AB, that is,

$$\frac{k_{rec}}{k_{diss}} = K_{eq}\,.$$

This equation is a fundamental relationship known as the principle of microscopic reversibility. It can readily be derived for the above two-step mechanism by solving for the pressure dependence of k_{diss} using the same line of reasoning that led us to the pressure dependence of k_{rec}. Because K_{eq} is a thermochemical quantity dependent on temperature only, the "falloff" pressure range over which k_{rec} and k_{diss} decrease from their "high-pressure limit" to "low-pressure limit" values is identical for both and similarly depends only on temperature. In terms of the kinetics of the two-step mechanism, one can understand the sense of the temperature dependence by observing that the dissociation rate coefficient k_{-a} for the energized complex AB* has to increase with temperature. The stabilization rate coefficient k_b has a much weaker temperature dependence, and so the value of [M] (and thus the pressure) needed to maintain high-pressure limiting behavior must be higher at higher temperature. This proves to be a rather large effect over a wide temperature range, and explicit account must be taken of it when constructing combustion models (cf. Warnatz 1983).

Neither the pressure falloff nor its temperature dependence, however, are accurately accounted for just by writing down the pair of reactions (a) and (b) as we have done above. This is because considering only a single energized intermediate AB* is an oversimplification of the real process. In order to account for the experimentally observed pressure falloff and its temperature dependence, one must recognize that a wide range of energization, and thus a wide range of k_a and k_{-a} values, really characterizes the observed reaction. The consequences of this situation have been thoroughly explored theoretically, primarily by means of what is termed "RRKM Theory", and it is readily possible to account for observed falloff behavior in terms of plausible molecular models for recombination and dissociation reactions (Forst 1972; Robinson and Holbrook 1972; Gilbert and Smith 1990).

Several better approximations have been introduced to allow more realistic descriptions of falloff behavior to be incorporated into kinetic models. One introduced by Troe and co-workers (Gilbert et al. 1983) involves adding a "broadening factor" F to the pressure-dependent k_{rec} formula

$$k_{rec} = k_\infty \left(\frac{P_r}{1 + P_r} \right) F$$

where F is a smooth but complicated function of temperature given by the set of equations

$$\begin{aligned}
\log F &= \left[1 + \left(\frac{\log P_r + c}{n - d(\log P_r + c)} \right)^2 \right]^{-1} \\
c &= -0.4 - 0.67 \log F_c \\
n &= 0.75 - 1.27 \log F_c \\
d &= 0.14 \\
F_c &= (1 - a) \exp(-T/T^{***}) + a \exp(-T/T^*) + \exp(-T^{**}/T) .
\end{aligned}$$

The idea is to fit F_c parameters a, T^{***}, T^*, and T^{**} to computed or experimental falloff data once, and then use these equations to compute $k_{rec}(P, T)$ for dynamic modeling. An alternative representation of $F(T)$ was introduced by Stewart et al. (1989)

$$\begin{aligned}
F &= \left[a \exp(-b/T) + \exp(-T/c) \right]^X \\
X &= 1/\left(1 + \log^2 P_r \right)
\end{aligned}$$

with the fitting parameters a, b, and c. Either of these approximations can be used with the Chemkin modeling package (Kee et al. 1989).

An alternative representation of falloff behavior is to compute theoretical k_{diss} or k_{rec} values for one fixed pressure at a number of temperatures, then fit these to an Arrhenius-like temperature function (see, for example, Tsang 1991). The pressure-dependent rate coefficients discussed in later sections are described in this way for pressures of 1/10, 1, and 10 atm. More compact and more accurate interpolation formulas have been discussed by various authors (Gardiner 1988; Oref 1989; Pawlowska et al. 1993; Wang and Frenklach 1993; Kazakov et al. 1994; Prezhdo 1995; Venkatesh et al. 1997).

2.3.2 Pressure-dependent bimolecular reactions

In recent years it has been recognized that similar falloff behavior is to be expected for any reaction that proceeds via a "chemically activated" energized complex, including reactions that involve only transient adduct formation. A variety of reaction types, including additions, recombinations, and insertions, as well as reactions whose overall stoichiometry might suggest a simple bimolecular reaction model, have been analyzed in this way using RRKM theory (for example, by Berman and Lin 1983; Larson et al. 1984; Golden and Larson 1985; Wagner et al. 1990; and Just 1994).

A special consideration that arises in analyzing reactions that proceed via energized complexes is the possibility of multiple dissociation channels. Radical addition, recombination and insertion reactions can manifest complex temperature and pressure behavior as stabilization channels compete with multiple decomposition channels. Significant errors may result if experimental measurements of rate coefficients for reactions of this type are extrapolated to different regions of temperature and pressure without accounting for the competition between unimolecular reactions of the energized complexes and their bimolecular collisional stabilization, as described in the following section.

2.3.3 Quantum Rice–Ramsperger–Kassel theory

A theoretical approach to chemical activation systems, intended for initial evaluations and modeling applications, was developed by Dean (1985). In its original formulation, it was based on a quantum version of Rice–Ramsperger–Kassel theory (QRRK) and used a geometric mean frequency to represent the internal dynamics of the reacting molecule. Robinson and Holbrook (1972) suggest that the quantum version is an improvement over the more commonly used classical version of Kassel theory because the quantum version considers the full set of internal vibrational modes of the reacting molecule, while in using the classical RRK version one typically reduces their number by a factor of one-half to one-third in order to fit experimental falloff data. In essence, QRRK theory uses statistical mechanics to calculate the probability that sufficient energy will be localized in a given vibrational mode for reaction to occur. It requires Arrhenius A-factors, either from the literature or from "Transition State Theory" (hereafter TST; cf. Benson 1976 and Zellner 1984) estimates. In combination with a modified "strong collision" model due to Troe (1979) to describe collisional stabilization, it was applied to the analysis of a variety of chemical activation reactions (Westmoreland et al. 1986; Dean and Westmoreland 1987; Bozzelli and Dean 1989, 1990, 1993) where there were sufficient experimental results in the literature to evaluate its predictions. Its success in these applications shows that QRRK theory provides a simple framework with which the effects of temperature and pressure can be readily understood and evaluated.

QRRK theory is a complementary technique to RRKM theory. In applying RRKM theory, one calculates an Arrhenius *A*-factor for k_∞ from computed properties of the transition state and the reactant molecule, that is, their vibrational frequencies and their internal rotor barriers, moments of inertia and symmetries, if applicable, and their moments of inertia for overall rotation. The value computed for this A-factor is generally determined by the values used for the low frequency vibrations of the transition state, which for complex molecules are sufficiently uncertain that adjustments are often made to obtain agreement with experiment. Thus, although RRKM theory is in principle a more comprehensive description of unimolecular reaction, uncertainties in the necessary input parameters severely restrict its predictive value even when significant computational effort is invested in computing these parameters. While use of RRKM theory may be appropriate in some contexts, it does not offer substantial advantages for the combustion modeler. Most of the unimolecular and chemical activation reactions of combustion chemistry are of such complexity that computing the molecular properties of the transition state at the required level of accuracy is infeasible.

2.3.4 Implementation of QRRK theory

The information required for a QRRK analysis is available from the literature, from thermodynamic parameter estimation codes (Ritter and Bozzelli 1991; Bozzelli et al. 1997), TST estimation codes (Bozzelli and Ritter 1993), and transport property estimation methods (Reid et al. 1987). The required input data include the Arrhenius A-factors for the high pressure limit rate coefficients k_∞ for dissociation of the adduct. It can be derived from measured values or estimated from rate coefficients of isoelectronic or structurally similar reactions with known kinetic parameters. Our preference is to use the equilibrium constant for the recombination reaction in conjunction with the recombination reaction rate coefficient and the equation $k_{diss} = k_{rec}/K_{eq}$. More data is usually available for the reverse (bimolecular) reaction; alternatively, in the case of a theoretical estimation, it is more reliable to estimate recombination than dissociation rate coefficients. This approach avoids the use of adjustable parameters for the unimolecular dissociation process and guarantees thermochemical consistency. Typically, we fit equilibrium constant values over the desired temperature range with appropriate algebraic expressions that account with good accuracy for changes in heat capacity. Equilibrium constants needed for all QRRK calculations reported here can be computed using the thermochemical database summarized in Tables 2.1 and 2.2, which were assembled from a variety of sources including the Sandia database (Kee and Miller 1987) and computations made with the group-additivity code THERM (Ritter and Bozzelli 1991). The reader is referred to a review (Dean et al. 1991) for further discussion of the QRRK method and guidance in selection of input parameters.

A shortcoming of the original QRRK approach was the use of a single frequency to represent the $3n - 6$ vibrational frequencies of a molecule. Weston (1986) compared two QRRK approaches (geometric and arithmetic mean), classical RRK theory, and a quantized version of classical RRK by Schranz et al. (1982) with

TABLE 2.1. Thermochemical data for elements and binary species.

Species	$\Delta_f H^\circ_{298}$ kJ/mol	S°_{298} $\mathrm{J\,mol^{-1}\,K^{-1}}$	$C^\circ_p(T)/\mathrm{J\,mol^{-1}\,K^{-1}}$ at T/K = 300	400	600	800	1000	1500
Ar	0.0	154.7	20.8	20.8	20.8	20.8	20.8	20.8
O	249.2	161.0	21.9	21.5	21.1	21.0	20.9	20.8
O_2	0.0	205.0	29.3	30.3	32.1	33.6	34.7	36.5
H	218.0	114.6	20.8	20.8	20.8	20.8	20.8	20.8
H_2	0.0	130.6	28.9	29.0	29.3	29.8	30.4	32.1
C	716.8	158.0	20.8	20.8	20.8	20.8	20.8	20.8
N	472.7	153.2	20.8	20.8	20.8	20.8	20.8	20.8
N_2	0.0	191.5	28.7	29.3	30.4	31.5	32.6	34.6
OH	39.7	183.6	29.1	29.2	29.6	30.2	30.8	32.6
HOO	15.9	229.0	35.3	37.6	41.6	44.8	47.6	51.8
H_2O	-241.8	188.7	33.9	34.5	36.3	38.7	41.2	46.8
HOOH	-136.1	232.9	43.6	48.0	54.8	59.9	63.6	70.5
CO	-110.5	197.6	28.7	29.4	30.8	31.9	33.0	35.0
CO_2	-393.5	213.7	37.2	41.2	47.3	51.5	54.3	58.3
CN	435.1	202.5	29.2	29.4	30.7	32.1	33.5	35.6
NCCN	309.1	241.6	57.1	61.5	68.3	73.1	76.3	81.1
NO	90.3	210.7	29.3	30.1	31.5	32.8	33.8	35.6
N_2O	82.1	219.8	38.7	42.6	48.4	52.3	54.8	58.3
NO_2	33.1	240.0	37.3	40.7	45.7	49.3	51.6	54.7
NO_3	71.1	252.6	47.6	55.9	67.1	73.2	76.4	79.9
CH	594.1	182.9	29.2	29.2	29.8	31.0	32.5	36.6
$^3CH_2$	386.4	193.8	34.6	36.1	39.2	42.5	45.5	51.1
$^1CH_2$	424.5	184.7	34.8	36.5	40.1	43.4	46.4	51.9
CH_3	145.7	194.1	38.8	42.2	48.5	54.2	59.3	68.2
CH_4	-74.9	186.2	35.9	41.2	52.0	62.2	71.1	86.1
C_2H	562.6	215.5	42.0	43.6	45.9	48.0	50.0	54.1
C_2H_2	225.4	200.9	44.2	50.1	57.5	62.5	66.6	74.1
C_2H_3	299.7	236.9	41.9	50.1	63.1	72.5	79.7	91.4
C_2H_4	52.4	219.5	42.7	53.2	71.1	84.3	94.3	110.4
C_2H_5	119.7	250.5	48.6	59.9	80.2	95.7	107.9	127.6
C_2H_6	-85.3	230.5	51.8	65.6	90.3	108.9	123.6	147.1
$HC{\equiv}C\dot{C}H_2$	341.3	249.2	57.6	67.5	82.8	93.7	101.8	114.9
C_4H_2	439.6	250.1	73.8	84.1	96.3	104.3	110.4	120.7
$HC{\equiv}C\dot{C}{=}CH_2$	491.9	282.3	73.6	87.5	107.5	119.4	128.3	142.5
^3NH	358.8	181.1	29.2	29.1	29.5	30.3	31.2	33.6
^1NH	505.8	174.4	29.2	29.3	30.4	31.8	33.1	35.3
NH_2	190.3	194.6	34.0	34.9	37.2	40.0	42.7	48.2
NH_3	-45.9	192.6	35.7	38.8	45.1	51.1	56.4	66.0
NNH	249.2	224.4	34.8	36.9	41.3	45.4	48.2	52.0
N_2H_2	209.2	218.5	36.6	41.2	49.6	56.5	62.1	70.5
N_2H_3	201.2	228.6	43.8	51.5	63.6	72.5	79.2	90.3
N_2H_4	95.4	238.6	51.1	61.7	77.5	88.4	96.2	110.1

See text for explanations of data sources. All values refer to the ideal gas state. Adjacent chemical symbols for the same element indicate atoms bonded to one another in the manner of peroxides. Unpaired electrons are denoted by dot symbols only where needed to resolve ambiguity.

TABLE 2.2. Thermochemical data for ternary and quaternary species.

Species	$\Delta_f H^\circ_{298}$ kJ/mol	S°_{298} $\mathrm{J\,mol^{-1}\,K^{-1}}$	$C^\circ_p(T)/\mathrm{J\,mol^{-1}\,K^{-1}}$ at T/K = 300	400	600	800	1000	1500
HCO	43.5	224.5	34.6	36.7	40.9	44.9	48.2	52.5
CH_2O	–108.8	218.9	35.4	39.6	48.1	55.8	62.2	70.9
^{1}HCOH	98.3	239.3	45.3	49.8	56.9	62.0	65.9	71.8
H_2COH	–15.1	247.0	49.2	55.5	64.7	72.0	77.9	87.2
HOCO	–207.9	251.6	44.6	50.6	60.2	67.0	71.6	77.0
HCO_2•	–160.0	242.4	41.2	48.4	59.0	67.0	72.0	77.4
CH_3OH	–200.8	239.7	43.9	51.6	67.0	79.7	89.5	105.1
CH_3O	16.6	233.6	39.8	46.2	59.1	69.7	77.7	89.7
HC≡COH	85.5	245.6	55.3	61.9	74.2	80.2	85.0	96.0
$H\dot{C}$=C=O	173.1	253.1	51.3	56.9	65.5	69.1	72.7	77.2
H_2C=C=O	–49.1	241.9	53.1	61.3	74.5	81.6	87.8	96.4
HNO	106.3	220.6	34.6	36.9	41.2	45.0	48.0	52.3
^{1}HON	255.2	214.4	35.0	37.2	41.3	44.8	47.7	52.2
^{3}HON	209.2	226.0	35.0	37.2	41.3	44.8	47.7	52.2
HNNO	231.8	253.3	44.8	50.7	59.7	65.8	69.9	75.9
NNOH	243.1	257.1	43.7	47.8	54.8	59.9	63.4	69.8
HONO	–77.8	254.0	45.1	51.1	59.9	65.4	68.9	74.9
^{1}HNOO	235.5	249.8	42.9	49.8	60.3	67.3	71.8	77.7
HNO_2	–59.2	237.4	37.7	43.5	53.2	60.5	66.0	73.5
HNOH	90.4	241.8	43.1	47.0	54.1	59.8	63.6	70.5
NH_2O	66.5	233.0	38.9	43.5	51.1	57.2	62.1	70.2
NH_2NO	74.9	252.3	50.9	59.9	74.6	85.3	92.5	100.5
c-HNNOH	89.9	259.6	49.4	59.2	74.7	85.4	92.4	100.4
t-HNNOH	80.7	258.2	49.1	61.1	77.8	87.7	93.3	100.5
HCN	135.2	201.7	36.0	39.0	43.7	47.5	50.3	55.1
HNC	189.1	205.9	36.0	39.1	43.9	47.4	50.1	55.2
HCNN	456.1	248.3	45.1	51.0	59.8	65.6	69.3	74.1
HCN_2	448.0	240.7	46.0	50.3	57.8	64.0	67.3	73.6
H_2CN	247.3	224.2	38.3	43.2	52.2	59.6	64.5	71.7
HCNH	276.6	234.0	38.2	43.2	52.2	59.5	64.1	72.3
CH_2NN	286.4	242.5	49.1	58.0	71.3	80.2	86.5	96.0
$^3CH_3N$	329.3	231.3	41.9	49.2	62.4	72.6	80.4	92.3
$^1CH_3N$	470.6	225.5	41.9	49.2	62.4	72.6	80.4	92.3
H_2C=NNH_2	190.4	264.0	59.8	72.0	82.8	105.4	121.8	135.3
H_2C=NH	91.4	234.1	38.2	44.7	57.9	69.5	77.7	89.5
CH_3N=NH	179.1	256.9	57.3	67.8	78.2	102.1	113.4	131.0
CH_3N=N	215.5	262.3	48.7	58.4	74.5	87.1	96.8	112.2
CH_3NH	181.0	249.7	46.6	55.0	70.2	81.2	89.7	105.3
CH_2NH_2	151.7	242.4	54.5	62.5	76.4	86.8	94.7	108.8
CH_3NH_2	–23.0	242.6	50.2	60.2	78.9	93.9	105.7	124.9
NCO	132.2	232.1	40.2	43.8	49.5	53.4	55.9	58.9
HCNO	176.3	242.7	46.6	53.0	61.3	66.9	71.1	77.7
HOCN	–9.5	242.1	46.1	50.5	57.3	62.4	66.3	72.7
HNCO	–101.7	240.7	46.6	51.2	59.0	64.5	68.1	73.7
CH_2NO	173.4	255.3	49.1	58.0	71.3	80.2	86.5	96.0
•N=CHOH	58.0	263.2	52.3	59.6	71.8	80.2	85.6	93.7
CH_3NO	77.4	255.2	52.7	62.7	79.1	89.7	98.7	111.3
NH_2CH_2O	9.2	274.5	54.8	66.1	76.1	99.6	110.0	127.7

See footnote to Table 2.1.

RRKM theory for a number of unimolecular dissociation reactions. He found that use of a geometric mean QRRK approach tends to overestimate $k(E)$, the energy-dependent dissociation rate coefficient, for typical molecules at energies within 40 kJ/mol of the critical energy for dissociation. The overestimate is greater when an arithmetic mean frequency is used. Comparisons of k/k_∞ between RRKM and QRRK theories in Weston's study showed that while the results were reaction specific, overall the agreement for the eight molecular systems analyzed was good. Significant deviations of QRRK from RRKM theory did appear at lower temperatures for several molecules. This is precisely the region where one would expect to find such differences, since the effectively coarser energy graining with a single frequency will have its most pronounced effect when it has to approximate the relatively compressed low-temperature thermal distribution function.

We addressed this issue by developing a three-frequency QRRK model. Three frequencies and associated degeneracies are computed from fits to heat capacity estimates (Bozzelli et al. 1997). Such six-parameter sets, together with the appropriate statistical thermodynamics formulas, were shown by Ritter (1989) to reproduce molecular heat capacities and vibrational state densities accurately. This approach avoids specification of the complete frequency distribution of the adduct, which is unavailable for most of the molecules of interest here. Comparisons of k/k_∞ for the eight molecular dissociations considered by Weston show that this three-frequency QRRK treatment yields significant improvement over the single frequency version. We believe that it is suitable for analyzing both unimolecular falloff and bimolecular chemical activation reactions.

In this chapter we report results of using QRRK theory to analyze a variety of reactions of nitrogen-containing species. The QRRK calculations we report were done with a computational procedure that is a significant extension of the one used in our earlier work (Dean et al. 1991). In addition to incorporating the three-frequency model, it was generalized to allow multiple intermediate species (potential energy wells) to be connected in many ways. The collisional stabilization model has also been improved in two ways. First, the collisional efficiency β is now calculated from the average energy α transferred in a down collision following the proposal of Gilbert et al. (1983)

$$\beta = \frac{\left(\dfrac{\alpha}{\alpha + F_E \mathrm{k_B} T}\right)^2}{\displaystyle\int_0^{E_0} f(E)\left[1 - \frac{F_E \mathrm{k_B} T}{\alpha + F_E \mathrm{k_B} T}\exp\left(\frac{E - E_0}{F_E \mathrm{k_B} T}\right)\right] \mathrm{d}E}.$$

As discussed by Gilbert et al., this approximation provides a more realistic description of weak collision effects at high temperatures than the suggestion of Troe (1979); in particular, β calculated in this manner does not decrease to unrealistically small values at high temperatures. Second, the collisional deactivation rate coefficient is now computed from reduced collisional integrals $\Omega^{(2,2)}$ (Reid et al. 1987), rather than set equal to the kinetic theory of gases formula for the hard-sphere collision frequency; this change has the effect of modifying slightly the $\sqrt{T}$ temperature dependence of collision frequency given by the kinetic theory formula.

2.4 ANALYSIS OF HYDROGEN ATOM ABSTRACTION REACTIONS

Elementary reactions in which a hydrogen atom is removed from a stable species in reaction with an atom or radical, called "abstraction reactions", play important roles in the combustion chemistry of nitrogen. Experimental rate data are available for only a small fraction of them, which forces us to introduce a number of estimated abstraction rate coefficients. In order to deal with them self-consistently, we developed the following estimation procedure. Discussion is limited to abstractions by H, O, OH, NH_2, and CH_3, the radicals likely to be in highest concentrations under combustion conditions. We also estimate rate coefficients for HO_2 abstractions for a limited number of reactions.

Rate coefficients were estimated by relating the parameters for a specific reaction type, say abstractions by OH, to those of a well-characterized hydrocarbon analog such as $OH + C_2H_6$. Consistent use of such prototypical reference reactions avoids propagating whatever inconsistencies are present in compilations. Moreover, it takes into account the different temperature dependences found for different types of hydrogen abstractions. It is well-recognized (Tsang and Hampson 1986; Baulch et al. 1992) that many hydrogen abstraction reactions show upward curvature in their Arrhenius plots, leading to appreciably higher rate coefficients at high temperatures than one would estimate from linear Arrhenius extrapolation of low-temperature measurements. Such temperature dependence has to be included in order to predict reliable rate coefficients under combustion conditions. A convenient way to include this dependence is to express the rate coefficient in the form $A\,T^m \exp(-E/RT)$. Zellner (1984) shows that m in this equation is expected to have maximum values of 1.5 for abstraction by atoms, 2 for abstraction by diatomic radicals, and 3 for abstraction by polyatomic radicals.

Our procedure is as follows:

(1) The values of A and m for all reactions in a homologous series are taken from a single reference reaction. For abstractions by H, O, and OH, we use the expressions recommended by Baulch et al. (1992) for the corresponding abstractions from C_2H_6 and assume that A scales in proportion to the number of equivalent abstractable hydrogen atoms. The preexponential factors thus take the form $A_{C_2H_6}(n_H/6)$, where n_H is the number of equivalent hydrogen atoms that can be abstracted from the molecule of interest.

CH_3, NH_2, and HO_2 have to be treated differently. The Baulch et al. rate coefficient expression for $CH_3 + C_2H_6 \longrightarrow CH_4 + C_2H_5$ sets m to 6, much larger than the theoretical expectation. A substantial difference in m for the abstractions from C_2H_6 by OH, for which $m = 2$, and by CH_3 is inconsistent with the temperature dependence of the equilibrium constant for the reaction $OH + CH_4 \longrightarrow CH_3 + H_2O$.[1] It is equally reasonable and more consistent to base the assignment of A and m for H abstractions by CH_3 on the basis of the rate coefficient for its reaction with H_2O implied by the rate coefficient and equilibrium constant of $OH + CH_4 \rightleftharpoons H_2O + CH_3$. The result is $m = 1.87$, much more consistent with theory (Zellner 1984). With A and m thus assigned for abstractions by CH_3, a value of E for abstractions by CH_3 is derived by fitting the Baulch expression for the rate coefficient of CH_3 + C_2H_6. Although the resulting fit is imperfect, it is adequate for our purpose and ensures thermodynamic consistency for the abstraction reactions of OH and CH_3.

The m value for NH_2 is similarly derived from the forward rate coefficient and the equilibrium constant of $OH + NH_3 \rightleftharpoons H_2O + NH_2$, and an E value is derived from the reaction of NH_2 with C_2H_6. For HO_2, the reverse of OH + H_2O_2 was used. Values for A and m are listed in Table 2.3.

(2) The value of E is estimated using the "Evans–Polanyi relationship", which states that the activation energy E for a set of reactions of similar type but different enthalpy change of reaction is proportional to the enthalpy change of reaction. This can be written as

$$E = E_{\text{ref}} - f(\Delta_r H^\circ_{\text{ref},298} - \Delta_r H^\circ_{298}) ,$$

where the ref subscript denotes the reaction in the set chosen as a reference and the proportionality constant f is known as the Evans–Polanyi factor. For abstractions from hydrocarbons, we estimated f using ethane as the reference molecule. For each reaction class, we selected the f values that gave the overall best fits to representative data listed in the NIST compilation (Mallard et al. 1993) and other available sources. Values of f for abstractions forming alkyl radicals are included in Table 2.3.

For formation of nitrogen-centered radicals by hydrogen abstraction, ammonia is used as the reference. The scarcity of data for these reaction types forced us to select f in a more arbitrary manner. For H, O, and OH, our assignments are based on the differences in reported E values for abstractions from NH_3 and HNO. The data for abstraction from NH_3 are well-characterized experimentally (cf. Sections 2.5.6 through 2.5.8); transition state theory results were adopted for the H-abstraction reactions from HNO by OH and H (cf. Section 2.7.11). Combining these data gives values for f_H and f_{OH}. Combining experimental data for $O + NH_3$ with the assumption that O + HNO is comparable to H + HNO (cf. Section 2.7.11) yields f_O. We obtained f_{NH_2} from the rate coefficients for $NH_2 + C_2H_6$ and $NH_2 + H_2O$. The data for $CH_3 + NH_3$ and $CH_3 + N_2H_4$ were combined to yield the f_{CH_3} values listed in Table 2.3.

---

[1] This reaction is a linear combination of the abstraction reactions of OH and CH_3 from C_2H_6; for this reason the temperature dependence of its equilibrium constant imposes a consistency requirement on the temperature dependence of the rate coefficients for abstraction of H atoms from C_2H_6 by OH and CH_3.

TABLE 2.3. Estimation parameters for hydrogen atom abstraction reactions.[1,2]

Radical	A	m	E_{ref}/kJ	$\Delta_r H^\circ_{ref,298}$/kJ	f
Formation of carbon-centered radicals; ethane reference.					
H	2.4×10^8	1.5	31.0	−13.0	0.65
O	1.7×10^8	1.5	24.3	−4.6	0.75
OH	1.2×10^6	2.0	3.8	−76.6	0.50
NH_2	9.2×10^5	1.94	30.1	−31.4	0.23
CH_3	8.1×10^5	1.87	44.4	−15.5	0.65
HO_2	1.4×10^4	2.69	79.1	53.1	0.60
Formation of nitrogen-centered radicals; ammonia reference.					
H	2.4×10^8	1.5	47.3	18.0	0.21
O	1.7×10^8	1.5	31.4	26.8	0.15
OH	1.2×10^6	2.0	3.3	−45.2	0.05
CH_3	8.1×10^5	1.87	41.8	15.9	0.15

[1] Only limited high-temperature rate data are available for the unusually exothermic abstraction reactions that produce resonance-stabilized radicals. The rate coefficient expressions reported for them are generally in the Arrhenius form, i.e., without a temperature-dependent preexponential factor. Using such expressions together with three-parameter expressions for reactions that produce alkyl-type radicals would lead to significant underestimates in rate coefficient assignments at combustion temperatures for reactions involving resonance-stabilized radicals. Although these abstractions are typically about 50 kJ/mol more exothermic than those that form normal radicals, the limited data available (Tsang 1991) suggest that their abstraction reactions are only slightly faster than reactions of normal exothermicity. Using the usual Evans–Polanyi factors would overestimate their rate coefficients. One can either employ different, i.e., smaller, correction factors for these reactions, or one can assume that these abstractions will have an activation energy about 4 kJ/mol lower than their non-resonant analogs but with the same m-exponent.

Additional complications arise in estimating A values for these reactions. Formation of the resonantly stabilized radical could restrict rotation about the radical center, lowering the entropy of the transition state and correspondingly decreasing the A-factor. Hydrogen atom abstractions from aldehydes and from carbon atoms adjacent to carbonyl groups also require special treatment.

[2] A complication that can arise when using the Evans–Polanyi relationship for reactions with very large exothermicities is that the resulting activation energy can become negative, which is physically meaningless for an abstraction reaction. To avoid this, a lower limit of $-300m$R is given to E. The activation energy in the two-parameter Arrhenius form must then be positive at 300 K.

A distinguishing feature of nitrogen chemistry is participation of long-lived radicals, for example NH_2O. These molecules require substantially more energy than their hydrocarbon analogs to break the "β" bond (the one on the far side of the atom adjacent to a radical site) to form an unsaturated molecule and a smaller radical. The increased endothermicity for these "β-scission" reactions in nitrogen chemistry results in much lower rate coefficients for dissociation. As a result, radical–radical reactions are more important in nitrogen than in carbon chemistry. Recombination reactions of radicals occur via chemically activated adducts, whose pressure-dependent reactivity can be described as outlined in the preceding section. However, estimates of the rate coefficients for "disproportionation" reactions, in which a hydrogen atom is transferred from one radical to another to form two stable products, are less reliable, because there is little information available about this type of reaction for nitrogen species.

Rate coefficients for disproportionation reactions of hydrocarbon radicals have been estimated in the past by collision theory models (for example, by Benson 1985). We prefer to estimate rate coefficients for these reactions as analogs of abstractions from stable molecules. Since radical–radical hydrogen transfer reactions are generally more exothermic than those involving a radical and a stable molecule, E must frequently be assigned the lower limit value $-300m\mathrm{R}$ (cf. footnote to Table 2.3). One would expect the transition state for hydrogen transfer between two radicals to be more extended than for an abstraction from a stable molecule, since bond strengths of hydrogens being transferred are usually weaker in radicals than in stable molecules. This means that our rate coefficient estimates for disproportionation reactions are lower limits. Comparisons of rate coefficients estimated with this method to values in the NIST database (Mallard et al. 1993) for hydrocarbon reactions confirm, however, that this method does provide close estimates, especially at the higher temperatures of interest in combustion. Specific comparisons for the few instances where information is available for reactions involving nitrogen are discussed later in this chapter.

In later sections we refer to rate coefficients estimated by the approach described here as DHT (for "Direct Hydrogen Transfer") estimates.

2.5 UPDATED RATE COEFFICIENTS FOR THE H/N/O SYSTEM

In this section we review reactions considered by Hanson and Salimian (1984) where there have been significant later measurements but where extensive theoretical analysis is not necessary.

2.5.1 $O + N_2 \longrightarrow N + NO$

There have been several new measurements, in quite remarkable agreement, as shown in Figure 2.1. Thielen and Roth (1984) used atomic resonance absorption spectroscopy to monitor N and O atoms produced from shock-heated N_2–N_2O–Ar mixtures over the temperature range 2400 to 4100 K. Their rate coefficient expression, $1.8 \times 10^{14} \exp(-38300/T)$ $\mathrm{cm^3 mol^{-1} s^{-1}}$, agrees well with the recommendation of Hanson and Salimian. Davidson and Hanson (1990a) measured

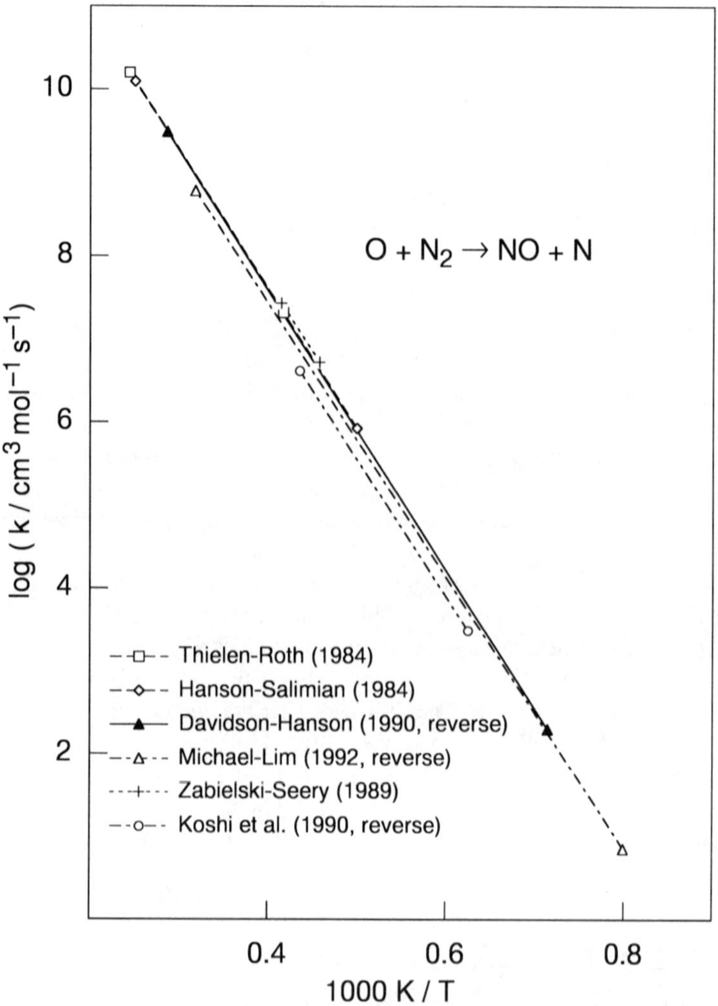

FIGURE 2.1. Arrhenius plot for $O + N_2 \longrightarrow NO + N$. The recommended expression is based on the measurements by Davidson and Hanson (1990) for the reverse reaction.

the reverse reaction rate by measuring NO disappearance behind shock waves from 1400 to 3500 K. Their result and the well-established thermochemistry of this reaction led to the forward reaction rate coefficient expression $1.95 \times 10^{14} \exp(-38660/T)$ $cm^3 mol^{-1} s^{-1}$, in agreement with the other measurements. Koshi et al. (1990), using NO as a source for N atoms, reported a slightly lower value for the reverse reaction rate over a narrower temperature range. Michael and Lim (1992) also studied the reverse reaction and derived the forward reaction rate coefficient expression $1.0 \times 10^{14} \exp(-37988/T)$ $cm^3 mol^{-1} s^{-1}$. Zabielski and Seery (1989) reported $k = 1.99 \times 10^{14} \exp(-38133/T)$ $cm^3 mol^{-1} s^{-1}$ from measurements of NO production in oxygen-enriched methane flames. Thus this rate coefficient for production of NO via the Zeldovich mechanism seems to be well-characterized. We suggest adoption of the Davidson and Hanson expression

$$k_1 = 1.95 \times 10^{14} \exp(-38660/T) \ \mathrm{cm^3 mol^{-1} s^{-1}} \ .$$

For the 1500 to 2500 K regime most relevant to combustion it appears that the recommended expression should be accurate to at least $\pm 25\%$. Such a spread encompasses most of the recent measurements with the exception of the higher temperature data of Michael and Lim and the data of Koshi et al., both of which are lower than given by the recommended expression.

Michaud et al. (1992) suggest that this reaction proceeds by attack of singlet oxygen atoms, $O(^1\Delta)$, rather than ground state $O(^3P)$ atoms. They report that this mechanism is consistent with the literature data if it is assumed that the $^1\Delta$ state is equilibrated with the ^3P ground state, but point out that it is not consistent with measurements of the reverse reaction rate by Felder and Young (1972), where $O(^1\Delta)$ production accounted for less than 10% of the overall reaction. It would appear that although the rate of this reaction is well-characterized, additional work would be helpful to establish its mechanism.

2.5.2 NO + Ar $\longrightarrow$ N + O + Ar

At the time of the Hanson–Salimian review, the reported values showed wide variations and no recommendation was made. The subsequent measurements of Thielen and Roth (1984) are quite close to the earlier work of Wray and Teare (1962) (Figure 2.2). While the higher values of Koshi et al. (1978) and the lower value of Myerson (1974) result in a substantial spread in measured values, we feel that the measurements of Thielen and Roth are to be preferred, essentially because the same type of analysis yielded good agreement for reaction 1. Accordingly, we recommend

$$k_2 = 9.6 \times 10^{14} \exp(-74700/T) \ \mathrm{cm^3 mol^{-1} s^{-1}}$$

over the temperature range 2400 to 6200 K. This recommendation is consistent with that of Tsang and Herron (1991) .

2.5.3 N_2O + Ar $\longrightarrow$ N_2 + O + Ar

There have been several new studies of this reaction in shock tubes (Roth and Just 1984; Frank and Just 1985; Fujii et al. 1989; Zuev and Starikovskii 1991a; Michael and Lim 1992) and in flow reactors (Johnsson et al. 1992). These are compared to the Hanson and Salimian recommendation in Figure 2.3. The remarkable agreement testifies that accurate rate coefficient determinations can indeed be made in well-chosen systems. The extended temperature range now

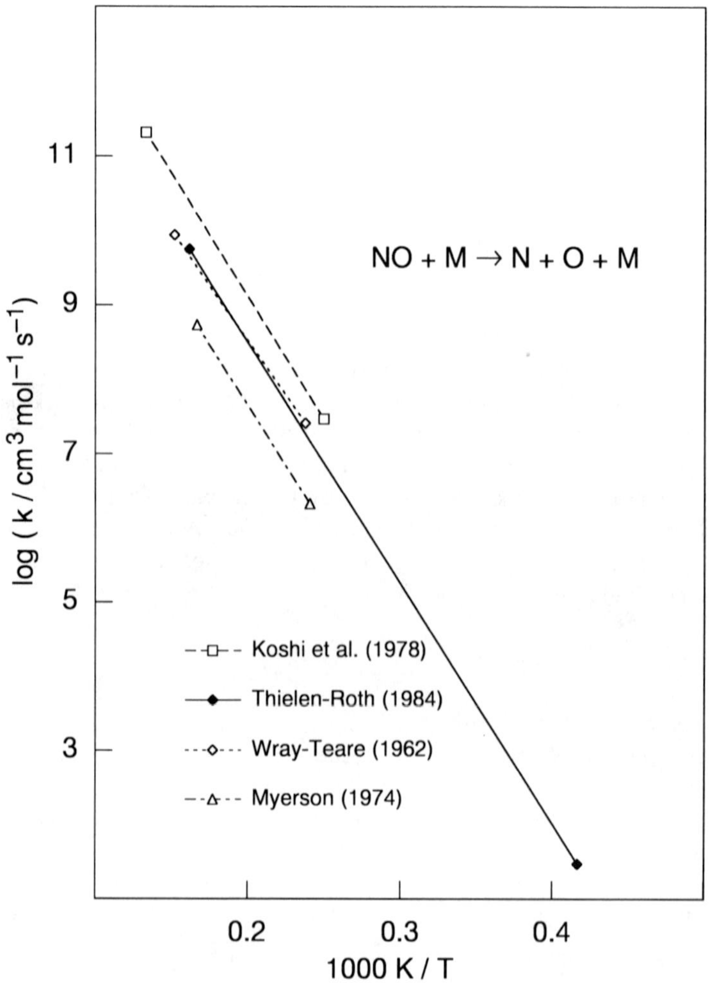

FIGURE 2.2. Arrhenius plot for NO dissociation. The recommended expression is that reported by Thielen and Roth (1984).

covered by experiments allows a recommendation to be made for a larger range. We recommend the expression suggested by Johnsson et al.

$$k_3 = 4.0 \times 10^{14} \exp(-28230/T) \ \text{cm}^3\text{mol}^{-1}\text{s}^{-1}$$

for M = Ar at temperatures from 1000 to 3000 K. Subsequent measurements by Röhrig et al. (1996) for temperatures from 1700 to 3000 K and pressures from 0.3 to 6 atm agree with this recommendation. These authors also report measurements at pressures up to 450 atm which confirm that the reaction is near the low-pressure limit for pressures up to 10 atm at temperatures over 1500 K. Falloff parameters are given by Tsang and Herron (1991) for N_2 and CO_2 bath gases.

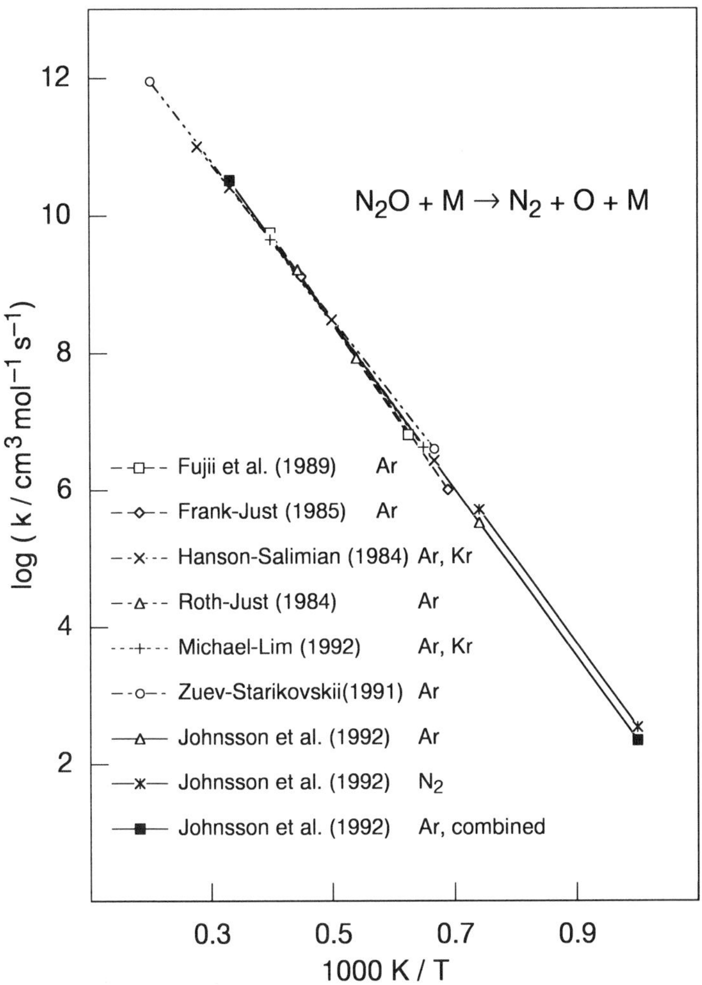

FIGURE 2.3. Arrhenius plot for N_2O dissociation. The recommended expression is the combined one reported by Johnsson et al. (1992).

2.5.4 $O + N_2O \longrightarrow$ Products

This reaction has two channels:

$$O + N_2O \longrightarrow N_2 + O_2 \tag{4a}$$

$$O + N_2O \longrightarrow 2\,NO\,. \tag{4b}$$

The recent measurements are summarized in Figures 2.4 and 2.5. Hidaka et al. (1985) reported rate coefficients for these reactions from a modeling analysis of N_2O/Ar and N_2O/H_2/Ar mixtures heated in shock waves. Their assessment is

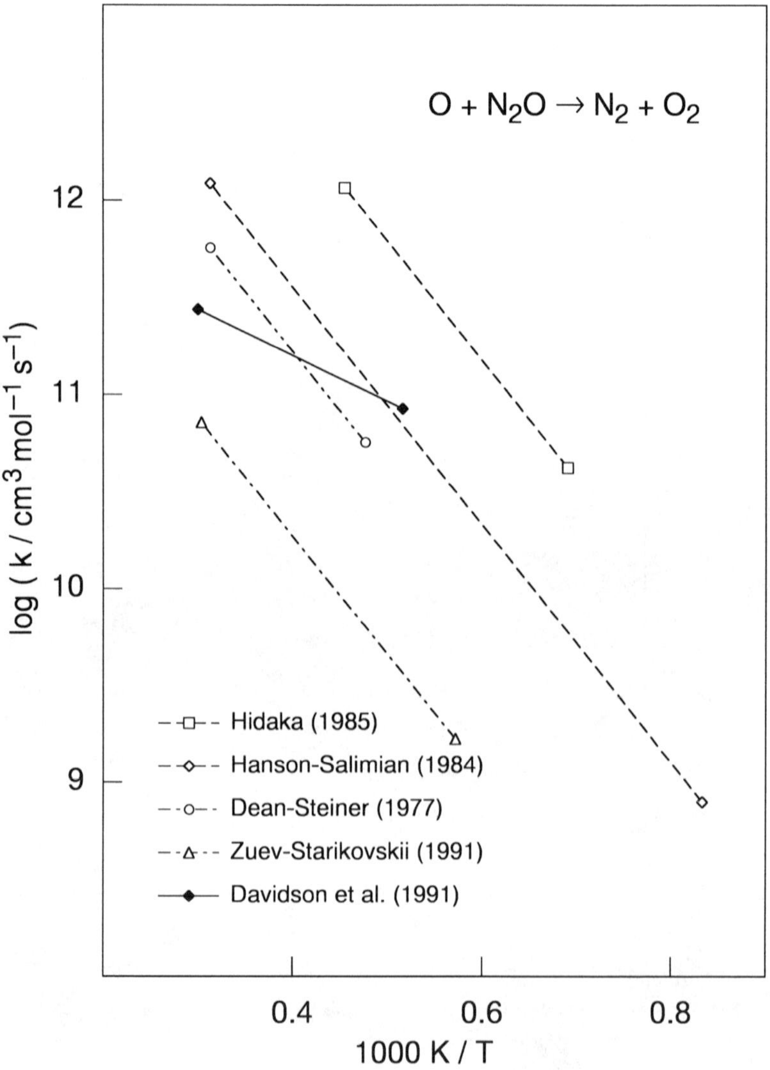

FIGURE 2.4. Arrhenius plot for $O + N_2O \longrightarrow N_2 + O_2$. The recommended expression is that reported by Davidson et al. (1991).

similar to that of Hanson and Salimian in that these reactions have rate coefficients of comparable magnitude, but Hidaka et al. assigned larger values than those suggested earlier. Zuev and Starikovskii (1991b) reported values of k_{4a} much lower than recommended by Hanson and Salimian, but similar values of k_{4b}. These reactions were reinvestigated by Davidson et al. (1991a) , who reported k_{4b} values similar to earlier ones but found that k_{4a} is noticeably smaller at higher temperatures. Earlier work (Dean and Steiner 1977) had also reported

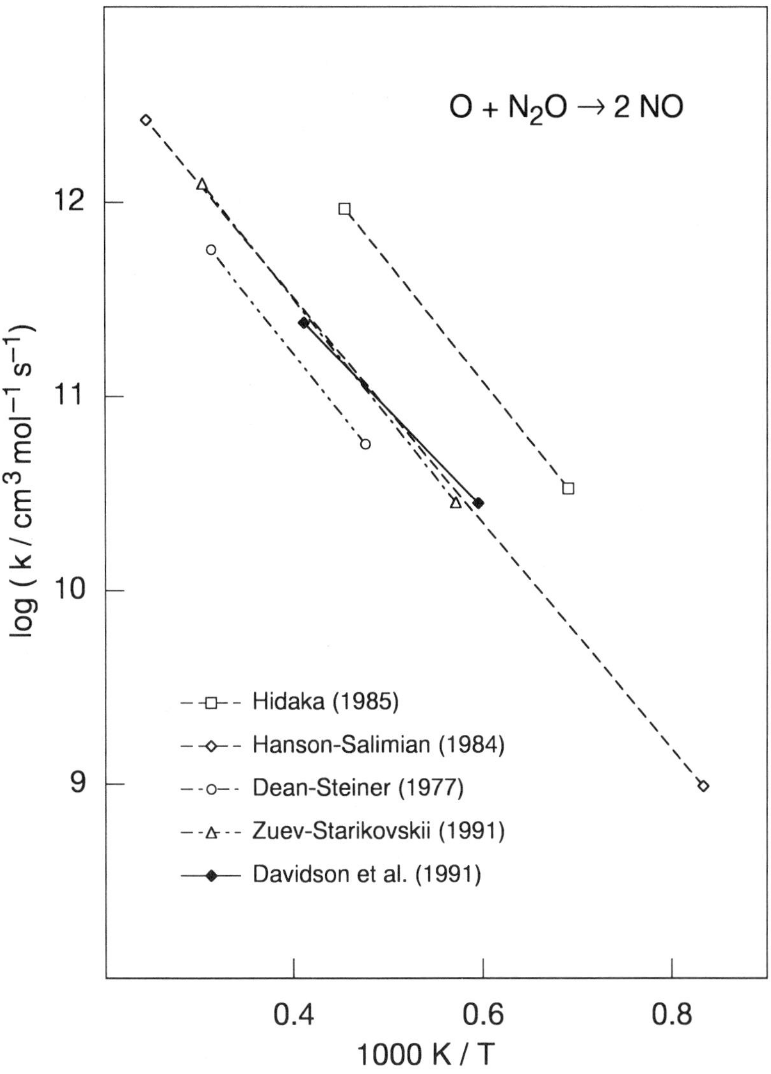

FIGURE 2.5. Arrhenius plot for $O + N_2O \longrightarrow 2\ NO$. The recommended expression is that reported by Davidson et al. (1991).

values somewhat lower than those recommended by Hanson and Salimian for both channels. The Davidson et al. (1991a) experiments have the advantage of direct observation of products from each channel (O_2 from 4a and NO from 4b) and for this reason we recommend their expressions

$$k_{4a} = 1.4 \times 10^{12} \exp(-5440/T)\ \text{cm}^3\text{mol}^{-1}\text{s}^{-1}$$
$$k_{4b} = 2.9 \times 10^{13} \exp(-11650/T)\ \text{cm}^3\text{mol}^{-1}\text{s}^{-1}$$

realizing that k_{4a} is still uncertain. This uncertainty is further amplified by the reports of Röhrig et al. (1996), Ross et al. (1997), and Meagher et al. (1997). Röhrig et al. found that using the Hanson and Salimian expression for k_{4a} gave better fits to their N_2O profiles than the Davidson et al. expression, Ross et al. report that the overall rate coefficient required to fit their O_2 profiles near 1400 K is much lower than even the Hanson and Salimian expression, while the Meagher et al. direct measurements from 1076 to 1276 K are significantly larger than those given by the Hanson and Salimian expression.

2.5.5 $NH_3 + Ar \longrightarrow NH_2 + H + Ar$

The measurements of Davidson et al. (1990b) agree with the expression recommended by Hanson and Salimian. Accordingly, that recommendation,

$$k_5 = 2.5 \times 10^{16} \exp(-47200/T) \ \mathrm{cm^3 mol^{-1} s^{-1}}$$

for M = Ar over the temperature range 1740 to 3450 K, is confirmed.

2.5.6 $NH_3 + H \longrightarrow NH_2 + H_2$

Hanson and Salimian (1984) noted the appreciable scatter in the high-temperature rate coefficients reported for this reaction. Since 1984, there have been several direct measurements of this reaction rate (Hack et al. 1986; Marshall and Fontijn 1986; Michael et al. 1986; Sutherland and Klemm 1987; and Ko et al. 1990), all in agreement with one another and all substantially faster than the expression recommended by Hanson and Salimian.

These newer results and the Hanson and Salimian recommendation are shown in Figure 2.6. We recommend that the combined expression developed by Ko et al. (1990),

$$k_6 = 5.4 \times 10^5 T^{2.40} \exp(-4990/T) \ \mathrm{cm^3 mol^{-1} s^{-1}}$$

be used over the temperature range 490 to 1780 K. This expression combines their measurements from 490 to 960 K with those at higher temperatures by Michael et al. (1986) and Hack et al. (1986).

2.5.7 $NH_3 + OH \longrightarrow NH_2 + H_2O$

Recent experiments have confirmed the marked non-Arrhenius behavior of this rate coefficient (Zabielski and Seery 1985; Jeffries and Smith 1986). Both groups combined their results with earlier work and developed non-Arrhenius fits, as shown in Figure 2.7. At high temperatures, their expressions are in good agreement with the expression recommended by Hanson and Salimian, but the newer expressions are much higher at lower temperatures. The significantly lower high-temperature measurement by Fujii et al. (1986), also indicated in Figure 2.7,

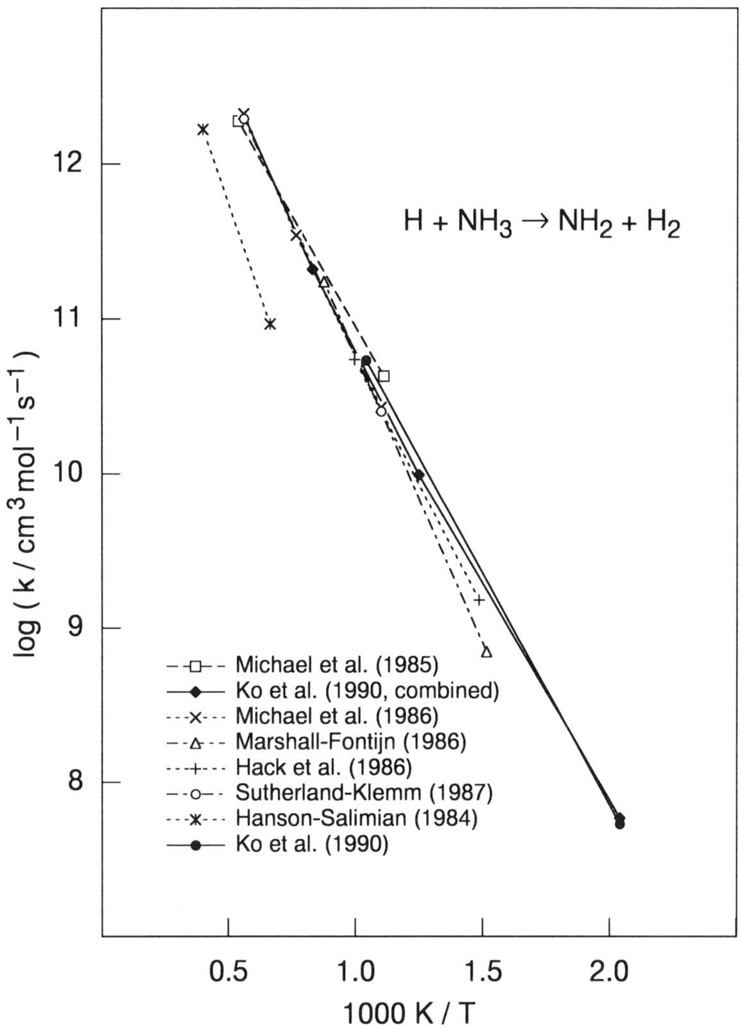

FIGURE 2.6. Arrhenius plot for $H + NH_3 \longrightarrow NH_2 + H_2$. The recommended expression is the combined one reported by Ko et al. (1990).

was derived by fitting a computer model to data taken using shock-heated NH_3–N_2O–Ar mixtures and is thus more indirect. This reaction was critically evaluated by Cohen and Westberg (1991). Their recommendation,

$$k_7 = 5.0 \times 10^7 T^{1.6} \exp(-480/T) \ \text{cm}^3\text{mol}^{-1}\text{s}^{-1},$$

also displayed in Figure 2.7, describes the data well over the temperature range from 225 to 3000 K and we accordingly recommend it.

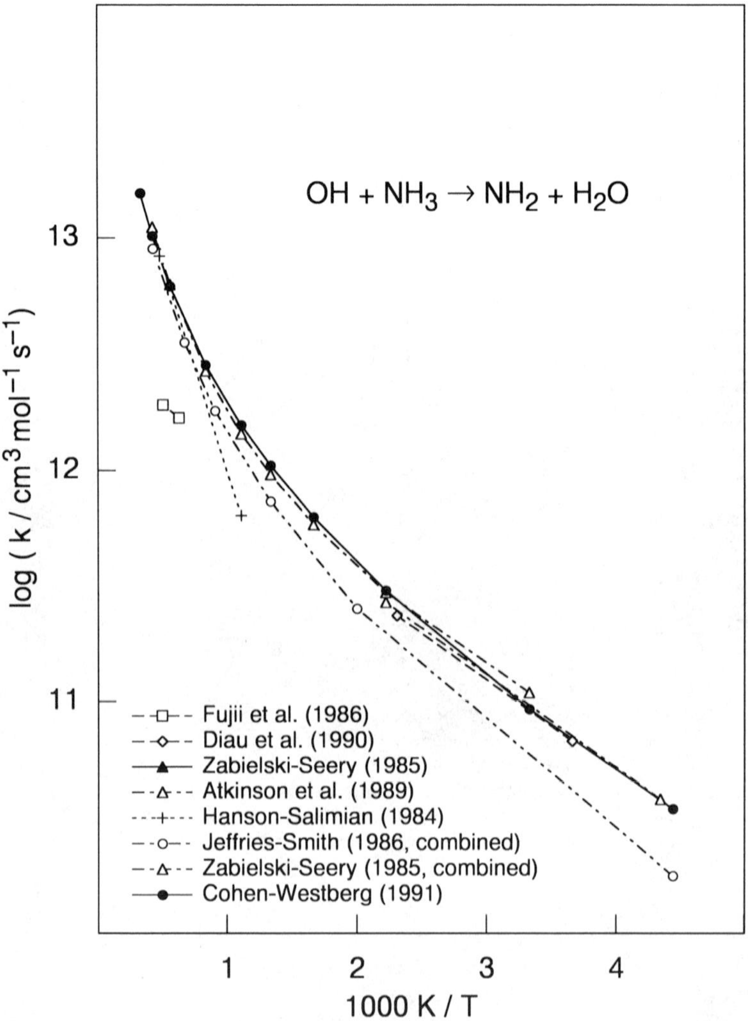

FIGURE 2.7. Arrhenius plot for $OH + NH_3 \longrightarrow NH_2 + H_2O$. The recommended expression is that of Cohen and Westberg (1991).

2.5.8 $NH_3 + O \longrightarrow NH_2 + OH$

Here again there has been significant progress since the Hanson–Salimian review of 1984. In addition to two new high-temperature measurements (Fujii et al. 1986; Sutherland et al. 1990) and one low-temperature study (Perry 1984), there have been an exhaustive review (Cohen 1987) and a critical evaluation (Baulch et al. 1992).

This information is summarized in Figure 2.8. Given the direct nature of both the Sutherland et al. and the Perry experiments, we suggest use of the Sutherland

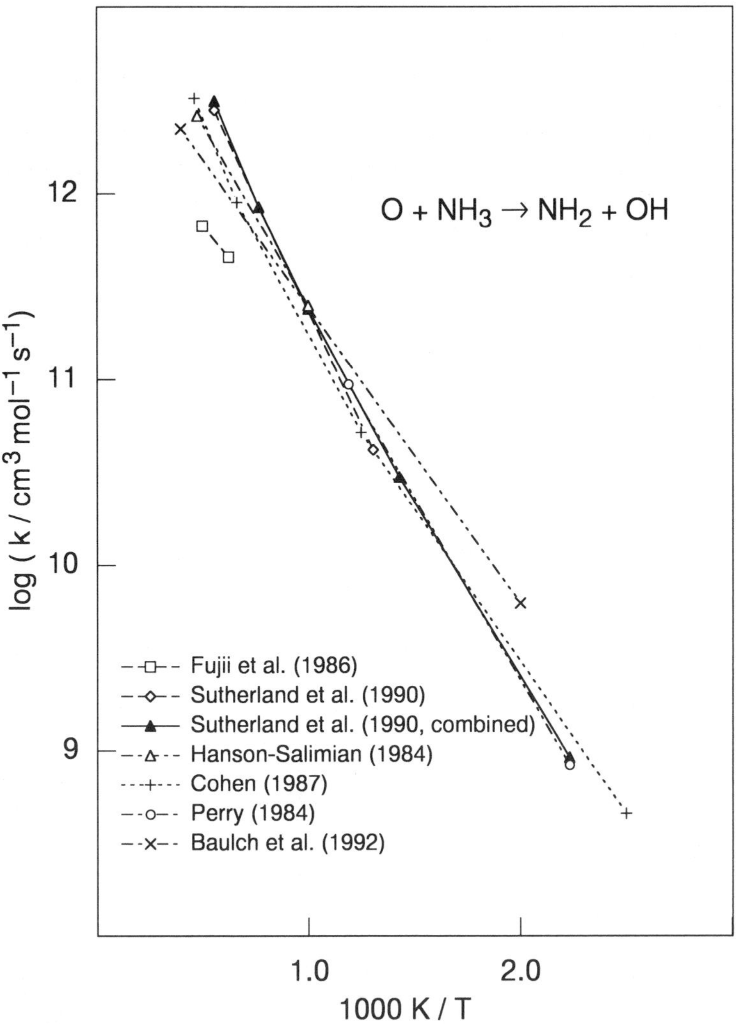

FIGURE 2.8. Arrhenius plot for $O + NH_3 \longrightarrow NH_2 + OH$. The recommended expression is the combined one reported by Sutherland et al. (1990).

et al. expression

$$k_8 = 9.4 \times 10^6 T^{1.94} \exp(-3250/T) \ \text{cm}^3\text{mol}^{-1}\text{s}^{-1},$$

which represents both data sets over the temperature range 450 to 1800 K. We prefer this expression at high temperatures to the slightly lower one suggested by Cohen since the latter is based upon a transition state theory extrapolation of Perry's value rather than measurements. These data also suggest that the recommendation of Baulch et al. (1992) is not an adequate representation of the more recent experiments.

Summarizing the data in Figures 2.6–2.8, it appears that the rate of NH_3 consumption in flames should be accurately described by the now thoroughly investigated elementary reactions with the principal reactive species in flame gases. The most prominent consequence of this is that the production of NH_2, the key intermediate in selective noncatalytic reduction processes for NO suppression, can be regarded as reasonably well understood. This allows us to focus on the subsequent reactions of NH_2 (and later chemistry) to develop improved understanding of these processes. We will address this topic several times in later sections.

2.6 QRRK TREATMENTS

2.6.1 $H + NH_2 \longrightarrow NH + H_2$

There is one new measurement (Davidson et al. 1990) to supplement the sparse database on this reaction, which has also been reviewed by Baulch et al. (1992). These results are compared to the tentative Hanson–Salimian recommendation in Figure 2.9. For completeness, the earlier measurements of Yumura and Asaba (1980) are also included. The Davidson et al. expression is consistent with the recommendation of Hanson and Salimian. The experimental scatter, however, suggests that additional study of this reaction is needed. The mechanism of this radical–radical reaction could be either direct hydrogen transfer or a complex recombination–elimination path. Spin conservation on the ground state NH_3 surface would require that the combination–elimination pathway produce $^1NH + H_2$, a channel appreciably higher in energy than the one producing ground state $^3NH + H_2$. A potential energy diagram for this pathway on the $^1NH_3$ surface is shown in Figure 2.10; the QRRK parameters are listed in Table 2.4.

TABLE 2.4. $H + NH_2 \longrightarrow$ Products.

	Reaction	A/s^{-1} or $cm^3mol^{-1}s^{-1}$	E_a/kJ mol^{-1}
(1)	$H + NH_2 \longrightarrow NH_3$	$1.2\times10^{15}\,T^{-0.4}$	0
(−1)	$NH_3 \longrightarrow H + NH_2$	2.6×10^{15}	454
(2)	$NH_3 \longrightarrow {}^3NH + H_2$	6.0×10^{13}	444
(3)	$NH_3 \longrightarrow {}^1NH + H_2$	1.8×10^{15}	550

Molecular constants of NH_3 adduct: $\nu_1 = 1101$ cm^{-1} (degen=2.10) $\nu_2 = 2430$ cm^{-1}(2.33) $\nu_3 = 3973$ cm^{-1}(1.57); $\sigma = 3.49$ Å and $\varepsilon/K = 310$.

Rate coefficients: k_1 based on $H + CH_3$ rate coefficient expression of Tsang et al. (1986); k_{-1} from microscopic reversibility (MR); k_2 A from TST estimate; E_a estimated to give a 42 kJ/mol barrier for the reverse reaction; k_3 via k_{-3} and MR based on the $^1CH_2 + H_2$ reaction.

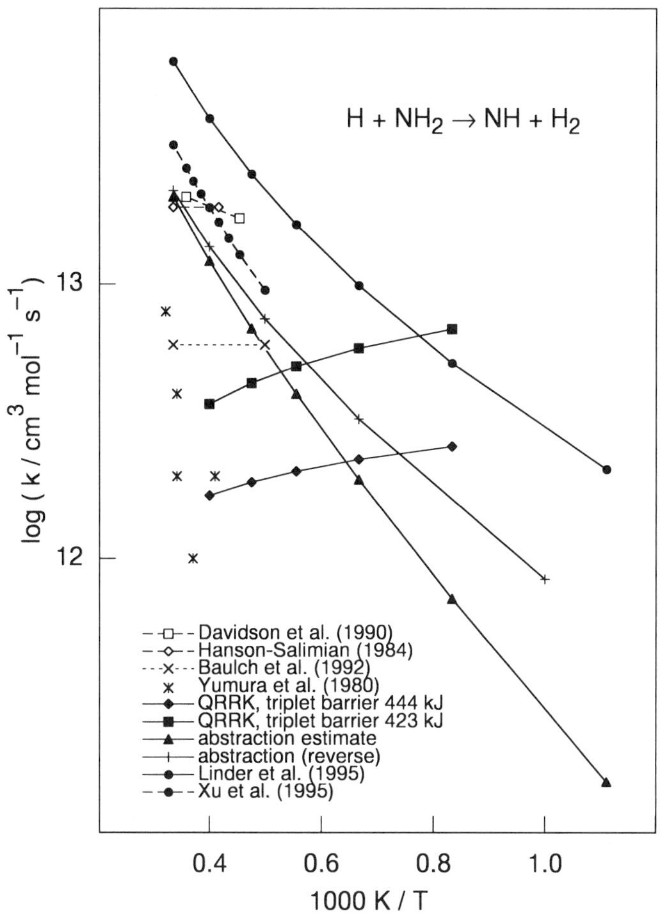

FIGURE 2.9. Arrhenius plot for $H + NH_2 \longrightarrow NH + H_2$. The QRRK predictions of reaction via combination–elimination are apparently somewhat low even with the triplet channel barrier reduced by 21 kJ/mol. Both the radical–radical hydrogen transfer estimate and the one based on the reverse reaction are close to the measurements of Davidson et al. Two sets of ab initio calculations (Linder et al. 1995; Xu et al. 1995) for the direct abstraction reaction are also included.

We are not aware of any quantum-mechanical computations for elimination via the triplet channel. Comparison with other non-spin-conserving processes (Yarkony 1993) suggests that the barrier is about 40 kJ/mol above the enthalpy change for this channel. One would anticipate little or no barrier above the energetic constraint for singlet NH elimination, because the reverse reaction of ^1NH insertion should be analogous to the isoelectronic $^1CH_2$ insertions, where barriers are observed to be very low. Even assuming the barrier to $^3NH + H_2$ to be 10 kJ/mol lower than that for dissociation of the complex back to the reactants $H + NH_2$, the much higher A-factor for the simple fission pathway results in

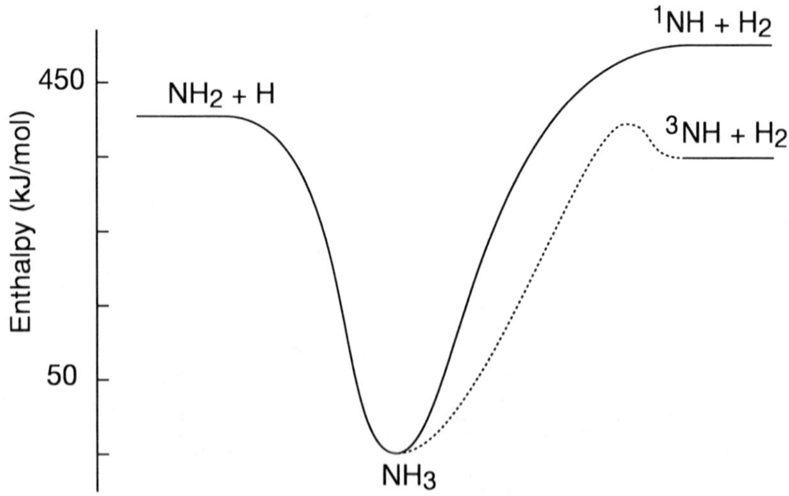

FIGURE 2.10. Potential energy diagram for the recombination–elimination pathway for $NH_2 + H$ on the ground state surface of NH_3. The dashed line shows the triplet pathway.

dissociation of the adduct back to reactants being the dominant channel. Thus the calculated rate coefficient for the $^3NH + H_2$ channel is lower than the expression reported by Davidson et al. These results are shown in Figure 2.9. Lowering the barrier to the ^3NH channel by 21 kJ/mol, so that the reverse barrier is now only 21 kJ/mol, serves to increase the rate coefficient slightly, but, as seen in Figure 2.9, this expression is still well below the Davidson et al. measurements. This leads us to conclude that the rate coefficient expression measured by Davidson et al. cannot be rationalized in terms of a recombination–elimination mechanism.

Another possible mechanism for this reaction is direct hydrogen transfer without transient formation of an intermediate that can be stabilized. The estimation procedures discussed in Section 2.4 were used for both directions of this reaction, with the k_{-9} estimate derived on the assumption that hydrogen abstractions by NH can be expected to have A-factors similar to those for hydrogen abstractions by OH. The activation energy for k_{-9} was estimated from reactions with similar enthalpies of reaction. Both approaches lead to expressions with similar values for k_9, but with somewhat greater temperature dependence than was found by Davidson et al. (Figure 2.9). The agreement does support direct hydrogen abstraction, and no pressure dependence of the rate coefficient is to be expected.

Similar conclusions were drawn from theoretical investigations of the direct abstraction mechanism by Linder et al. (1995) and Xu et al. (1995), both of whom found a temperature dependence similar to that found from our estimation procedure. We suggest use of the expression we derived for the forward reaction rate coefficient

$$k_9 = 4.8 \times 10^8 T^{1.5} \exp(-3995/T) \quad \mathrm{cm^3 mol^{-1} s^{-1}},$$

which is in reasonable agreement with the measurements of Davidson et al.

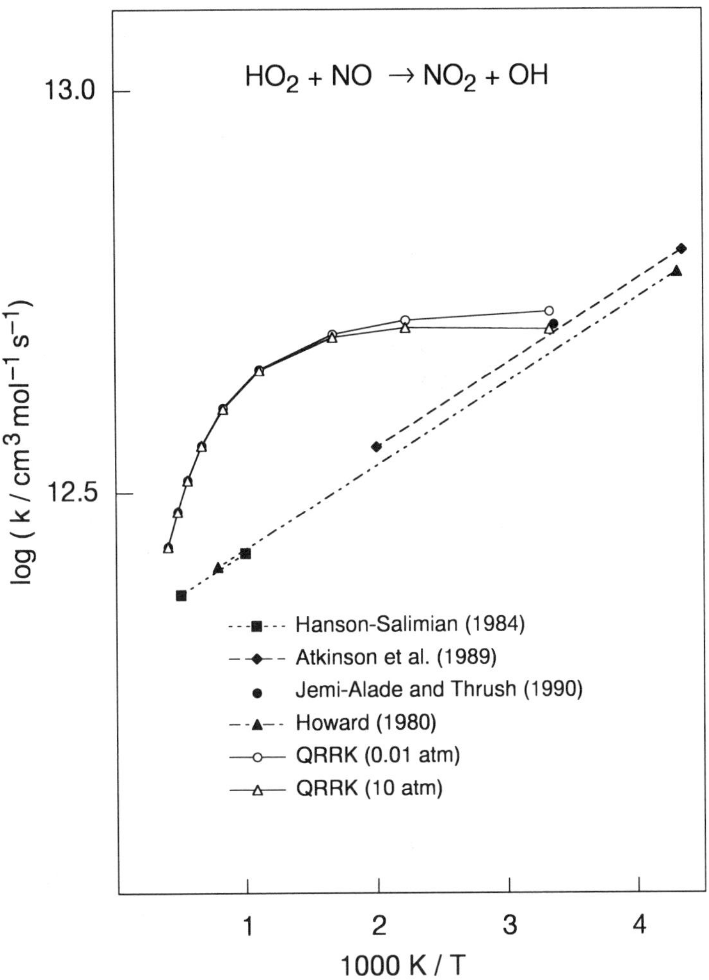

FIGURE 2.11. Arrhenius plot for $HO_2 + NO \longrightarrow NO_2 + OH$. The recommended expression is that of Atkinson et al. (1989) extended to 1250 K. Note that the QRRK calculations indicate virtually no dependence on pressure.

2.6.2 $HO_2 + NO \longrightarrow NO_2 + OH$

A low-temperature measurement (Jemi-Alade and Thrush 1990) appeared after the Hanson–Salimian review. It is consistent with the recommendation of Atkinson et al. (1989), which bases its recommended temperature dependence on the measurements of Howard (1980) over the range 423 to 1271 K with a slight difference in the preexponential term. The Hanson–Salimian recommendation was also based on the measurements of Howard. These expressions are compared in Figure 2.11.

We suggest using the expression recommended by Atkinson et al. with the temperature range extended to 1250 K:

$$k_{10} = 2.2 \times 10^{12} \exp(+240/T) \ \mathrm{cm^3 mol^{-1} s^{-1}}.$$

TABLE 2.5. $HO_2 + NO \longrightarrow$ Products.

	Reaction		$A/\mathrm{s^{-1}}$ or $\mathrm{cm^3mol^{-1}s^{-1}}$	E_a/kJ mol^{-1}
(1)	$HO_2 + NO$	$\longrightarrow HOONO$	5.6×10^{12}	0
(−1)	HOONO	$\longrightarrow HO_2 + NO$	8.5×10^{14}	120
(2)	HOONO	$\longrightarrow OH + NO_2$	2.0×10^{14}	87
(3)	HOONO	$\longrightarrow HONO_2$	7.1×10^{11}	105
(−3)	$HONO_2$	$\longrightarrow$ HOONO	1.4×10^{13}	218
(4)	$HONO_2$	$\longrightarrow OH + NO_2$	1.3×10^{15}	206

Molecular constants of HOONO adduct: $\nu_1 = 338$ cm^{-1}(degen=3.56) $\nu_2 = 1271$ cm^{-1}(4.30) $\nu_3 = 3994$ cm^{-1}(0.14); molecular constants of $HONO_2$ (nitric acid): $\nu_1 = 622$ cm^{-1}(4.55) $\nu_2 = 1422$ cm^{-1}(3.52) $\nu_3 = 3868$ cm^{-1}(0.93); $\sigma = 4.78$ Å and $\varepsilon/K = 486$ for both adducts.

Rate coefficients: k_1 this evaluation—see references in text; k_{-1} from microscopic reversibility (MR); k_2 via k_{-2} and MR; k_{-2} set equal to the OH + NO rate coefficient of Atkinson et al. (1989); k_3 TST with loss of 2 rotors and optical isomer, $\Delta S^\ddagger = -35.6$ J/K; E_a is an estimate for the lower limit [$E_a = 128$ kJ/mol for $NH_2NO \longrightarrow$ HNNOH (Melius and Binkley 1984b)]; k_{-3} MR; k_4 via k_{-4} and MR; k_{-4} same as k_{-2}.

One might expect this reaction to have an addition–elimination mechanism via the HOONO adduct, as shown in the potential energy diagram of Figure 2.12. The products $OH + NO_2$ are about 33 kJ/mol, and the HOONO adduct about 120 kJ/mol, more stable than the reactants (Melius 1988). Treatment of this system as a chemically activated reaction with the kinetic and thermodynamic assignments shown in Table 2.5 gives the rate coefficients shown in Figure 2.11. Both dissociation channels of the adduct are expected to have loose transition states and correspondingly high Arrhenius A-factors for unimolecular dissociation. The lower energy of the $OH + NO_2$ channel results in rapid dissociation of the adduct via this channel, rather than dissociation to the original reactants. Furthermore, dissociation is much more rapid than collisional stabilization. Thus one expects to see little pressure dependence even though the reaction proceeds via a transient complex. The predicted rate coefficient for the $NO_2 + OH$ channel is essentially invariant over a factor of 1000 in pressure; even at 10 atm, the stabilization channel contributes only a very small fraction to the total rate (Figure 2.11).

Another possible product channel not shown on Figure 2.11 would be HNO + O_2, which can occur via isomerization of the HOONO adduct to HN(O)OO followed by dissociation. However, this isomerization would be expected to have a relatively low A-factor, because of the loss of internal rotors in the transition

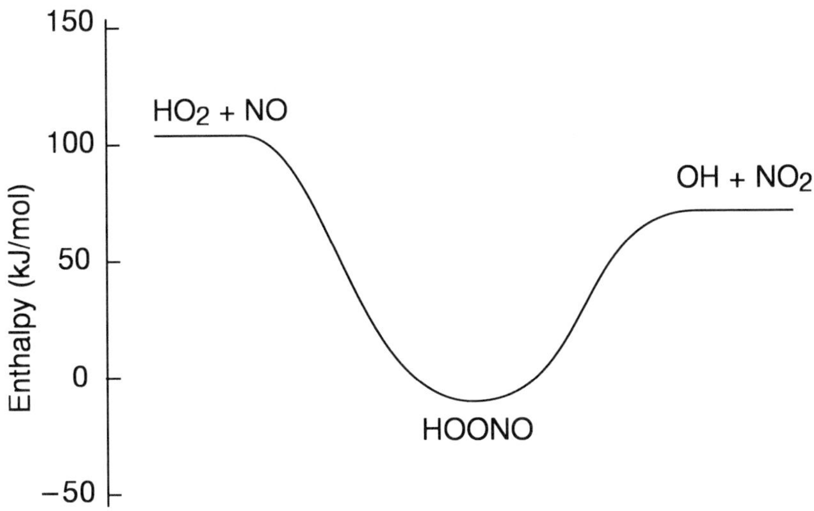

FIGURE 2.12. Potential energy diagram for the recombination–elimination pathway for HO_2 + NO. The curve forming $HONO_2$ (nitric acid, at −134 kJ/mol) is not shown.

state and also a high barrier to form a diradical, which eliminates this route from consideration. A third possible channel is isomerization to nitric acid, $HONO_2$, which would be stabilized at low temperatures and dissociate to OH + NO_2 at high temperatures. This isomerization would also be expected to have a lower *A*-factor. To our knowledge the barrier has not been calculated; by analogy to the NH_2NO system one can estimate that it would be greater than the barrier to OH + NO_2. Results from the QRRK analysis with the barrier set at the reasonable lower limit of 105 kJ/mol indicated that the $HONO_2$ channels, both stabilization and subsequent reaction to OH + NO_2, are about 1000 times less important than the OH + NO_2 channel via the HOONO adduct.[1]

The observed negative temperature dependence can be attributed to falloff effects, that is, the growing importance of the dissociation of the addition complex back to reactants. At higher temperatures, the larger *A*-factor for adduct dissociation back to reactants leads to increasing contributions from this channel and a predicted decrease in the observed rate coefficient. The somewhat lower *A*-factor for dissociation of the complex to OH + NO_2 (cf. Table 2.5) is based upon the measured expression for the reverse reaction rate coefficient.

Another possible explanation for the observed negative temperature dependence of the rate coefficient would be that the high-pressure addition rate coefficient has a slight negative temperature dependence. If this is assumed to be the case, then the computed QRRK rate coefficient expressions could be brought into better

[1] The reverse reaction OH + NO_2, however, would be expected to form nitric acid at a rate comparable to the rate of formation of the HOONO adduct. It is not an important reaction in combustion for the reasons discussed in Chapter 1.

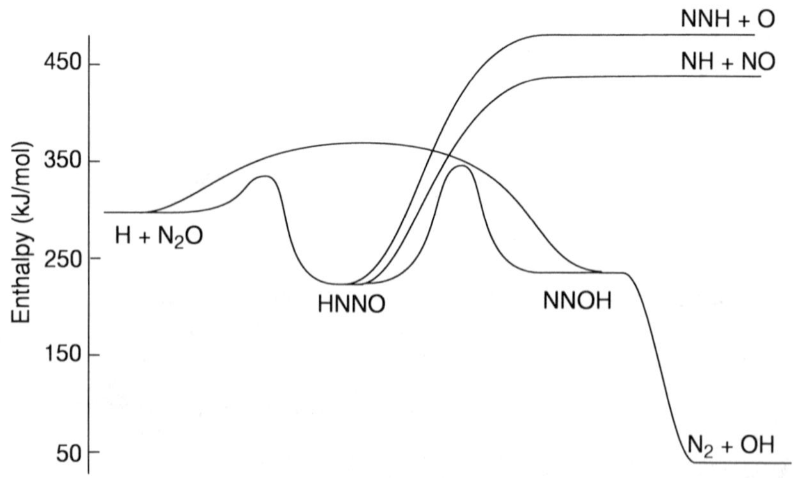

FIGURE 2.13. Potential energy diagram for H + N_2O. Note that both adducts can be formed directly from the reactants.

agreement with experiment in the middle of the temperature range. We did not pursue this assumption because the purpose of our analysis was not to replicate the experimental rate coefficient but only to show that the observed behavior is consistent with an addition–elimination reaction. The analysis strongly suggests that one does not expect any pressure effects in this system even though it appears not to be a direct reaction. The rate coefficient expression suggested by Atkinson should therefore be applicable at all pressures of interest in combustion.

2.6.3 H + $N_2O \longrightarrow$ Products

A potential energy diagram, following Marshall et al. (1987), for this reaction is shown in Figure 2.13. Aspects of it have been addressed in several investigations (Miller and Melius 1992a; Durant 1994; Melius 1993; Walch 1993a). Although a distinction was made in some of these studies between *cis*- and *trans*-HNNO, the overall features are in each case similar to those shown in Figure 2.13. We selected a barrier height of 38 kJ/mol for addition to form HNNO, representative of the values reported by the cited authors, and a barrier of 77 kJ/mol for direct addition to form NNOH, again close to the reported values. Following Melius (1993), we use a barrier of 123 kJ/mol for the isomerization of HNNO to NNOH.

There are four product channels:

$$\mathrm{H + N_2O \longrightarrow HNNO} \quad (11a)$$
$$\longrightarrow \mathrm{N_2 + OH} \quad (11b)$$
$$\longrightarrow \mathrm{NH + NO} \quad (11c)$$
$$\longrightarrow \mathrm{NNH + O}. \quad (11d)$$

Figure 2.13 suggests that the major channels, particularly at lower temperatures, would be HNNO or N_2 + OH. At higher temperatures, the initially formed adduct could have sufficient energy to dissociate to one of the higher energy channels.

Reaction 11b has been reasonably well characterized. Hidaka et al. (1985) used a computer model in conjunction with their shock tube observations in N_2O–Ar and N_2O–H_2–Ar mixtures, while Fontijn and coworkers (Marshall et al. 1987, 1989) measured the reaction directly in a high-temperature photochemical experiment. There is a difference in the higher temperature rate coefficients in the two photochemical studies. The authors point out that their earlier work employed a capped inlet which might have led to premature dissociation of some of the N_2O, thus introducing some error. As a result, we only consider the later data. These rate coefficients, as well as an earlier modeling analysis by Dean et al. (1980), are compared to the recommended expressions given by Hanson and Salimian (1984) and Tsang and Herron (1991) in Figure 2.14.

Data for the NH + NO channel (Reaction 11c), which has usually been studied in the reverse direction, as described below, are more scattered. The more recent information is shown in Figure 2.15. Cattolica et al. (1982) used measured NH and OH flame profiles in N_2O/H_2 flames to obtain estimates for this rate coefficient. The two shock tube studies of the reverse reaction (Roose 1981; Mertens et al. 1991a) report markedly different expressions. Also included in this plot is the expression obtained by using the recommended expression given by Miller and Bowman (1989) for the reverse reaction; their assignment is based primarily upon comparison of computed and experimental N_2O profiles in ammonia flames.

A confusing aspect of the NH + NO reaction is its multiple products:

$$\mathrm{NH + NO \longrightarrow N_2 + OH} \qquad (11e)$$

$$\mathrm{\longrightarrow H + N_2O} \qquad (11f)$$

$$\mathrm{\longrightarrow NNH + O\,.} \qquad (11g)$$

Mertens et al. (1991a) report a branching fraction for the N_2 channel $k_{11e}/k_{tot} = 0.19 \pm 0.10$ at 3000 K from measurements of NH and OH. Shock tube measurements of emissions from electronically excited OH and NH in HNCO–NO–Ar mixtures (Yokoyama et al. 1991a, 1991b) were used with a complex model to infer a branching fraction of 0.32 ± 0.07 for the N_2 channel at 3500 K. However, an analysis of ammonia flames (Vandooren et al. 1991) suggested that the N_2 channel is four to six times faster than the N_2O channel at 2000 K. The more direct nature of the Mertens et al. study suggests that their data are the most reliable.

At low temperatures, the measurements of the overall rate coefficient exhibit quite good agreement, but there is controversy with respect to the products. Harrison et al. (1986) suggest that their measurements are best interpreted assuming that the OH + N_2 channel is the dominant one. Yamasaki et al. (1991) report direct measurements of OH at 300 K and claim OH + N_2 is the exclusive channel. They also reported failure to detect H, supporting this assignment. However, later work from that laboratory (Okada et al. 1994) reversed this assignment, now reporting that the H + N_2O channel contributes 65% and OH + N_2 contributes 30%. These

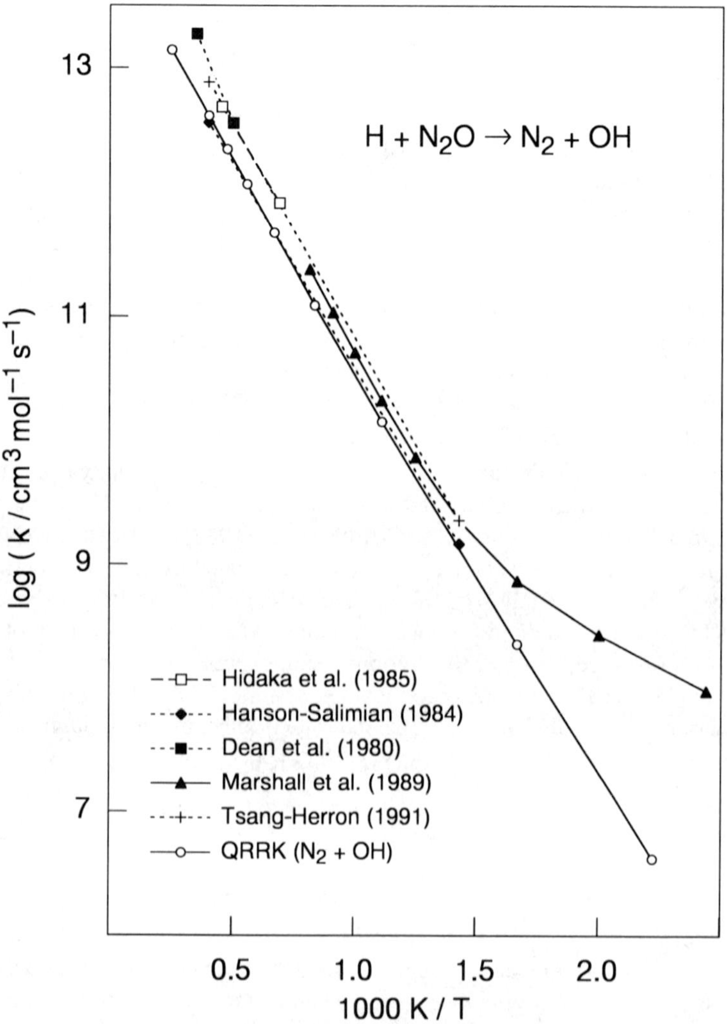

FIGURE 2.14. Arrhenius plot for $H + N_2O \longrightarrow N_2 + OH$. The recommended expression is that of Marshall et al. (1989) for temperatures above 1000 K. The QRRK calculations include contributions to this channel from both the HNNO and the NNOH adducts.

new assignments are consistent with a study by Durant (1994), who reported that $H + N_2O$ is the dominant channel at room temperature and the branching ratio is 0.8 ± 0.4. Lillich et al. (1994) report that the OH channel contributes $15 \pm 5\%$ at 300 K. Calculations by Harrison and Maclagan (1990) have been used to support the claim of $OH + N_2$ as the dominant channel. However, other calculations by Melius and Binkley (1984b) and by Miller and Melius (1992a) indicate that the N_2O channel is the dominant one over the entire temperature range. Modeling efforts in low pressure, stoichiometric H_2–N_2O–Ar flames (Sausa et al. 1993) have

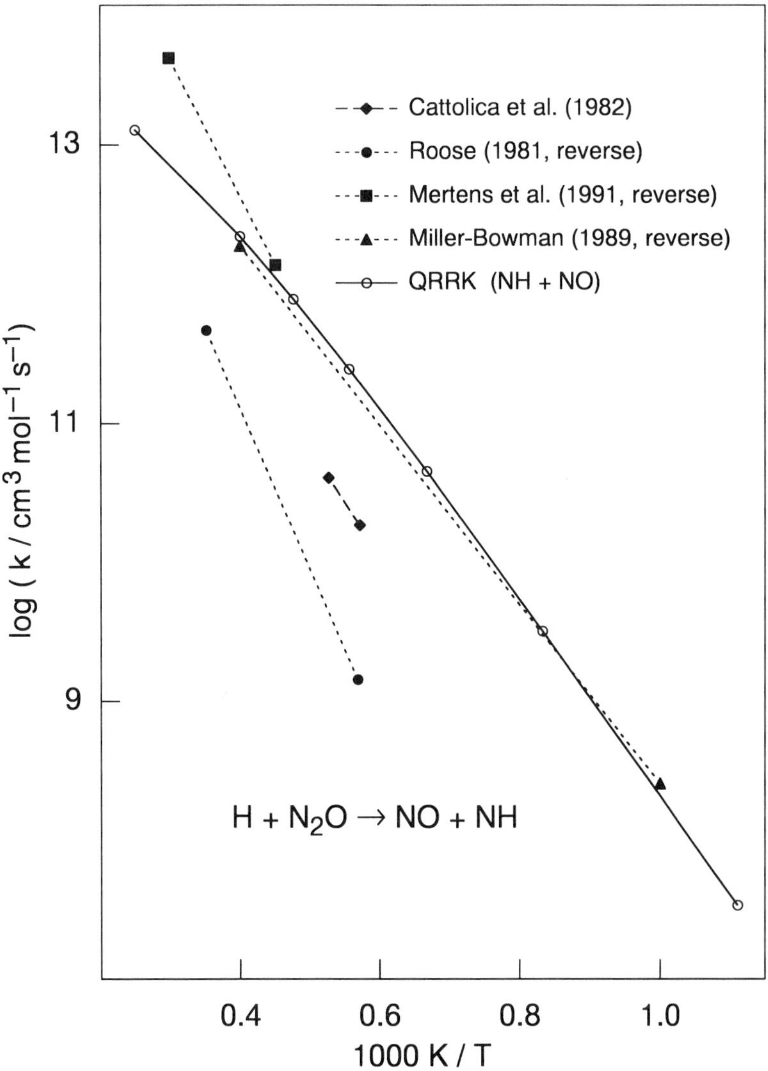

FIGURE 2.15. Arrhenius plot for $H + N_2O \longrightarrow NO + NH$. The "reverse" notation signifies that the rate coefficient was obtained via the equilibrium constant from measurements of the reverse reaction rate coefficient. The QRRK calculations are in agreement with the expression recommended by Miller and Bowman (1989).

used the branching fraction reported by Miller and Melius.

Given this ambiguity concerning the low-temperature branching fractions, we have plotted the various reported expressions for the total rate coefficient for NH + NO in Figure 2.16. Although the reported rate coefficients span a relatively narrow range around 1×10^{13} $cm^3 mol^{-1} s^{-1}$, there is a large variation in reported temperature dependence. The high-temperature shock tube results exhibit

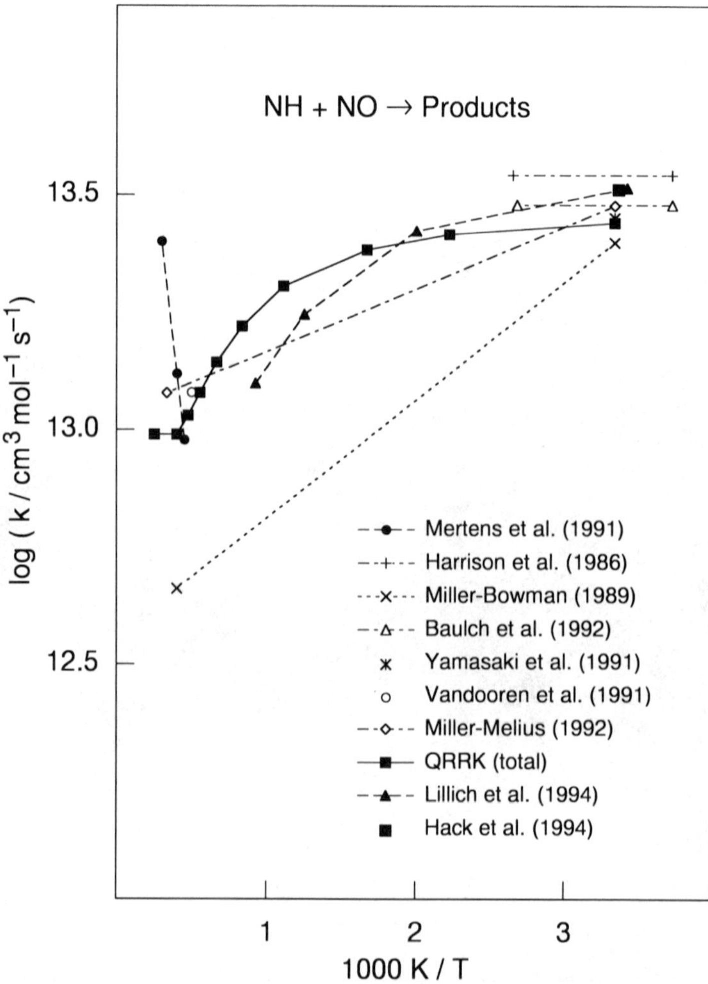

FIGURE 2.16. Arrhenius plot for NH + NO ⟶ Products. The QRRK predictions are seen to be in good agreement with the measured overall rate coefficients. Note the change in the predicted temperature dependence above 2000 K.

a substantially different temperature dependence than the lower temperature data and the calculations of Miller and Melius. Römming and Wagner (1996) reported shock tube measurements that gave a rate coefficient of about 1.0×10^{13} $cm^3 mol^{-1} s^{-1}$ over the temperature range 1200 to 2200 K. This result, together with those of Mertens et al. (1991) and Lillich et al. (1994), suggests a minimum rate coefficient near 1400 K.

As evident from the potential energy diagram, this reaction is not a simple one. We attempted to gain understanding of it by means of QRRK analysis. The input parameters are listed in Table 2.6.

TABLE 2.6. $H + NNO \longrightarrow$ Products.

	Reaction	A/s^{-1} or $cm^3mol^{-1}s^{-1}$	E_a/kJ mol^{-1}
(1a)	$H + NNO \longrightarrow HNNO$	8.5×10^{13}	38
(−1a)	$HNNO \longrightarrow H + NNO$	2.6×10^{13}	104
(1b)	$H + NNO \longrightarrow NNOH$	1.3×10^{14}	77
(2)	$HNNO \longrightarrow NH + NO$	4.0×10^{15}	209
(3)	$HNNO \longrightarrow O + NNH$	4.9×10^{15}	258
(4)	$HNNO \longrightarrow NNOH \longrightarrow N_2 + OH$	6.8×10^{12}	123

Molecular constants of HNNO adduct: $\nu_1 = 567\ cm^{-1}$ (degen=1.99) $\nu_2 = 1174\ cm^{-1}$ (2.49) $\nu_3 = 2517\ cm^{-1}$ (1.53); σ = 3.77 Å and ε/K = 280.

Rate coefficients: k_{1a} this study with E_a being a consensus value, cf. text; k_{-1a} microscopic reversibility (MR); k_{1b} this study, E_a consensus value, cf. text; A based upon H + NO_2 from Ko et al. (1991); this channel, effectively a direct route to N_2 + OH, is added to the chemical activation component as described in the text; k_2 via k_{-2} and MR with A_{-2} = 3.46×10^{13} and $E_a(-2) = 0$ from Harrison et al. (1986); k_3 via k_{-3} and MR with k_{-3} = 7.5×10^{13} [based upon O + C_2H_5, Tsang and Hampson (1986)]; k_4 TST, $\Delta S^{\ddagger} = -13.8$ J/K, E_a from Melius (1993)—this is the rate-controlling step for the N_2 + OH reaction.

We also explored use of a more complex reaction system, accounting separately for the *cis*- and *trans*-isomers of HNNO, as described by both Walch (1993a) and Durant (1994); the results were very similar to those presented below. We prefer the simpler analysis, since there remains considerable ambiguity as to the relative energies of the isomers, see Walch (1993a). This reaction system is unusual in that the addition reaction to form HNNO has a higher A-factor than the reverse unimolecular dissociation. Over the temperature range 300 to 2500 K, the average ratio of $A_{forward}/A_{reverse}$ for $H + NNO \longrightarrow HNNO$ is 3.2. Usually this ratio is much less than unity, strongly favoring the reverse reaction. For example, the ratio for $NH + NO \longrightarrow HNNO$ is 0.0086. This unusual case of a higher A-factor for addition than for dissociation is a result of the unusually low standard entropy of the reactants H + N_2O combined with loss of a rotor in the dissociation. The HNNO adduct will have a significantly higher probability for stabilization or reaction to other channels (as opposed to redissociating to H + N_2O) relative to many of the other systems discussed elsewhere in this chapter. In particular, this includes formation of the higher energy products NH + NO, since this channel has a more typical A-factor.

2.6.4 $H + N_2O \longrightarrow N_2 + OH$ and $H + N_2O \longrightarrow HNNO$

The products N_2 + OH can be formed from two adducts. Addition of H to form NNOH will lead immediately to N_2 + OH, and the rate coefficient for production of N_2 + OH via this channel is simply the rate of formation of the complex; no chemical activation calculation is needed. This rate coefficient is added to that obtained from the chemical activation analysis for addition of H to form

HNNO to obtain the total for this channel. The high temperature predictions, above 1000 K, are in good agreement with the data for the N_2 + OH channel (cf. Figure 2.14). This is particularly encouraging, since no parameters were adjusted in the calculation. However, this prediction does not account for the observed rate coefficient at lower temperatures. This strongly non-Arrhenius temperature dependence has been attributed to tunneling by Marshall et al. (1987). They point out that tunneling is consistent with the observed isotope effect. Although these researchers rule out stabilization as a major pathway on the basis of their observed lack of pressure dependence, our calculations suggest that stabilization may well account for the lower temperature data.

In Figure 2.17 we compare our predictions for all channels to the data of Marshall et al. The calculations were done for a pressure of 0.5 atm and with Ar as the buffer gas, conditions typical of those used in the experiments. Below 900 K, stabilization of the HNNO adduct is predicted to be important. (As discussed above, the low A-factor for dissociation of this adduct back to reactants makes this channel particularly susceptible to stabilization.) The temperature range over which stabilization dominates coincides with the region of significant non-Arrhenius behavior seen in the experiments of Marshall et al., and might account for the unusual behavior. The strong temperature dependence of the data, as well as the limited number of experiments at constant temperature and various pressures by Marshall et al., might obscure the predicted dependence on pressure. Additional experiments are needed to identify the source of the higher observed rate at lower temperatures. In the interim, we cautiously suggest use of a calculated stabilization rate coefficient. At 1 atm N_2

$$k_{11a} = 1.3 \times 10^{25} T^{-4.48} \exp(-5420/T) \ \mathrm{cm^3 mol^{-1} s^{-1}}$$

for the temperature range from 300 to 1200 K. Over the 600 to 1200 K temperature interval, the rate coefficient for this channel is close to the low pressure limit, meaning that it can be scaled linearly with pressure.

Given the ambiguity concerning the lower temperature behavior, we are reluctant to conclude that all of the observed loss of reactant in this regime can be attributed to production of N_2 + OH. Above 1000 K, however, it appears that the measurements of Marshall et al. describe this reaction well (Figure 2.17). We recommend use of the high-temperature component of their rate expression

$$k_{11b} = 2.2 \times 10^{14} \exp(-8430/T) \ \mathrm{cm^3 mol^{-1} s^{-1}}$$

for the temperature range from 1000 to 2000 K. (We have extended the high temperature range since this expression is consistent with the shock tube data.) The QRRK analysis suggests there should be no effect of pressure on the rate coefficient for this channel at higher temperatures.

2.6.5 H + N_2O $\longrightarrow$ NH + NO

The predicted rate coefficient for the NH + NO channel is compared to experimental measurements in Figure 2.15. We predict the rate coefficient for this channel

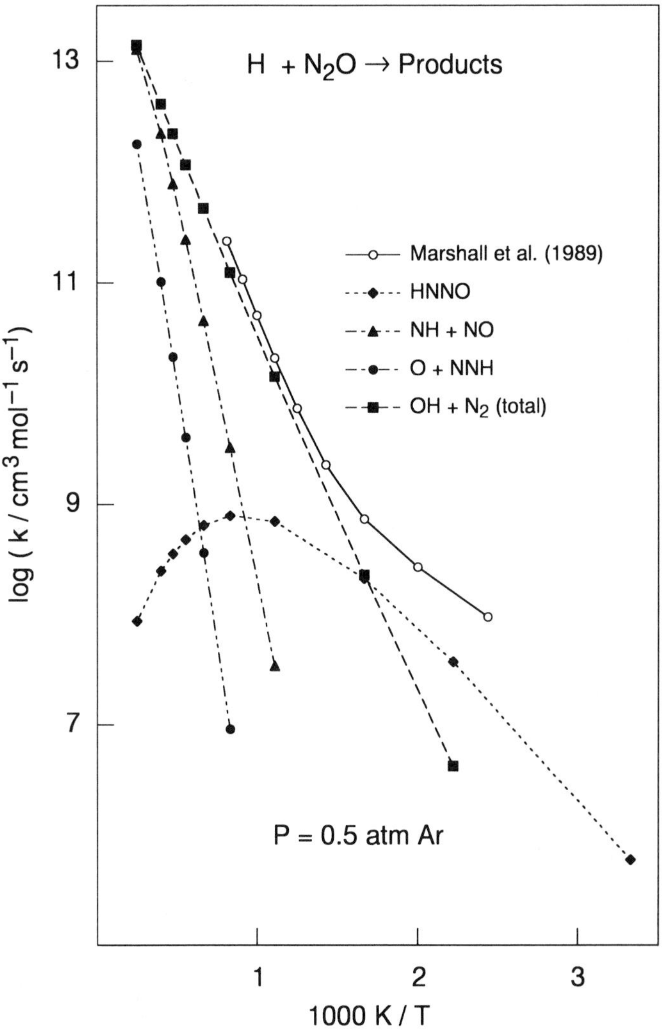

FIGURE 2.17. Arrhenius plot for the various channels of H + $N_2O \longrightarrow$ Products as predicted by QRRK calculations for 0.5 atm Ar. Note the importance of the HNNO stabilization channel at lower temperatures. Near 2500 K, the NH + NO channel has grown to become almost as important as the OH + N_2 channel.

to be independent of pressure over the range 10^{-3} to 100 atm. The calculations indicate that at higher temperatures the NH + NO channel becomes increasingly important, and is comparable to the N_2 + OH channel near 2500 K (cf. Figure 2.17). Selection of a preferred rate coefficient is difficult. The data of Mertens et al. were obtained from measurements of the reverse reaction rate; we suspect ambiguity in assigning product channels (see below), which would lead to a lower activation energy for k_{11c} and bring that expression closer to the one obtained from detailed modeling (of the reverse reaction) by Miller and Bowman, as well as that calculated in this work. Our analysis leads to rate coefficients close to those given by the Miller–Bowman expression. We suggest using our calculated expression

$$k_{11c} = 8.5 \times 10^{20} T^{-1.62} \exp(-17800/T) \ \mathrm{cm^3 mol^{-1} s^{-1}}$$

for the temperature range from 300 to 4000 K.

2.6.6 $H + N_2O \longrightarrow NNH + O$

Very little information is available on the NNH + O channel. The results in Figure 2.17 suggest it could become important at higher temperatures. Its smaller rate coefficient is primarily due to a higher barrier. Our results can be described by

$$k_{11d} = 2.4 \times 10^{19} T^{-1.26} \exp(-23700/T) \ \mathrm{cm^3 mol^{-1} s^{-1}}$$

for the temperature range from 300 to 4000 K. This expression is much lower than one obtains using the Miller–Bowman estimate for the reverse reaction rate coefficient. The calculations indicate that this channel has no pressure dependence.

2.6.7 $NH + NO \longrightarrow$ Products

The potential energy surface shown in Figure 2.13, with the same QRRK parameter values used for the $H + N_2O$ calculations, was also used to study the reaction NH + NO $\longrightarrow$ Products. The results (Figure 2.16) show that the features reported in the experiments, including the increase in rate at high temperature, are accounted for.

Predictions for the various channels are shown in Figure 2.18. Stabilization channels are predicted to be unimportant even at the lowest temperatures. The major product channel (except at the highest temperatures) is predicted to be $H + N_2O$, and our predicted rate coefficient expression agrees well with the calculations of Miller and Melius. Our analysis leads to a somewhat less important $N_2 + OH$ product channel than predicted by Miller and Melius. At 3000 K, we predict that $k_{11e}/k_{tot} = 9\%$, as compared to $19 \pm 10\%$ measured by Mertens et al.; Miller and Melius predict 29%. At room temperature, we predict 9%, compared to the Lillich et al. measurement of $15 \pm 5\%$ and the Miller–Melius prediction of 19%. We note that the reports of $N_2 + OH$ as the dominant channel at lower temperatures would suggest that the barrier to the HNNO $\rightleftharpoons$ NNOH isomerization used in both this work and by Miller and Melius is too high. While a detailed evaluation of the effect of this barrier on the product distribution is beyond the scope of this chapter, we note that changes of the magnitude needed would require a decrease in the barrier height well beyond the expected uncertainty in the Melius calculation. Such a conclusion is consistent with the data supporting that the N_2O channel is dominant at room temperature. It is also reinforced by the theoretical analysis of

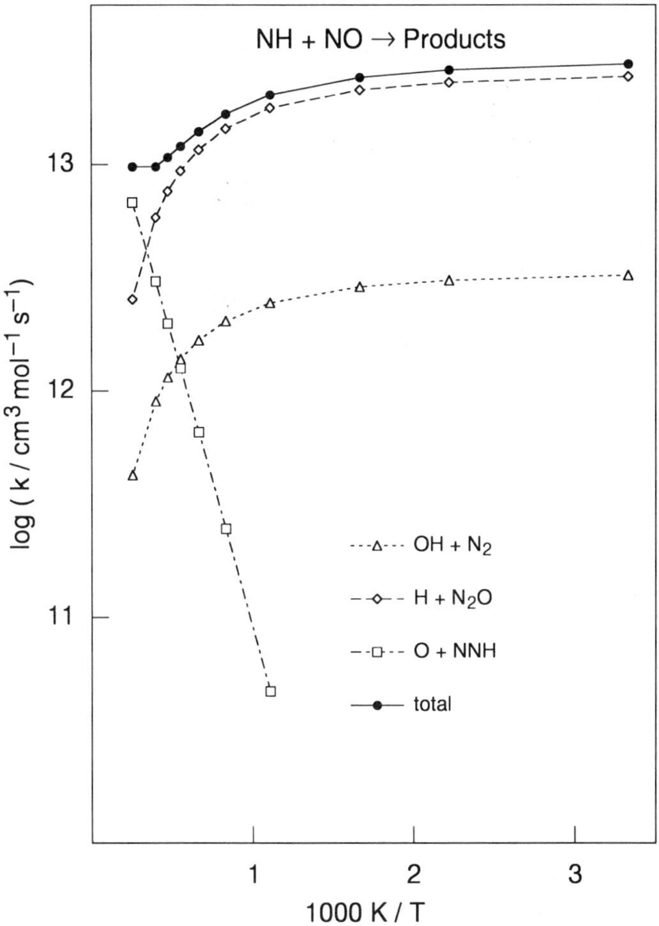

FIGURE 2.18. Arrhenius plot for the various channels of NH + NO ⟶ Products as predicted by QRRK calculations for 1 atm N_2. The change in temperature dependence near 2000 K is due to the onset of the O + NNH channel; at lower temperatures, H + N_2O is predicted to be dominant. As discussed in the text, the isomerization barrier between HNNO and NNOH would need to be significantly lower to have N_2 + OH dominate at lower temperature.

Bradley et al. (1995), who report a branching fraction of $13 \pm 3\%$ for the N_2 + OH channel and show that decreasing the isomerization barrier by 40 kJ/mol increases the branching fraction by less than a factor of two.

An additional feature of interest in Fig. 18 is the predicted importance of the NNH + O channel at the highest temperatures. The results suggest that NNH + O is a major product in the Mertens et al. shock tube experiments. This channel is more important for the NH + NO reaction than the H + N_2O reaction since the available energy is appreciably higher for the NH + NO reaction, thus allowing easier access to the endothermic NNH + O channel. This seems to be the most plausible explanation for the observed temperature dependence in the shock tube experiments. This reaction should be included in high temperature models; we show in Section 2.7.3 that it can have significant implications for NO production.

None of these channels are predicted to depend on pressure. Given the amount of energy in the adducts, as well as the low-energy dissociation channels, this behavior is expected—the unimolecular dissociation events are much too rapid to be intercepted by collisions. Our predicted rate coefficients are given by

$$k_{11e} = 1.4 \times 10^{17} T^{-1.49} \exp(-660/T)\ \text{cm}^3\text{mol}^{-1}\text{s}^{-1}$$
$$k_{11f} = 3.0 \times 10^{18} T^{-1.65} \exp(-720/T)\ \text{cm}^3\text{mol}^{-1}\text{s}^{-1}$$
$$k_{11g} = 1.7 \times 10^{14} T^{-0.20} \exp(-6140/T)\ \text{cm}^3\text{mol}^{-1}\text{s}^{-1}$$

for the temperature range from 300 to 4000 K.[1]

2.6.8 $NH + O_2 \longrightarrow$ Products

Most of the available high-temperature rate coefficients for this reaction have been inferred from applications of kinetic models to flames (Miller et al. 1983; Dean et al. 1984; Bian et al. 1990; Tsang and Herron 1991; Vandooren et al. 1991). The total rate coefficient was measured in a direct experiment behind shock waves (Mertens et al. 1991a) and a direct measurement at lower temperatures was made by Hack et al. (1985b), who concluded that OH + NO is the dominant pathway below 573 K. Baulch et al. (1992) recommend the Hack expression for the lower temperature regime. Later data (Hennig et al. 1993; Lillich et al. 1994) in the intermediate-temperature regime capture the marked curvature in the Arrhenius plot. Miller and Melius (1992a) calculated rate coefficients for two channels over a wide temperature range, showing that HNO + O is the dominant high-temperature pathway while OH and NO are the major products at low temperatures. The experimental data are summarized in Figure 2.19. The results inferred from the modeling studies are notably higher than the direct observations of Mertens et al.

A potential energy diagram for this system, based on the calculations of Miller and Melius (1992a), is shown in Figure 2.20.[2] Input parameters for the QRRK calculations are listed in Table 2.7, and the results are compared to literature expressions in Figure 2.21. The low A-factor for HONO formation suppresses OH + NO production, since this is the only path to these products. Although one

---

[1] The expression for k_{11f} is included only for illustration; since reaction (11f) is the reverse of reaction (11c), it should not be added to a reaction mechanism that already includes reaction (11c).

[2] For completeness, the additional pathway of HONO dissociation to $H + NO_2$ shown by the dashed line in Figure 2.20 was considered; it is not expected to compete with direct dissociation from HNO_2, because the A-factor for isomerization to HONO is appreciably smaller than the A-factor for direct dissociation.

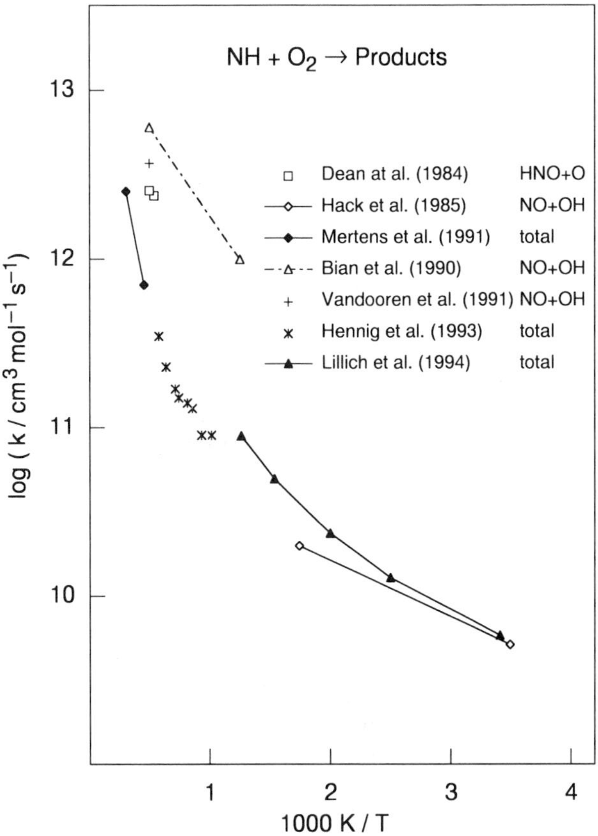

FIGURE 2.19. Arrhenius plot for $NH + O_2 \longrightarrow$ Products. The direct measurements of Mertens et al. (1991) and Hennig et al. (1993) suggest that the higher rate coefficient values derived from more complex systems are in error. The NO + OH channel appears to dominate at lower temperatures.

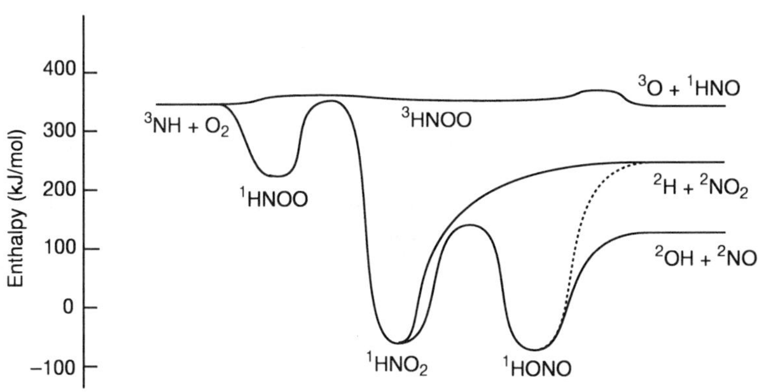

FIGURE 2.20. Potential energy diagram for $NH + O_2$. The dashed line indicates a reaction pathway not considered by Miller and Melius (1992a).

might expect a route to OH + NO via an H-shift in HNOO (to form NOOH, which rapidly dissociates), the estimated heat of formation of NOOH places it about 96 kJ/mol higher than HNOO; thus the barrier for an H-shift to form NOOH is too high for this pathway to be important.

TABLE 2.7. $NH + O_2 \longrightarrow$ Products.

	Reaction			A/s^{-1} or $cm^3mol^{-1}s^{-1}$	$E_a/kJ\ mol^{-1}$
(1)	$^3NH + O_2$	$\longrightarrow$	^1HNOO	2.0×10^{12}	0
(−1)	^1HNOO	$\longrightarrow$	$^3NH + O_2$	2.0×10^{14}	112
(2)	^1HNOO	$\longrightarrow$	$HNO + O$	spin forbidden; cf. triplet input	
(3)	^1HNOO	$\longrightarrow$	$^1HNO_2$	4.9×10^{12}	120
(−3)	$^1HNO_2$	$\longrightarrow$	HNOO	4.9×10^{13}	418
(4)	$^1HNO_2$	$\longrightarrow$	$H + NO_2$	5.1×10^{14}	319
(5)	$^1HNO_2$	$\longrightarrow$	$HNO + O$	spin forbidden; cf. triplet input	
(6)	$^1HNO_2$	$\longrightarrow$	^1HONO	7.7×10^{13}	205
(−6)	^1HONO	$\longrightarrow$	HNO_2	4.8×10^{12}	221
(7)	^1HONO	$\longrightarrow$	$H + NO_2$	3.0×10^{14}	327
(8)	^1HONO	$\longrightarrow$	$OH + NO$	8.5×10^{15}	203
(9)	$^3NH + O_2$	$\longrightarrow$	^3HNOO	6.0×10^{12}	21
(−9)	^3HNOO	$\longrightarrow$	$NH + O_2$	2.2×10^{14}	2
(10)	^3HNOO	$\longrightarrow$	$HNO + O$	5.0×10^{14}	28

Molecular constants of ^{1}HNOO adduct: ν_1 = 798 cm^{-1} (degen=1.83) ν_2 = 801 cm^{-1} (1.85) ν_3 = 1838 cm^{-1} (2.33); molecular constants of $^1HNO_2$ adduct: ν_1 = 1099 cm^{-1} (degen=1.97) ν_2 = 1107 cm^{-1} (1.91) ν_3 = 2553 cm^{-1} (2.12); molecular constants of ^{1}HONO: ν_1 = 548 cm^{-1} (degen=1.99) ν_2 = 1219 cm^{-1} (2.86) ν_3 = 3450 cm^{-1} (1.15); σ = 3.49 Å and ε/K = 350 for all adducts.

Rate coefficients: k_1 from $^3CH_2 + O_2$, Darwin et al. (1989) and Böhland et al. (1984); k_{-1} by microscopic reversibility (MR); k_3 TST, loss of 1 rotor, N–H stretch, gain of N–HO and O–OH bends, $E_a(3)$ set to be 8 kJ/mol above entrance channel (from Miller and Melius 1992), (variation of barrier height was checked, see text); k_{-3} MR; k_4 via k_{-4} and MR with $k_{-4} = 1.3\times10^{13}\exp(-1100/T)$ based on $H + C_2H_4$; k_6 TST, $E_a(6)$ from Miller and Melius (1992), degeneracy = 2; k_{-6} MR; k_7 via k_{-7} and MR with $k_{-7} = 1.3\times10^{14}\exp(-180/T)$ (Ko et al. 1991)—included for completeness (not considered by Miller and Melius); k_8 via k_{-8} and MR with $k_{-8} = 1.9\times10^{13}$, Atkinson et al. (1992); k_9 $3\times k_1$ because of triplet surface, barrier estimated; k_{-9} MR; k_{10} A_{10} via MR using $A_{-10} = 2.0\times10^{13}$ for O + CH_3O (Herron 1988) and barrier from the adjusted barrier of Miller and Melius (1992).

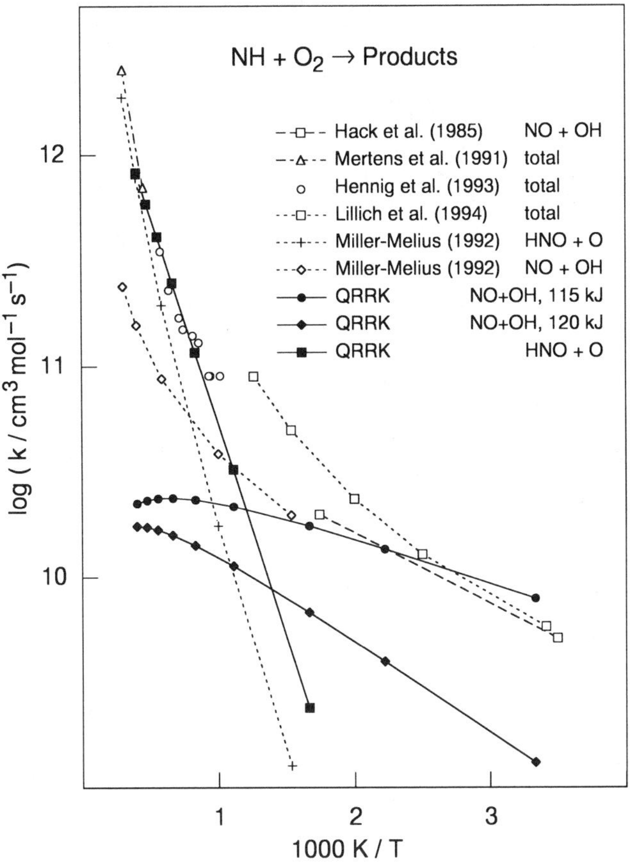

FIGURE 2.21. Comparison of QRRK predictions to measurements and calculations for $NH + O_2$. The QRRK calculations bracket the lower temperature data for the NO + OH channel. Small changes in the assumed isomerization barrier lead to the large changes in the QRRK rate coefficients, as generally found when the isomerization barrier is comparable to the entrance channel energy.

Possible product channels include

$$
\begin{aligned}
{}^3NH + O_2 &\longrightarrow {}^1HNOO && (12a)\\
&\longrightarrow {}^1HNO_2 && (12b)\\
&\longrightarrow {}^1HONO && (12c)\\
&\longrightarrow OH + NO && (12d)\\
&\longrightarrow H + NO_2 && (12e)\\
&\longrightarrow {}^3HNOO && (12f)\\
&\longrightarrow HNO + O\,. && (12g)
\end{aligned}
$$

Figure 2.21 shows that the QRRK prediction for production of O + HNO on the triplet surface agrees with the calculations of Miller and Melius and the data of Mertens et al. The dominant singlet channel produces OH + NO, in accord with the observations, but the predicted rate coefficient is below the data. Because the barrier height for isomerization of HNOO to HNO_2 is comparable to the well depth, however, the QRRK results are extremely sensitive to its value. A decrease of only 5 kJ/mol is sufficient to bring calculations and data into agreement.

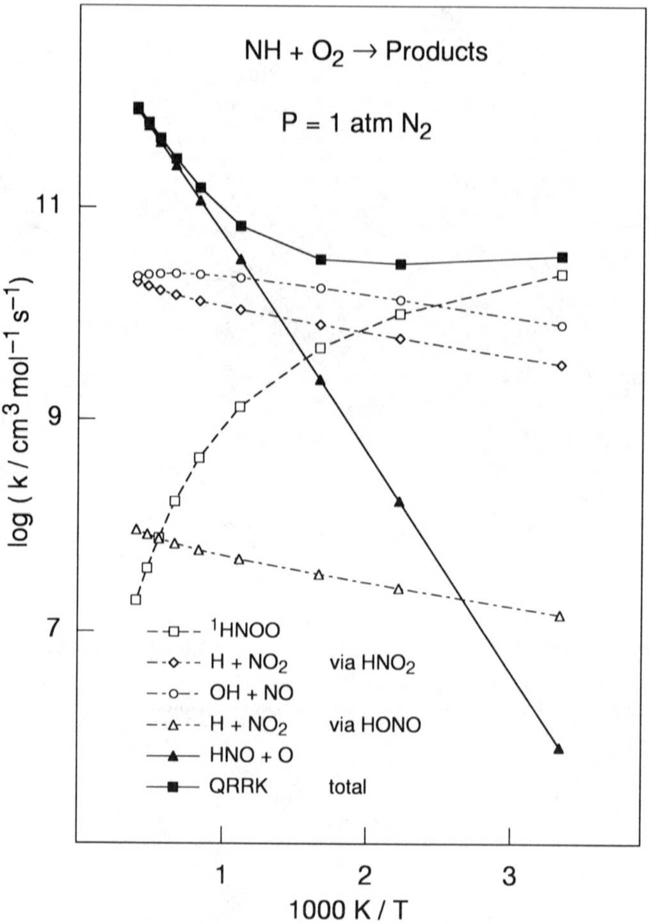

FIGURE 2.22. QRRK predictions for NH + $O_2 \longrightarrow$ Products at 1 atm N_2. Stabilization is predicted to be important at 1 atm and lower temperatures, while the HNO + O channel dominates at high temperatures.

The QRRK results can be used to explore the effects of temperature and pressure on the rate coefficient of this reaction. Figure 2.22 summarizes the predicted effect of temperature at 1 atm N_2. Near room temperature, substantial production of the ^{1}HNOO adduct is predicted. This channel would be two to three orders of magnitude slower in the low-pressure experiments of Hack et al., making the OH + NO channel dominant, as observed. Production of ^{1}NHOO falls off with increasing temperature, as typically observed for radical addition reactions without low energy exit channels. In addition to formation of OH + NO, one expects that smaller amounts of H + NO_2 would be produced. As expected, the production of H + NO_2 via HONO is much less than via HNO_2, for the reasons discussed above. At higher temperatures, production of HNO + O via the triplet surface dominates. Figure 2.23 shows that most of the reaction channels are independent of pressure and that at 1500 K stabilization is unimportant up to pressures of 100 atm.

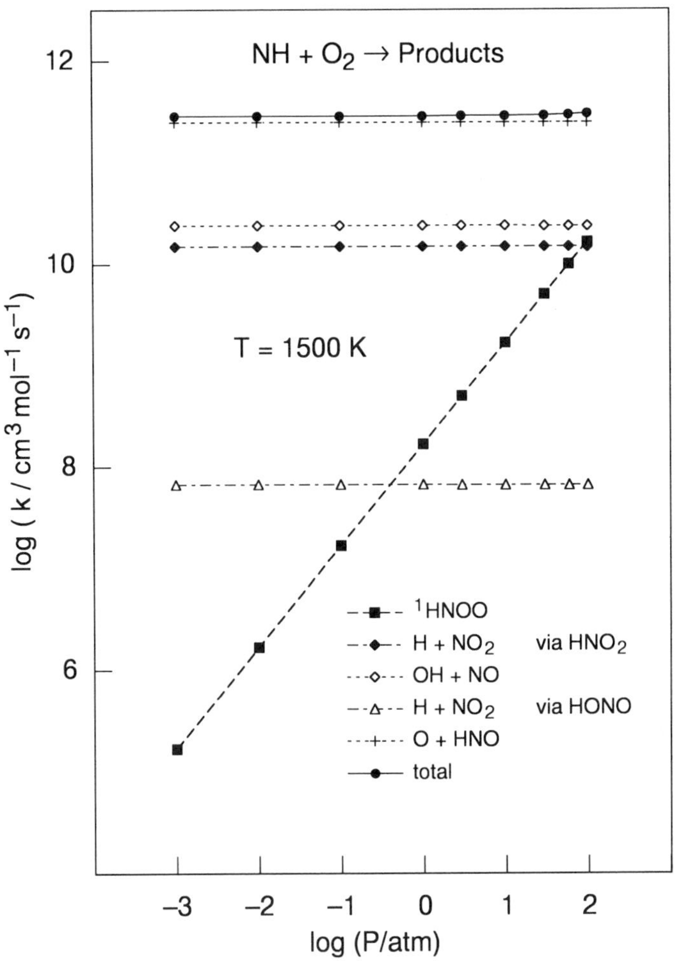

FIGURE 2.23. QRRK predictions of the effect of pressure at 1500 K for NH + $O_2 \longrightarrow$ Products. Rate coefficients for all major channels are independent of pressure, so stabilization should be unimportant for combustion applications.

The predicted absence of a pressure effect for the major channels at high temperatures allows one to use the literature expressions for rate coefficients for virtually all conditions of interest in combustion. We suggest use of the Hack et al. expression for the OH + NO channel

$$k_{12d} = 7.6 \times 10^{10} \exp(-770/T) \ \mathrm{cm^3mol^{-1}s^{-1}}$$

and the Miller–Melius recommendation for the HNO + O channel

$$k_{12g} = 4.6 \times 10^{5} T^{2} \exp(-3270/T) \ \mathrm{cm^3mol^{-1}s^{-1}}.$$

For completeness, we suggest use of the QRRK estimate for stabilization of the ^{1}HNOO adduct

$$k_{12a} = 3.7 \times 10^{24} T^{-5} \exp(-1155/T) \ \mathrm{cm^3mol^{-1}s^{-1}},$$

which is valid for N_2 at 1 atm. One can readily scale this expression to other pressures—since the calculations show this channel is in the low-pressure limit, its rate coefficient is directly proportional to pressure. The QRRK estimate for the H + NO_2 channel is

$$k_{12e} = 2.3 \times 10^{10} \exp(-1250/T) \ \mathrm{cm^3mol^{-1}s^{-1}}.$$

The QRRK calculations show that the other channels are not important.

2.6.9 $NH_2 + O_2 \longrightarrow$ Products

The NH_2 + O_2 reaction has received a considerable amount of attention and has been included in several mechanisms of ammonia oxidation. It is known to be relatively slow in comparison to other reactions of NH_2, with some authors (Melius and Binkley 1984a) favoring an endothermic pathway to NH_2O + O, while others, modeling ammonia oxidation and Thermal DeNOx (Fujii et al. 1981; Branch et al. 1982; Dean et al. 1982; Kimball-Linne and Hanson 1986), suggest that the products are HNO + OH, formed by isomerization and decomposition of the NH_2O_2 adduct. (Melius and Binkley concluded that the adduct did not have a potential well sufficiently deep to allow a rapid isomerization path.) Miller and Bowman (1989) followed earlier modeling studies in assuming that the products are HNO + OH. The literature was sufficiently ambiguous in 1984 for Hanson and Salimian to be unable to make a recommendation. Since then there have been additional experimental studies as well as a detailed QRRK analysis (Bozzelli and Dean 1989).

The experimental studies included a discordant variety of low-temperature measurements. Lesclaux (1984), Patrick and Golden (1984), Michael et al. (1985), and Tyndall et al. (1991) all reported a slow rate, with only upper limits measured, while Hack and Kurzke (1985a) reported rate coefficients some two

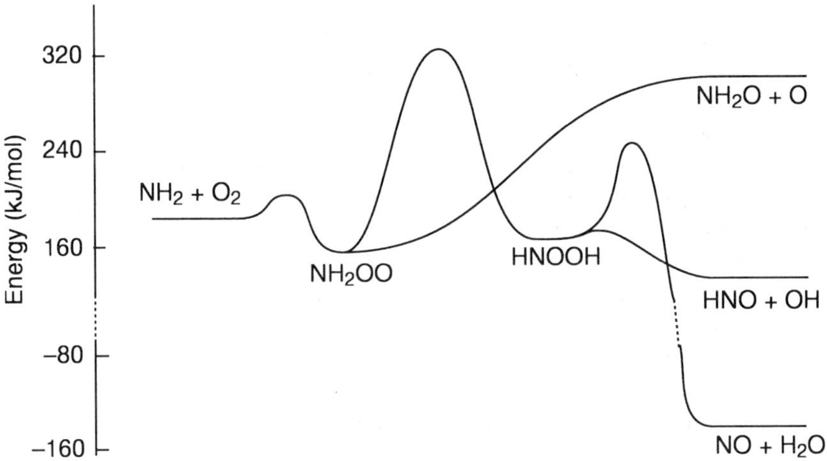

FIGURE 2.24. Potential energy diagram for $NH_2 + O_2$. The high barrier for isomerization of NH_2OO to HNOOH, coupled with the relatively low A–factor, makes the HNO + OH pathway unlikely. The additional barrier for H_2O elimination makes the NO + H_2O channel even less likely.

orders of magnitude higher. Baulch et al. (1992) recommend the lower value at room temperature. Hennig et al. (1995) measured NH_2 decay as well as OH and O production behind incident shock waves in N_2H_4–O_2–Ar mixtures between 1450 and 2300 K. They report that the radical formation channels contribute less than 15% of the overall reaction and that the main products of the reaction are NO and H_2O.

Figure 2.24 shows a potential energy diagram for this reaction. There are five possible channels:

$$NH_2 + O_2 \longrightarrow NH_2OO \quad (13a)$$
$$\longrightarrow NH_2O + O \quad (13b)$$
$$\longrightarrow HNOOH \quad (13c)$$
$$\longrightarrow HNO + OH \quad (13d)$$
$$\longrightarrow NO + H_2O\,. \quad (13e)$$

The combination of an extremely shallow well for the NH_2OO adduct and the high barrier (171 kJ/mol) calculated by Melius and Binkley for the intramolecular hydrogen shift implies that the energy required for simple bond fission to form NH_2O + O (151 kJ/mol) is actually lower than that required for isomerization. This isomerization barrier is consistent with what one would estimate assuming that the barrier height is the sum of the energy for hydrogen abstraction and the ring strain energy in the cyclic transition state. This assumption suggests a barrier of about 54 + 109 = 163 kJ/mol. The lower barrier for bond fission, coupled with the higher A-factor for this channel, would suggest that the fission channel dominates. Similarly, one expects formation of the molecular products NO + H_2O

to be unlikely since any HNOOH formed would decompose easily to HNO + OH. The shallow wells for both adducts suggest that stabilization is unlikely, especially at high temperatures.

We repeated our earlier QRRK analysis (Bozzelli and Dean 1989) using updated thermodynamic parameters and barrier estimates. The results were similar to our earlier ones, suggesting that the reaction is in the low-pressure limit for pressures below 10 atm and that the major channel for all conditions below 1000 K is stabilization, a non-reaction because dissociation of the NH_2OO adduct, with only a 25 kJ/mol potential well, is rapid. These results suggest negligible net stabilization, in agreement with most of the reported experimental results. We could not reconcile our results with data reported by Hack and Kurzke. The QRRK calculations suggest instead that the major high-temperature channel is NH_2O + O, with virtually no effect of pressure. Under typical conditions of temperature and pressure, one can describe this channel with

$$k_{13b} = 2.5 \times 10^{11} T^{0.48} \exp(-14900/T) \quad \mathrm{cm^3 mol^{-1} s^{-1}}.$$

The QRRK rate coefficient for the channel to HNO + OH is

$$k_{13d} = 6.2 \times 10^{7} T^{1.23} \exp(-17700/T) \quad \mathrm{cm^3 mol^{-1} s^{-1}},$$

substantially smaller than k_{13b}. The calculated rate coefficient for the NO + H_2O channel is several orders of magnitude lower than that for HNO + OH. Neither the NH_2O + O nor the HNO + OH channel is affected by change in bath gas over the temperature or pressure ranges considered. Furthermore, the stabilization and the NH_2O + O channels are not affected by plausible changes in the barrier to HNOOH. The barrier for the intramolecular H-shift would need to be lowered by over 40 kJ/mol for the HNO + OH channel to compete with the NH_2O + O channel. Such a drastic change seems unwarranted. Thus the QRRK analysis, confirming the earlier arguments of Melius and Binkley, seems to rule out dominance of the HNO + OH channel. This conclusion is also consistent with the observations of Hennig et al. that the NH_2O + O channel is more important than the HNO + OH channel, although their analysis suggests the difference between the two channels is less than we calculate. In any event, considering that HNO + OH was assumed to be the only channel in earlier ammonia oxidation modeling, for example that of Miller and Bowman (1989), it is clear that ammonia oxidation modeling needs to be revisited. The above results, as well as those on NH_2 + HO_2 discussed in the following section, indicate that NH_2O should be included in ammonia oxidation mechanisms.

A major unresolved issue is the importance of the NO + H_2O channel. The QRRK analysis suggesting that this channel is unimportant is in conflict with the interpretation of Hennig et al. that this is the dominant pathway. Their conclusion was derived from their observation that NH_2 decay is fast relative to production of H and OH. We cannot offer a satisfactory alternate explanation of their observations. However, our analysis of that experimental system suggests that it is surprisingly complex. For example, a sensitivity analysis for NH_2 decay in their experiments indicates that the most important reaction is $NH_2 + NH_2 \longrightarrow NH_3 + NH$, not any reactions of $NH_2 + O_2$. Dominance of the $NO + H_2O$ channel would have far-reaching consequences for ammonia oxidation, as discussed in Sections 2.8.1 and 2.8.3. Inclusion of this channel in the mechanism causes substantial chain termination and forces the reaction to proceed more slowly than observed.

2.6.10 $NH_2 + HO_2 \longrightarrow$ Products

The NH_2 + HO_2 reaction has received relatively little attention. The potential energy surface for recombination–elimination (Figure 2.25) shows three possible channels:

$$NH_2 + HO_2 \longrightarrow NH_2OOH \qquad (14a)$$

$$\longrightarrow NH_2O + OH \qquad (14b)$$

$$\longrightarrow HNO + H_2O\,. \qquad (14c)$$

Another possible pathway is the disproportionation reaction:

$$NH_2 + HO_2 \longrightarrow NH_3 + O_2\,. \qquad (14d)$$

There are no direct experimental studies. Two photochemical studies (Cheskis and Sarkisov 1979; Lesclaux 1984) on the NH_3–O_2 system led to estimates of the overall rate coefficient for this reaction of 3.0 and 1.5 $\times 10^{13}$ cm^{3}mol^{-1}s^{-1}, respectively. One theoretical study (Pouchan et al. 1987) was reported wherein the principal focus was the isomerization channel (14c); the results indicated that it is not significant. Lesclaux, however, concluded the opposite, that the HNO + H_2O product channel is probably the most important—he neglected the NH_2O + OH channel, arguing that it is not energetically accessible. This latter conclusion is surprising, as our thermodynamics (cf. Figure 2.25) suggest that these products are appreciably lower in energy than the reactants. Baulch et al. (1992) recommend 1.6 $\times 10^{13}$ cm^{3}mol^{-1}s^{-1} over the range 300 to 400 K for loss of reactant. Using the DHT method of Section 2.4 to estimate the disproportionation rate coefficient leads to

$$k_{14d} = 9.2 \times 10^5\, T^{1.94} \exp(+580/T)\ \text{cm}^3\text{mol}^{-1}\text{s}^{-1}\,.$$

At 300 K, this expression gives only about 4 $\times$ 10^{11} cm^{3}mol^{-1}s^{-1}, much slower than observed, suggesting that reaction 14d is not the pathway for this reaction at low temperature.

This system was studied using the QRRK approach by Bozzelli and Dean (1989). The most important feature of the reaction is that the $NH_2O + OH$ product channel is 92 kJ/mol below the lowest initial energy of the NH_2OOH adduct. While the $H_2O + HNO$ products are lower in energy than $NH_2O + OH$, the barrier to H_2O formation is high, as discussed above, so that this reaction is unimportant.

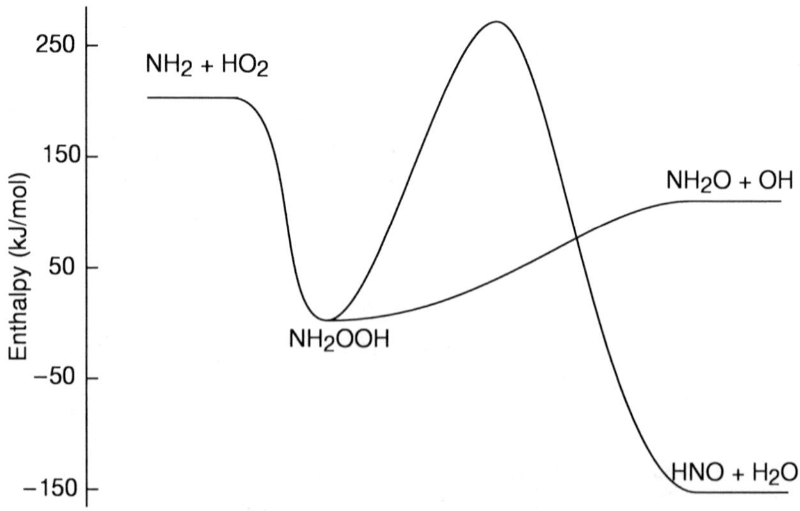

FIGURE 2.25. Potential energy diagram for NH_2 + HO_2. The high barrier for isomerization of NH_2OOH, coupled with the relatively low A-factor, makes the HNO + H_2O pathway unlikely in spite of its greater exothermicity.

The calculations indicate that production of NH_2O + OH is the only important reaction channel. The rate coefficient for forming these products is nearly constant over the temperature range from 200 to 1900 K and at pressures to above 10 atm. The rate coefficient for these conditions can be described by

$$k_{14b} = 2.5 \times 10^{13}\ \text{cm}^3\text{mol}^{-1}\text{s}^{-1}\,.$$

The results are insensitive to the thermochemistry used; increasing the heat of formation of NH_2O by 42 kJ/mol only increases the contribution of the stabilization channel from 0.1 to 1 percent of the collisions at 200 K and 1 atm. The barrier for the H-atom shift, 65 kJ/mol above the initial reactant energies, combined with the presence of lower energy dissociation channels (to products and back to reactants), essentially eliminates the HNO + H_2O channel from consideration. Another reason for this is the lower A-factor expected for the tight transition state for the shift relative to the simple bond fission transition states. Thus, the isomerization channel can only become favorable if the thermodynamics are changed to lower the shift barrier by about 210 kJ/mol, which is inconsistent with other four-member ring transition states.

Our calculations are consistent with the limited experimental measurements and provide an explanation for a high rate coefficient, that is, a large A-factor and zero barrier is expected for radical–radical recombination. However, we predict a completely different product set on the basis of the low energy exit channel. The largest error in determination of this rate coefficient is estimating the high-pressure limit A-factor for the NH_2 + HO_2 channel. The expression we chose for it, 2.5 $\times 10^{13}$ cm^{3}mol^{-1}s^{-1}, gives rate coefficients approximately halfway between the experimental results of Lesclaux and of Cheskis and Sarkisov.

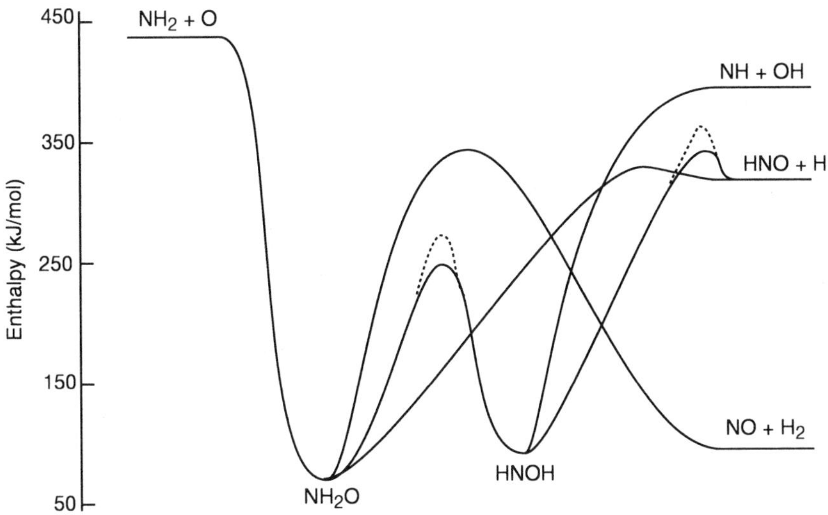

FIGURE 2.26. Potential energy diagram for NH_2 + O. The dashed lines illustrate adjustments to barrier heights that affect the branching ratio of products.

2.6.11 $NH_2 + O \longrightarrow$ Products

The NH_2 + O reaction system has been analyzed theoretically by Melius and Binkley (1984a), who calculated energies for the various possible molecular adducts (NH_2O and HNOH) and for the barriers leading to the various dissociation channels. Their potential energy surface is shown in Figure 2.26. The following channels are possible:

$$\begin{aligned} NH_2 + O &\longrightarrow NH_2O && (15a)\\ &\longrightarrow HNOH && (15b)\\ &\longrightarrow HNO + H && (15c)\\ &\longrightarrow NH + OH && (15d)\\ &\longrightarrow NO + H_2\,. && (15e) \end{aligned}$$

Sufficient energy is available in the initial NH_2O adduct that it can readily dissociate to HNO + H, unimolecularly isomerize by an H-atom shift and dissociate to NH + OH, or undergo concerted H_2 elimination to form NO + H_2.

Rate coefficients for channels (15c) and (15d) have been measured at room temperature by Dransfeld et al. (1984) using a flow reactor; NH and OH were monitored by laser magnetic resonance and HNO by laser induced fluorescence. The rate coefficient for the overall reaction was reported to be 5.3×10^{13} $cm^3 mol^{-1} s^{-1}$, with 87% of the reaction found to proceed via the HNO + H channel, in agreement with the predictions of Melius and Binkley. The product internal energy distribution was probed in a molecular beam study by Patel-Misra and Dagdigian (1991). The observed low level of rotational excitation of HNO was taken as evidence of a barrier to formation of HNO from the adduct. The Cohen–Westberg review (1991) recommends the Dransfeld et al. expression for the total rate coefficient and the 87% branching fraction to HNO + H. They include NO + H_2 as a minor channel with a rate coefficient slightly smaller than that for the NH + OH channel.

This reaction was also studied by Bozzelli and Dean (1989). The potential energy surface suggests that stabilization should be unimportant at typical pressures, since low energy exit channels are available. The QRRK results are consistent with this expectation; stabilization is calculated to contribute less than 1% at pressures below 10 atm. The QRRK analysis predicted that the HNO + H channel dominates, with the HNOH isomer contributing approximately twice as much as the NH_2O isomer at low temperatures. The HNOH contribution shifts slightly (about 20%) toward the NH + OH channel at higher temperatures because of the higher A-factor for dissociation of HNOH to NH + OH and the higher barrier to this channel. Using the barrier of Melius and Binkley for formation of HNO + H from HNOH dissociation (the dashed curve in Figure 2.26), the NH + OH channel was found to account for about 30% of the total reaction products, higher than observed by Dransfeld et al. The Melius and Binkley barrier for the HNO + H channel, however, implies a 42 kJ/mol barrier for the reverse reaction of H-atom addition to HNO, which is higher than most H-atom additions to unsaturated bond systems. In H + C_2H_4, for example, the observed barrier is in the 4 to 12 kJ/mol range (Kerr and Moss 1981) and reactions of H with NO and HCO have even lower barriers. Upon lowering the barrier to the HNO + H channel so that the reverse reaction had a barrier of 15 kJ/mol (the solid curve in Figure 2.26), the QRRK analysis gave the branching ratio measured by Dransfeld et al.

Bozzelli and Dean (1989) also pointed out other barrier adjustments that would give results consistent with the observations. In spite of these uncertainties concerning the potential energy surface, the qualitative picture is clear. The overall reaction is fast and the room temperature value of Dransfeld et al. is also applicable at higher temperatures. Thus we recommend

$$k_{15}(\text{total}) = 5.3 \times 10^{13} \text{ cm}^3\text{mol}^{-1}\text{s}^{-1}.$$

The measurements at 295 K by Adamson et al. (1994), who found k_{15}(total) to be 3.9×10^{13} cm^{3}mol^{-1}s^{-1}, with about 5 to 8% branching into the NH + OH channel, are in reasonable agreement with the Dransfeld et al. measurements. The branching ratio would be expected to have only a modest temperature dependence and (15c) would be expected to be the dominant channel at all temperatures. We recommend use of the Dransfeld et al. values

$$k_{15c} = 4.6 \times 10^{13} \text{ cm}^3\text{mol}^{-1}\text{s}^{-1}$$
$$k_{15d} = 7 \times 10^{12} \text{ cm}^3\text{mol}^{-1}\text{s}^{-1}.$$

In spite of the fact that the NO + H_2 channel is very exothermic, the high barrier for concerted elimination and the correspondingly low A-factor make this channel unimportant.

The potential energy surface computed by Wolf et al. (1994), who distinguished between *cis*- and *trans*-HNOH, agrees in most of its features with the one computed by Melius and Binkley.

Direct abstraction from NH_2 by O would also produce NH + OH. Using the DHT method described in Section 2.4 leads to

$$k_{15d2} = 3.3 \times 10^8\, T^{1.5} \exp(-2555/T)\ \ \mathrm{cm^3mol^{-1}s^{-1}}.$$

The contribution of this channel is negligible at low temperatures, but exceeds that for the addition pathway at 2000 K. A similar conclusion was reached by Duan and Page (1995) on the basis of theoretical calculations for the abstraction channel. Their expression

$$k = 0.87T^{4.01} \exp(842/T)\ \ \mathrm{cm^3mol^{-1}s^{-1}}$$

is within a factor of two of our expression from 300 to 2500 K.

2.6.12 $NH_2 + OH \longrightarrow$ Products

Little is known experimentally about this reaction. The expressions of Hanson and Salimian (1984) and Miller and Bowman (1989) are compared to earlier work of Cheskis and Sarkisov (1979) and an upper limit measurement by Diau et al. (1990) in Figure 2.27. Although all measurements cluster near 1×10^{13} $\mathrm{cm^3mol^{-1}s^{-1}}$, there is disagreement about the temperature dependence, suggesting that multiple channels may be involved.

As with the H + NH_2 reaction considered earlier, this reaction could be either direct hydrogen transfer or an addition–elimination reaction. For the direct route, the expected lowest energy products would be ^3NH + H_2O, a channel that is spin-forbidden for the addition–elimination route via a singlet adduct. The potential energy diagram is shown in Figure 2.28. Possible product channels for the addition–elimination route include

$$\begin{aligned} NH_2 + OH &\longrightarrow NH_2OH && (16a)\\ &\longrightarrow {}^3NH + H_2O && (16b)\\ &\longrightarrow {}^1NH + H_2O && (16c)\\ &\longrightarrow NH_2O + H && (16d)\\ &\longrightarrow HNOH + H. && (16e) \end{aligned}$$

The input parameters for the analysis are shown in Table 2.8. If one assumes the barrier to the spin-forbidden lowest energy products (^3NH + H_2O) is about 255 kJ/mol, one obtains the total rate coefficient predictions for the addition–elimination pathway shown in Figure 2.27. (Use of a lower barrier results in even higher total rate coefficients at the lower temperatures, clearly higher than the upper limit reported by Diau et al. in 300 Torr helium.) The QRRK predictions at higher temperatures are considerably lower than the literature values.

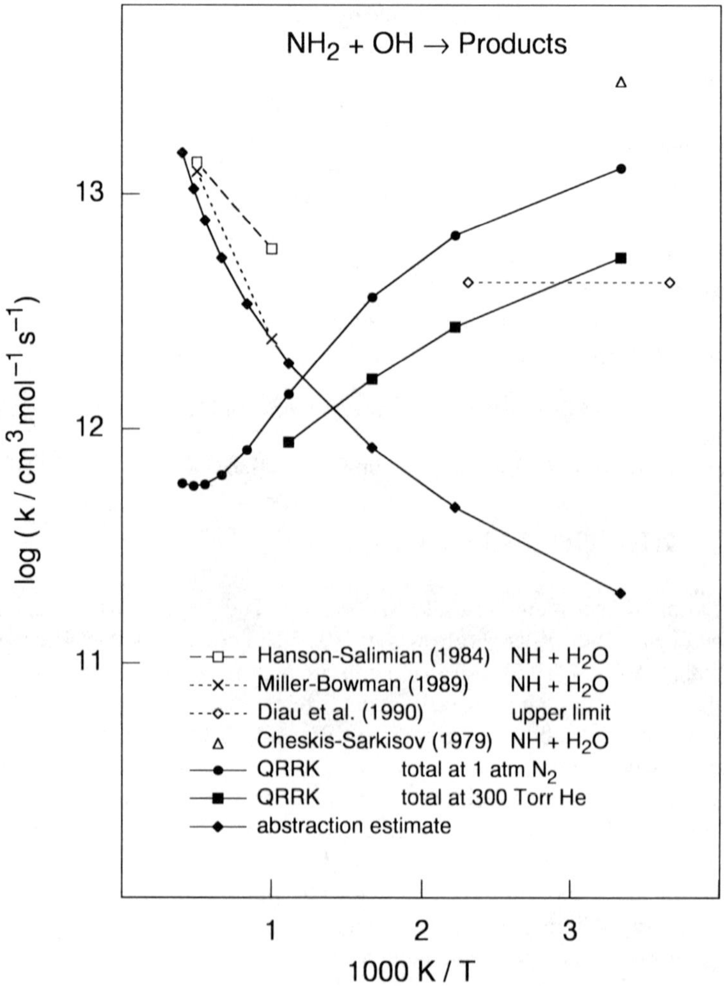

FIGURE 2.27. Comparison of QRRK predictions to rate coefficient measurements for NH_2 + OH. Also included is our estimate for the rate coefficient of the abstraction pathway to form NH + H_2O. The results suggest a change in mechanism from adduct formation at low temperature to hydrogen abstraction at high temperature.

Our calculations suggest that the dominant low-temperature pathway is formation of the NH_2OH adduct, while ^3NH and H_2O are the high-temperature products. Above 1000 K, adduct formation is at the low-pressure limit to at least 10 atm. The QRRK expressions for adduct formation in N_2 are

$$\begin{aligned} k_{16a} &= 1.8 \times 10^{32}\, T^{-6.91} \exp(-2070/T)\ \text{cm}^3\text{mol}^{-1}\text{s}^{-1} \quad && 0.1 \text{ atm} \\ &= 3.9 \times 10^{33}\, T^{-7.00} \exp(-2235/T)\ \text{cm}^3\text{mol}^{-1}\text{s}^{-1} && 1.0 \text{ atm} \\ &= 5.6 \times 10^{34}\, T^{-7.02} \exp(-2700/T)\ \text{cm}^3\text{mol}^{-1}\text{s}^{-1} && 10 \text{ atm.} \end{aligned}$$

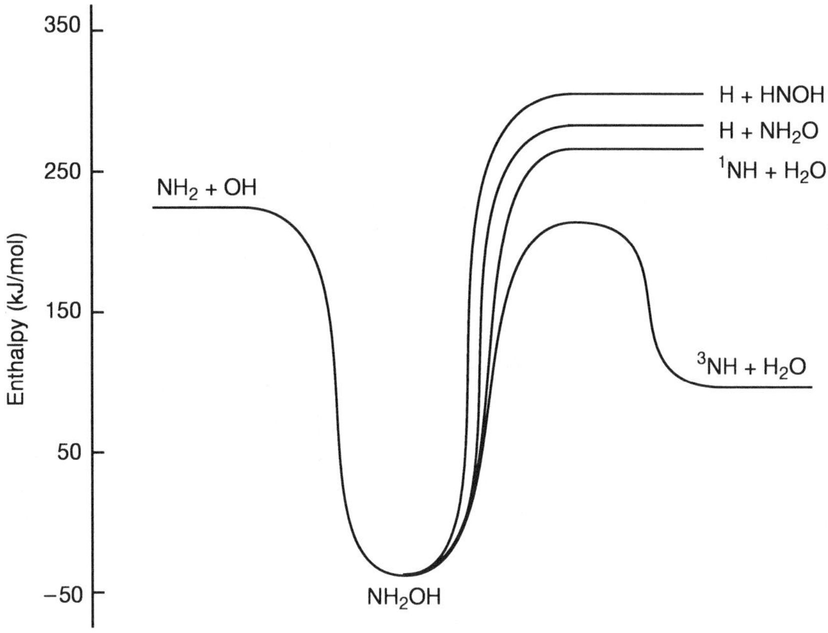

FIGURE 2.28. Potential energy diagram for NH_2 + OH. The barrier for ^{3}NH formation was adjusted to achieve consistency with the low temperature results of Diau et al. (1990).

TABLE 2.8. NH_2 + OH ⟶ Products.

	Reaction	A/s^{-1} or $cm^3 mol^{-1} s^{-1}$	E_a/kJ mol^{-1}
(1)	NH_2+ OH ⟶ NH_2OH	4.5×10^{13}	0
(−1)	NH_2OH ⟶ NH_2+ OH	3.6×10^{15}	263
(2)	NH_2OH ⟶ NH_2O + H	1.1×10^{14}	318
(3)	NH_2OH ⟶ ^{1}NH + H_2O	1.7×10^{13}	297
(4)	NH_2OH ⟶ H + HNOH	1.4×10^{15}	345
(5)	NH_2OH ⟶ ^{3}NH + H_2O	2.5×10^{13}	≥ 255

Molecular constants of NH_2OH adduct: $\nu_1 = 524\ cm^{-1}$(degen=2.78) $\nu_2 = 1468\ cm^{-1}$(3.60) $\nu_3 = 3167\ cm^{-1}$(2.61); $\sigma = 3.49$ Å and $\varepsilon/K = 370$.

Rate coefficients: k_1 based on CH_3 + OH, Washida (1980); k_{-1} microscopic reversibility (MR); k_2 via k_{-2} and MR with $A_{-2} = 3.6\times10^{13}$, $E_a(-2) = 0$, based on H + C_2H_5 (Tsang and Hampson, 1986); k_3 TST, $\Delta S^{\ddagger} = -14.6$ J/K, includes loss of 1 rotor, degeneracy = 2, $E_a = \Delta_r H^{\circ}$; k_4 via k_{-4} and MR with k_{-4} same as k_{-2}; k_5 TST for A_5, $E_a \geq 255$ kJ/mol, to be in agreement with data of Diau et al. (1990) for this channel.

Our conclusion that adduct formation dominates at lower temperatures differs from that of Cheskis and Sarkisov, who report NH and H_2O to be the products. We estimated the rate coefficient for direct hydrogen transfer using the DHT procedure described in Section 2.4 to be

$$k_{16b} = 2.4 \times 10^6\, T^2 \exp(-25/T)\ \text{cm}^3\text{mol}^{-1}\text{s}^{-1}.$$

This result, which is close to the expression used by Miller and Bowman (1989), has been included in Figure 2.27.

Our analysis suggests that the high-temperature reaction is direct H-atom transfer; thus no pressure dependence of the rate coefficient is expected at high temperatures, as concluded in Section 2.6.1 for the H + NH_2 reaction.

2.6.13 $NH_2 + NH_2 \longrightarrow$ Products

There are limited high-temperature measurements relevant to this reaction (Davidson et al. 1990; Dean et al. 1984). Low-temperature measurements by Stothard et al. (1995) indicate that H_2 production dominates. Because there must be a low energy barrier for this channel, formation of H_2 and singlet H_2NN are indicated; formation of H_2 and N_2H_2 would be expected to have a much higher barrier, as shown in Figure 2.29.

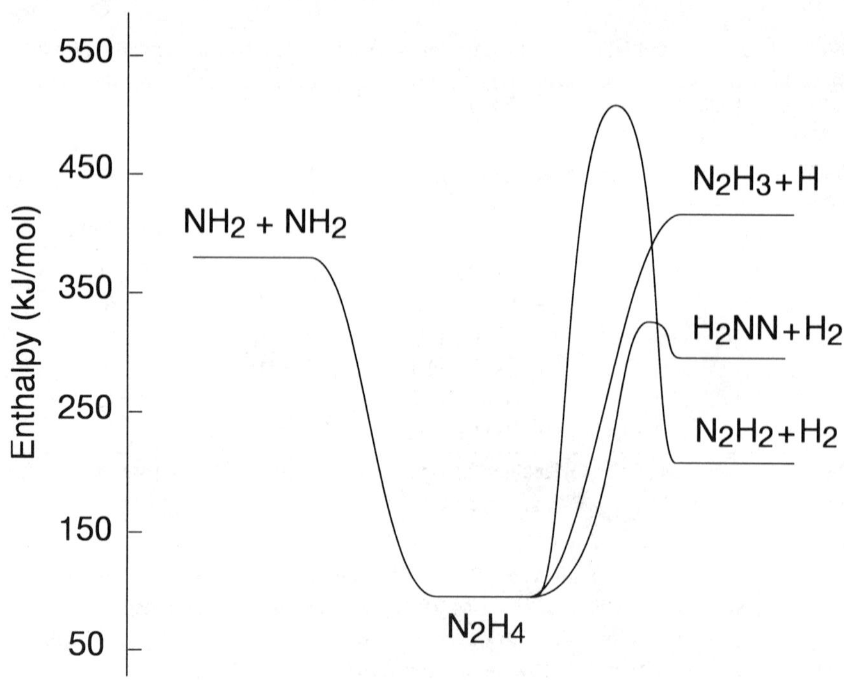

FIGURE 2.29. Potential energy diagram for the molecular and radical product channels of the $NH_2 + NH_2$ reaction.

Possible product channels include

$$NH_2 + NH_2 \longrightarrow N_2H_4 \quad (17a)$$
$$\longrightarrow H_2NN + H_2 \quad (17b)$$
$$\longrightarrow N_2H_2 + H_2 \quad (17c)$$
$$\longrightarrow N_2H_3 + H. \quad (17d)$$

Davidson et al. derived a rate coefficient from a model used to describe their shock tube measurements of ammonia pyrolysis and proposed

$$NH_2 + NH_2 \longrightarrow NH_3 + NH. \quad (17e)$$

Dean et al. proposed that the recombination reaction has two sets of products, N_2H_3 + H and N_2H_4, on the basis of radical profiles observed in rich ammonia flames. Miller and Bowman (1989) suggested that the high temperature products are $N_2H_2 + H_2$. These results are summarized in Figure 2.30. With the notable exception of Stothard et al. (1995), most of the lower temperature studies report that the product is N_2H_4. However, the reported rate coefficients vary widely (Mallard et al. 1993).

This reaction was analyzed using the QRRK method. The input parameters are listed in Table 2.9 and the results are compared to the reported rate coefficients in Figure 2.30. The barrier for channel 17b was adjusted to 238 kJ/mol so as to agree with the low-temperature, low-pressure data of Stothard et al. (1995), which is consistent with the reverse reaction, a singlet radical inserting into H_2, having a low barrier. The calculations indicate that the dominant product channel for the conditions of the Stothard et al. experiments is indeed 17b, for about 80% of the total rate, consistent with their finding that H_2 is the major product. The pressure of the experiments (about 1 mbar) was too low for N_2H_4 stabilization to be important.

TABLE 2.9. $NH_2 + NH_2 \longrightarrow$ Products.

	Reaction	A/s^{-1} or $cm^3 mol^{-1} s^{-1}$	E_a/kJ mol^{-1}
(1)	$NH_2 + NH_2 \longrightarrow NH_2NH_2$	3.1×10^{13}	0
(−1)	$NH_2NH_2 \longrightarrow NH_2 + NH_2$	1.5×10^{16}	279
(2)	$NH_2NH_2 \longrightarrow N_2H_3 + H$	4.3×10^{14}	321
(3)	$NH_2NH_2 \longrightarrow N_2H_2 + H_2$	8.7×10^{12}	414
(4)	$NH_2NH_2 \longrightarrow H_2NN + H_2$	1.5×10^{12}	238

Molecular constants of NH_2NH_2 adduct: ν_1 = 653 cm^{-1} (degen = 3.88) ν_2 = 1315 cm^{-1} (4.52) ν_3 = 3495 cm^{-1} (3.60); σ = 4.2 Å and ε/K = 205.

Rate coefficients: k_1 Back and Yokada (1973) and Sarkisov et al. (1984); k_{-1} microscopic reversibility (MR); k_2 via k_{-2} and MR with $k_{-2} = 1.0\times10^{14}$ from H + C_2H_3 of Duran et al. (1988); k_3 estimated using $E_a = \Delta_r H^\circ$ + 300 kJ/mol, based upon $C_2H_4 \longrightarrow C_2H_2 + H_2$ 1,2 elimination, A-factor from TST, loss of 1 rotor, degeneracy = 4; k_4 A-factor from Laufer et al. (1983) for $H_2C_2 + H_2$, E_a adjusted to fit data of Stothard et al. (1995).

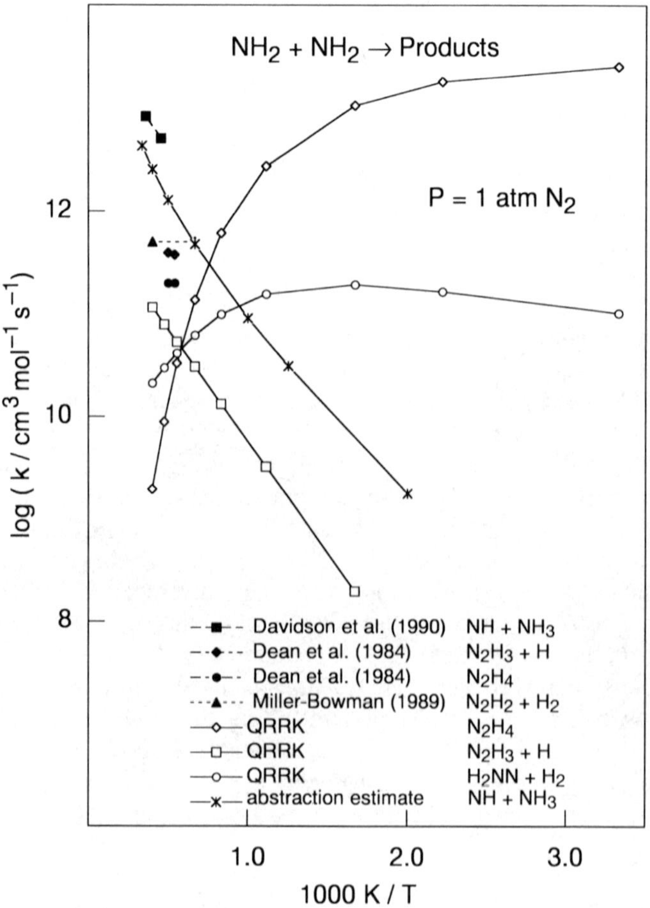

FIGURE 2.30. Comparison of QRRK predictions to high-temperature data for $NH_2 + NH_2$. Also included is the DHT estimate for abstraction to form $NH + NH_3$.

Several conclusions emerge from these comparisons. The channels that arise from adduct formation, 17a–d, are predicted to have rate coefficients appreciably lower than the observations of Davidson et al. However, the rate coefficient they reported is consistent with a hydrogen atom abstraction reaction. An estimate for this rate coefficient obtained with the DHT method of Section 2.4 is seen in Figure 2.30 to be similar in magnitude and temperature dependence to their data. It is also consistent with the rate coefficient for the reverse reaction reported by Röhrig et al. (1994). We recommend that the Davidson et al. expression

$$k_{17e} = 5.0 \times 10^{13} \exp(-5000/T) \ \text{cm}^3\text{mol}^{-1}\text{s}^{-1}$$

be used for this rate coefficient.

Our calculations also suggest that recombination to produce hydrazine (17a) is important up to quite high temperatures and should be explicitly considered in kinetic models. Figure 2.31 shows the predicted rate coefficients at 1500 K in N_2 as functions of pressure. We suggest the following expressions be used for temperatures from 600 to 2500 K at the specified pressures in N_2 bath gas:

$$k_{17a} = 2.0 \times 10^{46} T^{-10.93} \exp(-5030/T) \text{ cm}^3\text{mol}^{-1}\text{s}^{-1} \quad 0.1 \text{ atm}$$
$$= 5.6 \times 10^{48} T^{-11.30} \exp(-5980/T) \text{ cm}^3\text{mol}^{-1}\text{s}^{-1} \quad 1.0 \text{ atm}$$
$$= 3.2 \times 10^{49} T^{-11.18} \exp(-7040/T) \text{ cm}^3\text{mol}^{-1}\text{s}^{-1} \quad 10 \text{ atm.}$$

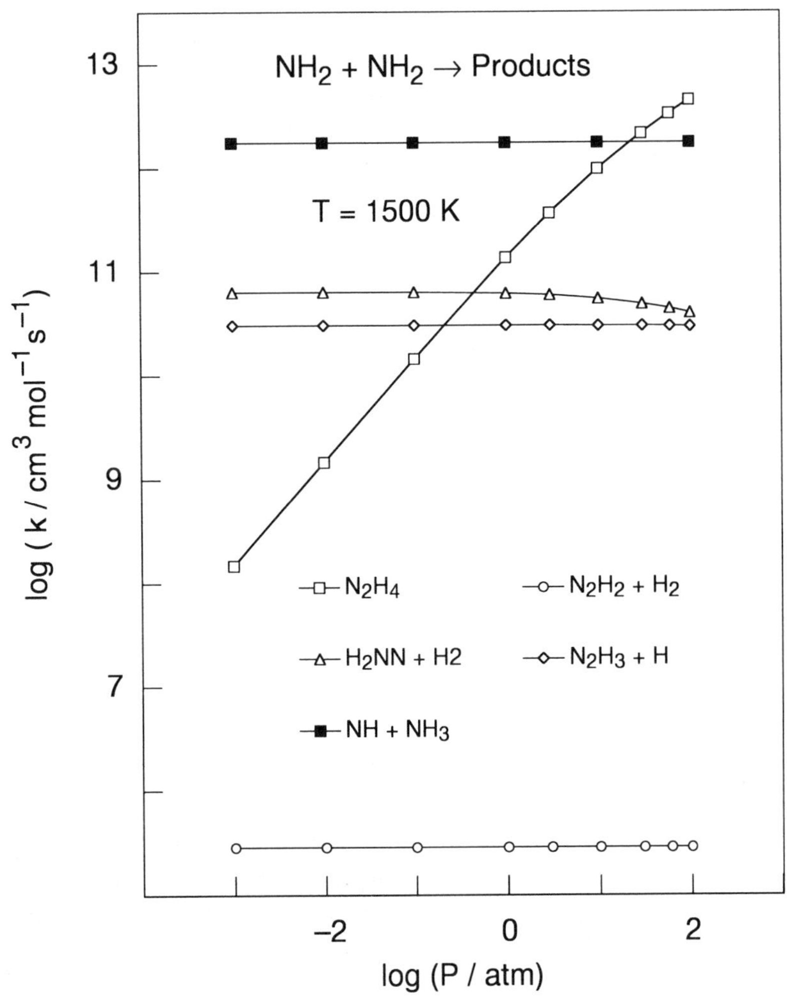

FIGURE 2.31. Predicted effect of pressure at 1500 K from the QRRK calculations for $NH_2 + NH_2 \longrightarrow$ Products. At pressures above 1 atm, stabilization is predicted to be important even at high temperatures.

These rate coefficient expressions cannot be extrapolated to room temperature. Our QRRK calculations at room temperature are in good agreement with the results of Patrick and Golden (1984). The next most important channel at lower temperatures is 17b, which implies that one has to account for the subsequent reactions of singlet H_2NN. This channel is also significant at lower pressures and higher temperatures (Figure 2.31). QRRK calculations for N_2 bath gas give

$$\begin{aligned} k_{17b} &= 2.4 \times 10^{20} T^{-2.91} \exp(-1075/T)\ \mathrm{cm^3 mol^{-1} s^{-1}} && 0.1\ \mathrm{atm} \\ &= 1.2 \times 10^{21} T^{-3.08} \exp(-1695/T)\ \mathrm{cm^3 mol^{-1} s^{-1}} && 1.0\ \mathrm{atm} \\ &= 2.3 \times 10^{19} T^{-2.54} \exp(-2105/T)\ \mathrm{cm^3 mol^{-1} s^{-1}} && 10\ \mathrm{atm.} \end{aligned}$$

At higher temperatures, production of $N_2H_3 + H$ also contributes. Because the N_2H_3 product will decompose rapidly to produce a second H atom, this channel can become a doubly significant H-atom source. Its rate coefficient is nearly independent of pressure. At 1 atm in N_2 the QRRK result is

$$k_{17d} = 1.2 \times 10^{12} T^{-0.03} \exp(-5075/T)\ \mathrm{cm^3 mol^{-1} s^{-1}}.$$

Figure 2.31 shows that the molecular channel forming N_2H_2 and H_2 is unimportant for all conditions, as expected from the high barrier for concerted elimination. Because this channel would have a tight transition state and correspondingly low A-factor, it is unlikely that the QRRK estimate could be sufficiently in error to make this channel important.

2.6.14 $NH_2 + NO \longrightarrow$ Products

This pivotal SNCR reaction has been the subject of extensive analysis since the Hanson and Salimian review. The product distribution is of critical importance and has been intensively studied. Most studies have focused on the two channels

$$\begin{aligned} NH_2 + NO &\longrightarrow N_2 + H_2O && (18a) \\ &\longrightarrow NNH + OH. && (18b) \end{aligned}$$

Modeling results are particularly sensitive to the branching ratio, since 18a is a chain termination reaction while reaction 18b, a source of H (from NNH dissociation) as well as OH, is a chain branching reaction. There is now substantial agreement on the overall rate coefficient (Atakan et al. 1989; Atkinson et al. 1989; Bulatov et al. 1989; Pagsberg et al. 1991; Baulch et al. 1992; Diau et al. 1994; Wolf et al. 1994) and growing evidence that the branching ratio β, that is, $k_{18b}/(k_{18a} + k_{18b})$, is smaller than previously believed, particularly at lower temperatures. For example, six investigations subsequent to the Hanson and Salimian review (Dolson 1986; Hall et al. 1986; Silver and Kolb 1987; Atakan et al. 1989; Bulatov et al. 1989; Pagsberg et al. 1991) uniformly reported β values in the range from 0.1 to 0.15 at room temperature; Baulch et al. (1992) recommend $\beta = 0.12$ at 298 K. The temperature dependence of β is controversial. Atakan et al. and Bulatov et al. report that it increases slightly with temperature, while Park and Lin (1996) and Halbgewachs et al. (1996) report stronger dependences. This information is summarized in Figure 2.32, where for completeness the recommendation of Atkinson et al. (1989) has also been included. The rate coefficients for the channel producing OH were obtained by multiplying the total rate coefficient and branching ratio from the individual studies cited.

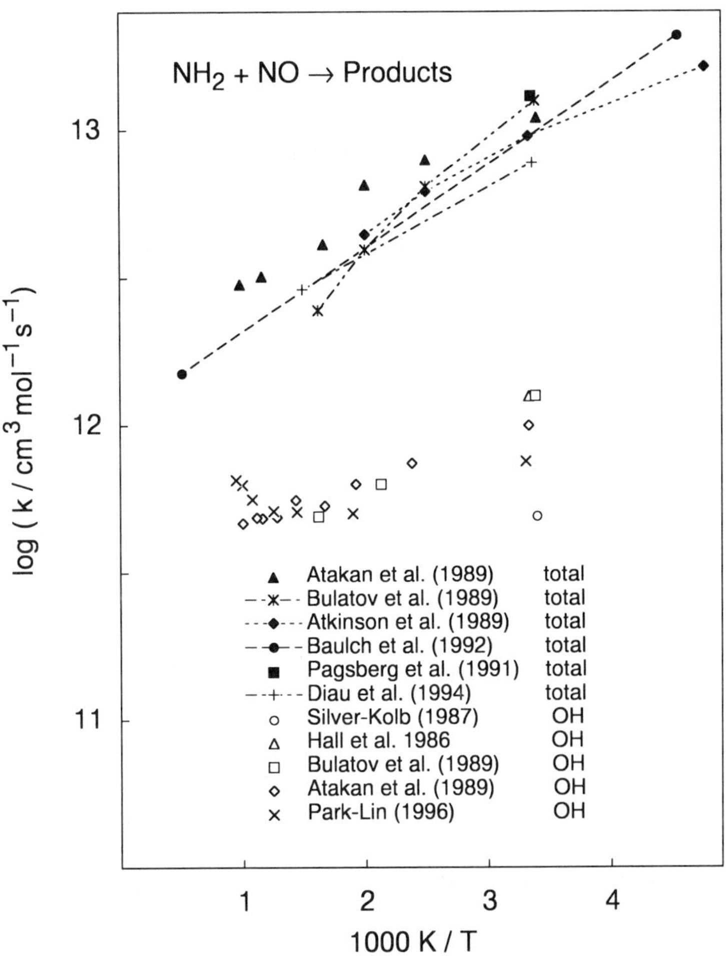

FIGURE 2.32. Arrhenius plot of experimental measurements of the total and OH production rate coefficients for NH_2 + NO $\longrightarrow$ Products.

In addition to the uncertainty regarding the temperature dependence of β, a second concern about this reaction is that H-atom production has not been observed. Because of the expected short lifetime of NNH produced in reaction 18b, H-atom production coincident with OH should be observable. No H has been seen, however, and experiments by Unfried et al. (1990) show that this cannot be attributed to reactions of NNH with NO, at least at room temperature. It is possible that some of the observed OH might come from sources other than reaction 18b (cf. Unfried et al., loc. cit.). Stephens et al. (1993) found the value of β to increase with temperature, but also reported that the total contribution of the two channels appeared to decrease at higher temperatures, possibly indicating the onset of an additional reaction channel. However, the modeling study by Diau et al. (1994) suggests that this phenomenon may result from secondary reactions.

A potential energy diagram for this system, based primarily on the calculations of Melius and Binkley (1984b), is shown in Figure 2.33. The low barrier for H-atom shift, through which the initial NH_2NO adduct can form *trans*-HNNOH, suggests that stabilization is relatively unimportant in this system. *Trans*-HNNOH can isomerize to *cis*-HNNOH or dissociate directly to NNH + OH. A significant fraction of the NNH + OH yield comes from the *trans* form, with the *cis* form decomposing primarily to N_2 and H_2O. The branching ratio is controlled by the differences in both A-factors and barriers for these channels. Walch (1993b), using a more extensive basis set, obtained results similar to those of Melius and Binkley. Diau and Smith (1996) used Walch's calculation as the basis for a microcanonical variational TST analysis of the reaction.

The input parameters for the QRRK calculation are listed in Table 2.10. The results at 0.01 atm, similar to the pressure in most of the experiments, are compared to experimental observations in Figure 2.34. Also shown in Figure 2.34 are the calculated values of Diau and Smith (1996). Much of the difference between the two predictions for the OH channel can be traced to differences in the total rate coefficient, as the values computed for the branching ratio as a function of temperature agree with one another. The agreement with respect to the overall rate coefficient is reasonable, while the calculations overpredict the amount of OH produced at higher temperatures. A small increase in the standard enthalpy of formation of NNH would improve the high-temperature fit but lead to a lower branching ratio than observed at room temperature. Nonetheless, this comparison suggests that a straightforward QRRK analysis does capture the intrinsic nature of this system. In particular, it supports that the QRRK predictions as functions of pressure should serve to indicate how the rate coefficients for the various channels are affected by pressure. Figure 2.35 illustrates the predicted effect of pressure at 1200 K, a typical temperature for Thermal DeNOx applications. The total rate coefficient for NH_2 + NO disappearance is insensitive to pressure over a wide range. Only at pressures approaching 10 atm is the effect of the stabilization channel noticeable. At room temperature, the same transition appears at pressures closer to 1 atm. Thus one should be able to use the measured low-pressure expressions for the rate coefficients for most applications of interest. For completeness, rate coefficients for formation of NH_2NO are included in Table 2.19.

Our recommendation for the rate coefficient of the OH channel

$$k_{18b} = 3.5 \times 10^{10} T^{0.335} \exp(+385/T)\ \mathrm{cm^3 mol^{-1} s^{-1}}$$

is based on a fit to the average of the results of Atakan et al. and Park and Lin.

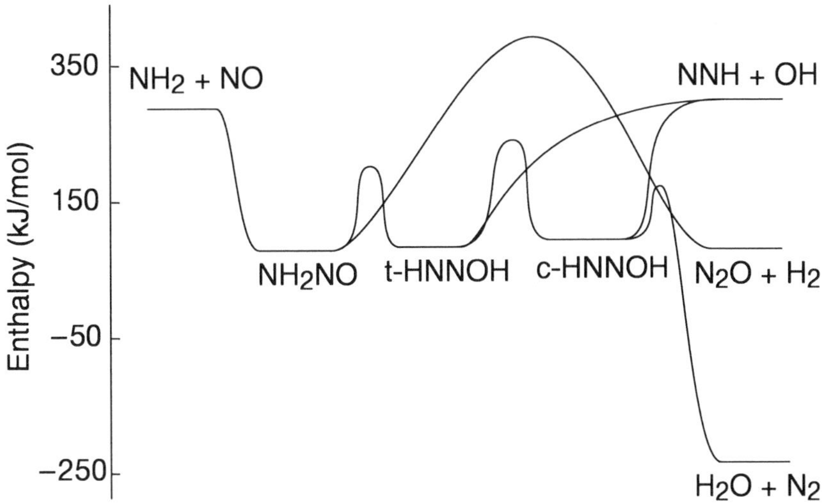

FIGURE 2.33. Potential energy diagram for NH_2 + NO.

TABLE 2.10. NH_2 + NO $\longrightarrow$ Products.

	Reaction	A/s^{-1} or $cm^3mol^{-1}s^{-1}$	E_a/kJ mol^{-1}
(1)	NH_2 + NO $\longrightarrow$ NH_2NO	1.0×10^{13}	0
(−1)	NH_2NO $\longrightarrow$ NH_2 + NO	1.9×10^{15}	194
(2)	NH_2NO $\longrightarrow$ $H_2 + N_2O$	6.3×10^{13}	315
(3)	NH_2NO $\longrightarrow$ HNNOH (trans)	1.5×10^{13}	128
(−3)	HNNOH(trans) $\longrightarrow$ NH_2NO	6.0×10^{12}	122
(4)	HNNOH(trans) $\longrightarrow$ NNH + OH	5.0×10^{15}	193
(5)	HNNOH(trans) $\longrightarrow$ HNNOH(cis)	1.3×10^{14}	151
(−5)	HNNOH(cis) $\longrightarrow$ HNNOH(trans)	1.4×10^{14}	144
(6)	HNNOH(cis) $\longrightarrow$ $H_2O + N_2$	5.5×10^{13}	91
(7)	HNNOH(cis) $\longrightarrow$ NNH + OH	5.6×10^{15}	185

Molecular constants of NH_2NO adduct: ν_1 = 506 cm^{-1} (degen=2.96) ν_2 = 1435 cm^{-1} (1.43) ν_3 = 1457 cm^{-1} (4.61); molecular constants of HNNOH(cis) adduct: ν_1 = 634 cm^{-1} (degen=3.29) ν_2 = 802 cm^{-1} (0.41) ν_3 = 1506 cm^{-1} (5.30); molecular constants of HNNOH(trans) adduct: ν_1 = 789 cm^{-1} (degen=4.42) ν_2 = 1213 cm^{-1} (3.96) ν_3 = 2914 cm^{-1} (0.62); σ = 3.97 Å and ε/K = 436 for all adducts.

Rate coefficients: k_1 review of literature, this study; k_{-1} microscopic reversibility (MR); k_2 nonsymmetric TST, E_a from C.F. Melius—average of two TST structures calculated; k_3 TST, Radicalc estimate (Bozzelli and Ritter 1993), E_a from Melius et al. (1984b); k_{-3} MR; k_4 via k_{-4} and MR with k_{-4} as for OH + CH_3, $A(-4) = 4.5\times10^{13}$, $E_a(-4)$ = 0, Washida (1980); k_5 TST, Radicalc estimate (Bozzelli and Ritter 1993) with E_a from C.F. Melius; k_{-5} MR; k_6 TST, Radicalc estimate (Bozzelli and Ritter 1993) with E_a from C.F. Melius; k_7 from k_{-7} by MR—see k_4.

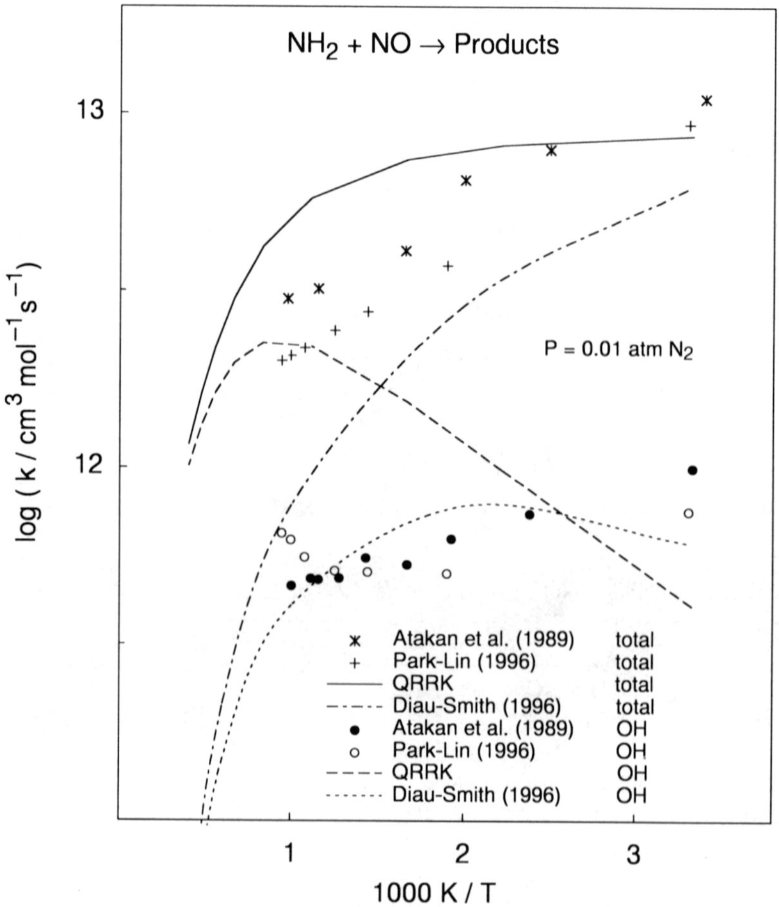

FIGURE 2.34. Comparison of QRRK predictions to the measurements of Atakan et al. (1989) for NH_2 + NO.

The rate coefficient for the N_2 channel

$$k_{18a} = 4.7 \times 10^{12} T^{-0.247} \exp(+605/T) \ \mathrm{cm^3 mol^{-1} s^{-1}}$$

is a fit to the difference between k_{18b} and the expression for the total rate coefficient recommended by Baulch et al. (1992) for the temperature interval from 300 to 1000 K. Because of the uncertainty in the temperature dependence of β, extrapolation to higher temperatures should be done with caution. At 1250 K, these expressions give $\beta = 0.28$, significantly lower than the 1273 K value (0.58) reported by Halbgewachs et al. and the temperature-independent value (0.51) used by Miller and Bowman (1989) for modeling Thermal DeNOx. As discussed in Section 2.8.2, such high β values are actually not needed to account for Thermal DeNOx kinetics; a β value near 0.3 at 1250 K is sufficient.

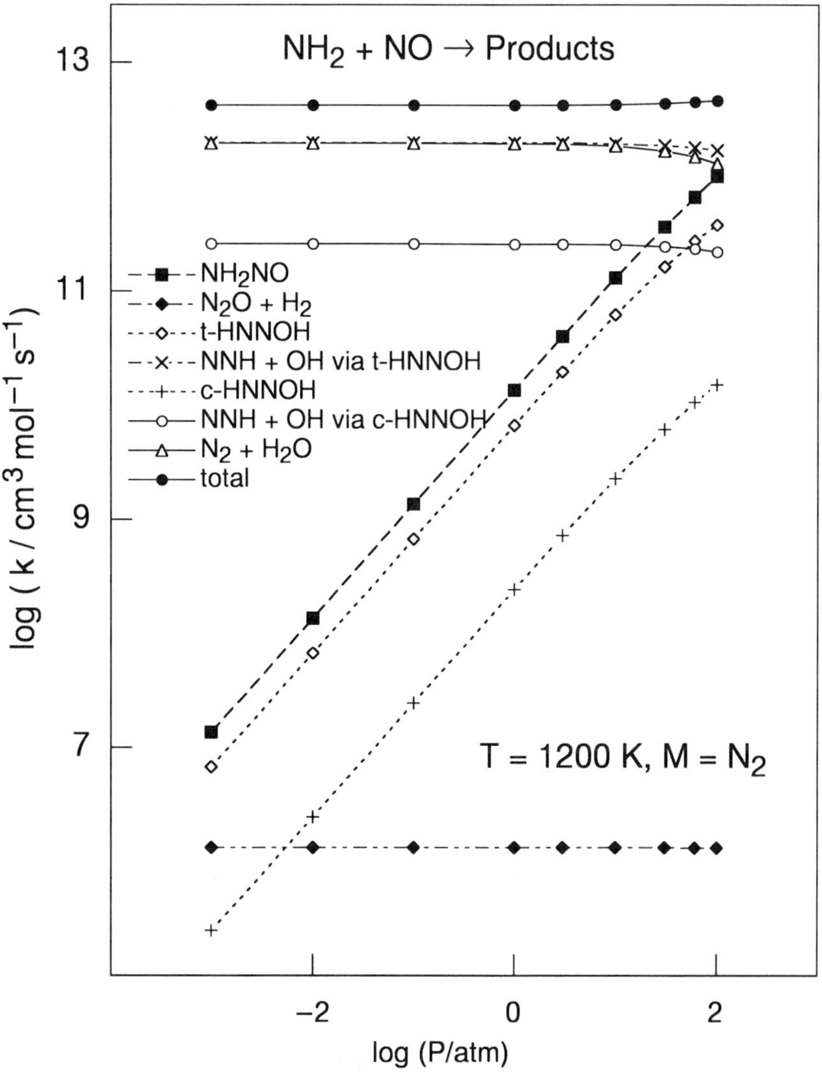

FIGURE 2.35. Predicted effect of pressure at 1200 K from QRRK calculations for NH_2 + NO ⟶ Products. The dominant pathways are virtually independent of pressure until above 10 atm.

2.6.15 CH_3 + NO ⟶ Products

As CH_3 is usually present in flames in much larger concentrations than other hydrocarbon radicals, this reaction may provide an important route for the removal of NO (Haynes 1978). There is quite good agreement among low-temperature measurements of the recombination channel rate coefficient (Washida 1980;

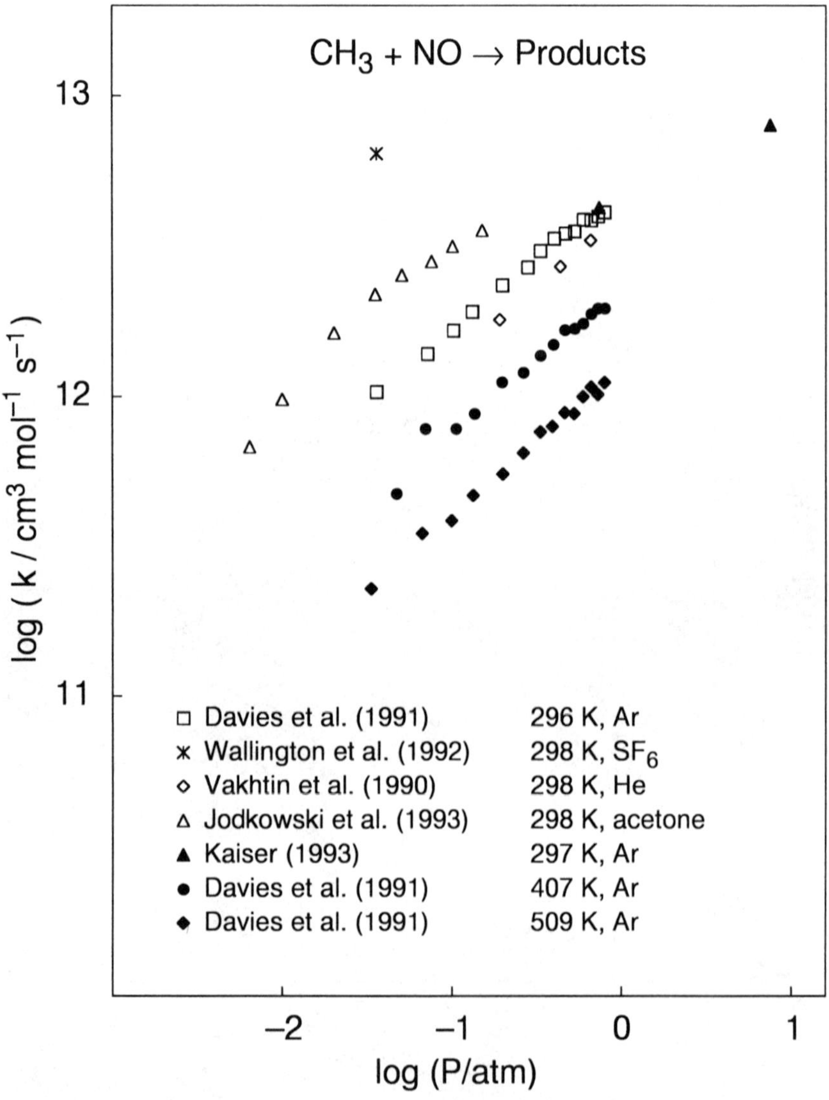

FIGURE 2.36. Low-temperature measurements of CH_3 + NO rate coefficients.

Vakhtin and Petrov 1990; Davies et al. 1991; Wallington et al. 1992; Jodkowski et al. 1993; Kaiser 1993) (Figure 2.36). The temperature and pressure dependence and the effect of buffer gases is what one expects for a simple recombination. At higher temperatures, there have been several shock tube studies (Wolff and Wagner 1988; Hoffmann et al. 1990; Yang et al. 1993; Lifshitz et al. 1993; Hennig and Wagner 1994). These data and earlier measurements are summarized in Figure 2.37. Hoffmann et al. measured production of OH and, using earlier measurements for the total rate coefficient, reported that OH production accounts for about 40% of the total methyl radical loss. They also presented evidence that the observed methyl loss at high temperatures cannot be attributed to formation of the adduct. Lifshitz et al. interpreted their single-pulse shock tube experiments in terms of $HCN + H_2O$ as the major products. Yang et al. measured loss of CH_3 and inferred a rate coefficient expression for the total reaction rate. Their expression extrapolates to a value close to that preported by Lifshitz et al., and they suggest that HCN and H_2O may be the major products in both studies. Hennig et al. measured H and HCN profiles and report that the contribution of the OH channel is 20–25% of that of the main channel. Overall, the data show appreciable scatter in both total rate and product distribution.

A potential energy level diagram for CH_3NO based on the calculations of Melius (1993) is shown in Figure 2.38. Possible reaction channels include

$$\begin{aligned} CH_3 + NO &\longrightarrow CH_3NO && (19a)\\ &\longrightarrow CH_2NHO && (19b)\\ &\longrightarrow CH_2NOH && (19c)\\ &\longrightarrow H_2CN + OH && (19d)\\ &\longrightarrow HCN + H_2O && (19e) \end{aligned}$$

This system is similar in some respects to $NH_2 + NO$, with which it is isoelectronic. In both cases, isomerization via H-atom shifts is followed by a two-channel dissociation, one into radical fragments and one into stable species. However, the higher barriers for isomerization of CH_3NO make stabilization and dissociation of the adduct back to reactants much more likely for this case. Isomerization to CH_2NOH can occur either directly or stepwise via CH_2NHO. The input parameters for the QRRK calculations are listed in Table 2.11. Transition state geometries and entropies were calculated with semi-empirical methods,[1] and the rate coefficients, including temperature-dependent preexponential factors, were derived from transition state theory. The similarity with the NH_2 + NO system continues in that the rate coefficient for production of the initial adduct was taken to be the same.

A comparison of these calculations to the room temperature data for Ar bath gas is shown in Figure 2.39. The QRRK rate coefficient predictions obtained using a normal value for collisional energy transfer ($\Delta E = 2.6$ kJ/mol) are

[1] The calculations were done with the MOPAC 6.0 program and PM3 parameterization. Transition states were located with the TS keyword; RHF was used for singlet species and UHF for doublet species. Transition states were confirmed by having one imaginary vibration frequency, by the atom positions, and by animation of the structure as it changed along the reaction coordinate. Entropy and heat capacity values were computed in MOPAC with the THERMO option for vibrations and overall rotation; contributions from frequencies corresponding to internal rotations were replaced by Pitzer–Gwinn estimates of the rotational entropy and heat capacity.

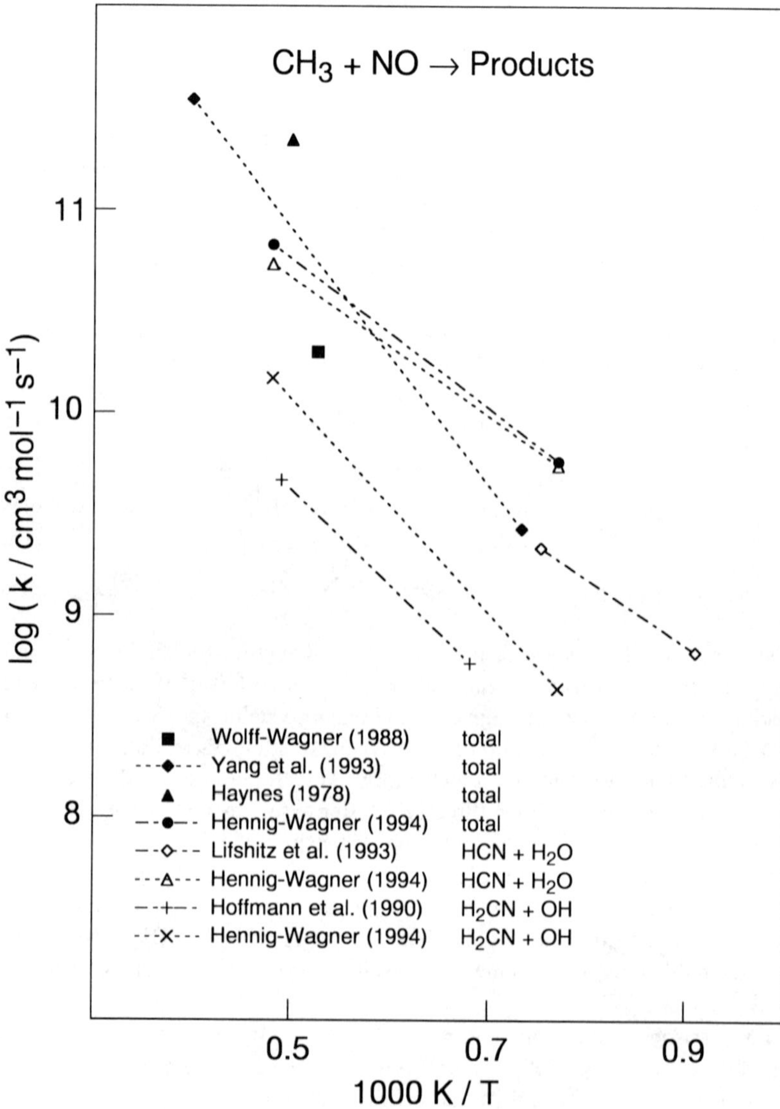

FIGURE 2.37. High-temperature rate coefficient measurements for CH_3 + NO.

significantly lower than observed. Davies et al. (1991) encountered a similar difficulty in analyzing the falloff regime and had to invoke unusually efficient energy transfer to account for their data. They suggested that the falloff data could be accounted for by recombination to form a triplet, followed by rapid intersystem crossing to form the stabilized singlet adduct. Figure 2.39 also illustrates that even the strong collision case (β = 1) does not fit the observed falloff. Somewhat better agreement can be forced by assuming greater stability of

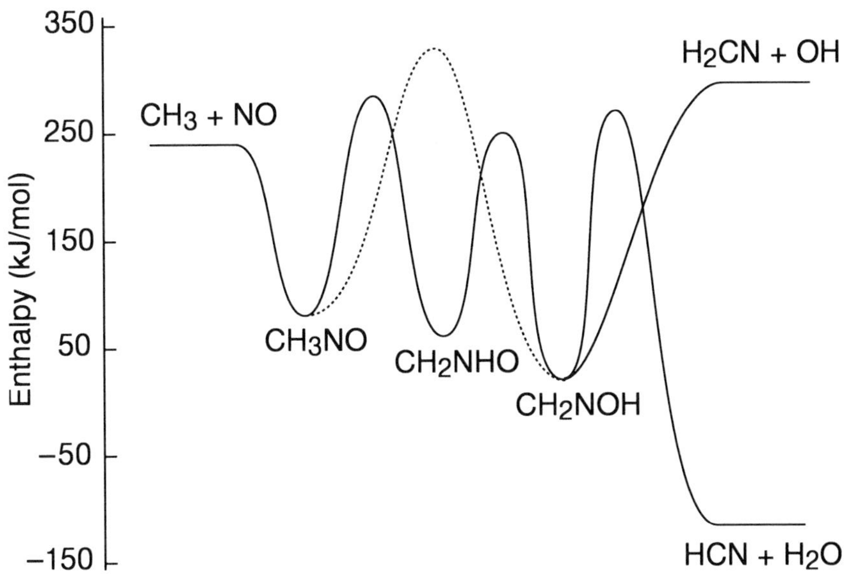

FIGURE 2.38. Potential energy diagram for CH_3 + NO ⟶ Products.

TABLE 2.11. CH_3 + NO ⟶ Products.

	Reaction			A/s^{-1} or cm^{3}mol^{-1}s^{-1}	m	E_a/kJ mol^{-1}
(1)	CH_3 + NO	⟶	CH_3NO	1.0×10^{13}		
(–1)	CH_3NO	⟶	CH_3+ NO	1.8×10^{15}		154
(2)	CH_3NO	⟶	CH_2NHO	6.8×10^{9}	1.25	198
(–2)	CH_2NHO	⟶	CH_3NO	2.2×10^{11}	0.98	238
(3)	CH_2NHO	⟶	CH_2NOH	3.1×10^{10}	1.22	184
(–3)	CH_2NOH	⟶	CH_2NHO	9.7×10^{10}	0.96	222
(4)	CH_3NO	⟶	CH_2NOH	2.9×10^{10}	1.00	243
(–4)	CH_2NOH	⟶	CH_3NO	2.9×10^{12}	0.47	321
(5)	CH_2NOH	⟶	HCN + H_2O	2.3×10^{11}	0.94	238
(6)	CH_2NOH	⟶	CH_2N + OH	4.3×10^{16}		259

Molecular constants of CH_3NO adduct: ν_1 = 658 cm^{-1} (degen=4.62) ν_2 = 1867 cm^{-1} (5.84) ν_3 = 3999 cm^{-1} (1.54); molecular constants of CH_2NHO adduct: ν_1 = 250 cm^{-1} (degen=2.22) ν_2 = 1119 cm^{-1} (6.29) ν_3 = 3278 cm^{-1} (3.49); molecular constants of CH_2NOH adduct: ν_1 = 384 cm^{-1} (degen=3.37) ν_2 = 1542 cm^{-1} (6.26) ν_3 = 3999 cm^{-1} (2.37); σ = 4.25 Å and ε/K = 410 for all adducts, an average of values for C_2H_5OH and CH_3CN.

Rate coefficients: k_1 Kaiser (1993); k_{-1} microscopic reversibility (MR); k_2 TST; k_{-2} MR; k_3 TST; k_{-3} MR; k_4 TST; k_{-4} MR; k_5 TST; k_6 via k_{-6} and MR with k_{-6} based on OH + CH_3 with $A(-6) = 4.5\times10^{13}$ and $E_a(-6) = 0$, Washida (1980).

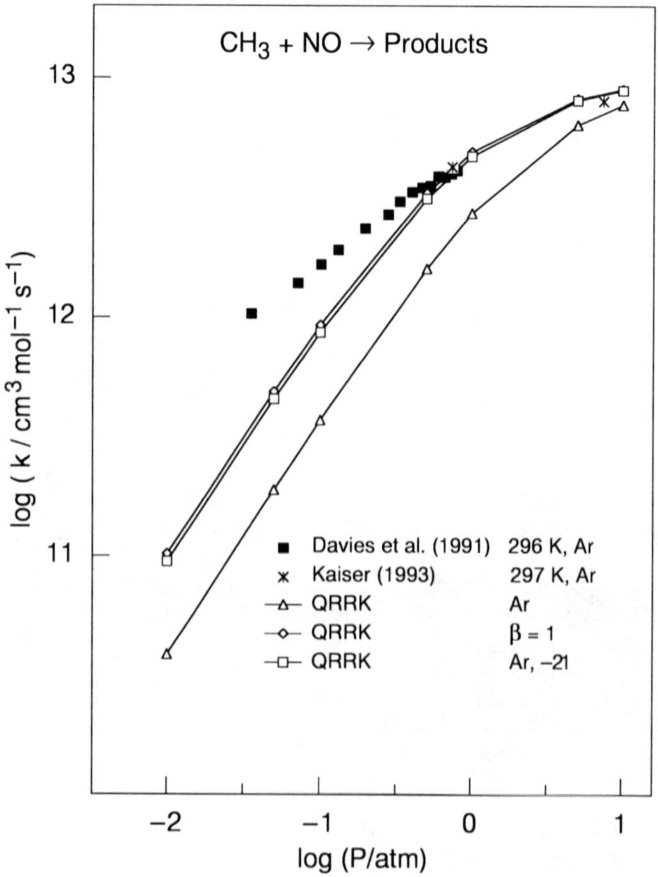

FIGURE 2.39. Comparison of QRRK predictions for CH_3 + NO to experimental falloff measurements in Ar buffer gas near room temperature. The line marked QRRK [Ar] used the normal energy transfer parameter for Ar, $\Delta E = 2.6$ kJ/mol. The line marked QRRK [$\beta = 1$] was calculated using a strong collision assumption. Increasing the CH_3NO well depth by 21 kJ/mol while using the usual Ar parameters, QRRK [Ar, −21], results in stabilization comparable to invoking the strong collision assumption.

CH_3NO. Figure 2.39 includes a calculation in which the well depth for CH_3NO was increased by 21 kJ/mol while using the usual collisional stabilization parameters; the QRRK results are similar to those for the strong collision case. Use of an even deeper well results in an approximately horizontal displacement of the calculated curves toward lower pressure; the slope remains steeper than observed. Similar difficulties arise in comparisons to the data at 407 and 509 K. Tunneling through the isomerization barrier would also increase the low-temperature stabilization rate coefficient computed using the Melius (1993) potential energy surface parameters. While ambiguities remain in understanding the low-temperature rate coefficient, the data do show that adduct formation is important under the conditions studied.

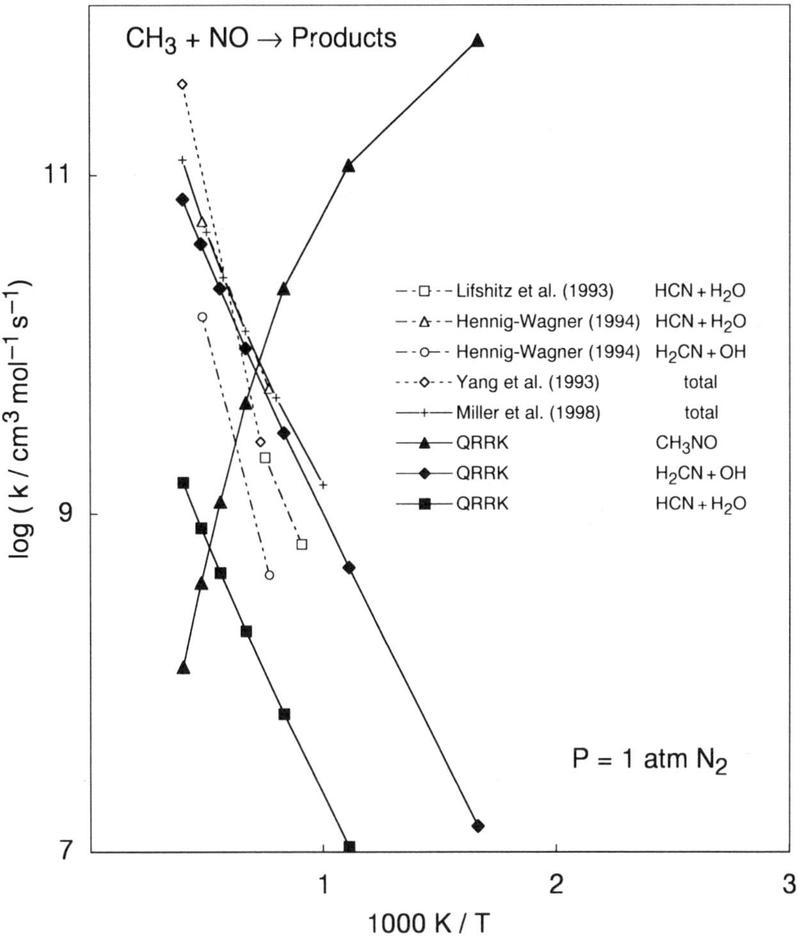

FIGURE 2.40. Arrhenius plot for the various channels of CH_3 + NO $\longrightarrow$ Products as predicted by QRRK calculations for 1 atm N_2. Formation of the stabilized adduct CH_3NO is predicted to dominate at lower temperatures, CH_2N + OH formation to dominate at higher temperatures, and the HCN + H_2O channel to be unimportant at all temperatures. The Miller et al. (1998) calculation considers only the HCN + H_2O and H_2CN + OH channels.

Figure 2.40 illustrates the QRRK predictions for the various reaction channels as a function of temperature at 1 atm N_2. Stabilization of the adduct dominates at lower temperatures as expected, while H_2CN + OH is predicted to be the major high-temperature channel. The transition between CH_3NO formation and H_2CN + OH formation suggested by Figure 2.40, however, is misleading. The computed bond dissociation energy of CH_3NO is only 160 kJ/mol, which implies that dissociation of CH_3NO back to CH_3 and NO is rapid. For this reason, the CH_3NO formation channel can be neglected at the high-temperature conditions of the studies shown in Figure 2.37. Even at 1000 K and 10 atm, where Figure 2.40 would suggest that CH_3NO formation would be the dominant channel, modeling studies for a mixture initially containing 20 ppm CH_3 and 2% NO confirmed that the predicted profiles for CH_3 loss and H_2CN formation are but slightly affected if the CH_3NO formation channel is omitted from the model.

The calculations predict in particular that the dominant high-temperature products are H_2CN and OH rather than HCN and H_2O. This prediction was supported by additional calculations in which the barrier to formation of HCN and H_2O was reduced by 42 kJ/mol. Even then, the H_2CN + OH channel was found to be the most important one above 1000 K. The main reason is the much higher A-factor for dissociation of CH_2NOH to H_2CN and OH than for isomerization of CH_2NOH and dissociation to HCN and H_2O. Using the somewhat higher A-factor for isomerization proposed by Lifshitz et al. would not change this result.

Some of the high-temperature results are included in Figure 2.40 for comparison. Several difficulties are apparent, especially with respect to the importance of the HCN + H_2O channel. While the QRRK calculations suggest that it is unimportant, it is reported by Lifshitz et al. to be the only channel and by Hennig and Wagner to be the dominant one. One way to reconcile this situation would be to assume that secondary reactions lead to the observed HCN. In the longer time scale Lifshitz et al. experiments, subsequent reactions of H_2CN (cf. Section 2.7.20) would readily account for the observed HCN production. The faster Hennig and Wagner experiments entailed observations of H-atom profiles and were interpreted as measures of the rate of OH + H_2CN production on the assumption of rapid H_2CN decomposition. There may be complications with this assumption. For example, using the H_2CN rate coefficient expressions discussed in Section 2.7.20 leads to an H_2CN dissociation lifetime of about 1 μs for the Hennig and Wagner conditions, sufficient for loss through secondary reactions such as recombination with NO, which is present in high concentrations. Any such reactions would reduce the appearance rate of H atoms and lead to an underestimation of the rate of H_2CN and OH appearance. The rate measurements based on HCN formation may be complicated by presence of vibrationally excited HCN, as the authors point out. Additional experiments and more detailed modeling are needed to resolve these complications. In any event, the QRRK results are clear in their support of OH and H_2CN being the main products of the CH_3 + NO reaction at high temperatures, and the predicted rate coefficient is in reasonable agreement with the observed reaction rate.

The conclusions of Miller et al. (1998) are in agreement with these predictions, as seen in Figure 2.40. Their calculated total rate coefficient for the sum of the two bimolecular channels is close to the combined QRRK rate coefficients; they infer that the branching ratio should favor the H_2CN + OH channel.

(Nguyen et al. (1996) report a QRRK analysis based upon ab initio molecular electronic structure calculations that predicts HCN + H_2O to be the major product channel; however, these workers did not investigate the H_2CN + OH channel and report a total rate constant an order of magnitude smaller than the experimental observations at high temperatures.)

QRRK calculations over an extended pressure range show that both the HCN + H_2O and the H_2CN + OH channel are insensitive to pressure at the higher temperatures, while the stabilization channels scale linearly with pressure. (At pressures above 10 atm, there is some falloff in all channels except CH_3NO.) Under all conditions the major channels are either formation of CH_3NO, which dominates completely at lower temperatures, or formation of H_2CN + OH. For low temperatures, we adopt the limiting rate coefficients derived by Kaiser for the stabilization channel in Ar bath gas

$$k_{0,19a} = 1.3 \times 10^{18} \mathrm{cm^6 mol^{-2} s^{-1}}$$
$$k_{\infty,19a} = 1.0 \times 10^{13} \ \mathrm{cm^3 mol^{-1} s^{-1}} .$$

Using these values, we tested the Gilbert et al. (1983) and the Stewart et al. (1989) fitting formulas against the data shown in Figure 2.36. The Stewart et al. formula (Section 2.3.1), with $a = 0.03$, $b = -790$, and $c = 1.0$, was found to give significantly better fits for the temperature range of the data, 300 to 500 K. At higher temperatures, where the other channels begin to have an influence, we suggest using the QRRK estimates, in $\mathrm{cm^3 mol^{-1} s^{-1}}$ units, for the stabilization channel in N_2:

$$\begin{aligned} k_{19a} &= 3.6 \times 10^{35} T^{-8.25} \exp(-2420/T) && 0.1 \text{ atm} \\ &= 1.0 \times 10^{37} T^{-8.38} \exp(-2630/T) && 1.0 \text{ atm} \\ &= 4.6 \times 10^{41} T^{-9.39} \exp(-4160/T) && 10 \text{ atm.} \end{aligned}$$

While these expressions underpredict the observed low-temperature stabilization rate coefficient, they should account properly for the temperature and pressure dependence over the whole range of these variables. As discussed earlier, rapid dissociation of CH_3NO makes modeling results insensitive to the stabilization rate coefficient at temperatures over 1000 K.

The QRRK estimate for the primary high-temperature reaction channel is

$$k_{19d} = 2.2 \times 10^9 T^{0.75} \exp(-5900/T) \ \mathrm{cm^3 mol^{-1} s^{-1}}.$$

The QRRK estimate for the HCN + H_2O channel is

$$k_{19e} = 4.9 \times 10^8 T^{0.46} \exp(-6240/T) \ \mathrm{cm^3 mol^{-1} s^{-1}}.$$

2.6.16 CH_3 + N ⟶ Products

This reaction provides a pathway for production of HCN and ultimately NO. It has been extensively studied at low temperatures (Stief et al. 1988; Marston et al. 1989a, 1989b), including measurement of the branching ratio between the major product channels

$$\begin{aligned} CH_3 + N &\longrightarrow H_2CN + H && (20a) \\ &\longrightarrow HCN + H_2 . && (20b) \end{aligned}$$

The reaction is observed to be very fast, with about 90% of the reaction producing $H_2CN + H$ and the remaining 10% producing $HCN + H_2$ via a spin-forbidden channel. The total rate has also been measured in shock tube experiments between 1600 and 2000 K by Davidson and Hanson (1990b). The data are summarized in Figure 2.41. The high-temperature rate coefficients are significantly lower than one would expect from a linear Arrhenius extrapolation of the lower temperature experimental results.

Figure 2.42 shows the relevant potential energy level diagram. Other channels considered besides those discussed in the reports on the experiments include

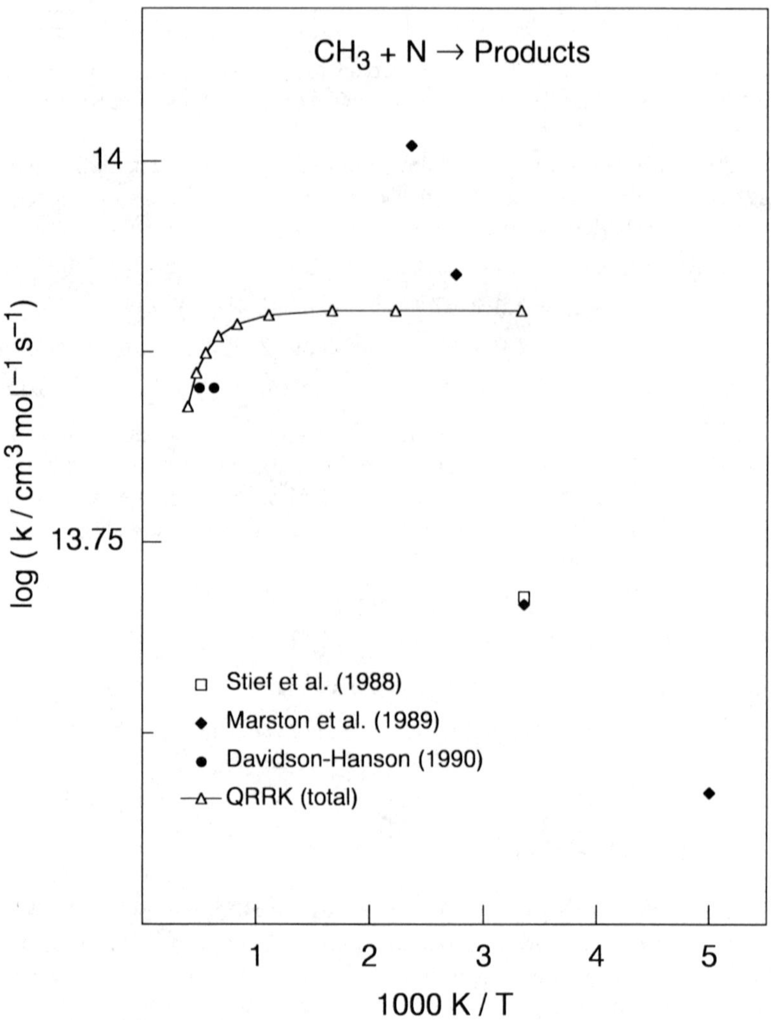

FIGURE 2.41. Comparison of QRRK predictions with experimental results for the reaction $CH_3 + N \longrightarrow$ Products.

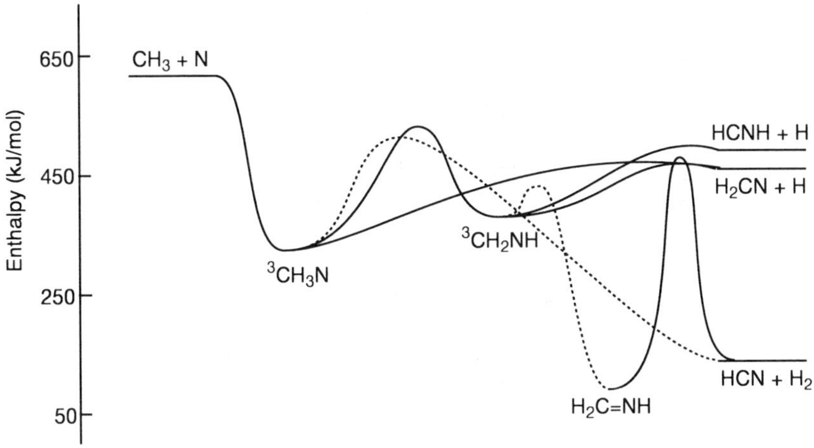

FIGURE 2.42. Potential energy diagram for CH_3 + N ⟶ Products. Spin-forbidden pathways are indicated by dashed lines.

$$CH_3 + N \longrightarrow {}^3CH_3N \tag{20c}$$

$$\longrightarrow {}^3CH_2NH \tag{20d}$$

$$\longrightarrow {}^1CH_2NH \tag{20e}$$

$$\longrightarrow HCNH + H\,. \tag{20f}$$

It can be seen that the possible reaction pathways are rather complex. To improve clarity, solid lines were used to indicate paths to allowed reaction product channels and isomerizations; dashed lines show the spin-forbidden reactions. The $^3CH_3N$ formed can dissociate to H_2CN + H, isomerize to $^3CH_2NH$, undergo a spin-forbidden molecular elimination of H_2, or dissociate back to reactants. The $^3CH_2NH$ can dissociate to H_2CN + H or to HC=NH + H, or undergo intersystem crossing to $^1CH_2NH$. The adduct well is sufficiently deep, about 290 kJ/mol, to put the barriers for isomerization and dissociation below the entrance channel. Thus one expects rapid reactions. Input parameters for the QRRK calculations are listed in Table 2.12.

These calculations indicate that H_2CN + H is the dominant channel at all temperatures and pressures, with very slight rate decreases as the temperature is increased due to an increased reverse reaction rate, that is, dissociation back to reactants. The QRRK predictions are compared to the observations in Figure 2.41. Although the temperature dependence of the low-temperature data is not accounted for, the overall agreement is acceptable. One possible explanation of the low-temperature results, suggested by Steif and co-workers, is the participation of a separate channel, formation of $^1CH_3N$ via an energy barrier, followed by rapid isomerization to $^1CH_2NH$, yielding an adduct (with about 530 kJ/mol excess energy) that could readily dissociate via a loose transition state to H_2CN + H or HC=NH + H. The isomerization would have to have a higher Arrhenius A-factor than the reverse dissociation, and adduct formation would still require an unusually high rate coefficient to explain the rapid increase in rate with temperature. This pathway was not included in the QRRK calculations.

TABLE 2.12. $CH_3 + N \longrightarrow$ Products.

	Reaction			A/s^{-1} or $cm^3mol^{-1}s^{-1}$	$E_a/kJ\ mol^{-1}$
(1)	$CH_3 + N$	$\longrightarrow$	$^3CH_3N$	8.0×10^{13}	0
(–1)	$^3CH_3N$	$\longrightarrow$	$CH_3 + N$	1.9×10^{15}	293
(2)	$^3CH_3N$	$\longrightarrow$	$CH_2{=}N + H$	9.3×10^{13}	148
(3)	$^3CH_3N$	$\longrightarrow$	$HCN + H_2$	7.0×10^{13}	188
(4)	$^3CH_3N$	$\longrightarrow$	$^3CH_2{=}NH$	1.4×10^{14}	205
(–4)	$^3CH_2{=}NH$	$\longrightarrow$	$^3CH_3N$	7.0×10^{13}	163
(5)	$^3CH_2{=}NH$	$\longrightarrow$	$CH_2{=}N + H$	2.9×10^{13}	89
(6)	$^3CH_2{=}NH$	$\longrightarrow$	$HC{=}NH{+}H$	8.7×10^{13}	118
(7)	$^3CH_2{=}NH$	$\longrightarrow$	$^1CH_2{=}NH$	1.0×10^{12}	51
(–7)	$^1CH_2{=}NH$	$\longrightarrow$	$^3CH_2{=}NH$	2.0×10^{12}	348
(8)	$^1CH_2{=}NH$	$\longrightarrow$	$HCN + H_2$	8.0×10^{13}	389

Molecular constants of $^3CH_3N$ adduct: ν_1 = 799 cm^{-1} (degen=2.61) ν_2 = 1426 cm^{-1} (3.39) ν_3 = 2772 cm^{-1} (3.00); molecular constants of $^3CH_2{=}NH$ adduct: ν_1 = 779 cm^{-1} (degen=3.27) ν_2 = 1487 cm^{-1} (3.03) ν_3 = 3419 cm^{-1} (2.70); molecular constants of $^1CH_2{=}NH$ adduct: ν_1 = 884 cm^{-1} (degen=2.81) ν_2 = 2046 cm^{-1} (5.59) ν_3 = 3986 cm^{-1} (0.60); σ = 3.49 Å and ε/K = 350 for all adducts.

Rate coefficients: k_1 was estimated by assuming that the dynamics of CH_3N adduct formation resemble those of the reaction O + CH_3, for which Baulch et al. (1992) adopted the rate coefficient 8.4×10^{13} of Slagle et al. (1987); k_{-1} microscopic reversibility (MR); k_2 from k_{-2} and MR with $A(-2) = 9.35\times10^{12}$, ($0.5\times A$-factor for H + C_2H_4) and $E_a(-2)$ = 8 kJ/mol from Mallard et al. (1993); k_3 TST, E_a from Gordon et. al (1986); k_4 TST, loss of symmetry, gain of rotor, E_a from Gordon et al. (1986)—cf. Walia and Kakker (1990, 1991) (Melius calculates $\Delta_r H^\circ$= 59 kJ/mol); k_{-4} MR; k_5 via k_{-5} and MR with k_{-5} taken to be the same as k_{-2}; k_6 via k_{-6} and MR with k_{-6} taken to be the same as k_{-2}; k_7 triplet–singlet conversion, E_a from Gordon et al. (1986) and our estimate for A; k_{-7} MR; k_8 TST, E_a based on 1,2-H_2 elimination from C_2H_4 (Yoshimine 1989).

Figure 2.43 illustrates the effect of pressure on the various pathways at 1500 K. The rate coefficients for the major channels are virtually independent of pressure, indicating that there is no need to assign pressure-dependent rate coefficients for this system. The most important channel is H_2CN + H, giving about 80% of the total reaction. Our results indicate that HCN + H_2 is the next important channel with about 13% of the total rate, followed by HC=NH + H at about 5% of the total rate. These results are consistent with the experimental observations that about 90% of the reaction produces H-atoms. As expected, most of the H_2CN + H is produced from the initially formed $^3CH_3N$, since isomerization to $^3CH_2NH$ has both a lower A-factor and a higher barrier than dissociation. The slower isomerization also accounts for the much slower rate of production of HCNH +

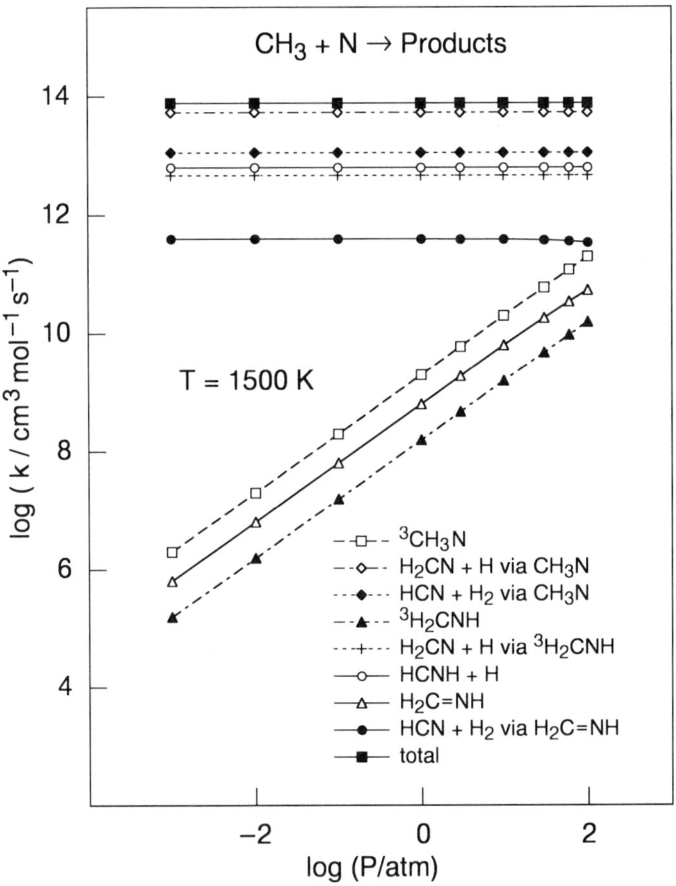

FIGURE 2.43. Effect of pressure at 1500 K for N + CH_3 ⟶ Products. The faster rates are virtually independent of pressure until above 100 atm.

H. In much the same way, most of the HCN + H_2 is produced directly from the $^3CH_3N$ via the spin-forbidden dissociation. The HCN and the HC=NH channels increase slightly with increasing temperature, as shown in Figure 2.44. Given the extremely rapid rate, we suggest the QRRK results be used to estimate the branching ratios at the high temperatures of interest in combustion. These results, in $cm^3mol^{-1}s^{-1}$ units, are

$$k_{20a} = 6.1 \times 10^{14} T^{-0.31} \exp(-145/T)$$
$$k_{20b} = 3.7 \times 10^{12} T^{0.15} \exp(+45/T)$$
$$k_{20f} = 1.2 \times 10^{11} T^{0.52} \exp(+185/T).$$

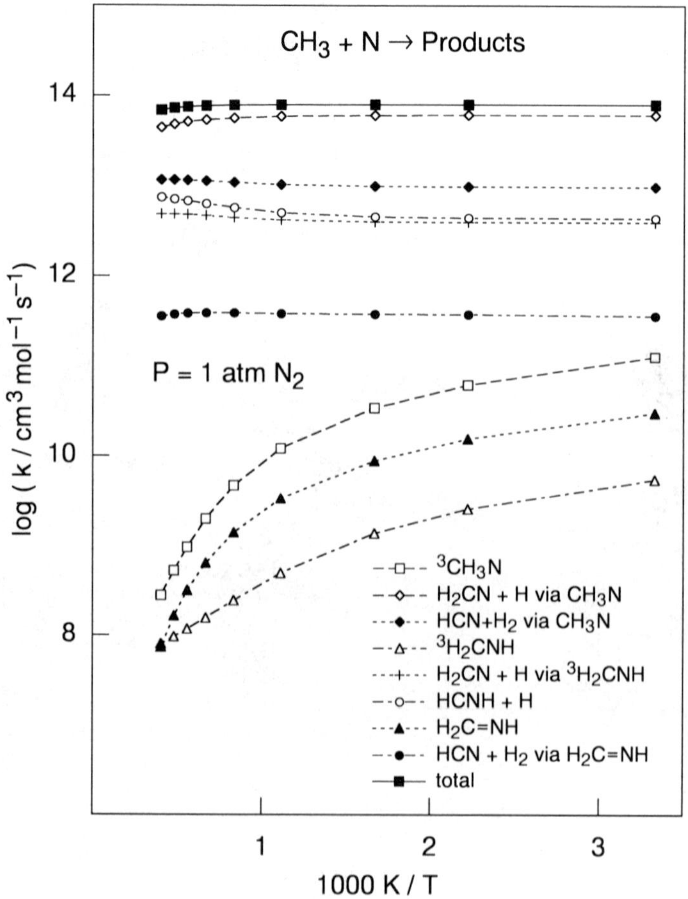

FIGURE 2.44. Arrhenius plot for the $CH_3 + N \longrightarrow$ Products reaction predicted by QRRK calculations for 1 atm N_2 buffer gas. The rate coefficients for the dominant channels are all predicted to be essentially independent of temperature.

2.6.17 $CH_3 + NH_2 \longrightarrow$ Products

This reaction can lead to recombination and other products

$$
\begin{aligned}
CH_3 + NH_2 &\longrightarrow CH_3NH_2 && (21a)\\
&\longrightarrow CH_2NH_2 + H && (21b)\\
&\longrightarrow CH_3NH + H && (21c)\\
&\longrightarrow H_2C{=}NH + H_2 . && (21d)
\end{aligned}
$$

Rate coefficients were estimated using the QRRK method and the parameters given in Table 2.13.

The results at 1 atm N_2 are shown in Figure 2.45. The dominant channel is formation of the stabilized adduct up to very high temperatures. Of the various elimination channels, the formation of CH_2NH_2 + H is favored. Stabilization is predicted to dominate at pressures above 0.1 atm even at 1500 K. (Figure 2.46)

TABLE 2.13. $CH_3 + NH_2 \longrightarrow$ Products.

	Reaction	A/s^{-1} or $cm^3mol^{-1}s^{-1}$	E_a/kJ mol^{-1}
(1)	$CH_3 + NH_2 \longrightarrow CH_3NH_2$	2.3×10^{13}	0
(−1)	$CH_3NH_2 \longrightarrow CH_3 + NH_2$	1.2×10^{16}	354
(2)	$CH_3NH_2 \longrightarrow \dot{C}H_2NH_2 + H$	2.0×10^{15}	395
(3)	$CH_3NH_2 \longrightarrow CH_3\dot{N}H + H$	1.8×10^{15}	420
(4)	$CH_3NH_2 \longrightarrow H_2C{=}NH + H_2$	4.9×10^{13}	433

Molecular constants of CH_3NH_2 adduct: $\nu_1 = 576\ cm^{-1}$(degen=3.12) $\nu_2 = 1416\ cm^{-1}$(6.46) $\nu_3 = 3121\ cm^{-1}$(4.92); σ = 3.77 Å and ε/K = 364.

Rate coefficients: k_1 geometric mean of $CH_3 + CH_3$ and $NH_2 + NH_2$ from Mallard et al. (1993); k_{-1} microscopic reversibility (MR); k_2 via k_{-2} and MR with $k_{-2} = 3.6\times10^{13}$ based on H + C_2H_5 from Baulch et al. (1992); k_3 same as k_2; k_4 TST with loss of 1 rotor, reaction degeneracy = 6, barrier based on $C_2H_6 \longrightarrow C_2H_4 + H_2$.

The calculations suggest a slight pressure dependence for the elimination channels, but at the temperatures where these are important this effect is small enough to be neglected. Because of the importance of stabilization, one must consider subsequent reactions of the monomethylamine adduct. The QRRK analysis suggests the rate coefficient expressions, in $cm^3mol^{-1}s^{-1}$ units:

$$k_{21a} = 1.3\times10^{54}T^{-12.72}\exp(-7855/T),\ \ 0.1\text{ atm N}_2,\ 600\text{–}2500\text{ K}$$

$$k_{21a} = 5.1\times10^{52}T^{-11.99}\exp(-8450/T),\ \ 1.0\text{ atm N}_2,\ 600\text{–}2500\text{ K}$$

$$k_{21a} = 1.6\times10^{47}T^{-10.15}\exp(-7895/T),\ \ 10\text{ atm N}_2,\ 600\text{–}2500\text{ K}$$

$$k_{21b} = 1.4\times10^{14}T^{-0.43}\exp(-5590/T)$$

$$k_{21c} = 4.4\times10^{13}T^{-0.31}\exp(-8375/T)$$

$$k_{21d} = 4.8\times10^{11}T^{-0.20}\exp(-9765/T)$$

The only experimental work on this system that we are aware of (Higashihara et al. 1987; Hwang et al. 1990; Lifshitz et al. 1991) used extensive computer simulations to derive elementary reaction rate coefficients applicable to monomethylamine pyrolysis and oxidation. For typical experimental conditions used by these authors to derive the CH_3NH_2 thermal dissociation rate coefficient (1% CH_3NH_2 in Ar at 1 atm and 1400 K), the QRRK analysis gives a falloff from the high-pressure limit of about a factor of 10, somewhat less than half of the amount they derived from their experiments. Because the thermochemistry used

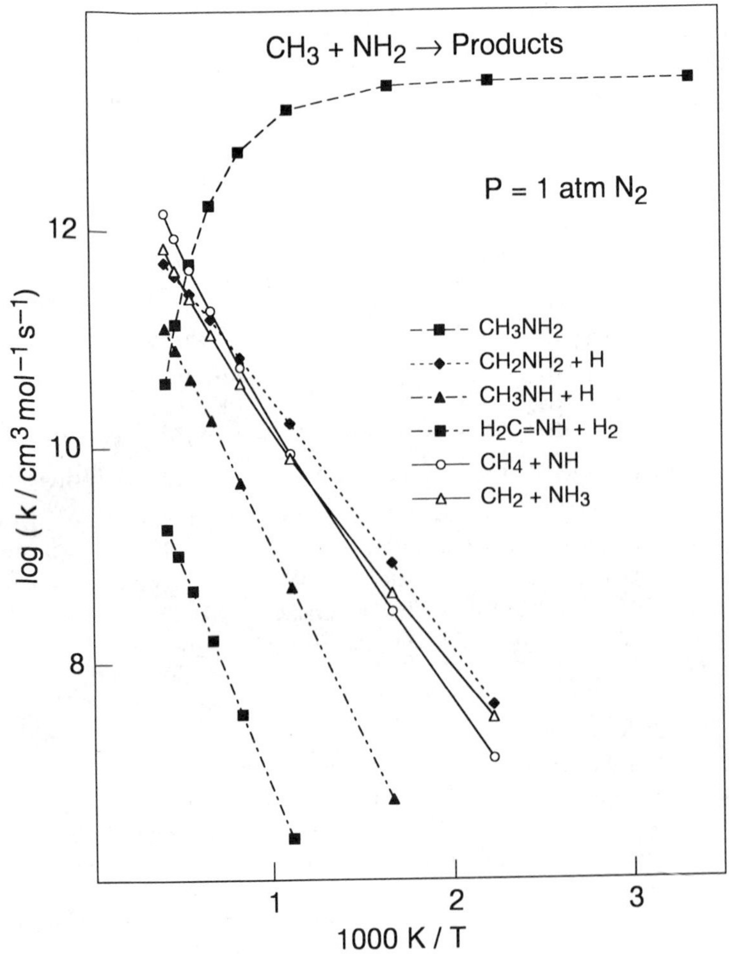

FIGURE 2.45. Arrhenius plot for the various channels of $CH_3 + NH_2 \longrightarrow$ Products predicted by QRRK (filled symbols) and DHT (open symbols) calculations for 1 atm N_2 buffer gas. Stabilization is predicted to be the dominant channel up to quite high temperatures; the rate of the $CH_2NH_2 + H$ channel does not equal the stabilization rate until 1800 K.

by these authors is different from ours and because of the complicated chemistry that they had to describe, extensive reanalysis of their data would be required to determine how well our QRRK rate coefficients for this and the other unimolecular reactions involved account for their data.

In addition to adduct formation followed by stabilization or dissociation in reactions 21b–21d, H-atom transfer can be significant:

$$CH_3 + NH_2 \longrightarrow CH_4 + NH \qquad (21e)$$

$$\longrightarrow CH_2 + NH_3. \qquad (21f)$$

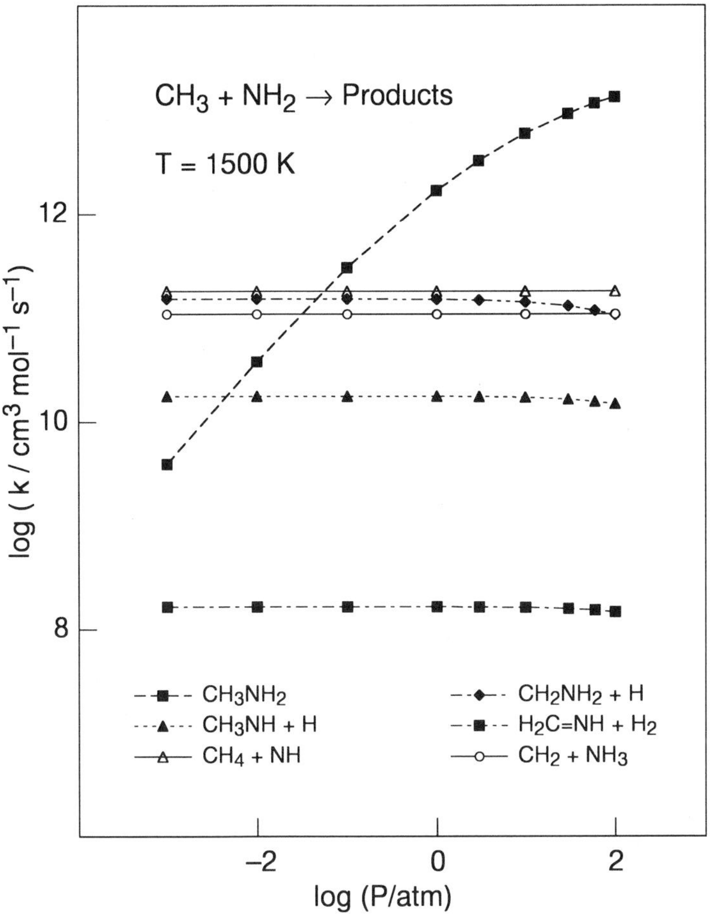

FIGURE 2.46. Effect of pressure at 1500 K predicted by QRRK calculations for CH_3 + $NH_2 \longrightarrow$ Products. Even at this relatively high temperature, stabilization dominates at N_2 buffer gas pressures above 0.1 atm.

The DHT method described in Section 2.4 leads to the rate coefficient expressions, in $cm^3mol^{-1}s^{-1}$ units,

$$k_{21e} = 2.8 \times 10^6\, T^{1.94} \exp(-4635/T)$$
$$k_{21f} = 1.6 \times 10^6\, T^{1.87} \exp(-3810/T)\,.$$

These estimates are included in Figures 2.45 and 2.46. It can be seen that these channels, together with the chemical activation channel producing CH_2NH_2 and H, dominate at high temperatures.

2.6.18 $CH_2 + N_2 \longrightarrow$ Products

The methylene radical reacts differently in its singlet $^1CH_2$ and triplet $^3CH_2$ forms. The more stable (by 38 kJ/mol) triplet form is commonly designated just as CH_2.

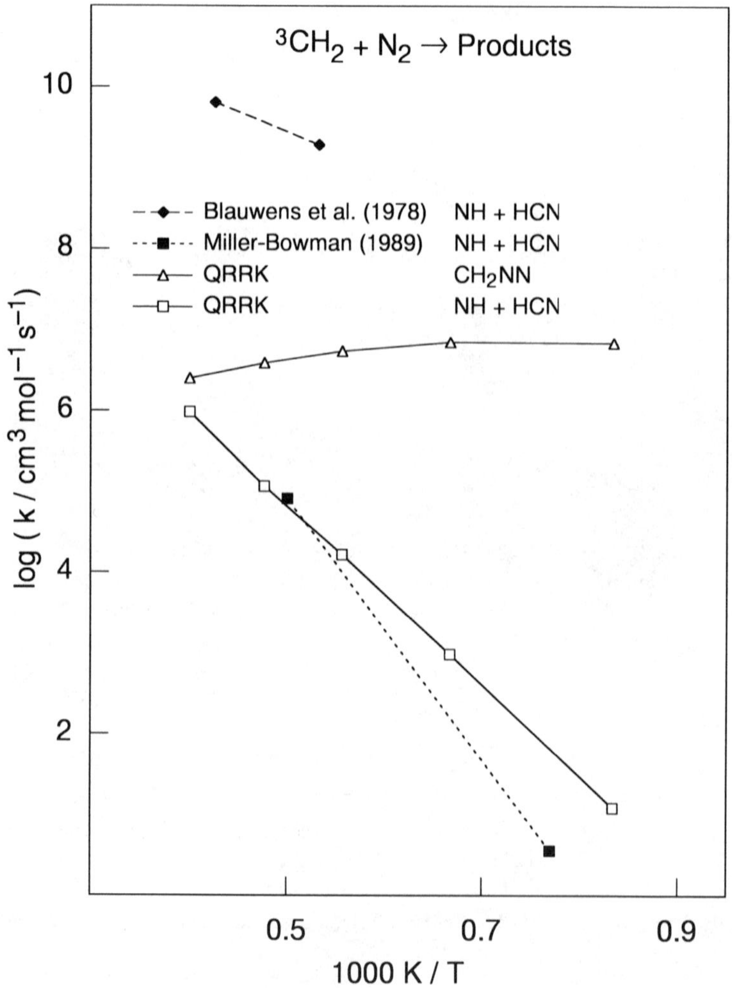

FIGURE 2.47. Comparison of QRRK predictions for $^3CH_2 + N_2 \longrightarrow$ Products to earlier rate coefficient estimates. The bath gas was assumed to be 1 atm N_2.

Attack of triplet methylene on molecular nitrogen was postulated as one of the sources of Prompt NO in rich hydrocarbon flames by Blauwens et al. (1978) and Hayhurst and Vince (1980). However, the kinetics and energetics of such flames make it difficult to distinguish experimentally between this reaction and CH + N_2. Miller and Bowman (1989) elected to use a much smaller rate coefficient than Blauwens et al., as shown in Figure 2.47. Figure 2.48 illustrates the

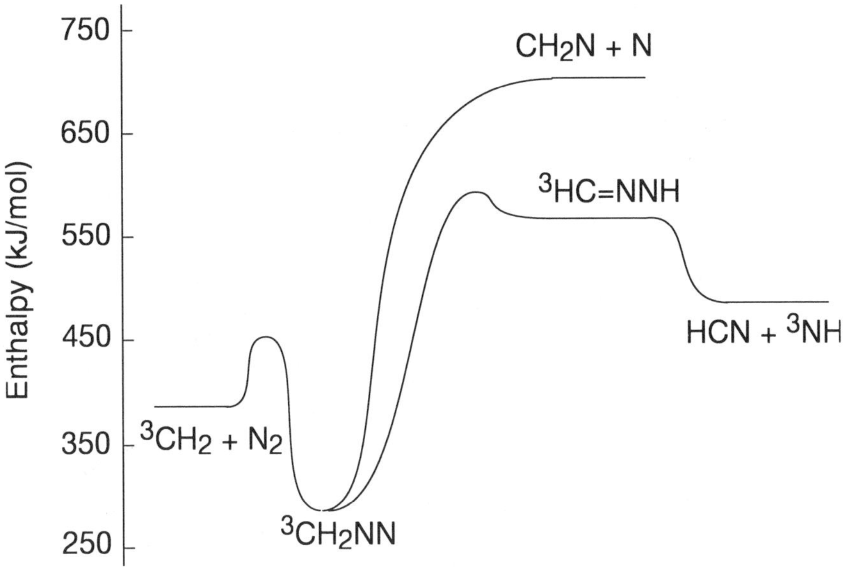

FIGURE 2.48. Potential energy diagram for $^3CH_2 + N_2$. All products are separated from the reactants by large energy barriers.

potential energy level diagram for the reaction based upon the parameters listed in Table 2.14. There are four channels for dissociation of the initially formed adduct in addition to dissociation back to the initial reactants. They include:

$$
\begin{aligned}
CH_2 + N_2 &\longrightarrow CH_2N_2 && (22a)\\
&\longrightarrow H_2CN + N && (22b)\\
&\longrightarrow HC{=}NNH && (22c)\\
&\longrightarrow HCN + NH\,. && (22d)
\end{aligned}
$$

There is an initial barrier of 66 kJ/mol for addition and a relatively shallow well, about 160 kJ/mol, below the top of the barrier to CH_2N_2. Subsequent isomerization of CH_2N_2 requires about 220 kJ/mol relative to the reactants. Although isomerization by H atom migration would rapidly lead to dissociation of the isomerization product and formation of HCN and ^3NH, the high barrier to isomerization would be expected to block this channel. Thus one expects the most likely fates of the initially formed adduct to be stabilization and/or dissociation back to reactants.

TABLE 2.14. $^3CH_2 + N_2 \longrightarrow$ Products.

	Reaction			A/s^{-1} or cm^{3}mol^{-1}s^{-1}	E_a/kJ mol^{-1}
(1)	$^3CH_2 + N_2$	$\longrightarrow$	$^3CH_2NN$	2.0×10^{12}	66
(−1)	$^3CH_2NN$	$\longrightarrow$	$^3CH_2 + N_2$	4.6×10^{14}	159
(2)	$^3CH_2NN$	$\longrightarrow$	$^3CH_2N + N$	7.0×10^{15}	429
(3)	$^3CH_2NN$	$\longrightarrow$	^{3}HC=N–NH	1.1×10^{13}	314
(−3)	^{3}HC=N–NH	$\longrightarrow$	$^3CH_2NN$	4.3×10^{11}	25
(4)	^{3}HC=N–NH	$\longrightarrow$	HCN + ^{3}NH	1.8×10^{13}	7

Molecular constants of $^3CH_2NN$ adduct: $\nu_1 = 616$ cm^{-1} (degen=2.92) $\nu_2 = 1163$ cm^{-1} (3.64) $\nu_3 = 2706$ cm^{-1} (2.45); molecular constants of ^{3}HC=N–NH adduct: $\nu_1 = 342$ cm^{-1} (degen=2.45) $\nu_2 = 401$ cm^{-1} (0.59) $\nu_3 = 1300$ cm^{-1} (5.46); $\sigma = 3.49$ Å and $\varepsilon/K = 298$, from CH_2CO, for both adducts.

Rate coefficients: k_1 from $^3CH_2 + O_2$, Böhland et al. (1984) and Darwin et al. (1989), E_a from Melius (1988); k_{-1} from microscopic reversibility (MR); k_2 via k_{-2} and MR with $k_{-2} = 7.1\times10^{13}$, based on N + NH_2 of Whyte and Phillips (1983); k_3 TST, with loss of hindered rotor, degeneracy = 2, and $E_a = \Delta_r H^\circ + 25$ kJ/mol; k_{-3} MR; k_4 from k_{-4} and MR with $k_{-4} = k_1$.

The QRRK parameters listed in Table 2.14 lead to the results compared to the Blauwens et al. and Miller and Bowman expressions in Figure 2.47. As expected, formation of the stabilized adduct CH_2N_2 is predicted to be the dominant channel, but even that channel is well below the expression obtained by Blauwens et al. Moreover, the stabilization rate is deceiving, because this adduct dissociates primarily by reverse reaction back to $CH_2 + N_2$, which must be considered in any mechanism where the forward reaction is included. The equilibrium constant shifts this system toward the dissociation channel at higher temperature, suppressing the contribution of the stabilization channel. The NH + HCN channel is much slower, close to the expression proposed by Miller and Bowman. The QRRK analysis indicates that the results at other pressures (other than for the stabilization channels) are similar to those shown in Figure 2.47 for 1 atm N_2. Thus the QRRK calculation is in agreement with the proposal of Miller and Bowman that this rate coefficient is much smaller than would be needed to explain Prompt NO formation as a result of triplet methylene reacting with nitrogen. As a consequence, the data of Blauwens et al. are probably best interpreted in terms of Prompt NO being formed in the CH + N_2 reaction. We recommend using the QRRK results for the stabilization channel and using the Miller–Bowman expression for reaction (22d), giving in cm^{3}mol^{-1}s^{-1} units for N_2 bath gas

$$\begin{aligned} k_{22a} &= 9.3\times10^{30}T^{-7.01}\exp(-9935/T) && 0.1\text{ atm} \\ &= 1.6\times10^{32}T^{-7.07}\exp(-10050/T) && 1.0\text{ atm} \\ &= 4.3\times10^{33}T^{-7.18}\exp(-10500/T) && 10\text{ atm} \\ k_{22d} &= 1.00\times10^{13}\exp(-37240/T)\,. \end{aligned}$$

An energy diagram for the reaction of $^1CH_2$ with N_2 is shown in Figure 2.49.

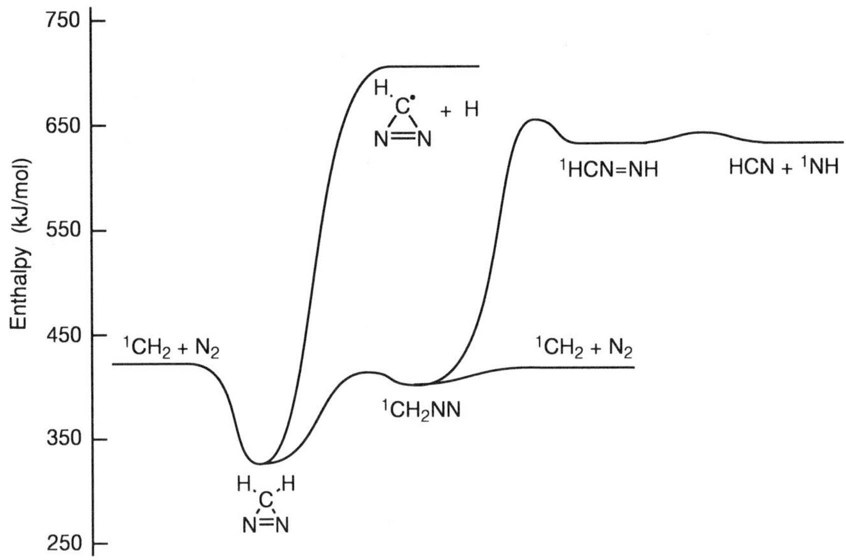

FIGURE 2.49. Potential energy diagram for $^1CH_2 + N_2$. Formation of all distinguishable products is very endothermic.

It forms a cyclic intermediate complex having a shallow well, about 100 kJ/mol, and dissociation back to reactants is its dominant unimolecular reaction. There is some stabilized adduct production at higher pressures, but the shallow well and the high A-factor for dissociation, about 10^{16} s^{-1}, imply a half-life of less than 10 ps at 1000 K. Considering the probable concentrations of radical species such as OH, CH_3, O, or H which could attack this adduct, further important bimolecular reactions of this intermediate are unlikely.

2.6.19 $^3CH_2 + NO \longrightarrow$ Products

Until recently, rate coefficient measurements for this reaction were confined to low temperatures (Laufer and Bass 1974; Vinckier and DeBruyn 1979; Darwin et al. 1989; Seidler et al. 1989). In these experiments the product distributions were not determined, although Seidler et al. concluded that OH and H were not the main products, that is, less than 10% of the total. In later work, Atakan et al. (1992) extended the temperature range up to 1025 K and also measured the rate of production of some of the reaction products. In particular, they reported that the HCN + OH channel contributes virtually nothing at 300 K but increases to 63 ±25% of the total products at 1025 K. In contrast to the conclusions of Seidler et al., Grussdorf et al. (1997) report that the dominant channel produces HCNO + H (about 84%), with HCN + OH accounting for about 15%. Bauerle et al. (1995) measured H, O, and OH appearance rates in shock waves over the temperature range from 1100 to 2600 K and concluded that HCNO + H and HCN + OH are

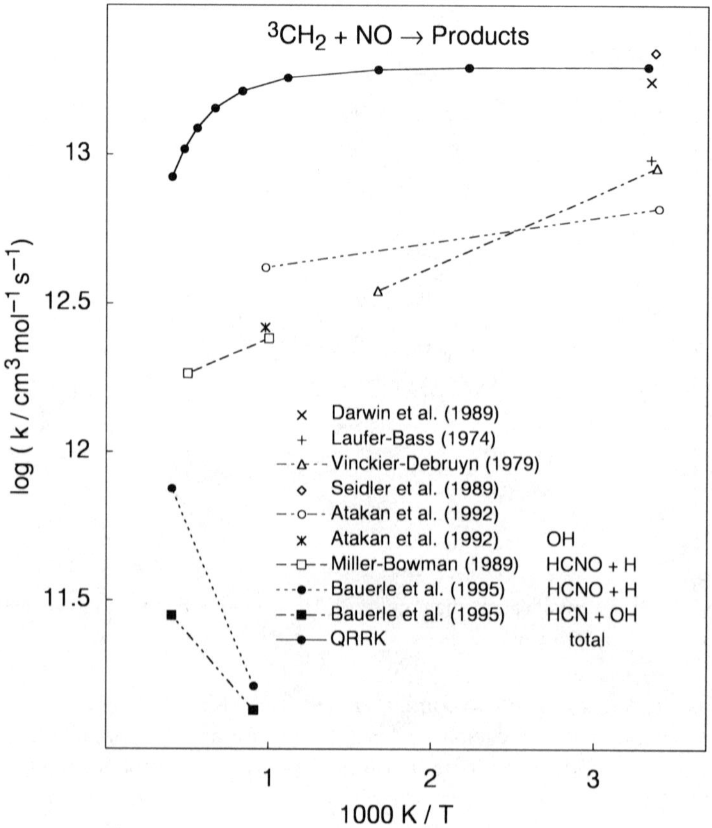

FIGURE 2.50. Comparison of QRRK predictions for the total rate coefficient to measurements for $^3CH_2 + NO \longrightarrow$ Products.

major product channels, with both rate coefficients increasing with temperature. These results, along with the expression used by Miller and Bowman in their flame modeling, are shown in Figure 2.50. A potential energy diagram for the reaction is shown in Figure 2.51. Possible channels include:

$$^3CH_2 + NO \longrightarrow CH_2NO \quad (23a)$$
$$\longrightarrow CyH_2CON \quad (23b)$$
$$\longrightarrow HNCHO \quad (23c)$$
$$\longrightarrow NCHOH \quad (23d)$$
$$\longrightarrow H_2NCO \quad (23e)$$
$$\longrightarrow HCNOH \quad (23f)$$
$$\longrightarrow HCNO + H \quad (23g)$$
$$\longrightarrow HOCN + H \quad (23h)$$

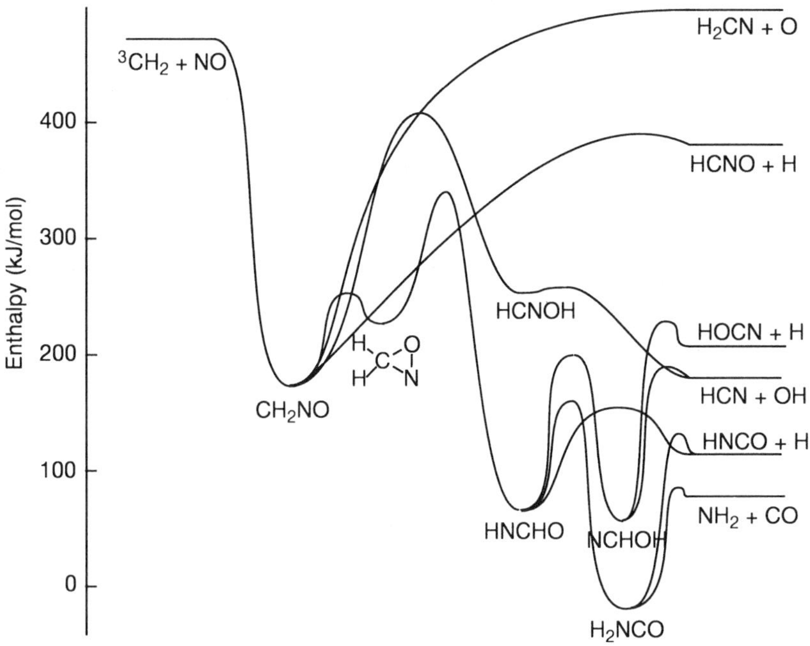

FIGURE 2.51. Potential energy diagram for $^3CH_2 + NO$.

$$\longrightarrow HCN + OH \tag{23i}$$

$$\longrightarrow HNCO + H \tag{23j}$$

$$\longrightarrow NH_2 + CO \tag{23k}$$

$$\longrightarrow H_2CN + O\,. \tag{23l}$$

The adduct CH_2NO well is quite deep, about 300 kJ/mol, and all reaction barriers except the one to $H_2CN + O$ are lower in energy than the entrance channel. Thus, one expects that the unimolecular reactions of the initially formed adduct will be rapid and stabilization will be minimal. The CH_2NO adduct can dissociate to HCNO + H, dissociate to $H_2CN + O$, or isomerize. There is a high-energy pathway to HCN + OH via HCNOH. Alternatively, there is a lower energy pathway via the cyclic intermediate CyH_2CON to HNCHO. This in turn can isomerize further to NCHOH or H_2NCO, or dissociate to HNCO + H. Finally, these isomers can dissociate to the indicated products. The QRRK parameters used are listed in Table 2.15.

The predicted rate coefficient for loss of reactant is compared to the observations in Figure 2.50. The predictions show less of a negative temperature dependence than observed at low temperatures, but a sharper falloff near combustion temperatures. The magnitude of the rate coefficient is fixed by the rate of adduct formation, and we have chosen this to match the measurements of Seidler et al. and Darwin et al.

Figure 2.52 shows QRRK predictions for the various channels at 1 atm pressure.

TABLE 2.15. $^3CH_2 + NO \longrightarrow$ Products.

	Reaction			A/s^{-1} or $cm^3mol^{-1}s^{-1}$	E_a/kJ mol^{-1}
(1)	$CH_2 + NO$	$\longrightarrow$	CH_2NO	2.0×10^{13}	0
(-1)	CH_2NO	$\longrightarrow$	$CH_2 + NO$	7.7×10^{15}	297
(2)	CH_2NO	$\longrightarrow$	$H + HCNO$	9.3×10^{13}	229
(3)	CH_2NO	$\longrightarrow$	CyH_2CON	1.5×10^{13}	79
(-3)	CyH_2CON	$\longrightarrow$	CH_2NO	1.9×10^{13}	25
(4)	CyH_2CON	$\longrightarrow$	HNCHO	9.0×10^{13}	121
(-4)	HNCHO	$\longrightarrow$	CyH_2CON	5.2×10^{13}	279
(5)	HNCHO	$\longrightarrow$	HNCO + H	7.2×10^{13}	84
(6)	HNCHO	$\longrightarrow$	NCHOH	1.7×10^{13}	130
(-6)	NCHOH	$\longrightarrow$	HNCHO	7.3×10^{12}	141
(7)	HNCHO	$\longrightarrow$	H_2NCO	1.5×10^{13}	88
(-7)	H_2NCO	$\longrightarrow$	HNCHO	2.6×10^{12}	182
(8)	NCHOH	$\longrightarrow$	HCN + OH	1.8×10^{14}	120
(9)	NCHOH	$\longrightarrow$	HOCN + H	1.8×10^{13}	157
(10)	H_2NCO	$\longrightarrow$	$NH_2 + CO$	3.5×10^{12}	123
(11)	H_2NCO	$\longrightarrow$	HNCO + H	1.2×10^{13}	149
(12)	CH_2NO	$\longrightarrow$	HCNOH	4.4×10^{13}	237
(-12)	HCNOH	$\longrightarrow$	CH_2NO	8.4×10^{12}	151
(13)	HCNOH	$\longrightarrow$	HCN + OH	1.1×10^{14}	15
(14)	CH_2NO	$\longrightarrow$	$H_2CN + O$	3.8×10^{15}	318

The QRRK frequencies in cm^{-1} units (ν_1,ν_2,ν_3) and degeneracies for the adducts are CH_2NO: (616,2.92; 1163,3.64; 2706, 2.45); CyH_2CON: (819,3.76; 823,2.45; 2868,2.79); HNCHO: (251,0.92; 910,5.67; 3450,2.40); NCHOH: (250,0.88; 903,5.70; 3448,2.41); H_2NCO: (251,1.68; 1016,4.36; 2744,2.46); HCNOH: (350,3.32; 1473,4.43; 3996,1.24); σ = 3.97 Å and ε/K = 436.

Rate coefficients: k_1 average of Darwin et al. (1989) and Siedler et al. (1989); k_{-1} from microscopic reversibility (MR); k_2 MR with $A_{-2} = 1.5\times10^{13}$ from H + C_3H_7, Allara et al. (1980), $E_a(-2)$ = 9 kJ/mol from H + C_2H_4; k_3 TST with loss of hindered rotor, $\Delta S^\ddagger$ = –9.3 J/K, $E_a = \Delta_r H^\circ$ + 25 kJ/mol; k_{-3} MR; k_4 TST with 1,2 H-shift, $\Delta S^\ddagger = 0$, degeneracy of 2, E_a = ring strain of 113 kJ/mol + E_a(abstraction) of 8 kJ/mol; k_{-4} MR; k_5 MR with $A_{-5} = 1.5\times10^{13}$ from 0.5(H + C_2H_4) and barrier from Melius (1988)—cf. text; k_6 TST, $\Delta S^\ddagger$ = –16.7 J/K and E_a from Melius (1988)—cf. text; k_{-6} MR; k_7 TST, $\Delta S^\ddagger$ = –9.2 J/K, E_a from Melius (1988)—cf. text; k_{-7} MR; k_8 MR with $k_{-8} = 1.2\times10^{13}\exp(-1060/T)$ from Miller et al. 1986; k_9 MR with $k_{-9} = 1.5\times10^{13}\exp(-1060/T)$, from 0.5(H + C_2H_4); k_{10} MR with $k_{-10} = 5.0\times10^{11}\exp(-3300/T)$, from CH_3 + CO (Anastasi et al. 1982); k_{11} MR with $k_{-11} = 1.5\times10^{13}\exp(-1060/T)$, from 0.5(H + C_2H_4); k_{12} TST with $\Delta S^\ddagger$ = –6.2 J/K, degeneracy of 2, E_a from Melius (1988)—cf. text; k_{-12} MR; k_{13} MR with $A_{-13} = 1.5\times10^{13}$ (Miller et al. 1986) and barrier from Melius (1988); k_{14} MR with k_{-14} = 8.0×10^{13} (Tsang et al. 1986).

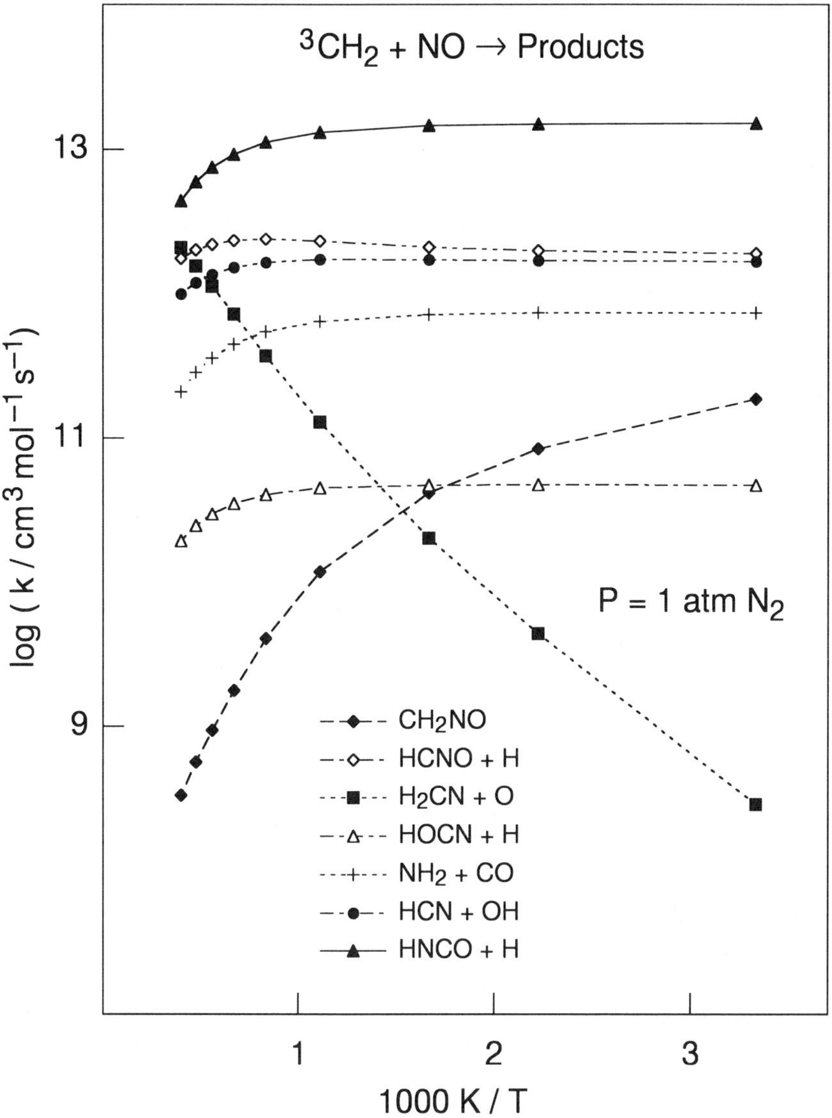

FIGURE 2.52. Arrhenius plot for the rate coefficients of various channels of $^3CH_2$ + NO ⟶ Products as predicted by QRRK calculations for 1 atm N_2 buffer gas. Stabilization is relatively unimportant and many different products are expected.

The major one at all temperatures is predicted to be HNCO + H, with most of it coming directly from HNCHO rather than through isomerization to H_2NCO followed by dissociation. The next most important channel is HCNO + H, followed by HCN + OH and, at high temperatures, by H_2CN + O.

Grussdorf et al., however, report that HCNO is the dominant channel. The potential energy curves shown in Figure 2.51 suggest that decreased production of HNCO relative to HCNO would be favored by an increased isomerization barrier for the conversion of the cyclic H_2CON to HNCHO, that is, decreasing the rate of reaction (4) in Table 2.15. For HCNO production to dominate, however, this barrier would have to be increased by more than 40 kJ/mol.

It is even more difficult to reconcile a non-H or non-OH channel dominating at low temperature, as reported by Seidler et al.; the only likely such channel at low temperature would be formation of NH_2 and CO. For this pathway to dominate, the barrier between HNCHO and HNCO + H must be substantially higher, or the one between HNCHO and H_2NCO much lower, than computed. The A-factor for dissociation is expected to be substantially larger than for isomerization, so the only plausible way to increase the selectivity to favor NH_2 and CO production would be by adjusting these barriers. Adjustments of other barriers were also tested. For example, increasing the barrier for isomerization of HNCHO to H_2NCO by 50 kJ/mol decreases the production rate of NH_2 and CO by only about a factor of two, with little effect on the other channels. Similarly, increasing the barrier for dissociation of HNCHO to HNCO and H by 20 kJ/mol has virtually no effect at all. Given the structure of the Figure 2.51 potential energy diagram, such small changes are to be expected. In all cases, the energy of the initially formed CH_2NO adduct is so much higher than the barriers that only small changes can be effected by adjusting them.

Near 1000 K, we find that the branching ratio to OH + HCN is only about 14%, consistent with the report of Grussdorf et al., and well below that reported by Atakan et al. One would need very large changes in the barriers to allow OH + HCN to be the major channel. For example, lowering the barriers for both $CH_2NO \longrightarrow$ HCNOH and HNCHO $\longrightarrow$ NCHOH (these are the rate-limiting steps for both the OH + HCN channels) by 21 kJ/mol increases the importance of this channel by about a factor of two, but HNCO + H is still dominant. At higher temperatures, O + H_2CN becomes important. It is thus difficult to reconcile the QRRK calculations with the high-temperature observations of Bauerle et al.; the only channel with a positive temperature dependence was found in the computations to be the one leading to H_2CN and O. The channels producing H and OH all become slower with increasing temperature.

Reaction exothermicity is not the controlling factor for any of the channels. For example, one sees a substantial contribution from HCNO + H, a relatively high energy channel. This can be traced to the fact that this channel is directly formed from the initially formed adduct, and the resulting large A-factor (relative to isomerization) offsets the larger energy barrier. Similar considerations apply to O + H_2CN. Conversely, the lowest energy channel, NH_2 + CO, is only accessible via a series of isomerizations, each with a relatively tight transition state, making it relatively unimportant. The most important stabilization channel is insignificant compared to the dissociation channels, even at 1 atm pressure and low temperature.

Because stabilization is found to be relatively unimportant, the effect of pressure is small. The calculations indicate that pressures well above 10 atm are required for stabilization to be significant even at low temperatures. The negative temperature dependence is also problematic. Our analysis (F2.ig. 50) suggests a more gradual decline at lower temperatures—followed by a steeper drop as one gets into the falloff region—than was observed by Vinckier et al. The work of Atakan et al., however, shows a temperature dependence closer to our calculations. We are reluctant to use QRRK analysis to revise a measured temperature dependence. It would be helpful if this rate coefficient could be remeasured over a wider temperature range, particularly at temperatures above 1000 K. In the interim, we suggest use of the measured temperature dependence of Atakan et al., but with a larger A-factor to reflect the faster room-temperature measurements of Siedler et al. and Darwin et al.; that is,

$$k_{23}(\text{total}) = 1.0 \times 10^{13} \exp(+190/T) \ \text{cm}^3\text{mol}^{-1}\text{s}^{-1}$$

for moderately high temperatures and pressures of less than 10 atm, where stabilization is unimportant. In view of the differences between the calculations and the available data, we reluctantly recommend, in $\text{cm}^3\text{mol}^{-1}\text{s}^{-1}$ units,

$$\begin{aligned}
k_{23\text{g}} &= 3.8 \times 10^{13} T^{-.36} \exp(-290/T) \\
k_{23\text{i}} &= 2.9 \times 10^{14} T^{-.69} \exp(-380/T) \\
k_{23\text{j}} &= 3.1 \times 10^{17} T^{-1.38} \exp(-640/T) \\
k_{23\text{k}} &= 2.3 \times 10^{16} T^{-1.43} \exp(-670/T) \\
k_{23\text{l}} &= 8.1 \times 10^{7} T^{1.42} \exp(-2070/T) .
\end{aligned}$$

There is a clear need for further experimental and theoretical work on the $^3CH_2$ + NO reactions.

2.6.20 CH + N_2 ⟶ Products

The reaction of CH with N_2, in flames and in other high-temperature oxidation systems, is important in coupling hydrocarbon oxidation chemistry with nitrogen chemistry. By forming HCN + N it generates two species that can both react further to form NO. As discussed in Section 2.6.18, it is unlikely that the CH_2 + N_2 reaction plays a significant role in NO formation, leaving CH + N_2 as the main route to Prompt NO. Because CH tends to react via insertion, forming two new bonds, one has to account for chemical activation effects in these very exothermic reactions.

The experimental rate coefficient measurements are shown in Figure 2.53. Berman and Lin (1983) studied the pressure and temperature dependence and interpreted their results in terms of adduct formation. An RRKM analysis was used to describe the observed falloff in rate with increasing temperature. Becker

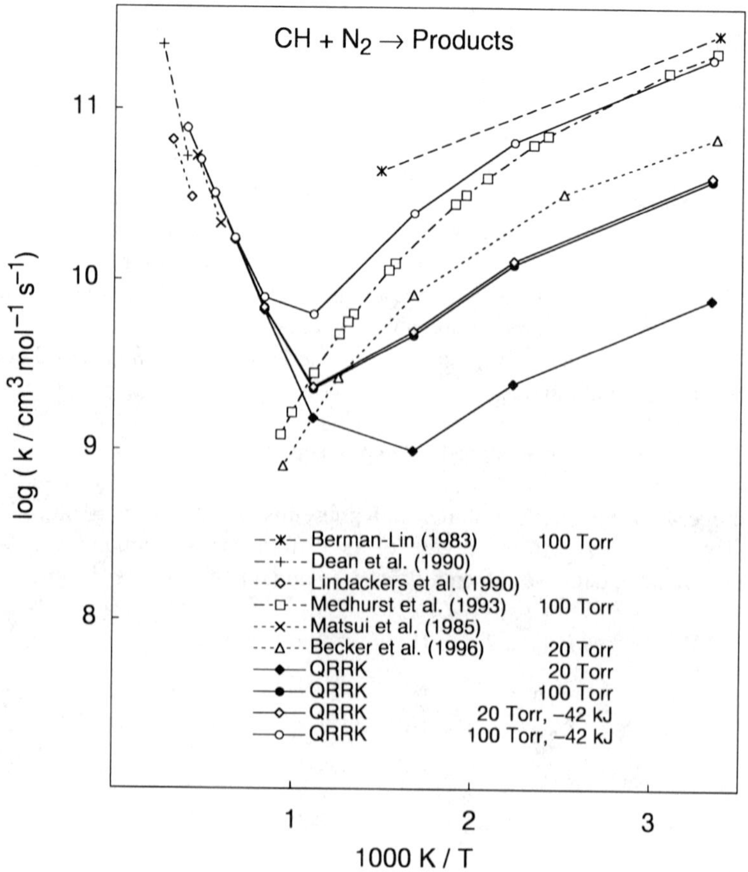

FIGURE 2.53. Comparison of QRRK predictions to rate coefficient measurements for CH + $N_2 \longrightarrow$ Products in Ar buffer gas. Two sets of QRRK calculations, differing in the well depths assigned to the adducts, are shown. Assigning the adduct well depths calculated by Manaa and Yarkony (1991, 1992) substantially underpredicts the low temperature data. Assigning wells 42 kJ/mol deeper results in rate coefficient values only slightly lower than observed at low temperature. Both sets correctly predict the increasing rate coefficient at higher temperatures.

et al. (1989) measured the CH decay rate at room temperature and 2 Torr. The measurements were later extended to higher temperatures (Becker et al. 1992a; Medhurst et al. 1993; Becker et al. 1996). The data are consistent, showing both the temperature and the pressure dependences expected for an adduct formation mechanism. Although the first report of Becker et al. (1992a) showed a definite minimum near 700 K, their 1996 data (included in Figure 2.53) do not; the authors consider the earlier data at temperatures over 560 K to be in error.

The higher temperature measurements (Matsui and Yuuki 1985; Dean et al. 1990; Lindackers et al. 1990) all show increases in rate coefficient with increasing

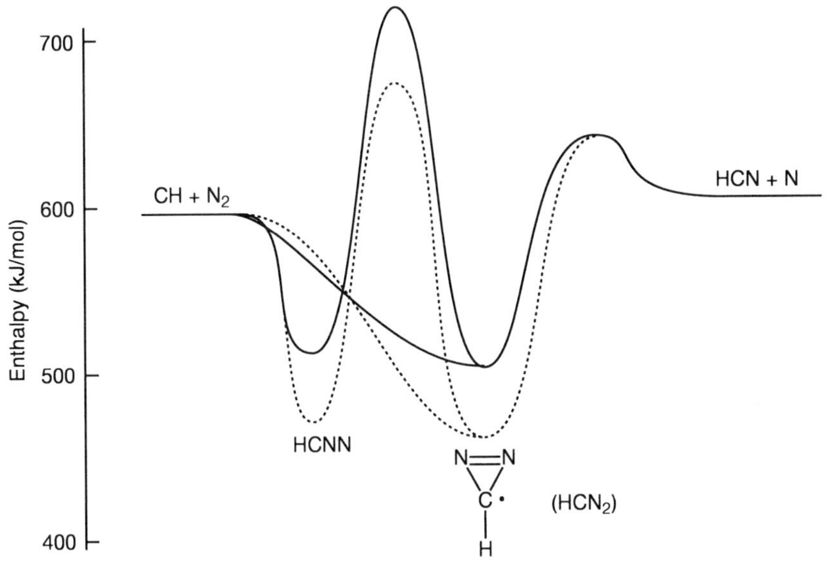

FIGURE 2.54. Potential energy diagram for CH + N_2. The solid lines are based on Manaa and Yarkony (1991, 1992), and a barrier between adducts from Martin and Taylor (1993). The dashed lines show the result of making both adducts more stable by 42 kJ/mol. For QRRK calculations it was assumed that both adducts can be reached directly from the reactant channels.

temperature. As pointed out by Berman and Lin, this behavior is consistent with formation of an energized adduct HCNN that can be collisionally stabilized, dissociate back to reactants (causing the falloff in rate at intermediate temperatures), or at higher temperatures dissociate via an endothermic channel.

The initially formed adduct has a doublet electronic state, and so there must be curve crossing to a quartet surface to form HCN + N. This crossing was studied theoretically by Manaa and Yarkony (1991; 1992), who report that the curve-crossing barrier is about 40 kJ/mol above the entrance channel. They also conclude that there are two adduct isomers, an acyclic HCNN and a cyclic insertion species HCN_2, which have well depths of about 92 and 100 kJ/mol respectively (Figure 2.54). Calculations by Martin and Taylor (1993) show a substantial (about 235 kJ/mol) barrier between the isomers. Others (Walch 1993c; Seideman and Walch 1994; Seideman 1994) indicate a significant entrance channel barrier for HCN_2 formation. We did not attempt to account for this feature. An additional complication is that Manaa and Yarkony report that the probability for crossing from the doublet to the quartet surface is much less than unity. Rodgers and Smith (1996) analyzed the system in terms of two separate reactions, a low-temperature one forming the HCNN adduct with no entrance channel barrier and a high-temperature one forming HCN and N through the cyclic HCN_2 structure. They assumed that the barrier to form HCN_2 is rate limiting.

Earlier attempts to characterize the well depth gave widely different results; Miller and Bowman (1989) use 75 kJ/mol, while Berman and Lin use 222 kJ/mol. Medhurst et al. (1993) found that a 120 kJ/mol well was consistent with their data. Rodgers and Smith found that a well depth in the range from 105 to 145 kJ/mol, depending on the choice of collision efficiency, gave the best fits to experiment. The dashed lines in Figure 2.54 show adjustments of the well depths made in attempts to fit the falloff data (see below).

TABLE 2.16. $CH + N_2 \longrightarrow$ Products.

	Reaction	A/s^{-1} or $cm^3 mol^{-1} s^{-1}$	E_a/kJ mol^{-1}
(1a)	$CH + N_2 \longrightarrow HCNN$	1.3×10^{13}	0
(-1a)	$HCNN \longrightarrow CH + N_2$	1.6×10^{14}	82, 124
(1b)	$CH + N_2 \longrightarrow HCN_2$	1.3×10^{13}	0
(-1b)	$HCN_2 \longrightarrow CH + N_2$	1.4×10^{15}	92, 134
(2)	$HCNN \longrightarrow HCN_2$	1.1×10^{13}	209
(-2)	$HCN_2 \longrightarrow HCNN$	9.7×10^{13}	220
(3)	$HCN_2 \longrightarrow HCN + N$	8.0×10^{13}	140, 182

Molecular constants of HCNN: $\nu_1 = 388$ cm^{-1} (degen=3.02) $\nu_2 = 900$ cm^{-1} (0.19) $\nu_3 = 1895$ cm^{-1} (2.79); molecular constants of HCN_2: $\nu_1 = 414$ cm^{-1} (degen=2.02) $\nu_2 = 1612$ cm^{-1} (3.33) $\nu_3 = 3998$ cm^{-1} (0.65); $\sigma = 3.49$ Å and $\varepsilon/K = 300$ for both adducts.

Rate coefficients: k_{1a} Berman and Lin (1983); k_{-1a} from microscopic reversibility (MR), based on Manaa et al. (1992), varied to fit falloff data; k_{1b} same as k_{1a}; k_{-1b} MR based on Manaa et al. (1992), varied to fit falloff data; k_2 TST, $\Delta S^\ddagger = -11.6$ J/K, E_a from Martin and Taylor (1993); k_{-2} MR; k_3 MR with A_{-3} for O + HCN from Perry and Melius (1984) and barrier from Manaa and Yarkony (1991), varied to fit falloff data.

The input parameters for the chemical activation calculation are listed in Table 2.16. Possible product channels include

$$CH + N_2 \longrightarrow HCNN \quad (24a)$$
$$\longrightarrow HCN_2 \quad (24b)$$
$$\longrightarrow HCN + N\,. \quad (24c)$$

Although the exit channel is uphill, it is energetically much more accessible than in the analogous $CH_2 + N_2$ case. Both adducts can form directly from the reactants. The isomerization barrier is sufficiently high to inhibit production of HCN + N unless the insertion complex can be formed directly. However, since CH is normally considered to insert, direct reaction is reasonable. We adjusted the well depth to fit the falloff data of Berman and Lin and Becker et al. The results of the QRRK calculations for total loss of reactants are compared to the available data in Figure 2.53. The QRRK predictions reflect the experimental observations in showing both pressure and temperature dependence at lower temperature. The dominant product channel is formation of stabilized adduct(s). The QRRK results underpredict the extent of stabilization when the well depths of Manaa and Yarkony are used. Setting the wells 42 kJ/mol deeper substantially increased the amount of stabilization; the results are then only slightly lower than the data of Berman and Lin at 100 Torr and Becker et al. at 20 Torr. This well depth is consistent with that assigned by Medhurst et al. and by Rodgers and Smith.

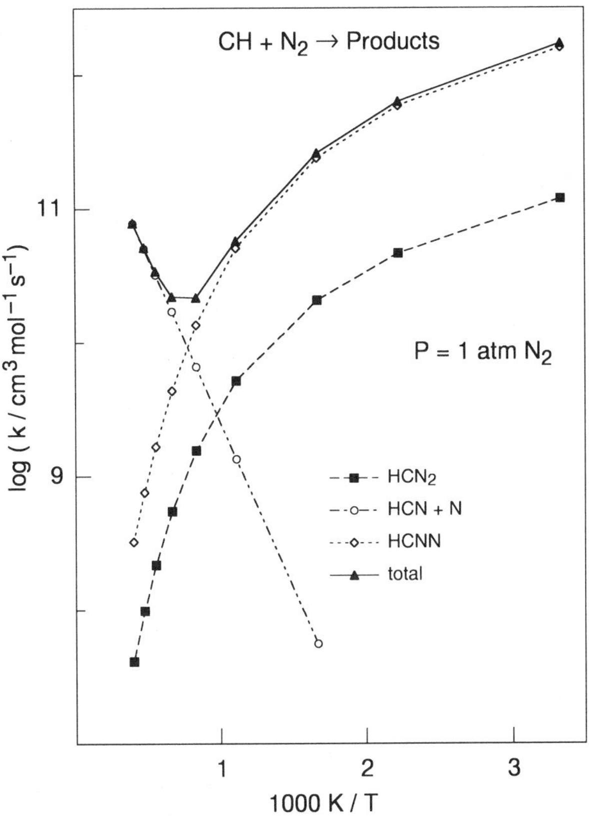

FIGURE 2.55. Arrhenius plot for the various channels of $CH + N_2 \longrightarrow$ Products in 1 atm N_2 as predicted by QRRK using the 42 kJ/mol deeper wells.

At higher temperatures, the calculations converge as the endothermic channel to HCN + N becomes accessible (Figure 2.55). This becomes the dominant channel, and the rate coefficient no longer exhibits a negative temperature dependence. The calculations predict this dominance to set in at a somewhat lower temperature than that observed in the experiments. This could be accommodated by adjusting the barrier for surface crossing. In view of the uncertainties in the surfaces and the probability for surface crossing, however, such adjustments are unwarranted. The observed kinetic behavior is in any event consistent with a general picture of adduct formation and stabilization at low temperature, an initial rate decrease with temperature as adduct dissociation becomes faster, and then a rate increase with

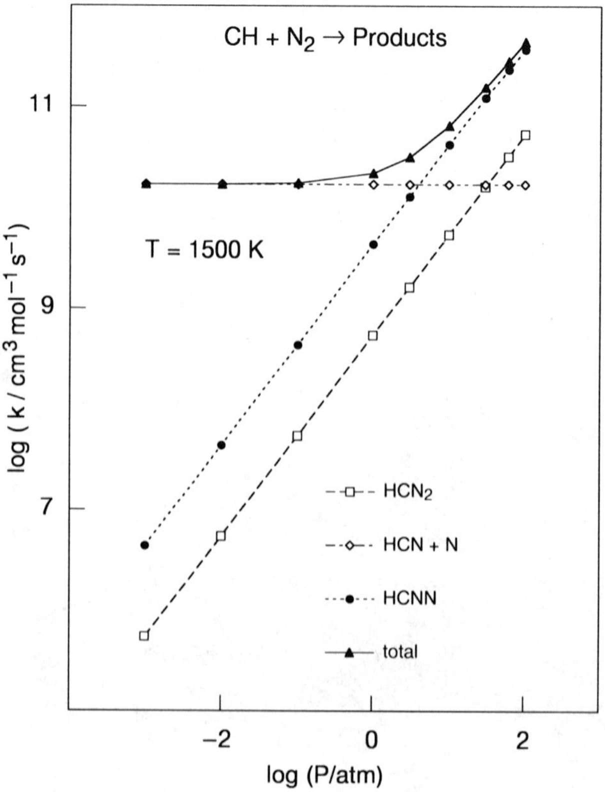

FIGURE 2.56. Predicted effect of pressure at 1500 K from QRRK calculations (using the 42 kJ/mol deeper wells) for CH + $N_2 \longrightarrow$ Products. Even at 1500 K, stabilization becomes more important than HCN + N above 3 atm.

temperature when the adduct has enough energy to make surface crossing and HCN + N formation significant. At very high pressures the stabilization channel is computed to continue to be important until significantly higher temperatures than those studied by Berman and Lin.

These effects are illustrated in Figures 2.55 and 2.56 for N_2 as bath gas. (The adjusted well depth was used for these calculations, as this gave much better agreement with the stabilization data.) Pressure effects could be important when including this reaction in kinetic models of high-pressure combustors; as shown in Figure 2.56, stabilization is important at pressures above 1 atm even up to 1500 K.

The disagreement between the two most direct high temperature measurements is disconcerting. Dean et al. (1991) reported a higher E_a value as well as a faster rate, a factor of 2 at 3150 K, than were found by Lindackers et al. As seen in Figure 2.56, one expects no pressure dependence for the HCN + N channel. We suggest using the expression of Dean et al.,

$$k_{24c} = 4.4 \times 10^{12} \exp(-11060/T) \ \text{cm}^3\text{mol}^{-1}\text{s}^{-1},$$

and the QRRK estimates for the stabilization channel, which are, in $\text{cm}^3\text{mol}^{-1}\text{s}^{-1}$

units, for 300 to 2500 K in N_2 bath gas:

$$\begin{aligned} k_{24a} &= 2.3 \times 10^{27} T^{-5.78} \exp(-1230/T) && 0.1 \text{ atm} \\ &= 3.6 \times 10^{28} T^{-5.84} \exp(-1320/T) && 1.0 \text{ atm} \\ &= 1.8 \times 10^{30} T^{-6.02} \exp(-1735/T) && 10 \text{ atm.} \end{aligned}$$

At temperatures above 1000 K, reaction 24a is close to the low pressure limit for pressures of 10 atm or less (cf. Figure 2.56 for 1500 K). The still-lower stabilization rate for formation of the cyclic complex can be ignored.

2.6.21 CH + NO $\longrightarrow$ Products

The reaction of CH with NO is important in flame chemistry because CH, along with other CH_x radicals, provides a path for NO destruction. Its contribution to NO destruction compared to that of other CH_x radicals is an important question that still needs to be resolved. Experimental data on this reaction are shown in Figure 2.57. Several groups have measured the rate at room temperature (Butler et al. 1981; Wagal et al. 1982; Lichtin et al. 1984; Okada et al. 1993), while two groups have measured the rate over extended ranges of low temperatures (Berman et al. 1982; Becker et al. 1993) . There is only one direct high-temperature measurement (Dean et al. 1991). All of these data present a consistent picture of a very rapid reaction with essentially no temperature dependence. Both Dean et al. and Okada et al. attempted to observe products of this reaction. In both cases, NH was confirmed to be a minor product (less than 10% and less than 15%, respectively); OH was observed to contribute less than 30% and less than 20%, respectively. These observations led Dean et al. to suggest that the major product is probably HCN + O, while Okada et al. were unable to rule out that NCO could be a major product. Miller and Bowman (1989) propose HCN + O as the sole products, but point out that yet another possible channel is CN + OH. Later work by Lambrecht and Herschberger (1994), using CD + NO, indicated that the major product channel is DCN + O (47.5±12.2%), with NCO + D at 18.8±5.5%, CN + OD less than 7.5%, and the sum of CO + DN and DCO + N at 33.7±13.8%.

Figure 2.58 shows the potential energy surface for reaction on the triplet surface. Addition of CH to NO forms ^{3}HCNO, which can form HCN + O via a very low exit barrier or isomerize to form ^{3}NCHO. NCHO has three dissociation routes, forming either N, O, or H and the corresponding triatomic partner. One expects these routes, with large A-factors and low-energy exit channels, to compete favorably with isomerization to HNCO or HOCN. If isomerization occurs, dissociation to H + NCO, NH + CO or OH + CN can follow. This plethora of products are all lower in energy than the reactants. The HCN + O channel is expected to dominate, because the direct route to these products from the first adduct has a low barrier and a high A-factor. The possible product channels encompass a rich variety of bond rearrangements:

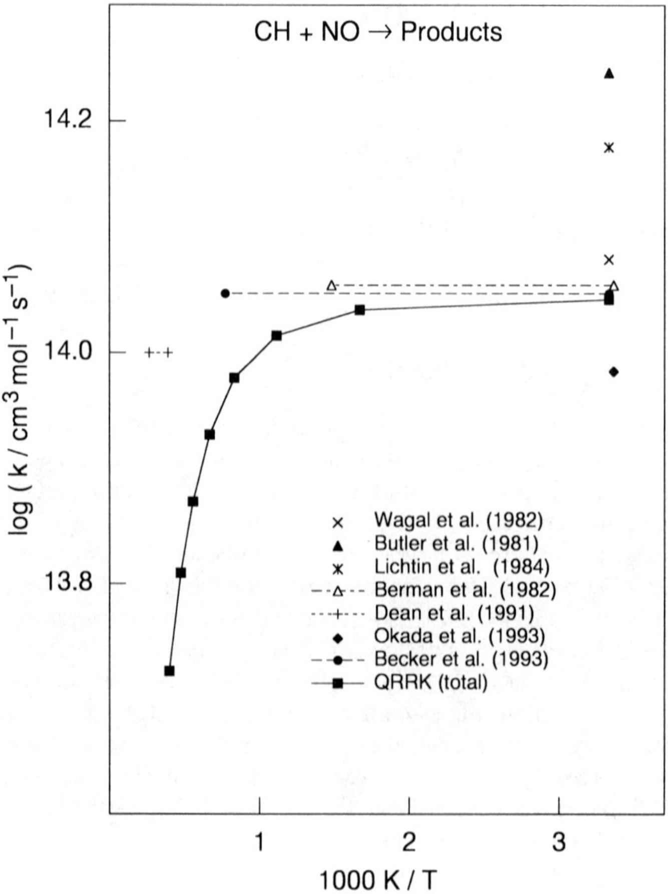

FIGURE 2.57. Comparison of QRRK predictions to rate coefficient measurements for the reaction CH + NO ⟶ Products.

$$
\begin{aligned}
CH + NO &\longrightarrow {}^{3}HCNO && (25a)\\
&\longrightarrow {}^{3}NCHO && (25b)\\
&\longrightarrow {}^{3}HNCO && (25c)\\
&\longrightarrow {}^{3}HOCN && (25d)\\
&\longrightarrow O + HCN && (25e)\\
&\longrightarrow H + NCO && (25f)\\
&\longrightarrow N + HCO && (25g)\\
&\longrightarrow NH + CO && (25h)\\
&\longrightarrow OH + CN\,. && (25i)
\end{aligned}
$$

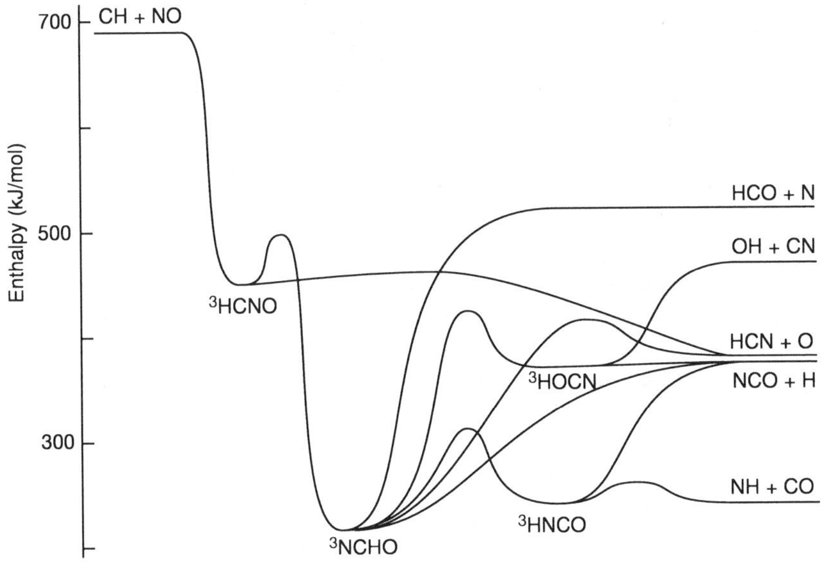

FIGURE 2.58. Potential energy diagram for CH + NO.

Table 2.17 describes the input parameters for this reaction on the triplet surface. In spite of the complexity of this system, several qualitative features are evident. First, the very low exit barriers suggest very little stabilization, implying little expected pressure dependence. Second, for the same reason, one expects the rate to be very fast. Third, one expects the distribution of products to be driven primarily by the A-factors for the unimolecular processes. Finally, one suspects the product distribution might be more complex than that used by Miller and Bowman.

These results are compared to the experimental measurements in Figure 2.57. Reasonable agreement is observed, although the calculations suggest more falloff than one would infer from the data of Dean et al. Figure 2.59 illustrates the predicted rate coefficients for the different pathways as functions of temperature.

For clarity only the sums of the three pathways leading to H + NCO and the two pathways to HCN + O are shown in Figure 2.59. (The major pathway for HCN + O is from HCNO; the major pathway for H + NCO is dissociation from NCHO, followed by HNCO and a small amount from HOCN.) In agreement with the observations of Lambrecht and Herschberger, the most important channel is predicted to be HCN + O, followed by HCO + N and NCO + H, with the diatomic product channels predicted to be least important. Overall, the agreement of these predictions with the detailed product measurements of Lambrecht and Herschberger is good. The only differences are that they observe slightly less O + HCN, 48% versus the QRRK prediction of 60%, and more H + NCO, 18% versus the QRRK prediction of 8%. Our predictions for the sum of the NH + CO and the HCO + N channels is 26% compared to the observed 34%, and our prediction of about 6% for the OH + CN channel is consistent with the upper limit measurement

TABLE 2.17. CH + NO $\longrightarrow$ Products (Triplet).

	Reaction	A/s^{-1} or $cm^3mol^{-1}s^{-1}$	E_a/kJ mol^{-1}
(1)	CH + NO $\longrightarrow$ ^{3}HCNO	1.1×10^{14}	0
(−1)	^{3}HCNO $\longrightarrow$ CH + NO	1.2×10^{16}	226
(2)	^{3}HCNO $\longrightarrow$ O + HCN	2.4×10^{13}	10
(3)	^{3}HCNO $\longrightarrow$ ^{3}NCHO	3.7×10^{13}	42
(−3)	^{3}NCHO $\longrightarrow$ ^{3}HCNO	5.0×10^{13}	272
(4)	^{3}NCHO $\longrightarrow$ O + HCN	4.7×10^{13}	192
(5)	^{3}NCHO $\longrightarrow$ H + NCO	3.6×10^{13}	170
(6)	^{3}NCHO $\longrightarrow$ ^{4}N + HCO	1.4×10^{15}	290
(7)	^{3}NCHO $\longrightarrow$ ^{3}HNCO	8.1×10^{13}	91
(−7)	^{3}HNCO $\longrightarrow$ ^{3}NCHO	7.8×10^{13}	67
(8)	^{3}HNCO $\longrightarrow$ NH + CO	1.4×10^{13}	21
(9)	^{3}HNCO $\longrightarrow$ H + NCO	3.4×10^{13}	143
(10)	^{3}NCHO $\longrightarrow$ ^{3}HOCN	8.1×10^{13}	205
(−10)	^{3}HOCN $\longrightarrow$ ^{3}NCHO	7.2×10^{13}	56
(11)	^{3}HOCN $\longrightarrow$ OH + CN	2.5×10^{15}	97
(12)	^{3}HOCN $\longrightarrow$ H + NCO	7.7×10^{13}	5

Molecular constants of ^{3}HCNO adduct: ν_1 = 417 cm^{-1} (degen=2.12) ν_2 = 1197 cm^{-1} (2.54) ν_3 = 2924 cm^{-1} (1.33); molecular constants of ^{3}NCHO adduct: ν_1 = 722 cm^{-1} (degen=2.75) ν_2 = 1724 cm^{-1} (2.86) ν_3 = 3990 cm^{-1} (0.39); molecular constants of ^{3}HNCO adduct: ν_1 = 396 cm^{-1} (degen=2.14) ν_2 = 1228 cm^{-1} (2.39) ν_3 = 3218 cm^{-1} (1.47); molecular constants of ^{3}HOCN adduct: ν_1 = 367 cm^{-1} (degen=1.78) ν_2 = 924 cm^{-1} (1.80) ν_3 = 2776 cm^{-1} (2.42); σ = 3.8 Å and ε/K = 232 for all adducts.

Rate coefficients: k_1 — 0.75 of the value measured by Berman and Lin (1983) and Becker et al. (1992b) to account for formation of the triplet product; k_{-1} from microscopic reversibilty (MR); k_2 from $A_{-2} = 6.0\times10^{12}$, Mallard et al. (1993), $E_a(2)$ assigned to be consistent with the reaction exothermicity; k_3 TST, $\Delta S^\ddagger$ = −4.1 J/K, E_a based on three-member ring closure; k_{-3} MR; k_4 from k_{-4} and MR with $k_{-4} = 6.0\times10^{12}$ exp(−3000/T) from Perry and Melius (1984); k_5 from k_{-5} and MR with k_{-5} based on H + C_2H_5 from Tsang and Hampson (1986); k_6 from k_{-6} and MR with k_{-6} taken as 4.0×10^{13}, half the rate coefficient used for N + CH_3 adduct formation (cf. Table 16); k_7 TST, $\Delta S^\ddagger$ = 4.1 J/K, partial gain of rotor, E_a for carbene H-shift; k_{-7} MR; k_8 from k_{-8} and MR with k_{-8} = 5.19×10^{11}exp(−3250/T), based on $^3CH_2$ + CO and CH_3 + CO, cf. Anastasi et al. (1982); E_a agrees with a BAC/MP4 calculation of C.F. Melius (1993); k_9 via k_{-9} and MR—cf. k_5; k_{10} TST, $\Delta S^\ddagger$ = 4.1 J/K, partial gain of rotor, E_a based on carbene H–shift; k_{-10} MR; k_{11} via k_{-11} and MR with k_{-11}= 4.0×10^{13} (Average of data reported by Mallard et al. 1993); k_{12} via k_{-12} and MR—cf. k_5.

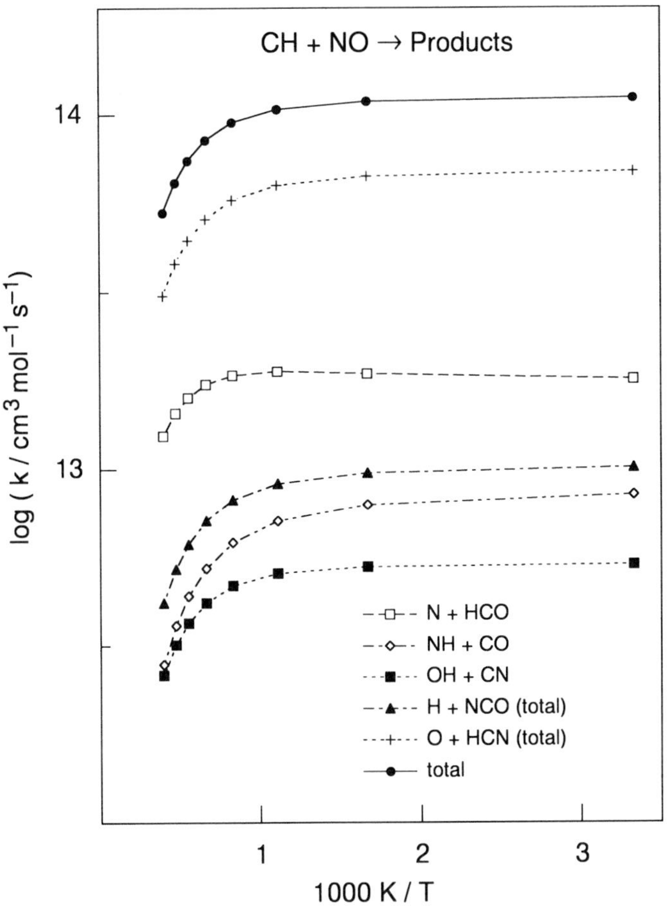

FIGURE 2.59. Arrhenius plot for the various channels of CH + NO ⟶ Products as predicted by QRRK calculations for 1 atm N_2 buffer gas. The low-energy exit channels make stabilization unimportant for this reaction; similarly, the effect of temperature is quite small.

of less than 7.5%. As discussed earlier, these results confirm that the energy of the exit channel is not necessarily the main factor influencing the rate. Here the highest-energy exit channel, HCO + N, is predicted to be much more important than the lowest, NH + CO. In this case, the higher barrier for HCO + N (272 kJ/mol higher than for NH + CO) is compensated by a much higher A-factor. The additional isomerization required to produce NH + CO lowers the rate along this path. The diminished role of energetic considerations in this reaction is made more pronounced by the low energies of all channels relative to the entrance channel.

Calculations over an extended pressure range confirm that pressure effects should be minimal for this system. The dissociation channels are found to dominate up to pressures over 100 atm even at 300 K. Thus one does not expect pressure to influence the branching ratios for this reaction.

Based on these considerations we suggest for the total rate coefficient

$$k_{25}(\text{total}) = 1.1 \times 10^{14}\ \text{cm}^3\text{mol}^{-1}\text{s}^{-1},\ \text{from 300 to 3800 K},$$

in agreement with all the measurements. Following the observations of Lambrecht and Herschberger we suggest that the following branching ratios be used:

$$\begin{aligned}
&(\text{O} + \text{HCN}) \quad k_{25\text{e}}/k_{25}(\text{total}) = 48\% \\
&(\text{H} + \text{NCO}) \quad k_{25\text{f}}/k_{25}(\text{total}) = 18\% \\
&(\text{N} + \text{HCO}) \quad k_{25\text{g}}/k_{25}(\text{total}) = 26\% \\
&(\text{NH} + \text{CO}) \quad k_{25\text{h}}/k_{25}(\text{total}) = 5\% \\
&(\text{OH} + \text{CN}) \quad k_{25\text{i}}/k_{25}(\text{total}) = 3\%.
\end{aligned}$$

The branching ratios vary only slightly with temperature, as can be seen from Figure 2.59, and so the ones listed can be applied at all temperatures.

We also examined reactions on the singlet surface. The slower overall rate of adduct formation leads to lower rate coefficients for all channels. Two new product channels become possible, namely ^{1}NH + CO and OH + CN; these are expected to contribute less than 10% to the overall reaction.

2.7 OTHER REACTIONS OF INTEREST

In this section we consider other reactions that are important in high-temperature nitrogen chemistry. While most of them have been included in earlier nitrogen chemistry mechanisms, we also discuss reactions of several species (in particular, NH_2O, H_2NN and H_2CN) that have not been considered before. The discussions are generally brief; appropriate references to the literature are given for those seeking further information. Our intent is to identify as many of the remaining potentially important reactions as possible and to collect their rate coefficients—for those reactions for which recommendations can be made—in one place for the convenience of the combustion modeling community.

2.7.1 Reactions of N atoms

One expects N atoms to participate in adduct formation reactions, as opposed to hydrogen abstractions, because the N–H bond dissociation energy is only about 340 kJ/mol, meaning that hydrogen abstractions by N atoms from molecules such as hydrocarbons would be endothermic.

(a) $N + O_2 \longrightarrow NO + O$.

This reaction was reviewed by Cohen (1992), who recommended

$$k_{26\text{a}} = 9.0 \times 10^9 T \exp(-3270/T)\ \text{cm}^3\text{mol}^{-1}\text{s}^{-1},$$

close to the expression obtained using the Hanson–Salimian expression for the reverse reaction rate coefficient.

(b) $N + OH \longrightarrow NH + O$.

Cohen and Westberg (1991) propose

$$k_{26b} = 6.4 \times 10^{12} T^{0.1} \exp(-10700/T) \ \text{cm}^3\text{mol}^{-1}\text{s}^{-1}$$

based upon an estimated expression for the reverse (abstraction) reaction 27e2. The high barrier assigned to this reaction is consistent with its 92 kJ/mol endothermicity. We suggest that this rate coefficient be calculated from the reverse reaction rate coefficient.

(c) $N + OH \longrightarrow NO + H$.

Natarajan et al. (1994) measured the reverse reaction rate behind shock waves over the range 2560 to 4040 K. A forward rate coefficient expression was obtained by fitting their data to an Arrhenius expression, invoking microscopic reversibility, and combining the result with the low-temperature recommendation of Atkinson et al. (1989) to obtain

$$k_{26c} = 1.1 \times 10^{14} \exp(-565/T) \ \text{cm}^3\text{mol}^{-1}\text{s}^{-1}.$$

Essentially the same result was obtained by Qin et al. (1997) in a similar investigation. This channel is much faster than the NH + O channel, which in view of its 205 kJ/mol exothermicity is not surprising.

(d) $N + CH \longrightarrow {}^3HCN \longrightarrow H + CN$.

This channel is exothermic by 415 kJ/mol, sufficient to convert the initially formed ^{3}HCN to the ground state singlet, which subsequently dissociates to H + CN. The rate coefficient at room temperature was reported by Messing et al. (1981) to be 1.3×10^{13} cm^{3}mol^{-1}s^{-1}. Brownsword et al. (1996) measured the rate from 216 to 584 K and reported

$$k_{26d} = 1.7 \times 10^{14} T^{-0.09} \ \text{cm}^3\text{mol}^{-1}\text{s}^{-1}.$$

(e) $N + CH_2 \longrightarrow H_2CN \longrightarrow HCN + H$.

This is another exothermic reaction having an exit channel with energy lower than that of the initially formed adduct. We are not aware of any measurements. Miller and Bowman use

$$k_{26e} = 5.0 \times 10^{13} \ \text{cm}^3\text{mol}^{-1}\text{s}^{-1},$$

which is close to the average of the rate coefficients for N + CH (reaction 26d) and $N + CH_3$ (cf. Section 2.6.16).

(f) $N + NH \longrightarrow N_2 + H$.

This reaction may be expected to proceed as formation of the NNH adduct followed by rapid dissociation to $N_2 + H$. Hack et al. (1994) studied it at 298 K and reported

$$k_{26f} = 1.5 \times 10^{13}\ \text{cm}^3\text{mol}^{-1}\text{s}^{-1}.$$

(g) $N + NH_2 \longrightarrow N_2 + H + H$.

Another exothermic adduct dissociation, perhaps with NNH as a second intermediate dissociating to $N_2 + H$, for which the rate coefficient is expected to be large. This expectation is confirmed by two studies (Whyte and Phillips 1983a; Dransfeld and Wagner 1987) giving at 298 K

$$k_{26g} = 7.1 \times 10^{13}\ \text{cm}^3\text{mol}^{-1}\text{s}^{-1}.$$

(h) $N + CN \longrightarrow C + N_2$.

Although Baulch et al. (1992) suggest 1.8×10^{14} $\text{cm}^3\text{mol}^{-1}\text{s}^{-1}$ for this reaction, the room temperature measurement was only 6×10^{13} $\text{cm}^3\text{mol}^{-1}\text{s}^{-1}$ (Whyte and Phillips 1983b). Dean et al. (1990) measured the reverse reaction rate in shock tube experiments and inferred the forward rate constant to be about 2.6×10^{13} $\text{cm}^3\text{mol}^{-1}\text{s}^{-1}$ with little temperature dependence. Combining the two values gives

$$k_{26h} = 2.4 \times 10^{13} \exp(+280/T)\ \text{cm}^3\text{mol}^{-1}\text{s}^{-1}.$$

2.7.2 Reactions of NH

Hydrogen abstraction from hydrocarbons by NH is expected to be slow, because the strength of the H–NH bond formed is only about 380 kJ/mol, meaning that most abstractions would be endothermic. As with N, adduct formation pathways are expected to be more important.

(a) $NH + NH \longrightarrow$ Products.

This reaction was studied behind shock waves by Mertens et al. (1989), who reported

$$k_{27a} = 5.1 \times 10^{13}\ \text{cm}^3\text{mol}^{-1}\text{s}^{-1},$$

consistent with an exothermic combination to form N_2H_2 followed by rapid dissociation to NNH + H, with likely final products $N_2 + H + H$.

(b) $NH + NH_2 \longrightarrow N_2H_2 + H$.

The most likely channel is formation of the N_2H_3 adduct followed by dissociation to form N_2H_2 + H. The overall process is exothermic by 113 kJ/mol. The shock tube experiments of Davidson and Hanson (1990b) were interpreted to yield

$$k_{27b1} = 1.5 \times 10^{15} T^{-0.5} \ \text{cm}^3\text{mol}^{-1}\text{s}^{-1},$$

consistent with the room-temperature measurement of Dransfeld et al. (1984) and with the expected rate for a process with an adduct formation mechanism. Adduct dissociation to form H_2NN + H is less likely, since this channel is only 33 kJ/mol exothermic. Another possibility would be the H abstraction to form N + NH_3, exothermic by 113 kJ/mol. The DHT estimation method of Section 2.4 gives

$$k_{27b2} = 9.2 \times 10^5 \, T^{1.94} \exp(-1230/T) \ \text{cm}^3\text{mol}^{-1}\text{s}^{-1}.$$

(c) $NH + OH \longrightarrow$ Products.

Cohen and Westberg (1991) reviewed this reaction. No data are available; they estimate rate coefficients by analogy to OH + OH and suggest

$$NH + OH \longrightarrow H + HNO, \ k_{27c1} = 2.0 \times 10^{13} \ \text{cm}^3\text{mol}^{-1}\text{s}^{-1},$$

consistent with the 71 kJ/mol exothermicity. Another pathway is direct hydrogen transfer

$$NH + OH \longrightarrow H_2O + N,$$

for which Cohen and Westberg suggest $2.0 \times 10^9 T^{1.2}$ $\text{cm}^3\text{mol}^{-1}\text{s}^{-1}$. This is somewhat larger than the DHT estimate

$$k_{27c2} = 1.2 \times 10^6 \, T^2 \exp(+245/T) \ \text{cm}^3\text{mol}^{-1}\text{s}^{-1}.$$

(d) $NH + H \longrightarrow H_2 + N$.

Baulch et al. (1992) suggest the value 1×10^{13} $\text{cm}^3\text{mol}^{-1}\text{s}^{-1}$, consistent with an exothermic direct abstraction reaction but not with an adduct formation pathway; adduct formation and dissociation would involve $^2NH_2$, from which dissociation to ^4N + $^1H_2$ is spin-forbidden. Davidson and Hanson (1990a) measured the reverse reaction rate from 1950 to 2850 K. Combining their result with the equilibrium constant gives

$$k_{27d} = 3.5 \times 10^{13} \exp(-870/T) \ \text{cm}^3\text{mol}^{-1}\text{s}^{-1},$$

which is higher than the Baulch et al. expression at combustion temperatures.

(e) $NH + O \longrightarrow$ Products.

Cohen and Westberg (1991) reviewed this reaction and suggested on the basis of collision theory arguments that $k = 3.0 \times 10^{13}$ $cm^3mol^{-1}s^{-1}$ for the H + NO channel, which is consistent with a very exothermic (–289 kJ/mol) adduct formation and dissociation process. For the hydrogen transfer channel to form OH + N they suggest $k = 3.0 \times 10^{12}$ $cm^3mol^{-1}s^{-1}$. Hack et al. (1994), however, report that the rate coefficient for this channel is less than 1.0×10^{11} at 298 K. The DHT estimate is

$$k_{27e2} = 1.7 \times 10^8\, T^{1.5} \exp(-1695/T)\ \ cm^3mol^{-1}s^{-1}.$$

This expression gives 3×10^9 $cm^3mol^{-1}s^{-1}$ at 298 K, consistent with the Hack et al. measurement, and 6×10^{12} $cm^3mol^{-1}s^{-1}$ at 2000 K, close to the Cohen and Westberg estimate. Sengupta and Chandra (1994) concluded on the basis of electronic structure and QRRK calculations that reaction 27e1 proceeds by intermediate adduct formation, while both direct abstraction and adduct formation are significant for 27e2.

There have been several measurements of the total rate coefficient. Dransfeld et al. (1984) reported $(5 \pm 2) \times 10^{13}$ $cm^3mol^{-1}s^{-1}$ and Adamson et al. (1994) reported 4.2×10^{13} $cm^3mol^{-1}s^{-1}$ at room temperature. Mertens et al. (1991) reported 9.2×10^{13} $cm^3mol^{-1}s^{-1}$ for temperatures between 2700 and 3400 K. Because the H-transfer channel is expected to be appreciably slower than the total for these measurements, we suggest that they should be assigned to the H + NO channel and recommend

$$k_{27e1} = 6 \times 10^{13}\ \ cm^3mol^{-1}s^{-1}.$$

(f) $NH + CH_3 \longrightarrow HNCH_3 \longrightarrow H_2CNH + H$.

This reaction is overall exothermic by 188 kJ/mol, suggesting that its rate coefficient should be similar to that for adduct formation. Using as an analog the Baulch et al. (1992) expression for $OH + CH_3$, we estimate

$$k_{27f} = 4.0 \times 10^{13}\ \ cm^3mol^{-1}s^{-1}.$$

Direct H-atom transfer is estimated to be at least two orders of magnitude slower.

2.7.3 Reactions of NNH

We discussed NNH in Section 2.6.14 as a dissociation product of the cis- or trans-HNNOH isomers of the initially-formed adduct of the $NH_2 + NO$ reaction.

(a) $NNH \longrightarrow N_2 + H$.

NNH is thought to be weakly bound and thus to dissociate rapidly; calculations by Koizumi et al. (1991) indicate a tunneling lifetime of only about 3 ns for the ground vibrational state. The pressure dependence of the dissociation rate can be estimated by means of a simplified QRRK analysis. The high pressure rate coefficient for NNH dissociation was estimated from the reverse reaction rate coefficient using an A-factor for H addition to N_2 comparable to H addition to C_2H_2 and a barrier that gives a room temperature value for the high pressure rate coefficient consistent with the calculated tunneling lifetime for the ground vibrational state. This barrier proved to be 20 kJ/mol. (It had to be lower than the 31 kJ/mol calculated by Koizumi, as no attempt had been made to correct the QRRK rate coefficient for tunneling.) The subsequent QRRK analysis suggested that this reaction is at the low pressure limit for conditions of interest in combustion, with a rate coefficient given by $k = 1.3 \times 10^{14} T^{-0.11} \exp(-2500/T)$ $\text{cm}^3\text{mol}^{-1}\text{s}^{-1}$for N_2 bath gas. At 1 atm the effective first-order rate coefficient is computed to be 6×10^7 s^{-1} at 1000 K and 1×10^8 s^{-1} at 2000 K, the small temperature effect being due to the unusually low barrier. These values are much lower than the estimated high-pressure limit of over 10^{11} s^{-1} at these temperatures and indicate that this reaction is deep in the falloff region. However, they are still much higher than the 10^4 s^{-1} value used by Miller and Bowman (1989). We are not able to reconcile the Miller and Bowman value theoretically.

The QRRK treatment of NNH dissociation assumed that dissociation occurs by thermal activation, the only consideration of tunneling being adoption of a lowered dissociation barrier. An alternative approach is to use the energy-dependent tunneling rate coefficients computed by Koizumi et al. to estimate the dissociation rate directly. Two contributions have to be included: (1) A thermal activation one with a 31 kJ/mol barrier height, corresponding to the calculation of Koizumi et al., and (2) a tunneling one at the rate calculated by Koizumi et al. The high pressure rate coefficient that results can be represented as $k_\infty = 4.1 \times 10^9 T^{1.13} \exp(-2610/T)$ s^{-1}, with the tunneling contribution dominating at all temperatures. At lower pressures, the analysis is complicated, because the pressure-independent term owing to tunneling from the lowest vibrational state should be the lower limit to the rate coefficient. An approximate analysis suggests that the dissociation rate coefficients from the higher levels are faster than collisional stabilization rates; in other words, the reaction is near its low-pressure limit. Assuming that the thermal activation rate is approximately $k^{coll}[\text{M}]$ leads to

$$k_{28a} \approx 3 \times 10^8 + 1 \times 10^{13} T^{1/2} \exp(-1540/T)[\text{M}]\ \text{s}^{-1},$$

where the collisional rate coefficient term represents activation to the first excited vibrational state, which gives a dissociation rate coefficient of about 1.2×10^9 s^{-1} at 1 atm and 1000 K and 1.6×10^9 s^{-1} at 2000 K, substantially larger than the QRRK estimate given above. We recommend this expression for combustion modeling. This assignment is consistent with the study of Thermal DeNO_x kinetics using isotopically labelled nitrogen by Duffy and Nelson (1996), who report the lifetime of NNH to be less than 10 ns.

The ambiguity in rate coefficient assignment for this reaction caused by the uncertain tunneling contribution is mitigated by the consideration that if tunneling is important it must be equally important for forward and reverse directions. Thus the net effect of tunneling is that this reaction is likely to be equilibrated under most conditions of interest in combustion, so that the concentration of NNH is essentially independent of the rate coefficient for dissociation. This is an important result—despite its large dissociation rate, NNH concentrations in flames prove not to be negligibly small, particularly in air, where the large N_2 concentration enhances the reverse reaction rate and can drive NNH concentrations to near-equilibrium levels. (For flames in air at 1800 K and 1 atm the equilibrium [NNH]/[H] ratio is about 2×10^{-6}; at 1200 K and 1 atm it is about 1×10^{-6}.) There are therefore conditions where the equilibrium NNH concentration is high enough for it to participate in bimolecular reactions. In general, one expects that NNH is more likely to be equilibrated in high-temperature systems in air, where the fast rate of NNH formation quickly leads to a near-equilibrium concentration.

One can estimate the importance of various reaction partners by estimating values of $k_x[\mathrm{X}]$, where k_x is the rate coefficient for species X reacting with NNH. We consider such reactions with O_2 and the dominant flame radicals H, OH, and O, which can attain mole fractions up to several percent in flame fronts, and with NH_2, HO_2, and NO, which are important under Thermal $DeNO_x$ conditions.

(b) $NNH + O_2 \longrightarrow$ Products.

The potential reactant at highest concentration is usually O_2. A QRRK approach can be used to estimate a rate coefficient for it. The HNNOO adduct formed can dissociate back to reactants or undergo intramolecular hydrogen transfer to form NNOOH, which will dissociate to either $HO_2 + N_2$ or $N_2O + OH$. The uncertain HNNOO well depth is probably somewhere in the range 70–90 kJ/mol. The A-factor for hydrogen transfer, where the transition state involves loss of the O_2 rotation degree of freedom, is estimated to be about 4×10^{12} cm^{3}mol^{-1}s^{-1}. We chose an isomerization barrier of 88 kJ/mol, as used by Lin et al. (1992), and a 92 kJ/mol well depth. (Assuming a shallower well significantly reduces the rate of product formation.) Partition of NNOOH to the two product channels was estimated at 4:1, favoring $HO_2 + N_2$, from estimated transition state entropies. These calculations give, in cm^{3}mol^{-1}s^{-1} units,

$$\mathrm{NNH} + \mathrm{O_2} \longrightarrow \mathrm{HO_2} + \mathrm{N_2}, \; k_{28b1} = 1.2 \times 10^{12} T^{-0.34} \exp(-75/T)$$
$$\mathrm{NNH} + \mathrm{O_2} \longrightarrow \mathrm{N_2O} + \mathrm{OH}, \; k_{28b2} = 2.9 \times 10^{11} T^{-0.34} \exp(-75/T)\,.$$

The value for the faster $HO_2 + N_2$ channel is only about 9×10^{10} cm^{3}mol^{-1}s^{-1} at 2000 K, suggesting a pseudo-first-order rate coefficient of about 1×10^5 s^{-1} in air at 1 atm for the reaction of NNH with O_2.

This analysis indicates that $NNH + O_2 \longrightarrow N_2 + HO_2$ is generally too slow to be of interest. At first glance this is surprising, given that the reaction appears to be a simple exothermic hydrogen abstraction. Extensive analysis of analogous reactions shows, however, that it is not direct hydrogen transfer but rather the result of an addition–isomerization–elimination sequence. Reactions of the form $R + O_2 \longrightarrow$ olefin + HO_2 have received considerable attention, and there is a growing consensus that they are not direct reactions. Kaiser (1995), for example, showed that the ethylene yield from $C_2H_5 + O_2$ decreases dramatically with increasing pressure. He interpreted this in terms of adduct formation, with the increasing pressure stabilizing the energized adduct prior to isomerization. In theoretical treatments of such reactions (Bozzelli and Dean 1990; Wagner et al. 1990; Bozzelli and Dean 1993) the available data were reconciled in terms of adduct formation. Thus for the NNH + O_2 reaction, even though the rate coefficient for formation of the HNNOO adduct is large and the pathway to $HO_2 + N_2$ is exothermic, the rate-limiting step is isomerization; the relatively shallow well for adduct formation puts the barrier for isomerization only 4 kJ/mol lower than the entrance channel. Since the A-factor for redissociation to reactants is higher than that for isomerization, dissociation to NNH + O_2 dominates and leads to a low rate of $N_2 + HO_2$ formation.

(c) NNH + H $\longrightarrow$ Products.

Recombination of H with NNH would form HNNH, while addition could form H_2NN. NH production by breaking the N–N bond in HNNH is too endothermic to be important. Formation of the HNNH adduct liberates about 250 kJ/mol, which is probably less than the barrier for H_2 elimination to form N_2 and H_2; in any event, the A-factor for this process should be much lower than for redissociation of the adduct to H + NNH. Stabilization of HNNH or H_2NN was also found to be relatively unimportant.

Another possibility is the direct reaction

$$\mathrm{NNH + H \longrightarrow H_2 + N_2}. \tag{28c}$$

for which the DHT estimate is

$$k_{28c} = 2.4 \times 10^8\, T^{1.5} \exp(+450/T)\ \mathrm{cm^3 mol^{-1} s^{-1}}.$$

For an H-atom mole fraction of 1% the pseudo-first-order rate coefficient of this reaction at 2000 K and 1 atm would be about $2\times10^6\,\mathrm{s^{-1}}$, substantially faster than reaction with 20% O_2.

(d) NNH + OH $\longrightarrow$ Products.

NNH reaction with OH to form HONNH is likely to form $N_2 + H_2O$ via concerted elimination of H_2O on the same potential energy surface considered earlier (Section 2.6.14) for NH_2 + NO. We analyzed this with the QRRK method using the parameters in Table 2.10. In the present case the surface is entered via OH + NNH to form trans HNNOH. The analysis leads, in $\mathrm{cm^3 mol^{-1} s^{-1}}$ units, to

$$\mathrm{NNH + OH \longrightarrow N_2 + H_2O}, \quad k_{28d1} = 2.4 \times 10^{22} T^{-2.88} \exp(-1230/T).$$

At 2000 K the value of this expression is 4×10^{12} $cm^3mol^{-1}s^{-1}$, significantly lower than the 5×10^{13} $cm^3mol^{-1}s^{-1}$ value used by Miller and Bowman (1989). At lower temperatures, its value is higher, the decline with temperature being due to normal fall-off effects, that is, the complex dissociates faster. For an OH mole fraction of 1% the pseudo-first-order rate coefficient at 2000 K for this reaction is about 2×10^5 s^{-1}.

Direct hydrogen transfer can also lead to these products. The DHT estimate for its rate coefficient is

$$k_{28d2} = 1.2 \times 10^6\, T^2 \exp(+600/T) \;\; cm^3mol^{-1}s^{-1}\,.$$

At 2000 K this expression is 6.5×10^{12} $cm^3mol^{-1}s^{-1}$. At 300 K it is much smaller than k_{28d1}. We suggest using both expressions together to reflect the change in mechanism with temperature. The sum of both rate coefficients is still about 5 times smaller than the Miller–Bowman value.

(e) $NNH + O \longrightarrow$ Products.

The NNH + O reaction has three product channels, two of which have already been discussed as reverse reactions. The ONNH complex can dissociate to form H + N_2O (the reverse of reaction 11d) or to NO + NH (the reverse of reaction 11g). On the basis of the discussion in Section 2.6.7 we expect both channels to be fast, because both are lower in energy than NNH + O; we expect a low A-factor for the H + N_2O channel to inhibit this pathway relative to NO + NH even though it is more exothermic. This is seen in the QRRK rate coefficients, k_{-11g} being about an order of magnitude larger than k_{-11d}. At 2000 K, k_{-11g} is calculated to be about 7×10^{13} $cm^3mol^{-1}s^{-1}$, leading to a pseudo-first-order rate coefficient of 4×10^6 s^{-1} for 1% O atoms in air. The QRRK estimate in $cm^3mol^{-1}s^{-1}$ units for the rate coefficient of the third (minor) channel producing OH + N_2

$$NNH + O \longrightarrow N_2 + OH,\; k_{28e1} = 1.7 \times 10^{16} T^{-1.23} \exp(-250/T)$$

is about 1×10^{12} $cm^3mol^{-1}s^{-1}$ at 2000 K, significantly less than that for the NH + NO channel.

Direct hydrogen transfer would give the same products. The DHT estimate for its rate coefficient is

$$k_{28e2} = 1.7 \times 10^8\, T^{1.5} \exp(+450/T) \;\; cm^3mol^{-1}s^{-1},$$

which is appreciably faster than the adduct formation pathway at high temperatures. We accordingly suggest again that both rate-coefficient expressions be used to reflect the change in mechanism with temperature.

In Section 2.6.7 we showed that NNH + O is a plausible high-temperature channel of the NH + NO reaction and that the calculated temperature dependence for this channel is consistent with the high-temperature measurements reported

in the shock tube study of Mertens et al. (1991a). Specifically, Mertens et al. reported an activation energy of 54 kJ/mol for the total rate coefficient for all channels of the NH + NO reaction over the temperature range 2220 to 3350 K. The NH + NO reaction has three likely product channels: H + N_2O, N_2 + OH and NNH + O. Only the latter is endothermic ($\Delta_r H = 52$ kJ/mol), and we expect the other two channels to have rate coefficients that decrease with temperature (cf. Figure 2.16). Thus the measured temperature dependence suggests that NNH + O is the dominant channel in the Mertens et al. experiments.
Accepting that NNH + O is the dominant product channel, the Mertens et al. rate coefficient expression implies, through microscopic reversibility, that the value of k_{-11g} should be about 1.3×10^{14} $\mathrm{cm^3mol^{-1}s^{-1}}$, a factor of two above the QRRK estimate.[1] The Mertens et al. experiments therefore provide indirect but convincing evidence that the reaction between NNH and O is very rapid. The QRRK rate coefficient estimate being in reasonable agreement with their data supports this conclusion.

(f) NNH + NH_2 $\longrightarrow$ Products.

The DHT estimate is

$$k_{28f} = 9.2 \times 10^5 T^{1.94} \exp(+580/T) \ \mathrm{cm^3mol^{-1}s^{-1}},$$

which is comparable to the H atom transfer reaction OH + NNH but slower than those of H and O.

(g) NNH + HO_2 $\longrightarrow$ Products.

This radical-radical reaction would be expected to have both adduct formation and direct hydrogen transfer channels. Because the N–H bond is much weaker than the O–H bond, formation of H_2O_2 and N_2 rather than NHNH and O_2 would be the expected molecular product channel. The DHT estimate for H_2O_2 production is, in $\mathrm{cm^3mol^{-1}s^{-1}}$ units,

$$\mathrm{NNH + HO_2 \longrightarrow N_2 + H_2O_2}, \ k_{28g1} = 1.4 \times 10^4 T^{2.69} \exp(+805/T).$$

Adduct formation followed by dissociation to HNNO + OH is endothermic by 8 kJ/mol. Assuming that the A-factor for this channel is comparable to that for redissociation to reactants gives, in $\mathrm{cm^3mol^{-1}s^{-1}}$ units,

$$\mathrm{NNH + HO_2 \longrightarrow HNNO + OH}, \ k_{28g2} = 2.4 \times 10^{13} \exp(-855/T).$$

[1] Miller and Melius (1992a, 1992b) have also noted that the endothermicity of this product channel may be responsible for the temperature dependence of the rate coefficient expression reported by Mertens et al.

(h) $NNH + NO \longrightarrow N_2 + HNO$.

We estimate

$$k_{28h} = 1.2 \times 10^6 T^2 \exp(+600/T)\ \ \mathrm{cm^3 mol^{-1} s^{-1}}$$

by assuming the reaction with NO to be analogous to that with OH. Even though the the H–NO bond is about 290 kJ/mol weaker than the H–OH bond, abstraction by NO is still exothermic by about 230 kJ/mol.

The $O + NNH \longrightarrow NH + NO$ reaction thus appears to be the most important of the bimolecular NNH reactions for combustion, with a pseudo-first-order rate coefficient at least an order of magnitude larger than the other bimolecular reactions and the only one that has a rate of product formation comparable to that for adduct formation. Roles of this reaction in NOx production in methane–air and hydrogen–air systems are discussed in Section 2.8 and by Bozzelli and Dean (1995).

No pressure dependence is expected for any of the radical reactions with NNH.

Our conclusions that NNH can be partially equilibrated during combustion processes and that its bimolecular reactions are important suggests that additional attention should be given to them. Especially important are the reactions with OH, H, and O_2. Our conclusion that the rate coefficients of these reactions are substantially smaller than those of the reaction with O is based on analogies to other reactions. These analogies should be tested by theoretical and experimental studies.

2.7.4 Reactions of N_2H_2 (HNNH)

N_2H_2 is unusual in that its H–N bond dissociation energy is only 258 kJ/mol, quite small for a stable molecule. One therefore expects it to undergo hydrogen transfer reactions, and the NNH abstraction product to dissociate quickly to N_2 + H. In some respects N_2H_2 might be expected to react like HNO, where the H–N bond dissociation energy is also low, 210 kJ/mol. Addition reactions to N_2H_2 are probably not important, as the adducts appear to have shallow potential wells. Consequently, the isomerization barriers are above the entrance channel, and the overall rates to non-adduct products are slow.

(a) $N_2H_2 + M \longrightarrow NNH + H + M$.

To our knowledge there are no measurements of the dissociation rate of N_2H_2. The same products are accessible both by direct dissociation and via isomerization to H_2NN followed by its dissociation. Because the isomerization barrier is not known, we estimated it by the G2 method (Foresman and Frisch 1996), which gave barriers in the 190 to 230 kJ/mol range. Based on this result it appears that both dissociation pathways contribute. The following QRRK estimates, in s^{-1} units, apply to dissociation in N_2 bath gas in the temperature range from 600 to 2500 K.

$$
\begin{aligned}
k_{29a1} &= 5.6 \times 10^{36}\,T^{-7.75}\exp(-35\,400/T) && 0.1 \text{ atm}\\
&= 1.8 \times 10^{40}\,T^{-8.41}\exp(-36\,900/T) && 1.0 \text{ atm}\\
&= 3.1 \times 10^{41}\,T^{-8.42}\exp(-38\,300/T) && 10 \text{ atm}\\
k_{29a2} &= 1.6 \times 10^{37}\,T^{-7.94}\exp(-35\,600/T) && 0.1 \text{ atm}\\
&= 2.6 \times 10^{40}\,T^{-8.53}\exp(-36\,700/T) && 1.0 \text{ atm}\\
&= 1.3 \times 10^{44}\,T^{-9.22}\exp(-38\,800/T) && 10 \text{ atm.}
\end{aligned}
$$

(b) $N_2H_2 \longrightarrow H_2NN$.

The isomerization rate coefficients estimated by QRRK, again for nitrogen bath gas and over the temperature range from 600 to 2500 K, are, in s^{-1} units,

$$
\begin{aligned}
k_{29b} &= 9.2 \times 10^{38}\,T^{-9.01}\exp(-34\,100/T) && 0.1 \text{ atm}\\
&= 2.0 \times 10^{41}\,T^{-9.38}\exp(-34\,400/T) && 1.0 \text{ atm}\\
&= 1.3 \times 10^{45}\,T^{-10.13}\exp(-35\,600/T) && 10 \text{ atm.}
\end{aligned}
$$

These results suggest that direct dissociation is faster than isomerization above about 700 K at 1 atm.

(c) $N_2H_2 + X \longrightarrow NNH + HX$, with X = H, OH, O, NH_2 or CH_3.

We are not aware of any experimental measurements for these reactions. DHT estimates are included in Table 2.19. Linder et al. (1995) reported variational transition state calculations of the rate coefficients for the reactions with H, OH, and NH_2. Their expression for abstraction by H agrees quite well with the DHT estimate for the entire 300–2000 K range, while their OH and NH_2 rate coefficients are generally lower.

(d) $N_2H_2 + NH \longrightarrow NNH + NH_2$.

Although appreciably less exothermic than the reaction with OH, the reaction with NH is still exothermic by about 160 kJ/mol. The same rate coefficient as we assigned to abstraction by OH is included in Table 2.19.

(e) $N_2H_2 + NO \longrightarrow N_2O + NH_2$.

This addition reaction can be considered to be similar to NO addition to HNO (Section 2.7.11). The *A*-factor is expected to be twice that for HNO addition. The adduct would be expected to undergo an H-shift with a lower barrier than that for HNO. This reasoning leads to

$$k_{29e} = 4.0 \times 10^{12}\exp(-6000/T)\ \text{cm}^3\text{mol}^{-1}\text{s}^{-1}.$$

2.7.5 Reactions of H_2NN

This species is expected to be a major product of the $NH_2 + NH_2$ reaction (Section 2.6.13). Like $^1CH_2$) and other singlet biradicals, H_2NN is expected to undergo insertion reactions. Its thermochemistry suggests that substituted hydrazines so formed will not have sufficient energy to dissociate into fragments.

^1H_2NN should not be considered as reactive as $^1CH_2$, because ^1H_2NN is stabilized by the overlap of the nitrogen lone pair orbitals. This is reflected in the shortening of the N–N bond length in the series N_2H_4 (1.43 Å), NH_2–NH (1.31 Å), ^1H_2NN (1.18 Å), and HN=NH (1.13 Å). This stabilization suggests that ^1H_2NN should be less reactive than $^1CH_2$, that is, more like pi-bonded species such as C_2H_4 or carbonyls. Molecular electronic structure calculations were carried out to characterize the transition states for reactions between radicals and ^1H_2NN and from them the corresponding rate parameters. Semi-empirical calculations with the PM3 method indicate transition states and A-factors similar to radical addition reactions.

Reaction of ^1H_2NN with radicals can occur at two sites, addition between the two N atoms or at the terminal N atom. The adducts formed often have an exothermic pathway in which beta-scission leads to formation of a double bond. Unimolecular reaction via this pathway is generally faster than stabilization under combustion conditions.

(a) $H_2NN \longrightarrow$ Products.

QRRK calculations indicate that isomerization to the lower energy isomer HNNH (N_2H_2) dominates below 1500 K. This reaction has already been discussed in the reverse direction. As with the HNNH isomer, isomerization may or may not precede dissociation. The QRRK results for the two pathways, valid for dissociation in N_2 bath gas from 600 to 2500 K, are in s^{-1} units,

$$\begin{aligned} k_{30a1} &= 5.9 \times 10^{32}\, T^{-6.99} \exp(-26\,100/T) && 0.1 \text{ atm} \\ &= 9.5 \times 10^{35}\, T^{-7.57} \exp(-27\,600/T) && 1.0 \text{ atm} \\ &= 5.0 \times 10^{36}\, T^{-7.43} \exp(-28\,800/T) && 10 \text{ atm} \\ k_{30a2} &= 7.2 \times 10^{28}\, T^{-5.77} \exp(-25\,500/T) && 0.1 \text{ atm} \\ &= 3.2 \times 10^{31}\, T^{-6.22} \exp(-26\,300/T) && 1.0 \text{ atm} \\ &= 5.1 \times 10^{33}\, T^{-6.52} \exp(-27\,300/T) && 10 \text{ atm}. \end{aligned}$$

(b) $H_2NN + O_2 \longrightarrow$ Products.

In this case adduct formation does provide sufficient energy to facilitate dissociation. Direct dissociation would produce $NH_2 + NO_2$, while isomerization would lead to HNN(O)OH, which could dissociate to H_2O and N_2O or to HNNO and OH. QRRK calculations suggest that the primary dissociation dominates

$$H_2NN + O_2 \longrightarrow NH_2 + NO_2 \tag{30b}$$

with a rate coefficient given by the insertion

$$k_{30b} = 1.5 \times 10^{12} \exp(-3000/T) \text{ cm}^3\text{mol}^{-1}\text{s}^{-1}.$$

The A-factor is that measured by Laufer et al. (1983) for the reaction of vinylidene with hydrogen. The dissociation barrier was estimated to be slightly lower than used for the $H_2NN + H_2$ channel of the $NH_2 + NH_2$ reaction. Although this barrier seems reasonable, additional experiments and theoretical calculations are needed. The rate of reaction 30b is greater than that of 30a in air at 1 atm for temperatures below 1900 K.

(c) $H_2NN + H \longrightarrow$ Products.

Both adduct formation and direct H-atom transfer have to be considered. Adduct formation followed by exothermic dissociation to HNNH + H has the effect of isomerization. The A-factor, in $\text{cm}^3\text{mol}^{-1}\text{s}^{-1}$ units, was estimated from calculated transition state geometries as described above.

$$H_2NN + H \longrightarrow N_2H_2 + H \quad k_{30c1} = 1.8 \times 10^{10}\, T^{0.97} \exp(-2250/T).$$

The DHT estimate for the direct reaction, again in $\text{cm}^3\text{mol}^{-1}\text{s}^{-1}$ units, is

$$H_2NN + H \longrightarrow H_2 + NNH \quad k_{30c2} = 4.8 \times 10^{8}\, T^{1.5} \exp(+450/T).$$

The direct reaction dominates.

(d) $H_2NN + O \longrightarrow$ Products.

The NH_2NO adduct formed in this reaction is highly energized. QRRK analysis based on the potential energy surface used for the NH_2 + NO reaction (Figure 2.33) shows that the H_2NN + O reaction leads almost exclusively to NH_2 and NO. The higher A-factor for this channel, relative to that for isomerization, governs the branching at such a high level of excess energy. The rate coefficient is therefore that of adduct formation, which we estimated from the calculated transition state geometries to be, in $\text{cm}^3\text{mol}^{-1}\text{s}^{-1}$ units,

$$H_2NN + O \longrightarrow NH_2 + NO \quad k_{30d1} = 3.2 \times 10^{9}\, T^{1.03} \exp(-1360/T).$$

The DHT estimate, likewise in $\text{cm}^3\text{mol}^{-1}\text{s}^{-1}$ units,

$$H_2NN + O \longrightarrow OH + NNH \quad k_{30d2} = 3.3 \times 10^{8}\, T^{1.5} \exp(+450/T),$$

is faster.

(e) $H_2NN + OH \longrightarrow$ Products.

Adduct formation followed by breaking of the O–H bond leads to NH_2NO + H with an exothermicity of 54 kJ/mol. We estimated its rate coefficient on the basis of a transition state calculation done with the PM3 semi-empirical method.

$$H_2NN + OH \longrightarrow NH_2NO + H \quad k_{30e1} = 2 \times 10^{12} \text{ cm}^3\text{mol}^{-1}\text{s}^{-1}.$$

The DHT estimate is, in $cm^3mol^{-1}s^{-1}$ units,

$$H_2NN + OH \longrightarrow H_2O + NNH \quad k_{30e2} = 2.4 \times 10^6\, T^2 \exp(+600/T)\,.$$

(f) $H_2NN + CH_3 \longrightarrow$ Products.

Adduct formation can be followed by dissociation into two exothermic channels, one breaking a C–H bond and one breaking an N–H bond. The transition state theory calculation leads to the rate coefficient expressions, in $cm^3mol^{-1}s^{-1}$ units,

$$\begin{aligned} H_2NN + CH_3 &\longrightarrow H_2NNCH_2 + H \quad k_{30f1} = 8.3 \times 10^5\, T^{1.93} \exp(-3270/T) \\ &\longrightarrow HNNCH_3 + H \quad k_{30f2} = 8.3 \times 10^5\, T^{1.93} \exp(-3270/T)\,. \end{aligned}$$

The DHT rate coefficient expression, also in $cm^3mol^{-1}s^{-1}$ units, is

$$H_2NN + CH_3 \longrightarrow CH_4 + NNH \quad k_{30f3} = 1.6 \times 10^6\, T^{1.87} \exp(-65/T)\,,$$

which is substantially faster than adduct formation.

(g) $H_2NN + NH_2 \longrightarrow$ Products.

Adduct formation followed by breaking an N–H bond is the only exothermic channel. The transition state rate coefficient expression, in $cm^3mol^{-1}s^{-1}$ units, is

$$H_2NN + NH_2 \longrightarrow HNNNH_2 + H \quad k_{30g1} = 7.9 \times 10^6\, T^{1.9} \exp(+670/T)\,,$$

significantly larger than the DHT rate coefficient expression

$$H_2NN + NH_2 \longrightarrow NH_3 + NNH \quad k_{30g2} = 1.8 \times 10^6\, T^{1.94} \exp(+580/T)\,.$$

(h) $H_2NN + HO_2 \longrightarrow$ Products.

Adduct formation followed by breaking an O–O bond is 192 kJ/mol exothermic. The TST result is, in $cm^3mol^{-1}s^{-1}$ units,

$$H_2NN + HO_2 \longrightarrow NH_2NO + OH \quad k_{30h1} = 6.6 \times 10^5\, T^{1.94} \exp(-3550/T)\,,$$

while the DHT method gives

$$H_2NN + HO_2 \longrightarrow NNH + H_2O_2 \quad k_{30h2} = 2.9 \times 10^4\, T^{2.69} \exp(+805/T)\,.$$

This channel is 88 kJ/mol more exothermic than formation of $N_2H_3 + O_2$.

2.7.6 Reactions of N_2H_3

Dissociation of N_2H_3 to N_2H_2 + H is endothermic by about 226 kJ/mol, significantly more than similar beta scission reactions, and so N_2H_3 is an unusually long-lived radical. We expect that its most important reactions are adduct formations and disproportionations; abstractions from hydrocarbons by N_2H_3 should be slow, as the strength of the H–N bond formed is only 322 kJ/mol.

(a) $N_2H_3 \longrightarrow N_2H_2 + H$.

QRRK analysis suggests that this reaction is in the falloff regime for typical combustion conditions. For N_2 bath gas and temperatures in the range 600 to 2500 K the QRRK analysis gives in s^{-1} units

$$\begin{aligned} k_{31a} &= 2.3 \times 10^{43} T^{-9.55} \exp(-32\,400/T) && 0.1 \text{ atm} \\ &= 3.6 \times 10^{47} T^{-10.38} \exp(-34\,700/T) && 1.0 \text{ atm} \\ &= 1.8 \times 10^{45} T^{-9.39} \exp(-35\,300/T) && 10 \text{ atm.} \end{aligned}$$

These expressions are shown in Figure 2.60.

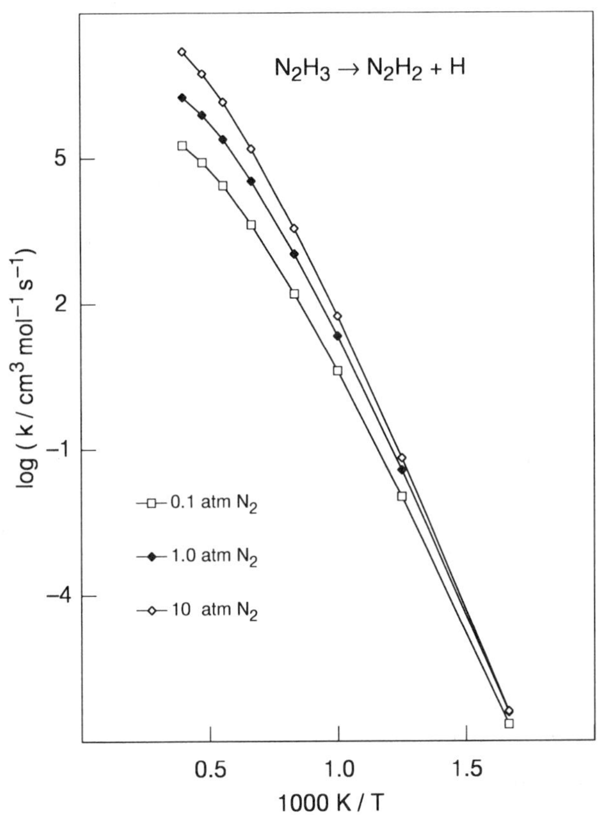

FIGURE 2.60. QRRK results for dissociation of N_2H_3.

(b) $N_2H_3 + H \longrightarrow$ Products.

The excited hydrazine formed by this association reaction can dissociate by the pathways shown in Figure 2.29. QRRK analysis indicates that the dominant channel is $NH_2 + NH_2$, the reverse of reaction 17d. Formation of $H_2NN + H_2$ is disfavored by its lower A-factor, which outweighs the somewhat lower barrier height. The DHT estimate for its rate coefficient is

$$N_2H_3 + H \longrightarrow N_2H_2 + H_2 \quad k_{31b} = 2.4 \times 10^8\, T^{1.5}\ \mathrm{cm^3 mol^{-1} s^{-1}}.$$

(c) $N_2H_3 + O \longrightarrow$ Products.

There are two exothermic channels

$$N_2H_3 + O \longrightarrow NH_2 + HNO, \quad \Delta_r H = -154\,\mathrm{kJ} \qquad (31c1)$$

$$NH_2NO + H, \quad \Delta_r H = -152\,\mathrm{kJ}. \qquad (31c2)$$

suggesting rapid reaction. This pair of reactions is analogous to the even more exothermic pair of $O + C_2H_5$ reactions, where the adduct can form either H + CH_3CHO or CH_3 + CH_2O. We adopt Baulch et al.'s recommendation for the total rate coefficient of $O + C_2H_5$ and assume that the two channels have equal rates.

$$k_{31c1} = 3 \times 10^{13}\ \mathrm{cm^3 mol^{-1} s^{-1}} \qquad k_{31c2} = 3 \times 10^{13}\ \mathrm{cm^3 mol^{-1} s^{-1}}.$$

For the hydrogen atom transfer reaction producing $N_2H_2 + OH$ the DHT estimate is

$$k_{31c3} = 1.7 \times 10^8\, T^{1.50} \exp(+325/T)\ \mathrm{cm^3 mol^{-1} s^{-1}}.$$

(d) $N_2H_3 + OH \longrightarrow$ Products.

The DHT estimate for the abstraction reaction producing $H_2O + N_2H_2$ is

$$k_{31d1} = 1.2 \times 10^6\, T^2 \exp(+600/T)\ \mathrm{cm^3 mol^{-1} s^{-1}}.$$

The N_2H_3OH adduct has an exothermic product channel forming $H_2NN + H_2O$. The barrier for this channel is low, because H_2NN is a singlet biradical; it is therefore analogous to $H_2NN + H_2$ formation from $NH_2 + NH_2$ (Section 2.6.13). QRRK analysis indicates that this pathway dominates and that the rate coefficient is that for adduct formation,

$$k_{31d2} = 3 \times 10^{13}\ \mathrm{cm^3 mol^{-1} s^{-1}}.$$

(e) $N_2H_3 + CH_3 \longrightarrow$ Products.

For the hydrogen transfer reaction producing $N_2H_2 + CH_4$ the DHT estimate is

$$k_{31e1} = 8.2 \times 10^5\, T^{1.87} \exp(-915/T)\ \mathrm{cm^3 mol^{-1} s^{-1}}.$$

The adduct has a low-energy exit channel to form $H_2NN + CH_4$. Despite the lower *A*-factor of this pathway it dominates the H-atom transfer pathway, the rate coefficient being approximately that of adduct formation,

$$k_{31e2} = 3 \times 10^{13} \ \text{cm}^3\text{mol}^{-1}\text{s}^{-1}.$$

(f) $N_2H_3 + NH_2 \longrightarrow$ Products.

The DHT estimate for the reaction to form $N_2H_2 + NH_3$ is

$$k_{31f1} = 9.2 \times 10^5 \, T^{1.94} \exp(+580/T) \ \text{cm}^3\text{mol}^{-1}\text{s}^{-1}.$$

The energetics for recombination channels are similar to those of the CH_3 + CH_3 reactions. QRRK analysis shows some falloff at high temperatures owing to dissociation of the adduct back to reactants, but formation of $H_2NN + NH_3$ dominates with a rate coefficient approximately that of adduct formation:

$$k_{31f2} = 3 \times 10^{13} \ \text{cm}^3\text{mol}^{-1}\text{s}^{-1}.$$

(g) $N_2H_3 + HO_2 \longrightarrow$ Products.

Adduct formation followed by rupture of the O–O bond to yield NH_2NHO + OH is 105 kJ/mol exothermic. The rate coefficient is expected to be that of adduct formation

$$k_{31g1} = 3 \times 10^{13} \ \text{cm}^3\text{mol}^{-1}\text{s}^{-1}.$$

The DHT estimate for the H-atom transfer reaction forming $N_2H_2 + H_2O_2$ is

$$k_{31g2} = 2.9 \times 10^4 \, T^{2.69} \exp(+805/T) \ \text{cm}^3\text{mol}^{-1}\text{s}^{-1}$$

and that for the one forming $N_2H_4 + O_2$ is

$$k_{31g3} = 9.2 \times 10^5 \, T^{1.94} \exp(-1070/T) \ \text{cm}^3\text{mol}^{-1}\text{s}^{-1}.$$

2.7.7 Reactions of N_2H_4

Hydrazine can be formed in ammonia flames by NH_2 + NH_2 association. Its dissociation to N_2H_3 + H can be neglected relative to dissociation to $NH_2 + NH_2$, the reverse of reaction 17a. Dissociation to form $H_2NN + H_2$ can be important at lower temperatures. The QRRK estimates for this reaction in N_2 bath gas in the temperature range from 600 to 2500 K are in s^{-1} units

$$k_{32} = 4.0 \times 10^{44} T^{-9.85} \exp(-35\,900/T) \qquad 0.1 \text{ atm}$$
$$k_{32} = 5.3 \times 10^{39} T^{-8.35} \exp(-34\,900/T) \qquad 1.0 \text{ atm}$$
$$k_{32} = 2.5 \times 10^{39} T^{-8.19} \exp(-35\,100/T) \qquad 10 \text{ atm.}$$

The N–H bond being relatively weak, about 322 kJ/mol, suggests that the hydrogen can be easily abstracted, but few data are available. Using the DHT method we estimated the following expressions for abstractions from hydrazine (in $cm^3mol^{-1}s^{-1}$ units),

$$
\begin{aligned}
N_2H_4 + H &\longrightarrow N_2H_3 + H_2 & k_{32a} &= 9.6 \times 10^8 T^{1.5} \exp(-2435/T) \\
+ O &\longrightarrow N_2H_3 + OH & k_{32b} &= 6.7 \times 10^8 T^{1.5} \exp(-1435/T) \\
+ OH &\longrightarrow N_2H_3 + H_2O & k_{32c} &= 4.8 \times 10^6 T^{2.0} \exp(+325/T) \\
+ CH_3 &\longrightarrow N_2H_3 + CH_4 & k_{32d} &= 3.3 \times 10^6 T^{1.87} \exp(-2680/T) \\
+ NH_2 &\longrightarrow N_2H_3 + NH_3 & k_{32e} &= 3.7 \times 10^6 T^{1.94} \exp(-820/T)\,.
\end{aligned}
$$

These estimates can be compared to the limited experimental data at low temperatures. The estimate for abstraction by hydrogen atoms gives $k = 1.5 \times 10^9$ $cm^3mol^{-1}s^{-1}$ at 300 K. This compares to experimental values ranging from 1 to 2 $\times 10^{11}$ $cm^3mol^{-1}s^{-1}$ at 300 K (Mallard et al. 1993). For abstraction by OH at room temperature, the above estimate gives 1.3×10^{12} $cm^3mol^{-1}s^{-1}$, whereas measurements of the total rate report values ranging from 1.3 to 3.9×10^{13} $cm^3mol^{-1}s^{-1}$ (Mallard et al. 1993). An experimental study of the abstraction by NH_2 (as a secondary reaction in the H + N_2H_4 system) by Gehring et al. (1971) gave $k = 3 \times 10^{11}$ $cm^3mol^{-1}s^{-1}$ at 300 K, compared to our estimate 1.5×10^{10} $cm^3mol^{-1}s^{-1}$. For CH_3 the DHT estimate is 2.9×10^8 $cm^3mol^{-1}s^{-1}$ at 400 K compared to the reported value of 1.8×10^8 $cm^3mol^{-1}s^{-1}$(Mallard et al. 1993). These comparisons are disappointing, especially with regard to the abstraction by H-atoms. The DHT estimates should be better at higher temperatures where the activation energy influence is diminished. There is some evidence that radical reactions with hydrazine may not be direct ones. For example, Vaghijiani (1996) reported that the OH yield in the reaction of O-atoms with hydrazine accounts for only 15% of the total reaction at 298 K. If all of the observed OH were due to the direct abstraction reaction it would nonetheless be about 20 times the DHT estimate.

2.7.8 Reactions of NO

Many reactions of NO have been considered earlier. For completeness we add several more here.

(a) $NO + C \longrightarrow$ Products.

There are two likely product channels:

$$NO + C \longrightarrow CO + N \qquad \Delta_r H = -444 \text{ kJ} \tag{33a1}$$

$$NO + C \longrightarrow CN + O \qquad \Delta_r H = -121 \text{ kJ.} \tag{33a2}$$

There have been two studies of these reactions behind shock waves, Dean et al. (1991) reporting a total rate coefficient of 4.8×10^{13} $cm^3mol^{-1}s^{-1}$

and Lindackers et al. (1990) reporting 2.0×10^{13} $cm^3mol^{-1}s^{-1}$. Dorthe et al. (1991) measured 1.6×10^{13} $cm^3mol^{-1}s^{-1}$ at room temperature. Dean et al. reported that CN + O contributes 40% of the total rate and CO + N the remaining 60%. We suggest use of a simple average of the overall rate measurements together with the reported branching ratio, that is,

$$k_{33a1} = 1.7 \times 10^{13} \ \ cm^3mol^{-1}s^{-1}$$
$$k_{33a2} = 1.1 \times 10^{13} \ \ cm^3mol^{-1}s^{-1}\,.$$

(b) $NO + OH + M \longrightarrow HONO + M$.

This reaction can serve as a sink for OH in the pre-ignition chemistry of internal combustion engines with exhaust gas recirculation, which could lead to effectively higher octane numbers. It is discussed as a reverse reaction in Section 2.7.15.

(c) $NO + HCCO \longrightarrow$ Products.

Modeling studies, for example by Lindstedt et al. (1995), identify this reaction as a major route for NO destruction in hydrocarbon flames. Boullart et al. (1994) reported a study of the rate over the temperature range from 290 to 670 K, leading to the rate coefficient expression $k_{33c} = 6 \times 10^{13} \exp(-350/T)$ $cm^3mol^{-1}s^{-1}$. The major channel at 700 K was found to be HCNO + CO, with HCN + CO_2 making up the balance. Assuming that the Boullart et al. branching ratio is temperature independent leads to the two expressions, in $cm^3mol^{-1}s^{-1}$ units,

$$\begin{aligned} NO + HCCO &\longrightarrow HCNO + CO \qquad & k_{33c1} &= 4.6 \times 10^{13} \exp(-350/T) \\ &\longrightarrow HCN + CO_2 \qquad & k_{33c2} &= 1.4 \times 10^{13} \exp(-350/T)\,. \end{aligned}$$

A theoretical study of the reaction was reported by Nguyen et al. (1994).

(d) $NO + C_2H \longrightarrow$ Products.

This reaction was proposed to influence NO_X chemistry in flames by Peeters et al. (1996), who reported an overall rate coefficient expression for the temperature range from 295 to 440 K

$$k_{33d} = 6.0 \times 10^{13} \exp(-287/T) \ \ cm^3mol^{-1}s^{-1}\,,$$

which is similar to the expression for the analogous reaction of NO with HCCO, 33(c). No information is available about the product distribution. Substantial rearrangement of the initially formed adduct would be required to reach the exothermic product channels forming HCN and CO (−630 kJ/mol) or CN and HCO (−170 kJ/mol). For this reason we hesitate to recommend likely products and rate coefficients. Additional experiments to measure branching ratios are needed for this reaction as well as for the reactions of NO with C_2H_3 and C_2H_5.

2.7.9 Reactions of NO_2

(a) $NO_2 + H \longrightarrow NO + OH$.

As shown in Figure 2.20, H + NO_2 can form HONO and HNO_2 adducts, with the likely final products from both being OH + NO. With this exothermic exit channel one expects fast reaction and little effect of pressure. The reaction was studied by Ko and Fontijn (1991) over the temperature range from 296 to 760 K. We suggest use of their rate coefficient expression,

$$k_{34a} = 1.3 \times 10^{14} \exp(-180/T)\ \mathrm{cm^3 mol^{-1} s^{-1}}.$$

An interesting feature of this reaction is that its reverse would be expected to produce primarily HONO, with some H + NO_2 at high temperatures, as the channel to H + NO_2 is 121 kJ/mol endothermic. Consequently, there could be high pressure, lower-temperature combustion conditions where one needs to account for reactions of HONO, as discussed in Section 2.7.15.

(b) $NO_2 + O \longrightarrow NO + O_2$.

This reaction is similar to reaction 34a in that two different adducts can be formed without barriers and there is a low-energy exit channel, here to NO + O_2, so that one could form either ONOO or the symmetrical structure NO_3. ONOO should rapidly dissociate to form NO + O_2, while there could be an appreciable barrier for isomerization–dissociation or concerted elimination from NO_3. The low-temperature data show a small decrease in rate coefficient with increasing temperature, suggesting adduct formation. This reaction was reviewed by Baulch et al. (1992), who recommend

$$k_{34b} = 3.9 \times 10^{12} \exp(+120/T)\ \mathrm{cm^3 mol^{-1} s^{-1}}$$

from 230 to 350 K. This expression gives $k = 4.2 \times 10^{12}\ \mathrm{cm^3 mol^{-1} s^{-1}}$ at 2000 K, which can be compared to the report by Zuev et al. (1991c) that $k = 6.0 \times 10^{12}\ \mathrm{cm^3 mol^{-1} s^{-1}}$ over the range 1780 to 2300 K.

(c) $NO_2 + M \longrightarrow NO + O + M$.

One expects thermal dissociation of this small molecule to be in the low-pressure limit over the pressure range of interest in combustion. Röhrig et al. (1997) studied the reaction in shock waves from 1350 to 2100 K and derived the low-pressure limit expression

$$k_{34c} = 4.0 \times 10^{15} \exp(-30200/T)\ \mathrm{cm^3 mol^{-1} s^{-1}}$$

for M = Ar. This expression is in reasonable agreement with the one derived by Tsang and Herron (1991) from the reverse reaction rate. Tsang and Herron present falloff parameters and third-body efficiencies as well.

(d) $NO_2 + NH_2 \longrightarrow$ Products.

Glarborg et al. (1995) reported as results of a modeling study of the NH_3–NO_2 system from 850 to 1350 K the expressions in $cm^3mol^{-1}s^{-1}$ units

$$\begin{aligned} NO_2 + NH_2 &\longrightarrow N_2O + H_2O \quad k = 3.2 \times 10^{18} T^{-2.2} \\ &\longrightarrow NH_2O + NO \quad k = 3.5 \times 10^{12}. \end{aligned}$$

Park and Lin (1997) studied the reaction over the temperature range from 300 to 900 K and reported

$$k_{total} = 8.1 \times 10^{16}\, T^{-1.44} \exp(-135/T) \;\; cm^3mol^{-1}s^{-1},$$

with the N_2O channel being 19% of the total, independent of temperature. This branching ratio is consistent with the results of Lindholm and Hershberger (1997), who reported a branching ratio of 24% to N_2O and H_2O, 76% to NH_2O and NO. In contrast, extrapolation of the Glarborg et al. result to 300 K predicts 77% for the N_2O channel. On the other hand, the branching ratio reported by Meunier et al. (1996), 59% to N_2O and H_2O, 40% to NH_2O and NO, are closer to the extrapolated Glarborg et al. expression. There remains ambiguity in the branching ratio of this reaction. For combustion modeling we suggest using the expression of Park and Lin together with their temperature-independent branching ratio, in $cm^3mol^{-1}s^{-1}$ units,

$$\begin{aligned} NO_2 + NH_2 &\longrightarrow N_2O + H_2O \quad k_{34d1} = 1.5 \times 10^{16}\, T^{-1.44} \exp(-135/T) \\ &\longrightarrow NH_2O + NO \quad k_{34d2} = 6.6 \times 10^{16}\, T^{-1.44} \exp(-135/T). \end{aligned}$$

(e) $NO_2 + CH_3 \longrightarrow CH_3O + NO$.

Biggs et al. (1993) report for this reaction

$$k_{34e} = 1.4 \times 10^{13} \;\; cm^3mol^{-1}s^{-1}$$

at room temperature. The reaction is 70 kJ exothermic and little temperature dependence is to be expected. At high pressure and low temperature, formation of the stabilized adducts CH_3NO_2 (nitromethane) and CH_3ONO is possible.

2.7.10 Reactions of N_2O

(a) $N_2O + OH \longrightarrow N_2 + HO_2$.

Little information is available on the rate of this potentially important reaction. It was reviewed by Tsang and Herron (1991). The available data are all from experiments at low temperatures, and most report only upper-limit rate coefficients. Tsang and Herron assign an activation energy of 42 kJ/mol to be consistent with the upper-limit measurements. Miller and Bowman (1989) use this activation energy together with an A-factor of 2×10^{12} $cm^3mol^{-1}s^{-1}$, four times lower than the Tsang-Herron upper-limit recommendation. The likely products, particularly at lower temperatures, are HO_2 and N_2. Use of either of these expressions predicts formation of N_2O at intermediate combustion temperatures, where HO_2 concentrations are large, and the reaction goes in reverse.

Sausa et al. (1993) reported that their flame data are best interpreted with an expression for this rate coefficient no larger than the one suggested by Miller and Bowman and that an even lower expression would also fit their data. A study of N_2O decomposition by Allen et al. (1994) suggests that use of the Miller-Bowman expression predicts too much NO_2 and too little NO. Their analysis showed that a better fit to the data could be obtained by lowering k_{35} by a factor of 160; this reduced the production of HO_2, which is responsible for converting NO to NO_2 under their conditions.

Mebel et al. (1996a, 1996b) studied this reaction with molecular electronic structure methods. An abstraction pathway to form $N_2 + HO_2$ as well as an addition pathway to form either a stabilized adduct HONNO or the dissociation products HNO + NO were included. Their transition state theory abstraction rate coefficient expression for the temperature range from 1000 to 5000 K,

$$N_2O + OH \rightarrow N_2 + HO_2 \quad k_{35} = 1.3 \times 10^{-2}\, T^{4.72} \exp(-18\,400/T)\,,$$

in $cm^3 mol^{-1} s^{-1}$ units, is consistent with the upper limit of Allen et al. (1994, 1995).

Mebel et al. conclude that above 1400 K and 1 atm pressure the HONNO adduct rapidly dissociates to HNO + NO, while stabilization dominates below 1000 K. The dissociation channel forming HNO + NO is accounted for as its reverse reaction in the following section.

The abstraction channel appears to dominate under combustion conditions. Even at 2000 K, however, k_{35} is only about 5×10^9 $cm^3 mol^{-1} s^{-1}$, so it is not likely that this reaction will be important under any combustion conditions.

2.7.11 Reactions of HNO

The H–NO bond strength in HNO is only 200 kJ/mol, intermediate between the low bond dissociation energies associated with unstable radicals and normal covalent bond strengths. Because it has only 3 atoms, its dissociation rate is close to the low pressure limit throughout the combustion temperature range. Unimolecular dissociation of HNO is sufficiently slow that its bimolecular reactions can be important, particularly at temperatures below 1500 K. Hydrogen abstractions should be especially important, as they will generally be exothermic. There is also the possibility of radical addition across the double bond.

As a relatively stable molecule with an easily abstractable hydrogen atom, HNO offers the following catalytic chain termination sequence, where X and X′

are atoms or radicals:

$$\begin{aligned} H + NO &\longrightarrow HNO \\ HNO + X &\longrightarrow HX + NO \\ \text{(overall)} \quad H + X &\longrightarrow HX\,. \end{aligned}$$

Other species, such as HONO or HNO_2, can function similarly. If we designate these species as XR, then the catalytic mechanism is

$$\begin{aligned} X + R &\longrightarrow XR \\ XR + X' &\longrightarrow XX' + R \\ \text{(overall)} \quad X' + X &\longrightarrow X'X\,. \end{aligned}$$

(a) $HNO + M \longrightarrow H + NO + M$.

Hanson and Salimian (1984) suggest for M = Ar

$$k_{36a} = 1.8 \times 10^{16} \exp(-24500/T) \quad \mathrm{cm^3 mol^{-1} s^{-1}}.$$

At 1500 K and 1 atm the effective first-order rate coefficient is $1.2 \times 10^4\ \mathrm{s^{-1}}$, over four orders of magnitude slower than NNH dissociation. A radical at a concentration of 1 nmol/cm^3 (about 100 ppm at 1 atm) reacting with HNO with a rate coefficient near $1 \times 10^{13}\ \mathrm{cm^3 mol^{-1} s^{-1}}$ would compete with HNO thermal dissociation.

(b) $HNO + HNO \longrightarrow N_2O + H_2O$.

This reaction was studied by He and Lin (1992), and a potential surface was reported by Lin et al. (1992a). The initially formed dimer undergoes several isomerizations prior to forming $N_2O + H_2O$. The barriers to isomerization are higher than that of the reverse reaction, and thus one sees rate falloff at higher temperatures. The observed loss of HNO is best explained by adduct formation rather than as formation of final products. He and Lin report

$$k_{36b} = 8.5 \times 10^{8} \exp(-1550/T) \quad \mathrm{cm^3 mol^{-1} s^{-1}}$$

for formation of $N_2O + H_2O$ over the temperature range 450 to 520 K. The rate coefficient is pressure dependent and was reported only at 1 atm. The He and Lin expression is orders of magnitude smaller than that used by Miller and Bowman.

(c) $HNO + OH \longrightarrow H_2O + NO$.

The abstraction channel is 293 kJ/mol exothermic and expected to be fast. Transition state calculations by Soto et al. (1991) led to

$$k_{36c} = 1.3 \times 10^{7} T^{1.88} \exp(+480/T) \quad \mathrm{cm^3 mol^{-1} s^{-1}},$$

in reasonable agreement with the limited data available. Soto et al. conclude that the abstraction channel is faster than the addition channel.

(d) $HNO + H \longrightarrow$ Products.

This reaction was analyzed by Soto and Page (1993), who used variational transition state theory to calculate a rate coefficient for the abstraction channel given in $cm^3mol^{-1}s^{-1}$ units by

$$H + HNO \longrightarrow H_2 + NO, \quad k_{36d1} = 4.5 \times 10^{11} T^{0.72} \exp(-330/T)$$

over the range 200 to 3000 K. Their expression, about an order of magnitude higher than earlier estimates, is supported by experiments on the reverse reaction by Natarajan et al. (1994), who monitored H-atom profiles in H_2–NO mixtures between 1760 and 2160 K. Their measurements imply $k_{36d1} \approx 5 \times 10^{13}$ $cm^3mol^{-1}s^{-1}$ at 1900 K, compared to the Soto and Page value of 8.7×10^{13} $cm^3mol^{-1}s^{-1}$.
In other work Page and Soto (1993) describe calculations which indicate that abstraction is faster than addition at all temperatures. For the addition–elimination channel Cohen and Westberg (1991) report in $cm^3mol^{-1}s^{-1}$ units

$$H + HNO \longrightarrow O + NH_2, \quad k_{36d2} = 3.5 \times 10^{15} T^{-0.3} \exp(-14730/T).$$

This expression gives rate coefficient values much lower than those of the abstraction channel, consistent with the Page–Soto analysis. As can be seen in Figure 2.26, the lower expression for the addition channel is primarily due to energetics. Not only is there a barrier in the entrance channel, but both $O + NH_2$ and NH + OH are endothermic exit channels. We cautiously recommend the rate coefficient calculated by Soto and Page for the abstraction channel. The abstraction is direct and so no pressure dependence is anticipated. The Cohen–Westberg expression for the addition–elimination route is consistent with an experimental measurement of the reverse reaction rate (cf. Section 2.6.11). We suggest adoption of the rate coefficient for the reverse reaction.

(e) $HNO + O \longrightarrow OH + NO$.

Campbell and Handy (1975) reported a lower limit of 1.1×10^{13} $cm^3mol^{-1}s^{-1}$ for the rate coefficient of this reaction at 425 K, consistent with other exothermic radical + HNO abstractions. (For example, the Soto–Page rate coefficient expression for H + HNO gives 1.6×10^{13} $cm^3mol^{-1}s^{-1}$ at 425 K.) We suggest use of the H + HNO rate coefficient expression for this reaction also.

(f) $HNO + NH_2 \longrightarrow NO + NH_3$.

We are not aware of any measurements or calculations for this reaction. Hanson and Salimian estimate $5 \times 10^{11} T^{0.5} \exp(-500/T)$ $cm^3mol^{-1}s^{-1}$, while Miller and Bowman use $2 \times 10^{13} \exp(-500/T)$ $cm^3mol^{-1}s^{-1}$. Again, from the nature of the species involved one expects fast exothermic abstraction. The DHT rate coefficient expression is

$$k_{36f} = 9.2 \times 10^5 T^{1.94} \exp(+580/T) \quad \text{cm}^3\text{mol}^{-1}\text{s}^{-1}.$$

This expression is 5 to 10 times less than the Hanson and Salimian and Miller and Bowman expressions. It is in much better agreement with the transition-state calculations of Mebel et al. (1996). As with the other HNO reactions, one expects abstraction to dominate over addition and no pressure dependence of the rate coefficient.

(g) $HNO + NO \longrightarrow N_2O + OH$.

Miller and Bowman use $k = 2 \times 10^{12} \exp(-13\,100/T)$ $\text{cm}^3\text{mol}^{-1}\text{s}^{-1}$, which is consistent with the reaction proceeding by addition followed by OH elimination to form the final products N_2O + OH. The A-factor is consistent with an addition reaction, and one expects a high barrier, caused by ring strain, for hydrogen transfer from the nitrogen to the oxygen on the same nitrogen, followed by rapid scission to give the final products. This rate coefficient is much smaller than that for H abstraction from HNO.

In a study of the NO–H_2 system, Diau et al. (1995) considered this reaction to be rate limiting. The rates of NO decay and CO_2 production from added CO were modeled to yield

$$k_{36g} = 8.5 \times 10^{12} \exp(-14\,890/T) \quad \text{cm}^3\text{mol}^{-1}\text{s}^{-1}.$$

This expression was shown to be consistent with an addition–isomerization–decomposition pathway. An RRKM analysis of the unimolecular isomerization step, which is rate-limiting, with a master equation treatment of the collisional stabilization process succeeded in matching the experimentally derived rate coefficient. Their expression is also reasonably consistent with the one used by Miller and Bowman (1989) and we recommend it for combustion modeling.

(h) $HNO + O_2 \longrightarrow NO + HO_2$.

Because of the weak H–N bond in HNO this reaction is nearly thermoneutral. By analogy to other RH + O_2 reactions we estimate

$$k_{36h} = 2 \times 10^{13} \exp(-8000/T) \quad \text{cm}^3\text{mol}^{-1}\text{s}^{-1}.$$

This expression gives rate coefficient values greater than the one proposed by Miller and Bowman (1991).

(i) $HNO + CH_3 \longrightarrow NO + CH_4$.

To provide an interaction with a hydrocarbon radical we estimated an expression for abstraction by CH_3, generally the most abundant radical in hydrocarbon flames, by the DHT method:

$$k_{36i} = 8.2 \times 10^5 \, T^{1.87} \exp(-480/T) \quad \text{cm}^3\text{mol}^{-1}\text{s}^{-1}.$$

2.7.12 Reactions of NH_2O

In Sections 2.6.9 and 2.6.10 we indicated that NH_2O can be an important intermediate in reactions of NH_2 with O_2 or HO_2. For a radical, NH_2O is unusually stable to thermal dissociation of the H atom, having an N–H bond strength of about 255 kJ/mol, and so isomerization to the more stable isomer HNOH is more likely than dissociation. As a consequence of the relatively strong N–H bond, one expects that the rates of disproportionation reactions, such as $NH_2O + CH_3 \longrightarrow HNO + CH_4$, would be somewhat smaller than usual since the exothermicity is less than in the typical case. Radical recombination reactions should be important; two of them, $NH_2O + OH$ and $NH_2O + O$, were introduced earlier as the reverses of $NH_2 + HO_2$ and $NH_2 + O_2$, respectively. (Cf. Sections 2.6.9 and 2.6.10.) One does not expect hydrogen transfer reactions of the form $NH_2O + RH \longrightarrow NH_2OH + R$, since the strength of the O–H bond formed being only about 322 kJ/mol means that such reactions would usually be very endothermic.

(a) $NH_2O + M \longrightarrow HNO + H + M$.

QRRK analysis predicts that the dissociation rate coefficient is close to its low-pressure limit for all conditions of interest in combustion, the rate coefficient expression for $M = N_2$ being given by

$$k_{37a} = 2.8 \times 10^{24} T^{-2.83} \exp(-32\,700/T) \quad \text{cm}^3\text{mol}^{-1}\text{s}^{-1}.$$

At 2000 K and 1 atm pressure the effective first-order rate coefficient is about 600 s^{-1}, indicating that under combustion conditions dissociation will be unimportant compared to reactions with radicals.

(b) $NH_2O + M \longrightarrow HNOH + M$.

QRRK calculations predict that the rate coefficient for isomerization is also near its low-pressure limit under combustion conditions. Near 1 atm pressure of N_2 the QRRK expression is

$$k_{37b} = 1.1 \times 10^{29} T^{-3.99} \exp(-22\,100/T) \ \text{cm}^3\text{mol}^{-1}\text{s}^{-1}.$$

Isomerization is much faster than dissociation; at 2000 K and 1 atm N_2, its effective first-order rate coefficient is about $7 \times 10^5 s^{-1}$, three orders of magnitude greater than the rate coefficient for dissociation.

(c) $NH_2O + H \longrightarrow H_2NOH \longrightarrow NH_2 + OH$
$\longrightarrow HNO + H_2$.

The adduct formation pathway, exothermic by about 60 kJ/mol, is analogous to $H + C_2H_5 \longrightarrow 2\ CH_3$. We estimate, based on the expression of Baulch et al. (1992) for the C_2H_5 analog,

$$k_{37c1} = 4.0 \times 10^{13} \ \text{cm}^3\text{mol}^{-1}\text{s}^{-1}.$$

The DHT estimate for the direct reaction producing $HNO + H_2$ is

$$k_{37c2} = 4.8 \times 10^8\, T^{1.5} \exp(-785/T)\ \text{cm}^3\text{mol}^{-1}\text{s}^{-1}.$$

The rate of the direct reaction approaches that of the adduct formation pathway at high temperatures.

(d) $NH_2O + O \longrightarrow$ Products.

The adduct dissociates quickly to NH_2 and O_2, which has been considered in Section 2.6.9 as the reverse reaction. The DHT estimate for the channel producing HNO and OH is, in $\text{cm}^3\text{mol}^{-1}\text{s}^{-1}$ units,

$$NH_2O + O \longrightarrow HNO + OH \quad k_{37d} = 3.3 \times 10^8\, T^{1.5} \exp(-245/T)\,.$$

(e) $NH_2O + OH \longrightarrow$ Products.

The channel forming $HNO + H_2O$, exothermic by 250 kJ/mol, is blocked from the adduct, the 115 kJ/mol adduct well depth being overwhelmed by the 250 kJ/mol exit barrier (Figure 2.25). The DHT estimate for the abstraction channel is, in $\text{cm}^3\text{mol}^{-1}\text{s}^{-1}$ units,

$$NH_2O + OH \longrightarrow HNO + H_2O \quad k_{37e} = 2.4 \times 10^6\, T^2 \exp(+600/T)\,.$$

(f) $NH_2O + CH_3 \longrightarrow$ Products.

The channel forming NH_2 and CH_3O is exothermic by about 8 kJ/mol. Consequently, the NH_2OCH_3 adduct is expected to dissociate both to reactants and to products. We estimate the rate coefficient to be half that of the high-pressure limit of the $CH_3 + CH_3$ reaction (Baulch et al. 1992),

$$NH_2O + CH_3 \longrightarrow CH_3O + NH_2 \quad k_{37f1} = 2 \times 10^{13}\ \text{cm}^3\text{mol}^{-1}\text{s}^{-1}.$$

The DHT method gives in the same units

$$NH_2O + CH_3 \longrightarrow HNO + CH_4 \quad k_{37f2} = 1.6 \times 10^6\, T^{1.87} \exp(-1490)\,.$$

The hydrogen transfer pathway to form NH_2OH and $^3CH_2$ is 130 kJ/mol endothermic and can be neglected. The adduct pathway is substantially faster than the direct one for all combustion conditions. The reaction is therefore chain propagating rather than chain terminating.

(g) $NH_2O + NH_2 \longrightarrow$ Products.

There are no accessible exothermic channels from the adduct, as the barrier to H_2O formation is high. The DHT estimate for the $HNO + NH_3$ channel is

$$k_{37g} = 1.8 \times 10^6\, T^{1.94} \exp(+580/T)\ \text{cm}^3\text{mol}^{-1}\text{s}^{-1}.$$

(h) $NH_2O + HO_2 \longrightarrow$ Products.

Transfer of an H-atom can proceed in two ways,

$$NH_2O + HO_2 \longrightarrow HNO + H_2O_2 \quad (37h1)$$
$$\longrightarrow NH_2OH + O_2\,, \quad (37h2)$$

with comparable exothermicity. The DHT estimate for abstraction by HO_2 gives

$$k_{37h1} = 2.9 \times 10^4 T^{2.69} \exp(+805/T)\ \text{cm}^3\text{mol}^{-1}\text{s}^{-1}\,.$$

Since there is insufficient data to estimate the rate coefficient for abstraction by NH_2O, we estimate $k_{37h2} = k_{37h1}$ based on the comparable exothermicities and expected similarity in A-factors.

2.7.13 Reactions of HNOH

Much of the discussion of NH_2O reactions in the previous section also applies to NHOH, as its relatively high O–H bond dissociation energy, about 230 kJ/mol, has similar implications. Dissociation will be relatively slow under combustion conditions, and so reactions with radicals should be considered. As for NH_2O, hydrogen abstraction reactions from stable molecules would be expected to be unimportant for HNOH, as the strength of the N–H bond formed would only be about 350 kJ/mol.

(a) $HNOH + M \longrightarrow HNO + H + M$.

QRRK analysis shows that the rate coefficient of this reaction is near its low-pressure limit for typical combustion conditions, with

$$k_{38a} = 2.0 \times 10^{24} T^{-2.84} \exp(-29\,700/T)\ \text{cm}^3\text{mol}^{-1}\text{s}^{-1}$$

for M = N_2. While faster than NH_2O dissociation (reaction 37a) it is still not fast enough to be important.

(b) $HNOH + H \longrightarrow H_2NOH \longrightarrow NH_2 + OH$
$\longrightarrow HNO + H_2$.

The adduct formation pathway is exothermic by about 80 kJ/mol and has no significant barrier to dissociation to products. We estimate

$$k_{38b} = 4 \times 10^{13}\ \text{cm}^3\text{mol}^{-1}\text{s}^{-1}.$$

from the Baulch et al. (1992) value for the rate coefficient of $H + C_2H_5 \longrightarrow CH_3 + CH_3$. The DHT estimate for the abstraction pathway is

$$k_{38b2} = 4.8 \times 10^8\, T^{1.5} \exp(-190/T)\ \text{cm}^3\text{mol}^{-1}\text{s}^{-1}.$$

(c) $HNOH + O \longrightarrow HN(O)OH \longrightarrow HNO + OH$
$\longrightarrow HNO + OH.$

Adduct formation followed by exothermic ($\Delta_r H = -200$ kJ/mol) dissociation is expected to be fast. We suggest

$$k_{38c1} = 7 \times 10^{13} \text{ cm}^3\text{mol}^{-1}\text{s}^{-1},$$

which is the Baulch et al. (1992) recommendation for the rate coefficient of the $O + C_2H_5$ reaction.

For the direct hydrogen abstraction path giving the same products the DHT estimate is

$$k_{38c2} = 3.3 \times 10^8 \, T^{1.5} \exp(+180/T) \text{ cm}^3\text{mol}^{-1}\text{s}^{-1}.$$

The two pathways have comparable rates at high temperatures.

(d) $HNOH + OH \longrightarrow HNO + H_2O.$

We expect the adduct formation pathway to be slow in this case because of the high barrier and a low A-factor to the product channel. The DHT estimate for direct reaction is

$$k_{38d} = 2.4 \times 10^6 \, T^2 \exp(+600/T) \text{ cm}^3\text{mol}^{-1}\text{s}^{-1}.$$

(e) $HNOH + CH_3 \longrightarrow CH_3NHOH \longrightarrow CH_3NH + OH$
$\longrightarrow HNO + CH_4.$

We estimate

$$k_{38e1} = 2 \times 10^{13} \text{ cm}^3\text{mol}^{-1}\text{s}^{-1}$$

for the adduct formation pathway on the basis that its 17 kJ/mol exothermicity makes it comparable to reaction 37f1. The DHT estimate for the abstraction pathway is

$$k_{38e2} = 1.6 \times 10^6 \, T^{1.87} \exp(-1055/T) \text{ cm}^3\text{mol}^{-1}\text{s}^{-1}.$$

The adduct formation pathway is faster.

(f) $HNOH + NH_2 \longrightarrow HN(OH)NH_2 \longrightarrow N_2H_3 + OH$
$\longrightarrow H_2NN + H_2O$
$\longrightarrow HNO + NH_3.$

The adduct formation pathway forming water, analogous to $H_2NN + H_2$ formation in the $NH_2 + NH_2$ reaction, has a low barrier, H_2NN being a singlet biradical. A barrier of 85 kJ/mol was assumed for its reverse reaction. QRRK analysis gave, in $\text{cm}^3\text{mol}^{-1}\text{s}^{-1}$ units,

$$\begin{aligned} HNOH + NH_2 &\longrightarrow N_2H_3 + OH, & k_{38f1} &= 6.7 \times 10^6 \, T^{1.82} \exp(-360/T) \\ &\longrightarrow H_2NN + H_2O, & k_{38f2} &= 4.6 \times 10^{19} \, T^{-1.94} \exp(-970/T). \end{aligned}$$

The DHT estimate for the abstraction pathway forming the molecular products HNO and NH_3 is

$$k_{38f3} = 1.8 \times 10^6\, T^{1.94} \exp(+580/T)\ \ \mathrm{cm^3mol^{-1}s^{-1}}.$$

$$\text{(g)}\ \ \mathrm{HNOH + HO_2 \longrightarrow HN(OH)OOH \longrightarrow HONHO + OH}$$
$$\mathrm{\longrightarrow HNO + H_2O_2}$$
$$\mathrm{\longrightarrow NH_2OH + O_2.}$$

The adduct formation pathway is exothermic by 70 kJ/mol and has no significant barrier. We estimate

$$k_{38g1} = 4 \times 10^{13}\ \ \mathrm{cm^3mol^{-1}s^{-1}}.$$

There are two molecular product channels that can be accessed by direct hydrogen abstraction. The DHT estimate for the channel forming HNO and H_2O_2 is

$$k_{38g2} = 2.9 \times 10^4\, T^{2.69} \exp(+805/T)\ \ \mathrm{cm^3mol^{-1}s^{-1}}.$$

In view of its comparable exothermicity and expected similar A-factor, we assume that $NH_2OH + O_2$ product channel has a similar rate and on this basis assign it the same rate coefficient expression as reaction 38g2.

$$k_{38g3} = 2.9 \times 10^4\, T^{2.69} \exp(+805/T)\ \ \mathrm{cm^3mol^{-1}s^{-1}}.$$

2.7.14 Reactions of ^{1}HNOO

Formed in the reaction between ^{3}NH and O_2, this adduct can dissociate quite rapidly (Figure 2.20), and reactions with other radicals can be neglected in combustion modeling. Reaction with O_2 is also expected to be too slow to be significant in combustion. The major reaction channels are thus dissociations. QRRK calculations using the parameters listed in Table 2.7 indicate that the dominant dissociation channel produces OH and NO, followed by dissociation to the reactants ^{3}NH and O_2. Dissociation to H and NO_2 is computed to be over 100 times slower than dissociation back to OH and NO. For combustion conditions this reaction is at the low-pressure limit. The QRRK estimate of its rate coefficient in N_2 is, in $\mathrm{cm^3mol^{-1}s^{-1}}$ units,

$$\mathrm{^1HNOO + M \longrightarrow OH + NO + M}\ \ k_{39} = 1.5 \times 10^{36}\, T^{-6.18} \exp(-15\,670/T)\,.$$

2.7.15 Reactions of HONO

In Section 2.7.8 we discussed the formation of HONO from OH and NO. Because of the relative stability of HONO its reactions with radical species have to be considered. The H–O bond strength is only 330 kJ/mol, so H-atom abstraction from HONO would be generally quite exothermic. Abstraction rate coefficients estimated by the DHT method are included in Table 2.19.

HONO thermal dissociation to form OH and NO was analyzed with the QRRK method. The dissociation rate coefficient for combustion conditions was found to be near its low-pressure limit even up to pressures characteristic of internal combustion engines and turbines. For N_2 bath gas we estimate

$$k_{40a} = 2.0 \times 10^{31} T^{-4.56} \exp(-25\,800/T)\ \text{cm}^3\text{mol}^{-1}\text{s}^{-1}.$$

Dissociation was found to dominate over isomerization to HNO_2, as expected from its much larger A-factor, that is, much looser transition state. Rate measurements on the reverse reaction OH + NO $\longrightarrow$ HONO by Zabarnick (1993) indicate that the high-pressure limit rate coefficient for this reaction is above 8×10^{12} $\text{cm}^3\text{mol}^{-1}\text{s}^{-1}$, and the study by Forster et al. (1995) at pressures up to 150 bar gave $k_\infty = 2 \times 10^{13}$ $\text{cm}^3\text{mol}^{-1}\text{s}^{-1}$ at 298 K. These values are consistent with the assumptions made for QRRK calculations (cf. Table 2.7).

Hsu et al. (1997) reported a molecular electronic structure study of the H + HONO reactions. Their calculations suggest that the adduct formation pathways are more important than direct abstraction, especially at lower temperatures. Their value for the abstraction channel, in $\text{cm}^3\text{mol}^{-1}\text{s}^{-1}$ units, is

$$\text{HONO} + \text{H} \rightarrow \text{H}_2 + \text{NO}_2\,, \quad k_{40b1} = 2.0 \times 10^{8}\, T^{1.55} \exp(-3330/T)\,,$$

which is lower than the DHT estimate at 300 K but within a factor of 2 of it at 1000 and 2000 K. Their results for the channels resulting from adduct formation are, in $\text{cm}^3\text{mol}^{-1}\text{s}^{-1}$ units

$$\begin{aligned} \text{HONO} + \text{H} &\longrightarrow \text{H}_2\text{O} + \text{NO} \quad & k_{40b2} &= 8.1 \times 10^{6}\ T^{1.89} \exp(-1935/T) \\ &\longrightarrow \text{OH} + \text{HNO} \quad & k_{40b3} &= 5.6 \times 10^{10}\, T^{0.86} \exp(-2500/T)\,. \end{aligned}$$

2.7.16 Reactions of HNO_2

Figure 2.20 illustrates some routes for the production of HNO_2. QRRK analysis suggests that the main unimolecular process is isomerization to HONO, which is in the falloff pressure regime for combustion conditions. For N_2 bath gas the QRRK result is in s^{-1} units

$$\begin{aligned} k_{41} &= 7.1 \times 10^{27} T^{-5.40} \exp(-26\,400/T) && 0.1\ \text{atm} \\ &= 1.3 \times 10^{29} T^{-5.47} \exp(-26\,600/T) && 1.0\ \text{atm} \\ &= 2.0 \times 10^{30} T^{-5.50} \exp(-27\,000/T) && 10\ \text{atm}. \end{aligned}$$

Because of their similar thermochemistry one expects abstractions from HNO_2 to be comparable to those from HONO. Rate coefficient expressions derived by the DHT method are listed in Table 2.19. A cause for concern is that the DHT expression for OH + $HNO_2 \longrightarrow H_2O + NO_2$ gives a room-temperature value appreciably lower than the one reported by Atkinson et al. (1989b). Although the room-temperature values estimated by the DHT method for abstractions by H and O-atoms are larger by factors of 20 and 30 than those estimated by Tsang and Herron (1991), the differences decrease to about a factor of 3 at temperatures above 1000 K.

2.7.17 Reactions of HCN

Because the C–H bond in HCN is very strong, about 520 kJ/mol, thermal dissociation is unimportant in combustion. The reactions which are most likely to be important are hydrogen abstraction by OH, addition to the triple bond, and isomerization to HNC:

$$\mathrm{HCN} \longrightarrow \mathrm{HNC} \quad \Delta_r H = 54\,\mathrm{kJ/mol}. \tag{42}$$

Isomerization was reported by Lin et al. (1992b) to be an important reaction of HCN at high temperatures. The isomerization barrier of about 200 kJ/mol permits isomerization to be a facile unimolecular reaction pathway for the relatively stable HCN molecule. Lin et al. computed the rate coefficient expression at infinite pressure to be

$$k_{42}^{\infty} = 3.5 \times 10^{13} \exp(-23,750/T)\,\mathrm{s}^{-1}.$$

One expects the actual isomerization rate coefficient of such a small molecule to be far from its high-pressure limit under combustion conditions. QRRK analysis showed that the reaction should usually be in the low-pressure limit at temperatures above 1000 K, the rate coefficient for N_2 being given by the expression

$$k_{42} = 1.6 \times 10^{26} T^{-3.23} \exp(-24950/T)\ \mathrm{cm^3 mol^{-1} s^{-1}}.$$

At 300 K the reaction is in the falloff regime at pressures near 10 atm. Rate coefficient expressions at 0.1, 1 and 10 atm are included in Table 2.19.

Wooldridge et al. (1995) reported that their experiments on OH + HCN kinetics were not affected by this isomerization. Our QRRK-derived falloff expression, however, suggests that the isomerization reaction proceeds on the same time scale as their experiments, with half-times ranging from 10 to 100 μs.

(a) HCN + OH $\longrightarrow$ CN + H_2O.

Baulch et al. (1992) recommend an expression similar to ones found for H-atom abstraction from hydrocarbons, $k = 9.0 \times 10^{12} \exp(-5400/T)\ \mathrm{cm^3 mol^{-1} s^{-1}}$, significantly greater than the measurements of Wooldridge et al. (1995) (Figure 2.61). We recommend their expression,

$$k_{42a} = 3.9 \times 10^{6}\, T^{1.83} \exp(-5180/T)\ \mathrm{cm^3 mol^{-1} s^{-1}}.$$

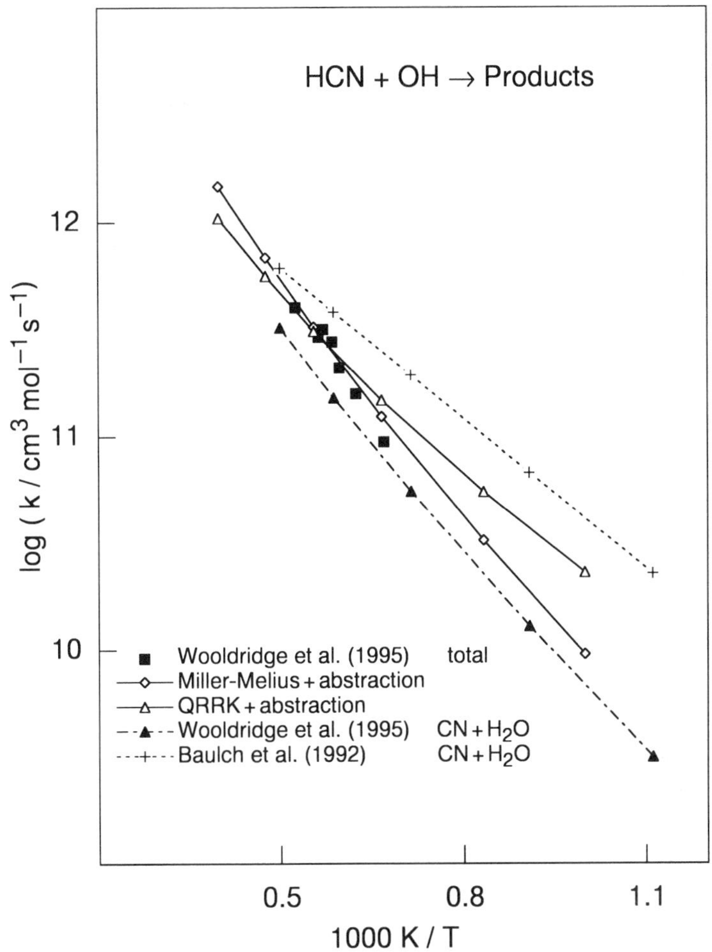

FIGURE 2.61. Rate coefficients for OH + HCN ⟶ Products. The results for the abstraction channel forming CN + H_2O reported by Wooldridge et al. (1995) are compared to the recommendation of Baulch et al. (1992). The total rate coefficient of Wooldridge et al. is also shown. The abstraction channel is dominant. The total is also compared to the sum of the rate coefficient measured for the abstraction channel and calculations of the total addition rate coefficient for all 3 addition channels. The comparisons indicate that both the Miller–Melius and the QRRK calculations give the correct magnitude for the addition channel rate coefficients, but that the temperature dependence of the Miller-Melius expression is more consistent with the data.

(b) HCN + OH ⟶ Products.

A potential energy diagram for the addition step is shown in Figure 2.51, where it appears as one of the product channels for the CH_2 + NO reaction. The originally formed NCHOH adduct (the higher-energy HCNOH can be safely ignored) has several possible dissociation paths, for which Miller and Melius (1986) have reported rate coefficients. Fitting their results over the temperature range from 1000 to 2500 K leads to the expressions in $cm^3mol^{-1}s^{-1}$ units

$$\begin{aligned} \text{HCN + OH} &\longrightarrow \text{HNCO + H} & k_{42b1} &= 4.4 \times 10^3 T^{2.26} \exp(-3220/T) \\ &\longrightarrow \text{HOCN + H} & k_{42b2} &= 1.1 \times 10^6 T^{2.03} \exp(-6730/T) \\ &\longrightarrow \text{CO + NH}_2 & k_{42b3} &= 1.6 \times 10^2 T^{2.56} \exp(-4530/T). \end{aligned}$$

QRRK analysis shows that at 1 atm pressure and temperatures less than 1000 K formation of the stabilized NCHOH adduct dominates, while at higher temperatures the major channels yield HNCO + H and HOCN + H. (As discussed in Section 2.6.19, the lowest energy NH_2 + CO channel is disfavored because of the isomerization step required to access it.) QRRK analysis for N_2 bath gas led to the rate coefficient expressions in $cm^3mol^{-1}s^{-1}$ units

$$\begin{aligned} \text{HCN + OH} &\longrightarrow \text{HNCO + H} & k &= 3.5 \times 10^{11} T^{0.02} \exp(-3560/T) \\ &\longrightarrow \text{HOCN + H} & k &= 9.9 \times 10^{10} T^{0.31} \exp(-5350/T) \\ &\longrightarrow \text{CO + NH}_2 & k &= 1.1 \times 10^{11} T^{-0.18} \exp(-3590/T) \end{aligned}$$

$$\begin{aligned} &\longrightarrow \text{NCHOH} & k_{42b4} &= 1.7 \times 10^{29} T^{-6.31} \exp(-2580/T) \quad 0.1 \text{ atm} \\ &\longrightarrow \text{NCHOH} & k_{42b4} &= 2.8 \times 10^{30} T^{-6.37} \exp(-2690/T) \quad 1.0 \text{ atm} \\ &\longrightarrow \text{NCHOH} & k_{42b4} &= 1.1 \times 10^{32} T^{-6.53} \exp(-3140/T) \quad 10 \text{ atm.} \end{aligned}$$

QRRK analysis generally indicates faster rates near 1000 K and a more gradual temperature dependence than were calculated by Miller and Melius. The differences could be due to small variations in the barrier heights. The predictions are compared to the measurements of Wooldridge et al. (1995) in Figure 2.61, where it can be seen that the abstraction channel (42a) dominates and that the sum of the calculated rate coefficients for the three adduct formation pathways and the measured rate coefficient for the abstraction channel agrees with the measured total rate coefficient. We suggest use of the Miller-Melius expressions for the non-adduct channels and the QRRK result for adduct formation.

(c) HCN + O $\longrightarrow$ Products.

A potential energy diagram for this addition reaction is shown in Figure 2.58, where it appears as one of the product channels for the CH + NO reaction. The NCHO adduct can react in a variety of ways. Perry and Melius (1984) analyzed this reaction and reported the rate coefficient expressions in $cm^3mol^{-1}s^{-1}$ units

$$\begin{aligned} \text{HCN + O} &\longrightarrow \text{NH + CO} & k_{42c1} &= 5.4 \times 10^8 T^{1.21} \exp(-3770/T) \\ &\longrightarrow \text{NCO + H} & k_{42c2} &= 2.0 \times 10^8 T^{1.47} \exp(-3820/T) \\ &\longrightarrow \text{CN + OH} & k &= 2.7 \times 10^9 T^{1.58} \exp(-13\,400/T). \end{aligned}$$

Formation of stabilized adducts is relatively unimportant. The significant channels are found by QRRK analysis to have the rate coefficient expressions in $cm^3mol^{-1}s^{-1}$ units

$$\begin{aligned} \mathrm{HCN + O} &\longrightarrow \mathrm{NH + CO} & k &= 4.0 \times 10^{14} T^{-0.65} \exp(-4285/T) \\ &\longrightarrow \mathrm{NCO + H} & k &= 7.6 \times 10^{10} T^{0.48} \exp(-3930/T) \\ &\longrightarrow \mathrm{CN + OH} & k_{42c3} &= 4.2 \times 10^{10} T^{0.40} \exp(-10\,400/T). \end{aligned}$$

In general, these QRRK expressions give somewhat smaller rate coefficients, essentially because Perry and Melius decreased all barriers by 11 kJ/mol in order to improve the fit to the high-temperature data. A major difference is that the QRRK results suggest a much lower rate for production of CN + OH. Since the Perry–Melius expression predicts rate coefficients for this channel about a factor of 3 higher than the measured upper limit at 2000 K of Szekely et al. (1984), we suggest use of the QRRK estimate for that channel.

2.7.18 Reactions of HNC

Although the strong H–N bond should make abstractions from HNC unimportant, radicals such as O and OH should react with it by other pathways. Lin et al. (1992b) obtained the rate coefficient expressions, in $\mathrm{cm^3 mol^{-1} s^{-1}}$ units,

$$\begin{aligned} \mathrm{HNC + O} &\longrightarrow \mathrm{HNCO} \longrightarrow \mathrm{NH + CO} & k_{43a} &= 4.6 \times 10^{12} \exp(-1100/T) \\ \mathrm{HNC + OH} &\longrightarrow \mathrm{HNCOH} \longrightarrow \mathrm{HNCO + H} & k_{43b} &= 2.8 \times 10^{13} \exp(-1860/T), \end{aligned}$$

by modeling CO formation during HCN oxidation by NO_2 over the 623–773 K temperature range.

HNC might also be expected to add to O_2. A QRRK analysis was done considering two product channels. The first was bond scission to form HNCO + O. Although this channel is 38 kJ/mol lower in energy than the reactants, the barrier between it and the adduct is expected to be 10 to 15 kJ/mol above the entrance channel (Melius 1993). The second pathway is formation of a 3-membered ring and subsequent formation of NH + CO_2. This barrier is expected to be lower than the entrance channel (Carpenter 1993), but a lower A-factor is to be expected because of the ring formation. The QRRK calculations led to the rate coefficient expressions in $\mathrm{cm^3 mol^{-1} s^{-1}}$ units

$$\begin{aligned} \mathrm{HNC + O_2} &\longrightarrow \mathrm{HNCO + O} & k_{43c1} &= 1.5 \times 10^{12} T^{0.01} \exp(-2070/T) \\ &\longrightarrow \mathrm{NH + CO_2} & k_{43c2} &= 1.6 \times 10^{19} T^{-2.25} \exp(-890/T). \end{aligned}$$

The higher A-factor for dissociation of the adduct to HNCO + O causes this channel to be more important at combustion temperatures, while NH + CO_2 dominates at lower temperatures.

2.7.19 Reactions of CN

The H–CN bond of hydrogen cyanide is so strong, about 520 kJ/mol, that one expects hydrogen transfer reactions to CN to be very important. Since CN itself is extremely stable, about 750 kJ/mol more stable than C + N, it can also react by radical addition and recombination.

(a) $CN + H_2 \longrightarrow HCN + H$.

The abstraction reaction from H_2 was studied in shock tubes by Natarajan et al. (1988) and Wooldridge et al. (1996a) and at lower temperatures by Atakan et al. (1989). Tsang (1992) has reviewed this reaction. We adopt his rate coefficient recommendation, which is consistent with these experimental results:

$$k_{44a} = 3.6 \times 10^8 T^{1.55} \exp(-1510/T) \ \text{cm}^3\text{mol}^{-1}\text{s}^{-1}.$$

(b) $CN + H_2O \longrightarrow HCN + OH$.

This is one of the few exothermic abstractions from water. Because of the large concentrations of water found in post-flame gases, it can be important in interconverting CN and HCN. Its rate coefficient can be inferred from the rate coefficient expression for the reverse reaction 42a.

(c) $CN + O \longrightarrow OCN \longrightarrow CO + N$.

This reaction is 322 kJ/mol exothermic overall. A likely mechanism is adduct formation by recombination followed by dissociation to low-energy products. It should be very fast, analogous to reaction 38b. This is consistent with the high-temperature measurements of Davidson et al. (1991b), who report

$$k_{44c} = 7.7 \times 10^{13} \ \text{cm}^3\text{mol}^{-1}\text{s}^{-1}.$$

(d) $CN + O_2 \longrightarrow NCOO \longrightarrow NCO + O$.

The overall reaction is exothermic by about 55 kJ/mol, suggesting that one should expect rate falloff at high temperatures as dissociation of the adduct back to the entrance channel becomes significant; a higher A-factor for entrance channel dissociation is expected because of the higher entropy for two diatomic products. Davidson et al. (1991b) reported a high-temperature measurement of

$$k_{44d} = 1.0 \times 10^{13} \ \text{cm}^3\text{mol}^{-1}\text{s}^{-1},$$

consistent with the Baulch et al. (1992) recommendation.

(e) $CN + OH \longrightarrow HOCN \longrightarrow H + NCO$.

This reaction is about 125 kJ/mol exothermic, suggesting that virtually all of the HOCN adduct dissociates to H + NCO. An estimate can be made by analogy to the recombination rate coefficient of $CH_3 + OH$. Using the Baulch et al. (1992) rate coefficient for $CH_3 + OH$, we infer for CN + OH

$$k_{44e} = 4 \times 10^{13} \ \text{cm}^3\text{mol}^{-1}\text{s}^{-1},$$

which is close to the value suggested by Tsang (1992) and identical to the value measured by Wooldridge et al. (1996a) over the temperature range 1250 to 1860 K.

(f) $CN + HCN \longrightarrow HC(CN)N \longrightarrow NCCN + H$.

This can be viewed as an addition–elimination reaction; overall it is exothermic by about 42 kJ/mol. One might again expect to see some falloff in rate at high temperatures owing to onset of the reverse reaction. This is not reflected, however, in the data summarized by Tsang (1992). We suggest adoption of his recommendation,

$$k_{44f} = 1.5 \times 10^7 T^{1.71} \exp(-770/T) \ \text{cm}^3\text{mol}^{-1}\text{s}^{-1},$$

which is consistent with the results of Wooldridge et al. (1996a).

(g) $CN + N_2O \longrightarrow$ Products.

CN might be expected to add at either end of the molecule, in a manner analogous to $H + N_2O$ (Sections 2.6.3 through 2.6.6). Addition to the O end would lead to NNOCN, which could dissociate to $NCO + N_2$. This overall reaction is exothermic by about 360 kJ/mol. With such a low-energy product channel one might expect the rate-limiting step to be the adduct formation. Addition to the N end would give NCNNO, which would have to isomerize first in order to access a low-energy product channel. Wang et al. (1991) report for the temperature range from 500 to 750 K $k = 9.5 \times 10^{11} \exp(-3560/T)$ $\text{cm}^3\text{mol}^{-1}\text{s}^{-1}$. They propose that the barrier to form NNOCN is sufficiently high that NCN + NO is the dominant reaction channel at their conditions. This analysis is consistent with the measurements of Williams et al. (1995), who report that these are the dominant products from 400 to 870 K, with the $NCO + N_2$ yield being less than 10%. There was some pressure dependence, the measurement at 100 Torr being expressed as $k = 6.0 \times 10^{13} \exp(-7730/T) + 1.9 \times 10^9 \exp(-730/T)$ $\text{cm}^3\text{mol}^{-1}\text{s}^{-1}$, about a factor of 3 smaller than the Wang et al. expression. Since this expression includes a contribution from the stabilization channel, we extracted an estimate for the NCN + NO channel from the low-pressure, high-temperature measurements alone:

$$k_{44g} = 4.2 \times 10^{11} \exp(-3610/T) \ \text{cm}^3\text{mol}^{-1}\text{s}^{-1}.$$

At higher temperatures additional reaction to form $NCO + N_2$ is to be expected. The total rate coefficient is nonetheless small enough that this reaction is not likely to be important in combustion.

(h) $CN + NO_2 \longrightarrow NCO + NO$.

The simplest way for this reaction to lead to these products is for CN to add to one of the oxygens to form NCONO, which would then dissociate to NCO + NO. Measurements by Park and Herschberger (1993b) indicate that this channel accounts for 87% of the total reaction. The rate coefficient has also been measured at higher temperatures in shock tube experiments by Wooldridge et al. (1994). Combining their data with that of Park and Herschberger and that of Wang et al. (1989), Wooldridge et al. report

$$k_{44h} = 6.2 \times 10^{15} T^{-0.752} \exp(-170/T) \ \text{cm}^3\text{mol}^{-1}\text{s}^{-1}$$

for temperatures from 300 to 1600 K. This very fast rate suggests that CN addition to NO_2 behaves as radical recombination, quite reasonable considering that NO_2 is indeed a radical.

(i) $CN + CH_4 \longrightarrow HCN + CH_3$.

We include this reaction as representative of CN abstractions that can take place in a hydrocarbon environment. Methane should not be taken as fully representative of all hydrocarbons, however, and this rate coefficient would not be appropriate for hydrogen abstraction from hydrocarbons having H–C bond strengths different from that of methane. There have been two rate measurements, in reasonable agreement with one another over extended temperature ranges, by Atakan and Wolfrum (1991) and Balla et al. (1991). We suggest adoption of the Atakan and Wolfrum expression

$$k_{44i} = 1.2 \times 10^{5} T^{2.64} \exp(+80/T) \ \text{cm}^3\text{mol}^{-1}\text{s}^{-1}.$$

(j) $CN + NH_3 \longrightarrow HCN + NH_2$.

Sims and Smith (1988) derived the rate coefficient expression

$$k_{44j} = 9.2 \times 10^{12} \exp(+180/T) \ \text{cm}^3\text{mol}^{-1}\text{s}^{-1}$$

from experimental measurements over the temperature range from 294 to 761 K. The dramatic difference in the temperature dependence of the reactions of CN with methane and ammonia bears some discussion. Sims and Smith studied the ammonia reaction over a wide temperature range, so the observed negative temperature dependence cannot be dismissed as based upon a too-limited investigation. Nonetheless, negative temperature dependence is very unusual for an abstraction reaction. In fact, it is really only consistent with an addition mechanism. Calculations by Meads et al. (1993) support formation of an H_3NCN complex and subsequent rearrangement to yield the same products one would expect from a simple abstraction. Meads et al. computed a rate coefficient and showed that it decreases with increasing temperature, in qualitative agreement with the experimental data.

2.7.20 Reactions of H_2CN

We have seen earlier that H_2CN is predicted to be a product of several reactions, including CH_3 + NO (Section 2.6.15). Dissociation to HCN + H is only 106 kJ/mol endothermic, so rapid dissociation at combustion temperatures should occur. At lower temperatures, one might expect H_2CN to transfer a weakly bonded H to radicals in a disproportionation reaction. Alternatively, radical–radical reactions could result in formation of an energized adduct and subsequent fragmentation. Most abstractions by H_2CN would be endothermic and therefore relatively unimportant, because the strength of the H–N bond formed in such a reaction would be only about 370 kJ/mol.

(a) $H_2CN \longrightarrow HCN + H$.

We are not aware of any measurements of the rate coefficient of this reaction. QRRK calculations indicate that it is in the falloff regime for combustion conditions, with rate coefficients for N_2 bath gas, in s^{-1} units,

$$\begin{aligned} k_{45a} &= 1.3 \times 10^{29} T^{-6.03} \exp(-15\,000/T) && 0.1 \text{ atm} \\ &= 6.0 \times 10^{31} T^{-6.46} \exp(-16\,200/T) && 1.0 \text{ atm} \\ &= 3.5 \times 10^{29} T^{-5.46} \exp(-16\,400/T) && 10 \text{ atm.} \end{aligned}$$

(b) $H_2CN + HO_2 \longrightarrow CH_2NO + OH$.

By analogy to hydrocarbon radical reactions with HO_2 one expects a rapid addition–dissociation reaction to break the weak O–O bond. On this basis we estimate

$$k_{45b} = 3 \times 10^{13} \text{ cm}^3\text{mol}^{-1}\text{s}^{-1},$$

which can be compared to the 2.4×10^{13} $\text{cm}^3\text{mol}^{-1}\text{s}^{-1}$ value suggested by Tsang and Hampson (1986) for the analogous channel of the C_2H_5 + HO_2 reaction. DHT estimates for the H transfer reactions are included in Table 2.19; these approach the adduct formation value at combustion temperatures.

(c) $H_2CN + O_2 \longrightarrow CH_2O + NO$.

Reaction by an addition pathway would be expected to result, by analogy to C_2H_3 + O_2, in these products, as formation of CH_2NO + O is endothermic by about 170 kJ/mol and therefore unimportant. The initially formed adduct should be appreciably less stable than C_2H_3OO, which implies a larger barrier for the cyclization to yield CH_2O + NO. From these considerations we estimate

$$k_{45c} = 3 \times 10^{12} \exp(-3000/T) \text{ cm}^3\text{mol}^{-1}\text{s}^{-1}.$$

(d) $H_2CN + CH_3 \longrightarrow HCN + CH_4$.

This is an example of H_2CN participating in radical disproportionation reactions. The DHT rate coefficient estimate is

$$k_{45d} = 8.1 \times 10^{5}\, T^{1.87} \exp(+560/T) \text{ cm}^3\text{mol}^{-1}\text{s}^{-1}.$$

(e) $H_2CN + OH \longrightarrow HCN + H_2O$.

This reaction is analogous to C_2H_5 + OH. For the reaction with C_2H_5 one can rationalize the observed very fast rate in terms of radical recombination followed by elimination of H_2O. A significant difference, however, is that the well depth of the H_2CNOH complex is not as deep as that of CH_3CH_2OH; consequently, the barrier for concerted H_2O elimination is closer to the entrance channel. For this reason one expects a falloff in the forward rate at higher temperatures as more of the adduct dissociates through the entrance channel. QRRK calculations for H_2CN reacting with OH in N_2 bath gas yield

$$\begin{aligned} k_{45e} &= 2.1 \times 10^{17} T^{-1.68} \exp(-160/T) \ \text{cm}^3\text{mol}^{-1}\text{s}^{-1} \quad 0.1 \text{ atm} \\ &= 1.5 \times 10^{19} T^{-2.18} \exp(-1090/T) \ \text{cm}^3\text{mol}^{-1}\text{s}^{-1} \quad 1.0 \text{ atm} \\ &= 9.5 \times 10^{21} T^{-2.91} \exp(-2830/T) \ \text{cm}^3\text{mol}^{-1}\text{s}^{-1} \quad 10 \text{ atm.} \end{aligned}$$

These expressions give appreciably smaller rate coefficients than are measured for OH + C_2H_5, and there is noticeable falloff for combustion conditions. The QRRK results also show an inverse pressure dependence, especially at low temperatures, where the adduct is in part stabilized by collisions.
The same products can also be formed directly. The DHT estimate is

$$k_{45e2} = 1.2 \times 10^6 \, T^2 \exp(+600/T) \ \text{cm}^3\text{mol}^{-1}\text{s}^{-1}.$$

At 2000 K direct pathway is calculated to be about 10 times faster than the adduct formation pathway. Both should be considered important under combustion conditions.

(f) $H_2CN + N \longrightarrow H_2CNN \longrightarrow CH_2 + N_2$.

This reaction is very exothermic and thus expected to be fast. Nesbitt et al. (1990) measured the rate of reactant disappearance over the temperature range from 200 to 363 K; assuming that the reaction has only the channel shown gives

$$k_{45f} = 6.0 \times 10^{13} \exp(-200/T) \ \text{cm}^3\text{mol}^{-1}\text{s}^{-1}.$$

(g) $H_2CN + H \longrightarrow HCN + H_2$.

The DHT estimate is

$$k_{45g} = 2.4 \times 10^8 \, T^{1.5} \exp(+450/T) \ \text{cm}^3\text{mol}^{-1}\text{s}^{-1}.$$

(h) $H_2CN + NH_2 \longrightarrow HCN + NH_3$.

The DHT estimate is

$$k_{45h} = 9.2 \times 10^5 \, T^{1.94} \exp(+580/T) \ \text{cm}^3\text{mol}^{-1}\text{s}^{-1}.$$

(i) $H_2CN + O \longrightarrow HCN + OH$

The DHT estimate is

$$k_{45i} = 1.7 \times 10^8\, T^{1.5} \exp(+450/T)\ \ \mathrm{cm^3mol^{-1}s^{-1}}.$$

Adduct formation leads to CH_2NO (Figure 2.51). QRRK analysis shows the dominant channels to be H atom loss:

$$\begin{aligned} H_2CN + O &\longrightarrow HNCO + H, \quad k_{45i2} = 6 \times 10^{13}\ \ \mathrm{cm^3mol^{-1}s^{-1}} \\ &\longrightarrow HCNO + H, \quad k_{45i2} = 2 \times 10^{13}\ \ \mathrm{cm^3mol^{-1}s^{-1}}. \end{aligned}$$

2.7.21 Reactions of HCNH

This species is formed in the CH_3 + N reaction (Section 2.6.16). Dissociation by loss of the weakly bound N-hydrogen,

$$HCNH \longrightarrow HCN + H \quad \Delta_r H = +75\ \mathrm{kJ/mol}, \tag{46}$$

should have a rate intermediate between those of NNH (Section 2.7.3) and HNO (Section 2.7.11) dissociation. The QRRK estimates for N_2 bath gas are

$$\begin{aligned} k_{46} &= 7.7 \times 10^{25} T^{-5.20} \exp(-11\,100/T)\ \mathrm{s^{-1}} && 0.1\ \mathrm{atm} \\ &= 6.1 \times 10^{28} T^{-5.69} \exp(-12\,200/T)\ \mathrm{s^{-1}} && 1.0\ \mathrm{atm} \\ &= 6.2 \times 10^{26} T^{-4.77} \exp(-12\,500/T)\ \mathrm{s^{-1}} && 10\ \mathrm{atm.} \end{aligned}$$

Under typical combustion conditions dissociation will be very rapid and other reactions of HCNH unimportant. However, as with NNH, important exceptions could arise in flame fronts where the mole fractions of H, OH, and O are high, approaching 1%, or concentrations of about 60 nmol/cm^3 at 1 atm. At such concentrations, a bimolecular rate coefficient for reaction with a flame radical over 10^{13} cm^{3}mol^{-1}s^{-1} will imply a first-order loss rate of HCNH above 10^6 s^{-1}, approaching the thermal dissociation loss rate, which at 2000 K and 1 atm is estimated to be about 2×10^7 s^{-1}. This is somewhat slower than for NNH dissociation, suggesting a greater chance for reactions of HCNH with flame radicals. We accordingly estimated these rate coefficients. Some of the estimates proved to be large enough to suggest that such bimolecular reactions can occur.

(a) HCNH + H $\longrightarrow$ Products.

Formation of H_2CNH liberates about 400 kJ/mol, providing ample energy for production of either CH_2N + H or HNC + H_2. Although the barrier for H_2 elimination is probably lower than that for production of CH_2N + H, we expect the latter, with its higher A-factor, will be most important in the high-temperature region where this reaction might occur. We estimate

$$HCNH + H \longrightarrow CH_2N + H \qquad k_{46a1} = 2 \times 10^{13}\ \ \mathrm{cm^3mol^{-1}s^{-1}}$$

from taking half the rate coefficient proposed by Baulch et al. (1992) for H + $C_2H_5 \longrightarrow 2\ CH_3$, which has similar energetics but would be expected to have a greater A-factor for adduct dissociation. The DHT estimate for the direct reaction is, in $cm^3mol^{-1}s^{-1}$ units,

$$HCNH + H \longrightarrow HCN + H_2 \quad k_{46a2} = 2.4 \times 10^8\, T^{1.5} \exp(+450/T)\,.$$

(b) HCNH + O $\longrightarrow$ Products.

The adduct HC(O)NH can dissociate rapidly to HNCO + H, which is overall 410 kJ/mol exothermic. We suggest that the rate coefficient be taken as the Baulch et al. (1992) recommendation for O + C_2H_5

$$HCNH + O \longrightarrow HNCO + H \qquad k_{46b1} = 7 \times 10^{13}\ cm^3mol^{-1}s^{-1}.$$

The DHT estimate for the direct reaction is, in $cm^3mol^{-1}s^{-1}$ units,

$$HCNH + O \longrightarrow HCN + OH \quad k_{46b2} = 1.7 \times 10^8\, T^{1.5} \exp(+450/T)\,.$$

(c) HCNH + OH $\longrightarrow$ Products.

Formation of H_2O from the HC(OH)NH adduct has a substantial barrier and a low A-factor. The same products can be formed by direct reaction, for which the DHT estimate is, in $cm^3mol^{-1}s^{-1}$ units

$$HCNH + OH \longrightarrow HCN + H_2O \quad k_{46c} = 1.2 \times 10^6\, T^2 \exp(+600/T)\,.$$

(d) HCNH + $CH_3 \longrightarrow$ HCN + CH_4.

The DHT estimate is

$$k_{46d} = 8.2 \times 10^5\, T^{1.87} \exp(+560/T)\ cm^3mol^{-1}s^{-1}.$$

2.7.22 Reactions of HCNN

This is the stable adduct formed in the reaction between CH and N_2 (Section 2.6.20). It redissociates quickly to reactants under combustion conditions, with a rate coefficient for dissociation that can be inferred from the formation rate coefficient. Because of the relatively weak C–N bond this channel will be so fast that reactions with other radicals should be unimportant. The reaction with O_2, however, should be considered. The adduct for this reaction can give two sets of products:

$$HCNN + O_2 \longrightarrow HCO_2 + N_2 \quad \Delta_r H = -615\,kJ/mol \qquad (47a)$$

$$\longrightarrow HCO + N_2O \quad \Delta_r H = -330\,kJ/mol. \qquad (47b)$$

The highly exothermic nature of these reactions suggests use of rate coefficients near that anticipated for formation of the initial adduct,

$$k_{47a} = k_{47b} = 4 \times 10^{12}\ cm^3mol^{-1}s^{-1}.$$

HCO_2 should rapidly dissociate to H + CO_2.

2.7.23 Reactions of H_2CNH

This species can be formed by the reaction of N with CH_3 (Section 2.6.16) and is also expected to play a role in methylamine pyrolysis and oxidation. Hydrogen abstractions from it can leave two different radicals:

$$H_2CNH + X \longrightarrow H_2CN + HX \quad (48a)$$
$$\longrightarrow HCNH + HX. \quad (48b)$$

The strength of the weaker H–N bond is about 370 kJ/mol. For estimating rate coefficients for abstracting from it we used the DHT method and obtained the rate coefficient expressions in $cm^3 mol^{-1} s^{-1}$ units:

$$\begin{aligned}
H_2CNH + H &\longrightarrow H_2CN + H_2 & k_{48a1} &= 2.4 \times 10^8 T^{1.5} \exp(-3685/T) \\
+ O &\longrightarrow H_2CN + OH & k_{48a2} &= 1.7 \times 10^8 T^{1.5} \exp(-2330/T) \\
+ OH &\longrightarrow H_2CN + H_2O & k_{48a3} &= 1.2 \times 10^6 T^2 \exp(+45/T) \\
+ CH_3 &\longrightarrow H_2CN + CH_4 & k_{48a4} &= 8.2 \times 10^5 T^{1.87} \exp(-3585/T) \\
+ NH_2 &\longrightarrow H_2CN + NH_3 & k_{48a5} &= 9.2 \times 10^5 T^{1.94} \exp(-2235/T).
\end{aligned}$$

For abstraction of an H attached to the C atom the bond strength is about 400 kJ/mol, compared to 465 kJ/mol in ethylene, reflecting resonance stabilization of the radical. The DHT estimate was corrected accordingly (cf. Section 2.4). The DHT estimates are, in $cm^3 mol^{-1} s^{-1}$ units,

$$\begin{aligned}
H_2CNH + H &\longrightarrow HCNH + H_2 & k_{48b1} &= 3.0 \times 10^8 T^{1.5} \exp(-3085/T) \\
+ O &\longrightarrow HCNH + OH & k_{48b2} &= 2.2 \times 10^8 T^{1.5} \exp(-2720/T) \\
+ OH &\longrightarrow HCNH + H_2O & k_{48b3} &= 2.4 \times 10^6 T^2 \exp(-230/T) \\
+ CH_3 &\longrightarrow HCNH + CH_4 & k_{48b4} &= 5.3 \times 10^5 T^{1.87} \exp(-4875/T) \\
+ NH_2 &\longrightarrow HCNH + NH_3 & k_{48b5} &= 1.8 \times 10^6 T^{1.94} \exp(-3065/T).
\end{aligned}$$

(a) $H_2CNH + O \longrightarrow CH_2O + NH$.

This reaction is exothermic by 100 kJ/mol and expected to be fast. We estimate it to have half of the addition rate of O atoms to ethylene and derive from the Baulch et al. (1992) expression

$$k_{48c} = 1.7 \times 10^6 T^{2.08} \ cm^3 mol^{-1} s^{-1}.$$

2.7.24 Reactions of CH_3NH

The weak C–H bond suggests that dissociation

$$CH_3NH \longrightarrow CH_2NH + H \quad \Delta_r H = 125 \text{ kJ/mol}$$

is the most important reaction of CH_3NH under combustion conditions. The QRRK method predicts for the temperature range from 600 to 2500 K in nitrogen

$$\begin{aligned}
k_{49} &= 1.6 \times 10^{36} T^{-7.92} \exp(-18\,300/T) \text{ s}^{-1} & & 0.1 \text{ atm} \\
&= 1.3 \times 10^{42} T^{-9.24} \exp(-20\,800/T) \text{ s}^{-1} & & 1.0 \text{ atm} \\
&= 2.3 \times 10^{44} T^{-9.51} \exp(-22\,800/T) \text{ s}^{-1} & & 10 \text{ atm.}
\end{aligned}$$

For combustion conditions this should be the only reaction of interest. We explored the possibility of addition to O_2 by QRRK analysis. The combination of a shallow potential well (estimated to be about 35 kJ/mol by analogy to the NH_2+O_2 addition reaction discussed in Section 2.6.9) and a high A-factor for dissociation of the adduct to reactants results in low rate coefficients for both product channels, that is, to $CH_3NHO + O$ and to $H_2CNH + HO_2$. At 2000 K the computed rate coefficients are less than 10^9 $cm^3mol^{-1}s^{-1}$, indicating pseudo-first-order rate coefficients in air on the order of 10^3 s^{-1}. In contrast, the dissociation rate coefficient for these conditions is about 1×10^7 s^{-1}. The shallow well also insures that reactions of the stabilized adduct can also be neglected. DHT estimates of the rate coefficients for direct hydrogen transfer to H, O, OH and CH_3 have been included in Table 2.19.

2.7.25 Reactions of CH_2NH_2

The weak N–H bond means that CH_2NH_2 is another species that can be expected to dissociate rapidly

$$CH_2NH_2 \longrightarrow CH_2NH + H \quad \Delta_r H = \; 160 \text{ kJ/mol}$$

at combustion temperatures. QRRK analysis gives for the temperature range from 600 to 2500 K and N_2 bath gas the rate coefficient expressions

$$\begin{aligned} k_{50} &= 1.1\times10^{45}T^{-10.24}\exp(-24\,100/T)\ \text{s}^{-1} && 0.1 \text{ atm} \\ &= 2.4\times10^{48}T^{-10.82}\exp(-26\,200/T)\ \text{s}^{-1} && 1.0 \text{ atm} \\ &= 3.2\times10^{46}T^{-9.95}\exp(-26\,900/T)\ \text{s}^{-1} && 10 \text{ atm.} \end{aligned}$$

Abstractions to form CH_3NH_2 are unlikely to compete with thermal dissociation under combustion conditions, as the strength of the C–H bond formed is only about 390 kJ/mol, so most abstractions would be endothermic. For most conditions radical concentrations are too small to allow reactions of CH_2NH_2 with radicals to compete with dissociation. For completeness we include the following estimates.

(a) $CH_2NH_2 + O_2$.

QRRK analysis of the addition of the carbon-centered radical to O_2 indicated that the major reaction channel is intramolecular hydrogen transfer and loss of HO_2, analogous to the $C_2H_5 + O_2$ reaction, and that the reaction is near its low pressure limit for combustion conditions. The QRRK analysis gave in $cm^3mol^{-1}s^{-1}$ units

$$CH_2NH_2 + O_2 \longrightarrow H_2CNH + HO_2 \quad k_{50a} = 1.0\times10^{22}T^{-3.09}\exp(-3400/T).$$

The QRRK estimate for the less important channel forming $O + NH_2CH_2O$ is also included in Table 2.19; because NH_2CH_2O should rapidly dissociate to $NH_2 + CH_2O$, these products are assumed to be produced directly.

(b) $CH_2NH_2 + H \longrightarrow$ Products.
The adduct formation pathway leading to $CH_3 + NH_2$ is the reverse of reaction 21b. The DHT estimate for the hydrogen transfer reaction forming $H_2CNH + H_2$ is

$$k_{50b} = 4.8 \times 10^8\, T^{1.5} \exp(+450/T)\ \text{cm}^3\text{mol}^{-1}\text{s}^{-1}.$$

(c) $CH_2NH_2 + O \longrightarrow$ Products.

Adduct formation followed by dissociation to $CH_2O + NH_2$ is 320 kJ/mol exothermic. We estimate its rate coefficient to be

$$k_{50c1} = 7 \times 10^{13}\ \text{cm}^3\text{mol}^{-1}\text{s}^{-1},$$

the Baulch et al. (1992) value for the rate coefficient of the association reaction of O atoms with C_2H_5. The DHT estimate gives, in $\text{cm}^3\text{mol}^{-1}\text{s}^{-1}$ units,

$$CH_2NH_2 + O \longrightarrow H_2CNH + OH \quad k_{50c2} = 3.3 \times 10^8\, T^{1.5} \exp(+450/T).$$

(d) $CH_2NH_2 + OH \longrightarrow$ Products.

The adduct formation pathway has a 92 kJ/mol exit channel producing $CH_2OH + NH_2$. We suggest

$$k_{50d1} = 4 \times 10^{13}\ \text{cm}^3\text{mol}^{-1}\text{s}^{-1},$$

the Baulch et al. (1992) rate coefficient for the $CH_3 + OH$ reaction. The DHT estimate for the direct reaction producing $H_2CNH + H_2O$ is

$$k_{50d2} = 2.4 \times 10^6\, T^2 \exp(+600/T)\ \text{cm}^3\text{mol}^{-1}\text{s}^{-1}.$$

(e) $CH_2NH_2 + CH_3 \longrightarrow$ Products.

Assuming that the adduct formation rate coefficient is $3 \times 10^{13}\ \text{cm}^3\text{mol}^{-1}\text{s}^{-1}$, the Baulch et al. (1992) recommendation for the high pressure limiting rate coefficient of $CH_3 + C_2H_5$, and that the A-factors for dissociation into the two possible exit channels are equal, leads to

$$k_{50e1} = 2 \times 10^{13} \exp(-1360/T)\ \text{cm}^3\text{mol}^{-1}\text{s}^{-1}$$

for the 12 kJ/mol endothermic reaction forming $C_2H_5 + NH_2$. At lower temperatures and higher pressures some degree of adduct stabilization is to be expected. The DHT estimate for the direct reaction producing $H_2CNH + CH_4$ is

$$k_{50e2} = 1.6 \times 10^6\, T^{1.87} \exp(+315/T)\ \text{cm}^3\text{mol}^{-1}\text{s}^{-1}.$$

2.7.26 Reactions of CH_3NH_2

This recombination product of CH_3 and NH_2 (Section 2.6.17) can undergo hydrogen transfer reactions as well as dissociation. The dissociation rate coefficient can be computed from the previously estimated recombination rate coefficient (Section 2.6.17). For flame front radical concentrations approaching 5×10^{-8} mol/cm^3, abstraction rate coefficients above 10^{12} cm^{3}mol^{-1}s^{-1} will imply abstraction rates that compete with thermal dissociation.

Hydrogen abstraction forms two products:

$$\begin{aligned} CH_3NH_2 + X &\longrightarrow CH_2NH_2 + HX \\ &\longrightarrow CH_3NH + HX\,. \end{aligned}$$

The site of radical attack is affected by the C–H and N–H bond dissociation energies as follows. The C–H bond dissociation energy of about 395 kJ/mol is some 30 kJ/mol less than typical primary C–H bond dissociation energies because of stabilization of the radical center formed by the adjacent nitrogen atom. The N–H bond dissociation energy is about 423 kJ/mol. Using these values and the DHT estimation method leads to the expressions listed in Table 2.19. (The Table 2.3 coefficients were adjusted to account for the partial resonance stabilization that accounts for the weakening of the C–H bond.) At 2000 K and 1 atm, with radical concentrations approaching 1%, these expressions for H, O, and OH imply that both abstraction rates are faster than thermal dissociation. Abstractions by NH_2 are about as fast as thermal dissociation while abstractions by CH_3 are somewhat slower.

2.7.27 Reactions of NCCN

Because the C–C bond in cyanogen is very strong, with $\Delta_r H^\circ = 560$ kJ/mol for dissociation, its dissociation rate is expected to be slow, as was confirmed in shock tube experiments by Natarajan et al. (1986). Their low pressure limit rate coefficient in argon test gas is given by

$$k_{52a} = 1.1 \times 10^{34}\, T^{-4.32} \exp(-65\,465/T)\ \text{cm}^3\text{mol}^{-1}\text{s}^{-1}.$$

At 2000 K and 1 atm the corresponding first-order rate coefficient is 2 s^{-1}, far too slow to be relevant in combustion. The most likely reactions would be addition–elimination sequences such as

$$\begin{aligned} NCCN + H &\longrightarrow NCHCN \longrightarrow HCN + CN \qquad \Delta_r H = +42\ \text{kJ/mol} \qquad (52b)\\ + O &\longrightarrow NCOCN \longrightarrow NCO + CN \qquad \Delta_r H = +8\ \text{kJ/mol} \qquad (52c)\\ + OH &\longrightarrow NCOHCN \longrightarrow HOCN + CN \qquad \Delta_r H = +75\ \text{kJ/mol} \qquad (52d) \end{aligned}$$

The reaction with H atoms is the reverse of reaction 44f. For the reaction with O atoms, Louge and Hanson (1984a,b) reported $k_{52c} = 5 \times 10^{11}$ cm^{3}mol^{-1}s^{-1} at 2000 K, which is consistent with the Miller and Bowman (1989) expression

$$k_{52c} = 4.6 \times 10^{12} \exp(-4470/T)\ \text{cm}^3\text{mol}^{-1}\text{s}^{-1}.$$

We are not aware of any high-temperature studies of reaction 52d. Miller and Bowman use an expression taken from low-temperature measurements that is appropriate for formation of the adduct but not for formation of the products HOCN + CN, which would not be accessible in the low-temperature experiments. Because of the relatively high endothermicity of reaction 52d it should be unimportant and can safely be omitted from combustion models. We estimate

$$k_{52d} = 2 \times 10^{12} \exp(-9500/T) \ \mathrm{cm^3 mol^{-1} s^{-1}}$$

by analogy to QRRK results with other OH addition reactions.

2.7.28 Reactions of NCO

Reactions of NCO are important when cyanuric acid is used as an NO_X abatement agent in the RAPRENOX process (Perry and Siebers 1986), wherein HNCO produced from thermal decomposition of cyanuric acid loses an H atom by abstraction. It can also be formed by the addition of O to HCN (cf. Section 2.7.17) or in reactions of OH or O_2 with CN (cf. Section 2.7.19). NCO is a relatively stable radical, about 230 kJ/mol more stable than N + CO, and thus has a relatively long lifetime. It can react with other radicals, with a high probability of breaking the N–C bond to form CO. Since the H–N bond in HNCO is so strong (about 470 kJ/mol), one expects that NCO can abstract hydrogen atoms from most species to produce HNCO.

(a) NCO + NO $\longrightarrow$ Products.

The formation of N_2 from reaction of NCO with NO is one of the reactions by which addition of cyanuric acid to combustion systems can remove nitric oxides. Other channels are also possible for this reaction, namely,

$$\begin{aligned} \mathrm{NCO + NO} &\longrightarrow \mathrm{CO_2 + N_2} & \Delta_r H &= -620 \ \mathrm{kJ/mol} & (53a1)\\ &\longrightarrow \mathrm{CO + N_2O} & \Delta_r H &= -250 \ \mathrm{kJ/mol} & (53a2)\\ &\longrightarrow \mathrm{CO + N_2 + O} & \Delta_r H &= -84 \ \mathrm{kJ/mol.} & (53a3) \end{aligned}$$

Mertens et al. (1992) combined their shock tube measurements of the overall rate coefficient with lower temperature data and reported

$$k_{total} = 1.4 \times 10^{18} T^{-1.73} \exp(-384/T) \ \mathrm{cm^3 mol^{-1} s^{-1}}$$

for the temperature range from 294 to 2660 K. The rate coefficient decreases by about an order of magnitude over this temperature range, consistent with intermediate formation of a chemically-activated adduct.

Product distributions were measured at room temperature by Cooper and Herschberger (1992) and Becker et al. (1992b). These groups report that the N_2O channel is 33 to 35% of the total; Cooper and Herschberger also report that the CO_2 channel is 44% of the total and the CO + O + N_2 is 22%. Measurements by Cooper et al. (1993) extended the temperature range for branching ratio measurements to 623 K and revised their 1992 measurements such that the CO_2 + N_2 pathway accounts for 56±7% and N_2O + CO for the remaining 44±7%, with no observed temperature dependence of the branching ratio. A theoretical analysis by Lin et al. (1993) can be made consistent with the data of Cooper et al. by a reduction in the barrier height for dissociation of the OCNNO complex to N_2O + CO. We suggest use of the above overall rate coefficient expression of Mertens et al. together with the temperature-independent branching ratios measured by Cooper et al. Measurements by Flatness and Kramlich (1996) gave a branching ratio for the N_2O channel in the 30 to 50% range from 1100 to 1400 K range, which is consistent with the lower temperature measurements.

(b) $NCO + M \longrightarrow N + CO + M$.

Baulch et al. (1992) recommend $k = 1.0 \times 10^{15} \exp(-23\,500/T)$ cm^{3}mol^{-1}s^{-1} for M = Ar. At 1500 K and 1 atm this corresponds to a pseudo-first-order rate coefficient of 0.5 s^{-1}. Using the enthalpy of formation value for NCO proposed by East and Allen (1993), however, implies that this reaction is 230 kJ/mol endothermic. The activation energy of Baulch et al. is thus probably too low. We recommend the expression of Mertens and Hanson (1996) for dissociation in Ar bath gas

$$k_{53b} = 2.2 \times 10^{14} \exp(-27\,200/T) \ \text{cm}^3\text{mol}^{-1}\text{s}^{-1}$$

for the temperature range from 2370 to 3050 K. This expression is more consistent with the East and Allen thermochemistry. This reaction is too slow to be important in combustion except at very high temperatures.

(c) $NCO + H_2 \longrightarrow HNCO + H$.

Miller and Melius (1992b) calculated the transition state theory expression

$$k_{53c} = 7.6 \times 10^2 T^3 \exp(-2000/T) \ \text{cm}^3\text{mol}^{-1}\text{s}^{-1},$$

which is in reasonable agreement with the limited data available.

(d) $NCO + O \longrightarrow ONCO \longrightarrow NO + CO$.

This reaction is expected to proceed by adduct formation followed by breaking of the weak N–C bond. Baulch et al. (1992) recommend

$$k_{53d1} = 4.2 \times 10^{13} \ \text{cm}^3\text{mol}^{-1}\text{s}^{-1},$$

which is consistent with this pathway. O atoms can also add to the C–O bond, and the adduct can dissociate exothermically to N and CO_2. We estimate the rate coefficient for this reaction to be half of that for O addition to ethylene:

$$k_{53d2} = 8 \times 10^{12} \exp(-1300/T) \ \text{cm}^3\text{mol}^{-1}\text{s}^{-1}.$$

(e) $NCO + H \longrightarrow HNCO \longrightarrow NH + CO$

Baulch et al. (1992) recommend

$$k_{53e} = 5.2 \times 10^{13} \ \text{cm}^3\text{mol}^{-1}\text{s}^{-1},$$

which is also consistent with a adduct formation–elimination mechanism.

(f) $NCO + N \longrightarrow NNCO \longrightarrow N_2 + CO$.

Brownsword et al. (1997) reported a room temperature rate coefficient

$$k_{53f} = 3.3 \times 10^{13} \ \text{cm}^3\text{mol}^{-1}\text{s}^{-1}.$$

While this reaction is spin-forbidden, it is also highly exothermic, by 710 kJ/mol, which explains the large rate coefficient value.

(g) $NCO + OH \longrightarrow$ Products.

Tsang (1992) considered this reaction to be an exothermic abstraction to form O + HNCO and recommended, based on the reverse reaction, the expression in $\text{cm}^3\text{mol}^{-1}\text{s}^{-1}$ units

$$NCO + OH \longrightarrow O + HNCO \qquad k_{53g1} = 7.8 \times 10^4 T^{2.27} \exp(+495/T).$$

These radicals can also recombine to form HONCO and then dissociate via two channels, one to HON + CO and the other to H + NO + CO. QRRK analysis gave the expressions in $\text{cm}^3\text{mol}^{-1}\text{s}^{-1}$ units

$$\begin{aligned} NCO + OH &\longrightarrow CO + HON & k_{53g2} &= 5.3 \times 10^{12} T^{-0.07} \exp(-2600/T) \\ &\longrightarrow H + CO + NO & k_{53g3} &= 8.3 \times 10^{12} T^{-0.05} \exp(-9080/T). \end{aligned}$$

(h) $NCO + NO_2 \longrightarrow$ Products.

This reaction is exothermic by 500 kJ/mol if the products are N_2O and CO_2. Other pathways, such as $CO + N_2 + O_2$, or $CO + 2\,NO$, are also thermodynamically accessible. Park and Herschberger (1993a) report a total rate coefficient and a product distribution over the temperature range from 298 to 500 K. We suggest using their expressions, which are, in $\text{cm}^3\text{mol}^{-1}\text{s}^{-1}$ units,

$$\begin{aligned} NCO + NO_2 &\longrightarrow CO_2 + N_2O & k_{53h1} &= 2.3 \times 10^{12} \exp(+440/T) \\ &\longrightarrow CO + 2\,NO & k_{53h2} &= 2.1 \times 10^{11} \exp(+440/T). \end{aligned}$$

At 1250 K these expressions yield 3.6×10^{12} $\text{cm}^3\text{mol}^{-1}\text{s}^{-1}$ for the total rate coefficient, in good agreement with the $4.5 \times 10^{12}\,\text{cm}^3\text{mol}^{-1}\text{s}^{-1}$ value measured by Wooldridge et al. (1994). The total rate coefficient into both channels was also measured by Juang et al. (1995) from 294 to 774 K, who report an expression about 60% higher than the Park and Herschberger result, with an indication of non-Arrhenius behavior, that extrapolates smoothly to the Wooldridge et al. value.

(i) $NCO + CH_4 \longrightarrow HNCO + CH_3$.

We include this reaction to represent abstractions from hydrocarbons. (As noted in Section 2.7.19 in connection with the CN reaction, methane is not a representative hydrocarbon; for other hydrocarbons this expression must be modified to account for different C–H bond strengths.) This reaction has been studied in a pump-probe experiment over the temperature range from 512 to 1113 K by Schuck et al. (1994), who report

$$k_{53i} = 9.8 \times 10^{12} \exp(-4090/T) \ \text{cm}^3\text{mol}^{-1}\text{s}^{-1}.$$

Studies of NCO reactions with other hydrocarbons have been reported (Park and Herschberger 1994; Becker et al. 1995a,b).

(j) $NCO + NH_3 \longrightarrow HNCO + NH_2$.

This reaction is approximately thermoneutral, while abstraction of an H atom by NCO from CH_4 is about 13 kJ/mol exothermic, and it has fewer H atoms to abstract than does CH_4. Nonetheless, its rate coefficient is reported by Becker et al. (1997) to be faster than that of the corresponding abstraction (53i) from methane.

$$k_{53j} = 2.8 \times 10^4 \, T^{2.48} \exp(-495/T) \ \text{cm}^3\text{mol}^{-1}\text{s}^{-1}.$$

2.7.29 Reactions of HCNO

This species has been reported to be the major product of the CH_2 + NO reaction (Section 2.6.19). Thus one needs to account for its subsequent reactions. Dissociation to HCN + O is endothermic by 210 kJ/mol, so this pathway will be relatively slow. QRRK analysis indicates that this reaction is in the falloff regime for combustion conditions and has the rate coefficient expressions in N_2 bath gas

$$\begin{aligned} k_{54} &= 2.0 \times 10^{30} T^{-6.03} \exp(-30\,600/T) \ \text{s}^{-1} && 0.1 \text{ atm} \\ &= 4.2 \times 10^{31} T^{-6.12} \exp(-30\,800/T) \ \text{s}^{-1} && 1.0 \text{ atm} \\ &= 5.9 \times 10^{31} T^{-5.85} \exp(-31\,200/T) \ \text{s}^{-1} && 10 \text{ atm.} \end{aligned}$$

Abstraction reactions by HCNO should be unimportant, because the strength of the C–H bond formed would only be about 220 kJ/mol. Recombination reactions could be important. QRRK calculations with the parameters shown in Table 2.15 gave, in $\text{cm}^3\text{mol}^{-1}\text{s}^{-1}$ units,

$$\begin{aligned} HCNO + H &\longrightarrow HNCO + H && k_{54a} = 2.1 \times 10^{15} T^{0.69} \exp(-1435/T) \\ &\longrightarrow HCN + OH && k_{54b} = 2.7 \times 10^{11} T^{0.18} \exp(-1065/T) \\ &\longrightarrow NH_2 + CO && k_{54c} = 1.7 \times 10^{14} T^{-0.75} \exp(-1455/T) \\ &\longrightarrow HOCN + H && k_{54d} = 1.4 \times 10^{11} T^{-0.19} \exp(-1250/T). \end{aligned}$$

Reaction 54a is the most important channel. Stabilization channels are not important under combustion conditions. Other radical recombinations, followed by dissociation, include

$$HCNO + O \longrightarrow HC(O)NO \longrightarrow HCO + NO \quad \Delta_r H = -290\ \text{kJ/mol} \quad (54e)$$
$$HCNO + OH \longrightarrow HC(OH)NO \longrightarrow HCOH + NO \quad \Delta_r H = -30\ \text{kJ/mol.} \quad (54f)$$

We expect these rate coefficients to reflect the recombination rates, as in the CH_2NH_2 reactions discussed in Section 2.7.25:

$$k_{54e} = 7 \times 10^{13}\ \text{cm}^3\text{mol}^{-1}\text{s}^{-1}$$
$$k_{54f} = 4 \times 10^{13}\ \text{cm}^3\text{mol}^{-1}\text{s}^{-1}.$$

The reaction with OH has a more exothermic exit channel, to CH_2O + NO, but QRRK analysis confirms that HCOH isomerization to CH_2O is slow, owing to the high barrier and the low A-factor caused by the entropy lost in forming a 3-member cyclic transition state. Decomposition of the HC(OH)NO adduct to HCOH and NO is expected to be the only output channel.

2.7.30 Reactions of HOCN

HOCN can be formed by the addition of OH to HCN, as discussed in Section 2.7.17. It can form NCO by loss of the hydrogen; the relatively weak, about 360 kJ/mol, O–H bond strength means that such abstractions would be exothermic and fast. Radicals could also add across the C–N bond. One such reaction is the addition of hydrogen atoms to form the adduct NCHOH (cf. Figure 2.51). QRRK analysis predicts for this adduct formation pathway, in $\text{cm}^3\text{mol}^{-1}\text{s}^{-1}$ units,

$$HOCN + H \longrightarrow HCN + OH \quad k_{55a} = 2.0 \times 10^{13} T^{-0.04} \exp(-1075/T)$$
$$\longrightarrow HNCO + H \quad k_{55b} = 3.1 \times 10^{8} T^{0.84} \exp(-965/T)$$
$$\longrightarrow NH_2 + CO \quad k_{55c} = 1.2 \times 10^{8} T^{0.61} \exp(-1045/T).$$

The first channel is the fastest. It is the reverse of reaction (42b2), and the same rate coefficient can also be inferred from the rate coefficient expression for k_{42b2}; accordingly, combustion modeling programs that automatically include reverse reactions should not have reactions (42b2) and (55a) both present.

Absent experimental data for the abstraction reactions, we estimated their direct pathway rate coefficients by the DHT method. The results are listed in Table 2.19. Abstractions by H, O, and OH compete with the H-atom addition channel at combustion temperatures.

2.7.31 Reactions of HNCO

HNCO is the initial decomposition product of cyanuric acid, the selective non-catalytic reducing agent of the Raprenox process for NO_X removal (Perry and

Siebers 1986). It is also predicted to be the major channel for the CH_2 + NO reaction (Section 2.6.19). Spin-forbidden thermal dissociation to the lowest energy products,

$$\mathrm{HNCO + M \longrightarrow {}^3NH + CO + M} \qquad \Delta_r H_{56a} = 350\,\mathrm{kJ/mol}\,,$$

has been studied in shock tubes by Mertens et al. (1989) and Wu et al. (1990) with concordant results. We suggest adoption of the Wu et al. expression,

$$k_{56a} = 8.4 \times 10^{15} \exp(-42\,640/T)\ \mathrm{cm^3mol^{-1}s^{-1}}\,,$$

which is valid for argon bath gas over the temperature range from 1830 to 3340 K. The measured activation energy suggests that the barrier for reverse reaction is small, which implies that there is no significant energy barrier on the surface for spin reversal.

HNCO participates in radical addition and abstraction reactions. QRRK analysis of hydrogen addition, which includes formation of either NCHOH or HNCHO (cf. Figure 2.51), predicts the dominant channel to be

$$\mathrm{HNCO + H \longrightarrow NH_2 + CO} \qquad (56b)$$

and a rate coefficient expression comparable to literature expressions, slightly higher at 1000 K than the expression suggested by Tsang (1992), but lower than his suggestion at 2000 K; and somewhat higher than the measurements of Mertens et al. (1991b) and the calculations of Miller and Melius (1992b). The Miller and Melius analysis confirms that addition dominates abstraction. Nguyen et al. (1996) computed a potential surface for reaction (56b) and from it QRRK rate coefficient expressions with tunneling corrections. We suggest adoption of their expression

$$k_{56b} = 3.6 \times 10^4\, T^{2.49} \exp(-1180/T)\ \mathrm{cm^3mol^{-1}s^{-1}}.$$

Reaction with O has two exothermic adduct formation pathways:

$$\begin{aligned} \mathrm{HNCO + O} &\longrightarrow \mathrm{HN(O)CO \longrightarrow HNO + CO} \quad \Delta_r H\,(56c1) = -160\ \mathrm{kJ/mol} \\ &\longrightarrow \mathrm{HNC(O)O \longrightarrow NH + CO_2} \quad \Delta_r H\,(56c2) = -190\ \mathrm{kJ/mol}. \end{aligned}$$

Assuming that adduct formation is rate limiting and that the rate coefficient is the same for oxygen atom addition to either N or C, we estimate

$$k_{56c1} = k_{56c2} = 1.7 \times 10^6 T^{2.08}\ \mathrm{cm^3mol^{-1}s^{-1}}$$

by taking half of the expression recommended for O–atom addition to ethylene by Baulch et al. (1922). This expression gives values significantly higher than the experimental results for k_{56c2} reported by Mertens et al. (1992a).

OH addition to the carbon leads to HNC(OH)O, which can form NH_2 + CO_2 after a hydrogen shift. Similarly, addition to nitrogen yields HN(OH)CO, which can form NCO + H_2O. QRRK analysis suggests the rate coefficient expressions in $cm^3mol^{-1}s^{-1}$ units

$$\begin{aligned} \mathrm{HNCO + OH} &\longrightarrow \mathrm{NH_2 + CO_2} \quad k_{56d1} = 6.3 \times 10^{10} T^{-0.06} \exp(-5860/T) \\ &\longrightarrow \mathrm{NCO + H_2O} \quad k_{56d2} = 5.2 \times 10^{10} T^{-0.03} \exp(-8840/T). \end{aligned}$$

These expressions are much lower than those for O–atom addition, reflecting the isomerizations required to reach the low energy exit channels in these reactions.

Direct abstractions from HNCO should be similar to abstractions from NH_3, except for there being but one abstractable hydrogen atom, as the bond strengths are the same. The reactions with H, O, CH_3 and NH_2 have appeared already as reverse reactions in Section 2.7.28. Wooldridge et al. (1996b) report an expression for the OH abstraction rate in $cm^3mol^{-1}s^{-1}$:

$$\mathrm{HNCO + OH \longrightarrow NCO + H_2O} \quad k_{56g} = 3.6 \times 10^7\, T^{1.5} \exp(-1810/T).$$

The abstraction products are the same as for one of the OH addition pathways (56d2). Direct abstraction is faster.

2.7.32 Reactions of CH_2NO

The endothermicity for loss of an H atom in direct dissociation of CH_2NO,

$$\mathrm{CH_2NO \longrightarrow HCNO + H} \quad \Delta_r H = +222 \text{ kJ/mol},$$

is comparable to that for the corresponding process in HNOH (cf. Section 2.7.13), and so we expect similar kinetics. In the CH_2NO case, however, there are other possibilities, as illustrated in Figure 2.51. From our earlier discussion of CH_2 + NO (Section 2.6.19) we expect

$$\mathrm{CH_2NO \longrightarrow HNCO + H} \quad \Delta_r H = -60 \text{kJ/mol}$$

to be important even though several isomerizations precede final bond-breaking. The long thermal dissociation lifetime of CH_2NO suggests that radical–radical reactions with it may be important.

(a) $CH_2NO \longrightarrow$ Products.

QRRK analysis indicates that CH_2NO thermal dissociation is in the falloff pressure regime under combustion conditions and that HNCO + H is the dominant product channel. A good fit to the rate coefficient is obtained by restricting the temperature range to 600 to 2500 K, for which the analysis gave in $cm^3mol^{-1}s^{-1}$ units for N_2 bath gas

$$\begin{aligned} k_{57a} &= 6.9 \times 10^{41} T^{-9.30} \exp(-26\,000/T) && 0.1 \text{ atm} \\ &= 2.3 \times 10^{42} T^{-9.11} \exp(-27\,100/T) && 1.0 \text{ atm} \\ &= 1.7 \times 10^{38} T^{-7.64} \exp(-27\,000/T) && 10 \text{ atm.} \end{aligned}$$

(b) $CH_2NO + O_2 \longrightarrow CH_2O + NO_2$.

This case is analogous to allyl radical addition to O_2 in that an O atom can add at the NO double bond and lead ultimately to $CH_2O + NO_2$. In the absence of experimental information we suggest using our QRRK expressions (Bozzelli and Dean 1993) for the allyl analog, in $cm^3 mol^{-1} s^{-1}$ units,

$$\begin{aligned} k_{57b} &= 2.9 \times 10^{12} T^{-0.31} \exp(-8900/T) \quad 300 \text{ to } 1000 \text{ K} \\ &= 1.2 \times 10^{15} T^{-1.01} \exp(-10\,100/T) \quad 1000 \text{ to } 2500 \text{ K}. \end{aligned}$$

(c) $CH_2NO + H \longrightarrow CH_3 + NO$.

Adduct formation followed by breaking the C–N bond is very exothermic, suggesting that the rate coefficient be taken as that of adduct formation,

$$k_{57c1} = 4 \times 10^{13} \ cm^3 mol^{-1} s^{-1},$$

using the Baulch et al. (1992) value for the $H + C_2H_5$ association reaction. The DHT method gives for the direct pathway

$$k_{57c2} = 4.8 \times 10^8 \, T^{1.5} \exp(+450/T) \ cm^3 mol^{-1} s^{-1}.$$

(d) $CH_2NO + O \longrightarrow$ Products.

Dissociation of the adduct to formaldehyde and NO is very exothermic, suggesting that the rate coefficient be taken as that of adduct formation,

$$k_{57d1} = 7 \times 10^{13} \ cm^3 mol^{-1} s^{-1},$$

the Baulch et al. (1992) value for the rate coefficient of the $O + C_2H_5$ reaction. Direct reaction forming $HCNO + OH$ is also very exothermic; the DHT estimate is

$$k_{57d2} = 3.3 \times 10^8 \, T^{1.5} \exp(+450/T) \ cm^3 mol^{-1} s^{-1}.$$

(e) $CH_2NO + OH \longrightarrow$ Products.

Formation of $H_2COH + NO$ from the adduct is 130 kJ/mol exothermic, suggesting that its rate coefficient be taken as that of adduct formation, for which we estimate

$$k_{57e1} = 4 \times 10^{13} \ cm^3 mol^{-1} s^{-1}$$

based on the Baulch et al. recommendation for $OH + CH_3$. The DHT rate coefficient estimate for direct hydrogen abstraction to form $HCNO + H_2O$ is

$$k_{57e2} = 2.4 \times 10^6 \, T^2 \exp(+600/T) \ cm^3 mol^{-1} s^{-1}.$$

(f) $CH_2NO + CH_3 \longrightarrow$ Products.

Formation of C_2H_5 + NO, which is 105 kJ/mol exothermic, is the only likely pathway for the adduct to decompose. The rate coefficient is expected to be that for adduct formation,

$$k_{57f1} = 3 \times 10^{13} \ \text{cm}^3\text{mol}^{-1}\text{s}^{-1},$$

the Baulch et al. (1992) recommendation for the rate coefficient of the CH_3 + C_2H_5 reaction. The DHT estimate for the rate coefficient for hydrogen abstraction to form HCNO + CH_4 is

$$k_{57f2} = 1.6 \times 10^6 \, T^{1.87} \exp(+560/T) \ \text{cm}^3\text{mol}^{-1}\text{s}^{-1}.$$

(g) $CH_2NO + NH_2 \longrightarrow$ Products.

Decomposition of the adduct to form CH_2NH_2 and NO is 115 kJ/mol exothermic; we suggest taking the same rate coefficient as for the reaction with CH_3,

$$k_{57g1} = 3 \times 10^{13} \ \text{cm}^3\text{mol}^{-1}\text{s}^{-1}.$$

The DHT estimate for the rate coefficient of the direct hydrogen transfer reaction forming HCNO + NH_3 is

$$k_{57g2} = 1.8 \times 10^6 \, T^{1.94} \exp(+580/T) \ \text{cm}^3\text{mol}^{-1}\text{s}^{-1}.$$

2.7.33 Reactions of CH_3NO

CH_3NO is an important intermediate in the reaction between CH_3 and NO at lower temperatures, as discussed in Section 2.6.15. Although rapid dissociation of this adduct is likely at higher temperatures, there is also the possibility of radical attack, because the C–H bond dissociation energy is only 315 kJ/mol. Rate coefficients for hydrogen abstraction (reactions 58a through 58e) computed by the DHT method are listed in Table 2.19.

Radicals can also add to CH_3NO and then eliminate CH_3. We estimated the corresponding rate coefficients as half of the corresponding rate coefficients for C_2H_4 (Baulch et al. 1992; Lightfoot and Pilling 1987: Tsang and Hampson 1986). The results, in $\text{cm}^3\text{mol}^{-1}\text{s}^{-1}$ units, are

$$\begin{aligned}
CH_3NO + H &\longrightarrow CH_3 + HNO & \quad k_{58f} &= 1.8 \times 10^{13} \exp(-1400/T) \\
+ O &\longrightarrow CH_3 + NO_2 & \quad k_{58g} &= 1.7 \times 10^6 T^{2.08} \\
+ OH &\longrightarrow CH_3 + HONO & \quad k_{58h} &= 2.5 \times 10^{12} \exp(-500/T).
\end{aligned}$$

2.7.34 Reactions of HON

HON appeared in Section 2.7.28 as a product of the recombination of OH with NCO. For completeness, we consider possible further reactions of this species. HON can dissociate to H + NO or isomerize to HNO. Melius (1993) estimated the barrier for isomerization to HNO to be about 85 kJ/mol, while the barrier for dissociation to H + NO should be only 5 to 10 kJ/mol above the endothermicity, that is, about 60 kJ/mol. The A-factor for dissociation should also be considerably higher, suggesting that dissociation is the dominant path. This conclusion was confirmed by QRRK analysis, which suggested that the dissociation channel will be at the low pressure limit for conditions of interest in combustion, with the rate coefficient expression in $cm^3mol^{-1}s^{-1}$ units for N_2 bath gas

$$\text{HON} + \text{M} \longrightarrow \text{H} + \text{NO} + \text{M}, \quad k_{59a} = 5.1 \times 10^{19} T^{-1.73} \exp(-8075/T)\,.$$

At 2000 K and 1 atm N_2, the effective first-order rate coefficient is about $1 \times 10^7\ s^{-1}$, similar to that for HCNH (Section 2.7.21). Thus one should consider the same types of reactions for HON—recombination with the flame radicals H, O, and OH, and addition to O_2. The adduct formation pathways all have exothermic dissociation channels. We expect them to have rate coefficients equal to those of adduct formation and estimate from the Baulch et al. (1992) recommendations for the analogous association reactions with C_2H_5, in $cm^3mol^{-1}s^{-1}$ units,

$$\begin{aligned} \text{HON} + \text{H} &\longrightarrow \text{HNO} + \text{H} & k_{59b1} &= 2 \times 10^{13} \\ + \text{H} &\longrightarrow \text{OH} + \text{NH} & k_{59b2} &= 2 \times 10^{13} \\ + \text{O} &\longrightarrow \text{OH} + \text{NO} & k_{59c} &= 7 \times 10^{13} \\ + \text{OH} &\longrightarrow \text{HONO} + \text{H} & k_{59d} &= 4 \times 10^{13}. \end{aligned}$$

The reaction with O_2 should be more important for HON, since the adduct can easily isomerize and then dissociate. However, the well depth of $HONO_2$ is expected to be quite shallow, so even the relatively low barrier for isomerization might be above the entrance channel. The adduct has an alternative low-energy exit channel in breaking the O–O bond. We expect the rate coefficient to be that for adduct formation,

$$^1\text{HON} + \text{O}_2 \longrightarrow \text{O} + \text{HONO}, \quad k_{59e} = 1.0 \times 10^{12} \exp(-2500/T)\ \text{cm}^3\text{mol}^{-1}\text{s}^{-1}.$$

2.7.35 Reactions of HCOH

HCOH can be produced in the reaction between HCNO and OH (reaction 54f). Its most likely reaction is isomerization to CH_2O, which is more stable by about 205 kJ/mol. We are not aware of any experimental or theoretical information available for this reaction. We estimate its rate coefficient from that of the analogous reaction in which CCH_2 isomerizes to HCCH. In both cases there is an exothermic hydrogen shift from a carbene-type carbon atom. The barrier for CCH_2 is about

20 kJ/mol; we estimate a more conservative value of 40 kJ/mol for HCOH. Using this value in a QRRK analysis gave for N_2 bath gas the expressions in s^{-1} units

$$
\begin{aligned}
k_{60} &= 3.5 \times 10^{17}\, T^{-2.86} \exp(-4470/T) && 0.1 \text{ atm} \\
&= 2.1 \times 10^{19}\, T^{-3.07} \exp(-4800/T) && 1.0 \text{ atm} \\
&= 1.8 \times 10^{21}\, T^{-3.32} \exp(-5460/T) && 10 \text{ atm.}
\end{aligned}
$$

This isomerization rate is comparable to that of NNH dissociation, and it is unlikely that bimolecular reactions would be sufficiently rapid to compete. We considered the reaction with O_2, but the energies of the various intermediates and the barriers for the various rearrangements were too uncertain to allow meaningful conclusions. The results suggest that reactions with O_2 may be about 20 times slower than isomerization for typical combustion conditions.

2.7.36 Reactions of NH_2OH

NH_2OH can be formed in the reaction between NH_2 and OH as well as in reactions of HO_2 with HNOH and NH_2O. Dissociation to NH_2 + OH, the reverse of reaction 16a (discussed in Section 2.6.12), is sufficiently slow that NH_2OH can also undergo radical attack and hydrogen abstraction at both ends of the molecule. Rate coefficient expressions derived by the DHT method are included in Table 2.19 for the direct abstraction reactions with H, O, OH, CH_3, NH_2, and HO_2.

2.7.37 Reactions of NH_2NO

NH_2NO is an adduct formed in the NH_2 + NO reaction and a product of the H_2NN + OH reaction. It dissociates mostly to N_2 and H_2O. (Dissociation to NH_2 and NO has appeared as the reverse of reaction 18c.) QRRK analysis indicates that most reaction proceeds via isomerization to HNNOH, the rest is direct. For combustion conditions the reaction is in the falloff pressure regime. The QRRK rate coefficient expressions for dissociation to N_2 and H_2O, derived using the data from Table 2.10, are, in $cm^3mol^{-1}s^{-1}$ units,

$$
\begin{aligned}
k_{62a} &= 4.1 \times 10^{33}\, T^{-7.18} \exp(-17\,700/T) && 0.1 \text{ atm} \\
&= 3.1 \times 10^{34}\, T^{-7.11} \exp(-18\,260/T) && 1.0 \text{ atm} \\
&= 2.9 \times 10^{31}\, T^{-5.91} \exp(-18\,205/T) && 10 \text{ atm.}
\end{aligned}
$$

Rate coefficient expressions derived by the DHT method are included in Table 2.19 for the direct abstraction reactions with H, O, OH, CH_3, NH_2, and HO_2.

2.7.38 Reactions of H_2NNHO

H_2NNHO is formed in the reaction between N_2H_3 and HO_2 (Section 2.7.6). A QRRK analysis gave for its most likely dissociation pathway to NH_2 and HNO the rate coefficient expressions for reaction in N_2, in $cm^3 mol^{-1} s^{-1}$ units,

$$\begin{aligned} k_{63a} &= 2.7 \times 10^{39}\, T^{-8.74} \exp(-20\,945/T) && 0.1 \text{ atm} \\ &= 2.4 \times 10^{40}\, T^{-8.73} \exp(-20\,940/T) && 1.0 \text{ atm} \\ &= 1.2 \times 10^{41}\, T^{-8.64} \exp(-20\,920/T) && 10 \text{ atm.} \end{aligned}$$

The reaction is close to the low-pressure limit at temperatures over 1000 K and pressures below 10 atm. A channel forming NH_2NO + H is also possible, but proved to be much slower and can be neglected.

Rate coefficient expressions derived by the DHT method are included in Table 2.19 for the direct abstraction reactions with H, O, OH, CH_3, NH_2, and HO_2.

2.7.39 Reactions of ClNO

Chemical bonds of chlorine are usually quite weak, the Cl–H bond in HCl being the most important exception. For this reason, chlorine atoms may be the radical species present at highest concentration in combustion and pyrolysis systems where chlorinated hydrocarbons or other sources of chlorine are present in significant quantities. Such combustion processes are usually operated under fuel-lean conditions, where hydrogen needed to form HCl is in limited supply. Because the O–H bonds in OH and H_2O are stronger than the H–Cl bond, hydrogen is generally not found to be bonded to chlorine under these conditions. Fuel lean operation favors high levels of Cl atoms, or at low temperatures Cl_2 or other Cl adducts. Thermochemical properties of the chlorine-containing species of interest in this section are shown in Table 2.18.

Several reactions involving chlorine have to be considered in connection with NO_x chemistry, specifically the addition reaction with NO and subsequent reactions of the ClNO adduct. Because the adduct has a relatively weak Cl–NO bond, 160 kJ/mol, NO can react with Cl much as with H and OH, as discussed previously, to provide a catalyst for combination reactions

$$\begin{aligned} \text{Cl} + \text{NO} &\longrightarrow \text{ClNO} \\ \text{ClNO} + \text{X} &\longrightarrow \text{XCl} + \text{NO} \\ \text{(overall)}\ \text{Cl} + \text{X} &\longrightarrow \text{XCl}, \end{aligned}$$

where X is an atom or radical present in the system. The rates of dissociation and addition reactions of triatomic molecules like ClNO are in or near the low-pressure limit under combustion conditions. Even for weakly bonded molecules like ClNO, however, the low-pressure dissociation rate at typical combustion conditions is slow enough that transient concentrations become significant. Thus ClNO can be an important species both in hazardous waste disposal processes and in general-purpose industrial or municipal waste incinerators. We consider several of the reactions of ClNO with radicals expected to be in the highest concentrations.

TABLE 2.18. Thermochemical data for chlorine species.

Species	$\Delta_f H^\circ_{298}$ kJ/mol	S°_{298} $\mathrm{J\,mol^{-1}\,K^{-1}}$	$C^\circ_p(T)/\mathrm{J\,mol^{-1}\,K^{-1}}$ at $T/\mathrm{K} =$ 300	400	600	800	1000	1500
Cl	120.9	165.3	21.8	22.3	22.6	22.4	22.2	22.0
Cl_2	0.0	223.0	33.9	35.1	36.6	37.3	37.6	38.1
HCl	-92.3	186.8	29.1	29.1	29.6	30.5	31.6	33.9
ClO	101.3	226.4	31.4	33.1	35.3	36.4	36.9	37.7
HOCl	-74.5	236.4	37.3	40.0	43.9	46.6	48.5	51.9
ClNO	51.7	261.5	44.6	47.0	50.7	53.1	54.9	57.3

All values refer to the ideal gas state.

(a) ClNO + M ⟶ NO + Cl + M.
We suggest using the Baulch et al. (1981) recommendation, which covers the temperature range 800 to 1500 K and is valid for Ar bath gas:

$$k_{64a} = 1.3 \times 10^{15} \exp(-16100/T) \quad \mathrm{cm^3 mol^{-1} s^{-1}}.$$

(b) O + ClNO ⟶ ClO + NO.
We suggest use of the expression of Abbatt et al. (1989) but caution that their data were collected at low temperatures (220–450 K).

$$k_{64b} = 5 \times 10^{12} \exp(-1520/T) \quad \mathrm{cm^3 mol^{-1} s^{-1}}.$$

(c) OH + ClNO.
This reaction has two exothermic pathways, one being abstraction of Cl, the other being addition of OH to the unsaturated bond in ClN=O followed by dissociation of the weaker Cl–N bond to release Cl. The addition–dissociation channel has, as one would expect, a slower rate than the abstraction channel. We suggest adoption of the expressions, in $\mathrm{cm^3 mol^{-1} s^{-1}}$ units, of Abbatt et al. (1989) for both the abstraction channel

$$\mathrm{OH + ClNO \longrightarrow HOCl + NO}, \; k_{64c1} = 5.4 \times 10^{12} \exp(-1130/T)$$

and the addition channel

$$\mathrm{OH + ClNO \longrightarrow HONO + Cl}, \; k_{64c2} = 5.5 \times 10^{10} \exp(+240/T)\,.$$

(d) Cl + ClNO ⟶ Cl_2 + NO.
We suggest adoption of the expression of Abbatt et al. (1989):

$$k_{64d} = 4.0 \times 10^{13} \exp(+128/T) \quad \mathrm{cm^3 mol^{-1} s^{-1}}.$$

(e) H + ClNO ⟶ HCl + NO.
The temperature dependence has been measured at low temperatures (240–460 K) by Wagner et al. (1976), who reported

$$k_{64e} = 4.6 \times 10^{13} \exp(-457/T) \quad \mathrm{cm^3 mol^{-1} s^{-1}}.$$

This expression is consistent with other measurements at room temperature. (Mallard et al. 1993).

2.8 ILLUSTRATIVE MODELING RESULTS

To illustrate the kinetics of nitrogen chemistry for representative combustion conditions, we assembled the elementary reactions discussed in the preceding sections into the full mechanism given as Table 2.19 at the end of this chapter. It is not complete for combustion modeling, however, because even if one restricts one's interest to ammonia oxidation it is still necessary to add additional reactions to describe the H–O chemistry that runs in parallel to the H–O–N chemistry. Exploring the combustion chemistry consequences of interactions between hydrocarbon radicals and nitrogen species requires one to include elementary reactions describing all of the appropriate hydrocarbon oxidation chemistry.

To facilitate a number of comparisons with the well-tested earlier nitrogen chemistry model of Miller and Bowman (1989), we combined the Table 2.19 reactions with the hydrogen and hydrocarbon reactions of the Miller and Bowman mechanism to compose a full C–H–O–N mechanism that we will designate as DBN/MBC. It comprises 522 reactions of 87 species (151 reactions from Miller and Bowman, with their third-body efficiencies, and the 371 reactions in Table 2.19). All elementary reactions were considered to be reversible, the reverse reaction rate coefficient expressions being derived from equilibrium constants computed using the thermodynamic data listed in Tables 2.1 and 2.2. With this mechanism we computed product distributions, reaction rates, and sensitivity spectra for representative reaction conditions that we found to be suitable for exploring the main kinetic features of the combustion chemistry of nitrogen.[†]

We also present results computed using the complete Miller–Bowman mechanism, that is, using their mechanism for both the hydrocarbon and the nitrogen chemistry, in order to compare predictions made using our nitrogen chemistry to those obtained with a mechanism commonly used in recent years. We call this mechanism MB; it includes the Miller and Bowman (1989) reactions, the updates described by Miller and Bowman (1991), and the $NH_3 + M \longrightarrow NH_2 + H + M$ reaction. Comparing MB profiles to DBN/MBC profiles allows us to uncover the differences that can result using alternate descriptions of nitrogen chemistry. Full discussion of the causes of these differences is beyond the scope of this chapter, but just showing them illustrates some important issues confronting how one models nitrogen chemistry in combustion today. Detailed study of these differences will reveal important topics for further research in this area.

[†] A number of rate coefficients for reactions that play key roles in the prototype situations described in this section have been derived from measurements that were made under conditions where Ar was the dominant collision partner. In the construction of Table 2.19 these were converted to expressions valid for N_2 as collision partner using the assumption that the ratio of Ar to N_2 collision efficiency is 0.7. Other collision efficiencies relative to N_2 were assigned to be $H_2O = 7.0$, $CO_2 = 2.0$ and $CH_4 = 2.0$. See Section 9.4 of Chapter 1 for further discussion of this topic.

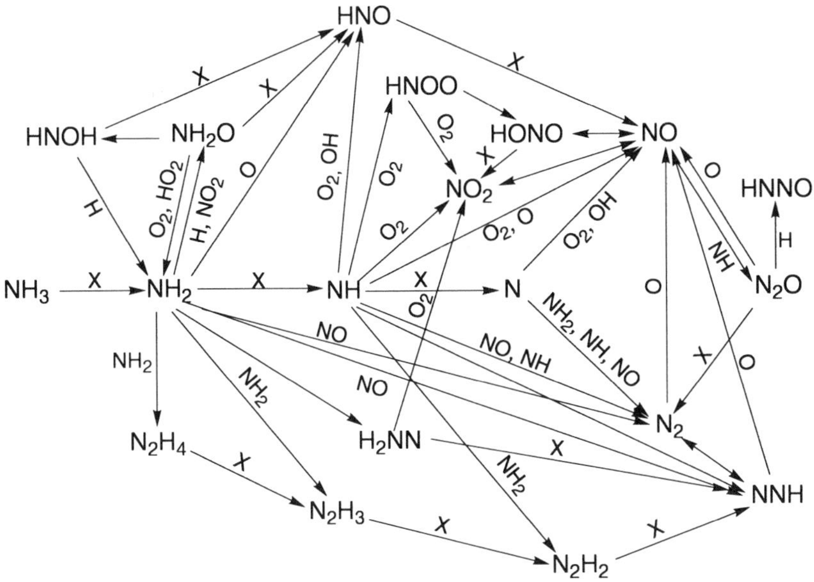

FIGURE 2.62. Overview of reactions connecting various species in the N/H/O system. The bimolecular partners for the major reactions are shown. An X designates that the pathway involves a series of hydrogen transfer reactions and sometimes unimolecular dissociation as well.

2.8.1 Ammonia oxidation

Figure 2.62 shows the reaction paths connecting the species of the H–O–N system. An interesting aspect of this diagram is that most species can be converted readily either into NO or into N_2. Small shifts in conditions can affect the balance of reactive flux in alternative directions. A significant difference between the Figure 2.62 description of the H–O–N system and earlier ones is the absence of a direct pathway from NH_2 to HNO via O_2. As discussed in Section 2.6.9, a better description is that this route to HNO proceeds via NH_2O and HNOH. Other differences are the role of NNH in NO production, as discussed in Section 2.7.3, and the participation of H_2NN. Figure 2.62 also shows the potential involvement of HONO as a route for NO loss. The reaction pathways shown in Figure 2.62 make clear that radical–radical reactions play important roles in nitrogen combustion chemistry—the large dissociation barriers of nitrogen-containing radicals, relative to their hydrocarbon counterparts, imply much longer radical lifetimes.

Kinetic model predictions for ammonia oxidation are shown in Figures 2.63–68. A comparison with the experimental data of Dean et al. (1982) at 1280 K and 1.2 atm is given in Figure 2.63. While there was experimental evidence that the observed induction time was influenced by surface reactions in the quartz reactor, other aspects of the kinetics appeared to be unaffected. The amount of NO produced was much less than the amount of NH_3 reacted—8 ppm of NO were

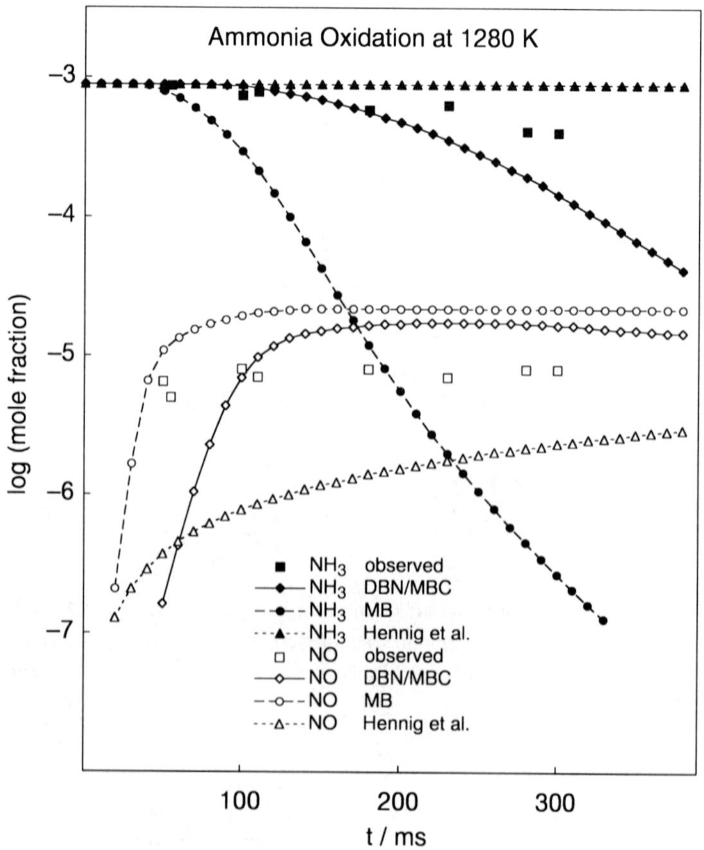

FIGURE 2.63. Comparison of predictions for NH_3 decay and NO formation to the data of Dean et al. (1982) for the oxidation of 900 ppm NH_3 by 4% O_2 with 1% H_2O in helium at 1279 K and 1.2 atm. The induction time was observed to be influenced by the reactor surface, suggesting heterogeneous initiation. Since this was not considered in the model, one expects to see somewhat longer induction times in the model predictions than were observed in the experiment. Taking this offset in induction time into account, the mechanism based on the rate coefficients recommended in this chapter (DBN/MBC) is seen to describe much of the observed kinetics. The Miller–Bowman mechanism (MB) predicts much faster decay of ammonia than observed as well as too much NO production. Predictions using the Hennig et al. (1995) rate coefficient for $NH_2 + O_2$ in the DBN/MBC mechanism are much too slow.

produced by loss of about 500 ppm NH_3. Most dynamic features are qualitatively captured by the DBN/MBC model. Given the heterogeneous character of the initiation, the early time mismatch for NO is to be expected. Aside from this offset, the model correctly predicts the "burst" of NO at early time, and the computed NO yield is somewhat larger than observed. The predicted rate of NH_3 decay is faster than observed.

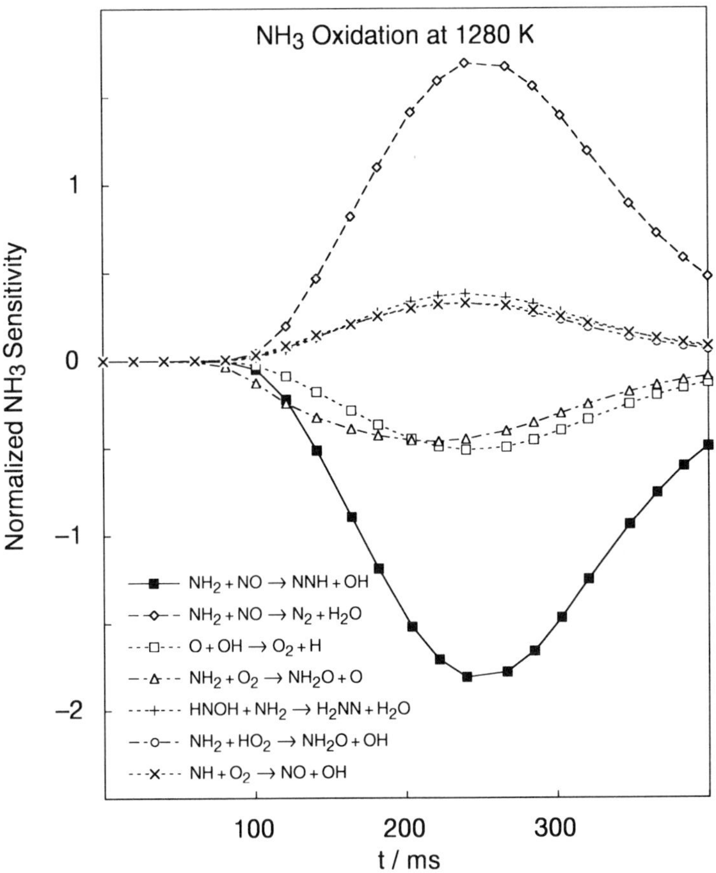

FIGURE 2.64. Normalized sensitivity coefficients computed by the SENKIN program (Lutz et al. 1991) for the reactions with the largest effects on ammonia concentrations during ammonia oxidation for the conditions of Figure 2.63. The clearly important reactions involving H_2NN, HNOH, and NH_2O have not been included in previous ammonia oxidation mechanisms.

It is encouraging that these features are reasonably reproduced with elementary reaction rate coefficient expressions assigned from an assessment of the experimental measurements and theoretical plausibility, with no adjustment to fit the Dean et al. data. The burst of NO, for example, was accounted for without invoking direct conversion of NH_2 to HNO as was done in the original analysis (cf. Section 2.6.9). For these conditions the MB mechanism substantially overpredicts the NH_3 decay rate and the NO yield. It can also be seen in Figure 2.63 that the Hennig et al. (1995) set of rate coefficients leads to predicted reaction rates that are much slower than found experimentally. The essential reason for this is that Hennig et al. considered the major channel of the $NH_2 + O_2$ reaction to be formation of NO and H_2O, which depletes the radical pool too rapidly.

In the DBN/MBC reaction mechanism, the species NH_2O, NHOH, and H_2NN play important roles. These species have not been considered in most previously proposed ammonia oxidation mechanisms; indeed, HNOH and H_2NN have not been considered before at all. Reaction flux analysis shows that most of the NH_2 is formed by

$$NH_3 + OH \longrightarrow NH_2 + H_2O. \tag{7}$$

Some of the NH_2 reacts with HO_2 to form NH_2O,

$$NH_2 + HO_2 \longrightarrow NH_2O + OH, \tag{14b}$$

which isomerizes rapidly

$$NH_2O \longrightarrow NHOH. \tag{37b}$$

The NHOH product reacts mostly with NH_2:

$$HNOH + NH_2 \longrightarrow H_2NN + H_2O, \tag{38f2}$$

and the H_2NN produced mostly reacts with O_2 to regenerate NH_2 and form NO_2,

$$H_2NN + O_2 \longrightarrow NH_2 + NO_2, \tag{30b}$$

which then react with one another in

$$NH_2 + NO_2 \longrightarrow NH_2O + NO \tag{34d2}$$

and then cycle through the 37b–34d2 sequence.

Figure 2.64 shows the evolution in time of the logarithmic sensitivity spectrum for NH_3 consumption under the ammonia oxidation conditions of Figure 2.63. The two most sensitive reactions, with opposing effects, are seen to be the two major channels of the NH_2 + NO reaction. Reactions involving NH_2O, H_2NN and HNOH are also important.

The rate of NH_3 decay is sensitive to several rate coefficients for which there is little experimental data. In particular, the rate coefficients of

$$NH_2 + O_2 \longrightarrow NH_2O + O \tag{13b}$$
$$HNOH + NH_2 \longrightarrow H_2NN + H_2O \tag{38f2}$$
$$NH_2 + HO_2 \longrightarrow NH_2O + OH, \tag{14b}$$

all of which are important for NH_3 decay (Figure 2.64). The fit to the NH_3 decay profile (Figure 2.63) could have been improved by adjusting one or more of these expressions, but experimental and theoretical investigations are to be preferred. The fact that the computed NH_3 profile is close to the observed one suggests that the mechanism and estimated rate coefficient expressions are reasonable, but even that suggestion needs to be tested by further studies of the relevant elementary reactions themselves.

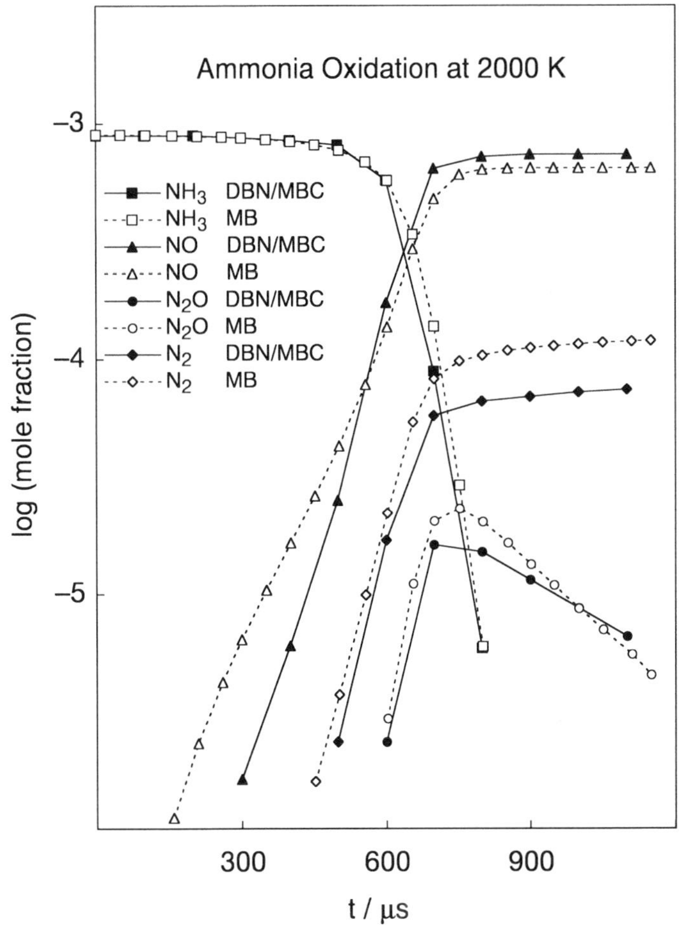

FIGURE 2.65. Comparison of predictions using a mechanism comprising the reactions and rate coefficient expressions recommended in this chapter (DBN/MBC) to those of the Miller–Bowman mechanism (MB) for oxidation of 900 ppm NH_3 by 4% O_2 at 2000 K and 1 atm. The rates and selectivities to major products computed by the two mechanisms are about the same. Ammonia is almost completely converted to NO at high temperatures.

The effects of temperature can be seen in Figure 2.65, which shows computed concentration profiles for 900 ppm NH_3 and 4% O_2 in Ar reacting at 2000 K. At this temperature the two models are in better agreement with one another. The NH_3 decay rate and the major product yields are similar for the two mechanisms. The DBN/MBC model predicts slightly more NO production, with less N_2 and N_2O. Unlike at lower temperatures, NH_3 is almost completely converted into NO.

To explore the reasons for the shift in selectivity toward NO at high temperatures we have to identify the major reactions affecting NO production at the two temperatures. The DBN/MBC reaction rates at 1280 and 2000 K shown in Figures

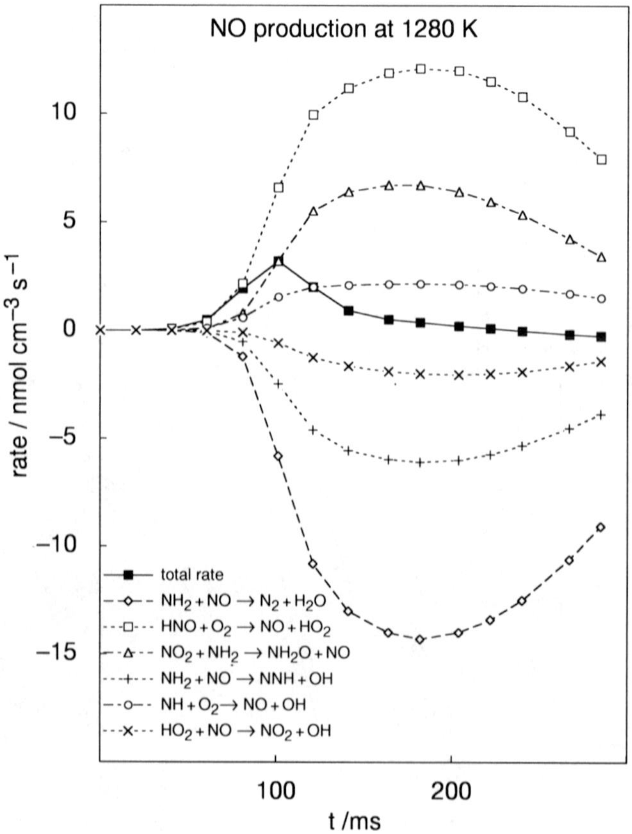

FIGURE 2.66. Major routes of NO production and decay in the DBN/MBC model for the conditions of Figure 2.63. The rapid burst of NO predicted near 100 ms is due to $HNO + O_2 \longrightarrow NO + HO_2$, with $NO_2 + NH_2 \longrightarrow NH_2O + NO$ arriving slightly later. The rate then decays to essentially zero as NO is consumed, mainly by reaction with NH_2.

2.66 and 2.67 suggest that the NO burst at low temperature is due mostly to

$$HNO + O_2 \longrightarrow NO + HO_2 \tag{36h}$$

and

$$NO_2 + NH_2 \longrightarrow NH_2O + NO. \tag{34d2}$$

After this burst NO decays mainly through reaction with NH_2, with the result that there is virtually no further increase of NO. In contrast, at 2000 K (Figure 2.67) none of the important reactions consume NO. The major source of NO is thermal dissociation of HNO, which is much faster at the higher temperature. The absence of reactions that consume NO at 2000 K can be attributed to the increased NH_2 concentration being offset by the decrease in the overall rate coefficient for NH_2 + NO so that this rate is comparable to that at 1280 K, about 20 nmol cm^{-3}s^{-1}, while the NO production rate at 2000 K peaks near 30 μmol cm^{-3}s^{-1}, overwhelming all NO consumption rates (cf. Figures 2.34 and 2.66).

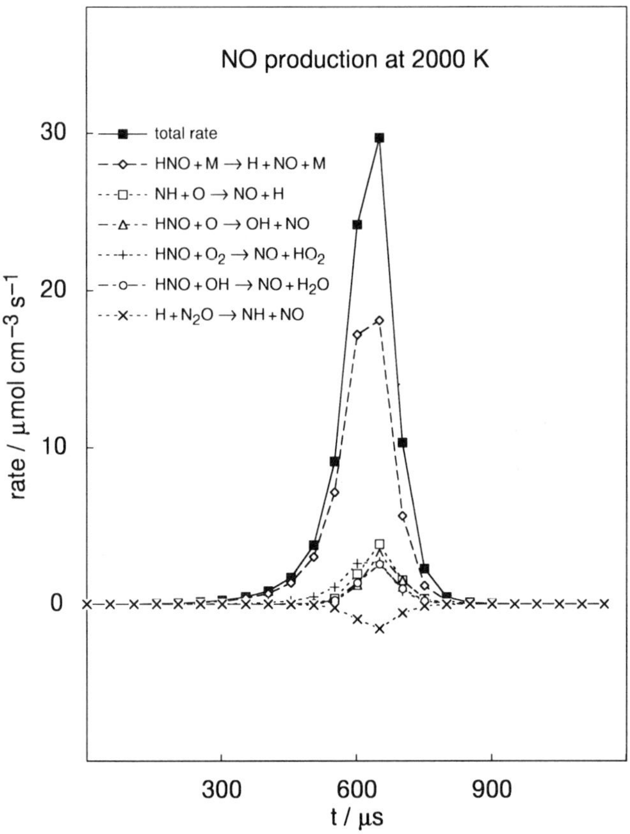

FIGURE 2.67. Major routes of NO production in the DBN/MBC model for the high temperature conditions of Figure 2.65. The higher temperature causes $HNO + M \longrightarrow H + NO + M$ to dominate NO production. In contrast to the lower temperature regime, reduction of NO by reaction with NH_2 is unimportant, and one sees almost quantitative conversion of ammonia to NO.

HNO is the major precursor of NO at both temperatures. At 1280 K, 5 reactions contribute to producing HNO (Figure 2.68). The long-lived radicals NH_2O and HNOH (which are also involved in the alternate path to NO going through the reaction sequence 37b–34d2) are prominent in this chemistry, primarily through losing an H atom in reactions with other radicals. Some HNO is also produced by

$$NH + O_2 \longrightarrow HNO + O \qquad (12g)$$

$$NH_2 + \longrightarrow HNO + H. \qquad (15c)$$

At 2000 K reaction 15c accounts for most of the HNO production, essentially because at this temperature the concentration of O atoms is much greater.

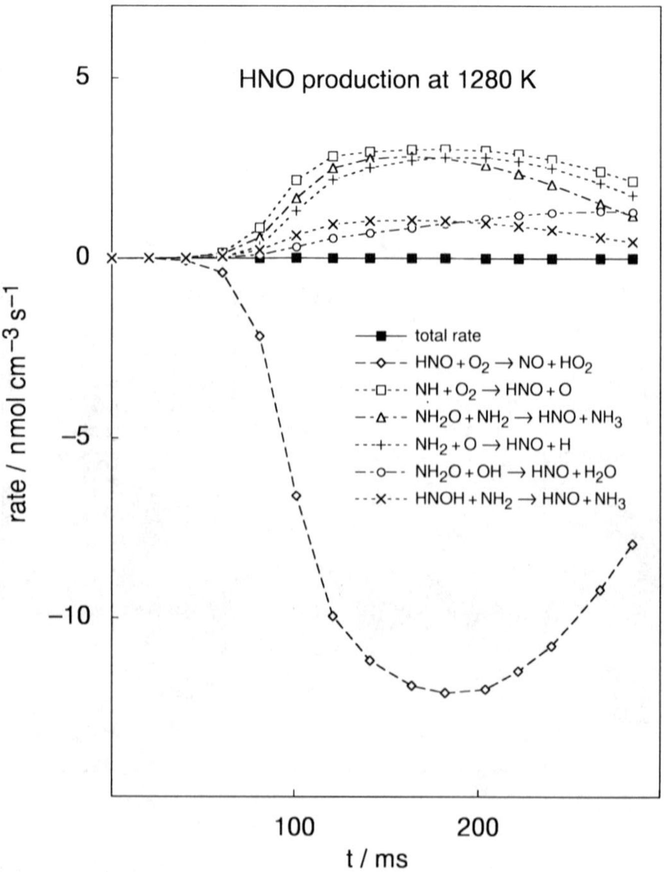

FIGURE 2.68. Rate analysis for the most important reactions contributing to HNO production for the conditions of Figure 2.63. Most of them involve hydrogen transfer between radicals.

2.8.2 Kinetics of selective noncatalytic reduction of NO

The elementary reactions involved in selective noncatalytic reduction of NO (SNCR), whereby NH_2 sources are injected into the post-combustion zone, have already been discussed in various sections of this chapter. The earliest SNCR realization, the Thermal DeNO$_x$ process invented by Lyon in 1975, uses NH_3 as the NH_2 source; it has been installed in over a hundred commercial furnaces and boilers (Lyon and Hardy 1986). The kinetics common to all SNCR processes are interesting in that each has a temperature window—for NH_3 addition between 1100 and 1400 K—where some of the NO is reduced to N_2. At lower temperatures the kinetics are too slow and at higher temperatures more NO is produced than consumed. The representative laboratory data (Lyon 1987) shown in Figure 2.69

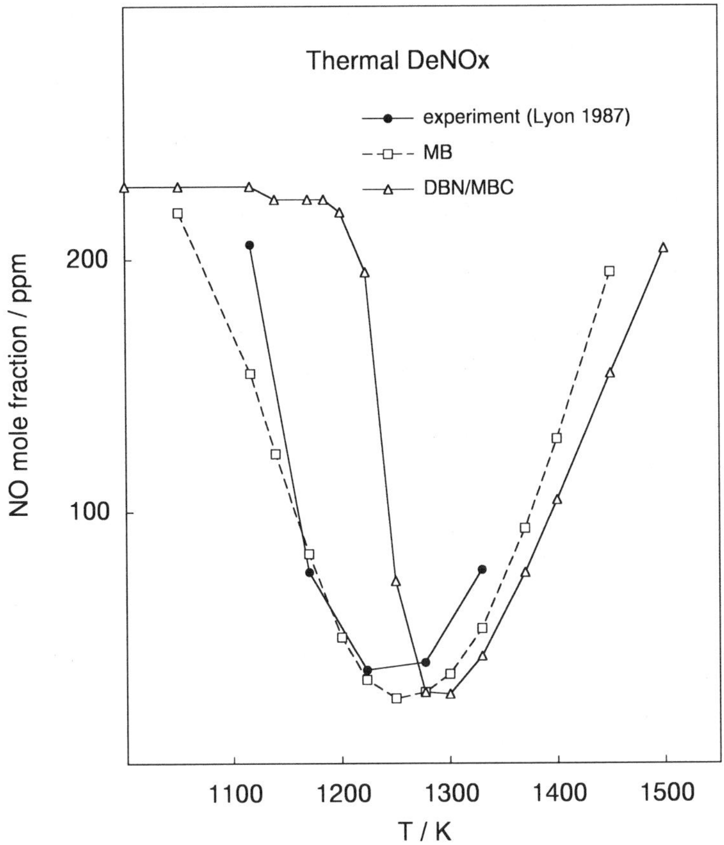

FIGURE 2.69. Comparison of predictions for NO removal at various temperatures to the experimental data of Lyon for typical Thermal DeNO$_X$ conditions: 380 ppm NH_3, 230 ppm NO, 4% O_2, 10% H_2O in helium at 1.1 atm with a residence time of 0.2 s. The Miller–Bowman mechanism (MB) is in good agreement with the data. Although the DBN/MBC mechanism requires higher temperatures for NO reduction than observed, the overall agreement is especially satisfying in that no rate coefficient adjustments were made to attain it.

illustrate the effectiveness of SNCR: at the center of the temperature window, near 1250 K, addition of 380 ppm NH_3 reduces a starting NO concentration of 230 ppm to about 35 ppm.

Many models describing SNCR chemistry have been advanced and tested against the available experimental data. Glarborg et al. (1994) summarize the earlier studies and discuss the remaining problems, in particular, two troublesome aspects of the MB mechanism. The first is their use of a larger branching ratio for OH production in the NH_2 + NO reaction (cf. Section 2.6.14). Glarborg et al. reduced the branching ratio from the 0.51 MB value to values near 0.3 for the SNCR temperature range. The second one is MB's use of an NNH lifetime much longer than expected from theory (cf. Section 2.7.3). Glarborg et al. reduced the assumed dissociation lifetime of NNH from the Miller and Bowman value of 100 μs to 1 μs and found that the combination of these two changes significantly diminished the agreement between model and experiment.

Figure 2.69 shows model predictions for typical SNCR conditions. The MB mechanism captures much of the detail of the Lyon and Hardy data. Use of the DBN/MBC mechanism leads to slower chemistry, with the result that the SNCR window is offset to higher temperatures. Further calculations showed that a slight increase in the branching ratio of the NH_2 + NO reaction (Section 2.6.14), from 0.28 to 0.32 at 1200 K,[1] suffices to shift the minimum of the DBN/MBC temperature window from 1300 to 1250 K. Thus the BDN/MBC model, despite the imperfect match with the data, captures the main features of the SNCR process. This is especially encouraging in that the model correctly addresses both of the problems noted by Glarborg et al. (1994), that is, the branching ratio of the NH_2 + NO reaction and the lifetime of NNH.

Figure 2.70 illustrates the effect of H_2O on SNCR kinetics. In both the experiments and the DBN/MBC model the SNCR window of the dry system shifts to lower temperature by about 25 to 50 K.

The overall agreement with experiment found for SNCR conditions using the DBN/MBC mechanism is encouraging. We feel that because of the complexity of the model it would not be appropriate to adjust the rate coefficient expressions of reactions 18a and 18b to achieve agreement with the experimental SNCR data. Further experiments offer a better alternative to deepen our understanding of nitrogen chemistry under these conditions.

2.8.3 Fuel-rich ammonia flames

Dean et al. (1984) concluded that it was necessary to include a number of reactions of various NH_i species that form N–N bonds in order satisfactorily to account for species profiles that they had measured in fuel-rich ammonia flames. We have compared the predictions of the DBN/MBC mechanism to the profiles measured in that work as a further test of our rate coefficient assignments. Such comparisons are stringent tests of the mechanism in that concentrations of the reactive intermediates NH, NH_2, and OH were measured in addition to NH_3 and NO in the flame experiments.

---

[1] For modified branching ratio calculations we adjusted the expressions listed in Table 2.19 for the two NH_2 + NO reactions, k_{18a} and k_{18b}, so as to keep their sum constant but their ratio greater by 0.04 over the temperature range of the SNCR window, that is, from 1000 to 1300 K.

Votsmeier et al. (1998) reported a direct measurement of the branching ratio between 1350 and 1750 K. Extrapolation of the formula they derive to lower temperature gives a branching ratio of 0.35 at 1250 K, similar to the value suggested by the above modeling exercise.

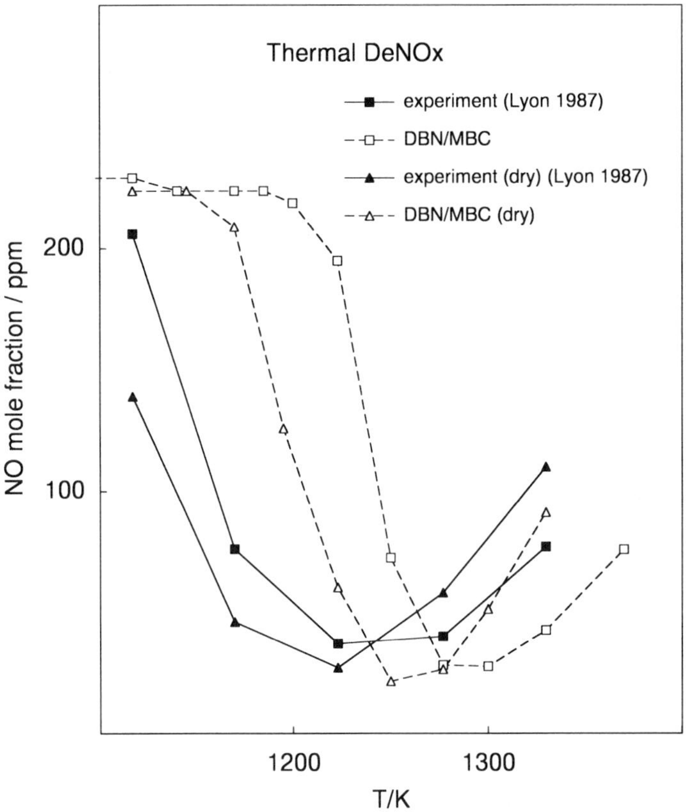

FIGURE 2.70. Effect of H_2O on NO reduction in the Thermal $DeNO_x$ process. Conditions as in Figure 2.69 except that no H_2O was added for the dry case. The predicted shift is similar to the one observed.

Figures 2.71 and 2.72 compare the predictions of the DBN/MBC and MB mechanisms to the observed profiles of NO and NH for an atmospheric pressure NH_3–O_2–N_2 flame with an equivalence ratio of 1.81. The observed peak flame temperature was 1875 K at a point 1.6 mm above the burner surface. Much of the NO initially produced in the flame front region is destroyed in the hot post-flame environment. The peak NH concentration is lower than that for NO, and its decay is more gradual. For both of these species, the DBN/MBC mechanism provides a better description than the MB mechanism. The extent of the agreement with the observed profiles is good, especially considering that no normalization was used and that no adjustments were made to the DBN/MBC mechanism.

Profiles were also computed using the Hennig et al. (1995) values for NH_2 + O_2 rate coefficients (which include NO + H_2O as the major product channel, cf. Section 2.6.9) in the DBN/MBC mechanism. The results are similar to the effect seen above for lower temperature ammonia oxidation (cf. Figure 2.63). Inclusion of this pathway not only overpredicts the NO concentration, but the

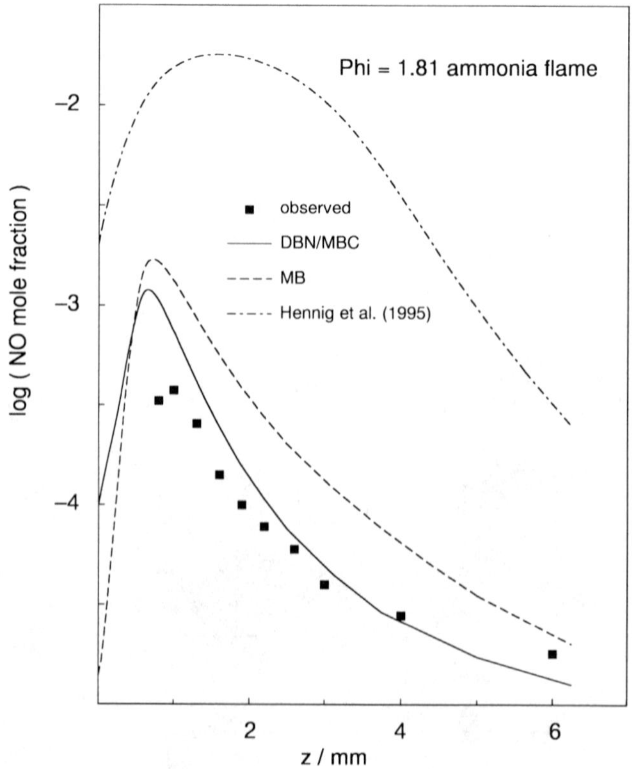

FIGURE 2.71. Comparison of observed and predicted NO profiles in an ammonia flame with an equivalence ratio of 1.81 (Dean et al. 1984). The predictions using the mechanism developed in this work (DBN/MBC) are closer to the experimental results than those obtained using the Miller/Bowman mechanism (MB). Use of the rate coefficients of Hennig et al. (1995) for $NH_2 + O_2$ with the DBN/MBC mechanism substantially overpredict the amount of NO formed and move the peak to larger distances from the burner surface.

chain termination nature of this reaction makes the chemistry much slower, such that the predicted peak NO concentration is too far above the burner surface. The NH predictions also show the peak much too high above the burner surface. In this case, the predicted peak concentration is lower than that observed.

Figures 2.73 and 2.74 compare the rates of NO production as a function of height above burner for the two mechanisms. The peak production rate is slightly lower with the DBN/MBC mechanism (Figure 2.73) than with the MB mechanism (Figure 2.74). In both, the dominant pathway for NO production is HNO dissociation, just as was found earlier for lean ammonia oxidation at 2000 K. However, there the similarities with lean ammonia oxidation cease; in the rich flame there are significant pathways for NO destruction, with the result that in the fuel-rich flame little of the NH_3 is converted to NO, whereas NH_3 is almost completely converted to NO under fuel-lean conditions.

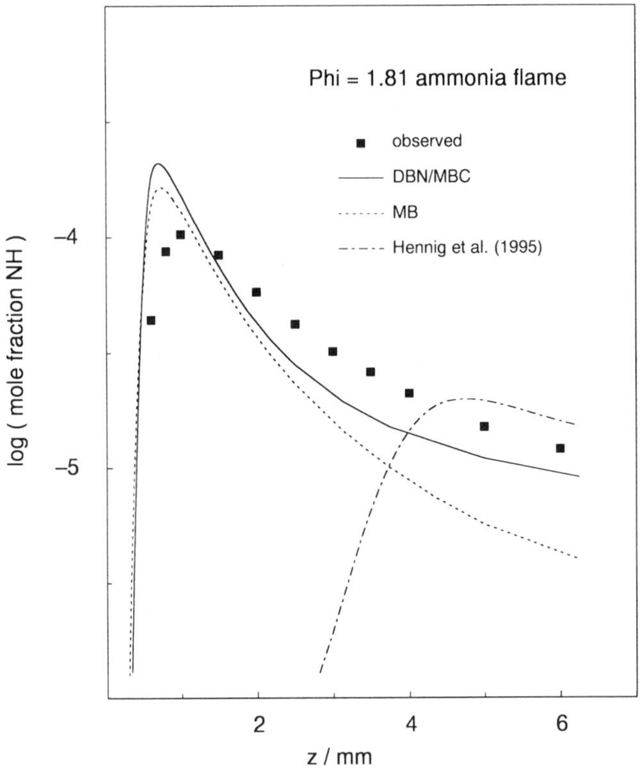

FIGURE 2.72. Comparison of observed and predicted NH profiles in an ammonia flame with an equivalence ratio of 1.81 (Dean et al. 1984). The predictions using the mechanism developed in this work (DBN/MBC) are somewhat closer to the experimental results than those obtained using the Miller/Bowman mechanism (MB). Use of the rate coefficients of Hennig et al. (1995) for $NH_2 + O_2$ with the DBN/MBC mechanism underpredict the amount of NH formed and move the peak to much larger distances from the burner surface.

Comparing the predictions of the MB and DBN/MBC models, we see that the major sink for NO is different. In the MB mechanism the dominant channel is

$$\mathrm{NNH + NO \longrightarrow N_2 + HNO,} \tag{28h}$$

while in the DBN/MBC mechanism it is

$$\mathrm{NH_2 + NO \longrightarrow N_2 + H_2O.} \tag{18a}$$

Much of this difference can be attributed to the use of a much lower rate coefficient for reaction 28h in DBN/MBC. There are other substantial differences; for example, only 3 of the fastest 6 reaction rates in the DBN/MBC model are among the fastest 6 in the MB model.

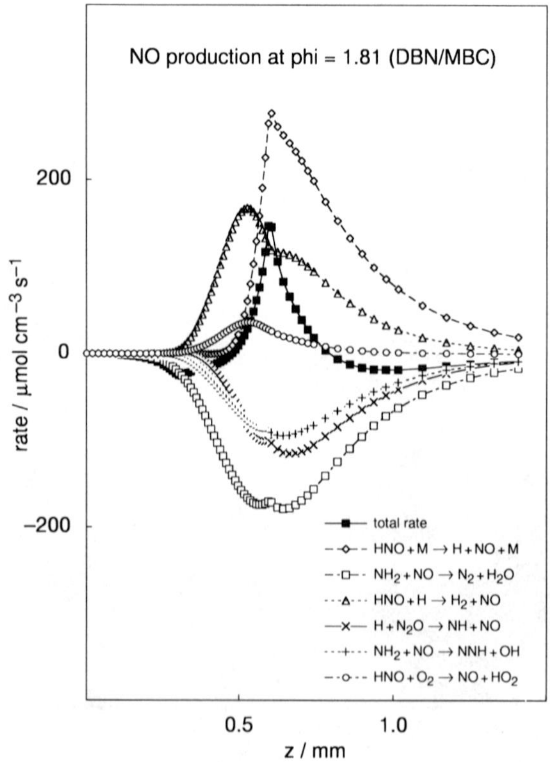

FIGURE 2.73. Rate analysis based upon the DBN/MBC mechanism for the most important reactions contributing to NO production and/or decay in an ammonia flame with an equivalence ratio of 1.81. Most of the NO is produced by reactions of HNO, while the most important sinks involve reactions with NH_2 and NH.

Figures 2.75 and 2.76 compare the rates of NH production. Again there are significant differences. The dominant source of NH in the DBN/MBC model is

$$2\,NH_2 \longrightarrow NH_3 + NH, \tag{17e}$$

while this reaction is not included in the MB model. In both models the main reaction consuming NH is

$$NH_2 + NH \longrightarrow N_2H_2 + H, \tag{27b1}$$

which is important in accounting for the low conversion of NH_3 to NO in this rich flame—it consumes NH_i species while forming an N-N bond. Thus, the NH_i species are no longer available to be oxidized to NO and much of the N_2H_2 goes on to form N_2. A second loss route for NH in the DBN/MBC mechanism,

$$2\,NH \longrightarrow N_2 + 2\,H, \tag{27a}$$

also converts NH_i to N_2.

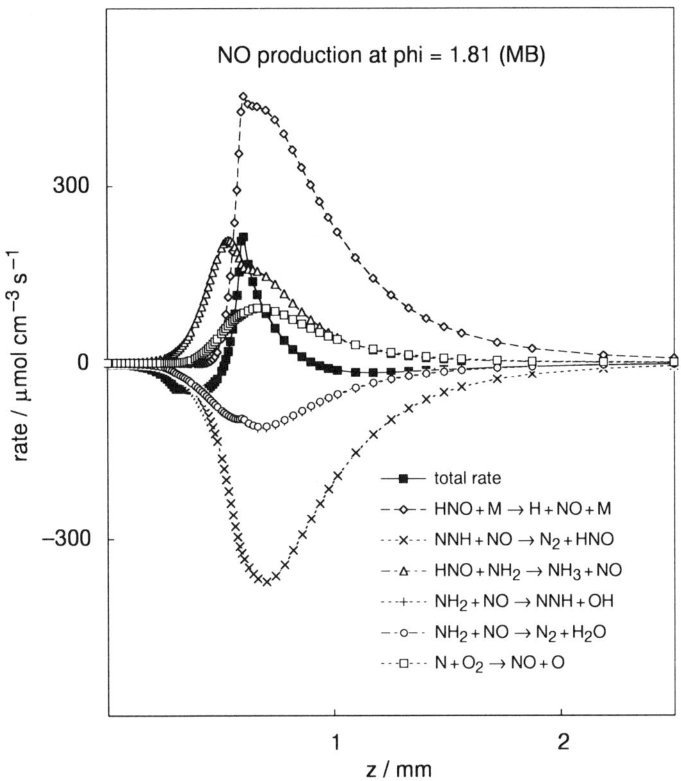

FIGURE 2.74. Rate analysis based upon the MB mechanism for the most important reactions contributing to NO production and/or decay in the ammonia flame with an equivalence ratio of 1.81. The production of NO from HNO dissociation is approximately twice as large as for the DBN/MBC mechanism, and the major sink with the MB mechanism is NNH + NO, which is much less important in the DBN/MBC mechanism.

2.8.4 Implications of the O + NNH reaction

We presented an overview of NO_X production mechanisms in Section 2.2. In the absence of hydrocarbons, NO production is governed by the concentration of oxygen atoms, as shown in Figure 2.62. In the course of our investigation of the combustion chemistry of nitrogen we identified a new pathway for NO_X formation, the reaction

$$NNH + O \longrightarrow NH + NO, \qquad (-11g)$$

which leads directly to one molecule of NO and indirectly, through subsequent reactions of NH, to still more. As discussed in Section 2.7.3, NNH can rapidly become equilibrated with H and N_2 in flames, such that bimolecular reactions of NNH become important. This equilibrium has consequences for NO_X chemistry,

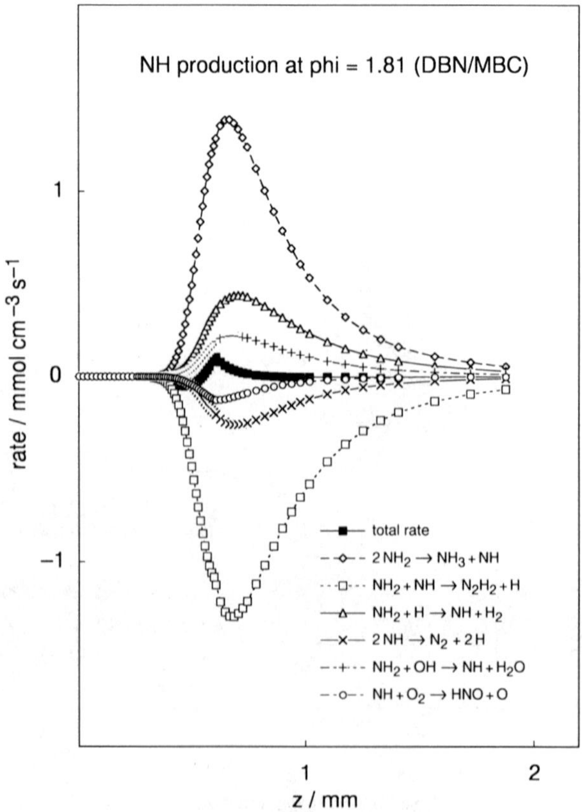

FIGURE 2.75. Rate analysis based upon the DBN/MBC mechanism for the most important reactions contributing to NH production and/or decay in a $\phi = 1.81$ ammonia flame. Most of the NH is produced and consumed by reactions of NH_2.

for example in hydrogen/air flames. Figure 2.77 compares the predictions of the DBN/MBC and MB models for a stoichiometric mixture of hydrogen and air at 1800 and 2400 K. At 1800 K the DBN/MBC model predicts much larger amounts of NO than the MB model, while at 2400 K, where most of the NO arises from the Zeldovich mechanism, the differences between the two models are small.

Harrington et al. (1995, 1996) measured NO production rates in low-pressure, low-temperature (near 1200 K) hydrogen–air flames. For these conditions, NO from other NO production routes is computed to be negligible. The amount of NO observed was consistent with the amount computed to be formed via the O + NNH channel, which was reported to be the only explanation for the observed NO.

Figure 2.78 shows the major pathways for NO production and decay computed using the DBN/MBC model for H_2 burning in air at 1800 K. The two most important reactions for NO production are

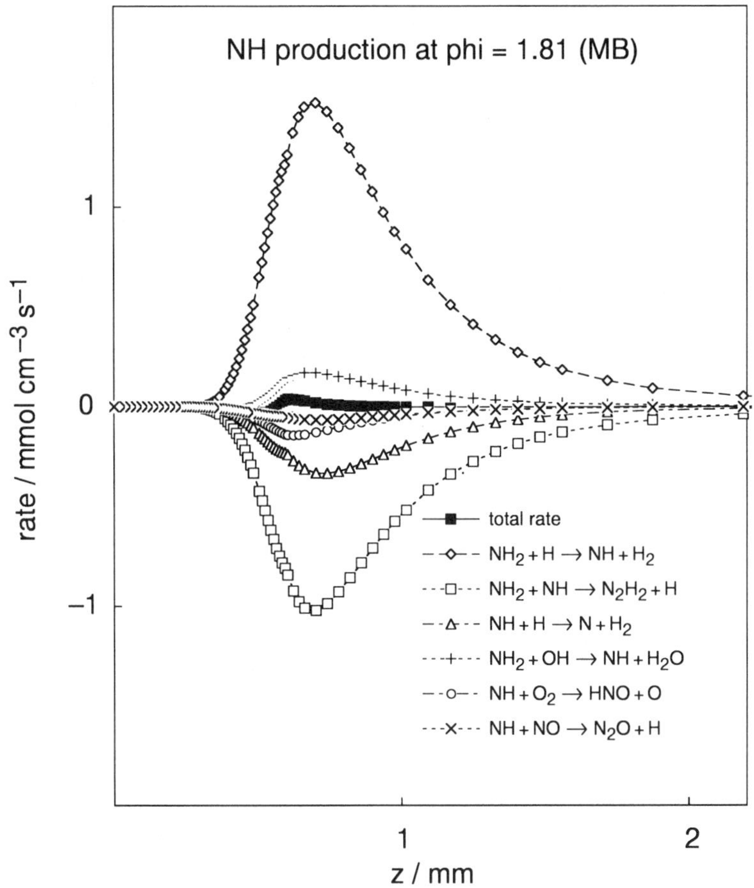

FIGURE 2.76. Rate analysis based upon the MB mechanism for the most important reactions contributing to NH production and/or decay in the ammonia flame with an equivalence ratio of 1.81. In contrast to the DBN/MBC mechanism, the major source of NH is $H + NH_2$, since the reaction $NH_2 + NH_2 \longrightarrow NH_3 + NH$ is not considered in this mechanism. The major sink is the same as in the DBN/MBC mechanism.

$$NNH + O \longrightarrow NH + NO \quad \text{and} \qquad (-11g)$$

$$HNO + H \longrightarrow H_2 + NO. \qquad (36d1)$$

This clarifies the difference between the DBN/MBC and MB predictions; the MB model does not include reaction 11g, and its rate coefficient for reaction 36d1 is over 10 times slower than the DBN/MBC expression (Section 2.7.11).

As discussed in Section 2.7.3, the flame front concentration of NNH established by equilibration with $H + N_2$ is high enough that NNH reacts there with the high (over 1% mole fraction) flame front concentration of atomic oxygen. The reaction

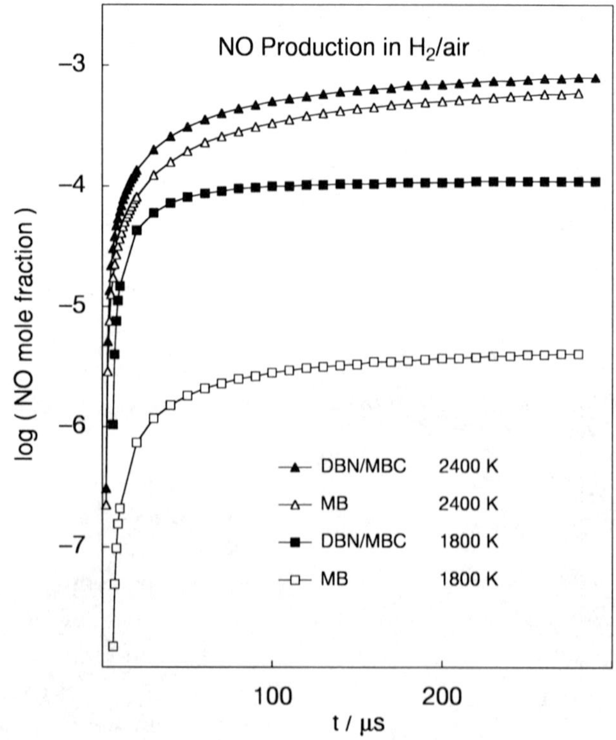

FIGURE 2.77. Predicted effect of temperature on NO production in a stoichiometric hydrogen/air mixture using the DBN/MBC and MB models. Results are shown for a plug flow model at constant temperatures of 1800 K and 2400 K and 1 atm.

NNH $\longrightarrow$ N_2 + H is effectively in equilibrium at the flame front even if a lower value for NNH dissociation rate coefficient is assumed, that is, if the tunneling contribution (cf. Section 2.7.3) is omitted. However, use of the Miller–Bowman value of $10^4\,s^{-1}$, five orders of magnitude smaller, does completely suppress this NO production route. With the MB rate coefficient, the (reverse) reaction between H and N_2 is not rapid enough and there is insufficient NNH to react with O atoms.

The theoretical evidence (Section 2.7.3) supports a dissociation rate coefficient for NNH much larger than the value assumed by Miller and Bowman, and therefore a much larger formation rate coefficient. One has to consider in this context the reliability of the equilibrium constant for this reaction. Our calculations used the NNH thermochemistry of Melius (1993), which is consistent with that implied by the potential energy surface of Koizumi et al. (1991). The uncertainty in the entropy of NNH can be estimated to be about 8 kJ $mol^{-1}K^{-1}$, and the heat of formation uncertainty to be about 12 kJ/mol. A worst case combination of these errors would shift the equilibrium constant by about a factor of 6 at 2000 K; a more realistic uncertainty estimate would be a factor of 3. Such a change in

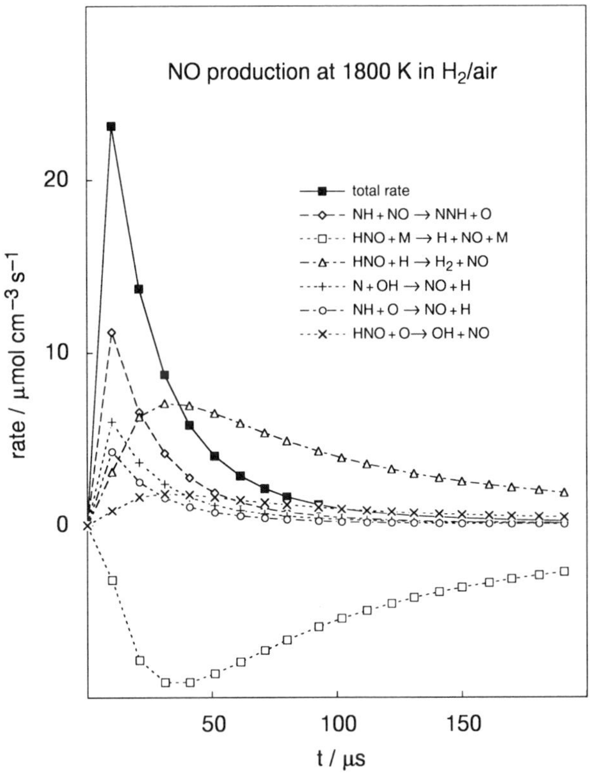

FIGURE 2.78. Rate analysis based on the DBN/MBC mechanism for NO production and decay in a stoichiometric H_2-air mixture at 1800 K and 1 atm. Most of the NO is formed by reactions of NNH, HNO and NH. The major sink is conversion back to HNO.

the equilibrium constant (and thus the concentration of NNH) translates into a comparable shift in the predicted rate of NO production and would also affect the time scale required for the reaction to become partially equilibrated. Since an order of magnitude increase in the time required to attain equilibrium did not appreciably affect the results, the main thermochemical influence would be to affect the rate of NO production; the NNH $\longrightarrow$ N_2 + H reaction will still be partially equilibrated in flame fronts.

Comparison of the NNH production rate in Figure 2.79 to that for NO production in Figure 2.78 shows that they are comparable to one another even though the concentration of NNH is orders of magnitude less than the concentration of NO. NNH is produced by both the collisional activation and the tunneling pathways. Even though the rate of NNH + O $\longrightarrow$ NH + NO is smaller than that for NNH + H $\longrightarrow$ $N_2 + H_2$, rapid resupply of NNH by the H + N_2 reaction maintains a near-equilibrium NNH concentration, and NO production continues to be significant until the concentrations of H and O begin to decay.

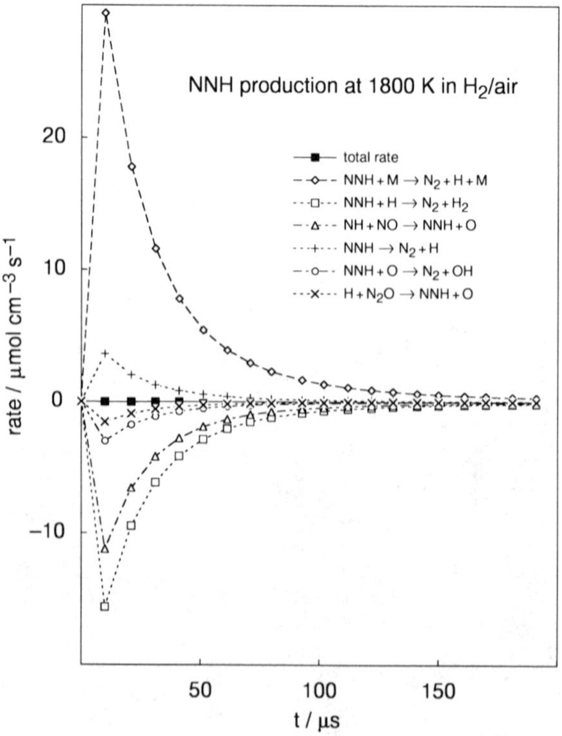

FIGURE 2.79. Rate analysis based on the DBN/MBC mechanism for NNH production and decay in a stoichiometric H_2–air mixture at 1800 K and 1 atm. NNH is produced by H addition to N_2 and mainly consumed in reactions with H and O. The reaction with H recycles NNH to N_2.

2.8.5 Nitrogen chemistry in hydrocarbon–air flames

Nitrogen chemistry is much more complex in the presence of hydrocarbons, specifically hydrocarbon radicals, as illustrated in Figure 2.80. Nitrogen-containing species can react with CH_i species to form a wide variety of compounds that may ultimately cascade back to nitrogen species, not necessarily those from wheRE they came. NO can be formed from N_2, via attack by CH (Prompt NO), or converted to N_2 by hydrocarbon radical attack, the essence of NO_x reburning strategies. Reliable prediction of the kinetics of these processes requires an accurate description of both the hydrocarbon and nitrogen chemistry. For purposes of this review, we can illustrate some of the interactions that occur in flames using a plug flow model. This neglect of diffusion does not introduce major errors at pressures of 1 atm or higher, and the calculation is appreciably simpler.

Figure 2.81 illustrates the effect of temperature on NO production during the combustion of a stoichiometric mixture of methane and air as predicted by the DBN/MBC and MB models. The results are for a pressure of 1 atm and temperature held fixed at either 1800 or 2400 K. As with the H_2/air case

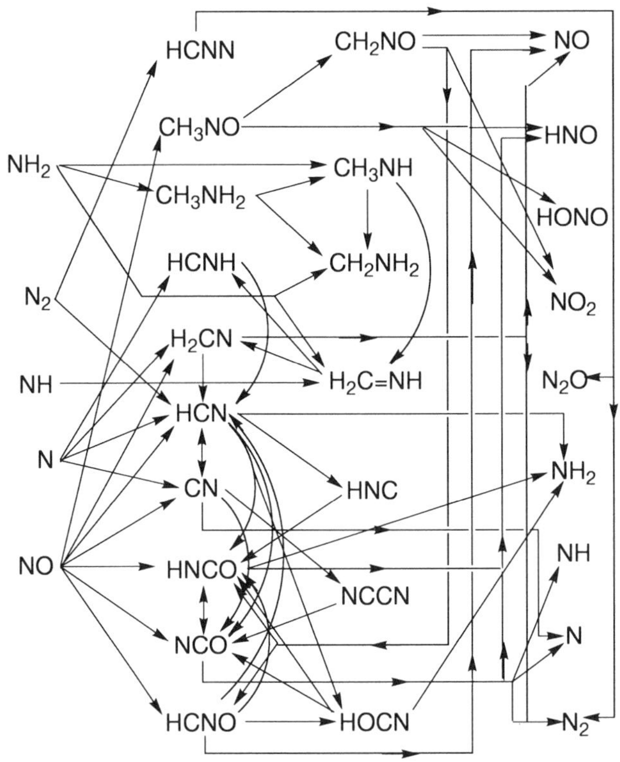

FIGURE 2.80. Overview of the reactions connecting various species in the C/N/H/O system. The N-containing species in the left column can react with the hydrocarbon radicals CH, CH_2, and CH_3 to produce the various $CH_xN_yO_z$ species that undergo extensive rearrangements while cascading back to the N-containing species in the right column. No attempt was made to denote the co-reactants for the transformations. The N/H/O submodel is outlined in Figure 2.62.

considered in the previous section, the results of the two models are close at 2400 K, while at 1800 K the DBN/MBC model predicts an order of magnitude higher final concentration of NO than the MB model. The DBN/MBC results indicate that the increase in temperature from 1800 to 2400 K results in an order of magnitude increase in NO formed. The NO profile increases much more steeply at 2400 K, following the more rapid increase in oxygen atom concentration at this temperature. The difference in the NO profiles is greater than the difference in the O-atom profiles because the rate coefficient for NO production (reaction 1) has a high activation energy.

The reasons for the difference in NO production at 1800 K can be seen in the analysis of the NO production rates in the two models, as seen in Figures 2.82 and

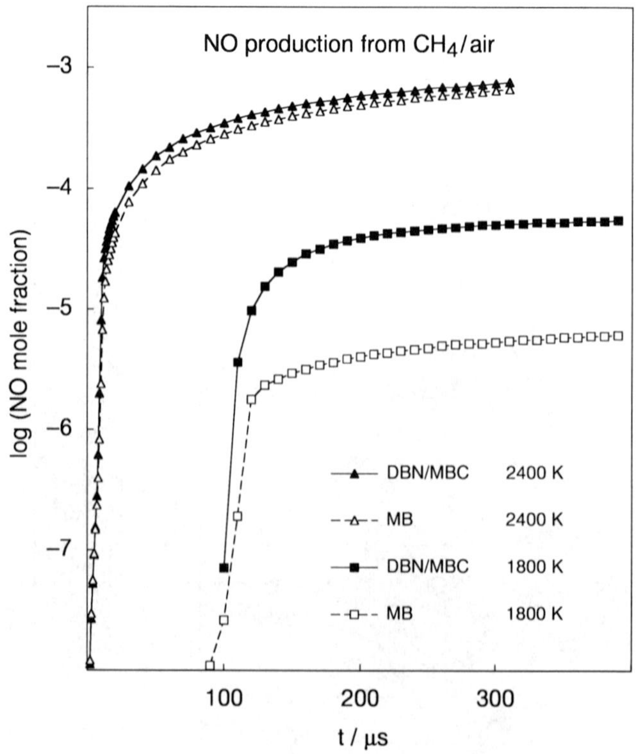

FIGURE 2.81. Predicted effect of temperature on NO production in a stoichiometric methane/air mixture using the DBN/MBC and MB models. Results are shown for a plug flow model at 1 atm and constant temperatures of 1800 K and 2400 K.

2.83. The major contributors to NO production in the DBN/MBC model are

$$O + NNH \longrightarrow NH + NO \qquad (-11g)$$

$$NH + O \longrightarrow NO + H \qquad (27e1)$$

$$HNO + H \longrightarrow H_2 + NO. \qquad (36d1)$$

In the DBN/MBC model, most NO production at 1800 K is non-Zeldovich NO. In contrast, the MB model attributes most NO production to the Zeldovich pair

$$N + OH \longrightarrow NO + H \qquad (26c)$$

$$N + O_2 \longrightarrow NO + O. \qquad (26a)$$

At 2400 K essentially the same amount of NO is predicted for stoichiometric methane and hydrogen flames (cf. Figures 2.77 and 2.81), while at 1800 K the NO concentration predicted for a hydrogen flame is about twice that predicted for a methane flame. Comparison of Figures 2.78 and 2.82 reveals that the reactions producing most of the NO are the same in both flames: The early sources of NO are reactions −11g and 27e1, while reaction 36d1 generates NO at later times. The rates of all three reactions are lower in the methane flame.

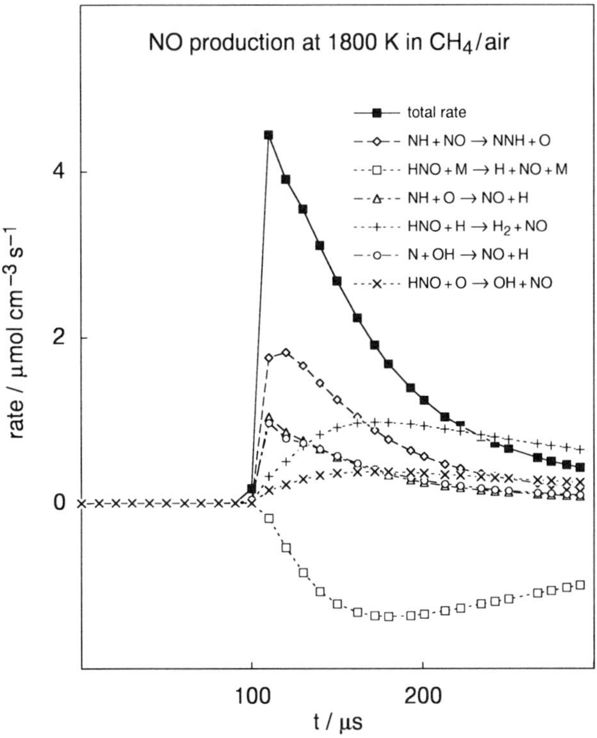

FIGURE 2.82. Rate analysis based on the DBN/MBC mechanism for NO production and decay in a stoichiometric CH_4–air mixture at 1800 K and 1 atm. The major sources of NO are NNH, HNO, and NH as in the hydrogen/air system shown in Figure 2.78. HNO is the major sink, as in the hydrogen system. The peak rate of NO production, however, is about five times lower for methane than for hydrogen.

2.8.6 General conclusions from modeling tests

The modeling results described in the foregoing sections show that the Table 2.19 mechanism captures the essential features of several high-temperature nitrogen chemistry systems. The agreement found in the instances where predictions can be compared to data is especially gratifying in that the rate coefficient assignments were not adjusted to fit data taken in such complex chemical situations. An important aspect of these tests is the emergence of hitherto unfamiliar species and reaction pathways that appear to play central roles in high-temperature nitrogen chemistry. A number of such systems were shown to warrant further study.

A generalization emerging from the results of these modeling tests is that radical–radical reactions are important, essentially because nitrogen radicals usually have slower unimolecular decay rates than their hydrocarbon analogs. This is traceable to the underlying thermochemistry. With hydrocarbon radicals, beta-scission reactions usually have barriers ranging from 120 to 170 kJ/mol.

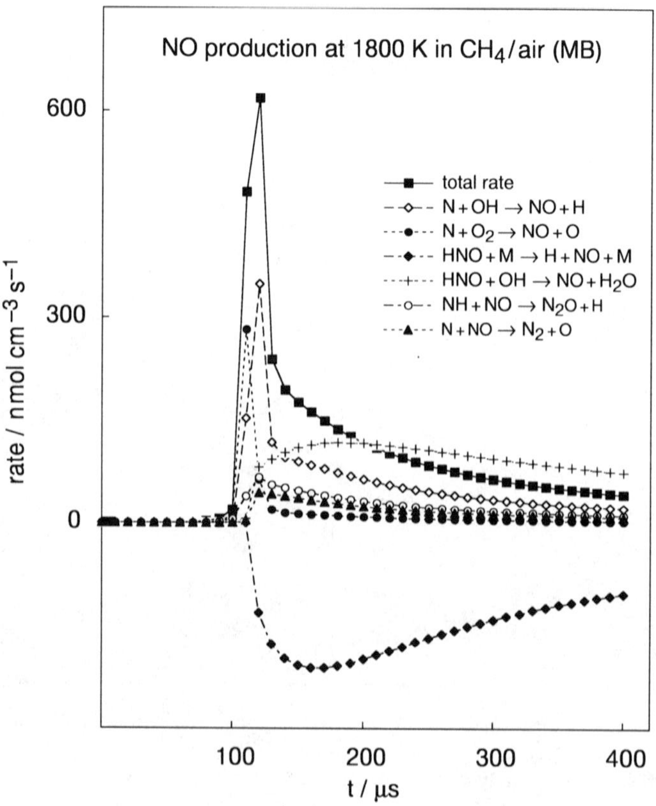

FIGURE 2.83. Rate analysis based on the MB mechanism for NO production and decay in a stoichiometric methane–air mixture at 1800 K and 1 atm. N atoms are the major source of NO and HNO the major sink. The peak rate of NO production is about seven times lower than predicted using the DBN/MBC mechanism (Figure 2.82).

Most of the energy needed to break a C–C or C–H single bond is compensated by energy gained in formation of a carbon–carbon or carbon–oxygen double bond. With nitrogen radicals, the energy compensation by double bond formation is substantially lower, leading to higher barriers for beta scission. Further refinement of rate coefficient estimation procedures, for example, our DHT procedure for radical–radical hydrogen transfer rate coefficients, could lead to improved understanding of the effects of these radical–radical reactions.

The literature contains many flame studies which suggest that important aspects of hydrocarbon–nitrogen interactions remain to be discovered. Examples include the studies by Etzkorn et al. (1992) and Williams and Fleming (1994) of NO-doped methane flames, of Garo et al. (1992) on production of NO and HCN in flames, and the analysis of reburn chemistry by Staph and Leuckel (1996). Reinvestigation of the relevant reaction pathways using the Table 2.19 mechanism, together with an appropriate hydrocarbon oxidation mechanism, for the conditions studied by these authors would be a useful undertaking.

2.9 SUMMARY

Table 2.19 summarizes our recommended rate coefficient expressions for the elementary reactions considered in this chapter. They are based upon measurements as far as possible, and consistency with thermochemical and theoretical expectations has been enforced. For important elementary reactions for which experimental data are not yet available, QRRK analysis was used to derive interim rate coefficient expressions and to compute the effects of temperature and pressure on rate coefficients that are expected to be pressure-dependent.

An important caveat about our rate coefficient recommendations is that using them as part of a dynamic model will not resolve all basic ambiguities. There are questions in C–H–O–N chemistry for which experimental answers are still needed before rate coefficients and/or product channels can be assigned.

We urge acceptance of a modeling philosophy in which one makes maximum use of the best available expressions for the elementary reaction rate coefficients, preferably obtained from studies where the reaction of interest is well isolated, as the basis set for the model. For reactions for which such information is unavailable, we suggest adoption of techniques like those used in this chapter to obtain plausible theoretical estimates. If using such models leads to predictions significantly different from the observations, the most likely explanation is that important elementary reactions have been overlooked. Sensitivity analysis will suggest where to look for missing chemistry, where to adjust estimates of reaction barriers, and so on. It is generally more productive, when faced with such circumstances, to improve one's understanding of the chemistry than to adjust rate coefficient expressions to improve a fit to data. Such adjustments may only serve to obscure a significant scientific problem.

We hope that our analysis has clarified which C–H–O–N reactions have rate coefficient expressions based on reliable measurements and shown how plausible estimates can be made for the rate coefficients of reactions lacking such measurements. We hope that combustion researchers will use our material extensively for analysis of experimental data on the combustion chemistry of nitrogen. Where our mechanism does not satisfactorily rationalize new data, we hope that it will help reveal the C–H–O–N combustion chemistry that remains to be discovered.

2.10 ACKNOWLEDGMENTS

We are grateful to C.F. Melius for providing results of BAC-MP4 calculations of standard enthalpies of formation for many species and transition state energies for many reactions. We also thank the many researchers who gave us pre-publication copies of their manuscripts.

2.11 REFERENCES

Abbatt, J.P.D., Toohey, B.W., Fenter, F.F., Stevens, P.S., Brune, W.H., & Anderson, J.G. (1989). J. Phys. Chem. **93** 1022.

Adamson, J.D., Farhat, S.K., Morter, C.L., Glass, G.P., Curl, R.F., & Phillips, L.F. (1994). J. Phys. Chem., **98,** 5665.

Allara, D.L., & Shaw, R. (1980). J. Chem. Phys. Ref. Data, **9,** 523.

Allen, M.T., Yetter, R.A., & Dryer, F.L. (1994). *The Decomposition of Nitrous Oxide at Elevated Pressures,* Spring Western States Section Meeting of the Combustion Institute, Paper No. 94–012.

Allen, M.T., Yetter, R.A., & Dryer, F.L. (1995). Int. J. Chem. Kinet., **27,** 883.

Anastasi, C., & Maw, P.R. (1982). J. Chem. Soc. Faraday Trans. I, **78,** 2423.

Atakan, B., & Wolfrum, J. (1991). Chem. Phys. Lett., **186,** 547.

Atakan, B., Jacobs, A., Wahl, M., Weller, R., & Wolfrum, J. (1989). Chem. Phys. Lett., **155,** 609.

Atakan, B., Kocis, D., Wolfrum, J., & Nelson, P. (1992). 24th Symposium (International) on Combustion, p. 691.

Atkinson, R. (1989). *Kinetics and Mechanisms of Gas-Phase Reactions of the Hydroxyl Radical with Organic Compounds.,* J. Phys. Chem. Ref. Data Monograph No. 1.

Atkinson, R., Baulch, D.L., Cox, R.A., Hampson, R.F., Jr., Kerr, J.A., & Troe, J. (1989). J. Phys. Chem. Ref. Data, **18,** 881.

Atkinson, R., Baulch, D.L., Cox, R.A., Hampson, R.F., Jr., Kerr, J.A. & Troe, J. (1992). J. Phys. Chem. Ref. Data, **21,** 1125.

Back, R.A., & Yokota, T. (1973). Int. J. Chem. Kinet., **5,** 1039.

Baldwin, A.C., & Golden, D.M. (1978). Chem. Phys. Lett., **55,** 350.

Balla, R.J., Casleton, K.H., Adams, J.S., & Pasternack, L. (1991). J. Phys. Chem., **95,** 8694.

Bauerle, S., Klatt, M., & Wagner, H.Gg. (1995). Ber. Bunsenges. Phys. Chem., **99,** 97.

Baulch, D.L., Cobos, C.J., Cox, R. A., Esser, C., Frank, P., Just, T., Kerr, J.A., Pilling, M.J., Troe, J., Walker, R.W., & Warnatz, J. (1992). J. Phys. Chem. Ref. Data, **21,** 411.

Baulch, D.L., Cobos, C.J., Cox, R.A., Frank, P., Hayman, G., Just, T., Kerr, J.A., Murrells, T., Pilling, M.J., Troe, J., Walker, R.W. & Warnatz, J. (1994). Combust. Flame, **98,** 59.

Baulch, D.L., Duxbury, J., Grant, S.J., & Montague, D.C. (1981) J. Phys. Chem. Ref. Data, **10,** Suppl. 1

Becker, K.H., Engelhardt, B., Geiger, H., Kurtenbach, R., Schrey, G., & Wiesen, P. (1992a). Chem. Phys. Lett., **195,** 322.

Becker, K.H., Engelhardt, B., Geiger, H., Kurtenbach, R. & Wiesen, P. (1993). Chem. Phys. Lett., **210,** 135.

Becker, K.H., Engelhardt, B., Wiesen, P., & Bayes, K.D. (1989). Chem. Phys. Lett., **154,** 342.

Becker, K.H., Geiger, H., & Wiesen, P. (1996). Int. J. Chem. Kinet., **28,** 115.

Becker, K.H., Kurtenbach, R., Schmidt, F., & Wiesen, P. (1995a). Chem. Phys. Lett., **235,** 230.

Becker, K.H., Kurtenbach, R., Schmidt, F., & Wiesen, P. (1997). Ber. Bunsenges., **101,** 128.

Becker, K.H., Kurtenbach, R., & Wiesen, P. (1992b). Chem. Phys. Lett., **198,** 424.

Becker, K.H., Kurtenbach, R., & Wiesen, P. (1995b). J. Phys. Chem., **99,** 5986.

Benson, S.W. (1976). *Thermochemical Kinetics.,* New York, John Wiley & Sons.

Benson, S.W. (1985). J. Phys. Chem., **89,** 4366.

Berman, M.R., Fleming, J.W., Harvey, A.B., & Lin, M.C. (1982). 19th Symposium (International) on Combustion, p. 73.

Berman, M.R., & Lin, M C. (1983). J. Phys. Chem., **87,** 3933.

Bian, J., Vandooren, J., & Van Tiggelen, P.J. (1990). 23rd Symposium (International) on Combustion, p. 379.

Biggs, P., Canosa-Mas, C.E., Fracheboud, J.-M., Parr, A.D., Shallcross, D.E., Wayne, R.P., & Caralp, F. (1993). J. Chem. Soc. Faraday Trans., **89,** 4163.

Blauwens, J., Smets, B., & Peeters, J. (1978). 16th Symposium (International) on Combustion, p. 1055.

Böhland, T., Temps, F., & Wagner, H.Gg. (1984). Ber. Bunsenges. Phys. Chem., **88,** 455.

Boullart, W., Nguyen, M.T., & Peeters, J. (1994). J. Phys. Chem., **98,** 8036.

Bowman, C.T. (1992). 24th Symposium (International) on Combustion, p. 859.

Bozzelli, J.W., Chang, A.Y., & Dean, A.M. (1997). Int. J. Chem. Kinet., **29,** 161.

Bozzelli, J.W., & Dean, A.M. (1989). J. Phys. Chem., **93,** 1058.

Bozzelli, J.W., & Dean, A.M. (1990). J. Phys. Chem., **94,** 3313.

Bozzelli, J.W., & Dean, A.M. (1993). J. Phys. Chem., **97,** 4427.

Bozzelli, J.W., & Dean, A.M. (1995). Int. J. Chem. Kinet., **27,** 1097.

Bozzelli, J.W., & Ritter, E.R. (1993). *Radicalc---Computer code to calculate entropy & heat capacity contributions to transition states & radical species from changes in vibrational frequencies, barriers, moments of inertia, and internal rotations.* Proceedings of the Eastern States Section of the Combustion Institute, **103,** 459.

Bradley, K.S., McCabe, P., Schatz, G.C., & Walch, S.P. (1995). J. Chem. Phys., **102,** 6696.

Branch, M.C., Miller, J.A., & Kee, R.J. (1982). Combust. Sci. Technol., **29,** 147.

Brownsword, R.A., Gatenby, S.D., Herert, L.B., Smith, I.W.M., Stewart, D.W.A., & Symonds, A. C.(1996). J. Chem. Soc. Faraday Trans., **92,** 723.

Brownsword, R.A., Hancock, G., & Heard, D.E. (1997). J. Chem. Soc. Farad. Trans., **93,** 2473.

Bulatov, V.P., Ioffe, A.A., Lozovsky, V.A., & Sarkisov, O.M. (1989). Chem. Phys. Lett., **161,** 141.

Butler, J.E., Fleming, J.W., Goss, L.P., & Lin, M.C. (1981). Chem. Phys., **56,** 355.

Campbell, I.M., & Handy, B.J. (1975). J. Chem. Soc. Faraday Trans. I, **71,** 2097.

Carpenter, B.K. (1993). J. Am. Chem. Soc., **115,** 9806.

Cattolica, R.J., Smooke, M.D., & Dean, A.M. (1982). *A Hydrogen--Nitrous Oxide Flame Study,* Sandia National Laboratories Report SAND82.

Cheskis, S.G., & Sarkisov, O.M. (1979). Chem. Phys. Lett., **67,** 72.

Cohen, N. (1987). Int. J. Chem. Kinet., **19,** 319.

Cohen, N. (1990). *A Revised Model for Transition State Theory Calculations for Rate Coefficients of OH with Alkanes,* Aerospace Corporation Report ATR–90(7179)–1.

Cohen, N. (1992). *Chemical Kinetic Data Sheets for High--Temperature Chemical Reactions, Vol. III.,* Aerospace Corporation Report ATR–91(7189)–2.

Cohen, N., & Westberg, K.R. (1991). J. Phys. Chem. Ref. Data, **20,** 1211.

Cooper, W.F., & Hershberger, J.F. (1992). J. Phys. Chem., **96,** 771.

Cooper, W.F., Park, J., & Hershberger, J.F. (1993). J. Phys. Chem., **97,** 3283.

Darwin, D.C., Young, A.T., Johnston, H.S., & Moore, C.B. (1989). J. Phys. Chem., **93,** 1074.

Davidson, D.F., & Hanson, R.K. (1990a). Int. J. Chem. Kinet., **22,** 843.

Davidson, D.F., & Hanson, R.K. (1990b). 23rd Symposium (International) on Combustion, p. 267.

Davidson, D.F., Dean, A.J., DiRosa, M.D., & Hanson, R.K. (1991b). Int. J. Chem. Kinet., **23,** 1035.

Davidson, D.F., DiRosa, M.D., Chang, A.Y., & Hanson, R.K. (1991a). *18th International Symposium on Shock Waves,* Sendai, p. 813.

Davidson, D.F., Kohse–Höinghaus, K., Chang, A.Y., & Hanson, R.K. (1990). Int. J. Chem. Kinetics, **22,** 513.

Davies, J.W., Green, N.J.B., & Pilling, M.J. (1991). J. Chem. Soc. Faraday Trans., **87,** 2317.

Dean, A.J., Hanson, R.K., & Bowman, C.T. (1990). 23rd Symposium (International) on Combustion, p. 259.

Dean, A.J., Hanson, R.K., & Bowman, C.T. (1991). J. Phys. Chem., **95,** 3180.

Dean, A.M. (1985). J. Phys. Chem., **89,** 4600.

Dean, A.M., & Steiner, D.C. (1977). J. Chem. Phys., **66,** 598.

Dean, A.M., & Westmoreland, P.R. (1987). Int. J. Chem. Kinet., **19,** 207.

Dean, A.M., Bozzelli, J.W., & Ritter, E.R. (1991). Combust. Sci. Tech., **80,** 63.

Dean, A.M., Chou, M.S., & Stern, D. (1984). Int. J. Chem. Kinet., **16,** 633.

Dean, A.M., Hardy, J.E., & Lyon, R.K. (1982). 19th Symposium (International) on Combustion, p. 97.

Dean, A.M., Johnson, R.L., & Steiner, D.C. (1980). Combust. Flame, **37,** 41.

Diau, E.W., Halbgewachs, M.J., Smith, A.R., & Lin, M.C. (1995). Int. J. Chem. Kinet., **27,** 867.

Diau, E.W.G., & Smith, S.C. (1996). J. Phys. Chem., **100,** 12349.

Diau, E.W.-G., Tso, T.-Y., & Lee, Y.-P. (1990). J. Phys. Chem., **94,** 5261.

Diau, E.W., Yu, T., Wagner, M.A.G., & Lin, M.C. (1994). J. Phys. Chem., **98,** 4034.

Dolson, D.A. (1986). J. Phys. Chem, **90,** 6714.

Dorthe, G., Caubet, P., Vias, T., Barrere, B., & Marchais, J. (1991) J. Phys. Chem., **95,** 5109.

Drake, M.C., & Blint, R.J. (1991a). Combust. Sci. Tech., **75,** 261.

Drake, M.C., & Blint, R.J. (1991b). Combust. Flame, **83,** 185.

Drake, M.C., Ratcliffe, J.W., Blint, R.J., Carter, C.D., & Laurendeau, N.M. (1990). 23rd Symposium (International) on Combustion, p. 387.

Dransfeld, P., & Wagner, H.Gg. (1987). Z. Phys. Chem. (Munich), **153,** 89.

Dransfeld, P., Hack, W., Kurzke, H., Temps, F., & Wagner, H.Gg. (1984). 20th Symposium (International) on Combustion, p. 655.

Duan, X., & Page, M. (1995). J. Chem. Phys., **102,** 6121.

Duffy, B.L., & Nelson, P.F. (1996). 26th Symposium (International) on Combustion, p. 2099.

Duran, R.P., Amorebieta, V.T., & Colussi, A.J. (1988). J. Phys. Chem., **92,** 636.

Durant, J.L. (1994). J. Phys. Chem., **98,** 518.

East, A.L.L., & Allen, W.D. (1993). J. Chem. Phys., **99,** 4638.

Etzkorn, T., Muris, S., Wolfrum, J., Dembny, C., Bockhorn, H., Nelson, P.F., Attia–Shanin, A., & Warnatz, J. (1992). 24th Symposium (International) on Combustion, p. 925.

Felder, W., & Young, R.A. (1972). J. Chem. Phys., **57,** 572.

Fenimore, C.P. (1970). 13th Symposium (International) on Combustion, p. 373.

Fenimore, C.P. (1976). Comb. Flame, **26,** 249.

Fenimore, C.P. (1980). Comb. Flame, **37,** 245.

Flatness, S.A., & Kramlich, J.C. (1996). 26th Symposium (International) on Combustion, p. 567.

Foresman, J.B., & Frisch, Æ, (1996) *Exploring Chemistry with Electronic Structure Methods, Second Edition,,* Gaussian, Pittsburgh.

Forst, W. (1972). *Theory of Unimolecular Reactions,* Academic Press, New York.

Forster, R., Frost, M., Fulle, D., Hamann, H.F., Hippler, H., Schlepegrell, A., & Troe, J. (1995). J. Chem. Phys., **103,** 2949-2957.

Frank, P., & Just, T. (1985). Ber. Bunsenges. Phys. Chem., **89,** 181.

Frerichs, H., Koch, R., Schliephake, V., Tappe, M., & Wagner, H.Gg. (1990). Z. Phys. Chem., **166,** 145.

Fujii, N., Chiba, K., Uchida, S., & Miyama, H. (1986). Chem. Phys. Lett., **127,** 141.

Fujii, N., Sagawai, S., Sato, T., Nosaka, Y., & Miyama, H. (1989). J. Phys. Chem., **93,** 5474.

Fujii, N., Miyama, H., Koshi, M., & Asaba, T. (1981). 18th Symposium (International) on Combustion, p. 873.

Gardiner, W.C., Jr. (1977) Accts. Chem. Res., **10,** 326.

Gardiner, W.C., In *The 12th IMACS World Congress on Scientific Computation,* Vichnevetsky, R., Borne, P., & Vignes, J., Eds., IMACS, Paris, 1988, p. 58.

Garo, A., Hilaire, C., & Puechberty, D. (1992). Combust. Sci. Tech., **86,** 87.

Gehring, V.M., Hoyermann, K., Wagner, H.Gg., & Wolfrum, J. (1971). Ber. Bunsenges. Phys. Chem., **75,** 1287.

Gilbert, R.G., & Smith, S.C. (1990). *Theory of Unimolecular and Recombination Reactions.,* Oxford, Blackwell Scientific Publications.

Gilbert, R.G., Luther, K., & Troe, J. (1983) Ber. Bunsenges. Phys. Chem., **87,** 169.

Glarborg, P., Dam-Johansen, K., & Miller, J.A. (1995). Int. J. Chem. Kinet., **27,** 1207.

Glarborg, P., Dam-Johansen, K., Miller, J.A., Kee, R.J., & Coltrin, M.E. (1994). Int. J. Chem. Kinet., **26,** 421-436.

Golden, D.M., & Larson, C.W. (1985). 20th Symposium (International) on Combustion, p. 595.

Gordon, M.S., Truong, T.N., & Pople, J.A. (1986). Chem. Phys. Lett., **130,** 245.

Grussdorf, J., Temps, F., & Wagner, H.Gg. (1997). Ber. Bunsenges. Physik. Chemie, **101,** 134.

Hack, W., & Kurzke, H. (1985a). Ber. Bunsenges. Phys. Chem., **89,** 86.

Hack, W., Kurzke, H., & Wagner, H.Gg. (1985b). J. Chem. Soc. Faraday Trans. 2, **81,** 949.

Hack, W., Rouverirolles, P., & Wagner, H.Gg. (1986). J. Phys. Chem., **90,** 2505.

Hack, W., Wagner, H.Gg., & Zasypkin, A. (1994). Ber. Bunsenges. Phys. Chem., **98,** 156.

Halbgewachs, M.J., Diau, E.W.G., Mebel, A.M., Lin, M.C., & Melius, C. F. (1996). 26th Symposium (International) on Combustion, p. 2109.

Hall, J.L., Zeitz, D., Stephens, J.W., Kasper, J.V.V., Glass, G.P., Curl, R.F., & Tittel, F.K. (1986). J. Phys. Chem., **90,** 2501.

Hanson, R.K., & Salimian, S. (1984). In *Combustion Chemistry,* W.C. Gardiner, Jr., Ed., Springer–Verlag, New York, 361.

Harrington, J.E., Noble, A.R., Smith, G.P., Jeffries, J.B., & Crosley, D.R. (1995). *Evidence for a New NO Production Mechanism in Flames.* Fall Meeting of the Western States Section of the Combustion Institute, Stanford, California, Paper 95F-193.

Harrington, J.E., Smith, G.P., Berg, P.A., Noble, A.R., Jeffries, J.B., & Crosley, D.R. (1996). 26th Symposium (International) on Combustion, p. 2133.

Harrison, J.A., & Maclagan, R.G.A.R. (1990). J. Chem. Soc. Faraday Trans., **86,** 3519.

Harrison, J.A., Whyte, A.R., & Phillips, L.F. (1986). Chem. Phys. Lett., **129,** 346.

Hayhurst, A.N., & Vince, I.M. (1980). Prog. Energy Combust. Sci., **6,** 35.

Haynes, B.S. (1978). Astronaut. Aeronaut., **62,** 359.

He, Y., & Lin, M.C. (1992). Int. J. Chem. Kinet., **24,** 743.

Hennig, G., & Wagner, H.Gg. (1994). Ber. Bunsenges. Phys. Chem., **98,** 749.

Hennig, G., & Wagner, H.Gg. (1995). Ber. Bunsenges. Phys. Chem., **99,** 863.

Hennig, G., Klatt, M., Spindler, B. &Wagner, H.Gg. (1995). Ber. Bunsenges. Phys. Chem., **99,** 651.

Hennig, G., Röhrig, M., & Wagner, H.Gg. (1993). Ber. Bunsenges. Phys. Chem., **97,** 830.

Herron, J.T. (1988). J. Phys. Chem. Ref. Data, **17,** 967.

Hidaka, Y., Taskuma, H., & Suga, M. (1985). Bull. Chem. Soc. Jpn., **58,** 2911.

Higashihara, T., Gardiner, W.C., Jr., & Hwang, S.M. (1987). J. Phys. Chem., **91,** 1900.

Hoffmann, A., Wagner, H.Gg., Wolff, T., & Hwang, S.M. (1990). Ber. Bunsenges. Phys. Chem., **94,** 1407.

Howard, C.J. (1980). J. Amer. Chem. Soc., **102,** 6937.

Hsu, C.C., Lin, M.C., Mebel, A.M., & Melius, C.F. (1997). J. Phys. Chem. A, **101,** 60.

Hwang, S.M., Higashihara, T., Shin, K.S., & Gardiner, W.C., Jr. (1990) J. Phys. Chem., **94,** 2883.

Jeffries, J.B., & Smith, G.P. (1986). J. Phys. Chem., **90,** 487.

Jemi-Alade, A.A., & Thrush, B.A. (1990). J. Chem. Soc. Faraday Trans., **86,** 3355.

Jodkowski, J.T., Ratajczak, E., Sillesen, A., & Pagsberg, P. (1993). Chem. Phys. Lett., **203,** 490.

Johnsson, J.E., Glarborg, P., & Dam–Johansen, K. (1992). 24th Symposium (International) on Combustion, p. 917.

Juang, D.Y., Lee, J.-S., & Wang, N.S. (1995). Int. J. Chem. Kinet., **27,** 1111.

Just, T. (1994). 25th Symposium (International) on Combustion, p. 687.

Kaiser, E.W. (1993). J. Phys. Chem., **97,** 11681.

Kaiser, E.W. (1995). J. Phys. Chem., **99,** 707.

Kaiser, E.W., & Japar, S.M. (1978). J. Phys. Chem., **82,** 2753.

Kazakov, A., Wang, H., & Frenklach, M. (1994). J. Phys. Chem., **98,** 10598.

Kee, R.J., & Miller, J.A. (1987). *The Chemkin Thermodynamic Data Base.* Sandia National Laboratory Report SAND87–8215 UC4.

Kee, R.J., Grcar, J.F., Smooke, M.D., & Miller, J.A. (1985). *A Fortran Program for Modeling Steady Laminar One-Dimensional Premixed Flames.* Sandia National Laboratory Report SAND85–8240 UC401.

Kee, R.J., Rupley, F.M., & Miller, J.A. (1989). *Chemkin-II. A Fortran Chemical Kinetics Package for the Analysis of Gas-Phase Chemical Kinetics.* Sandia National Laboratory Report SAND89–8009 UC-401.

Kerr, J.A., & Moss, S.J., Ed. (1981). *Handbook of Bimolecular and Termolecular Gas Phase Reactions.* CRC Press, Boca Raton.

Khe, P.V., Soulignac, J.C., & Lesclaux, R. (1977). J. Phys. Chem., **81,** 210.

Kimball-Linne, M. A., & Hanson, R. K. (1986). Combust. Flame, **64,** 337.

Knispel, R., Koch, R., Siese, M., & Zetzsch, C. (1990). Ber. Bunsenges. Phys. Chem., **94,** 1375.

Ko, T., & Fontijn, A. (1991). J. Phys. Chem., **95,** 3984.

Ko, T., Marshall, P., & Fontijn, A. (1990). J. Phys. Chem., **94,** 1401.

Koizumi, H., Schatz, G.C., & Walch, S.P. (1991). J. Chem. Phys., **95,** 4130.

Koshi, M., Bando, S., Saito, M., & Asaba, T. (1978). 17th Symposium (International) on Combustion, p. 553.

Koshi, M., Yoshimura, M., Fukuda, K., Matsui, H., Saito, K., Watanabe, M., Imamura, A., & Chen, C. (1990). J. Chem. Phys., **93,** 8703.

Lambrecht, R., & Herschberger, J. F. (1994). J. Phys. Chem., **98,** 8406.

Lang, V.I. (1992). J. Phys. Chem., **96,** 3047.

Larson, C.W., Patrick, R., & Golden, D.M. (1984). Combust. Flame, **58,** 229.

Laufer, A., & Bass, A.M. (1974). J. Phys. Chem., **78,** 1344.

Laufer, A.H., Gardner, E.P., Kwok, T.L., & Yung, Y.L. (1983). Icarus, **56,** 560.

Lesclaux, R. (1984). Rev. Chem. Intermed., **5,** 347.

Lichtin, D.A., Berman, M.R., & Lin, M.C. (1984). Chem. Phys. Lett., **108,** 18.

Lifshitz, A., Bidani, M., Carroll, H.F., Hwang, S.M., Fu, P.-Y., Shin, K.S., & Gardiner, W.C., Jr. (1991). Comb. Flame, **86,** 229.

Lifshitz, A., Tamburu, C., Frank, P., & Just, T. (1993). J. Phys. Chem., **97,** 4085.

Lightfoot, P.D., & Pilling, M.J. (1987.) J. Phys. Chem., **91,** 3373.

Lillich, H., Schuck, A., Volpp, H.-R., & Wolfrum, J. (1994). 25th Symposium (International) on Combustion, p. 993.

Lin, M.C., He, Y., & Melius, C.F. (1992a). Int. J. Chem. Kinet., **24,** 489.

Lin, M.C., He, Y., & Melius, C.F. (1992b). Int. J. Chem. Kinet., **24,** 1103.

Lin, M.C., He, Y., & Melius, C.F. (1993). J. Phys. Chem., **97,** 9124.

Lindackers, D., Burmeister, M., & Roth, P. (1990). 23rd Symposium (International) on Combustion, p. 251.

Linder, D.P., Duan, X., & Page, M. (1995). J. Phys. Chem., **99,** 11458.

Lindholm, N., & Hershberger, J.F. (1997). J. Phys. Chem. A, **101,** 4991.

Lindstedt, R.P., Lockwood, F.C., & Selim, M.A. (1995). Combust. Sci. Technol., **108,** 231.

Louge, M.Y., & Hanson, R.K. (1984a). Combust. Flame, **58,** 291.

Louge, M.Y., & Hanson, R.K. (1984b). Int. J. Chem. Kinet., **16,** 231.

Lutz, A.E., Kee, R.J., & Miller, J.A. (1991) *SENKIN: A Fortran Program for Predicting Homogeneous Gas Phase Chemical Kinetics with Sensitivity Analysis.* Sandia National Laboratory Report SAND87-8248.

Lyon, R. K. (1975). U. S. Patent No. 3,900,554.

Lyon, R.K. (1987). Div. Fuel Chem., **32,** 433.

Lyon, R.K., & Hardy, J.E. (1986). Ind. Eng. Fundam., **25,** 19.

Mackie, J.C., Colket, M.B., & Nelson, P.F. (1990). J. Phys. Chem, **94,** 4099.

Mallard, W.G., Westley, F., Herron, J.T., Hampson, R.F., & Frizzell, D.H. (1993). *NIST Chemical Kinetics Data Base: Version 5.0.* National Institute of Standards and Technology, Gaithersburg.

Malte, P.C., & Pratt, D.T. (1974). Combust. Sci. Tech., **9,** 221.

Manaa, M.R., & Yarkony, D.R. (1991). J. Chem. Phys., **95,** 1808.

Manaa, M.R., & Yarkony, D.R. (1992). Chem. Phys. Lett., **188,** 352.

Marshall, P., & Fontijn, A. (1986). J. Chem. Phys., **85,** 2637.

Marshall, P., Fontijn, A., & Melius, C. F. (1987). J. Chem. Phys., **86,** 5540.

Marshall, P., Ko, T., & Fontijn, A. (1989). J. Phys. Chem., **93,** 1922.

Marston, G., Nesbitt, F.L., & Stief, L.J. (1989a). J. Chem. Phys., **91,** 3483.

Marston, G., Nesbitt, F.L., Nava, D.F., Payne, W.A., & Stief, L.J. (1989b). J. Phys. Chem., **93,** 5769.

Martin, J.M.L., & Taylor, P.R. (1993). Chem. Phys. Lett., **209,** 143.

Matsui, Y., & Yuuki, A. (1985). Japan J. Appl. Phys, **24,** 598.

Meads, R.F., Machigan, G.A.R., & Phillips, L.F. (1993). J. Phys. Chem., **97,** 3257.

Meagher, N.E., Anderson, W.R., Goumri, A., & Fontijn, A. (1997). *Kinetics of the $O(^3P)$ + N_2O Reaction at Intermediate Temperatures.*, 4th International Conference on Chemical Kinetics, Gaithersburg, Maryland.

Mebel, A.M., Diau, E.W.G., Lin, M.C., & Morokuma, K. (1996a). J. Phys. Chem., **100,** 7517.

Mebel, A.M., Lin, M.C., Morokuma, K., & Melius, C.F. (1996b). Int. J. Chem. Kinet., **28,** 693.

Medhurst, L.J., Garland, N.L., & Nelson, H.H. (1993). J. Phys. Chem., **97,** 12275.

Melius, C.F. (1988). *BAC--MP4 Heats of Formation & Free Energies.* Privately circulated report.

Melius, C.F. (1993). *BAC--MP4 Heats of Formation & Free Energies.* Privately circulated report.

Melius, C.F., & Binkley, J.S. (1984a). In *The Chemistry of Combustion Processes,* T.M. Sloane , Ed., ACS Symposium Series, 103.

Melius, C.F., & Binkley, J.S. (1984b). 20th Symposium (International) on Combustion, p. 575.

Mertens, J.D., & Hanson, R.K. (1996). 26th Symposium (International) on Combustion, p. 551.

Mertens, J.D., Chang, A.Y., Hanson, R.K., & Bowman, C.T. (1989). Int. J. Chem. Kinet., **21,** 1049.

Mertens, J.D., Chang, A.Y., Hanson, R.K., & Bowman, C.T. (1991a). Int. J. Chem. Kinet., **23,** 173.

Mertens, J.D., Chang, A.Y., Hanson, R.K., & Bowman, C.T. (1992a). Int. J. Chem. Kinet., **24,** 279.

Mertens, J.D., Dean, A.J., Hanson, R.K., & Bowman, C.T. (1992b). 24th Symposium (International) on Combustion, p. 701.

Mertens, J.D., Kohse–Höinghaus, K., Hanson, R.K., & Bowman, C.T. (1991b). Int. J. Chem. Kinet., **23,** 655.

Messing, I., Filseth, S.V., Sadowski, C.M., & Carrington, T. (1981). J. Chem. Phys., **74,** 3874.

Meunier, H., Pagsberg, P., & Sillesen, A. (1996). Chem. Phys. Lett., **261,** 277.

Michael, J.V., & Lim, K.P. (1992). J. Chem. Phys., **97,** 3228.

Michael, J.V., Klemm, R.B., Brobst, W.D., Bosco, S.R., & Nava, D.F. (1985). J. Phys. Chem., **89,** 3335.

Michael, J.V., Sutherland, J.W., & Klemm, R.B. (1986). J. Phys. Chem., **90,** 497.

Michaud, M.G., Westmoreland, P.R., & Feitelberg, A.S. (1992). 24th Symposium (International) on Combustion, p. 879.

Miller, J.A., & Bowman, C.T. (1989). Prog. Energy Combust. Sci., **15,** 287.

Miller, J A., & Bowman, C.T. (1991). Int. J. Chem. Kinet., **23,** 289.

Miller, J.A., & Melius, C.F. (1986). 21st Symposium (International) on Combustion, p. 919.

Miller, J.A., & Melius, C.F. (1992a). 24th Symposium (International) on Combustion, p. 719.

Miller, J.A., & Melius, C.F. (1992b). Int. J. Chem. Kinet., **24,** 421.

Miller, J.A., Branch, M.C., McLean, W.J., Chandler, D.W., Smooke, M.D., & Kee, R.J. (1984). 20th Symposium (International) on Combustion, p. 673.

Miller, J.A., Melius, C.F., & Glarborg, P. (1998). Int. J. Chem. Kinet., **30,** 223.

Miller, J.A., Smooke, M.D., Green, R.M., & Kee, R.J. (1983). Combust. Sci. Technol., **34,** 149.

Morley, C. (1980). 18th Symposium (International) on Combustion, p. 23.

Myerson, A.L. (1974). 14th Symposium (International) on Combustion, p. 219.

Natarajan, K., & Roth, P. (1988). 21st Symposium (International) on Combustion, p. 729.

Natarajan, K., Mick, H.J., Woiki, D., & Roth, P. (1994). Combust. Flame, **99,** 610.

Natarajan, K., Thielen, K., Hermanns, H.D., & Roth, P. (1986). Ber. Bunsenges. Phys. Chem., **90,** 533.

Nesbitt, F.L., Marston, G., & Stief, L.J. (1990). J. Phys. Chem., **94,** 4946.

Nguyen, M.T., Boullart, W., & Peeters, J. (1994). J. Phys. Chem., **98,** 8030.

Nguyen, M.T., Sengupta, D., & Vanquickenborne, L.G. (1996). J. Phys. Chem., **100,** 10956.

Nguyen, M.T., Sengupta, D., Vereecken, L., Peeters, J., & Vanquickenborne, L.G. (1996). J. Phys. Chem., **100,** 1615.

Okada, S., Tezaki, A., Miyoshi, A., & Matsui, H. (1994). J. Chem. Phys, **101,** 9582.

Okada, S., Yamasaki, K., Matsui, H., Saito, K., & Okada, K. (1993). Bull. Chem. Soc. Japan, **66,** 1004.

Oref, I. (1989). J. Phys. Chem., **87,** 3465.

Page, M., & Soto, M.R. (1993). J. Chem. Phys., **99,** 7709.

Pagsberg, P., Sztuba, B., Ratajczak, E., & Sillesen, A. (1991). Acta Chem. Scandinavica, **45,** 329.

Park, J., & Hershberger, J.F. (1993a). J. Phys. Chem., **97,** 13647.

Park, J., & Hershberger, J.F. (1993b). J. Chem. Phys., **99,** 3488.

Park, J., & Hershberger, J.F. (1994). Chem. Phys. Lett., **218,** 537.

Park, J., & Lin, M.C. (1996). J. Phys. Chem., **100,** 3317.

Park, J., & Lin, M.C. (1997). J. Phys. Chem. A, **101,** 2643.

Patel–Misra, D., & Dagdigian, P.J. (1991). Chem. Phys. Lett., **185,** 387.

Patrick, R., & Golden, D.M. (1984). J. Phys. Chem., **88,** 491.

Pawlowska, Z., Gardiner, W.C., & Oref, I. (1993). J. Phys. Chem., **97,** 5024.

Peeters, J., Van Look, H., & Ceursters, B. (1996). J. Phys. Chem., **100,** 15124.

Perry, R.A. (1984). Chem. Phys. Lett., **106,** 223.

Perry, R.A., & Melius, C.F. (1984). 20th Symposium (International) on Combustion, p. 639.

Perry, R.A., & Siebers, D.L. (1986). Nature, **324,** 657.

Pouchan, C., Lam, B., & Bishop, D.M. (1987). J. Phys. Chem., **91,** 4809.

Prezhdo, O. (1995). J. Phys. Chem., **99,** 8633.

Quandt, R.W., & Hershberger, J.F. (1996). J. Phys. Chem., **100,** 9407.

Qin, Z., Yang, H., Lissianski, V., Gardiner, W.C., & Shin, K.S. (1997). *Chem. Phys. Lett.,*, 110.

Reid, R.C., Prausnitz, J.M., & Poling, B.E. (1987). *Properties of Gases and Liquids,* McGraw Hill, New York.

Ritter, E. R. (1989). *Ph.D. Thesis,* New Jersey Institute of Technology.

Ritter, E.R., & Bozzelli, J.W. (1991). Int. J. Chem. Kinet., **23,** 767.

Robinson, P.J., & Holbrook, K.A. (1972). *Unimolecular Reactions,* Wiley Interscience, London.

Rodgers, A.S., & Smith, G.P. (1996). Chem. Phys. Lett., **253,** 313.

Röhrig, M., Petersen, E.L., Davidson, D.F., & Hanson, R.K. (1996). Int. J. Chem. Kin., **28,** 599.

Röhrig, M., Petersen, E.L., Davidson, D.F., & Hanson, R.K. (1997). Int. J. Chem. Kinet., **29,** 483.

Röhrig, M., Römming, H.-J., & Wagner, H.Gg. (1994). Ber. Bunsenges. Phys. Chem., **98,** 1332.

Römming, H.-J., & Wagner, H.G. (1996). 26th Symposium (International) on Combustion, p. 559.

Roose, T.R. (1981). Ph.D. Thesis, Stanford University.

Ross, S.K., Sutherland, J.W., Kuo, S.-C., & Klemm, R.B. (1997). J. Phys. Chem. A, **101,** 1104.

Roth, P., & Just, T. (1984). 20th Symposium (International) on Combustion, p. 807.

Sarkisov, O.M., Cheskis, S.G., Nadtochenko, V.A., Sviridenkov, E.A., & Vedeneev, V.I. (1984). Arch. Combust., **4,** 111.

Sausa, R.C., Anderson, W.R., Dayton, D.C., Faust, C.M., & Howard, S. L. (1993). Combust. Flame, **94,** 407.

Schranz, H.W., Nordholm, S., & Hamer, N.D. (1982). Int. J. Chem. Kinet., **14,** 543.

Schuck, A., Volpp, H.-R., & Wolfrum, J. (1994). Combust. Flame, **99,** 491.

Seideman, T. (1994). J. Chem. Phys., **101,** 3662.

Seideman, T., & Walch, S.P. (1994). J. Chem. Phys., **101,** 3656.

Seidler, V., Temps, F., Wagner, H.Gg., & Wolf, M. (1989). J. Phys. Chem., **93,** 1070.

Sengupta, D., & Chandra, A.K. (1994). J. Chem. Phys., **101,** 3906-3915.

Silver, J.A., & Kolb, C.E. (1987). J. Phys. Chem., **91,** 3713.

Sims, I.R., & Smith, I.W.M. (1988). J. Chem. Soc. Faraday Trans. 2, **84,** 527.

Slagle, I.R., Sarzynski, D., & Gutman, D. (1987). J. Phys. Chem., **91,** 4375.

Soto, M.R., & Page, M. (1993). J. Chem. Phys., **97,** 7287.

Soto, M.R., Page, M.J., & McKee, M.L. (1991). Chem. Phys. Lett., **153,** 415.

Stapf, D., & Leuckel, W. (1996). 26th Symposium (International) on Combustion, p. 2083.

Stephens, J.W., Morter, C.L., Farhat, S.K., Glass, G.P., & Curl, R.F. (1993). J. Phys. Chem., **97,** 8944.

Stewart, P.H., Larson, C.W., & Golden, D.M. (1989). Combust. Flame, **75,** 25-31.

Stief, L.J., Marston, G., Nava, D.F., Payne, W.A., & Nesbitt, F.L. (1988). Chem. Phys. Lett., **147,** 570.

Stothard, N., Humpfer, R., & Grotheer, H.-H. (1995). Chem. Phys. Lett., **240,** 474.

Sutherland, J.W., & Klemm, R.B. (1987). 16th International Symposium on Shock Tubes and Waves, p. 395.

Sutherland, J.W., Patterson, P.M., & Klemm, R.B. (1990). J. Phys. Chem., **94,** 2471.

Szekely, A., Hanson, R.K., & Bowman, C.T. (1984). 20th Symposium (International) on Combustion, p. 647.

Thielen, K., & Roth, P. (1984). 20th Symposium (International) on Combustion, p. 685.

Troe, J. (1979). J. Phys. Chem., **83,** 114

Tsang, W. (1991). J. Phys. Chem. Ref. Data, **20,** 221.

Tsang, W. (1992). J. Phys. Chem. Ref. Data, **21,** 753.

Tsang, W., & Hampson, R.F. (1986). J. Phys. Chem. Ref. Data, **15,** 1087.

Tsang, W., & Herron, J.T. (1991). J. Phys. Chem. Ref. Data, **20,** 609.

Tyndall, G.S., Orlando, J.J., Nickerson, K.E., Cantrell, C.A., & Calvert, J. G. (1991). J. Geophys. Res., **96,** 20761.

Unfried, K.G., Glass, G.P., & Curl, R.F. (1990). Chem. Phys. Lett., **173,** 337.

Vaghjiani, G.L. (1996). J. Chem. Phys., **104,** 5479.

Vakhtin, A.B., & Petrov, A.K. (1990). Spectrochimica Acta., **46A,** 603.

Vandooren, O.M., Sarkisov, O.M., Balakhnin, V.P., & Van Tiggelen, P.J. (1991). Chem. Phys. Lett., **184,** 294.

Venkatesh, P.K., Chang, A.Y., Dean, A.M., Cohen, M.H., & Carr, R.W. (1997). A.I.Ch.E. Journal, **43,** 1331.

Vinckier, C., & Debruyn, W. (1979). J. Phys. Chem., **83,** 2057.

Votsmeier, M., Song, S., Hanson, R.K., & Bowman, C.T. (1998). *Sensitive Detection of NH_2 in Shock Tube Kinetics Experiments Using Frequency Modulation Spectroscopy,* Work-in-Progress Poster Session W4F06, 27th Symposium (International) on Combustion, Boulder, to be published.

Wagal, S.S., Carrington, T., Filseth, S.V., & Sadowski, C. M. (1982). Chem. Phys., **69,** 61.

Wagner, A.F., Slagle, I.R., Sarzynski, D., & Gutman, D. (1990). J. Phys. Chem., **94,** 1853.

Wagner, H.Gg., Welzbacher, U., & Zellner, R. (1976). Ber. Bunsenges. Phys. Chem., **80,** 1023

Walch, S.P. (1993a). J. Chem. Phys., **98,** 1170.

Walch, S.P. (1993b). J. Chem. Phys., **99,** 5295.

Walch, S.P. (1993c). Chem. Phys. Lett., **208,** 214.

Walch, S.P., Duchovic, R.J., & Rohlfing, C.M. (1989). J. Chem. Phys., **90,** 3230.

Walia, N., & Kakker, R. (1990). Indian J. Chem., **29B,** 1007.

Walia, N., & Kakker, R. (1991). Indian J. Chem., **30A,** 308.

Wallington, T.J., Maricq, M.M., Ellermann, T., & Nielsen, O.J. (1992). J. Phys. Chem., **96,** 982.

Wang, H., & Frenklach, M. (1993). Chem. Phys. Lett., **205,** 271.

Wang, N.S., Yang, D.L., & Lin, M.C. (1989). Chem. Phys. Lett., **163,** 480.

Wang, N.S., Yang, D.L., & Lin, M.C. (1991). Int. J. Chem. Kinet., **23,** 151.

Warnatz, J. (1983). Combust. Sci. Technol., **34,** 177.

Warnatz, J. (1984). In *Combustion Chemistry,* W.C. Gardiner., Jr., Ed., Springer–Verlag, New York.

Washida, N. (1980). J. Chem. Phys., **73,** 1665.

Westmoreland, P.R., Howard, J.B., Longwell, J.P., & Dean, A.M. (1986). AIChE Journal, **32,** 1971.

Weston, R.E. (1986). Int. J. Chem. Kinet., **18,** 1259.

Whyte, A.R., & Phillips, L.F. (1983a). Chem. Phys. Lett., **102,** 451.

Whyte, A.R., & Phillips, L.F. (1983b). Bull. Soc. Chim. Belg., **92,** 635.

Williams, B.A., & Fleming, J.W. (1994). Combust. Flame, **98,** 93.

Williams, B.A., Papas, P., & Nelson, H.H. (1995). J. Phys. Chem., **99,** 13471.

Wolf, M., Yang, D.L., & Durant, J.L. (1994). J. Photochem. Photobiol. A Chem., **80,** 85-93.

Wolff, T., & Wagner, H.Gg. (1988). Ber. Bunsenges. Phys. Chem., **92,** 678.

Wolfrum, J. (1972). Chemie Ingenieur Technik, **44,** 656.

Wooldridge, S.T., Hanson, R.K., & Bowman, C.T. (1995). Int. J. Chem. Kinet., **27,** 1075.

Wooldridge, S.T., Hanson, R.K., & Bowman, C.T. (1996a). Int. J. Chem. Kinet., **28,** 245.

Wooldridge, M.S., Hanson, R.K., & Bowman, C.T. (1996b). Int. J. Chem. Kinet., **28,** 361.

Wooldridge, S.T., Mertens, J.D., Hanson, R.K., & Bowman, C.T. (1994). 25th Symposium (International) on Combustion, p. 983.

Wray, K.L., & Teare, J.D. (1962). J. Chem. Phys., **36,** 2582.

Wu, C.H., Wang, H.T., Lin, M.C., & Fifer, R.A. (1990). J. Phys. Chem., **94,** 3344.

Xu, Z.-F., Fang, D.-C., & Fu, X.-Y. (1995). J. Phys. Chem., **99,** 5889.

Yamasaki, K., Okada, S., Koshi, M., & Matsui, H. (1991). J. Chem. Phys., **95,** 5087.

Yang, H.–Y., Lissianski, V., Okoroanyanwu, J.U., Gardiner, W.C., & Shin, K.S. (1993). J. Phys. Chem., **97,** 10042.

Yarkony, D.R. (1993). J. Phys. Chem., **97,** 4407.

Yokoyama, K., Sakane, Y., & Fueno, T. (1991a). Bull. Chem. Soc. Jpn., **64,** 1738.

Yokoyama, K., Takane, S., & Fueno, T. (1991b). Bull. Chem. Soc. Jpn., **64,** 2230.

Yoshimine, M.J. (1989) J. Chem. Phys., **90,** 378

Yumura, M., & Asaba, T. (1980). 18th Symposium (International) on Combustion, p. 863.

Zabarnick, S. (1993). Chem. Phys., **171,** 265.

Zabarnick, S., & Lin, M.C. (1989). Chem. Phys., **134,** 185.

Zabielski, M.F., & Seery, D.J. (1985). Int. J. Chem. Kinet., **17,** 1191.

Zabielski, M.F., & Seery, D.J. (1989). In *Fossil Fuels Combustion Symposium,* Vol. 25, American Society of Mechanical Engineers, 91.

Zeldovich, Y.B. (1946). Acta Physicochim. URSS, **21,** 577.

Zellner, R. (1984). "Bimolecular Reaction Rate Coefficients", In *Combustion Chemistry,* W.C. Gardiner, Jr., Ed., Springer-Verlag, New York, 127.

Zuev, A.P., & Starikovskii, A.Y. (1991a). Khim. Fiz., **10,** 52.

Zuev, A.P., & Starikovskii, A.Y. (1991b). Khim. Fiz., **10,** 179.

Zuev, A P., & Starikovskii, A.Y. (1991c). Khim. Fiz., **10,** 190.

TABLE 2.19(a) Summary of Rate Coefficient Parameters[1] Part I.

	Reaction	A (Note 1)	m	E_a/R	Notes
1	$O + N_2 \longrightarrow N + NO$	2.0×10^{14}		38 660	
2	$NO + M \longrightarrow N + O + M$	1.4×10^{15}		74 700	
3	$N_2O + M \longrightarrow N_2 + O + M$	5.7×10^{14}		28 230	
4a	$N_2O + O \longrightarrow N_2 + O_2$	1.4×10^{12}		5440	
4b	$N_2O + O \longrightarrow 2\,NO$	2.9×10^{13}		11 650	
5	$NH_3 + M \longrightarrow NH_2 + H + M$	3.6×10^{16}		47 200	
6	$NH_3 + H \longrightarrow NH_2 + H_2$	5.4×10^{5}	2.40	4990	
7	$NH_3 + OH \longrightarrow NH_2 + H_2O$	5.0×10^{7}	1.60	480	
8	$NH_3 + O \longrightarrow NH_2 + OH$	9.4×10^{6}	1.94	3250	
9	$NH_2 + H \longrightarrow NH + H_2$	4.8×10^{8}	1.50	3995	
10	$HO_2 + NO \longrightarrow NO_2 + OH$	2.2×10^{12}		−240	
11a	$N_2O + H \longrightarrow HNNO$	1.2×10^{24}	−4.46	5385	0.1 atm
11a		1.3×10^{25}	−4.48	5420	1.0 atm
11a		3.2×10^{26}	−4.58	5650	10 atm
11a	$N_2O{+}H{+}M \longrightarrow HNNO + M$	1.1×10^{27}	−3.48	5420	Note 2.
11b	$N_2O + H \longrightarrow N_2 + OH$	2.2×10^{14}		8430	Note 3.
11c	$H + N_2O \longrightarrow NH + NO$	8.5×10^{20}	−1.62	17 800	
11d	$H + N_2O \longrightarrow NNH + O$	2.4×10^{19}	−1.26	23 700	
11e	$NH + NO \longrightarrow N_2 + OH$	1.4×10^{17}	−1.49	660	
11f	$NH + NO \longrightarrow N_2O + H$	3.0×10^{18}	−1.65	720	Note 4.
11g	$NH + NO \longrightarrow NNH + O$	1.7×10^{14}	−0.20	6140	
12a	$NH + O_2 \longrightarrow HNOO$	3.5×10^{23}	−5.00	1145	0.1 atm
12a		3.7×10^{24}	−5.00	1155	1.0 atm
12a		5.4×10^{25}	−5.05	1235	10 atm
12a	$NH{+}O_2{+}M \longrightarrow HNOO + M$	3.0×10^{26}	−4.00	1155	Note 5.
12d	$NH + O_2 \longrightarrow NO + OH$	7.6×10^{10}		770	
12e	$NH + O_2 \longrightarrow H + NO_2$	2.3×10^{10}		1250	
12g	$NH + O_2 \longrightarrow HNO + O$	4.6×10^{5}	2.00	3270	
13b	$NH_2 + O_2 \longrightarrow NH_2O + O$	2.5×10^{11}	0.48	14 890	
13d	$NH_2 + O_2 \longrightarrow HNO + OH$	6.2×10^{7}	1.23	17 665	
14b	$NH_2 + HO_2 \longrightarrow NH_2O + OH$	2.5×10^{13}			
14d	$NH_2 + HO_2 \longrightarrow NH_3 + O_2$	9.2×10^{5}	1.94	−580	
15c	$NH_2 + O \longrightarrow HNO + H$	4.6×10^{13}		0	
15d1	$NH_2 + O \longrightarrow NH + OH$	7.0×10^{12}		0	
15d2	$NH_2 + O \longrightarrow NH + OH$	3.3×10^{8}	1.50	2555	
16a	$NH_2 + OH \longrightarrow NH_2OH$	1.8×10^{32}	−6.91	2070	0.1 atm
16a		3.9×10^{33}	−7.00	2235	1.0 atm
16a		5.6×10^{34}	−7.02	2700	10 atm
16b	$NH_2 + OH \longrightarrow NH + H_2O$	2.4×10^{6}	2.00	25	

All pressures refer to N_2 buffer gas. Notes: (1) Rate coefficients are expressed in the form $k = A\,T^m \exp(-E_a/RT)$ for concentration units mol/cm^3 and time in s. (2) P = 10 atm, T > 1000 K or P = 1 atm, T > 300 K. (3) For 1000 to 2000 K. (4) Reverse of reaction 11c. (5) Slight falloff at T = 300 K, P = 10 atm.

TABLE 2.19(b) Summary of Rate Coefficient Parameters[1] Part II.

	Reaction	A (Note 1)	m	E_a/R	Notes
17a	$NH_2 + NH_2 \longrightarrow N_2H_4$	2.0×10^{46}	−10.93	5030	Note 2.
17a		5.6×10^{48}	−11.30	5980	Note 3.
17a		3.2×10^{49}	−11.18	7040	Note 4.
17b	$NH_2 + NH_2 \longrightarrow H_2NN + H_2$	2.4×10^{20}	−2.91	1075	0.1 atm
17b		1.2×10^{21}	−3.08	1695	1.0 atm
17b		2.3×10^{19}	−2.54	2105	10 atm
17d	$NH_2 + NH_2 \longrightarrow N_2H_3 + H$	9.2×10^{11}	−0.01	5040	0.1 atm
17d		1.2×10^{12}	−0.03	5075	1.0 atm
17d		4.7×10^{12}	−0.20	5345	10 atm
17e	$NH_2 + NH_2 \longrightarrow NH_3 + NH$	5.0×10^{13}		5000	
18a	$NH_2 + NO \longrightarrow N_2 + H_2O$	4.7×10^{12}	−0.25	−605	
18b	$NH_2 + NO \longrightarrow NNH + OH$	3.5×10^{10}	0.34	−385	
18c	$NH_2 + NO \longrightarrow NH_2NO$	1.9×10^{30}	−6.67	1760	0.1 atm
18c		3.5×10^{31}	−6.75	1875	1.0 atm
18c		1.7×10^{33}	−6.92	2320	10 atm
19a	$CH_3 + NO(+N_2) \longrightarrow CH_3NO\ (+N_2)$	1.0×10^{13}		0	k_∞
		1.9×10^{18}		0	k_0
	SRI falloff coefficients: a=0.03, b=−790, c=1.0				Note 5.
19a	$CH_3 + NO \longrightarrow CH_3NO$	3.6×10^{35}	−8.25	2420	0.1 atm
19a		1.0×10^{37}	−8.38	2630	1.0 atm
19a		4.6×10^{41}	−9.39	4160	10 atm
19d	$CH_3 + NO \longrightarrow H_2CN + OH$	2.2×10^{9}	0.75	5900	
19e	$CH_3 + NO \longrightarrow HCN + H_2O$	4.9×10^{8}	0.46	6240	
20a	$CH_3 + N \longrightarrow H_2CN + H$	6.1×10^{14}	−0.31	145	
20b	$CH_3 + N \longrightarrow HCN + H_2$	3.7×10^{12}	0.15	−45	
20f	$CH_3 + N \longrightarrow HCNH + H$	1.2×10^{11}	0.52	−185	
21a	$CH_3 + NH_2 \longrightarrow CH_3NH_2$	1.3×10^{54}	−12.72	7855	Note 2.
21a		5.1×10^{52}	−11.99	8450	Note 3.
21a		1.6×10^{47}	−10.15	7895	Note 4.
21b	$CH_3 + NH_2 \longrightarrow CH_2NH_2 + H$	1.1×10^{13}	−0.13	4985	0.1 atm
21b		1.4×10^{14}	−0.43	5590	1.0 atm
21b		7.4×10^{12}		6075	10 atm
21c	$CH_3 + NH_2 \longrightarrow CH_3NH + H$	1.2×10^{13}	−0.15	8125	0.1 atm
21c		4.4×10^{13}	−0.31	8375	1.0 atm
21c		1.4×10^{14}	−0.42	8990	10 atm
21d	$CH_3 + NH_2 \longrightarrow H_2C{=}NH + H_2$	2.1×10^{11}	−0.10	9610	0.1 atm
21d		4.8×10^{11}	−0.20	9765	1.0 atm
21d		2.9×10^{12}	−0.40	10 320	10 atm
21e	$CH_3 + NH_2 \longrightarrow CH_4 + NH$	2.8×10^{6}	1.94	4635	
21f	$CH_3 + NH_2 \longrightarrow CH_2 + NH_3$	1.6×10^{6}	1.87	3810	

All pressures refer to N_2 buffer gas. Notes: (1) Rate coefficients are expressed in the form $k = A\,T^m \exp(-E_a/RT)$ for concentration units mol/cm^3 and time in s. (2) Valid for 0.1 atm, 600 to 2500 K. (3) Valid for 1 atm, 600 to 2500 K. (4) Valid for 10 atm, 600 to 2500 K. (5) Stewart et al. (1989). The $k_\circ$ expression of Kaiser (1993) has been increased by a factor of 1.4 to account for the greater collision efficiency of N_2. These expressions are valid for the temperature range 300 to 500 K. The following expressions are suitable for temperatures over 500 K. (See text Section 6.15.)

TABLE 2.19(c) Summary of Rate Coefficient Parameters[1] Part III.

	Reaction	A (Note 1)	m	E_a/R	Notes
22a	$CH_2 + N_2 \longrightarrow CH_2NN$	9.3×10^{30}	−7.01	9935	0.1 atm
22a		1.6×10^{32}	−7.07	10 050	1.0 atm
22a		4.3×10^{33}	−7.18	10 500	10 atm
22d	$CH_2 + N_2 \longrightarrow HCN + NH$	1.0×10^{13}		37 240	
23Σ	$CH_2 + NO \longrightarrow$ Products	1.0×10^{13}		−190	Note 2.
23g	$CH_2 + NO \longrightarrow HCNO + H$	3.8×10^{13}	−0.36	290	
23i	$CH_2 + NO \longrightarrow HCN + OH$	2.9×10^{14}	−0.69	380	
23j	$CH_2 + NO \longrightarrow HNCO + H$	3.1×10^{17}	−1.38	640	
23k	$CH_2 + NO \longrightarrow NH_2 + CO$	2.3×10^{16}	−1.43	670	
23l	$CH_2 + NO \longrightarrow H_2CN + O$	8.1×10^{7}	1.42	2070	
24a	$CH + N_2 \longrightarrow HCNN$	2.3×10^{27}	−5.78	1230	0.1 atm
24a		3.6×10^{28}	−5.84	1320	1.0 atm
24a		1.8×10^{30}	−6.02	1735	10 atm
24c	$CH + N_2 \longrightarrow HCN + N$	4.4×10^{12}		11 060	
25e	$CH + NO \longrightarrow HCN + O$	5.3×10^{13}		0	
25f	$CH + NO \longrightarrow H + NCO$	2.0×10^{13}		0	
25g	$CH + NO \longrightarrow N + HCO$	2.9×10^{13}		0	
25h	$CH + NO \longrightarrow NH{+}CO$	5.5×10^{12}		0	
25i	$CH + NO \longrightarrow OH{+}CN$	3.3×10^{12}		0	
26a	$N + O_2 \longrightarrow NO + O$	9.0×10^{9}	1.00	3270	
26b	$N + OH \longrightarrow NH + O$	6.4×10^{12}	0.10	10 700	Note 3.
26c	$N + OH \longrightarrow NO + H$	1.1×10^{14}		565	
26d	$CH + N \longrightarrow CN + H$	1.7×10^{14}	−0.09	0	
26e	$CH_2 + N \longrightarrow HCN + H$	5.0×10^{13}		0	
26f	$NH + N \longrightarrow N_2 + H$	1.5×10^{13}		0	
26g	$NH_2 + N \longrightarrow N_2 + 2H$	7.1×10^{13}		0	
26h	$CN + N \longrightarrow C + N_2$	2.4×10^{13}		−280	
27a	$NH + NH \longrightarrow N_2 + 2H$	5.1×10^{13}		0	
27b1	$NH_2 + NH \longrightarrow N_2H_2 + H$	1.5×10^{15}	−0.50	0	
27b2	$NH_2 + NH \longrightarrow NH_3 + N$	9.2×10^{5}	1.94	1230	
27c1	$NH + OH \longrightarrow HNO + H$	2.0×10^{13}		0	
27c2	$NH + OH \longrightarrow N + H_2O$	1.2×10^{6}	2.00	−245	
27d	$NH + H \longrightarrow N + H_2$	3.5×10^{13}		870	
27e1	$NH + O \longrightarrow NO + H$	6.0×10^{13}		0	
27e2	$NH + O \longrightarrow N + OH$	1.7×10^{8}	1.50	1695	
27f1	$NH + CH_3 \longrightarrow H_2C{=}NH + H$	4.0×10^{13}		0	
27f2	$NH + CH_3 \longrightarrow CH_4 + N$	8.2×10^{5}	1.87	2945	

All pressures refer to N_2 buffer gas. Notes: (1) Rate coefficients are expressed in the form $k = A\,T^m \exp(-E_a/RT)$ for concentration units mol/cm^3 and time in s. (2) Total rate coefficient for all channels of CH_2 + NO. (3) Reverse of reaction 27e2.

TABLE 2.19(d) Summary of Rate Coefficient Parameters[1] Part IV.

	Reaction	A (Note 1)	m	E_a/R	Notes
28a	$NNH \longrightarrow N_2 + H$				
		$3.0\times10^{8} + 1.0\times10^{13}$[M]	0.5	1540	
28b1	$NNH + O_2 \longrightarrow N_2 + HO_2$	1.2×10^{12}	−0.34	75	
28b2	$NNH + O_2 \longrightarrow N_2O + OH$	2.9×10^{11}	−0.34	75	
28c	$NNH + H \longrightarrow N_2 + H_2$	2.4×10^{8}	1.50	−450	
28d1	$NNH + OH \longrightarrow N_2 + H_2O$	2.4×10^{22}	−2.88	1235	
28d2	$NNH + OH \longrightarrow N_2 + H_2O$	1.2×10^{6}	2.00	−600	
28e1	$NNH + O \longrightarrow N_2 + OH$	1.7×10^{16}	−1.23	250	
28e2	$NNH + O \longrightarrow N_2 + OH$	1.7×10^{8}	1.50	−450	
28f	$NNH + NH_2 \longrightarrow N_2 + NH_3$	9.2×10^{5}	1.94	−580	
28g1	$NNH + HO_2 \longrightarrow N_2 + H_2O_2$	1.4×10^{4}	2.69	−805	
28g2	$NNH + HO_2 \longrightarrow HNNO + OH$	2.4×10^{13}		855	
28h	$NNH + NO \longrightarrow N_2 + HNO$	1.2×10^{6}	2.00	−600	
29a1	$N_2H_2 \longrightarrow NNH + H$	5.6×10^{36}	−7.75	35 355	Note 2.
29a1		1.8×10^{40}	−8.41	36 935	Note 3.
29a1		3.1×10^{41}	−8.42	38 270	Note 4.
29a2	$N_2H_2 \longrightarrow NNH + H$	1.6×10^{37}	−7.94	35 610	Note 2.
29a2		2.6×10^{40}	−8.53	36 700	Note 3.
29a2		1.3×10^{44}	−9.22	38 790	Note 4.
29b1	$N_2H_2 \longrightarrow H_2NN$	9.2×10^{38}	−9.01	34 085	Note 2.
29b1		2.0×10^{41}	−9.38	34 450	Note 3.
29b1		1.3×10^{45}	−10.13	35 610	Note 4.
29c1	$N_2H_2 + H \longrightarrow NNH + H_2$	4.8×10^{8}	1.50	795	
29c2	$N_2H_2 + O \longrightarrow NNH + OH$	3.3×10^{8}	1.50	250	
29c3	$N_2H_2 + OH \longrightarrow NNH + H_2O$	2.4×10^{6}	2.00	−600	
29c4	$N_2H_2 + NH_2 \longrightarrow NH_3 + NNH$	1.8×10^{6}	1.94	−580	
29c5	$N_2H_2 + CH_3 \longrightarrow NNH + CH_4$	1.6×10^{6}	1.87	1495	
29d	$N_2H_2 + NH \longrightarrow NNH + NH_2$	2.4×10^{6}	2.00	−600	
29e	$N_2H_2 + NO \longrightarrow N_2O + NH_2$	4.0×10^{12}		6000	
30a1	$H_2NN \longrightarrow NNH + H$	5.9×10^{32}	−6.99	26 065	Note 2.
30a1		9.6×10^{35}	−7.57	27 600	Note 3.
30a1		5.0×10^{36}	−7.43	28 835	Note 4.
30a2	$H_2NN \longrightarrow NNH + H$	7.2×10^{28}	−7.77	25 545	Note 2.
30a2		3.2×10^{31}	−6.22	26 330	Note 3.
30a2		5.1×10^{33}	−6.52	27 285	Note 4.
30b	$H_2NN + O_2 \longrightarrow NH_2 + NO_2$	1.5×10^{12}		3000	
30c1	$H_2NN + H \longrightarrow N_2H_2 + H$	1.8×10^{10}	0.97	2250	
30c2	$H_2NN + H \longrightarrow NNH + H_2$	4.8×10^{8}	1.50	−450	
30d1	$H_2NN + O \longrightarrow NH_2 + NO$	3.2×10^{9}	1.03	1360	
30d2	$H_2NN + O \longrightarrow OH + NNH$	3.3×10^{8}	1.50	−450	
30e1	$H_2NN + OH \longrightarrow NH_2NO + H$	2.0×10^{12}		0	
30e2	$H_2NN + OH \longrightarrow NNH + H_2O$	2.4×10^{6}	2.00	−600	Note 5.

All pressures refer to N_2 buffer gas. Notes: (1) Rate coefficients are expressed in the form $k = A\,T^m \exp(-E_a/RT)$ for concentration units mol/cm^3 and time in s. (2) Valid for 0.1 atm, 600 to 2500 K. (3) Valid for 1 atm, 600 to 2500 K. (4) Valid for 10 atm, 600 to 2500 K. (5) Reaction 30 entries continue in Part V.

TABLE 2.19(e) Summary of Rate Coefficient Parameters[1] Part V.

	Reaction	A (Note 1)	m	E_a/R	Notes
30f1	$H_2NN + CH_3 \longrightarrow H_2C{=}NNH_2 + H$	8.3×10^5	1.93	3270	
30f2	$H_2NN + CH_3 \longrightarrow CH_3N{=}NH + H$	8.3×10^5	1.93	3270	
30f3	$H_2NN + CH_3 \longrightarrow CH_4 + NNH$	1.6×10^6	1.87	65	
30g1	$H_2NN + NH_2 \longrightarrow HNNNH_2 + H$	7.9×10^6	1.90	−670	
30g2	$H_2NN + NH_2 \longrightarrow NH_3 + NNH$	1.8×10^6	1.94	−580	
30h1	$H_2NN + HO_2 \longrightarrow NH_2NO + OH$	6.6×10^5	1.94	3550	
30h2	$H_2NN + HO_2 \longrightarrow NNH + H_2O_2$	2.9×10^4	2.69	−805	
31a	$N_2H_3 \longrightarrow N_2H_2 + H$	2.3×10^{43}	−9.55	32 445	Note 2.
31a		3.6×10^{47}	−10.38	34 730	Note 3.
31a		1.8×10^{45}	−9.39	35 300	Note 4.
31b	$N_2H_3 + H \longrightarrow N_2H_2 + H_2$	2.4×10^8	1.50	0	
31c1	$N_2H_3 + O \longrightarrow NH_2 + HNO$	3.0×10^{13}		0	
31c2	$N_2H_3 + O \longrightarrow NH_2NO + H$	3.0×10^{13}		0	
31c3	$N_2H_3 + O \longrightarrow N_2H_2 + OH$	1.7×10^8	1.50	−325	
31d1	$N_2H_3 + OH \longrightarrow N_2H_2 + H_2O$	1.2×10^6	2.00	−600	
31d2	$N_2H_3 + OH \longrightarrow H_2NN + H_2O$	3.0×10^{13}		0	
31e1	$N_2H_3 + CH_3 \longrightarrow N_2H_2 + CH_4$	8.2×10^5	1.87	915	
31e2	$N_2H_3 + CH_3 \longrightarrow H_2NN + CH_4$	3.0×10^{13}		0	
31f1	$N_2H_3 + NH_2 \longrightarrow N_2H_2 + NH_3$	9.2×10^5	1.94	−580	
31f2	$N_2H_3 + NH_2 \longrightarrow H_2NN + NH_3$	3.0×10^{13}		0	
31g1	$N_2H_3 + HO_2 \longrightarrow NH_2NHO + OH$	3.0×10^{13}		0	
31g2	$N_2H_3 + HO_2 \longrightarrow N_2H_2 + H_2O_2$	2.9×10^4	2.69	−805	
31g3	$N_2H_3 + HO_2 \longrightarrow N_2H_4 + O_2$	9.2×10^5	1.94	1070	
32	$N_2H_4 \longrightarrow H_2NN + H_2$	4.0×10^{44}	−9.85	35 910	Note 2.
32		5.3×10^{39}	−8.35	34 880	Note 3.
32		2.5×10^{39}	−8.19	35 060	Note 4.
32a	$N_2H_4 + H \longrightarrow N_2H_3 + H_2$	9.6×10^8	1.50	2435	
32b	$N_2H_4 + O \longrightarrow N_2H_3 + OH$	6.7×10^8	1.50	1435	
32c	$N_2H_4 + OH \longrightarrow N_2H_3 + H_2O$	4.8×10^6	2.00	−325	
32d	$N_2H_4 + CH_3 \longrightarrow N_2H_3 + CH_4$	3.3×10^6	1.87	2680	
32e	$N_2H_4 + NH_2 \longrightarrow N_2H_3 + NH_3$	3.7×10^6	1.94	820	
33a1	$NO + C \longrightarrow CO + N$	1.7×10^{13}		0	
33a2	$NO + C \longrightarrow CN + O$	1.1×10^{13}		0	
33c1	$NO + HCCO \longrightarrow HCNO + CO$	4.6×10^{13}		350	
33c2	$NO + HCCO \longrightarrow HCN + CO_2$	1.4×10^{13}		350	
34a	$NO_2 + H \longrightarrow NO + OH$	1.3×10^{14}		180	
34b	$NO_2 + O \longrightarrow NO + O_2$	3.9×10^{12}		−120	
34c	$NO_2 + M \longrightarrow NO + O + M$	5.7×10^{15}		30 190	
34d1	$NO_2 + NH_2 \longrightarrow N_2O + H_2O$	1.5×10^{16}	−1.44	135	
34d2	$NO_2 + NH_2 \longrightarrow NH_2O + NO$	6.6×10^{16}	−1.44	135	
34e	$NO_2 + CH_3 \longrightarrow NO + CH_3O$	1.4×10^{13}		0	
35	$N_2O + OH \longrightarrow N_2 + HO_2$	1.3×10^{-2}	4.72	18400	

All pressures refer to N_2 buffer gas. Notes: (1) Rate coefficients are expressed in the form $k = A\,T^m \exp(-E_a/RT)$ for concentration units mol/cm^3 and time in s. (2) Valid for 0.1 atm, 600 to 2500 K. (3) Valid for 1 atm, 600 to 2500 K. (4) Valid for 10 atm, 600 to 2500 K.

TABLE 2.19(f) Summary of Rate Coefficient Parameters[1] Part VI.

	Reaction	A (Note 1)	m	E_a/R	Notes
36a	$HNO + M \longrightarrow H + NO + M$	2.6×10^{16}		24 500	
36b	$2HNO \longrightarrow N_2O + H_2O$	8.5×10^{8}		1550	1.0 atm
36c	$HNO + OH \longrightarrow NO + H_2O$	1.3×10^{7}	1.88	−480	
36d1	$HNO + H \longrightarrow H_2 + NO$	4.5×10^{11}	0.72	330	
36d2	$HNO + H \longrightarrow O + NH_2$	3.5×10^{15}	−0.30	14 730	Note 2.
36e	$HNO + O \longrightarrow OH + NO$	4.5×10^{11}	0.72	330	
36f	$HNO + NH_2 \longrightarrow NH_3 + NO$	9.2×10^{5}	1.94	−580	
36g	$HNO + NO \longrightarrow N_2O + OH$	8.5×10^{12}		14 890	
36h	$HNO + O_2 \longrightarrow NO + HO_2$	2.0×10^{13}		8000	
36i	$HNO + CH_3 \longrightarrow NO + CH_4$	8.2×10^{5}	1.87	480	
37a	$NH_2O + M \longrightarrow HNO + H + M$	2.8×10^{24}	−2.83	32 670	
37b	$NH_2O \longrightarrow HNOH$	8.2×10^{25}	−4.94	22 040	0.1 atm
37b		1.3×10^{27}	−4.99	22 135	1.0 atm
37b		2.6×10^{28}	−5.06	22 530	10 atm
37b	$NH_2O + M \longrightarrow HNOH + M$	1.1×10^{29}	−3.99	22 135	Note 3.
37c1	$NH_2O + H \longrightarrow NH_2 + OH$	4.0×10^{13}		0	
37c2	$NH_2O + H \longrightarrow HNO + H_2$	4.8×10^{8}	1.50	785	
37d	$NH_2O + O \longrightarrow HNO + OH$	3.3×10^{8}	1.50	245	
37e	$NH_2O + OH \longrightarrow HNO + H_2O$	2.4×10^{6}	2.00	−600	
37f1	$NH_2O + CH_3 \longrightarrow CH_3O + NH_2$	2.0×10^{13}		0	
37f2	$NH_2O + CH_3 \longrightarrow CH_4 + HNO$	1.6×10^{6}	1.87	1490	
37g	$NH_2O + NH_2 \longrightarrow HNO + NH_3$	1.8×10^{6}	1.94	−580	
37h1	$NH_2O + HO_2 \longrightarrow HNO + H_2O_2$	2.9×10^{4}	2.69	−805	
37h2	$NH_2O + HO_2 \longrightarrow O_2 + NH_2OH$	2.9×10^{4}	2.69	−805	
38a	$HNOH + M \longrightarrow H + HNO + M$	2.0×10^{24}	−2.84	29 660	
38b1	$HNOH + H \longrightarrow NH_2 + OH$	4.0×10^{13}		0	
38b2	$HNOH + H \longrightarrow HNO + H_2$	4.8×10^{8}	1.50	190	
38c1	$HNOH + O \longrightarrow HNO + OH$	7.0×10^{13}		0	
38c2	$HNOH + O \longrightarrow HNO + OH$	3.3×10^{8}	1.50	−180	
38d	$HNOH + OH \longrightarrow HNO + H_2O$	2.4×10^{6}	2.00	−600	
38e1	$HNOH + CH_3 \longrightarrow CH_3N.H + OH$	2.0×10^{13}		0	
38e2	$HNOH + CH_3 \longrightarrow CH_4 + HNO$	1.6×10^{6}	1.87	1055	
38f1	$HNOH + NH_2 \longrightarrow N_2H_3 + OH$	6.7×10^{6}	1.82	360	
38f2	$HNOH + NH_2 \longrightarrow H_2NN + H_2O$	4.6×10^{19}	−1.94	970	
38f3	$HNOH + NH_2 \longrightarrow HNO + NH_3$	1.8×10^{6}	1.94	−580	
38g1	$HNOH + HO_2 \longrightarrow HONHO + OH$	4.0×10^{13}		0	
38g2	$HNOH + HO_2 \longrightarrow HNO + H_2O_2$	2.9×10^{4}	2.69	−805	
38g3	$HNOH + HO_2 \longrightarrow NH_2OH + O_2$	2.9×10^{4}	2.69	−805	
39	$^{1}HNOO + M \longrightarrow OH + NO + M$	1.5×10^{36}	−6.18	15 670	

All pressures refer to N_2 buffer gas. Notes: (1) Rate coefficients are expressed in the form $k = A\,T^m \exp(-E_a/RT)$ for concentration units mol/cm^3 and time in s. (2) Reverse of reaction 15c. (3) T > 1000 K.

TABLE 2.19(g) Summary of Rate Coefficient Parameters[1] Part VII.

	Reaction	A (Note 1)	m	E_a/R	Notes
40a	$HONO + M \longrightarrow OH + NO + M$	2.0×10^{31}	−4.56	25 755	
40b1	$HONO + H \longrightarrow H_2 + NO_2$	2.0×10^{8}	1.55	3330	
40b2	$HONO + H \longrightarrow H_2O + NO$	8.1×10^{6}	1.89	1935	
40b3	$HONO + H \longrightarrow OH + HNO$	5.6×10^{10}	0.86	2500	
40c	$HONO + O \longrightarrow OH + NO_2$	1.7×10^{8}	1.50	1525	
40d	$HONO + OH \longrightarrow H_2O + NO_2$	1.2×10^{6}	2.00	−300	
40e	$HONO + CH_3 \longrightarrow NO_2 + CH_4$	8.1×10^{5}	1.87	2770	
40f	$HONO + NH_2 \longrightarrow NO_2 + NH_3$	9.2×10^{5}	1.94	965	
41	$HNO_2 \longrightarrow HONO$	7.1×10^{27}	−5.40	26 440	0.1 atm
41		1.3×10^{29}	−5.47	26 580	1.0 atm
41		2.0×10^{30}	−5.50	27 020	10 atm
41a	$HNO_2 + H \longrightarrow H_2 + NO_2$	2.4×10^{8}	1.50	2095	
41b	$HNO_2 + O \longrightarrow OH + NO_2$	1.7×10^{8}	1.50	1190	
41c	$HNO_2 + OH \longrightarrow H_2O + NO_2$	1.2×10^{6}	2.00	−400	
41d	$HNO_2 + CH_3 \longrightarrow NO_2 + CH_4$	8.1×10^{5}	1.87	2435	
41e	$HNO_2 + NH_2 \longrightarrow NO_2 + NH_3$	9.2×10^{5}	1.94	440	
42	$HCN \longrightarrow HNC$	1.5×10^{23}	−4.20	24 890	0.1 atm
42		1.9×10^{24}	−4.23	24 950	1.0 atm
42		5.3×10^{25}	−4.34	25 260	10 atm
42	$HCN + M \longrightarrow HNC + M$	1.6×10^{26}	−3.23	24 950	Note 2.
42a	$HCN + OH \longrightarrow CN + H_2O$	3.9×10^{6}	1.83	5180	
42b1	$OH + HCN \longrightarrow HNCO + H$	4.4×10^{3}	2.26	3220	
42b2	$OH + HCN \longrightarrow HOCN + H$	1.1×10^{6}	2.03	6730	
42b3	$OH + HCN \longrightarrow NH_2 + CO$	1.6×10^{2}	2.56	4530	
42b4	$OH + HCN \longrightarrow \cdot N{=}CHOH$	1.7×10^{29}	−6.31	2580	0.1 atm
42b4		2.8×10^{30}	−6.37	2690	1.0 atm
42b4		1.1×10^{32}	−6.53	3140	10 atm
42c1	$HCN + O \longrightarrow NH + CO$	5.4×10^{8}	1.21	3770	
42c2	$HCN + O \longrightarrow NCO + H$	2.0×10^{8}	1.47	3820	
42c3	$HCN + O \longrightarrow CN + OH$	4.2×10^{10}	0.40	10 405	
43a	$O + HNC \longrightarrow NH + CO$	4.6×10^{12}		1100	
43b	$OH + HNC \longrightarrow HNCO + H$	2.8×10^{13}		1860	
43c1	$HNC + O_2 \longrightarrow HNCO + O$	1.5×10^{12}	0.01	2070	
43c2	$HNC + O_2 \longrightarrow NH + CO_2$	1.6×10^{19}	−2.25	895	
44a	$CN + H_2 \longrightarrow HCN + H$	3.6×10^{8}	1.55	1510	
44b	$CN + H_2O \longrightarrow HCN + OH$	7.8×10^{12}		3750	Note 3.
44c	$CN + O \longrightarrow CO + N$	7.7×10^{13}		0	
44d	$CN + O_2 \longrightarrow NCO + O$	1.0×10^{13}		0	
44e	$CN + OH \longrightarrow NCO + H$	4.0×10^{13}		0	
44f	$CN + HCN \longrightarrow NCCN + H$	1.5×10^{7}	1.71	770	
44g	$CN + N_2O \longrightarrow NCN + NO$	4.2×10^{11}		3610	
44h	$CN + NO_2 \longrightarrow NCO + NO$	6.2×10^{15}	−0.75	175	
44i	$CN + CH_4 \longrightarrow HCN + CH_3$	1.2×10^{5}	2.64	−80	
44j	$CN + NH_3 \longrightarrow HCN + NH_2$	9.2×10^{12}		−180	

All pressures refer to N_2 buffer gas. Notes: (1) Rate coefficients are expressed in the form $k = A\,T^m \exp(-E_a/RT)$ for concentration units mol/cm^3 and time in s. (2) T > 1000 K. (3) Reverse of reaction 42a.

TABLE 2.19(h) Summary of Rate Coefficient Parameters[1] Part VIII.

	Reaction	A (Note 1)	m	E_a/R	Notes
45a	$H_2CN \longrightarrow HCN + H$	1.3×10^{29}	−6.03	15 045	0.1 atm
45a		6.0×10^{31}	−6.46	16 160	1.0 atm
45a		3.5×10^{29}	−5.46	16 380	10 atm
45b	$H_2CN + HO_2 \longrightarrow CH_2NO + OH$	3.0×10^{13}		0	
45b1	$H_2CN + HO_2 \longrightarrow HCN + H_2O_2$	1.4×10^{4}	2.69	−810	
45b2	$H_2CN + HO_2 \longrightarrow H_2CNH + O_2$	1.4×10^{4}	2.69	−810	
45c	$H_2CN + O_2 \longrightarrow CH_2O + NO$	3.0×10^{12}		3000	
45d	$H_2CN + CH_3 \longrightarrow HCN + CH_4$	8.1×10^{5}	1.87	−560	
45e	$H_2CN + OH \longrightarrow HCN + H_2O$	2.1×10^{17}	−1.68	160	0.1 atm
45e		1.5×10^{19}	−2.18	1090	1.0 atm
45e		9.5×10^{21}	−2.91	2835	10 atm
45e2	$H_2CN + OH \longrightarrow HCN + H_2O$	1.2×10^{6}	2.00	−600	
45f	$H_2CN + N \longrightarrow N_2 + CH_2$	6.0×10^{13}		200	
45g	$H_2CN + H \longrightarrow HCN + H_2$	2.4×10^{8}	1.50	−450	
45h	$H_2CN + NH_2 \longrightarrow HCN + NH_3$	9.2×10^{5}	1.94	−580	
45i1	$H_2CN + O \longrightarrow HCN + OH$	1.7×10^{8}	1.50	−450	
45i2	$H_2CN + O \longrightarrow HNCO + H$	6.0×10^{13}		0	
45i3	$H_2CN + O \longrightarrow HCNO + H$	2.0×10^{13}		0	
46	$HCNH \longrightarrow HCN + H$	7.7×10^{25}	−5.20	11 065	0.1 atm
46		6.1×10^{28}	−5.69	12 215	1.0 atm
46		6.2×10^{26}	−4.77	12 490	10 atm
46a1	$HCNH + H \longrightarrow H_2CN + H$	2.0×10^{13}		0	
46a2	$HCNH + H \longrightarrow HCN + H_2$	2.4×10^{8}	1.50	−450	
46b1	$HCNH + O \longrightarrow HNCO + H$	7.0×10^{13}		0	
46b2	$HCNH + O \longrightarrow HCN + OH$	1.7×10^{8}	1.50	−450	
46c	$HCNH + OH \longrightarrow HCN + H_2O$	1.2×10^{6}	2.00	−600	
46d	$HCNH + CH_3 \longrightarrow HCN + CH_4$	8.2×10^{5}	1.87	−560	
47a	$HCNN + O_2 \longrightarrow H + CO_2 + N_2$	4.0×10^{12}		0	
47b	$HCNN + O_2 \longrightarrow HCO + N_2O$	4.0×10^{12}		0	
48a1	$H_2C{=}NH + H \longrightarrow H_2CN + H_2$	2.4×10^{8}	1.50	3685	
48a2	$H_2C{=}NH + O \longrightarrow H_2CN + OH$	1.7×10^{8}	1.50	2330	
48a3	$H_2C{=}NH + OH \longrightarrow H_2CN + H_2O$	1.2×10^{6}	2.00	−45	
48a4	$H_2C{=}NH + CH_3 \longrightarrow H_2CN + CH_4$	8.2×10^{5}	1.87	3585	
48a5	$H_2C{=}NH + NH_2 \longrightarrow H_2CN + NH_3$	9.2×10^{5}	1.94	2235	
48b1	$H_2C{=}NH + H \longrightarrow HCNH + H_2$	3.0×10^{8}	1.50	3085	
48b2	$H_2C{=}NH + O \longrightarrow HCNH + OH$	2.2×10^{8}	1.50	2720	
48b3	$H_2C{=}NH + OH \longrightarrow HCNH + H_2O$	2.4×10^{6}	2.00	230	
48b4	$H_2C{=}NH + CH_3 \longrightarrow HCNH + CH_4$	5.3×10^{5}	1.87	4875	
48b5	$H_2C{=}NH + NH_2 \longrightarrow HCNH + NH_3$	1.8×10^{6}	1.94	3065	
48c	$H_2C{=}NH + O \longrightarrow CH_2O{+}NH$	1.7×10^{6}	2.08	0	

All pressures refer to N_2 buffer gas. Note: (1) Rate coefficients are expressed in the form $k = A\,T^m \exp(-E_a/RT)$ for concentration units mol/cm^3 and time in s.

TABLE 2.19(i) Summary of Rate Coefficient Parameters[1] Part IX.

	Reaction	A (Note 1)	m	E_a/R	Notes
49a	$CH_3NH \longrightarrow H_2C{=}NH + H$	1.6×10^{36}	−7.92	18 290	Note 2.
49a		1.3×10^{42}	−9.24	20 805	Note 3.
49a		2.3×10^{44}	−9.51	22 770	Note 4.
49b	$CH_3NH + H \longrightarrow H_2C{=}NH + H_2$	7.2×10^{8}	1.50	−450	
49c	$CH_3NH + O \longrightarrow H_2C{=}NH + OH$	5.0×10^{8}	1.50	−450	
49d	$CH_3NH + OH \longrightarrow H_2C{=}NH + H_2O$	3.6×10^{6}	2.00	−600	
49e	$CH_3NH + CH_3 \longrightarrow H_2C{=}NH + CH_4$	2.4×10^{6}	1.87	−560	
50	$CH_2NH_2 \longrightarrow H_2C{=}NH + H$	1.1×10^{45}	−10.24	24 065	Note 2.
50		2.4×10^{48}	−10.82	26 190	Note 3.
50		3.2×10^{46}	−9.95	26 940	Note 4.
50a1	$CH_2NH_2 + O_2 \longrightarrow H_2C{=}NH + HO_2$	1.0×10^{22}	−3.09	3400	
50a2	$CH_2NH_2 + O_2 \longrightarrow NH_2 + CH_2O + O$	6.0×10^{18}	−1.59	15 195	
50b	$CH_2NH_2 + H \longrightarrow H_2C{=}NH + H_2$	4.0×10^{8}	1.50	−450	
50c1	$CH_2NH_2 + O \longrightarrow CH_2O + NH_2$	7.0×10^{13}		0	
50c2	$CH_2NH_2 + O \longrightarrow H_2C{=}NH + OH$	3.3×10^{8}	1.50	−450	
50d1	$CH_2NH_2 + OH \longrightarrow CH_2OH + NH_2$	4.0×10^{13}		0	
50d2	$CH_2NH_2 + OH \longrightarrow H_2C{=}NH + H_2O$	2.4×10^{6}	2.00	−600	
50e1	$CH_2NH_2 + CH_3 \longrightarrow C_2H_5 + NH_2$	2.0×10^{13}		1360	
50e2	$CH_2NH_2 + CH_3 \longrightarrow H_2C{=}NH + CH_4$	1.6×10^{6}	1.87	−315	
51a1	$CH_3NH_2 + H \longrightarrow CH_2NH_2 + H_2$	5.6×10^{8}	1.50	2750	
51a2	$CH_3NH_2 + O \longrightarrow CH_2NH_2 + OH$	4.0×10^{8}	1.50	2615	
51a3	$CH_3NH_2 + OH \longrightarrow CH_2NH_2 + H_2O$	3.6×10^{6}	2.00	120	
51a4	$CH_3NH_2 + CH_3 \longrightarrow CH_2NH_2 + CH_4$	1.5×10^{6}	1.87	4615	
51a5	$CH_3NH_2 + NH_2 \longrightarrow CH_2NH_2 + NH_3$	2.8×10^{6}	1.94	2765	
51b1	$CH_3NH_2 + H \longrightarrow CH_3NH + H_2$	4.8×10^{8}	1.50	4885	
51b2	$CH_3NH_2 + O \longrightarrow CH_3NH + OH$	3.3×10^{8}	1.50	3195	
51b3	$CH_3NH_2 + OH \longrightarrow CH_3NH + H_2O$	2.4×10^{6}	2.00	225	
51b4	$CH_3NH_2 + CH_3 \longrightarrow CH_3NH + CH_4$	1.6×10^{6}	1.87	4450	
51b5	$CH_3NH_2 + NH_2 \longrightarrow CH_3NH + NH_3$	1.8×10^{6}	1.94	3595	
52a	$NCCN + M \longrightarrow CN + CN + M$	1.6×10^{34}	−4.32	65 465	
52b	$NCCN + H \longrightarrow HCN + CN$	1.4×10^{14}		4000	Note 5.
52c	$NCCN + O \longrightarrow NCO + CN$	4.6×10^{12}		4470	
52d	$NCCN + OH \longrightarrow HOCN + CN$	2.0×10^{12}		9560	
53aΣ	$NCO + NO \longrightarrow$ Products	1.4×10^{18}	−1.73	385	Note 6.
53a1	$NCO + NO \longrightarrow CO_2 + N_2$	7.8×10^{17}	−1.73	385	
53a2	$NCO + NO \longrightarrow N_2O + CO$	6.2×10^{17}	−1.73	385	
53b	$NCO + M \longrightarrow N + CO + M$	3.3×10^{14}		27 200	
53c	$NCO + H_2 \longrightarrow HNCO + H$	7.6×10^{2}	3.00	2000	
53d1	$NCO + O \longrightarrow NO + CO$	4.2×10^{13}		0	
53d2	$NCO + O \longrightarrow N + CO_2$	8.0×10^{12}		1260	
53e	$NCO + H \longrightarrow NH + CO$	5.2×10^{13}		0	
53f	$NCO + N \longrightarrow N_2 + CO$	3.3×10^{13}		0	Note 7.

All pressures refer to N_2 buffer gas. Notes: (1) Rate coefficients are expressed in the form $k = A\,T^m \exp(-E_a/RT)$ for concentration units mol/cm³ and time in s. (2) Valid for 0.1 atm, 600 to 2500 K. (3) Valid for 1 atm, 600 to 2500 K. (4) Valid for 10 atm, 600 to 2500 K. (5) Reverse of reaction 44f. (6) Total for all channels of reaction 53a. (7) Reaction 53 entries continue in Part X.

TABLE 2.19(j) Summary of Rate Coefficient Parameters[1] Part X.

	Reaction	A (Note 1)	m	E_a/R	Notes
53g1	$NCO + OH \longrightarrow HNCO + O$	7.8×10^{4}	2.27	−500	
53g2	$NCO + OH \longrightarrow HON + CO$	5.3×10^{12}	−0.07	2580	
53g3	$NCO + OH \longrightarrow H + CO + NO$	8.3×10^{12}	−0.05	9080	
53h1	$NCO + NO_2 \longrightarrow CO_2 + N_2O$	2.3×10^{12}		−440	
53h2	$NCO + NO_2 \longrightarrow CO + 2NO$	2.1×10^{11}		−440	
53i	$NCO + CH_4 \longrightarrow HNCO + CH_3$	9.8×10^{12}		4090	
53j	$NCO + NH_3 \longrightarrow HNCO + NH_2$	2.8×10^{4}	2.48	495	
54	$HCNO \longrightarrow HCN + O$	2.0×10^{30}	−6.03	30 565	0.1 atm
54		4.2×10^{31}	−6.12	30 805	1.0 atm
54		5.9×10^{31}	−5.85	31 170	10 atm
54a	$HCNO + H \longrightarrow HNCO + H$	2.1×10^{15}	−0.69	1435	
54b	$HCNO + H \longrightarrow HCN + OH$	2.7×10^{11}	0.18	1065	
54c	$HCNO + H \longrightarrow NH_2 + CO$	1.7×10^{14}	−0.75	1455	
54d	$HCNO + H \longrightarrow HOCN + H$	1.4×10^{11}	−0.19	1250	
54e	$HCNO + O \longrightarrow HCO + NO$	7.0×10^{13}		0	
54f	$HCNO + OH \longrightarrow HCOH + NO$	4.0×10^{13}		0	
55a	$HOCN + H \longrightarrow HCN + OH$	2.0×10^{13}	−0.04	1075	Note 2.
55b	$HOCN + H \longrightarrow HNCO + H$	3.1×10^{8}	0.84	965	
55c	$HOCN + H \longrightarrow NH_2 + CO$	1.2×10^{8}	0.61	1045	
55d	$HOCN + H \longrightarrow H_2 + NCO$	2.4×10^{8}	1.50	3330	
55e	$HOCN + O \longrightarrow OH + NCO$	1.7×10^{8}	1.50	2080	
55f	$HOCN + OH \longrightarrow H_2O + NCO$	1.2×10^{6}	2.00	−125	
55g	$HOCN + CH_3 \longrightarrow CH_4 + NCO$	8.2×10^{5}	1.87	3330	
55h	$HOCN + NH_2 \longrightarrow NCO + NH_3$	9.2×10^{5}	1.94	1835	
56a	$HNCO + M \longrightarrow NH + CO + M$	1.3×10^{16}		42 460	
56b	$HNCO + H \longrightarrow NH_2 + CO$	3.6×10^{4}	2.49	1180	
56c1	$HNCO + O \longrightarrow HNO + CO$	1.7×10^{6}	2.08	0	
56c2	$HNCO + O \longrightarrow NH + CO_2$	1.7×10^{6}	2.08	0	
56d1	$HNCO + OH \longrightarrow NH_2 + CO_2$	6.3×10^{10}	−0.06	5860	
56d2	$HNCO + OH \longrightarrow NCO + H_2O$	5.2×10^{10}	−0.03	8840	
56e	$HNCO + H \longrightarrow NCO + H_2$	1.8×10^{5}	2.40	4990	Note 3.
56f	$HNCO + O \longrightarrow NCO + OH$	3.1×10^{6}	1.94	3250	Note 4.
56g	$HNCO + OH \longrightarrow NCO + H_2O$	3.6×10^{7}	1.50	1810	
56h	$HNCO + CH_3 \longrightarrow NCO + CH_4$	1.0×10^{12}		5000	Note 5.
56i	$HNCO + NH_2 \longrightarrow NCO + NH_3$	1.0×10^{12}		4500	Note 6.

All pressures refer to N_2 buffer gas. Notes: (1) Rate coefficients are expressed in the form $k = A\,T^m \exp(-E_a/RT)$ for concentration units mol/cm^3 and time in s. (2) Reverse of reaction 42b2. (3) Reverse of reaction 53c. (4) Reverse of reaction 53g1. (5) Reverse of reaction 53i. (6) Reverse of reaction 53j.

TABLE 2.19(k) Summary of Rate Coefficient Parameters[1] Part XI.

	Reaction	A (Note 1)	m	E_a/R	Notes
57a	$CH_2NO \longrightarrow HNCO + H$	6.9×10^{41}	−9.30	26 020	Note 2.
57a		2.3×10^{42}	−9.11	27 095	Note 3.
57a		1.7×10^{38}	−7.64	26 965	Note 4.
57b	$CH_2NO + O_2 \longrightarrow CH_2O + NO_2$	1.2×10^{15}	−1.01	10 130	Note 5.
57b	$CH_2NO + O_2 \longrightarrow CH_2O + NO_2$	2.9×10^{12}	−0.31	8910	Note 6.
57c1	$CH_2NO + H \longrightarrow CH_3 + NO$	4.0×10^{13}		0	
57c2	$CH_2NO + H \longrightarrow HCNO + H_2$	4.8×10^{8}	1.50	−450	
57d1	$CH_2NO + O \longrightarrow CH_2O + NO$	7.0×10^{13}		0	
57d2	$CH_2NO + O \longrightarrow HCNO + OH$	3.3×10^{8}	1.50	−450	
57e1	$CH_2NO + OH \longrightarrow CH_2OH + NO$	4.0×10^{13}		0	
57e2	$CH_2NO + OH \longrightarrow HCNO + H_2O$	2.4×10^{6}	2.00	−600	
57f1	$CH_2NO + CH_3 \longrightarrow C_2H_5 + NO$	3.0×10^{13}		0	
57f2	$CH_2NO + CH_3 \longrightarrow HCNO + CH_4$	1.6×10^{6}	1.87	−560	
57g1	$CH_2NO + NH_2 \longrightarrow CH_2NH_2 + NO$	3.0×10^{13}		0	
57g2	$CH_2NO + NH_2 \longrightarrow HCNO + NH_3$	1.8×10^{6}	1.94	−580	
58a	$CH_3NO + H \longrightarrow CH_2NO + H_2$	4.4×10^{8}	1.50	190	
58b	$CH_3NO + O \longrightarrow CH_2NO + OH$	3.3×10^{8}	1.50	1820	
58c	$CH_3NO + OH \longrightarrow CH_2NO + H_2O$	3.6×10^{6}	2.00	−600	
58d	$CH_3NO + CH_3 \longrightarrow CH_2NO + CH_4$	7.9×10^{5}	1.87	2725	
58e	$CH_3NO + NH_2 \longrightarrow CH_2NO + NH_3$	2.8×10^{6}	1.94	540	
58f	$CH_3NO + H \longrightarrow CH_3 + HNO$	1.8×10^{13}		1400	
58g	$CH_3NO + O \longrightarrow CH_3 + NO_2$	1.7×10^{6}	2.08	0	
58h	$CH_3NO + OH \longrightarrow CH_3 + HONO$	2.5×10^{12}		500	
59a	$HON + M \longrightarrow NO + H + M$	5.1×10^{19}	−1.73	8075	
59b1	$HON + H \longrightarrow HNO + H$	2.0×10^{13}		0	
59b2	$HON + H \longrightarrow OH + NH$	2.0×10^{13}		0	
59c	$HON + O \longrightarrow OH + NO$	7.0×10^{13}		0	
59d	$HON + OH \longrightarrow HONO + H$	4.0×10^{13}		0	
59e	$HON + O_2 \longrightarrow HONO + O$	1.0×10^{12}		2500	
60	$HCOH \longrightarrow CH_2O$	3.5×10^{17}	−2.86	4470	0.1 atm
60		2.1×10^{19}	−3.07	4800	1.0 atm
60		1.8×10^{21}	−3.32	5465	10 atm

Notes: All pressures refer to N_2 buffer gas. Notes: (1) Rate coefficients are expressed in the form $k = A\,T^m \exp(-E_a/RT)$ for concentration units mol/cm^3 and time in s. (2) Valid for 0.1 atm, 600 to 2500 K. (3) Valid for 1 atm, 600 to 2500 K. (4) Valid for 10 atm, 600 to 2500 K. (5) Valid for 1000 to 2500 K. (6) Valid for 300 to 1000 K.

TABLE 2.19(l) Summary of Rate Coefficient Parameters[1] Part XII.

	Reaction	A (Note 1)	m	E_a/R	Notes
61b1	$NH_2OH + H \longrightarrow HNOH + H_2$	4.8×10^{8}	1.50	3145	
61b2	$NH_2OH + H \longrightarrow NH_2O + H_2$	2.4×10^{8}	1.50	2550	
61c1	$NH_2OH + O \longrightarrow HNOH + OH$	3.3×10^{8}	1.50	1945	
61c2	$NH_2OH + O \longrightarrow NH_2O + OH$	1.7×10^{8}	1.50	1515	
61d1	$NH_2OH + OH \longrightarrow HNOH + H_2O$	2.4×10^{6}	2.00	−165	
61d2	$NH_2OH + OH \longrightarrow NH_2O + H_2O$	1.2×10^{6}	2.00	−300	
61e1	$NH_2OH + CH_3 \longrightarrow HNOH + CH_4$	1.6×10^{6}	1.87	3195	
61e2	$NH_2OH + CH_3 \longrightarrow NH_2O + CH_4$	8.2×10^{5}	1.87	2765	
61f1	$NH_2OH + NH_2 \longrightarrow HNOH + NH_3$	1.8×10^{6}	1.94	1625	
61f2	$NH_2OH + NH_2 \longrightarrow NH_2O + NH_3$	9.2×10^{5}	1.94	950	
61g1	$NH_2OH + HO_2 \longrightarrow HNOH + H_2O_2$	2.9×10^{4}	2.69	4810	
61g2	$NH_2OH + HO_2 \longrightarrow NH_2O + H_2O_2$	1.4×10^{4}	2.69	3230	
62a	$NH_2NO \longrightarrow N_2 + H_2O$	4.1×10^{33}	−7.18	17 700	0.1 atm
62a		3.1×10^{34}	−7.11	18 260	1.0 atm
62a		2.9×10^{31}	−5.91	18 205	10 atm
62b	$NH_2NO + H \longrightarrow HNNO + H_2$	4.8×10^{8}	1.50	3730	
62c	$NH_2NO + O \longrightarrow HNNO + OH$	3.3×10^{8}	1.50	2365	
62d	$NH_2NO + OH \longrightarrow HNNO + H_2O$	2.4×10^{6}	2.00	−35	
62e	$NH_2NO + CH_3 \longrightarrow HNNO + CH_4$	1.6×10^{6}	1.87	3615	
62f	$NH_2NO + NH_2 \longrightarrow HNNO + NH_3$	1.8×10^{6}	1.94	2285	
62g	$NH_2NO + HO_2 \longrightarrow HNNO + H_2O_2$	2.9×10^{4}	2.69	6355	
63a	$H_2NNHO \longrightarrow NH_2 + HNO$	2.7×10^{39}	−8.74	20 945	0.1 atm
63a		2.4×10^{40}	−8.73	20 940	1.0 atm
63a		1.2×10^{41}	−8.64	20 925	10 atm
63b	$H_2NNHO + H \longrightarrow HNNHO + H_2$	4.8×10^{8}	1.50	−450	
63c	$H_2NNHO + O \longrightarrow HNNHO + OH$	3.3×10^{8}	1.50	−450	
63d	$H_2NNHO + OH \longrightarrow HNNHO + H_2O$	2.4×10^{6}	2.00	−600	
63e	$H_2NNHO + CH_3 \longrightarrow HNNHO + CH_4$	1.6×10^{6}	1.87	190	
63f	$H_2NNHO + NH_2 \longrightarrow HNNHO + NH_3$	1.8×10^{6}	1.94	−580	
63g	$H_2NNHO + HO_2 \longrightarrow HNNHO + H_2O_2$	2.9×10^{4}	2.69	−805	
64a	$ClNO + M \longrightarrow NO + Cl + M$	1.9×10^{15}		16 100	
64b	$O + ClNO \longrightarrow ClO + NO$	5.0×10^{12}		1520	
64c1	$OH + ClNO \longrightarrow HOCl + NO$	5.4×10^{12}		1130	
64c2	$OH + ClNO \longrightarrow HONO + Cl$	5.5×10^{10}		−240	
64d	$Cl + ClNO \longrightarrow Cl_2 + NO$	4.0×10^{13}		−130	
64e	$H + ClNO \longrightarrow HCl + NO$	4.6×10^{13}		455	

All pressures refer to N_2 buffer gas. Notes: (1) Rate coefficients are expressed in the form $k = A\,T^m \exp(-E_a/RT)$ for concentration units mol/cm^3 and time in s.

3

Kinetics and Mechanisms of the Oxidation of Gaseous Sulfur Compounds

Anthony J. Hynes[1]

Paul H. Wine[2]

3.1 INTRODUCTION

The problems associated with global climate change in general, and acid rain in particular, have led to a great deal of research on the atmospheric and combustion chemistry of sulfur. Developments over the last decade have led to significant progress in our understanding of the kinetics and mechanisms of the atmospheric oxidation chemistry of natural and anthropogenic sulfur. Rather less effort, however, has been placed on developing an understanding of sulfur combustion kinetics; the emphasis of mitigation research has instead been placed on removal of sulfur from fuels or development of scrubbing techniques to remove SO_2 from stack gases.

The last major review of the oxidation and combustion chemistry of sulfur, by Cullis and Mulcahy in 1972, still provides an excellent review of the field. It is striking that although their review is a quarter-century old it is only the section on elementary reactions, which deals mostly with low-temperature kinetics studies, that is outdated. In this chapter we update the Cullis and

[1]Division of Marine and Atmospheric Chemistry, Rosenstiel School of Marine and Atmospheric Sciences, University of Miami, 4600 Rickenbacker Causeway, Miami, FL 33149

[2]School of Chemistry and Biochemistry, School of Earth and Atmospheric Sciences, and Georgia Tech Research Institute, Georgia Institute of Technology, Atlanta, GA 30332

Mulcahy review with emphasis on areas that have developed significantly since then, focusing particularly on progress in our understanding of the elementary reactions, diagnostic techniques, and NO_x–SO_x interactions in flames.

3.2 SULFUR EMISSIONS

Estimates of natural and anthropogenic sulfur emissions vary widely, reflecting the difficulty of establishing accurate inventories (Cullis and Hirschler 1980; Möller 1984; Aneja and Cooper 1989; Aneja 1990; Bates et al. 1992; Spiro et al. 1992). There is general agreement, however, that combustion of fossil fuels accounts for the bulk of the anthropogenic emissions, of which coal combustion accounts for about 60%, petroleum combustion about 30%, and metal smelting the remainder (Spiro et al. 1992). Anthropogenic emissions have been estimated to account for approximately 75% of the total atmospheric sulfur burden in two detailed studies (Bates et al. 1992; Spiro et al. 1992). The anthropogenic emissions, almost exclusively SO_2, are concentrated in the northern hemisphere. In contrast, biogenic emissions, always in the form of reduced sulfur, consist of a variety of organic and inorganic species, such as H_2S, CS_2, COS, dimethyl sulfide (DMS, CH_3SCH_3), dimethyl disulfide (DMDS, CH_3SSCH_3), and methyl mercaptan (CH_3SH). Approximately 50% of the biogenic sulfur is emitted in the marine environment. These emissions are dominated by DMS, produced by marine planktonic algae and thought to be the precursor of the sulfate aerosols that act as cloud condensation nuclei in the marine environment and thus play a role in global climate regulation (Charlson et al. 1987).

While coal and petroleum combustion has a single sulfur-containing end product, SO_2, the chemical nature of bound sulfur in each medium is highly variable, which has important consequences for the detailed mechanism of combustion. Attar and Corcoran (1977) reviewed the literature on the nature of bound sulfur in coal. They noted that sulfur constitutes 0.2 to 10 wt% of coal and includes both organic and inorganically bound species, with an inorganic:organic ratio that is typically 2:1. The nature of the bound organic sulfur is difficult to determine; Attar and Corcoran conclude that most is in the form of thiophenic, aromatic, and aliphatic structures. During pyrolysis a complex mixture of gases is produced, including H_2S, CS_2, and thiophene and its derivatives.

3.3 ELEMENTARY REACTIONS

There has been a transformation of our understanding of the detailed elementary reaction kinetics of sulfur since the review of Cullis and Mulcahy (1972). At that time, with the exception of reactions of excited atomic sulfur, $S(^1D)$, little was known about the elementary reactions of sulfur species; Cullis and Mulcahy presented rate data on only about twenty elementary reactions other than those of $S(^1D)$. A 1976 review of gas phase kinetics data contained results for 63 elementary reactions of sulfur (Baulch et al. 1976), while the 1992 NIST database includes 447 reactions (Mallard et al. 1992). The increase in both the quantity and quality of elementary reaction rate data has been driven largely by the desire to understand the elementary chemical processes taking place in the troposphere and stratosphere. By and large, therefore, kinetic studies have focused on low-temperature rate studies as required to get rate coefficient data for atmospheric modeling purposes. In cases where extensive temperature dependencies have been examined, the limitations of the most commonly used measurement techniques, flash photolysis and flow discharge, mean that data were taken at temperatures below 1000 K and extrapolations are required for flame temperatures. If the temperature dependence of the rate coefficient is well described by a simple Arrhenius expression, then this procedure can provide reliable results. In general, however, even direct abstraction reactions follow Arrhenius behavior only approximately (Zellner 1984) and, as will be seen below, this type of behavior is the exception rather than the rule for elementary reactions of sulfur species. Elementary reactions that proceed through formation of an addition complex show complex temperature and pressure dependence, and Arrhenius extrapolation of low-temperature rate data for such reactions is inappropriate. The limited high-temperature data on sulfur reactions have been derived from shock tube or flame studies; detailed assumptions about complex chemistry were required to extract values of elementary reaction rate coefficients.

It would be impractical to review the complete elementary gas phase kinetic database on sulfur reactions for the purposes of this chapter. A feeling for the development of our understanding in this area can be obtained from the compilations and evaluations of Baulch et al. (1976), Heicklen (1985), Singleton and Cvetanovic (1988), Plane (1989), Tyndall and Ravishankara (1991), Mallard et al. (1992), DeMore et al. (1992), and Atkinson et al. (1992). These sources will guide the reader to information of interest which is not covered here. We review instead just the subset of these reactions for which significant progress has been made and which are involved in the oxidation mechanism of sulfur compounds that are important in combustion and atmospheric chemistry. Our intention is to indicate general features, such as the predominance of reactions that proceed via addition to the sulfur atom and the importance of weakly bound intermediates. Describing sulfur combustion chemistry requires elementary kinetic data obtained under pressure and temperature conditions appropriate to the environment which is being modeled, in contrast to hydrocarbon combustion chemistry, where weakly-bound intermediates appear only occasionally.

Table 3.1 lists the reactions discussed in this and later sections. In some cases evaluated rate expressions and estimates of the temperature range over which we consider them to be reliable have been included. In a few cases we offer our own estimates or list previously reported values (often estimates) which together with a citation at least provide a starting point for further investigation. Experimentally derived rate coefficients for many of the radical–radical reactions required for comprehensive kinetic modeling of sulfur chemistry are unavailable.

TABLE 3.1, Part 1. Elementary reaction rate coefficients.[a]

	Reaction	$\Delta_r H^\circ_{298}$	A	E_a	ΔT	Notes
1	$SO_2 + M \longrightarrow SO + O + M$	551	1.7×10^{16}	469	2900–5200	Kiefer 1975
2	$SO_2 + O + M \longrightarrow SO_3 + M$	−348	1.5×10^{16}	8.3	200–400	Atkinson et al. 1992
			1.1×10^{13}	−65.4	1700–2500	Astholz et al. 1979
3	$SO_3 + O \longrightarrow SO_2 + O_2$	−150	1.3×10^{12}	25.5	1435–1850	Smith et al. 1982
4	$SO + O_2 \longrightarrow SO_2 + O$	−53	8.4×10^{10}	19	230–420	Atkinson et al. 1992
5	$S + O_2 \longrightarrow SO + O$	−23	1.3×10^{12}		230–400	Atkinson et al. 1992
			1.5×10^{13}	15.3	1000–3000	Miyoshi et al. 1998
6	$O + S_2 \longrightarrow SO + S$	−96	1.0×10^{13}		298	Singleton & Cvetanovich 1988
7	$H_2S + M \longrightarrow SH + H + M$	381	2.0×10^{14}	310	2700–3800	Bowman & Dodge 1977
8	$SH + H \longrightarrow S + H_2$	−84	1.5×10^{13}		295	Cupitt & Glass 1975
	Reverse reaction		6.0×10^{14}	100	1237–3137	Woiki & Roth 1995
9	$SH + SH \longrightarrow S + H_2S$	−30	2.4×10^{13}		298	Stachnik & Molina 1987
10	$H_2S + H \longrightarrow SH + H_2$	−54	$1.2 \times 10^7 T^{2.1}$	2.9	250–2500	Yoshimura et al. 1992
11	$S + OH \longrightarrow SO + H$	−123	4.0×10^{13}		298	Jourdain et al. 1979
12	$S + SH \longrightarrow S_2 + H$	−93	2.7×10^{13}		300	Von Mihelcic & Schindler 1970
13	$SH + O \longrightarrow SO + H$	−169	1.4×10^{14}		298	Singleton & Cvetanovich 1988
14	$SH + O_2 \longrightarrow$ Products		$< 2.4 \times 10^5$		298	Stachnik & Molina 1987
	$\longrightarrow SO_2 + H$	−222	1×10^{13}	84	1000–2000	Frenklach et al. 1981
	$\longrightarrow SO + OH$	−99	5×10^{13}	84	1000–2000	Frenklach et al. 1981
15	$S_2 + OH \longrightarrow SH + SO$	−20				Note *b*.
16	$SH + HO_2 \longrightarrow H_2S + O_2$	−186	1×10^{12}		1000–2000	Frenklach et al. 1981

a Concentration units mol/cm^3; time in s; $\Delta_r H^\circ_{298}$ and E_a in kJ/mol. *b* Not an elementary reaction. See text.

TABLE 3.1, Part 2. Elementary reaction rate coefficients.[a]

	Reaction	$\Delta_r H^\circ_{298}$	A	E_a	ΔT	Notes
17	$SO + OH \longrightarrow SO_2 + H$	−123	5×10^{13}		298	Jourdain et al. 1979
18	$SO_2 + H + M \longrightarrow HSO_2 + M$	−109	$< 5 \times 10^{14}$		298	Kallend 1972
19	$SO_2 + OH + M \longrightarrow HOSO_2 + M$	−127	$2.2 \times 10^{25} T^{-3.3}$		200–400	Atkinson et al. 1992
20	$HOSO_2 + O_2 \longrightarrow HO_2 + SO_3$	+2	8×10^{11}	2.7	290–420	
21	$H_2S + O \longrightarrow SH + OH$	−46	5.5×10^{12}	15	200–600	DeMore et al. 1992
	$SO + H_2$	−222	1×10^{13}	42	1000–2000	Frenklach et al. 1981
	$HSO + H$	41	1×10^{13}	42	1000–2000	Frenklach et al. 1981
22	$H_2S + OH \longrightarrow SH + H_2O$	−117	2.7×10^{12}		300	Atkinson et al. 1992
23	$HSO_2 + H \longrightarrow SO_2 + H_2$	−327	3.0×10^{12}		1940–1990	Halstead & Jenkins 1968
24	$HSO_2 + OH \longrightarrow SO_2 + H_2O$	−390	$< 4.8 \times 10^{12}$		1940–1990	Halstead & Jenkins 1968
25	$HOSO_2 + H \longrightarrow SO_2 + H_2O$	−372				Note *b*.
26	$SO_3 + H_2O + M \longrightarrow H_2SO_4 + M$	−97	$8.2 \times 10^4 T \exp(6544/T)[H_2O]^2$			Note *c*.
27	$CH_3SH + H \longrightarrow CH_3S + H_2$	−70	2.1×10^{13}	7.0	250–410	Wine et al. 1986
	$\longrightarrow H_2S + CH_3$	−69	Possible minor channel			Martin et al. 1988
28	$CH_3SH + O \longrightarrow$ Products		$9 \times 10^{11} + 1.8 \times 10^{13} \exp(1260/T)$			Note *d*.
	$HSO + CH_3$ 60%	−83				Cvetanovic et al. 1981
	$\longrightarrow CH_3SO + H$ 30%	−39				Cvetanovic et al. 1981
	$\longrightarrow CH_3S + OH$ 10%	−62				Cvetanovic et al. 1981
29	$CH_3SH + OH \longrightarrow CH_3S + H_2O$	−133	2×10^{12}			Atkinson et al. 1992
30	$CH_3SCH_3 + H \longrightarrow CH_3SH + CH_3$	−56	1.6×10^{13}	10.0	200–500	See text.
31	$CH_3SCH_3 + O \longrightarrow CH_3SO + CH_3$	−126	8.2×10^{12}	−3.4	250–660	Note *d*

a Concentration units mol/cm^3; time in s; $\Delta_r H^\circ_{298}$ and E_a in kJ/mol. *b* No data available. *c* Lovejoy et al. (1996) report this expression as an effective first-order representation of the rate coefficient for SO_3 loss in the presence of water from 250 to 360 K. The negative activation energy suggests formation of an adduct that subsequently reacts with water to form sulfuric acid. *d* From Singleton and Cvetanovic (1988).

TABLE 3.1, Part 3. Elementary reaction rate coefficients.[a]

	Reaction	$\Delta_r H^\circ_{298}$	A	E_a	ΔT	Notes
32	$CH_3SCH_3 + OH \longrightarrow CH_3SCH_2 + H_2O$	−106	5.8×10^{12}	2.0	350–400	Hynes et al. 1986
33	$CH_3SCH_3 + OH + M \longrightarrow CH_3S(OH)CH_3 + M$	−60	$6 \times \times 10^{12}$			Note *b*.
34	$CH_3S(OH)CH_3 + O_2 \longrightarrow$ Products		4.8×10^{11}	0	250–300	Hynes et al. 1995
	$\longrightarrow (CH_3)_2SO + HO_2$	−200	Branching ratio 0.5			Hynes et al. 1993
35	$CS_2 + O_2 \longrightarrow CS + SO_2$	−134	1×10^{12}	134	1400–2400	Saito et al. 1986
36	$CS_2 + O \longrightarrow$ Products		1.9×10^{13}	5.4	200–500	Singleton & Cvetanovich 1988
	$\longrightarrow CS + SO$ (88.5%)	−81				Cooper & Herschberger 1992
	$\longrightarrow CO + S_2$ (8.5%)	−348				Cooper & Herschberger 1992
	$\longrightarrow COS + S$ (3%)	−228				Cooper & Herschberger 1992
37	$CS_2 + OH \longrightarrow$ Products		9×10^{11}			Note *c*.
38	$CS_2 + OH + M \longrightarrow CS_2OH + M$	−50	2.9×10^{17}		270–230	Atkinson et al. 1992
39	$CS_2OH + O_2 \longrightarrow COS + SO_2 + H$	−332	1.7×10^{10}		250–350	Atkinson et al. 1992
40	$COS + S \longrightarrow CO + S_2$	−313	9.1×10^{11}	15.2	233–445	Klemm & Davis 1974
			6.0×10^{13}	37.9	1100–2500	Woiki et al. 1993
41	$S + HNO \longrightarrow SO + NH$	−17				Note *d*.
42	$SO + N \longrightarrow S + NO$	−110	4.8×10^{11}		298	Jacob & Winkler 1970
43	$SO + CN \longrightarrow S + NCO$	−33	5×10^{12}		2000	Note *e*.
44	$NS + O \longrightarrow S + NO$	−144	$1 \times 10^{12} T^{1/2} \exp(-17500/T)$			Note *f*.
45	$NS + O \longrightarrow SO + N$	−34	$6.3 \times 10^{11} T^{1/2} \exp(-4030/T)$			Wendt et al. 1983
46	$NS + NO \longrightarrow SO + N_2$	−349	1×10^{13}		2000	Pfefferle & Churchill 1989
47	$SH + NH \longrightarrow NS + H_2$	−237	1×10^{13}		2000	Pfefferle & Churchill 1989
48	$CS + NO \longrightarrow NS + CO$	−219	1×10^{13}		2000	Pfefferle & Churchill 1989

a Concentration units mol/cm^3; time in s; E_a and $\Delta_r H^\circ_{298}$ in kJ. *b* At 1 atm and 260 K; from Hynes et al. (1986). *c* Effective rate coefficient for a complex CS_2 loss mechanism at 295 K and 700 Torr; from Hynes et al. (1988). *d* No data are available. *e* Estimated by analogy to the parameters for the CN + O_2 reaction reported by Thielen and Roth (1987). *f* Computed from the reverse rate coefficient of Wendt et al. (1983).

In addition to the compilations cited above, several experimental and modeling studies discussed later list extensive reaction mechanisms and suggest estimates of needed rate coefficients where experimental data are not available. These include the papers by Muller et al. (1979), Frenklach et al. (1981), Wendt et al. (1983), Smith et al. (1983), Zachariah and Smith (1987), Chernysheva et al. (1990, 1992), Pauwels et al. (1992), Basevich et al. (1994), and Glarborg et al. (1996).

In the following sections we discuss a number of the elementary reactions thought to important play roles in the combustion chemistry of sulfur. In contrast to Table 3.1, where elementary reactions are grouped by the elements involved, the following discussions address first reactions of atoms and radicals with stable sulfur-containing molecules and then reactions of sulfur-containing radicals. Most of the experimental work cited has been done at temperatures well below those of flames. In a final section we summarize some of the elementary reaction results that have been derived from shock tube experiments, where it is possible to study chemical reactions at pressures and temperatures like those of flames, but under better controlled conditions.

3.3.1 Reactions of atoms and radicals with sulfur-containing molecules

The OH radical plays central roles in atmospheric and combustion chemistry. The atmosphere can, in fact, be characterized as a cool flame, and many of the elementary reactions important in combustion also appear in atmospheric chemistry. Consideration of the similarities and differences between the two situations gives a perspective on the dominant processes in each. The major differences between ordinary combustion and a lean, cool atmospheric flame are these:

In ordinary flames OH, O, and H radicals are all important, their relative concentrations depending on equivalence ratio—OH is always significant, while H and O are important in rich and lean flames, respectively; radical concentrations are high; and there is dynamic interplay between kinetics and thermodynamics. Typically, at 2500 K the relative concentrations of most flame radicals are controlled by thermodynamics, while below 2000 K kinetics plays an increasingly important role. In the atmospheric flame, far from thermodynamic equilibrium in all respects, kinetics dominates; radical concentrations are low; and the chemistry is dominated by OH, with O atom chemistry becoming important in the upper atmosphere, above around 15 km.

$O + SO_2 + M \longrightarrow SO_3 + M$ (2)

This extensively studied reaction is thought to be responsible for the sulfur-catalyzed recombination of flame radicals in lean flames via the sequence (Smith et al. 1983):

$$O + SO_2 + M \longrightarrow SO_3 + M \qquad (2)$$

$$O + SO_3 \longrightarrow SO_2 + O_2 . \qquad (3)$$

The rate coefficient has a positive temperature dependence at low temperature but a negative one at high temperature. Astholtz et al. (1979) showed that this is consistent with spin-forbidden recombination on a potential energy surface with a recombination barrier of about 14 kJ/mol. The rate coefficients they derived from measurements of the dissociation rate of SO_3 in shock waves from 1700 to 2500 K and the equilibrium constant are about a factor of three lower than those obtained from two indirect measurements in flames (Smith et al. 1982).

Formation of SO_3 in practical combustion systems is important because of its implications for corrosion. Thermodynamic considerations favor a shift from SO_2 towards SO_3 as the temperature is lowered. The possibility that routes other than the relatively slow three body recombination (reaction 20, for example) could lead to SO_3 formation has not been fully considered so far.

$H + SO_2 + M \longrightarrow HSO_2 + M$ (18)

This reaction is widely accepted as the main initiator of sulfur-catalyzed recombination of flame radicals, as discussed in the following section. Webster and Walsh (1965) appear to have been the first to suggest that it could play a significant role in combustion systems. They examined the effect of SO_2 on the second explosion limit of hydrogen–oxygen mixtures and attributed the significant inhibiting effect that they found to H-atom loss in this reaction. They derived $k_{18} = 1.5 \times 10^{16}$ cm^{6}mol^{-2}s^{-1} at 800 K. Later rate coefficient values inferred from flame studies are discussed in detail in the evaluation by Baulch et al. (1976), who recommend $k_{18} = 5.1 \times 10^{16}$ cm^{6}mol^{-2}s^{-1}.

Direct studies at low temperature, however, find this reaction to be slow and suggest only upper limits to the rate coefficient. Binns and Marshall (1991) reported an experimental upper limit $k_{18} < 5 \times 10^{14}$ cm^{6}mol^{-2}s^{-1} at 500 K and calculated rate coefficients of 4×10^9 cm^{6}mol^{-2}s^{-1} at 500 K and 4×10^{12} cm^{6}mol^{-2}s^{-1} at 800 K with Ar as the third body. Their ab initio calculations also predict that only the HO–S=O isomer is stable, with an H–OSO bond energy of 109 kJ/mol; formation of the H–SO_2 isomer is calculated to be 25 kJ/mol endothermic. This contrasts with earlier estimates by Benson (1978) which suggested that both isomers would be stable. He suggested that HO–S=O would be the more stable isomer, with a bond dissociation energy of 160 kJ/mol. The ab initio study by Boyd et al. (1980) gave an H–SO_2 bond energy of 142 kJ/mol. Binns and Marshall calculate a substantial barrier to recombination, about 95 kJ/mol, leading to the unusually large predicted temperature dependence for recombination. Their calculated rate coefficient of 4×10^{14} cm^{6}mol^{-2}s^{-1} at 2000 K is one to two orders of magnitude lower than values derived from flame measurements (Fenimore and Jones, 1965; Kallend, 1972).

It appears that there is a fundamental inconsistency between the interpretation of Webster and Walsh of their explosion inhibition results and the more recent work of Binns and Marshall. This is important in the development of theories of sulfur catalysis in flames since Fenimore and Jones (1965) assumed that reaction (18) was the initial and rate determining step in sulfur catalyzed recombination in their flames and determined a rate coefficient that they considered consistent with the value determined by Webster and Walsh (1965). Subsequent work has accepted this, largely discounting the proposal by Wheeler (1968) that recombination of OH with SO_2 is the initial step. For example, Webster and Walsh did not consider the effects of reaction 19, which could potentially inhibit explosion by competing with the chain propagation step $OH + H_2 \longrightarrow H + H_2O$; $HOSO_2$ formed in reaction (19) would then react with O_2 or H via reactions 20 or 25, generating HO_2 and SO_3 or H_2O and SO_2. The SO_3 would then be expected to react rapidly with H atoms. Unfortunately, as noted above, the rate coefficient for reaction 17 at 800 K has not been measured. Kinetic modeling of the H_2–O_2 system [e.g., as undertaken by Dougherty and Rabitz (1980)] extended to include sulfur chemistry for the conditions used by Webster and Walsh could be enlightening.

$OH + SO_2 + M \longrightarrow HOSO_2 + M$ (19)

This reaction is the initial step in the gas phase oxidation of SO_2 in the atmosphere and has been suggested as a potential initial step in the sulfur catalyzed recombination of flame radicals.

In the atmosphere reaction (19) is followed by the sequence (Wang and Howard 1990):

$$HOSO_2 + O_2 \longrightarrow HO_2 + SO_3 \quad (20)$$

$$SO_3 + H_2O \longrightarrow H_2SO_4 \quad \text{(multiple steps probable).} \quad (26)$$

Both liquid and gas phase processes are important in the atmospheric oxidation of SO_2 to H_2SO_4 (Schwartz 1989).

Wheeler (1968) proposed the catalytic recombination cycle

$$OH + SO_2 + M \longrightarrow HOSO_2 + M \quad (19)$$

$$HOSO_2 + H \longrightarrow H_2O + SO_2 \quad (25)$$

as a potential route for catalytic recombination of flame radicals. Reaction 20, however, has not been considered as a source of SO_3 in combustion systems, possibly as a result of Benson's (1978) estimate that it is endothermic by 25 kJ/mol. Current values for the standard enthalpy of formation of $HOSO_2$, however, imply that the standard enthalpy change of reaction 20 is only 1 ± 2 kJ/mol, so that it could be a normal H-atom transfer reaction under combustion conditions.

The OH + SO_2 association reaction has been studied extensively at low temperature, and the rate coefficients appropriate for use in atmospheric models are considered to be reasonably well defined. Wine et al. (1984b) studied the reaction as a function of temperature and pressure from 260 to 420 K using a laser photolysis-resonance fluorescence technique. They used a Troe formalism to characterize the pressure and temperature dependence of the rate coefficient, deriving temperature-dependent as well as temperature-independent values for the collision-broadening factor F_c and the low-pressure limiting rate coefficient k_0. Evaluations from NASA (DeMore et al. 1992) and CODATA (Atkinson et al. 1992) are also available and can be used to extrapolate to higher temperatures. Unfortunately, none of these expressions should be considered reliable outside the ranges over which the data were fit. In fact, the high-temperature rate coefficients derived using the temperature-dependent F_c and k_0 are substantially smaller than those which would be derived from temperature-independent expressions recommended by the NASA or CODATA panels for modeling atmospheric processes. For instance, extrapolation of the Wine et al. expression to one atm at 2000 K would give an effective second-order rate coefficient of 4.7×10^7 $cm^3mol^{-1}s^{-1}$. This is considerably smaller than values obtained from flame modeling and would eliminate the OH + SO_2 reaction as a significant reaction pathway for catalytic recombination in ordinary flames. The fitting procedure of Wine et al., however, used an estimated HO–SO_2 bond dissociation energy of 151 kJ/mol, which is now known to be too large. Accurate thermochemical data on $HOSO_2$ would allow this data to be refit and provide an estimate for k_{19} in the 200 to 1000 K regime.

The OH + SO_2 reaction may be significant in sulfur-catalyzed radical recombination in flames and in the inhibition of the hydrogen-oxygen second explosion limit by SO_2. The reaction of the $HOSO_2$ adduct with oxygen provides a mechanism for converting SO_2 to SO_3 that could be important in lean combustion environments. More accurate kinetic data for appropriate third bodies at high temperatures and accurate thermodynamic data for $HOSO_2$ are needed to evaluate this possibility.

$H + H_2S \longrightarrow H_2 + SH$ (10)

This reaction is important in sulfur combustion under rich conditions. The data prior to 1976 were reviewed by Baulch et al. (1976). A recent study by Yoshimura et al. (1992) suggests that it displays non-Arrhenius behavior. Arrhenius extrapolation of low-temperature data obtained using flash photolysis techniques at 293 K (Yoshimura et al. 1992) and over the temperature ranges from 190 to 464 K (Kurylo et al. 1971) and 243 to 368 K (von Mihelic and Schindler 1970) agrees with measurements obtained from shock-tube experiments from 1965 to 2560 K (Roth et al. 1982). Results on the thermal decomposition of H_2S from 808 to 937 K by Pratt and Rodgers (1977), however, and the high temperature shock tube studies of Yoshimura et al. (1992) suggest upward curvature in the Arrhenius plot. Figure 3.1 shows the Arrhenius expressions of Yoshimura et al., Kurylo et al., Roth et al., and Pratt and Rodgers. The Baulch et al. review recommended the Kurylo et al. expression, and the study of Roth et al. appeared to provide confirmation that this was one of the few sulfur reactions which displays Arrhenius behavior over a wide temperature range. As a consequence the expression has been widely used in subsequent studies of sulfur combustion. The work of Yoshimura et al. combined shock tube experiments utilizing laser flash photolysis and atomic resonance absorption spectroscopy with a transition state theory analysis based upon ab initio characterization of the

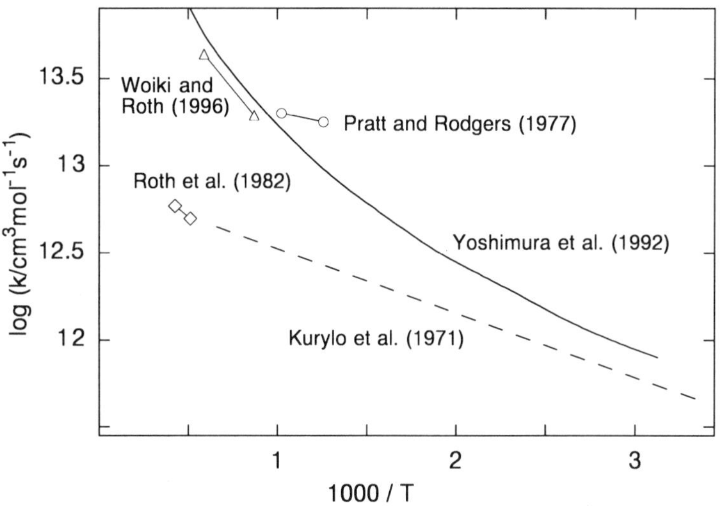

FIGURE 3.1. Rate coefficients for H + H_2S from the Arrhenius expressions of several laboratories. The expression of Kurylo et al. (1971) is an extrapolation of an expression based on data taken between 190 and 460 K.

transition state. The agreement they found between experiment and theory and the good agreement with the study of Pratt and Rodgers and the result of Woiki and Roth (1996) for the temperature range 1200 to 1700 K lead us to recommend the expression of Yoshimura et al. Further confirmation of this result, especially in the 500 to 1500 K temperature range, would be valuable.

$O + H_2S \longrightarrow$ Products (21)

Relatively good agreement exists on the low-temperature rate coefficient of this reaction; the products are assumed to be OH and SH. Shock tube studies indicate that channels leading to SO and H_2 and HSO and H may be important at higher temperatures (Frenklach et al. 1981).

$OH + H_2S \longrightarrow SH + H_2O$ (22)

This reaction is important in both the atmospheric and combustion chemistry of sulfur and has been well studied at low temperatures, where it was found to be fast and to have little or no activation energy. Early work by Westenberg and de Haas (1973) had indicated a substantial positive activation energy, but most subsequent studies are not consistent with this. Nevertheless, the Arrhenius expression they derived has been widely used for extrapolation to higher temperatures. Although the rate coefficient is well established at low temperature, there have been no mechanistic studies and it is unclear whether the reaction proceeds by abstraction or by addition. Detailed studies of the temperature dependence, kinetic isotope effect, and reaction products in the 200 to 1000 K range would be useful in clarifying the mechanism. As noted by Tyndall and Ravishankara (1991), it has not been demonstrated that SH and H_2O are the products.

$$H + CH_3SH \longrightarrow \text{Products} \qquad (27)$$

Our recommended rate coefficient is the average of three direct studies (Wine et al. 1986; Martin et al. 1988; Vaghjiani and Ravishankara, 1990). In the low pressure study of Martin et al. both product channels were observed, i.e., $CH_3S + H_2$ and CH_3+ H_2S, indicating that, as in the case of H + CH_3SCH_3, the reaction mechanism involves addition of H to the sulfur atom followed by adduct decomposition to products.

$$O + CH_3SH \longrightarrow \text{Products} \qquad (28)$$

Four studies of this reaction between 250 and 660 K are in agreement and show a strongly curved Arrhenius plot (Singleton and Cvetanovic 1988). The results were interpreted in terms of addition to the sulfur atom followed by rapid decomposition of the adduct.

$$OH + CH_3SH \longrightarrow CH_3S + H_2O \qquad (29)$$

At low temperature this reaction appears to proceed via addition to the sulfur atom followed by rapid decomposition of the adduct to products (Wine et al. 1984a). There is no evidence for decomposition of the thermalized adduct back to reactants in a manner analogous to the $CH_3S(OH)CH_3$ adduct. Arrhenius analysis of the rate coefficients obtained in the low temperature studies indicates a negative activation energy. At high temperatures an abstraction channel, probably involving the weaker S–H bond, would be expected to become significant, and hence a curved Arrhenius plot is possible.

$$H + CH_3SCH_3 \longrightarrow \text{Products} \qquad (30)$$

The one direct and one relative rate study of this reaction are in agreement (Lee et al. 1981; Yokota and Strausz 1979); the rate coefficient recommended in Table 3.1 is the average of the two measurements. The product determination in the relative rate study was consistent with reaction proceeding by H atom addition to the sulfur followed by decomposition to CH_3SH and CH_3.

$$O + CH_3SCH_3 \longrightarrow \text{Products} \qquad (31)$$

This reaction has been well studied at low temperature. The results are consistent with reaction proceeding by addition of the oxygen to the sulfur atom followed by decomposition to CH_3SO and CH_3(Singleton and Cvetanovic 1988). The reaction has a small negative activation energy. Extrapolations to higher temperatures should be viewed with caution.

$OH + CH_3SCH_3 \longrightarrow$ Products (32–34)

This is an important reaction in the atmospheric sulfur cycle. Hynes et al. (1986) showed that atmospheric oxidation proceeds via a two-channel mechanism in which addition and abstraction routes are both important.

$$OH + CH_3SCH_3 \longrightarrow CH_3SCH_2 + H_2O \tag{32}$$
$$OH + CH_3SCH_3 + M \longrightarrow CH_3S(OH)CH_3 + M \tag{33}$$
$$O_2 + CH_3S(OH)CH_3 \longrightarrow \text{products}. \tag{34}$$

They directly measured rate coefficients for reaction 32 and, based on the assumed mechanism, derived rates for (33), (–33) and (34). Although unable to observe adduct formation directly, they could estimate an adduct binding energy. In subsequent work Hynes et al. (1993, 1995) were able to observe adduct formation directly, confirming the mechanism proposed earlier and deriving an adduct binding energy of 59 ± 12 kJ/mol. Barone et al. (1997) also observed adduct formation directly, deriving an adduct binding energy of 43 ± 6 kJ/mol. As in the case of CS_2OH, the addition route is only important at low temperature. The observed rate coefficient is 3.7×10^{12} cm^{3}mol^{-1}s^{-1}at 300 K in one atm of air. The rate coefficient for the reaction of the $(CH_3)_2SOH$ adduct with oxygen is 5.4×10^{11} cm^{3}mol^{-1}s^{-1}at 300 K. Extrapolation of the Arrhenius parameters for a variety of sulfides to high temperatures suggests they react rapidly with OH via hydrogen abstraction with a rate coefficient of about 10^{13} cm^{3}mol^{-1}s^{-1}at 2000 K.

$O + CS_2 \longrightarrow$ Products (36)

This reaction has been extensively studied over the temperature range from 220 to 1000 K and shows simple Arrhenius dependence with a positive activation energy (DeMore et al. 1992; Atkinson et al. 1992). Three channels have been observed in low-temperature studies:

$$O + CS_2 \longrightarrow CS + SO \tag{36a}$$
$$\longrightarrow COS + S \tag{36b}$$
$$\longrightarrow CO + S_2. \tag{36c}$$

with the simple abstraction channel (36a) accounting for 90% of the reaction. Cooper and Hershberger (1992) were able to assign branching ratios by time-resolved IR diode laser absorption-spectroscopic detection of the CO and COS products of the minor channels.

$OH + CS_2 \longrightarrow$ Products (37)

The initial step in the atmospheric degradation of CS_2 (the low-temperature oxidation route) is the addition of OH to CS_2 to form an adduct that then reacts further with O_2 to form products.

$$OH + CS_2 \longrightarrow CS_2OH \quad (38)$$
$$CS_2OH + O_2 \longrightarrow COS + SO_2 + H. \quad (39)$$

Hynes et al. (1988) observed this equilibration process directly using a pulsed laser photolysis–pulsed laser induced fluorescence technique. By measuring the forward and reverse rates as a function of temperature they obtained the temperature dependence of the equilibrium constant and hence the binding energy of the adduct, 41 $\pm$ 3 kJ/mol. From an analysis of the variation in the effective rate of the reaction as a function of oxygen pressure they were able to determine the rate coefficient of the adduct's reaction with O_2. The observed second-order rate coefficient for the reaction of OH with CS_2 in 700 Torr of air was found to be 9×10^{11} cm^{3}mol^{-1}s^{-1}at 300 K and 8.9×10^{10} cm^{3}mol^{-1}s^{-1}at 350 K. The rate coefficient for the reaction of the CS_2OH adduct with oxygen was found to be $k_{39} = 1.7 \times 10^{10}$ cm^{3}mol^{-1}s^{-1}at 300 K. Subsequent studies (Murrells et al. 1990; Diau and Lee 1991a,b) confirmed these observations. The stability of the adduct is so weak, however, that this route shuts down at high temperatures. Any reaction would then have to occur via a bimolecular route, which appears to be slow (Friedl 1984; Murrells et al. 1990). Stickel et al. (1993) have shown that COS and SO_2 are the major products of reaction 39 and that CO is also produced. Details of the overall mechanism remain uncertain.

3.3.2 Sulfur radical reactions

Our knowledge on the elementary reactions of sulfur radicals ranks as SH > SO > CS > NS, with no kinetic data at all reported for NS. Several shock tube studies of S atom reactions using atomic resonance absorption spectroscopy (ARAS) or laser flash photolysis in conjunction with ARAS have been reported.

$S + O_2 \longrightarrow O + SO$ (5)

This reaction has been well studied at low temperature, and the temperature-independent rate coefficient 1.4×10^{12} cm^{3}mol^{-1}s^{-1} has been recommended for the temperatures range from 400 to 230 K. Several high-temperature studies suggest that the rate coefficient rises steeply above 1200 K. Saito et al. (1986) monitored O-atom concentrations in COS–O_2–Ar shock waves and reported $k_5 = 3.7 \times 10^{12}$ cm^{3}mol^{-1}s^{-1} for the temperature range from 1900 to 2300 K. Two later studies (Woiki and Roth 1995; Miyoshi et al. 1996) suggest an acceleration in the rate coefficient starting at about 1000 K, reaching the value of Saito et al. near 1500 K. Miyoshi et al. showed that in the pressure range of their experiments O atoms are produced in unit yield at high and low temperatures and found no evidence of recombination to form SO_2. The reason for the extreme curvature in the Arrhenius plot is unclear. The temperature dependence of the preexponential factor predicted by transition state theory is insufficient to explain the extreme curvature. Miyoshi et al. speculate that multiple channels on a variety of excited surfaces exist. If these surfaces have substantial barriers they will not affect the low-temperature rate but will become increasingly important at elevated temperatures.

$H + SH \longrightarrow S + H_2$ (8)

Woiki and Roth (1995) reported a high-temperature study of the reverse reaction (−8) using shock-initiated pyrolysis of COS–H_2–Ar mixtures and laser photolysis of shock-heated CS_2–H_2–Ar mixtures, monitoring S atoms by ARAS in both cases. The two approaches gave concordant results and led to an Arrhenius expression with an activation energy of 100 kJ/mol, about 20 kJ/mol above the reaction enthalpy, suggesting a small barrier to reaction.

$SH + O_2 \longrightarrow$ Products (14)

This reaction is important in lean combustion and potentially important in atmospheric chemistry. The upper limit of 6×10^5 cm^{3}mol^{-1}s^{-1}established for the rate coefficient by Stachnik and Molina (1987) is still high enough to make it the dominant sink for SH under clean tropospheric conditions. Arrhenius expressions for two reaction channels have been inferred from a shock-tube study (Frenklach et al. 1981):

$$SH + O_2 \longrightarrow OH + SO \qquad (14a)$$

$$\longrightarrow H + SO_2 \qquad (14b)$$

Arrhenius expressions were derived by modeling a complex mechanism, and, as discussed below, using a rate coefficient for reaction (10) much slower than given by the Table 3.1 expression. The authors noted that their expression for k_{10} gave slower values than other experimentally derived expressions and cautioned that using faster rate coefficient expressions would require changes in the expressions they derived for k_{14a} and k_{14b}. Extrapolation to room temperature gives $k_{14a} = 5.4 \times 10^5$ cm^{3}mol^{-1}s^{-1} with a large uncertainty. It does suggest, however, that the reaction could be important in the atmosphere and that measurements over the temperature range 300 to 1000 K would be useful in establishing an appropriate atmospheric rate coefficient.

$OH + SO \longrightarrow H + SO_2$ (17)

Jourdain et al. (1979) found a large rate coefficient for this reaction at room temperature, $k_{23} = 5 \times 10^{13}$ cm^{3}mol^{-1}s^{-1}. It was reported by Muller et al. (1979) to be balanced in rich flames. Their equilibrium constant for this reaction (derived from JANAF data) at 2000 K suggests that the reverse reaction is about two orders of magnitude faster than recombination to form HSO_2. Zachariah and Smith (1987), however, found this fast rate to be inconsistent with their observations of SO:SO_2 ratios in rich flames at temperatures between 1450 and 1800 K.

$S + COS \longrightarrow CO + S_2 \quad (40)$

This reaction may interfere with inferences drawn in shock tube experiments in which COS is used as a precursor of S atoms. Roth and coworkers have studied this reaction in shock heated COS–Ar mixtures using thermal decomposition (Woiki and Roth 1992) and laser photolysis (Woiki et al. 1993) methods, monitoring S atoms by ARAS in both cases. The higher-temperature studies, which used decomposition as a source of S atoms, gave results in agreement with the lower-temperature flash photolysis studies. This is the only high-temperature study of this reaction.

$SO + N \longrightarrow S + NO \quad (42)$

Jacob and Winkler (1970) reported the only low-temperature study of this reaction. Three later shock-tube studies of the reverse reaction have been reported. Oya et al. (1987, 1991) used recombination emission to monitor S-atom decay in either H_2S–NO–Ar and COS–NO–Ar mixtures shock-heated to temperatures from 1800 to 5000 K. Woiki and Roth used ARAS to measure S, O, and N atom profiles in COS–NO–Ar mixtures shock heated to temperatures from 2400 to 3900 K. Both the Woiki and Roth and the 1991 Oya et al. study gave a similar activation energy, although there is a factor of about two difference in the absolute rate coefficient expressions. We recommend the Arrhenius expression reported by Woiki and Roth, whose N atom measurements allowed them to identify reaction (−42) as the main channel, with a small contribution from reaction (−44).

3.3.3 Sulfuric acid formation

It has long been recognized that while SO_3 is indeed the anhydride of sulfuric acid, the chemical reaction

$$SO_3 + H_2O + M \longrightarrow H_2SO_4 + M \qquad (26)$$

is very slow in the gas phase. It is nonetheless thought to be the elementary reaction primarily responsible for conversion of sulfur trioxide to sulfuric acid in both the troposphere and the stratosphere. It has been the subject of several experimental studies, which have been reviewed by Jayne et al. (1997). The more recent studies (Lovejoy et al. 1996; Jayne et al. 1997) indicate that it has a second-order dependence on water vapor concentration, a strong negative temperature dependence, and a significant isotope effect. Over the pressure range from 80 to 760 Torr it shows no pressure dependence.

The observed behavior would appear to be consistent with formation of an $SO_3{\cdot}H_2O$ adduct that subsequently reacts with another water molecule, or, alternatively, reaction of SO_3 with a water dimer. Uncertainties in the energetics of adduct and dimer formation preclude distinguishing between the two possibilities. The rate parameters in Table 3.1, from Lovejoy et al., give the effective SO_3 first-order loss rate as a function of temperature and water vapor concentration.

They are appropriate for the temperature range from 250 to 370 K and must be used with caution outside this range. The rate coefficient expression given by Lovejoy et al. is in full agreement with the later study of Jayne et al. Theoretical investigations by Hofman and Schleyer (1994) and Morokuma and Murgurama (1994) confirm that a large energy barrier must be overcome in the isomerization of the H_2OSO_3 adduct to sulfuric acid. As a consequence, the simple termolecular process is very slow at atmospheric temperatures. Conversely, at combustion temperatures the concentrations of the weakly bound adduct and the water dimer would be very small, and a simple termolecular mechanism could provide the major route to sulfuric acid formation.

3.4 BASIC CHEMISTRY OF SULFUR IN COMBUSTION ENVIRONMENTS

3.4.1 Hydrogen–oxygen flames

Much of our current understanding of sulfur reactions in hydrogen–oxygen flames, that is, in absence of carbon and nitrogen, has not changed since the 1972 review of Cullis and Mulcahy. One of the characteristics of hydrogen–oxygen flames is the generation of superequilibrium radical concentrations early in the flame (Bulewicz et al. 1956). The fast chain branching and propagation reactions

$$\begin{aligned} O + H_2 &\longrightarrow OH + H \\ H + O_2 &\longrightarrow OH + O \\ OH + H_2 &\longrightarrow H_2O + H \end{aligned}$$

produce radical concentrations that can be an order of magnitude greater than those predicted from equilibrium considerations. This set of three reactions running their course leads to the radicals attaining a partially equilibrated state such that in a rich flame H and OH are related by

$$K = \frac{[H][H_2O]}{[H_2][OH]} .$$

Because K, $[H_2]$, and $[H_2O]$ are essentially constant throughout the post-flame gas, the OH and H-atom concentrations have a constant ratio; as this equilibration is established on a microsecond time scale, anything that affects one radical will also be reflected in the concentration of the other. The radicals decay to their equilibrium concentrations through slow termolecular reactions such as

$$\begin{aligned} H + H + M &\longrightarrow H_2 + M \\ O + O + M &\longrightarrow O_2 + M \\ OH + H + M &\longrightarrow H_2O + M , \end{aligned}$$

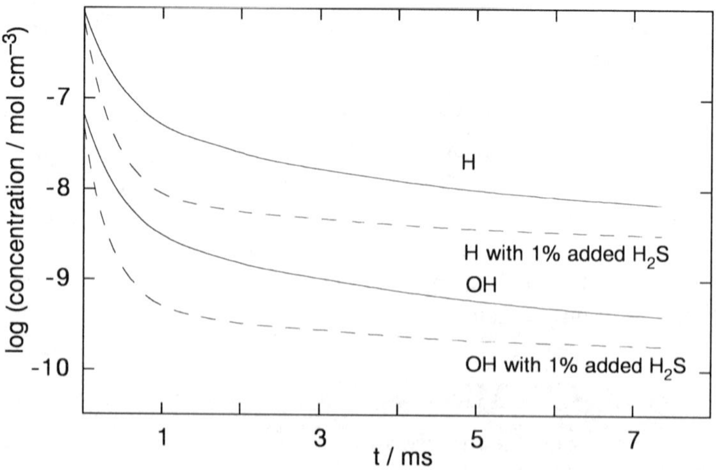

FIGURE 3.2. Calculated profiles of H and OH in a gas mixture initially consisting of H_2 and O_2 in a molar ratio of 4:1 at 1760 K. Profiles are also shown for the same mixture containing 1% added H_2S.

Because of this partial equilibrium, a process which speeds up any of these recombination reactions brings the whole radical pool into equilibrium more rapidly. The essential features of this chemistry are apparent even in a simple model of the H_2–O_2 system. Figure 3.2 shows concentration profiles for H and OH in a gas mixture initially consisting of H_2 and O_2 in a molar ratio of 4:1 at 1760 K, the adiabatic flame temperature of a 4:1:6 H_2:O_2:N_2 flame, calculated using the ACUCHEM program (Braun et al. 1988) with a reaction scheme comprising the above 6 hydrogen–oxygen reactions. The scheme included both forward and reverse reactions and used the rate coefficient expressions of Baulch et al. (1992). It can be seen that the radical profiles greatly overshoot their equilibrium values at the end of a brief induction period. At this temperature, the time to reach partial equilibrium is approximately 10 μs, while full equilibrium is approached on a timescale of ms. Figure 3.2 also shows calculated profiles that demonstrate the effects of sulfur on radical recombination discussed below.

A series of reactions involving sulfur species analogous to reactions in the H_2–O_2 system is possible (Sugden et al. 1962; Demerdache and Sugden 1963):

$$H_2S + H \longrightarrow SH + H_2 \tag{10}$$

$$SH + H \longrightarrow S + H_2 \tag{8}$$

$$SO + SH \longrightarrow S_2 + OH \tag{15}$$

$$S + H_2O \longrightarrow SO + H_2 \tag{49}$$

$$SO + H_2O \longrightarrow SO_2 + H_2 . \tag{50}$$

While the four-center reactions (15, 49, and 50) would appear to be slow ones, these authors note that because of the rapid reactions interconverting the hydrogen and oxygen species, the same interconversions cannot be distinguished from summations of alternate radical-molecule reactions. (For example, reaction 50 is the sum of $SO + OH \longrightarrow SO_2 + H$ [17] and $H + H_2O \longrightarrow OH + H_2$.)

The fact that fast bimolecular processes like these rapidly interconvert sulfur species means that observed distributions are a result of their interactions with the superequilibrium radical distributions. This explains the observation that in rich flames, where reduced sulfur might be expected to be dominant, the superequilibrium H atom concentrations depress SH, S_2, and H_2S concentrations to levels below their equilibrium values, and SO_2 and SO make up most of the sulfur mass balance. It also explains the observation, first noted by Fenimore and Jones in 1965, that fuel SO_2 and H_2S profiles are essentially indistinguishable in these flames.

The catalytic effects of sulfur addition are illustrated in Figure 3.3, taken from the work of Muller et al. (1980), which shows a series of concentration profiles for OH, obtained using laser induced fluorescence, in an $H_2{:}O_2{:}N_2 = 4{:}1{:}6$ flame doped with 0, 0.25, 0.5 and 1% H_2S. The enhanced rate of radical recombination in the presence of H_2S is clear. McAndrew and Wheeler (1962) first observed that additions of small amounts of SO_2 to a propane–air flame strongly accelerated radical recombination and assumed that the effect was due to enhanced three body recombination of OH with SO_2,

$$OH + OH + SO_2 \longrightarrow H_2O + SO_3\,. \qquad (51)$$

Webster and Walsh (1965) examined the inhibition of the second explosion limit of hydrogen–oxygen mixtures and proposed that reaction 18 was primarily responsible,

$$H + SO_2 + M \longrightarrow HSO_2 + M\,. \qquad (18)$$

Fenimore and Jones (1965) suggested that reaction 18 was responsible for sulfur-catalyzed radical recombination via the reaction sequence

$$H + SO_2 + M \longrightarrow HSO_2 + M \qquad (18)$$
$$H + HSO_2 \longrightarrow SO_2 + H_2 \qquad (23)$$
$$OH + HSO_2 \longrightarrow SO_2 + H_2O\,. \qquad (24)$$

Subsequent studies appear to be consistent with these observations and proposals, and the catalytic cycle of reactions 18, 23, and 24 has been widely accepted as the explanation (Cullis and Mulcahy, 1972). Wheeler (1968) suggested that the OH + SO_2 recombination reaction (19) could be important, but this suggestion has been largely discounted as OH concentrations are significantly lower than those of H atoms in rich flames.

Three more recent studies have examined sulfur in fuel-rich flames and, although the observed behavior is qualitatively consistent with these mechanisms, HSO_2 has not been directly observed, and some contradictory results have been obtained. Kallend (1972) measured temperature and H-atom concentrations in

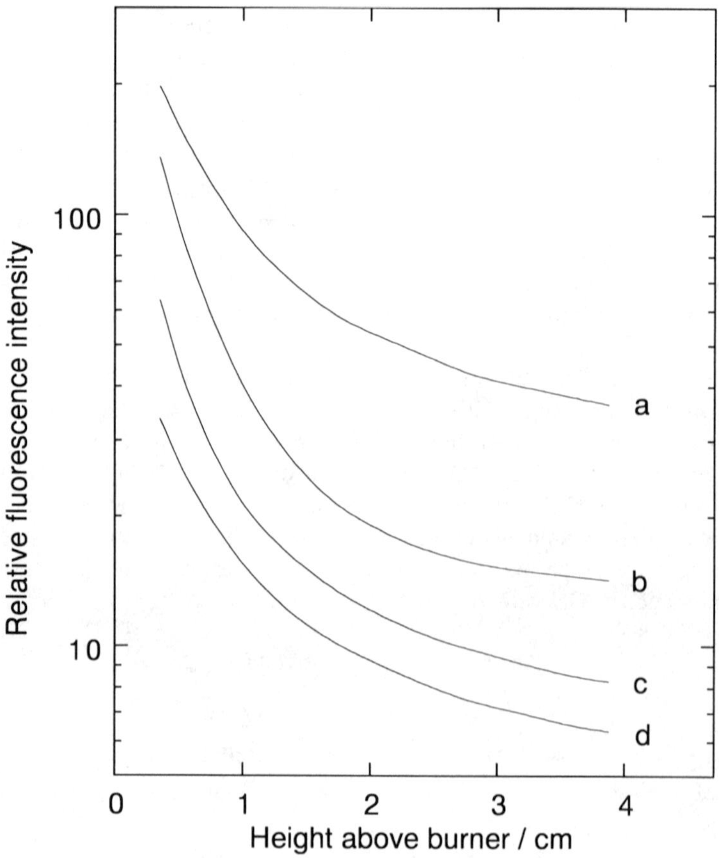

FIGURE 3.3. OH concentration profiles above a flat flame burner for an H_2:O_2:N_2= 4:1:5 flame doped with (a) 0%, (b) 0.25%, (c) 0.5%, and (d) 1.0% H_2S. The data, taken from the work of Muller et al. (1980), were obtained using laser induced fluorescence to monitor OH concentrations.

rich H_2–O_2–N_2 flames. From a detailed kinetic analysis he concluded that reaction 18 was the rate-limiting step and inferred its rate coefficient from the measured H and calculated SO_2 profiles. He also concluded that reaction 18 should be the rate determining step in the catalytic cycle under most conditions and that reaction 24 is unimportant in rich flames.

Muller et al. (1979, 1980) used laser-induced fluorescence (LIF) to monitor OH, SH, S_2, SO, and SO_2 in a series of rich flames over the temperature range 1700 to 2300 K. Their approach to determining absolute concentrations, discussed below, allowed them to test a variety of equilibration schemes. From the available kinetic data they concluded that equilibration of the following reactions would explain the observed distribution of sulfur-containing species:

$$SH + H \longrightarrow S + H_2 \qquad (8)$$

$$SH + SH \longrightarrow S + H_2S \tag{9}$$
$$H_2S + H \longrightarrow SH + H_2 \tag{10}$$
$$S + OH \longrightarrow H + SO \tag{11}$$
$$SH + S \longrightarrow H + S_2 \tag{12}$$
$$SH + O \longrightarrow SO + H \tag{13}$$
$$OH + SO \longrightarrow SO_2 + H \tag{17}$$
$$OH + H_2S \longrightarrow H_2O + SH \tag{22}$$

From a qualitative analysis of available kinetic data they concluded that reactions 12 and 17 play a dominant role in controlling the distribution of SO_2. Although flame profiles were obtained for many of the flame radical species, no attempts to model flame profiles quantitatively were reported. The measured profiles were used to test for equilibration of the summary reactions:

$$S_2 + H_2 \longrightarrow SH + SH \tag{I}$$
$$SH + OH \longrightarrow SO + H_2 \tag{II}$$
$$H_2 + SO_2 \longrightarrow SO + H_2O \tag{III}$$
$$SO_2 + 2\,H_2 \longrightarrow SH + OH + H_2O \tag{IV}$$

The equilibrium constant expressions as a function of temperature were then compared on van't Hoff plots with the behavior implied by currently accepted thermochemistry. In essence this tests for equilibration of different sets of elementary reactions; for example, observation of equilibration in I would be consistent with equilibration of several elementary reaction sequences, such as 8 and 12, or 9, 10 and 12. The results of this analysis for equilibria I and III are shown in Figures 3.4 and 3.5. The agreement is good for I and II, somewhat poorer for IV, and much poorer for III particularly at low temperature. Muller et al. (1980) considered the deviations in III and IV to be due to non-optimal measuring conditions and the difficult data reduction for SO_2, which has a complex laser-induced fluorescence spectrum.

Kramlich et al. (1981) studied H_2S production as a function of equivalence ratio for H_2–O_2–N_2–H_2S mixtures at 0.9 atm and 1700 K in a jet stirred reactor with probe sampling. Although SO_2 proved to be the dominant sulfur species, the H_2S levels exceeded calculated ones by two orders of magnitude in rich mixtures. As the measured levels agreed with SO levels predicted from an equilibrated sulfur system, they concluded that SO was reduced to H_2S in the sample probe.

Zachariah and Smith (1987) used molecular-beam mass spectrometric sampling to study a series of low pressure rich H_2–O_2–N_2–SO_2 flames. A specific objective of this work was direct observation of HSO_2. Profiles of the major flame species, H_2, O_2, and H_2O together with SO_2, SO, S_2, H_2S, and SH were obtained and compared with profiles calculated using a one-dimensional flame simulation program. Reasonable qualitative agreement was obtained for SO_2, SO, and

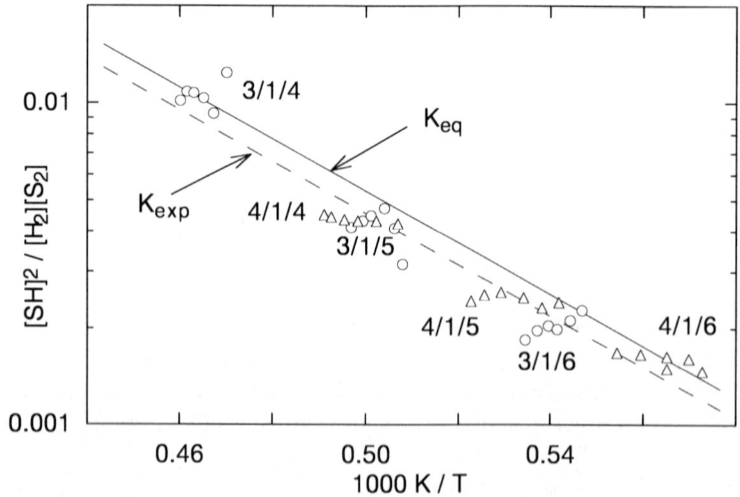

FIGURE 3.4. Test of equilibration of $S_2 + H_2 \rightleftharpoons SH + SH$ (I), taken from the work of Muller et al. (1979).

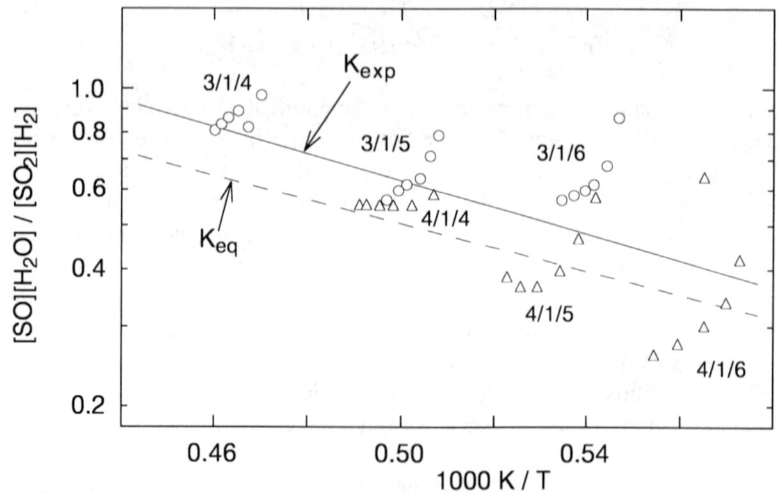

FIGURE 3.5. Test of equilibration of $SO_2 + H_2 \rightleftharpoons SO + H_2O$ (III), taken from the work of Muller et al. (1979).

S_2 profiles, but the agreement for H_2S and SH was poor. The authors noted that H_2S concentrations are sensitive to radical concentrations and concluded that the discrepancy reflected the difficulty in modeling the catalytic effects of HSO_2. It should be noted, however, that the rate constants used for reactions 10 and 22, which control H_2S consumption, may be significantly in error. Direct experimental observation of HSO_2 was not possible, which the authors concluded was a result of interferences from other species of the same mass. They also concluded that several of the fast sulfur radical reactions were not in partial

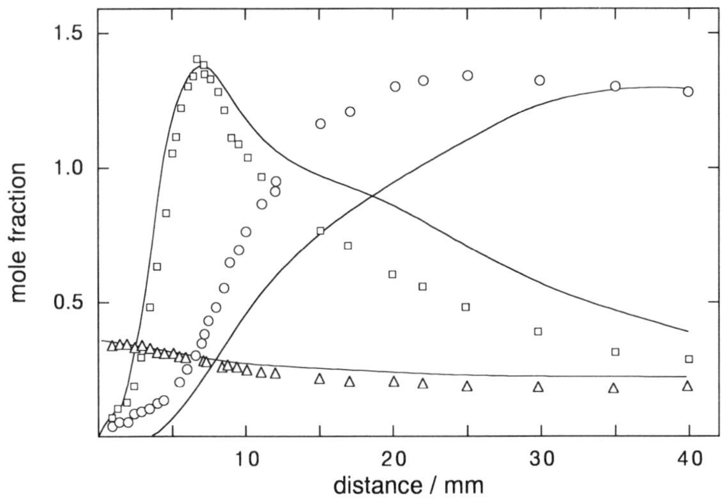

FIGURE 3.6. Calculated and experimental profiles for SO_2, SO, and S_2 above a low-pressure (150 Torr) H_2–O_2–Ar flame doped with SO_2. The data, taken from the work of Zachariah and Smith (1987), were obtained with a molecular beam-mass spectrometer sampling system. Triangles $SO_2 \times 10$, squares SO $\times$ 400, circles $S_2 \times 300$.

equilibrium. As noted above, partial equilibrium considerations imply that the concentration of H-atoms is proportional to the concentration of OH radicals. A balanced reaction (17) implies

$$K_{17} = \frac{[\mathrm{H}][\mathrm{SO_2}]}{[\mathrm{OH}][\mathrm{SO}]},$$

and therefore, at constant temperature, the [SO]:[SO_2] ratio should also remain constant in the post-flame gases. Figure 3.6, taken from Zachariah and Smith (1987), shows, however, that the [SO]:[SO_2] ratio varies dramatically through the flame, indicating that the above relation does not hold if the reported concentration profiles are accurate. Zachariah and Smith attribute this discrepancy to the difference in temperature between their flames and those of Muller et al. (1979). They note that at low temperature, the test of balance for equilibrium of reaction III, as shown in Figure 3.5, shows poor agreement with the assumption of an equilibrated reaction 17. To obtain agreement between their observed and calculated downstream drop in SO concentrations, Zachariah and Smith had to use a rather slow rate coefficient for reaction 17, 1.8×10^{13} cm^{3}mol^{-1}s^{-1}; the 298 K rate coefficient for this reaction has been reported to be 5.2×10^{13} cm^{3}mol^{-1}s^{-1}(Jourdain et al. 1979). They considered the competition between reactions 18 and 23 and concluded that reaction 23 is the rate determining step, which would appear to contradict the results of Kallend (1972). They also concluded that sulfur chemistry perturbs the H_2–O_2 partial equilibrium.

The qualitative features of sulfur addition can be illustrated using relatively simple chemistry. This is illustrated in Figures 3.2 and 3.7, which show

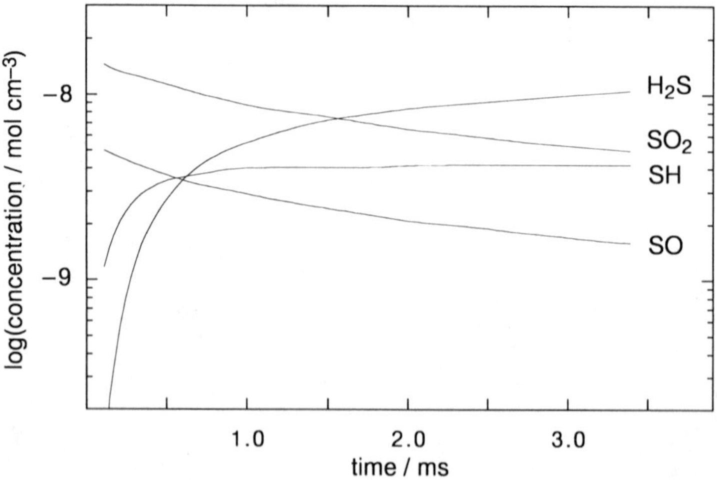

FIGURE 3.7. Calculated profiles of H_2S, SO_2, SH and SO in a gas mixture initially consisting of H_2 and O_2 in a molar ratio of 4:1 containing 1% added H_2S at 1760 K.

concentration profiles for a gas mixture initially consisting of H_2 and O_2 in molar ratio 4:1 with the addition of 1% H_2S at 1760 K. The reaction scheme consisted of the H_2–O_2 reactions together with the forward and reverses of reactions 8–13, 17, 18, and 23, the scheme suggested by Muller et al. (1979), whose estimated rate coefficients were used for all reactions except reaction 18. For reaction 18, the initial step in the catalytic recombination cycle, we used the faster rate coefficient reported by Fenimore and Jones (1965).

In this simple model the effectiveness of sulfur in catalyzing recombination is sensitive to the forward and reverse rates of reaction 18. Using the fast forward rate coefficient of Fenimore and Jones and the equilibrium constant of Halstead and Jenkins (1968), we calculated the OH- and H-atom profiles shown in Figure 3.2. The enhanced radical recombination rate is qualitatively consistent with experiment. Reaction 18 is indeed the rate-determining step in recombination, as concluded by Kallend (1972). It was also possible to obtain similar enhancements in the rate of radical recombination by eliminating reaction 18 and including instead fast OH + SO_2 recombination by reaction 1.

Figureure 3.7 shows calculated profiles of the four sulfur species H_2S, SO_2, SH, and SO, which qualitatively reproduce the experimental profiles of Muller et al. (1980). In fact the calculated concentrations are actually in quite reasonable quantitative agreement, given the simple "isothermal box model" chemistry, which takes no account of transport or adiabatic flow. Although sulfur is added as H_2S, as noted above, it is rapidly converted to SO_2. The observed qualitative agreement does not necessarily indicate that we understand the detailed kinetic mechanism, but rather that thermodynamics is dominant in determining species distributions. In this model sulfur addition never perturbs the H_2–O_2 partial equilibrium, and the sulfur chemistry is itself rapidly equilibrated.

3.4.2 Hydrocarbon flames

Given that our understanding of the simplest hydrogen–oxygen flames is quite incomplete, it is not surprising that the effects of sulfur addition to hydrocarbon flames are less well studied and less well understood. Fenimore and Jones (1965) found that the superequilibrium radical overshoots which characterize H_2–O_2 and H_2–O_2–CO flames are not present in hydrocarbon flames, so the effects of added sulfur might be thought to be less dramatic. However, the observation that, as with H_2–O_2 flames, fuel H_2S and SO_2 are indistinguishable, leads again to the conclusion that H_2S is rapidly converted to SO_2 in the flame front (Johnson et al. 1970).

Since the review of Cullis and Mulcahy (1972) only a few flame investigations of sulfur chemistry have appeared. Kramlich et al. (1981) studied a variety of hydrocarbon flames doped with H_2S, SO_2, and C_4H_4S (thiophene). Measurements were performed in a jet-stirred reactor at 0.9 atm at temperatures between 1600 and 1800 K and equivalence ratios between 0.7 and 1.7. OH, SH, and CS were measured in situ by absorption spectroscopy. Stable products were sampled with a probe and analyzed by gas chromatography. SO_2 was found to be the dominant sulfur product at all equivalence ratios. Measurements of H_2S were perturbed by probe reactions which, they concluded, could generate H_2S, COS and CS_2 from SO under rich conditions. CS_2 and COS found in the richer flames was attributed to reactions of sulfur radicals with unburnt hydrocarbons in the probe to generate CS that then reacted further to generate CS_2 and COS:

$$\mathrm{SO} + \mathrm{C_XH_Y} \longrightarrow \mathrm{CS} + \mathrm{products} \longrightarrow \mathrm{COS\ or\ CS_2}.$$

Evidence for the initial reaction was obtained from spectroscopic observation of CS in the reactor as the proportion of unburned hydrocarbons increased. Concentrations of CS increased dramatically with equivalence ratio for equivalence ratios greater than 1.2. For the H_2–O_2–N_2–H_2S flames discussed above Kramlich et al. found that partial equilibrium based on H–O–S chemistry described the observed concentrations after correction for H_2S production in the probe. They concluded the same for the hydrocarbon flames, again based on the assumption that SO was converted to H_2S, COS, and CS_2. SO_3 yields were found to vary from 3% of the total sulfur in lean flames to undetectable amounts at equivalence ratios greater than 1.2.

Three studies have appeared on sulfur effects on methanol combustion. Pauwels et al. (1986) studied low pressure stoichiometric methanol–air flames doped with H_2S. They used probe sampling coupled with ESR detection and corrected for loss of labile species via reaction and destruction on probe walls. They were able to spatially resolve H_2S loss and formation of SO_2 and SO. Elevated levels of H_2 in the sulfur-doped flame were attributed to reaction 10. They examined the departure of radical concentrations from partial and thermodynamic equilibrium and found significant departures from the latter. In particular, O- and H-atom concentrations were almost two orders of magnitude greater than expected in an equilibrated flame. The deviations from partial equilibrium increased as the percentage of H_2S in the mixture increased, essentially as a result of H and OH radical concentrations approached equilibrium values while O-atom concentrations remained unchanged.

In a similar study Taniguchi et al. (1987) used probe sampling coupled with ESR to study a stoichiometric, atmospheric pressure, methanol–air flame doped with CS_2. Addition of 0.46% CS_2 to the flame produced elevated levels of H_2 within the inner flame front and depressed the concentrations of OH, H, and O. They attributed the elevated H_2 levels to the decreased OH level reducing the rate of OH the H_2-consuming reaction OH + H_2 in the inner flame front. Although radical concentrations measured in the undoped flame were in good agreement with those calculated from partial equilibrium considerations, the concentration profiles in the doped flame were significantly lower than expected from partial equilibrium considerations. They considered their findings to be generally in accord with those of Pauwels et al. (1986).

Pauwels et al. (1992) reported further work on low-pressure methanol–air flames doped with H_2S, again using probe sampling and ESR detection. They studied three flames, all slightly rich and doped with 0, 1.2, and 2.4% H_2S, and computed flame profiles using the SANDIA flame code (Kee et al. 1985) with a reaction mechanism that did not include catalytic recombination via the reaction sequences initiated by reactions 18 or 19. They claimed good agreement between observations and model predictions for CH_3OH, O_2, H_2, H_2O, H_2S, SO_2, and SO, but the experimental and calculated profiles they show for SO do not support this conclusion. For CO, H, O, and OH they acknowledge that the predicted profiles are higher than the experimentally determined profiles, with particularly large discrepancies for H atom. They investigated the effects of the introduction of the catalytic recombination cycle involving HSO_2, that is, reaction sequence 18, 23, 24, and found that it did not improve the agreement between modeled and experimental H-atom profiles. A global sensitivity analysis to determine which reactions had the greatest effect on sulfur chemistry suggested that for their flames reactions 10 and 22 were most important. As noted above for the work of Zachariah and Smith (1987), the rate coefficients used for reactions 10 and 22 may be significantly in error.

Smith et al. (1983) investigated the effects of sulfur addition to a CO–O_2–Ar flame using mass spectrometric sampling to measure concentration profiles for O, SO_3, CO, CO_2, and O_2. At temperatures over 1500 K they found reactions 2 and 3 to be equilibrated and that addition of 1.2% SO_2 eliminated O-atom overshoot in their flames. These observations can be immediately understood as SO_2-catalyzed recombination of O atoms, because the sum of reactions 2 and 3 is just $O + O \longrightarrow O_2$.

Glarborg et al. (1996) examined the effects of SO_2 addition on the moist oxidation of CO in an isothermal flow reactor. They found that SO_2 inhibits CO oxidation, particularly for conditions favoring high O-atom concentrations, which can again be interpreted as catalytic recombination of O atoms through reactions 2 and 3. SO_2:SO_3 ratios computed using a detailed kinetic model agreed with the ones observed experimentally.

3.4.3 Sulfur–nitrogen interactions

A significant development in high-temperature studies of sulfur oxidation has been the attempt to understand sulfur–nitrogen interactions in flames. Since both SO_2 and NO_X contribute to atmospheric pollution, the synergistic effects of fuel sulfur on NO_X need to be understood in order to optimize pollution abatement strategies. For example, if the presence of fuel sulfur results in significant NO_X reduction, then an abatement strategy based on the removal of SO_2 from stack gases could be more cost effective than fuel desulfurization. The issue is complicated by the fact that at least three mechanisms of NO_X formation are important in flames (Chapters 1 and 2).

The early work on sulfur–nitrogen interactions emerged from the laboratory of J.O.L. Wendt at the University of Arizona. It was based on probe sampling of atmospheric-pressure flames. Profiles of stable species were obtained by chemiluminescence or gas chromatographic techniques. Wendt and Ekmann (1975) appear to have been the first to report that fuel sulfur can inhibit NO_X formation in flames. They studied the effects of H_2S and SO_2 addition to atmospheric pressure methane–air flames. NO and SO_2 profiles were obtained from probe sampling and chemiluminescence analysis. They were able to distinguish between Thermal and Prompt NO and found that addition of 5% SO_2 significantly reduced Thermal NO and led to a small decrease in Prompt NO. They found that H_2S was also effective in depressing NO levels in lean flames. The most dramatic reductions in NO were in slightly lean flames. The effect appears to be attributable to catalysis by sulfur of radical recombination lowering the peak O atom concentration.

Subsequently, Wendt et al. (1979) studied the influence of sulfur on fuel nitrogen by adding H_2S and SO_2 to methane–oxygen–argon flames doped with cyanogen (C_2N_2). They found both enhancement and inhibition of fuel NO production, depending on equivalence ratio and residence time. In typical flame profiles, sulfur addition produced a dramatic increase in NO close to the flame front, after which the NO profiles decayed to levels near the undoped ones. They suggested that sulfur addition enhances prompt NO by direct interaction of sulfur and nitrogen radicals via reactions such as

$$N + SO \longrightarrow NO + S \tag{42}$$

and that this effect increases with temperature. At the same time, sulfur-catalyzed recombination of the major flame radicals can reduce thermal NO formation. In an effort to simplify the flame chemistry, Wendt et al. (1983) studied moist, rich $CO–Ar–O_2$ flames doped with NO, C_2N_2, or NH_3 as fuel nitrogen and SO_2 as fuel sulfur. CO was chosen as a fuel because they considered its kinetics to be well understood and the reactions involved form a subset of those involved in hydrocarbon oxidation. Both the NO and N_2 formed in the flame were measured to test the nitrogen mass balance. From an analysis of a kinetic model of the flame they concluded that the only important reaction producing N_2 was the reverse

Zeldovich reaction:

$$N + NO \longrightarrow N_2 + O\,.$$

Because the effect of adding 1% fuel SO_2 to these flames was a small lowering of NO and enhancement of N_2 they concluded that SO_2 addition must enhance N atom levels. They added to their model a series of reactions to account for the effects of sulfur addition, including the catalytic recombination reactions and the direct interactions

$$N + SO \longrightarrow NO + S \tag{42}$$

$$NO + S \longrightarrow NS + O \tag{-44}$$

$$NS + O \longrightarrow SO + N. \tag{45}$$

They concluded that the direct interactions, rather than radical recombination effects, were responsible for the modifications of the NO and N_2 profiles.

Corley and Wendt (1984) extended this work to three rich CH_4–O_2–He flames doped with C_2H_2 and SO_2. The effects of sulfur addition on NO emission varied drastically as the equivalence ratio ϕ increased from 1.44 to 2.18. At the lowest equivalence ratio, 1.44, addition of 4% SO_2 had almost no effect. At $\phi = 1.71$ addition of 0.5% SO_2 produced a doubling in NO emissions combined with a significantly different flame profile, while at $\phi = 2.18$ addition of 0.9% SO_2 produced a dramatic reduction in NO. Addition of SO_2 increased HCN concentrations in all three flames examined. The increased complexity of the chemistry made it difficult to establish the role of competing chemical mechanisms. Catalytic recombination of flame radicals, interaction of sulfur with the cyanide and amine subsystem, or direct radical interactions via reactions 42, 44 and 45 or

$$CN + SO \longrightarrow NCO + S \tag{43}$$

$$NH + SO \longrightarrow HNO + S \tag{-41}$$

could play a role in the observed behavior. The authors noted that monitoring the concentrations of unstable nitrogenous species is needed to understand these interactions more fully.

In addition to the studies of Wendt et al., the problem of sulfur–nitrogen interactions has been studied in the laboratories of O.I. Smith at UCLA and P.C. Malte at the University of Washington. Tseregounis and Smith (1983) studied the effects of SO_2 addition on atmospheric pressure C_2H_2–O_2–Ar and H_2–O_2–Ar flames doped with C_2N_2. They measured temperatures and NO profiles by probe sampling and chemiluminescence analysis and found that sulfur addition produced a large increase in the concentration of fuel NO in the flame front, while downstream the profiles remained flat or increased slightly. Addition of up to 0.5% SO_2 led to approximately proportional NO enhancement. They concluded that a direct interaction such as reaction 42 could not be responsible for the observed increase, based on the assumption that the rate coefficient for reaction 42 is independent of temperature. Tseregounis and Smith later (1984) extended their work to low-pressure H_2–O_2–Ar flames using probe sampling with molecular beam mass spectrometric detection, which enabled them to obtain profiles of reactants, products, stable intermediates, and the major flame radicals. They studied three flames doped with C_2H_2 and measured the effects of adding of 0.5% SO_2 to each flame. Sulfur addition again resulted in enhancement of NO formation, although the effect was less pronounced than in atmospheric flames, and had little effect on measured HCN profiles at short residence times. They concluded that rapidly formed NO decays in the postflame gases through

$$H + NO + M \longrightarrow HNO + M$$

and that the enhancing effect of sulfur is to lower H-atom concentrations via catalytic recombination, making HNO formation less effective in removing NO.

Chen et al. (1984) used an atmospheric-pressure jet-stirred reactor to study hydrogen and propane flames of varying equivalence ratio doped with NO or CH_3CN as fuel nitrogen and SO_2 as fuel sulfur. Concentrations were measured by probe sampling or optical absorption spectroscopy. In the hydrogen flames, SO_2 addition produced an increase in NO concentration consistent with catalytic recombination reducing H atom concentrations and inhibiting H + NO + M. Direct radical interactions were not thought to be significant. In the propane flames addition of 1% SO_2 increased the NO concentration of rich flames and produced a slight decrease in flames with ϕ less than 1.2. Similarly, HCN and NH_3 levels increased in rich flames upon SO_2 addition. Catalytic recombination and consequent suppression of OH and H concentrations were again thought to be most important in explaining the observed behavior; direct interactions between nitrogen and sulfur radicals were considered to be "supporting instead of major".

Direct interactions of sulfur and nitrogen radicals were observed by Jeffries and Crosley (1986), who found a laser-induced fluorescence signal from NS in the flame front of a variety of methane–oxidant flames. The laser excitation spectrum provides strong, although not unambiguous, evidence for the presence of NS in the flame.[1] They examined a series of rich flames of methane with nitrous oxide and oxygen oxidants. As the detection system could not be calibrated directly, they derived, with an estimated accuracy of a factor of 3, absolute NS concentrations from known and estimated spectroscopic and kinetic parameters. In a rich CH_4–N_2O flame doped with H_2S or SF_6 the NS concentration was about 0.001 of the amount of added sulfur, which ranged from 0.01 to 5%. Addition of

[1] LIF at short wavelengths is susceptible to photochemical perturbations in which the exciting laser pulse photolyses a parent molecule to produce a photofragment that undergoes LIF. Photofragment LIF signals generally show a quadratic dependence on laser power; a true LIF signal is linearly dependent on laser power at low excitation powers and saturates at high powers. At the low excitation powers used by Jeffries and Crosley the possibility of photochemical perturbation is remote.

NH_3 lowered the NS concentration. In a similar flame using commercial natural gas, the 2 ppm CH_3SH dopant added by the gas supplier was sufficient to produce an observable NS signal. Similar results were obtained in a $\phi = 1.2$ CH_4–O_2 flame. In a $\phi = 2.1$ CH_4–air flame, which was presumably considerably cooler than the other flames, addition of H_2S and NH_3 again resulted in an NS signal; however, the concentration was estimated to be 0.02% of the added H_2S, considerably lower than in the hotter flames. Jeffries and Crosley demonstrated that significant NS concentrations were only observable in the flame front, suggesting that NS is removed rapidly; this implies that the production mechanism is rapid also. Because the observed NS concentrations represent only steady-state levels, the fraction of the initial sulfur dopant processed through NS may be much greater than its steady-state fraction. As their studies were done with a torch burner rather than a flat flame burner, conclusions about the flame chemistry producing NS could not be drawn even though SO_2, NO, and S_2 were also observed. Nonetheless, this important study clearly establishes the existence of direct sulfur–nitrogen interactions, at least in rich hydrocarbon flames.

Pfefferle and Churchill (1989) included direct sulfur–nitrogen interaction chemistry in their interpretation of fuel sulfur effects on NO production in a thermally stabilized plug-flow burner. They examined the effects of H_2S addition to ethane–air flames doped with ammonia by monitoring NO emissions as a function of equivalence ratio using a chemiluminescence analyzer. A small drop in NO emissions was found to result from doping ethane–air flames at equivalence ratios between 0.8 and 1.6 with 0.03 and 0.06 weight percent H_2S. These 5 to 10% reductions were attributed to reductions in thermal NO production as a result of catalytic recombination of flame radicals. Larger NO reductions were observed in flames doped with fuel nitrogen, where addition of H_2S produced reductions of 10 to 20%, the reduction increasing with increased sulfur doping. However, the reduction was proportionally less at higher dopant levels. They also examined the NO concentration as a function of residence time for a stoichiometric flame doped with 0.14 weight percent ammonia with and without 0.03 weight percent H_2S. The experimental results were compared with a detailed reaction model which included the reactions proposed by Wendt et al. (1983). The model underestimated the fuel NO reduction produced by H_2S doping. Addition of eight radical–radical reactions involving sulfur–nitrogen and sulfur–carbon interactions dramatically improved the agreement between the calculated and the observed NO reductions. Most of the improvement resulted from including reactions involving the NS radical:

$$\mathrm{SH + NH \longrightarrow NS + H_2} \tag{38}$$

$$\mathrm{NS + NO \longrightarrow N_2 + SO} \tag{39}$$

$$\mathrm{NO + CS \longrightarrow NS + CO.} \tag{40}$$

In the absence of experimental data, the above reactions were assigned rate coefficients of 9×10^{12} $cm^3 mol^{-1} s^{-1}$, rather high values for four center reactions.

Hampartsoumian and Nimo (1995) examined the effects of sulfur addition to liquid fuel spray flames that simulated the conditions of utility burners.

Although the behavior depended in a complex way on burner operating conditions, sulfur:nitrogen ratio, and fuel type, sulfur addition consistently resulted in increased NO_x emissions.

3.4.4 Sodium–sulfur interactions

Sodium and sulfur in flames contribute to corrosion of industrial combustion systems, as discussed by Cullis and Mulcahy (1972), who concluded that when sulfur is present in an excess of sodium, sodium sulfate formed in the gas phase condenses onto cool surfaces. Subsequent work, however, has not confirmed this view.

Stearns et al. (1977) used mass spectrometric sampling from a lean CH_4–O_2 flame heavily doped with sodium and reported direct observation of Na_2SO_4, $NaSO_2$, and $NaSO_3$ in the gas phase. Steinberg and Schofield (1990) and Schofield and Steinberg (1992) reported measurements of OH and Na concentrations in a series of lean and rich flames doped with SO_2. A partial equilibrium analysis developed from the earlier work of Muller et al. (1979, 1980) was used to calculate the concentration profiles of the sulfur radicals in the flame, while the sodium chemistry was based on previous work in their laboratory (Hynes et al. 1984, Steinberg and Schofield 1987). A comprehensive thermochemical analysis was carried out to estimate the rate coefficients of a series of 43 reactions involving the various sodium and sulfur species that could potentially be present in these flames; it was concluded that $NaSO_2$ is the only significant one in lean flames, but still constitutes only a small fraction of the total sodium. The levels of sodium depletion predicted by the model agreed with the experimentally observed depletion produced by addition of 1% and 2.76% SO_2 to a lean flame. In rich flames their model predicted that NaOS is the dominant molecular species but that again it is only a small fraction of the total sodium present. In this case the reported model and experimental sodium profiles are in poorer agreement. The authors concluded that Na_2SO_4 formation in the gas phase is unlikely and suggested that the results of Stearns et al. (1977) may be due to heterogeneous chemistry at the mass spectrometer sampling orifice. In situ diagnostic techniques for detecting gas-phase sodium–sulfur species are needed for further clarification of these interactions.

3.4.5 Sulfur reaction studies in shock tubes

Shock tubes provide better resolution of species profiles and more closely controlled reaction conditions than can be obtained in laboratory flames. They are therefore the main approach one has for recording kinetics data at higher temperatures than can be obtained in flow reactors. Unfortunately, little shock-tube work on sulfur-containing molecules has been reported other than the studies discussed in Section 3.3, where we noted a number of examples where elementary reaction rate coefficients were inferred from experiments in which a shock tube was used to generate hot test gas and laser flash photolysis or other methods were used to generate reactive atoms.

Conventional shock-tube studies of combustion reactions entail observation of the onset of ignition in a test gas, usually with optical diagnostics, and then fitting the experimental data using a complex kinetic model. Since the 1972 Cullis and Mulcahy review, three such studies of CS_2 and H_2S oxidation have appeared.

Hardy and Gardiner (1976) studied the shock-initiated onset of ignition in CS_2–O_2–Ar mixtures, monitoring chemiluminescence from electronically excited SO_2 at 490 nm and infrared emission from CS_2 and CO at 5 μm. They were able to account for the observed ignition delays, but found no model that was able to reproduce the observed chemiluminescence and infrared emission profiles. They identified high-temperature reactions of CS as a particular uncertainty in their mechanism and concluded that further studies are needed. Saito et al. (1986) studied O-atom production in shock-heated CS_2–O_2–Ar mixtures and interpreted the results using a complex mechanism.

Chernysheva et al. (1992) developed a 51-reaction mechanism for CS_2 oxidation, of which 20 reactions were considered to to have non-zero reverse reaction rate coefficients. Rate coefficients were estimated for the 23 reactions for which they found no information in the literature. The mechanism was used to interpret a number of experimental studies, with approximate agreement to the data being found in all cases. The experiments included a room-temperature, 1.4 Torr flow system study of a CS_2–O-atom flame reported by Slagle et al. (1974); a stirred-reactor study of CS_2–O_2 mixtures of variable composition at 573 K and 3 Torr reported by Kondratiev (1940); another stirred-reactor study of CS_2–O_2 mixtures at 623 K and 3 Torr reported by Sarkisyan et al. (1968), in which O-atom and SO-concentrations were measured; a flow reactor study of SO_2 reduction by CO at 1215 K and 50 Torr reported by Arutyunov et al. (1990); ignition delays in an 8% COS, 12% O_2 mixture in argon investigated by Lifshitz et al. (1975); and an O-atom profile in a 1615 K, 1 atm reflected shock wave experiment reported by Saito et al. (1986). Basevich et al. (1994) later used this mechanism to model the CS_2–air flames studied by Egerton and Sen (1953), obtaining reasonable agreement with the flame speeds but flame thicknesses about half of the experimental values.

Frenklach et al. (1981) studied ignition delays in moist and dry H_2S/air mixtures at temperatures between 950 and 1200 K. They obtained expressions for the temperature dependence of the induction time for the dry and moist mixtures. For the supporting modeling study they combined one of their own data sets with earlier data from Bradley and Dobson (1967). To model the observed ignition delays they included the following reactions in their mechanism:

$$H_2S + O \longrightarrow SO + H_2 \tag{21b}$$

$$H_2S + O \longrightarrow HSO + H \tag{21c}$$

$$HS + O_2 \longrightarrow SO + OH \tag{14a}$$

$$HS + O_2 \longrightarrow SO_2 + H \tag{14b}$$

$$HS + HO_2 \longrightarrow H_2S + O_2 . \tag{16}$$

They were able to model the experimentally observed delays and reported the corresponding Arrhenius parameters for these five reactions. Their model, however, did not include potential catalytic radical recombination cycles initiated by reactions 18 and 19, and it incorporated a temperature-independent rate coefficient 2.3×10^{11} $cm^3mol^{-1}s^{-1}$ for reaction (10), factors of 70 at 1050 K and 370 at 2000 K smaller than given by the recommended k_{10} expression in Table 3.1. The authors noted that their value for this rate coefficient was smaller than other experimental expressions and cautioned that using a larger one would require changes in their Arrhenius expressions for reactions 14a and 14b.

Chernysheva et al. (1990) developed an elaborate mechanism for the H_2S–O_2 system by estimating many of the needed rate coefficients. It comprised 44 H–O reactions, 2 S reactions, 27 H–S reactions, 25 S–O reactions, and 103 H–S–O reactions. It was used to interpret a number of previous experimental studies, including the ignition delay experiments of Frenklach et al. (1981), data on H_2S–O_2 explosion limits, and conversion of H_2S to sulfur in the Claus process (Tesner et al. 1984). Basevich et al. (1994) used the same mechanism to model data on two H_2S–O_2 flames reported by Levy and Merryman (1965). The agreement between computation and experiment was fair at best in all cases.

3.5 THERMOCHEMISTRY OF SULFUR-CONTAINING COMPOUNDS

The thermochemistry of sulfur-containing free radicals important in combustion and atmospheric chemistry is a subject of active research interest. Experiments including molecular beam studies (Nourbakhsh et al. 1990, 1991a,b; Continetti et al. 1991; Balucani et al. 1993), photoionization mass spectrometry studies (Ruscic and Berkowitz 1992, 1993), and Br + RH $\rightleftharpoons$ R + HBr (R = HS, CH_3S, CH_3SCH_2) kinetics studies (Nicovich et al. 1992; Jefferson et al. 1994), and theoretical studies (Fournier and DePristo 1992; Nobes and Radom 1992; Curtiss et al. 1992) have significantly improved our understanding of the thermochemistry of the sulfur radicals HS, CH_3S, CH_2SH, CH_3SCH_2, HCS, and HSO. Xantheas and Dunning (1993) report that HSO is 23 kJ/mol more stable than HOS; previous calculations predicted HOS to be the more stable species (Hinchliffe 1980; Luke and McLean 1985; Moore Plummer 1990).

Recommended standard enthalpies of formation at 298 K for over 100 sulfur species of interest in combustion are given in Table 3.2. As can be seen from perusal of the table, uncertainties in the heats of formation of a number of key oxygenated sulfur species remain unacceptably high. Examples include HOS, HSO_2, HOSO, $HOSO_2$, H_2SO_4(g), CH_3SO, CH_3SO_2, CH_3SO_3, CH_3SOH, $CH_3S(O)OH$, and CH_3SO_3H. Further experimental and theoretical research aimed at improving the accuracy of the thermochemistry of the above species should be a high priority for understanding sulfur oxidation.

We have attempted to make Table 3.2 comprehensive and thereby included many species not discussed in this chapter. A large number of halogenated

TABLE 3.2a. Standard Molar Enthalpies of Formation of Sulfur Compounds of Interest in Combustion.[a]

Species	$\Delta_f H^\circ_{298}$	Notes	Species	$\Delta_f H^\circ_{298}$	Notes
S	277.0 ± 0.3	1	S_2	128.6 ± 0.3	1
S_3	142 ± 8	1, 2	S_4	146 ± 8	1, 2
S_5	109 ± 8	1, 2	S_6	102 ± 8	1, 2
S_7	114 ± 8	1, 2	S_8	100.4 ± 0.6	1, 2
SO	5.0 ± 1.3	1	S_2O	-56 ± 34	1
SO_2	-296.8 ± 0.2	1	SO_3	-395.8 ± 0.7	1
SH	143 ± 3	3	HS_2	$27 \pm [20]$	4
H_2S	-20.5 ± 0.8	1	HSSH	$16 \pm [15]$	5
HSO	-4 ± 3	4, 6–9	HOS	$18 \pm [15]$	8
HSO_2	-54 ± 15	10	HOSO	-188 ± 15	10
$HOSO_2$	$-385 \pm [10]$	11, 12	H_2SO_4(g)	-735.1 ± 8.4	1
H_2SO_4(l)	-814.0 ± 0.7	1			
CS	280 ± 25	1	CS_2	117 ± 1	13
COS	-138.4 ± 0.5	13			
HCS	$295 \pm [10]$	14, 15	H_2CS	$115 \pm [10]$	14–16
CH_3S	125 ± 2	17	CH_2SH	214 ± 9	18
CH_3SH	-22.9 ± 0.6	13	C_2H_5SH	-46.3 ± 0.6	13
C_6H_5SH	112.4 ± 0.8	13	$H_2C{=}C{=}S$	$165 \pm [15]$	13
CH_3SCH_3	-37.5 ± 0.5	13	CH_3SCH_2	135 ± 3	19, 20
CH_3SSCH_3	-24.2 ± 1.0	13	CH_3SSSCH_3	$11 \pm [10]$	13
CH_3SS	72 ± 5	5	CH_3SSS	86 ± 5	5
c-CH_2CH_2S	82.1 ± 1.2	1, 13	$CH_3SC_2H_5$	-59.6 ± 1.1	13
$(C_2H_5)_2S$	-84 ± 1	13	$(C_6H_5)_2S$	231 ± 3	13
C_4H_4S	115.0 ± 0.4	13, 21	C_4H_8S	-34.1 ± 0.9	13, 21
CH_3SO	$-62 \pm [15]$	5, 22	CH_3SOO	76 ± 4	23
CH_3SO_2	$-238 \pm [15]$	7	CH_3SO_3	$-350 \pm [15]$	22
$(CH_3)_2SO$	-151.3 ± 0.8	13	$(CH_3)_2SO_2$	-373 ± 3	13
c-CH_2CH_2SO	$-30 \pm [15]$	13	$(C_2H_5)_2SO$	-205.6 ± 1.5	13
$(C_6H_5)_2SO$	107 ± 3	13	$(CH_3O)_2SO$	-483 ± 2	13
$CH_3C(O)SH$	-175 ± 8	13	CH_3SOH	$-90 \pm [25]$	22
$CH_3S(O)OH$	$-360 \pm [25]$	22	CH_3SO_3H	$-567 \pm [25]$	22
SCSOH	$110 \pm [10]$	7	$(CH_3)_2SOH$	$60 \pm [10]$	24

TABLE 3.2b. Standard Molar Enthalpies of Formation of Sulfur Compounds of Interest in Combustion.[a] (Concluded.)

Species	$\Delta_f H^\circ_{298}$	Notes	Species	$\Delta_f H^\circ_{298}$	Notes
HSNO	94 ± [20]	12	NS	263 ± 105	1
HNCS	126 ± [15]	5	CH_3NCS	131 ± [15]	13
CH_3SCN	160 ± [15]	13	$(SCN)_2$	350 ± 6	13
$(NH_2)_2CS$	−25 ± [15]	5	$CH_3C(S)NH_2$	10 ± 1	13
FS	13.0 ± 6.3	1	SF_2	−297 ± 17	1
ClS	156 ± 17	1	S_2Cl	78.6 ± 8.4	1
Cl_2S	17.6 ± 3.3	1	Br_2S	21 ± 17	5
SF_6	−1220.5 ± 0.8	1	SF_5Cl	−1039 ± 11	1
SF_4Cl_2	−858 ± 13	5	SF_5	−908 ± 15	1
SF_4Cl	−741 ± [20]	5	SF_4	−763 ± 21	1
SF_3	−503 ± 34	1	HSI	42 ± 3	25
F_2CS	−350 ± [15]	13	Cl_2CS	−27 ± [15]	13
CH_3SCl	−28 ± 6	5	CH_3SI	30 ± 3	26
C_6H_5SCl	106 ± 6	5	CH_3SCH_2Cl	−90 ± [5]	1
CH_3SCH_2Cl	26 ± 5	26			
FSSF	−335 ± 42	5	ClSSCl	−16.7 ± 4	1
BrSSBr	35 ± [10]	5, 13	CH_3SSCl	−21 ± 6	5
C_6H_5SSCl	113 ± 6	5			
F_2SO	−544 ± 21	5	Cl_2SO	−213 ± [20]	5
Br_2SO	−107 ± [20]	5, 13	F_2SO_2	−758.6 ± 8.4	1
Cl_2SO_2	−354.8 ± 2.1	1	$FClSO_2$	−557 ± 21	1

a Units are kJ/mol; uncertainties in brackets are our estimates. 1, Chase et al. 1985; 2, Cyclic; 3, Continetti et al. 1991; 4, Luke and McLean 1985; 5, Benson 1978; 6, Lovejoy et al. 1987; 7, DeMore et al. 1992; 8, Xantheas and Dunning 1993; 9, Balucani et al. 1993; 10, Binns and Marshall 1991; 11, Margitan 1984; 12, Atkinson et al. 1992; 13, Lias et al. 1988; 14, Curtiss et al. 1992; 15, Ruscic and Berkowitz 1993; 16, Nobis and Radom 1992; 17, Nicovich et al. 1992; 18, Ruscic and Berkowitz 1992; 19, Nourbakhsh et al. 1991b; 20, Jefferson et al. 1994; 21, Heterocyclic; 22, Yin et al. 1990; 23, Turnipseed et al. 1992; 24, Hynes et al. 1993; 25, Hwang and Benson 1979; 26, Shum and Benson 1983.

compounds are included; while little kinetics research has been reported in this area, interactions between halogen-containing and sulfur-containing species could play significant roles in hazardous waste incineration. In atmospheric chemistry, gas phase sulfur–halogen interactions may be most significant in the marine boundary layer, where there are large oceanic sources of CH_3SCH_3 and CH_3X (X

= Cl, Br, or I) and because heterogeneous reactions on sea salt aerosol particles are thought to be significant sources of ClO_x and BrO_x radicals. (Singh 1995.) Reactions on the surfaces of sulfate aerosol particles are thought to be significant sources of ClO_x and BrO_x radicals in the lower stratosphere. (Tolbert 1994) The kinetics of hydrogen transfer reactions involving Br, HBr, and sulfur species have provided routes to thermochemical information on the sulfur-containing radicals SH, CH_3S, and CH_3SCH_2. (Nicovich et al. 1992; Jefferson et al. 1994)

3.6 OBSERVATIONS AND CONCLUSIONS

3.6.1 Elementary reactions

The paucity of high-temperature kinetics data on elementary reactions of sulfur is a substantial roadblock to understanding sulfur combustion and sulfur–nitrogen interactions. More high-temperature studies are needed for almost all of the reactions of sulfur species discussed in this review. The mechanistic complexity of many of the reactions is such that even quite detailed studies over the 200 to 1000 K range accessible to flash photolysis experiments may not be adequate. The development of techniques which utilize flash photolysis studies on high-temperature gas mixtures behind reflected shock waves is promising in this regard, as demonstrated by the work of Yoshimura et al. (1992) and Michael and coworkers (Michael and Lim 1993).

Certain reactions are pivotal in establishing a "baseline" for our understanding. Until the H–O–S system is well understood it will be difficult to have confidence in the interpretation of more complex systems involving hydrocarbons and/or sulfur–nitrogen interactions. Critical needs include determination of the rate coefficients for reactions 18 and 19 with H_2, O_2, and H_2O as third bodies. The observed effects on the catalytic recombination of flame radicals are beyond dispute. If the conclusions of Binns and Marshall (1991) are correct, however, the role of reaction 18 in radical inhibition at medium temperatures, that is, near 800 K, is seriously in question. High-quality rate data for the recombination reactions are required to assess their relative importance in catalytic recombination. Among the radical–radical interactions, reaction 17 is particularly important since it controls the SO:SO_2 ratio in flames. Given the potential importance of the NS radical in sulfur–nitrogen interactions in flames it is clear that understanding its elementary reactions is required. At present the data on NS rate coefficients are sparse even at room temperature.

3.6.2 High-temperature studies

An attempt to review the present status and future prospects of high-temperature experiments on sulfur chemistry divides itself into two areas, diagnostics on the one hand and kinetics and modeling on the other.

Diagnostic techniques

Unfortunately, many of the techniques used in sulfur chemistry studies are highly intrusive and call into question the validity of the raw data obtained. Simply stated, probe techniques cannot be used to sample sulfur species reliably. To understand and model "probe chemistry" is as challenging as understanding and modeling the system being probed. Heterogeneous effects are notorious in sulfur chemistry and at present impossible to quantify and model. Molecular beam sampling, while clearly intrusive, does offer a chance of sampling wide ranges of free radical and stable species. The possibilities of heterogeneous chemistry at the nozzle, chemistry and clustering occurring during expansion through the nozzle, and the fact that weakly bound intermediates may dissociate during passage through the nozzle and skimmers or upon ionization, however, require that such experiments be independently validated.

Optical diagnostics are nonintrusive and provide unambiguous detection of stable and free radical species. The development of the tunable dye laser and the laser-induced fluorescence (LIF) technique have had impacts both on low-temperature reaction kinetics and on combustion diagnostics.

LIF is typically more sensitive than absorption spectroscopy but more difficult to calibrate in order to obtain absolute concentrations, primarily because of excited state quenching effects. To determine a rate coefficient for the reaction of a reactive radical with a stable molecule one ideally measures the reaction rate under pseudo-first order conditions with a high and therefore constant concentration of the stable molecule and a low radical concentration. The rate coefficient determination then requires knowledge only of the absolute concentration of the stable molecule and the first-order rate of loss of the radical, for which one needs only relative rather than absolute concentrations. In this respect LIF is an ideal technique and corrections are usually not required for quenching effects, because the quencher concentrations are also contant.

The requirements for flame diagnostics are rather different. Radical concentrations are typically large, radical–radical interactions are important, and absolute concentrations are required. In addition, the quenching environment can change drastically from flame to flame and, particularly for turbulent combustion, even within the same flame. Hence, acquisition of even relative concentration profiles for a series of flames or within a single turbulent flame is not trivial.

In LIF studies on sulfur-doped flames two approaches to the determination of absolute concentrations have been tested. Muller et al. (1979) used narrow bandwidth detection of fluorescence provided by a monochromator with narrow slits. By exciting and monitoring a single rovibrational transition they were able to assume that the quenching rate could be taken as being equal to the gas kinetic collision rate. Absolute concentrations were obtained by referencing the signals to those obtained in a part of a flame thought to be at equilibrium, for which all concentrations can be calculated. They validated this approach by using it to obtain H-atom and OH-radical concentration profiles in a series of rich flames and found that the inferred equilibrium constant for $OH + H_2 \rightleftharpoons H_2O + H$ agreed with the value computed from JANAF (Chase et al. 1985) thermochemistry. They used a similar detection approach for S_2, SH, SO, and SO_2, and, as discussed earlier, after testing various equilibration schemes concluded that the sulfur chemistry was balanced and that the observation of balance validated their technique. This strategy is most directly applicable to hydrides, which have a well-spaced rotational spectrum that allows excitation and monitoring of a single rotational transition. For a molecule such as SO_2, excited in a continuum that is not well defined spectroscopically, the validity of the approach is more questionable.

Jeffries and Crosley (1986), in their study of NS, attempted to calculate an absolute concentration from the observed fluorescence intensity, a complicated procedure requiring calibration of the optical collection system, calculation of excitation and fluorescence efficiencies, and estimating the detection volume. The combined measurement uncertainties lead to uncertainty in absolute concentrations of about a factor of three. The accuracy of this approach can be somewhat improved by referencing the fluorescence signal to a well characterized scattering process such as Rayleigh or Raman scattering, but that again is difficult because of polarization and the anisotropic nature of these processes.

While such difficulties in establishing absolute concentrations prevail, downstream relative concentration profiles in a flame are usually well defined by LIF, because the concentrations of the major quenching species N_2, H_2O, and H_2 are essentially constant in this region.

Two of the difficulties in evaluating the quality of the various experimental approaches are that there are no standard burners and flames that can be used to validate a technique and that most investigators shun attempting to reproduce results from other laboratories.

Much of the progress in elementary reaction kinetics has come as a result of the development of widely differing techniques for measurement of rate coefficients. Each new experimental approach has to be tested against rate coefficients that are known with reasonable accuracy. In our view there is a serious need for intercomparison of the two most promising diagnostic approaches, LIF and molecular beam sampling, on identical flames, ideally simultaneously, or at least on similar burners.

Kinetics and modeling studies

Modeling studies on sulfur chemistry under combustion conditions have been handicapped by the lack of a regularly upgraded, evaluated high temperature database similar to the NASA or CODATA compilations used for atmospheric modeling. The atmospheric compilations serve the research community by establishing readily available sources of data and references and provide a baseline allowing different models to be compared using the same data set. While reviews and evaluations of combustion rate coefficients are available, they typically focus only on a specific area and there is little uniformity in their use. This leads to a situation where experimental combustion chemistry data are fit using a variety of chemical models and a variety of rate coefficients. When discrepancies arise it is difficult if not impossible to pinpoint their sources. For similar reasons the uniform use of a reference flame code, such as that developed at Sandia National Laboratory (Kee et al. 1985), which computes flame profiles and also generates a sensitivity analysis, is desirable. An example of this approach is given by Kee et al. (1988).

Until the behavior of sulfur in the H_2–O_2 system is well understood little confidence can be placed in interpretation of results in more complex systems involving hydrocarbons or nitrogen–sulfur interactions. This is underscored by the remarkable fact that the two most detailed studies so far of the simplest sulfur doped flames draw opposing conclusions:

Muller et al. (1979) essentially concur with earlier studies that the system is well defined by partial equilibrium considerations and that measurement of one major flame radical, such as OH, allows the composition of the whole radical pool to be calculated. This approach makes no assumptions about the mechanism of catalytic recombination, because the one directly measured radical concentration allows the other radical concentrations to be calculated from partial equilibrium considerations. As discussed in Section 3.4.1, this approach has varying degrees of success and it is unclear whether problems, for instance in the test of equilibrium (III), lie in the measurement of SO_2 concentrations, or accurately reflect the fact that critical reactions no longer proceed sufficiently rapidly to maintain balanced chemistry. Its utility is nonetheless well demonstrated in the work of Schofield and Steinberg (1992) on sodium–sulfur interactions, where a limited number of measurements were sufficient to define the flame chemistry.

Zachariah and Smith (1987) attempted to compare experimental profiles with the predictions of a detailed flame code. Unfortunately, they were unable to measure either of the major flame radicals, OH or H, directly. Since the concentration profiles of the sulfur species are intimately linked to those of the major flame radicals, the success of this approach depends critically on the accuracy of the assumed mechanism of catalytic recombination. They indicate that discrepancies between their observed and calculated species profiles, particularly for H_2S, may reflect problems introduced by probe sampling. However, they may also reflect inadequacy of the basic mechanism. In their view neither SO nor S_2 is in partial equilibrium, and they conclude that in fact sulfur does perturb the pseudo-equilibrium of the H_2–O_2 system. This is not consistent with the work of Muller et al. at all, and it is unclear whether these results could reflect a transition to a regime in which radical distributions are controlled by kinetic rather than thermodynamic considerations.

Finally it seems appropriate to recall a basic principle of chemical kinetics which is often forgotten, apparently never more so than in the analysis of the effects of sulfur on catalytic recombination of radicals in combustion: Observations that demonstrate kinetic behavior of one or two species in good agreement with the predictions of a complex mechanism are only a confirmation of consistency. Such tests can only disprove, never prove, that a reaction mechanism is correct. Extensive direct observation of the profiles of stable and radical species using validated techniques, coupled with comparison against profiles calculated using flame models that have well validated rate coefficients, is the pathway to understanding sulfur combustion chemistry.

3.7 ACKNOWLEDGMENTS

The authors acknowledge support of their work on elementary reaction kinetics of sulfur species by the Atmospheric Chemistry Program of the National Science Foundation and the Office of Naval Research. A.J. Hynes acknowledges the hospitality of Professors D.L. Baulch and M.J. Pilling and the use of the High Temperature Kinetics Evaluation Facility at Leeds University, Leeds, UK, and the support of NSF-INT which made that use possible.

3.8 REFERENCES

Amano, A., Yamada, M., Hashimoto, K., & Sugiura, K. (1983). Nippon Kagaku Kaishi, 385.

Aneja, V.P. (1990). J. Air Waste Manage. Assoc., **40**, 469.

Aneja, V.P., & Cooper, W.J. (1989). In *Biogenic Sulfur in the Environment*, Saltzman and Cooper (Eds.), ACS Symp. Ser. No. **393**, pp. 2–13.

Arutyunov, V.S., Basevich, V.Ya., Vedeneev, V.I., Ushakov, V.A., & Chernysheva, A.V. (1990). Kinet. Katal., **31**, 1053.

Astholz, D.C., Glänzer, K., & Troe, J. (1979). J. Chem. Phys., **70**, 2409.

Atkinson, R., Baulch, D.L., Cox, R.A., Hampson, Jr., R.F., Kerr, J.A., & Troe, J. (1992). J. Phys. Chem. Ref. Data, **21**, 1125.

Attar, A., & Corcoran, W.H. (1977). Ind. Eng. Chem., Prod. Res. Dev., **16**, 168.

Balucani, N., Casavecchia, P., Stranges, D., & Volpi, G.G. (1993). Chem. Phys. Lett., **211**, 469.

Barone, S.B., Turnipseed, A.A., & Ravishankara, A.R. (1997). J. Chem. Phys. **100**, 14694.

Basevich, V.Ya., Vedeneev, V.I., & Arutyunov, V.S. (1994). Russian Journal of Chemical Physics, **13**, 137.

Bates, T.S., Lamb, B.K., Guenther, A., Dignon, J., & Stoiber, R.E. (1992). J. Atmos. Chem., **14**, 315.

Baulch, D.L., Drysdale, D.D., Duxbury, J., & Grant, S.J. (1976). Evaluated Kinetic Data for High Temperature Reactions, Vol. **3**, Butterworths, London.

Baulch, D.L., Cobos, C.J., Cox, R.A., Esser, C., Frank, P., Just, Th., Kerr, J.A., Pilling, M.J., Troe, J., Walker, R.W., & Warnatz, J. (1992). J. Phys. Chem. Ref. Data, **21**, 411.

Benson, S.W. (1978). Chem. Rev., **78**, 23.

Binns, D., & Marshall, P. (1991). J. Chem. Phys., **95**, 4940.

Bowman, C.T., & Dodge, L.G. (1977). 16th Symposium (International) on Combustion, p. 971.

Boyd, R.J., Gupta, A., Langler, R.F., Lownie, S.P., & Pincock, J.A. (1980). Can. J. Chem., **58**, 331.

Braun, W., Herron, J.T., & Kahaner, D. (1988). Int. J. Chem. Kinet., **20**, 51.

Bradley, J.N, & Dobson, D.C. (1967). J. Chem. Phys. **46**, 2872.

Bulewicz, E.M., James, C.G., & Sugden, T.M. (1956). Proc. Roy. Soc., **A235**, 89.

Charlson, R.J., Lovelock, J.E., Andreae, M.O., & Warren, S.G. (1987). Nature, **326**, 655.

Chase, Jr., M.W., Davies, C.A., Downey, Jr., J.R., Frurip, D.J., McDonald, R.A., & Syverud, A.N. (1985). J. Phys. Chem. Ref. Data, **14**, Supplement No. 1.

Chen, A.T., Malte, P.C., & Thornton, M.M. (1984). 20th Symposium (International) on Combustion, p. 769.

Chernysheva, A.V., Basevich, V.Ya., Vedeneev, V.I., & Arutyunov, V.S. (1990). Izv. Akad. Nauk SSSR, Ser. Khim., No. 9. 1956; English translation in Bull. Acad. Sci. USSR: Div. Chem. Sci. (1990), Consultants Bureau, New York, p. 1775.

Chernysheva, A.V., Basevich, V.Ya., Vedeneev, V.I., & Arutyunov, V.S. (1992). Izv. Akad. Nauk SSSR, Ser. Khim., No. 4. 812; English translation in Bull. Acad. Sci. USSR: Div. Chem. Sci. (1992), Consultants Bureau, New York, p. 629.

Continetti, R.E., Balko, B.A., & Lee, Y.T. (1991). Chem. Phys. Lett., **182**, 400.

Cooper, W.F., & Hershberger, J.F. (1992). J. Phys. Chem., **96**, 5405.

Corley, T.L., & Wendt, J.O.L. (1984). Combust. Flame,, **58**, 141.

Cullis, C.F., & Hirschler, M.M. (1980), Atmos. Environ., **14**, 1263.

Cullis, C.F., & Mulcahy, M.F.R. (1972). Combust. Flame, **18**, 225.

Cupitt, L.T., & Glass, G.P. (1975). Int. J. Chem. Kin., **7**, 39.

Curtiss, L.A., Nobes, R.H., Pople, J.A., & Radom, L. (1992). J. Chem. Phys., **97**, 6766.

Cvetanovic, R.J., Singleton, D.L., & Irwin, R.S. (1981). J. Am. Chem. Soc., **103**, 3530.

Demerdache, A., & Sugden, T.M. (1963). In *The Mechanism of Corrosion by Fuel Impurities,* H.R. Johnson & D.J. Littler, Eds., Butterworths, London, pp. 216–227.

DeMore, W.B., Sander, S.P., Golden, D.M., Hampson, R.F., Kurylo, M.J., Howard, C.J., Ravishankara, A.R., Kolb, C.E., & Molina, M.J. (1992). Chemical Kinetics and Photochemical Data for Use in Stratospheric Modeling, Evaluation No. 10, JPL Publication 92–20.

Diau, E.W.-G., & Lee, Y.-P. (1991a). J. Phys. Chem., **95**, 379.

Diau, E.W.-G., & Lee, Y.-P. (1991b). J. Phys. Chem., **95**, 7726.

Dougherty, E.P., & Rabitz, H. (1980). J. Chem. Phys., **72**, 6571.

Edgerton, A., & Sen, D. (1953). 5th Symposium (International) on Combustion, Reinhold Publishing Corp., New York, p. 321.

Fenimore, C.P., & Jones, G.W. (1965). J. Phys. Chem., **69**, 3593.

Fournier, R., & DePristo, A.E. (1992). J. Chem. Phys., **96**, 1183.

Frenklach, M., Lee, J.H., White, J.N., & Gardiner, W.C. (1981). Combust. Flame, **41**, 1.

Friedl, R.R. (1984). Spectroscopic & chemical kinetic studies of the SH and SD free radicals, dissertation, Harvard University.

Glarborg, P., Kubel, D., Dam-Johansen, K., Chiang, H.-M., & Bozzelli, J.W. (1996). Int. J. Chem. Kinet., **28**, 773.

Halstead, C.J., & Jenkins, D.R. (1968). Trans. Faraday Soc., **65**, 3013.

Hampartsoumian, E., & Nimmo, W. (1995). Combust. Sci. Technol., **110**, 487.

Hardy, J.E., & Gardiner, W.C. (1976). 16th Symposium (International) on Combustion, p. 373.

Heicklen, J. (1985). Rev. Chem. Int. , **6**, 175.

Hinchliffe, A. (1980). J. Mol. Struct., **66**, 235.

Hofman, M., & Schleyer, P. (1994). J. Am. Chem. Soc., **116**, 4947.

Hwang, R.J., & Benson, S.W. (1979). Int. J. Chem. Kinet., **11**, 579.

Hynes, A.J., Steinberg, M., & Schofield, K. (1984). J. Chem. Phys., **80**, 2585.

Hynes, A.J., Stickel, R.E., Pounds, A.J., Zhao, Z., McKay, T., Bradshaw, J.D., & Wine, P.H. (1993). In *Dimethylsulfide: Oceans, Atmosphere, and Climate*, Restelli and Angeletti, Eds., Kluwer Academic Publishers, Dordrecht, pp. 211–221.

Hynes, A.J., Stoker, R.B., Pounds, A.J., McKay, T., Bradshaw, J.D., Nicovich, J.M., & Wine, P.H. (1995). J. Phys. Chem., **99**, 16967.

Hynes, A.J., Wine, P.H., & Nicovich, J.M. (1988). J. Phys. Chem., **92**, 3846.

Hynes, A.J., Wine, P.H., & Semmes, D.H. (1986). J. Phys. Chem., **90**, 4148.

Jacob, A., & Winkler, C.A. (1970). Can. J. Chem., **48**, 1774.

Jayne, J.T., Poschl, U., Dai, D., Molina, L.T., Worsnop, D.R., Kolb, C.E., & Molina, M.J. (1997). J. Phys. Chem., **101**, 10000.

Jefferson, A., Nicovich, J.M., & Wine, P.H. (1994). J. Phys. Chem., **98**, 7128.

Jeffries, J.B., & Crosley, D.R. (1986). Combust. Flame, **64**, 55.

Jourdain, J.L., LeBras, G., & Combourieu, J. (1979). Int. J. Chem. Kinet., **11**, 569.

Kallend, A.S. (1972). Combust. Flame, **19**, 227.

Kee, R.J., Grcar, J.F., Smooke, M.D, & Miller, J.A. (1985). Report SAND85-8240, Sandia National Laboratories, Livermore CA.

Kee, R.J., Miller, J.A., Evans, G.H., & Dixon-Lewis, G. (1988). 13th Symposium (International) on Combustion, p. 1479.

Kiefer, J.H. (1975). J. Chem. Phys., **62**, 1354.

Klemm, R.B., & Davis, D.D. (1974). J. Phys. Chem., **78**, 1137.

Kondratiev, V.N. (1940). Zh. Fiz. Khim., **14**, 281.

Kramlich, J.C., Malte, P.C., & Grosshandler, W. (1981). 18th Symposium (International) on Combustion, p. 151.

Kurylo, M.J., Peterson, N.C., & Braun, W. (1971). J. Chem. Phys., **54**, 943.

Lee, J.H., Machen, R.C., Nava, D.F., & Stief, L.J. (1981). J. Chem. Phys., **74**, 2839.

Lias, S.G., Bartmess, J.E., Liebman, J.F., Holmes, J.L., Levin, R.D., & Mallard, W.G. (1988). J. Phys. Chem. Ref. Data, **17**, Supplement No. 1.

Lifshitz, A., Frenklach, M., Schechner, P., & Carroll, H.F. (1975). Int. J. Chem. Kinet., **7**, 753.

Lovejoy, E.R., Hanson, D.R., & Huey, G.L. (1996). J. Phys. Chem., **100**, 19911.

Lovejoy, E.R., Wang, N.S., & Howard, C.J. (1987), J. Phys. Chem., **91**, 5749.

Luke, B.T., & McLean, A.D. (1985). J. Phys. Chem., **89**, 4592.

Mallard, W.G., Westley, F., Herron, J.T., & Hampson, R.F. (1992). NIST Chemical Kinetics Database – Version 4.0, NIST Standard Reference Data, Gaithersburg, MD.

Margitan, J.J. (1984). J. Phys. Chem., **88**, 3314.

Martin, D., Jourdain, J.L., & LeBras, G. (1988). Int. J. Chem. Kinet., **20**, 897.

McAndrew, R., & Wheeler, R. (1962). J. Phys. Chem., **66**, 229.

Michael, J.V., & Lim, K.P. (1993). Ann. Rev. Phys. Chem., **44**, 429.

Miyoshi, A., Shiina, H., Tsuchiya, K., & Matsui, H. (1996) 26th Symposium (International) on Combustion, p. 535.

Möller, D. (1984). Atmos. Environ., **18**, 19.

Moore Plummer, P. L. (1990). J. Chem. Phys., **92**, 6627.

Morokuma, K., & Murguruma, C. (1994). J. Am. Chem. Soc., **116**, 10316.

Muller, C.H. III, Schofield, K., Steinberg, M., & Broida, H.P. (1979). 17th Symposium (International) on Combustion, p. 867.

Muller, C.H. III, Schofield, K., & Steinberg, M. (1980). In Laser Probes for Combustion Chemistry, ACS Symp. Ser. No. 134 pp. 103–130.

Murrells, T.P., Lovejoy, E.R., & Ravishankara, A.R. (1990). J. Phys. Chem., **94**, 2381.

Nicovich, J.M., Kreutter, K.D., van Dijk, C.A., & Wine, P.H. (1992). J. Phys. Chem., **96**, 2518.

Nobes, R.H., & Radom, L. (1992). Chem. Phys. Lett., **189**, 554.

Nourbakhsh, S., Liao, C.-L., & Ng, C.Y. (1990). J. Chem. Phys., **92**, 6587.

Nourbakhsh, S., Norwood, K., Yin, H.-M., Liao, C.-L., & Ng, C.Y. (1991a). J. Chem. Phys., **95**, 946.

Nourbakhsh, S., Norwood, K., Yin, H.-M., Liao, C.-L., & Ng, C.Y. (1991b). J. Chem. Phys., **95**, 5014

Oya, M., Tsuchiya, K., & Asaba, T. (1987). Proc. 16th Int. Symp. on Shock Tubes and Shock Waves, p. 421.

Oya, M., Tsuchiya, K., & Asaba, T. (1991). Proc. 17th Int. Symp. on Shock Tubes and Shock Waves, p. 799.

Pauwels, J.F., Carlier, M., Devolder, P., & Sochet, L.R. (1986). J. Phys. Chem., **90**, 4377.

Pauwels, J.F., Carlier, M., Devolder, P., & Sochet, L.R. (1992). Combust. Sci. Tech., **90**, 4377.

Pfefferle, L.D., & Churchill, S.W. (1989). Ind. Eng. Chem. Res., **28**, 1004.

Plane, J.M.C. (1989). In *Biogenic Sulfur in the Environment*, Saltzman & Cooper (Eds.), ACS Symp. Ser. No. 393 pp. 404–423.

Pratt, G., & Rodgers, D. (1977). J. Chem. Soc. Far. Trans. I, **73**, 54.

Roth, P., Löhr, R., & Barner, U. (1982). Combust. Flame, **45**, 273.

Ruscic, B., & Berkowitz, J. (1992). J. Chem. Phys., **97**, 1818.

Ruscic, B., & Berkowitz, J. (1993). J. Chem. Phys., **98**, 2568.

Saito, K., Ueda, Y., Kakumoto, T., & Imamura, A. (1986). Int. J. Chem. Kinet., **18**, 871.

Sarkisyan, E.N., Azatyan, V.A., & Nalbandyan, A.B. (1968). Dokl. Akad. Nauk. SSSR, **178**, No. 2, 389.

Schofield, K., & Steinberg, M. (1992). J. Phys. Chem., **96**, 715.

Schwartz, S.E. (1989). Science, **243**, 753.

Shum, L.G.S., & Benson, S.W. (1983). Int. J. Chem. Kinet., **15**, 433.

Shum, L.G.S., & Benson, S.W. (1985). Int. J. Chem. Kinet., **17**, 277; 749.

Singh, H.B. (1995). In *Composition, Chemistry and Climate of the Atmosphere*, Singh, H.B., Ed., Van Nostrand Reinhold, New York, NY, pp. 216–250.

Singleton, D.L., & Cvetanovic, R.J. (1988). J. Phys. Chem. Ref. Data, **17**, 1377.

Slagle, I.R., Gilbert, J.R., & Gutman, D.J. (1974). J. Chem. Phys., **61**, 704.

Smith, O.I., Tseregounis, S.I., & Wang, S. (1982). Int. J. Chem. Kinet., **14**, 679.

Smith, O.I., Wang, S., Tseregounis, S.I., & Westbrook, C.K. (1983).

Spiro, P.A., Jacob, D.J., & Logan, J.A. (1992). J. Geophys. Res., **97**, 6023.

Stachnik, R.A., & Molina, M.J. (1987). J. Phys. Chem., **91**, 4603.

Stearns, C.A., Miller, R. A., Kohl, F.J., & Fryburg, G.C. (1977). J. Electrochem. Soc., **124**, 1145.

Steinberg, M., & Schofield, K. (1987). Paper 3a-042 Joint meeting of the Western States and Japanese Sections of the Combustion Institute, Honolulu, HI.

Steinberg, M., & Schofield, K. (1990). Prog. Energy Combust. Sci., **16**, 311.

Stickel, R.E., Chin, M., Daykin, E.P., Hynes, A.J., & Wine, P.H. (1993). J. Phys. Chem., **97**, 13653.

Sugden, T.M., Bulewicz, E.M., & Demerdache, A. (1962). *Chemical Reactions in the Lower and Upper Atmosphere,* John Wiley & Sons, New York, NY, pp. 89–108.

Taniguchi, M., Hirasawa, N., & Yoshida, H. (1987). Bull. Chem. Soc. Jpn., **60,** 2349.

Tesner, P.A., Rubinov, R.Kh., & Sherov, I.Sh. (1984). Gaz. Prom., **9,** 28.

Thielen, K., & Roth, P. (1987). Combust. Flame, **69,** 141.

Tolbert, M.A. (1994). Science, **264,** 527.

Tsang, W., & Herron, J.T. (1991). J. Phys. Chem. Ref. Data **20,** 609.

Tseregounis, S.I., & Smith, O.I. (1983). Combust. Sci. Technol., **30,** 231.

Tseregounis, S.I., & Smith, O.I. (1984). 20th Symposium (International) on Combustion, p. 761.

Turnipseed, A.A., Barone, S.B., & Ravishankara, A.R. (1992). J. Phys. Chem., **96,** 7502.

Tyndall, G.S., & Ravishankara, A.R. (1991). Int. J. Chem. Kinet., **23,** 483.

Vaghjiani, G.L., & Ravishankara, A.R. (1990). J.°Chem. Phys., **92,** 996.

Von Mihelcic, D., & Schindler, R.N. (1970). Ber. Bunsenges. Phys. Chem., **74,** 1280.

Wang, N.S., & Howard, C.J. (1990). J. Phys. Chem., **94,** 8787.

Webster, P., & Walsh, A.D. (1965). 10th Symposium (International) on Combustion, p. 463.

Wendt, J.O.L., & Ekmann, J.M. (1975). Combust. Flame, **25,** 355.

Wendt, J.O.L., Morcomb, J.T., & Corley, T.L. (1979). 17th Symposium (International) on Combustion, p. 671.

Wendt, J.O.L., Wottan, E.C., & Corley, T.L. (1983). Combust. Flame, **49,** 261.

Westenberg, A.A., & de Haas, N.N. (1973). J. Chem. Phys., **59,** 6685.

Wheeler, R.J. (1968). J. Phys. Chem., **72,** 3359.

Wine, P.H., Thompson, R.J., & Semmes, D.H. (1984a). Int. J. Chem. Kinet., **16,** 1623.

Wine, P.H., Thompson, R.J., Ravishankara, A.R., Semmes, D.H., Gump, C.A., Torabi, A., & Nicovich, J.M. (1984b). J. Phys. Chem., **88,** 2095.

Wine, P.H., Nicovich, J.M., Hynes, A.J., & Wells, J.R. (1986). J. Phys. Chem., **90,** 4033.

Woiki, D., & Roth, P. (1995). Int. J. Chem. Kinet., **27,** 547.

Woiki, D., & Roth, P. (1996). Israel J. Chem., **36,** 279.

Woiki, D., & Roth, P. (1998). 26th Symposium (International) on Combustion, p. 583.

Woiki, D., Markus, M.W., & Roth, P. (1993). J. Phys. Chem., **1997,** 9682.

Xantheas, S.S., & Dunning, Jr., T.H. (1993). J. Phys. Chem., **97,** 6616.

Yin, F., Grosjean, D., & Seinfeld, J.H. (1990). J. Atmos. Chem., **11**, 309.

Yokota, T., & Strausz, O.P. (1979). J. Phys. Chem., **83**, 3196.

Yoshimura, Y., Koshi, M., Matsui, H., Kamiya, K., & Umeyama, H. (1992). Chem. Phys. Lett., **189**, 199.

Zachariah, M.R., & Smith, O.I. (1987). Combust. Flame, **69**, 125.

Zellner, R.E. (1984). In *Combustion Chemistry,* W.C. Gardiner, Ed., Springer-Verlag, New York.

4

Survey of Rate Coefficients in the C-H-Cl-O System

Selim M. Senkan[1]

4.1 INTRODUCTION

The reactions of halogens and their compounds, in particular of chlorine and chlorinated hydrocarbons (CHCs), are of growing interest to the combustion community because of the increasing role that incineration plays in the treatment of hazardous chemical wastes, which frequently contain these compounds (Senkan 1988, Oppelt 1987). Although a reasonably large body of knowledge appears to exist on the rates and mechanisms of reactions of chlorine-containing species, most of our present understanding is derived from experiments done at or near ambient temperatures. (See for example Atkinson et al. 1992, DeMore et al. 1992, and references therein.) Studies at high temperatures are limited; more are urgently needed for the development of quantitative detailed chemical kinetic mechanisms describing the oxidation and pyrolysis of CHCs at combustion conditions. To cope with the scarcity of high-temperature data, techniques have been developed to extrapolate rate information obtained at lower temperatures to combustion conditions (Atkinson 1986, Dean 1985, Dean and Westmoreland 1987, Cohen and Benson 1987, Senkan and Quam 1992), and these permit the available thermochemical and kinetic data to be used to develop mechanisms suitable for

---

[1] Department of Chemical Engineering, University of California, Los Angeles, CA 90024

modeling the combustion of CHCs. (See, for example, Karra et al. 1988, Chang and Senkan 1989, Fisher et al. 1990, and Ho et al. 1992.) For such modeling studies, however, it is required that the rate parameters of a large number of reactions be estimated. Although estimation in itself is not undesirable, as there will always be more reactions in such complex systems than resources available to study them individually in isolated experiments, experimental rate measurements on reactions of chlorinated species at combustion temperatures serve as vital references to assess the validity of predictions and to calibrate theoretical methods. Only with reliable experimental rate data together with robust theoretical methods will we be in a position to develop mechanisms with broad ranges of applicability and to address issues related to the formation and destruction of pollutants in the combustion of CHCs.

In developing this survey the following approach was taken. First, the most reliable information on the rates and mechanisms of chemical reactions related to the field of interest was compiled. For elementary reactions for which earlier evaluations were available, those evaluations were used as starting points for rate parameter reviews. When prior evaluations were not available, the relevant earlier studies in the original literature were reviewed and evaluated. Finally, since large data bases for elementary reactions involving chlorine at high temperatures are not available, relevant low temperature kinetic studies were reviewed, as such information can then be used to predict high-temperature kinetic behavior using appropriate mathematical models.

It is expected that over a broad temperature range the simple two-parameter Arrhenius-form rate expression $k = A\exp(-E_a/\mathrm{RT})$ is inadequate to represent experimental rate data, especially for bimolecular reactions that have small energy barriers. (Zellner 1984) Consequently, a three-parameter expression of the form $k = AT^m \exp(-E_a/\mathrm{RT})$ was adopted in this survey to represent recommended rate expressions. Values of m were determined from appropriate fits to the experimental data. The use of this rate coefficient expression brings about several issues that must be recognized. First, for bimolecular reactions the numerical value of the pre-exponential A factors may be considerably different from the collision frequency, generally in the range 10^{13} to 10^{14} $\mathrm{cm^3mol^{-1}s^{-1}}$, because of the presence of the T^m factor. Second, for endothermic reactions we can no longer assume that E_a should be equal to at least the heat of reaction, because the temperature dependence of the rate coefficient is also influenced by the T^m factor.

Before presenting results on reaction rates, the structure and properties of the Cl atom will be described and compared to those of H, C, and O atoms, because these properties strongly influence the structure, bonding, and reactions of chlorine-containing compounds. This discussion will then be followed by a review of the thermochemistry of CHCs and their associated radicals and a brief presentation of some of the distinguishing features of the reactions of CHCs and chlorine-containing radicals.

4.2 ELECTRONIC STRUCTURE AND THERMOCHEMISTRY

Chlorine is a Group VIIB element with 17 electrons and the electronic structure $1s^2 2s^2 2p^6 3s^2 3p^5$. That is, Cl has an unpaired electron in its $3p$ shell and consequently it reacts vigorously to complete this shell to 8 electrons so as to attain the electronic structure of argon. As a consequence of its electronic structure, Cl has a high electron affinity and electronegativity (Pauling 1960). For example, the electron affinity of Cl is 356 kJ/mol compared to 120, 142, and 67 kJ/mol for C, O, and H atoms, respectively. Similarly, the electronegativity of Cl is 3.16 on the Pauling scale compared to 2.55, 3.44 and 2.2 for C, O and H, respectively. In hydrocarbons and oxygenated hydrocarbons, Cl displaces H to form analogous chlorinated compounds. For example, upon chlorine substitution CH_4, C_2H_2, CH_2CHCH_3, and C_6H_6 form CHCs such as CH_3Cl, C_2HCl, CH_2CHCH_2Cl, and C_6H_5Cl. Similarly, the displacement of hydrogens in oxygenated hydrocarbons such as CH_2O yields CHClO and CCl_2O (phosgene), and CH_3CHO forms CH_3CClO, $CH_2ClCClO$, $CHCl_2CClO$, and CCl_3CClO. As is evident from these examples, Cl maintains a single bond structure in hydrocarbons because the electronegativity of C is considerably less than that of Cl. With hydrogen, chlorine even forms the very stable HCl. However, in combination with more electronegative oxygen, Cl exhibits multiple bonding as a consequence of the participation of its $3s$ and $3p$ electrons. Examples of well-characterized Cl_xO_y species include ClO, Cl_2O, Cl_2O_3, ClO_2, Cl_2O_4, Cl_2O_3, ClO_4, and Cl_2O_7: Thus chlorine can form as many as 7 bonds, corresponding to full utilization of its M shell electrons. These species are all unstable; they have large positive enthalpies of formation and thus tend to decompose to the elements. (DeMore et al. 1992)

The standard enthalpies of formation ($\Delta_f H^\circ_{298}$) of chlorinated C_1– and C_2–hydrocarbons and several other chlorinated species are presented in Table 4.1 together with C–H and C–Cl bond dissociation energies. Several important features should be noted. First, increasing chlorination tends to stabilize the hydrocarbon molecule and/or radical. For example, the enthalpies of formation of chlorinated methanes decrease from –75 for CH_4 to –84, –95, –98, and –85 kJ/mol for CH_3Cl, CH_2Cl_2, $CHCl_3$, and CCl_4, respectively. A parallel trend can also be seen for C_2 chlorinated alkanes, alkenes, and alkynes. Similarly, the standard enthalpies of formation of C_1 radicals decrease from 146 kJ/mol for CH_3 to 70 kJ/mol for CCl_3. Second, increasing chlorination decreases the C–H bond dissociation energies (BDE) of α hydrogens in alkanes (Benson 1976). For example, in the case of chlorinated methanes the C–H BDEs decrease from 435 kJ/mol in CH_4 to 414 kJ/mol in CH_3Cl, to 397 kJ/mol in CH_2Cl_2, and to 385 kJ/mol in $CHCl_3$. A similar trend is also seen in chlorinated ethanes, with one important additional feature, namely, that chlorine substitution increases the C–H BDE at the β positions. For example, although the BDE of an α hydrogen in C_2H_5Cl decreases to about 406 kJ/mol from 418 kJ/mol in C_2H_6, the BDEs of β hydrogens increase to about 427 kJ/mol. Further chlorination of the α carbon atom ultimately leads to the strengthening of the β C–H bond to 435 kJ/mol, the same as the BDE in

TABLE 4.1a. Standard enthalpies of formation at 298 K for chlorine-containing species and related hydrocarbons.[a]

Species	$\Delta_f H^\circ_{298}$	Species	$\Delta_f H^\circ$
H	218.0	$CHCl_2CH_2$	84.5
Cl	120.9	CH_3CCl_3	−140.6
HCl	−92.5	CH_3CCl_2	42.7
ClO	122.2 (102.1)	CCl_3CH_2	78.7
$ClOO$	140.6 (96.7)	CH_2ClCH_2Cl	−135.6
$ClOO_2$	(69.9)	$CH_2ClCHCl$	54.8
ClO_3	(217.6)	$CH_2ClCHCl_2$	−147.3
Cl_2O	(81.6)	CH_2ClCCl_2	25.3
Cl_2O_2	(129.7)	$CHCl_2CHCl$	43.1
Cl_2O_3	(142.3)	CH_2ClCCl_3	−148.1
$HOCl$	−75.3	CCl_3CHCl	46.4
$COCl_2$	−220.1	$CHCl_2CHCl_2$	−151.0
$COHCl$	−185.4	$CHCl_2CCl_2$	35.4
$COCl$	−27.1	C_2HCl_5	−147.3
CH_4	−74.9	C_2Cl_5	30.7
CH_3	145.6	C_2Cl_6	−136.0
CH_3Cl	−84.1	C_2H_4	52.3
CH_2Cl	113.4	C_2H_3	286.2
CH_2Cl_2	−95.4	C_2H_3Cl	19.5
$CHCl_2$	81.6	CH_2CCl	256.9
$CHCl_2O_2$	−5.6	$CHClCH$	271.1
$CHCl_3$	−97.5	$CHClCHCl$	−4.7
CCl_3	69.5	$CHClCCl$	231.8
CCl_3O_2	−3.0	CH_2CCl_2	3.0
$CHCl$	334.7	CH_2CCl	256.9
CCl_2	221.8	CCl_2CH	260.7
CCl_4	−84.5	C_2HCl_3	−17.2
C_2H_6	−84.1	C_2Cl_3	227.6
C_2H_5	118.0	C_2Cl_4	−24.1
C_2H_5Cl	−113.8	C_2H_2	226.8
CH_3CHCl	76.1	C_2H	476.6
CH_2ClCH_2	95.8	C_2HCl	213.8
CH_3CHCl_2	−134.3	C_2Cl	560.7
CH_3CCl_2	43.1	C_2Cl_2	235.1

[a] Units are kJ/mol. From unpublished calculations done in collaboration with C.F. Melius; values in parentheses are from DeMore et al. (1992).

CH_4. These results have important implications for the rates and mechanisms of elementary reactions involving CHCs. For example, the attack of electronegative radicals, such as Cl, O, and OH on chlorinated alkanes (C_2 and higher) results in

TABLE 4.1b. Bond dissociation energies of C_1 and C_2 chlorinated hydrocarbons.[a]

Species	α-C–H	β-C–H	α-C–Cl	β-C–Cl	C–C
CH_4	435				
CH_3Cl	414		351		
CH_2Cl_2	397		331		
$CHCl_3$	385		301		
CCl_4			285		
C_2H_6	418				372
C_2H_5Cl	406	427	351		372
CH_3CHCl_2	393	435	331		364
CH_3CCl_3		435	301		356
CH_2ClCH_2Cl	406		351		336
$CH_2ClCHCl_2$	389	406	322	334	333
CH_2ClCCl_3	410		289	343	298
$CHCl_2CHCl_2$	397		314		322
$CHCl_2CCl_3$	393		301	314	301
C_2Cl_6			285		276
C_2H_4	452				
C_2H_3Cl	452	469	389		
CHClCHCl	456		397		
CH_2CCl_2	473		372		
C_2HCl_3	460		372		
C_2Cl_4			372		
C_2H_2	548				
C_2HCl	469		452		
C_2Cl_2			351		

[a] Units are kJ/mol. The values listed here are derived from the Table 1a data.

preferential abstraction of α hydrogens, and this, in turn, leads to the production of chlorinated alkenes by subsequent reactions of the product via β–scission or by hydrogen atom abstraction by O_2. This issue will be revisited later.

A similar trend is also observed in the R–O_2 bond strengths in alkoxy radicals that are formed as a consequence of the combination of chlorinated alkyl radicals (R) with molecular oxygen. For example, the R–O_2 BDE decreases from 136 kJ/mol for CH_3O_2 (Slagle and Gutman 1985) to 121 and 106 kJ/mol for CH_2ClO_2 and $CHCl_2O_2$ (Russell et al. 1990a), and 83 kJ/mol for CCl_3O_2, respectively (Russell et al. 1990b). These results are significant as they help in the establishment of temperature levels at which changes in the mechanisms of R + O_2 reactions occur as a result of the thermal decomposition of the adduct RO_2. From the above results it is clear that the decomposition of CCl_3O_2 should occur at the lowest temperature among all the chlorinated methyl radicals. Consequently, CCl_3 radicals undergo recombination reactions, leading to the formation of higher molecular weight products at lower temperatures.

4.3 CHARACTERISTIC FEATURES OF ELEMENTARY REACTIONS OF CHLORINE

The kinetics of CHC reactions also exhibit distinguishing features when compared to reactions of analogous nonsubstituted hydrocarbons. First, as a consequence of the weaker C–Cl bond dissociation energies, CHCs decompose at temperatures that are significantly lower than the decomposition temperatures of corresponding hydrocarbons and result in the generation of highly reactive Cl radicals. For example, the C–Cl BDE in CH_3Cl is 351 kJ/mol compared to 414 kJ/mol for the C–H BDE (cf. Table 4.1). Since the C–Cl and C–H bond dissociation processes both proceed via a loose transition state (Benson 1976), their activation energies are expected to be close to the bond dissociation energies. In addition, one expects that the preexponential A factors should be close to one another. Consequently, at 1000 K the rate of C–Cl bond fissions should be greater than those of C–H bond fissions by a factor of about $\exp(-(414-351)/RT) \approx 2000$. At lower temperatures the differences in Cl and H generation rates are even greater.

Second, because of the high electronegativity of Cl relative to C and H and the BDE issues discussed above, 1,2– and 1,1–HCl eliminations from CHCs are major reaction channels. The relative ease of formation of charged transition state structures, which occur as a consequence of the interaction of highly electronegative Cl atoms with H atoms, lowers the activation energy barrier for the production of HCl. (Benson 1976) Because charged transition states cannot form between identical atoms, H_2 and Cl_2 elimination reactions are less important. However, Cl_2 elimination from CHCs does become an increasingly important reaction with increasing chlorination of the CHC molecule.

Third, Cl atoms abstract H atoms from hydrocarbon, for example, alkanes, very efficiently, with small energy barriers. In fact, with the exception of CH_4 ($E_a \approx 13$ kJ/mol) the activation energies for H-atom abstractions from alkanes by Cl atoms are virtually zero at temperatures of interest in combustion. (See Section 4.5.) Methane differs from other alkanes because of its unusually high (435 kJ/mol) C–H bond dissociation energy. In contrast, abstractions of Cl and H atoms from CHCs by H atom attack exhibit significant activation energies. For example, activation energies range from 8 to 17 kJ/mol for $H + CHC \longrightarrow HCl + R$ reactions and from 33 to 46 kJ/mol for $H + HC \longrightarrow H_2 + R$ reactions, in spite of the exothermic nature of all of these processes. In Figure 4.1, showing the rate coefficients for H atom abstraction from CH_4 by Cl, H, OH, and O, it can be seen that the rate coefficient of the Cl atom reaction is significantly higher than the others at temperatures below 2000 K and also exhibits the lowest activation energy. In contrast, H atom attack has the lowest rate coefficient and the highest activation energy. Consequently, rate coefficient of $H + CH_4 \longrightarrow CH_3 + H_2$ becomes the largest one only at very high temperatures. Abstractions of Cl from CHCs by attack of Cl, OH, or O are endothermic processes, because C–Cl bond dissociation energies are considerably larger than those of both Cl–Cl and O–Cl; therefore, these reactions have high activation energies and are less important in combustion processes.

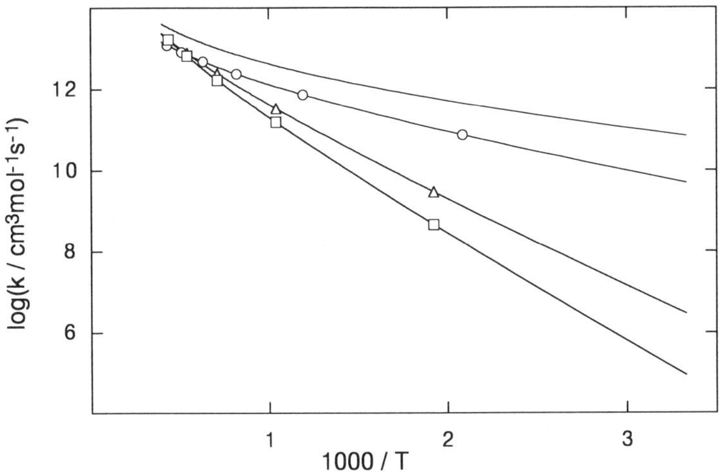

FIGURE 4.1. Comparison of rate coefficient for abstraction of an H atom from methane by Cl atoms with the rate coefficients for abstraction by OH (○), O (△), and H (□).

These kinetic features of the reactions of chlorine and chlorinated compounds have important implications for combustion modeling, in particular with regard to their effects on flame propagation and inhibition, flame structure, and on the formation of high molecular weight aromatic products and ultimately soot. For example, the high sooting tendency of CHCs (Senkan et al. 1983) can be explained, in part, as early formation of Cl and hydrocarbon radicals by unimolecular decomposition of CHCs and chlorinated alkoxy radicals and subsequent fast reactions of Cl with hydrocarbons, leading to increased production of more hydrocarbon radicals. Increased production of hydrocarbon radicals then enhances the formation of higher molecular weight species through recombination and/or addition to unsaturated compounds.

Fourth, in view of the bond dissociation energy considerations presented above, the attack of radicals such as Cl, OH, and O on chlorinated alkanes preferentially abstracts α H atoms, that is, H atoms next to the Cl atom in the CHC molecule. For example, as shown in Figure 4.2, the reaction of Cl with C_2H_5Cl results primarily in the production of CH_3CHCl radicals rather than of CH_2ClCH_2; the production rate of CH_3CHCl exceeds that for C_2ClCH_2 formation by at least a factor of 2 over the entire temperature range of interest for combustion in spite of the 3:2 greater abundance of β H atoms in C_2H_5Cl. The preferential production of CH_3CHCl over CH_2ClCH_2 has fundamental implications. Most importantly, β–scission of CH_3CHCl results in formation of C_2H_3Cl, a chlorinated hydrocarbon product. Formation of C_2H_3Cl from CH_2ClCH_2 is insignificant, because β–scission energetics favor Cl atom loss and C_2H_4 production. (In the presence of O_2, however, H abstraction from CH_3CHCl does occur, that is, $CH_3CHCl + O_2 \longrightarrow C_2H_3Cl + HO_2$, is an efficient pathway for C_2H_3Cl production [Senkan et al. 1992].) As a consequence of the reactions with O_2, increased levels of CHC intermediates are expected to be formed in flames with lower equivalence ratios,

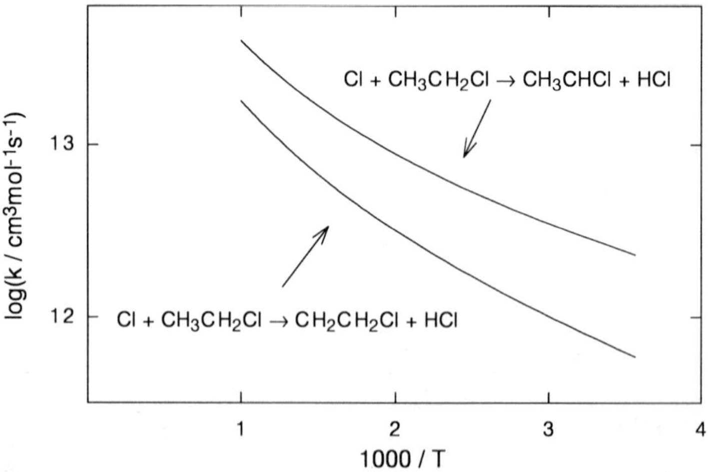

FIGURE 4.2. Rate coefficient expressions for abstraction of α and β hydrogens from ethyl chloride by Cl atoms.

as has indeed been observed experimentally in the CHC combustion. (See for example Kassem and Senkan 1991, Qun and Senkan 1994.)

Fifth, chlorine and chlorinated compounds compete with the chain branching reaction $H + O_2 \longrightarrow OH + O$ (E_a = 68 kJ/mol) for H atoms and thereby inhibit hydrocarbon combustion. This inhibition manifests itself as reduction in the laminar burning velocities of hydrocarbon flames (Valeiras et al. 1984) and by increased ignition delays (Chang and Senkan 1985). At the molecular level the inhibition of hydrocarbon combustion by chlorinated compounds (and halogens in general) is interpreted as being caused by the combination of the lower activation energy of H + CHC reactions, relative to the $H + O_2 \longrightarrow OH + O$ reaction, and the stability of the HCl product. That is, chlorine compounds not only compete for the H radicals, but the formation of highly stable product HCl also ties up the H atoms for a considerable period of time. Halogenated compounds have also been proposed to catalyze the recombination of H into H_2 and thereby exert yet another inhibition of hydrocarbon combustion (Westbrook 1982). In Figure 4.3 the rate coefficients of several H reactions with chlorinated species are presented and compared to the chain branching reaction; while the reactions of H with chlorinated compounds dominate at low temperatures, the rate of the higher activation energy $H + O_2 \longrightarrow OH + O$ reaction increases rapidly with increasing temperature, thereby diminishing the effectiveness of halogen inhibition reactions at high temperatures.

Because chlorinated compounds decompose more rapidly than hydrocarbons of similar size, because of the low dissociation energy of the C–Cl bond, and are effective in decreasing the concentrations of H and therefore OH and O radicals in the earlier, i.e., cooler, parts of premixed flames while having little influence at high temperatures, they impact the formation of pollutants such as NO_x, hydrocarbons and soot in a complex manner. In the case of NO_x, chlorinated compounds would

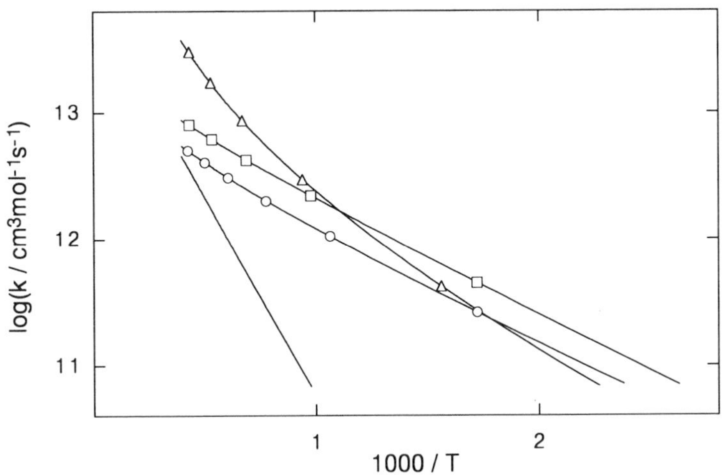

FIGURE 4.3. Comparison of rate coefficient expression for H-atom reaction with O_2 to H atom reactions with CCl_4 (○), CH_3Cl (△), and HCl (□).

on the one hand be expected to decrease the rates of formation of fuel and prompt NO, which are lower temperature processes in which small radicals such as H, OH, O play important roles (Miller and Bowman 1989). On the other hand, because chlorinated hydrocarbons also lead to early and more abundant production of CH, CH_2, CHCl and similar radicals, they may also act to promote the formation of prompt NO. However, chlorinated compounds should not affect the rate of thermal NO formation, which is important only at high flame temperatures. The presence of chlorinated compounds would only decrease the levels of thermal NO produced in combustion through the reduction of flame temperature caused by dilution. The reduction of OH and O radical concentrations in flames also suppresses CO and hydrocarbon oxidation, and this, in turn, promotes the formation of high molecular weight products and ultimately soot.

4.4 REACTION MECHANISMS

To illustrate the various kinetic issues discussed above let us consider the combustion of CH_3Cl, which has been studied extensively over the past several years in flames (Karra and Senkan 1988) and in flow reactors (Granada et al. 1987). Detailed chemical kinetic mechanisms have been proposed (Weissman and Benson 1984; Karra et al. 1988). For discussion purposes consider the reactions of 9% CH_3Cl in air, a mixture which corresponds to an equivalence ratio of about 0.7 and an H:Cl ratio of 3. Using this mixture composition as the starting point, the reaction mechanism of Karra et al. (1988) was integrated isothermally at 1300 K and 1 atm pressure. (The results are shown in Figures 4.4 and 4.5.) Aside from CH_2Cl and CH_3, which form from the parent fuel CH_3Cl, the radical abundances are HO_2 > Cl > H > OH > O during the induction period. As ignition approaches, the

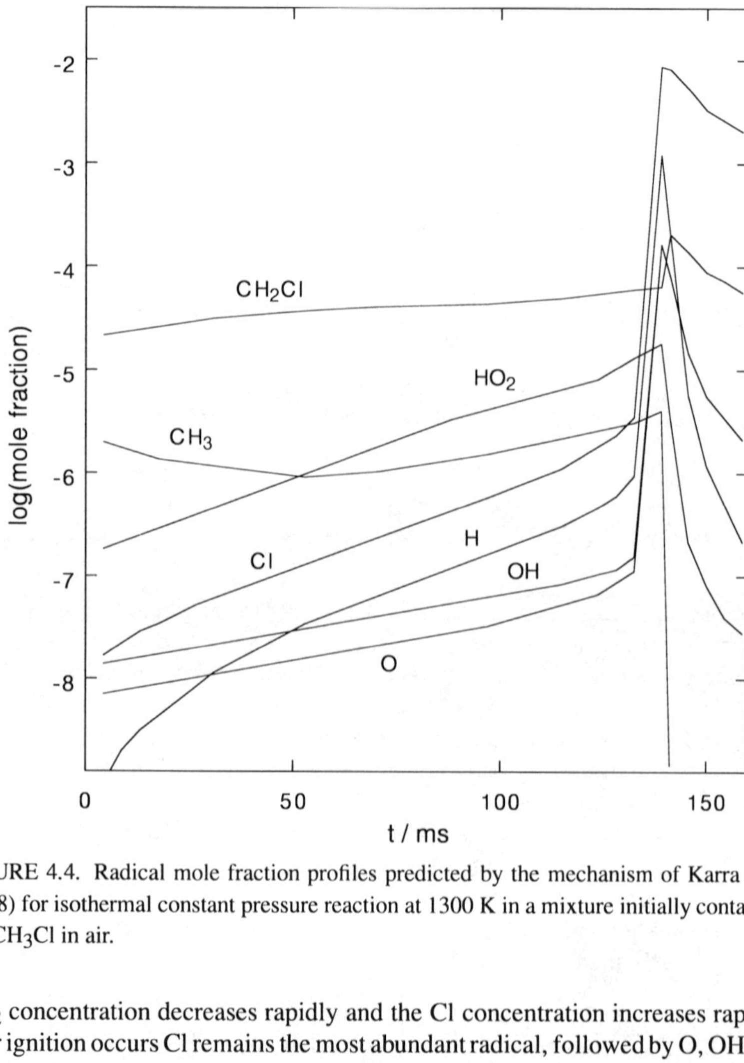

FIGURE 4.4. Radical mole fraction profiles predicted by the mechanism of Karra et al. (1988) for isothermal constant pressure reaction at 1300 K in a mixture initially containing 9% CH_3Cl in air.

HO_2 concentration decreases rapidly and the Cl concentration increases rapidly; after ignition occurs Cl remains the most abundant radical, followed by O, OH, and H. (Absolute concentrations, of course, do not directly rank relative importance in the reaction processes.)

The five largest reaction rates involving CH_3Cl are shown in Figure 4.5. Attack of Cl is the most significant loss pathway for CH_3Cl throughout the entire reaction period. In the early parts of the process, unimolecular decomposition and HO_2 attack are also important. Upon the increase of free radical concentrations the relative contribution of unimolecular decomposition decreases.

Increased production of highly chlorinated intermediates, compared to fuel-rich combustion, has been observed in fuel-lean combustion of CHCs (Kassem and Senkan 1991; Qun and Senkan 1990). Although this observation remains qualitative because of experimental difficulties associated with the sampling of fuel-lean flames, which have narrow reaction zones, the rate parameters of the

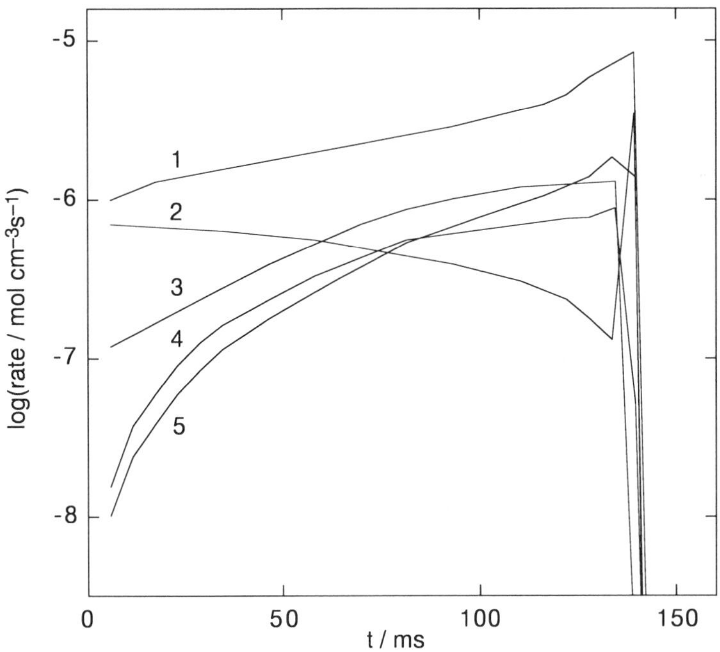

FIGURE 4.5. Induction zone rates of radical attack on CH_3Cl as computed by the mechanism of Karra et al. (1988) for the conditions of Figure 4.4.

1	$Cl + CH_3Cl$	$\longrightarrow$	$HCl + CH_2Cl$
2	CH_3Cl	$\longrightarrow$	$CH_3 + Cl$
3	$HO_2 + CH_3Cl$	$\longrightarrow$	$H_2O_2 + CH_2Cl$
4	$H + CH_3Cl$	$\longrightarrow$	$H_2 + CH_2Cl$
5	$H + CH_3Cl$	$\longrightarrow$	$HCl + CH_3$

C–H–Cl–O reactions presented in Section 4.5 offer plausible explanations. For example, at low equivalence ratios, the higher levels of electronegative radicals such as Cl, OH and O preferentially abstract H atoms from CHCs, because Cl abstraction and formation of products like Cl_2, HOCl, and ClO are endothermic. Moreover, these electronegative species preferentially abstract α H atoms from CHCs. Rapid β-scission of the radicals formed, as in $CH_3CHCl \longrightarrow C_2H_3Cl + H$, and/or reactions with O_2, as in $CH_3CHCl + O_2 \longrightarrow C_2H_3Cl + HO_2$, then lead to CHC intermediates. The high OH and O concentrations in fuel-lean combustion lead to HCl oxidation, that is, $HCl + (O,OH) \longrightarrow Cl + (OH, H_2O)$, and release of Cl atoms. Production of Cl_2 by Cl recombination can then further chlorinate the CHC radicals and lead to more highly chlorinated products. Thus production of increasingly chlorinated hydrocarbons is favored in fuel-lean combustion.

In the next section the kinetics and mechanisms of selected reactions involving Cl-containing species are discussed. Rate coefficient parameters for these and related elementary reactions are presented in tabular form at the end of this chapter. Standard enthalpy of reaction ($\Delta_r H$) values are also provided wherever

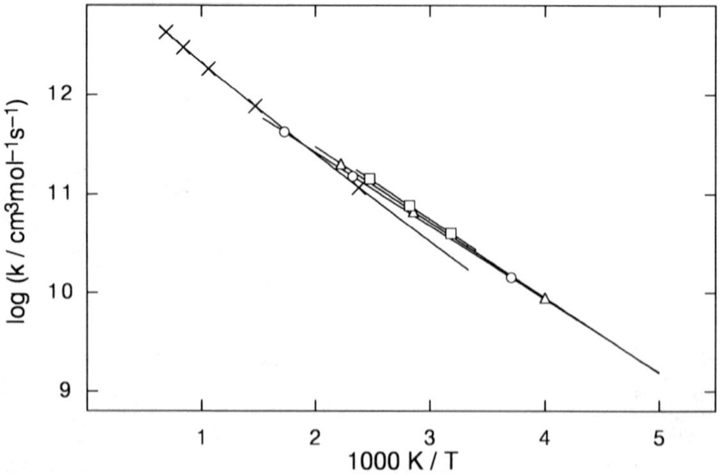

FIGURE 4.6. Rate coefficient expressions for H + HCl ⟶ H_2 + Cl. × Recommendation ○ Baulch et al. (1981) △ Miller and Gordon (1981) □ Kita and Stedman (1982).

sufficient thermochemical information is available. For reactions with multiple product sets, each pathway is identified and discussed separately. Because space limitations prevent discussing every tabulated elementary reaction in detail, we provide discussions only for reactions that are found to be important in combustion modeling and for which a reasonable body of kinetic data exists.

4.5 SURVEY OF ELEMENTARY REACTIONS

4.5.1 Reactions of H atoms

There have been few reliable rate measurements and mechanism studies on the reactions of H atoms with chlorinated compounds. We particularly need to know the branching ratios for H and Cl abstractions from CHCs by H-atom attack better, as both pathways are important in combustion mechanisms and need to be distinguished. Because the H-atom reactions compete directly with H + O_2 ⟶ OH + O, they are important in flame inhibition. Additional studies of the reactions of H atoms with CHCs are needed for the development of reaction mechanisms describing the combustion of CHCs.

H + HCl ⟶ H_2 + Cl

This reaction is an important inhibitor in later stages of CHC combustion, that is, after HCl has accumulated in the system. Since the evaluation by Baulch et al. (1981) it was studied by Miller and Gordon (1981), Kita and Stedman (1982),

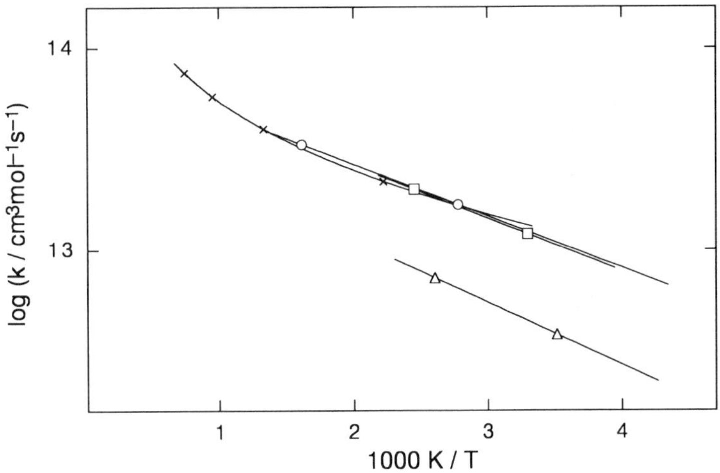

FIGURE 4.7. Rate coefficient expressions for $H + Cl_2 \longrightarrow HCl + Cl$. ○ Baulch et al. (1981) □ Wagner et al. (1977) ◇ Ambidge et al. (1976) × recommendation.

and Adusei and Fontijn (1993) over a broad temperature range. The latter authors also derived a rate expression for the reverse reaction. The measurements are in reasonable agreement from 200 to 700 K (Figure 4.6) and indicate slight Arrhenius graph curvature at combustion temperatures. Since no experimental data at very high temperatures are available, we use $m = 0.3$ to express the non-Arrhenius behavior. The rate coefficient expression of Adusei and Fontijn was taken as the basis for the recommended expression.

$H + Cl_2 \longrightarrow HCl + Cl$

This can be an important flame inhibition reaction, especially in the combustion of CHCs with Cl:H ratios larger than 1 and under fuel-lean conditions. The evaluation of Baulch et al. (1981) was taken as the basis for our recommendation. We modified their expression to account for the expected Arrhenius graph curvature by setting $m = 1$ and recalculating A and E_a (Figure 4.7).

$H + CH_3Cl \longrightarrow CH_3 + HCl$ (a)
$\longrightarrow CH_2Cl + H_2$ (b)

Attack of H on CH_3Cl can lead to either Cl or H abstraction at combustion temperatures. Both channels are exothermic. Experiments have led to the conclusion that HCl is the major product. The activation energy of 34–38 kJ/mol suggested by Jones et al. (1973) and Westenberg and DeHaas (1975) for the Cl abstraction path appears to be too high, as this level of barrier would be expected to be associated with removal of H from CH_3Cl in a manner similar to reaction of H with CH_4. Consequently, we used the results of Hart et al. (1974) and Aders et al. (1975) to derive the three-parameter rate coefficient recommendation shown in Figure 4.8.

$H + CCl_4 \longrightarrow CCl_3 + HCl$

As seen in Figure 4.9, the earlier measurements of the rate coefficient parameters for this reaction by Vance and Bauman (1938) are in agreement with the later measurements of Jones et al. (1973). Siedel et al. (1965) noted a higher activation energy at high temperatures, consistent with the three-parameter expression of Vance and Bauman. We recommend the parameters of Vance and Bauman.

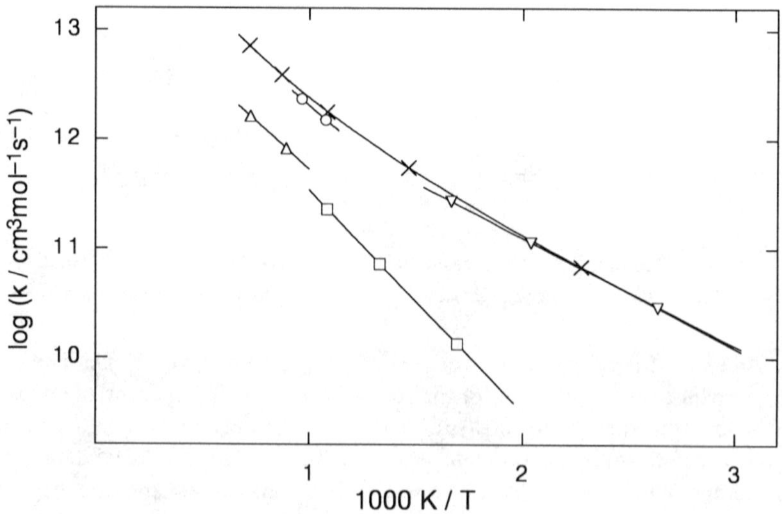

FIGURE 4.8. Rate coefficient of $H + CH_3Cl \longrightarrow CH_3 + HCl$. × Recommendation O Hart et al. (1974) Δ Jones et al. (1973) □ Westenberg and de Haas (1975) ∇ Aders et al. (1975).

Rate coefficients in the following tables are expressed in the form $k = AT^m \exp(E_a/RT)$ for K, mol/cm^3 and s units; values of $\Delta_r H$ and E_a are in kJ/mol units. Literature citations are abbreviated in the tables by the first three letters of the author's surname for single-author papers and by the authors' initials for multiple-author papers; thus Bed 1991b for the second 1991 citation of Bednar and WFA 1987 for Wang, Fu, and Anderson 1987. Measurement and calculation methods are denoted as follows: ARAS, atomic resonance absorption spectroscopy: CO_2, CO_2 laser; DFR, discharge flow reactor; DGA, differential gas analysis; DLA, diode laser absorption; FR, flow reactor; GA, gas analysis; GC, gas chromatography; HTP, high temperature photochemistry; IRC, infra red chemiluminescence; IRS, infra red spectroscopy; IS, internal standard, relative rate method; LFFB, laminar flat flame burner; LIF, laser induced fluorescence; LMR, laser magnetic resonance; LP, laser photolysis; MD, microwave discharge; MMS, molecular modulation spectroscopy; MS, mass spectrometry; OS, optical spectroscopy; RC, radiochemistry; RFS, resonance fluorescence spectroscopy; PH, photolysis; PR, pulse radiolysis; SAR, structure activity relationship; SR, static reactor; ST, shock tube; TRISP, time resolved infrared spectroscopy; TST, transition state theory; UVS, ultraviolet spectroscopy; VLPP, very low pressure pyrolysis.

TABLE 4.2. Reactions of H atoms.[a]

Reaction	$\Delta_r H$	A	m	E_a	ΔT	Method	Reference
$H + HCl \longrightarrow H_2 + Cl$	−4.4	1.7×10^{13}		17.3	298–1192	HTP, RFS	AF 1993
		1.0×10^{13}		14.6	199–502	FP, RFS	MG 1981
		1.2×10^{13}		14.9	294–424	DFR, RFS	KS 1982
		8.0×10^{12}		14.2	200–650	Review	BDGM 1981
Recommendation		1.8×10^{12}	0.3	15.9	298–1500		
$H + Cl_2 \longrightarrow HCl + Cl$	−189.1	8.6×10^{13}		4.9	250–730	Review	BDGM 1981
		8.7×10^{13}		5.0	252–458	DFR, MS	WWZ 1976
		4.6×10^{13}		5.9	292–434	DFR, MS	ABW 1976
Recommendation		6.0×10^{10}	1	0.8	298–1500		
$H + HOCl \longrightarrow Products$		3.0×10^{12}			298	DFR, MS	EB 1988
$H + Cl_2O \longrightarrow HCl + ClO$	−70.90	2.5×10^{13}			300	DFR, IR	WS 1989
$H + CH_4 \longrightarrow CH_3 + H_2$	2.6	1.3×10^{4}	3	33.6	1000–2500	Review	ABCHKT 1992
		2.2×10^{4}	3	36.6	1000–2500	Review	War 1984
$H + CH_3Cl \longrightarrow CH_3 + HCl$	−19.30	4.8×10^{8}			298	DFR, ESR	JM 1986
		3.1×10^{13}		33.8		Review	JMT 1973
		3.7×10^{13}		38.9	510–998	DFR, MS	WdeH 1975
		9.5×10^{13}		31.9	870–1088	LFFB, MS	HGF 1974
		1.3×10^{13}		19.3	326–652	FR, MS	APW 1975
Recommendation		1.1×10^{7}	2	12.6	326–1500		

[a] See footnote on page 402 for units and comments.

TABLE 4.2. Reactions of H atoms.[a] (Continued)

Reaction	$\Delta_r H$	A	m	E_a	ΔT	Method	Reference
$H + CH_3Cl \longrightarrow CH_2Cl + H_2$	–20.9						
$H + CH_2Cl_2 \longrightarrow CH_2Cl + HCl$	–101.3	1.1×10^{13}		25.5	298–462	DFR, MS	CLeBP 1973
$H + CH_2Cl_2 \longrightarrow CHCl_2 + H_2$	–37.2						
$H + CHCl_3 \longrightarrow CHCl_2 + HCl$	–127.6						
$H + CHCl_3 \longrightarrow CCl_3 + H_2$	–51.0						
$H + CCl_4 \longrightarrow CCl_3 + HCl$	–147.7	2.2×10^{11}	0.5	14.5	294–473	DFR	VB 1938
		5.9×10^{12}		16.3	294–473	Review	JMT 1973
				16.3	800–1000	LFFB, GC	D-LW 1955
				20.1	2137–2468	DFR	SMM 1965
Recommendation		2.2×10^{11}	0.5	14.5	298–1500		
$H + C_2H_6 \longrightarrow C_2H_5 + H_2$	–16.9	1.5×10^{9}	1.5	31.0	1000–2500	Review	ABCHKT 1992
		5.4×10^{2}	3.5	21.8	1000–2500	Review	War 1984
$H + C_2H_5Cl \longrightarrow C_2H_5 + HCl$	–79.1	6.3×10^{13}		18.8	300–1000	Estimate	SQ 1992
$H + C_2H_5Cl \longrightarrow CH_3CHCl + H_2$	–28.0	2.0×10^{9}	2	36.4	300-1000	Estimate	SQ 1992
$H + C_2H_5Cl \longrightarrow CH_2ClCH_2 + H_2$	–7.9	3.0×10^{9}	2	40.6	300–1000	Estimate	SQ 1992
$H + C_2H_4 \longrightarrow C_2H_3 + H_2$	15.8	5.4×10^{14}		62.3	1000–2500	Review	ABCHKT 1992
		1.5×10^{13}		42.7	1000–2500	Review	War 1984
$H + C_2H_4 \longrightarrow C_2H_5$	–153.2						
k_∞		1.0×10^{13}		6.6	1000–2500	Review	War 1984

[a] See footnote on page 402 for units and comments.

TABLE 4.2. Reactions of H atoms.[a] (Continued)

Reaction	$\Delta_r H$	A	E_a	ΔT	Method	Reference
$H + C_2H_3Cl \longrightarrow CH_2ClCH_2$	–141.4	1.3×10^{13}	14.6	850–1100	Estimate	ML 1988
$H + C_2H_3Cl \longrightarrow CH_3CHCl$	–161.5	1.4×10^{13}	6.3	850–1100	Estimate	ML 1988
$H + C_2H_3Cl \longrightarrow CHClCH + H_2$	33.5	5.0×10^{13}	62.8	850–1100	Estimate	ML 1988
$H + C_2H_3Cl \longrightarrow C_2H_3 + HCl$	–43.5	3.9×10^{13}	38.9	850–1100	Estimate	ML 1988
		1.0×10^{14}	18.8	300–1000	Review	KM 1981
$H + C_2H_3Cl \longrightarrow CH_2CCl + H_2$	19.2	1.0×10^{14}	48.1	300–1000	Estimate	SQ 1992
$H + C_2H_3Cl \longrightarrow CHClCH + H_2$	33.5	1.0×10^{14}	60.7	300–1000	Estimate	SQ 1992
$H + CH_2CCl_2 \longrightarrow$ Products		3.6×10^{12}		298	DFR, MS	AJ 1985
$H + C_2HCl_3 \longrightarrow$ Products		6.3×10^{11}		298	DRF, MS	AJ 1985
$H + C_2Cl_4 \longrightarrow C_2Cl_3 + HCl$	–58.6	1.2×10^{15}	58.9	950–1100	ST, IS	TW 1991
$H + C_2Cl_4 \longrightarrow C_2HCl_3 + Cl$	–90.4	2.0×10^{14}	38.7	950–1100	ST, IS	THMC 1988
$H + C_2Cl_4 \longrightarrow$ Products		3.0×10^{9}		500	SR, PH, UVS	Fabry1986
		1.4×10^{14}	38.2	950–1100	ST, IS, GC	TW 1991

[a] See footnote on page 402 for units and comments.

TABLE 4.2. Reactions of H atoms.[a] (Concluded)

Reaction	$\Delta_r H$	A	E_a	ΔT	Method	Reference
$H + C_6H_5Cl \longrightarrow C_6H_6 + Cl$	–48.5	2.2×10^{13}	37.7	1050–1150	ST, IS	CHT 1989
$H + o\text{-}C_6H_4Cl_2 \longrightarrow C_6H_5Cl + Cl$		9.1×10^{14}	43.1	1050–1150	ST, IS	CHT 1989
$H + p\text{-}C_6H_4Cl_2 \longrightarrow C_6H_5Cl + Cl$		4.6×10^{13}	38.6	1050–1150	ST, IS	CHT 1989
$H + 1,2,3\text{-}C_6H_3Cl_3 \longrightarrow o\text{-}C_6H_4Cl + Cl$		1.9×10^{14}	24.6	1050–1150	ST, IS	CHT 1989
$H + 1,2,3\text{-}C_6H_3Cl_3 \longrightarrow m\text{-}C_6H_4Cl + Cl$		3.6×10^{12}	15.4	1050–1150	ST, IS	CHT 1989
$H + o\text{-}C_6H_4ClOH \longrightarrow C_6H_5Cl + OH$		3.2×10^{14}	35.8	1000–1080	ST, IS	TCW 1989
$H + o\text{-}C_6H_4ClOH \longrightarrow C_6H_5OH + Cl$		3.3×10^{13}	35.8	1000–1080	ST, IS	TCW 1989
$H + p\text{-}C_6H_4ClCH_3 \longrightarrow p\text{-}C_6H_4ClCH_2 + H_2$		9.6×10^{13}	35.5	1010–1130	ST, IS	THMC 1988
$H + p\text{-}C_6H_4ClCH_3 \longrightarrow C_6H_5Cl + CH_3$		3.5×10^{13}	30.8	1010–1130	ST, IS	THMC 1988
$H + p\text{-}C_6H_4ClCH_3 \longrightarrow C_6H_5CH_3 + Cl$		4.8×10^{13}	39.9	1010–1130	ST, IS	THMC 1988
$H + o\text{-}C_6H_4ClCH_3 \longrightarrow o\text{-}C_6H_4ClCH_2 + H_2$		1.2×10^{14}	35.5	1010–1130	ST, IS	THMC 1988
$H + o\text{-}C_6H_4ClCH_3 \longrightarrow C_6H_5Cl + CH_3$		4.8×10^{13}	30.8	1010–1130	ST, IS	THMC 1988
$H + o\text{-}C_6H_4ClCH_3 \longrightarrow C_6H_5CH_3 + Cl$		5.4×10^{13}	39.9	1010–1130	ST, IS	THMC 1988

[a] See footnote on page 402 for units and comments.

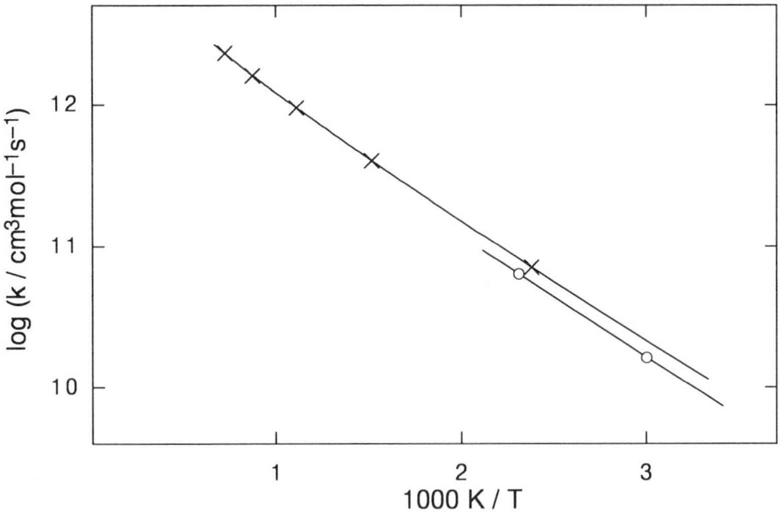

FIGURE 4.9. Rate coefficient expressions for H + $CCl_4 \longrightarrow HCl + CCl_3$. × Vance and Bauman (1938) and recommendation. ○ Jones et al. (1973).

H + chlorinated alkenes, alkynes, and aromatics

As can be seen in Table 4.2, a number of measurements and estimates exist for the rates of H-atom reactions with chlorinated olefins. With few exceptions, however, product identities were not reported, which limits the utility of the results for combustion modeling. Extensive studies of the reactions of H with chlorinated aromatics and olefins were made by Tsang and coworkers, who used a shock tube and internal standards (Tsang and Walker 1991; Cui et al. 1989; Tsang et al. 1988, 1990). Since these studies provide reliable measurements with proper accounting of isomeric reaction products, we recommend their use in combustion modeling.

Reactions of H with compounds having double and triple bonds are in principle complex, because they proceed via the formation of chemically activated adducts. Consequently, these reactions can become pressure dependent if fall-off conditions exist in the system. In addition, multiple reaction channels may be accessible to the energy-rich adduct. Therefore the determination of suitable rate coefficient parameters for individual reaction channels requires energy transfer calculations, which depend on the composition of the bath gas and the pressure. These calculations require the knowledge of the high-pressure reaction rate coefficients together with molecular properties of the adduct (Section 4.7). H easily replaces Cl upon addition to double and/or triple bonds because of the bond dissociation energy considerations noted above. For example, addition of H to C_2H_3Cl

$$H + C_2H_3Cl \longrightarrow [CH_2ClCH_2]^* \longrightarrow C_2H_4 + Cl$$

leads to an energy-rich adduct, the CH_2ClCH_2 radical, which has a low activation energy decomposition channel leading to the formation of C_2H_4 and Cl. Other radicals that bond to carbon more strongly than Cl lead to similar reactions.

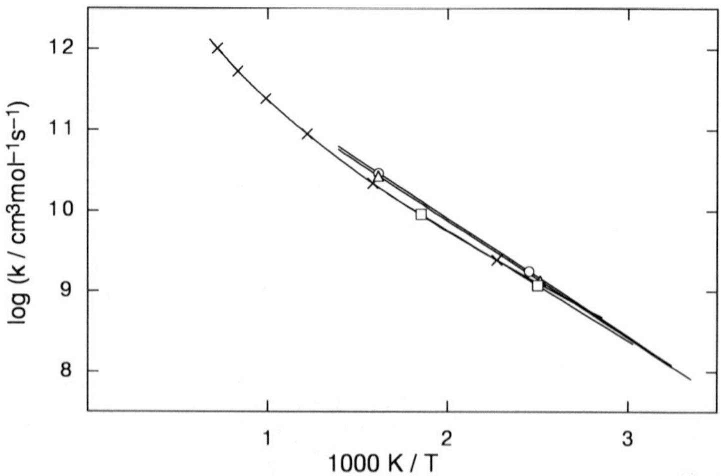

FIGURE 4.10. Rate coefficient expressions for O + HCl ⟶ OH + Cl. × Mahmud et al. (1990) and recommendation ○ Baulch et al. (1981) △ Atkinson et al. (1992) □ Singleton and Cvetanovich (1981).

4.5.2 Reactions of O atoms

Our current knowledge of O reactions with CHCs also needs improvement, for few experimental measurements exist. Unlike H reactions, attack of O on CHCs preferentially abstracts hydrogens, because abstraction of Cl and ClO production is endothermic, typically with $\Delta_r H > 80$ kJ/mol. There are also significant knowledge gaps in our understanding of the isomer-specific reaction paths of O reactions with C_2 and higher CHCs.

$O + HCl \longrightarrow OH + Cl$

This reaction, important in HCl oxidation under fuel-lean conditions, releases Cl into the system. Subsequent Cl recombination forms Cl_2, which in turn leads to chlorination of radicals and production of highly CHCs in fuel-lean flames of CHCs (Qun and Senkan 1989; Kassem and Senkan 1990; Cicek and Senkan 1993). Direct combination of Cl with hydrocarbon and chlorinated hydrocarbon radicals can also lead to production of increasingly more chlorinated CHCs (Figure 4.4). The most significant new investigation of this reaction since the reviews of Baulch et al. (1981), DeMore et al. (1992), and Atkinson et al. (1992), which deal with the temperature range from 293 to 718 K, was that of Mahmud et al. (1990), who studied it from 350 to 1480 K using a high-temperature flow reactor (Figure 4.10). All of the cited review recommendations and the results of Mahmud et al. are in agreement at low temperatures. However, Mahmud et al. also found significant curvature in the Arrhenius plots for this reaction, which necessitated the use of three-parameter rate expression. The Mahmud et al. rate expression is recommended for use in combustion modeling.

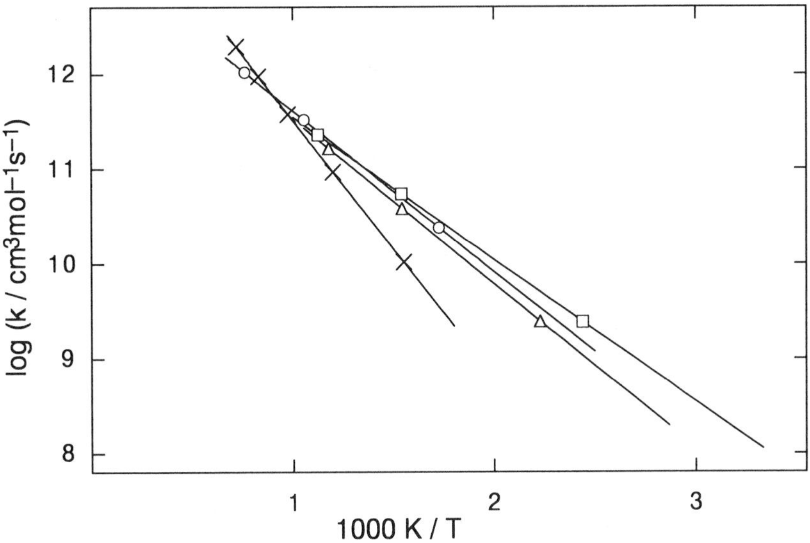

FIGURE 4.11. Rate coefficient expressions for O + CH_3Cl ⟶ CH_2Cl + OH. × Ko et al. (1992) and recommendation ○ Herron (1988) △ Wilson and O'Donovan (1967) □ Baulch et al. (1981).

O + Cl_2 ⟶ ClO + Cl

This reaction can be important at intermediate combustion temperatures. At high temperatures, where O may be abundant, Cl_2 readily decomposes, while at temperatures where Cl_2 remains intact the concentration of O is generally low. Since the evaluation of Baulch et al. (1981) it was studied by Wine et al. (1985). Collectively, the measurements indicate a slight upward Arrhenius plot curvature for this reaction.

O + CH_3Cl ⟶ CH_2Cl + OH

There were several studies of this reaction prior to 1988, summarized by Baulch et al. (1981) and Herron (1988). It was later studied by Ko et al. (1992) from 556 to 1485 K using high-temperature photochemistry and laser-flash photolysis shock tube techniques. The results obtained with the two techniques agree with one another. The rate coefficients showed significant non-Arrhenius behavior. At low temperatures the results of Ko et al. are lower than prior recommendations and significantly below the room temperature value of Force and Wiesenfeld (1981). Since the room temperature rate coefficient reported by Force and Wiesenfeld is close to the collision rate, without activation energy considerations, it appears to be unreliable. We recommend the rate coefficient expression of Ko et al. for combustion modeling (Figure 4.11).

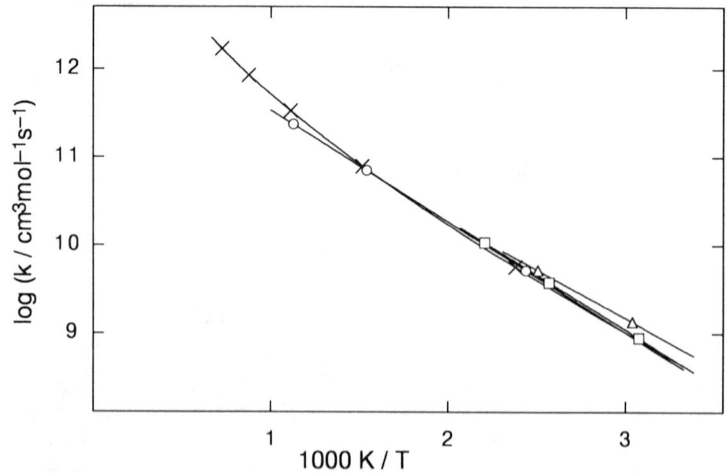

FIGURE 4.12. Rate coefficient expressions for $O + CH_2Cl_2 \longrightarrow CHCl_2 + OH$. × Su et al. (1994) and recommendation ○ Herron (1988) △ Barassin and Combourieu (1974) □ Barassin and Combourieu (1973).

$O + CH_2Cl_2 \longrightarrow CHCl_2 + OH$

The rate data for this reaction have been evaluated by Herron (1988). Barassin and Combourieu (1973, 1974), using a discharge flow reactor and mass spectrometry, obtained rate coefficients in the temperature range from 295 to 485 K that are significantly smaller than those reported by Force and Wiesenfeld (1981), who used laser photolysis and mass spectrometry. Again the results of Force and Wiesenfeld are apparently in error, as their room temperature rate is close to the collision rate without activation energy considerations. Su et al. (1994) investigated the reaction by high-temperature photochemistry and laser photolysis-shock tube methods over the temperature ranges 500 to 1090 K and 470 to 1350 K, respectively. The rate coefficients they determined by the two methods were consistent with one another but substantially lower than those reported by Barassin and Combourieu near 500 K. We recommend the rate coefficient expression of Su et al. for combustion modeling (Figure 4.12).

$O + CHCl_3 \longrightarrow CCl_3 + OH$

Barassin and Combourieu (1974) also studied this reaction in their discharge flow reactor. The later measurement of the rate of this reaction by Force and Weisenfeld (1981) at room temperature we again believe to be unreliable. The review by Herron (1988) supports the measurements of Barassin and Combourieu. Su et al. (1994) also investigated this reaction as they did the $O + CH_2Cl_2$ reaction and again obtained internally consistent results that imply lower rate coefficients than reported by Barassin and Combourieu near 500 K. We recommend the rate coefficient expression of Su et al. for combustion modeling.

$O + CCl_4 \longrightarrow CCl_3 + ClO$

This reaction is endothermic by 14 kJ/mol. The recommendation of DeMore et al. (1992) which proposes no activation energy, is therefore likely to be in error. For similar reasons the rate coefficient of Force and Weisenfeld (1981) appears too high. The measurements of Ung and Schiff (1962) and Brassin and Combourieu (1974), within a factor of 6 of one another, have similar activation energies. We prefer the results of Brassin and Combourieu.

O + higher hydrocarbons

Little information exists on the reactions of O with C_2 and higher molecular weight CHCs. The most comprehensive studies of O reactions with CHCs were done by Herron and Huie (1969) and Herron (1988). Although these investigators reported results on a large variety of CHCs, they did not derive branching ratios for abstraction of different hydrogens, which limits the utility of these measurements for modeling. Nevertheless, the information provided illustrates trends in total reaction rates. In Table 4.3 the overall rate parameters measured are presented without any recommendations. Fontijn and co-workers studied the high-temperature reactions of O with chlorinated ethylenes (Hranisavljevic et al. 1993) and found the rates of O + C_2H_3Cl and O + CH_2CCl_2 to be virtually identical, suggesting that O may preferentially attack the unsubstituted carbon. These measurements were consistent with theoretical calculations of addition of CH_3 to chlorinated ethylenes (Shi and Senkan 1991). Additional measurements of the kinetics of isomer-specific reactions of O with higher molecular weight CHCs would be useful for development of CHC combustion models. Again it should be noted that reactions of O with compounds having double and triple bonds are expected to proceed by adduct formation and may exhibit pressure-dependent rate coefficients and multiple product channels. These issues are discussed in Section 4.7.

4.5.3 OH radical reactions

Because of their importance in atmospheric chemistry (Atkinson 1985), a considerable body of knowledge exists on the rates and mechanisms of OH reactions with chlorinated compounds at near-ambient temperatures. Experimental data at higher temperatures are less abundant but steadily growing as a result of ongoing initiatives in this area. As for O reactions, attack of OH on CHCs primarily results in abstraction of H because Cl abstraction to form HOCl is quite endothermic, with $\Delta_r H$ values greater than 80 kJ/mol. Because most OH reaction rate measurements have been done at ambient and near-ambient temperatures, empirical methods based on transition state theory have been developed to extend them to elevated temperatures (Cohen and Benson 1987). In addition, techniques have been developed to estimate OH reaction rate parameters from structure–activity relationships (SAR) based on physical or chemical properties such as ionization potentials or bond dissociation energies. (See for example Atkinson 1986,

TABLE 4.3. Reactions of O atoms.[a]

Reaction	$\Delta_r H$	A	m	E_a	ΔT	Method	Reference
$O + H_2 \longrightarrow OH + H$	7.8	5.1×10^{4}	2.67	26.3	1000–2500	Review	ABCHKT 1992
		1.5×10^{7}	2	30.4	1000–2500	Review	War 1984
$O + HCl \longrightarrow OH + Cl$	3.4	4.5×10^{2}	3.13	13.0	350–1480	HTP, OS	MKF 1990
		6.0×10^{12}		27.8	293–718	Review	ABCHKT 1992
		3.1×10^{12}		26.2	330–600	PC, IRC	SC 1981
		6.8×10^{12}		28.0	293–718	Review	BDGM 1981
		6.0×10^{12}		27.4	298	Review	DSHKGHRM 1992
Recommendation		4.5×10^{2}	3.13	13.0	350–1480		
$O + Cl_2 \longrightarrow Cl + ClO$	–26.8	2.5×10^{12}		11.4	174–602	Review	BDGM 1981
		4.5×10^{12}		13.7	245–371	LP, RFS	WNR 1985
		3.7×10^{12}		13.6	306–424	SR, ESR	MYD 1979
		6.2×10^{12}		15.0	295–461	DFR, ESR	BWZ 1973
$O + ClO \longrightarrow Cl + O_2$	–230.1	2.3×10^{13}		0.0	200–300	Review	ABCHKT 1992
		5.7×10^{13}		1.5	200–500	Review	BDGM 1981
		3.6×10^{13}		0.8	220–450	DFR, RFS	Mar 1984
		3.0×10^{13}		0.8	236–422	DFR, RFS	Leu 1984
Recommendation		3.3×10^{8}	2	0.8	300–1200		
$O + OClO \longrightarrow O_2 + ClO$	–252.7	1.5×10^{12}		7.9	240–400	Review	ABCHKT 1992
$O + Cl_2O \longrightarrow ClO + ClO$	–134.7	1.8×10^{13}		5.2	235–300	Review	ABCHKT 1992
		7.8×10^{12}		4.2	230–400	LMR, DFR	SA 1992
		1.9×10^{13}		5.8	237–297	DFR, ESR	WJS 1982
$O + HOCl \longrightarrow OH + ClO$	–34.3	6.3×10^{12}		18.3	200–300	Review	DSHKGHRM 1992

[a] See footnote on page 402 for units and comments.

TABLE 4.3. Reactions of O atoms.[a] (Continued)

Reaction	$\Delta_r H$	A	m	E_a	ΔT	Method	Reference
$O + CH_4 \longrightarrow CH_3 + OH$	10.5	9.3×10^{8}	1.5	35.5	1000–2500	Review	ABCHKT 1992
		1.2×10^{8}	2.1	31.9	1000–2500	Review	War 1984
$O + CH_3Cl \longrightarrow CH_2Cl + OH$	−13.4	1.5×10^{13}	0.31	46.4	556–1485	HTP, LP, ST	KFLM 1992
		2.0×10^{13}		32.4	400–1500	Review	Her 1988
		1.1×10^{13}		28.7	300–1000	Review	BDGM 1981
		1.3×10^{14}			298	LP, MS	FW 1981
		1.8×10^{13}		33.2	353–949	DFR, MS	WO'D 1967
Recommendation		1.5×10^{13}	0.31	49.8	556–1485		
$O + CH_2Cl_2 \longrightarrow CHCl_2 + OH$	−29.3	6.1×10^{12}		24.1	300–1000	Review	Her 1988
		5.9×10^{12}		23.8	295–483	DFR, MS	BC 1973
		2.9×10^{12}		21.0	295–433	DFR, MS	BC 1974
		1.6×10^{14}			298	LP, MS	FW 1981
		6.6×10^{6}	1.99	23.7	468–1355	HTP, LP-ST	SLMHF 1994
Recommendation		6.6×10^{6}	1.99	23.7	298–1500		
$O + CHCl_3 \longrightarrow CCl_3 + OH$	−43.3	3.0×10^{12}		20.8	300–1000	Review	Her 1988
		1.8×10^{14}			298	LP, MS	FW 1981
		2.9×10^{12}		20.9	298–500	DFR, MS	BC 1974
		1.1×10^{13}	0.22	39.5	468–1355	HTP, LP-ST	SLMHF 1994
Recommendation		1.1×10^{13}	0.22	39.5	300–1500		
$O + CCl_4 \longrightarrow CCl_3 + ClO$	14.2	3.0×10^{11}		18.3	300–1000	Review	Her 1988
		1.6×10^{11}		17.9	296–502	DFR, MS	BC 1974
		2.1×10^{14}			298	LP, MS	FW 1981
		2.0×10^{14}			298–398	Review	DSHRGKHKM 1992
		2.5×10^{10}		18.9	277–373	DFR, MS	US 1962
Recommendation		8.0×10^{5}	2	16.7	300–1000		

[a] See footnote on page 402 for units and comments.

TABLE 4.3. Reactions of O atoms.[a] (Continued)

Reaction	$\Delta_r H$	A	m	E_a	ΔT	Method	Reference
$O + COCl_2 \longrightarrow COCl + ClO$	55.3	2.2×10^{14}		0.0	298–398	Review	DSHRGKHKM 1992
		6.0×10^{9}			292	Review	Her 1988
$O + C_2Cl_2 \longrightarrow$ Products		9.6×10^{10} [b]			823	FR, PH, MS	RSSG 1989
$O + C_2H_3Cl \longrightarrow$ Products		5.5×10^{11}	0.33	6.5	288–1335	HTP, LP, OS	HAXF 1993
		3.4×10^{12}		5.4	290–450	Review	Cve 1987
		4.1×10^{10}			297	SR, PH, RFS	WdeH 1977
		3.8×10^{10}			302	FR, LP, MS	SGG 1975
		5.2×10^{11}			307	Review	HH 1973
$O + CH_2CCl_2 \longrightarrow$ Products		3.9×10^{8}	1.32	2.2	295–1218	HTP, LP, OS	HAXF 1993
		5.9×10^{11}			298	Review	Cve 1987
O + t-CHClCHCl $\longrightarrow$ Products		3.4×10^{7}	1.56	3.6	326–1258	HTP, LP, OS	HAXF 1993
$O + C_2HCl_3 \longrightarrow$ Products		5.9×10^{10}			298	Review	Cve 1987
$O + C_2Cl_4 \longrightarrow C_2Cl_3 + ClO$	103.3	3.6×10^{10}			298	DFR, ESR	BR 1986
$O + C_2H_6 \longrightarrow C_2H_5 + OH$	−9.1	1.0×10^{9}	1.5	24.3	1000–2500	Review	ABCHKT 1992
		3.0×10^{7}	2	21.4	1000–2500	Review	War 1984
$O+C_2H_5Cl \longrightarrow CH_3CHCl+OH$ (a)	−20.1						
$O+C_2H_5Cl \longrightarrow CH_2ClCH_2+OH$ (b)	5.0						
		4.8×10^{13} [c]		27.7	340–650	Review	Her 1988
		4.8×10^{13} [c]		27.7	336–622	DFR, MS	HH 1969
$O+CH_2ClCH_2Cl \longrightarrow CH_2ClCHCl+OH$	−19.7						
		5.0×10^{12}		20.8	300–700	Review	Her 1988
		4.9×10^{12}		20.9	290–650	DFR, MS	BRC 1977

[a] See footnote on page 402 for units and comments.
[b] Upper limit.
[c] Sum of channels (a) and (b).

TABLE 4.3. Reactions of O atoms.[a] (Concluded)

Reaction	$\Delta_r H$	A	m	E_a	ΔT	Method	Reference
$O + CH_3CHCHCl \longrightarrow$ Products		1.3×10^{12}			298	DFR, MS	WH 1987
$O + CH_3CHCl_2 \longrightarrow CH_3CCl_2 + OH$ (a)	–32.6						
$O + CH_3CHCl_2 \longrightarrow CHCl_2CH_2 + OH$ (b)	8.6						
		3.0×10^{12} *b*		20.3	300–700	Review	Her 1988
		2.9×10^{12} *b*		20.3	290–650	DFR, MS	BRC 1977
$O + CH_2CHCH_2Cl \longrightarrow$ Products		6.1×10^{11}			298	Review	Cve 1987
		4.6×10^{11}			298	DFR, MS	WH 1987
$O+CH_3CHClCH_3 \longrightarrow CH_3CClCH_3+OH$ (a)							
$O+CH_3CHClCH_3 \longrightarrow CH_3CHClCH_2+OH$ (b)							
		2.4×10^{13} *b*		21.3	300–700	Review	Her 1988
		2.6×10^{13}		21.8	336–622	DFR, MS	HH 1969
$O + CH_2CCHCl \longrightarrow$ Products		Relative	rate		298	SR, LP, UVS	FUL 1980
$O + CH_3CCCl \longrightarrow$ Products		2.4×10^{13}		8.8	298–876	FR, LP, MS	RSSG 1989
$O + (CH_3)_3CCl \longrightarrow (CH_3)_2CClCH_2 + OH$		2.6×10^{13}		25.0	336–622	DFR, MS	HH 1969
$O + CH_3CH_2CHClCH_3 \longrightarrow$ Products		3.5×10^{13}		21.0	336–622	DFR, MS	HH 1969
$O + (CH_3)_3CHCH_2Cl \longrightarrow$ Products		3.6×10^{13}		21.0	336–622	DFR, MS	HH 1969
$O + C_6H_5Cl \longrightarrow$ Products		2.1×10^{13}		19.9	295–871	DRF, MS	FTW 1989
		3.0×10^{11}			298	Review	Cve 1987
$O + n\text{-}C_7H_{15} \longrightarrow$ Products		8.0×10^{10}		22.2	298–330	Review	HH 1973
		3.0×10^{11}			298	Review	Cve 1987

[a] See footnote on page 402 for units and comments.

[b] Sum of channels *a* and *b*.

1987 and references therein for a comprehensive discussion of various estimation techniques.)

Because of its simplicity, the SAR approach has grown into a predictive tool that can play a major role in modeling combustion processes. In this approach, overall rate coefficients of OH reactions are expressed as

$$k(\text{total}) = k(\text{H abstraction}) + k(\text{OH addition to double and triple bonds}) + k(\text{OH addition to aromatics}).$$

SAR calculations of rate coefficients for H abstraction from C–H and O–H are based on group rate coefficients depending on the identity of the substituent groups. The designations $(X)CH_3$, $(X)(Y)CH_2$, $(X)(Y)(Z)CH$ are used for primary, secondary and tertiary carbons, and correspondingly for XOH. For example, the total H atom abstraction rate coefficient can be represented by:

$$k(\text{H abstraction})/\text{cm}^3\text{mol}^{-1}\text{s}^{-1} = 2.69 \times 10^6 T^2 e^{(T_x-303)/T} + 2.60 \times 10^6 T^2 e^{(T_x+T_y+233)/T} + 1.14 \times 10^6 T^2 e^{(T_x+T_y+T_z+711)/T}.$$

Appropriate values for the fitting parameters, determined by Atkinson (1987) using a least squares fitting process, are given in Table 4.4.

TABLE 4.4. Structure–Activity Parameters for Abstraction of H Atoms by OH Radicals[a] (Atkinson 1987)

Species	T_x, T_y, T_z	Note[b]
CH_3	1	
Cl	–288	
CH_2Cl	–168	–280
$CHCl_2$	–168	–359
CCl_3	–718	
CClO	≈–207	
$-CH_2-$	76	
>CH	76	

[a] The quantities T_x, T_y and T_z are denoted by Atkinson (1987) as E_x, E_y, and E_z; in both of these sets the units are K.

[b] From Taylor et al. (1992).

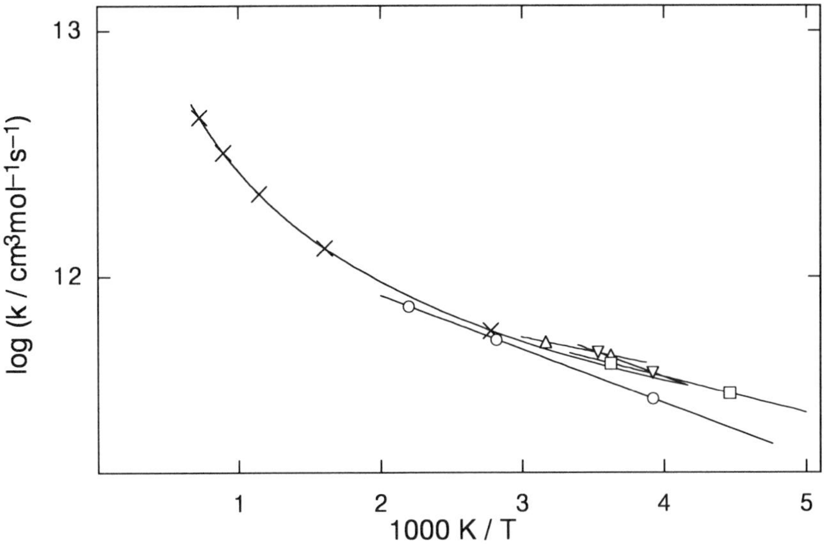

FIGURE 4.13. Rate coefficient expressions for $OH + HCl \longrightarrow H_2O + Cl$. × Ravishankara et al. (1985) and recommendation ○ Baulch et al. (1981) △ Keyser (1984) ▽ Molina et al. (1984) □ Atkinson et al. (1989).

The rates of OH addition to double and triple bonds depend on the number of unconjugated bonds and the degree, identity, and configuration of substitution around the bonds (Atkinson 1987). Although such addition processes are of major significance at or near ambient temperatures, they are expected to be less important at combustion conditions. The rates of addition processes are, in principle, pressure-dependent, implying a need to consider collisional energy transfer processes to and from chemically activated adducts (Section 4.6). We invoke SAR analysis in this section in discussing OH radical reactions with hydrocarbons and CHCs.

$OH + HCl \longrightarrow H_2O + Cl$

This is an important HCl oxidation reaction in fuel-lean combustion, releasing Cl to the system, and is partially responsible for the formation of highly CHCs in fuel-lean flames of CHCs (Qun and Senkan 1989, Kassem and Senkan 1990, Cicek and Senkan 1993). It is exothermic by about 67 kJ/mol, while the product channel that leads to the formation of HOCl + H is endothermic by 197 kJ/mol and thus unimportant. Several studies and evaluations of this reaction are in good agreement with one another at lower temperatures (Figure 4.13). Ravishankara et al. (1985) studied the reaction over the broadest temperature range, that is. from 240 to 1055 K, in a flow reactor using laser photolysis and laser fluorescence spectroscopy, and proposed a three-parameter rate expression to account for the observed non-Arrhenius behavior. We recommend their expression for combustion modeling.

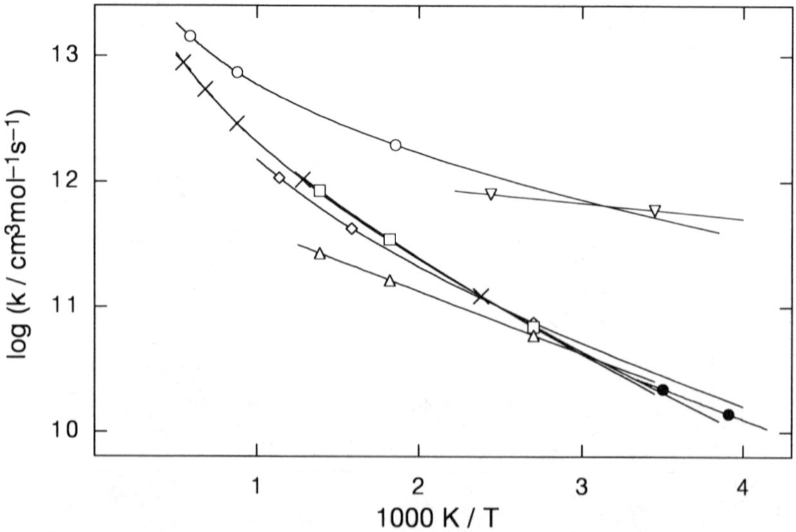

FIGURE 4.14. Rate coefficient expressions for $OH + CH_3Cl \longrightarrow CH_2Cl + H_2O$. × Cohen and Westberg (1991) and recommendation ○ Cohen and Benson (1987) ◇ Atkinson (1987) □ Taylor et al. (1989) △ DeMore et al. (1992) ▽ Baulch et al. (1981) ○ Atkinson et al. (1992).

$OH + Cl_2 \longrightarrow HOCl + Cl$

This reaction generally is important at intermediate temperatures in combustion processes. At high temperatures, where OH radicals are prevalent, the weakly bonded Cl_2 decomposes, while at lower temperatures where Cl_2 remains intact the concentration of OH is small.

$OH + CH_3Cl \longrightarrow CH_2Cl + H_2O$

This exothermic reaction ($\Delta_r H = -94$ kJ/mol) has been studied experimentally and theoretically by several investigators. Cohen and Benson (1987) analyzed existing kinetic information using transition state theory and derived an empirical correlation to predict the kinetic parameters. The SAR method developed by Atkinson (1987) also provided a framework to estimate the rate parameters for this reaction. Subsequent experimental studies by Taylor et al. (1989) used laser photolysis and laser-induced fluorescence spectroscopy. Re-evaluations have been reported by Atkinson et al. (1992), who recommended rate parameters for use in low-temperature studies, and by Cohen and Westberg (1991), who recommended rate parameters for a broader range of temperature. The experimental measurements, theoretical predictions and recommendations are in good agreement with one another, especially at low temperatures (Figure 4.14). We recommend the Cohen and Westberg three-parameter expression for combustion modeling.

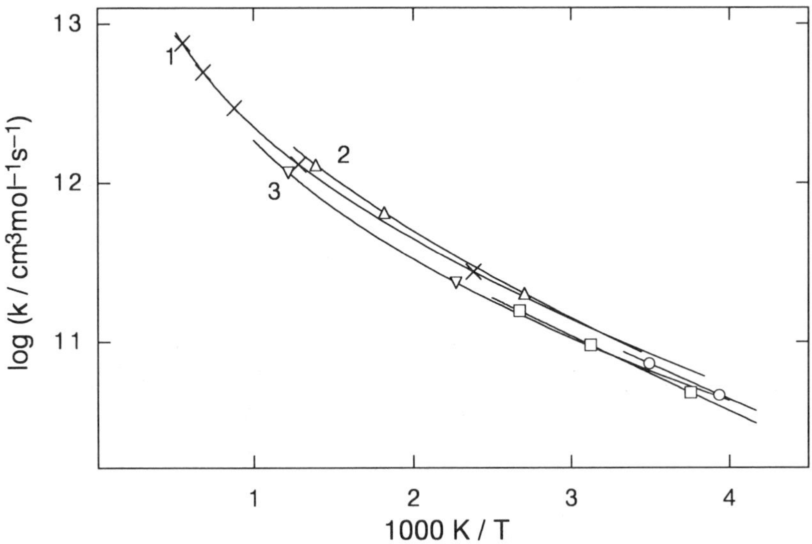

FIGURE 4.15. Rate coefficient expressions for $OH + CH_2Cl_2 \longrightarrow CHCl_2 + H_2O$. × Cohen and Benson (1987), Cohen and Westberg (1991) and recommendation △ Taylor et al. (1989) ▽ Atkinson (1987) □ Baulch et al. (1981) ○ Atkinson et al. (1992).

$OH + CH_2Cl_2 \longrightarrow CHCl_2 + H_2O$

This reaction is significantly more exothermic ($\Delta_r H = -101$ kJ/mol) than the $OH + CH_3Cl$ reaction because of the weakening of the C–H bond in CH_2Cl_2 (cf. Table 4.1). Consequently, the activation energy for this reaction would be expected to be smaller, which is supported by experimental data and SAR analysis (Atkinson 1987). The same groups who studied the $OH + CH_3Cl$ reaction also investigated this reaction, and, as seen in Figure 4.15, their results agree with one another. Again we prefer the rate expression proposed by Cohen and Westberg (1991).

$OH + CHCl_3 \longrightarrow CCl_3 + H_2O$

The C–H bond dissociation energy in $CHCl_3$ is the weakest of all chlorinated methanes (see Table 4.1), and this, as expected, leads to the lowest activation energy for Cl abstraction, as seen in Table 4.5. The same groups cited for the $OH + CH_3Cl$ reaction also studied this reaction and obtained results that are consistent with each other (Figure 4.16). The three-parameter rate coefficient expression recommended by Cohen and Westberg is preferred for combustion modeling.

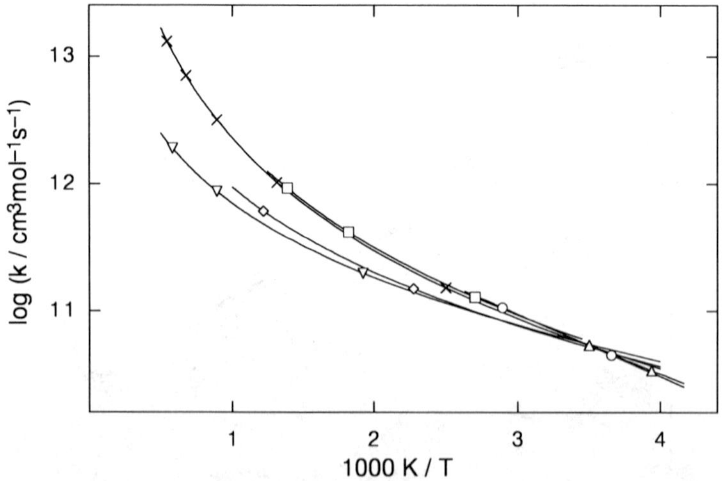

FIGURE 4.16. Rate coefficient expressions for $OH + CHCl_3 \longrightarrow CCl_3 + H_2O$. × Cohen and Westberg (1991) and recommendation ▽ Cohen and Benson (1987) ◇ Atkinson (1987) □ Taylor et al. (1989) ○ Baulch et al. (1981) △ Atkinson et al. (1992).

OH + higher chlorohydrocarbons

Rate data for reactions of OH with C_2 and higher CHCs is sparse. Nearly all of the available information comes from overall rate measurements of H abstraction by OH without regard to the site of abstraction. The structure–activity relationships developed by Atkinson (1986, 1987) provide the only means to estimate isomer-specific reaction rate parameters for CHC alkanes. Since the SAR approach has been shown to provide reliable rate predictions for chlorinated methanes, we used it extensively to generate rate coefficient recommendations for C_2 alkanes in this survey. For CHCs in which there is only one type of hydrogen, SAR predictions were compared to the available experimental data to establish better recommendations. Both the experimental data and the SAR relationships for chlorinated alkenes and alkynes are still provisional. We include them here and in Table 4.5 to support development of more definitive recommendations. The kinetics of many of these reactions are pressure-dependent. Additional studies are needed to generate an expanded data base for combustion modeling.

$$\begin{aligned} OH + C_2H_5Cl &\longrightarrow CH_3CHCl + H_2O \text{ (a)} \\ &\longrightarrow CH_2ClCH_2 + H_2O \text{ (b)} \end{aligned}$$

This reaction has two product channels, corresponding to abstraction of α (reaction a) and β (reaction b) hydrogens, respectively. As noted before, since the bond dissociation energies of α hydrogens are about 20 kJ/mol less than those of β hydrogens, the pathway leading to production of CH_2ClCH_2 would be expected to be slower based on Evans–Polanyi considerations and the SAR analysis of Atkinson

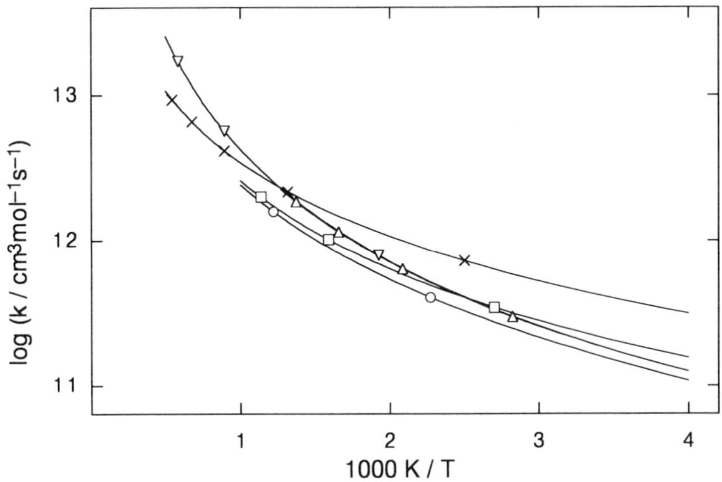

FIGURE 4.17. Rate coefficient expressions for reaction of OH with ethyl chloride. × Cohen and Benson (1987) and recommendation for total rate coefficient ∇ Cohen and Westberg (1991) recommendation for total rate coefficient △ Kasner et al. (1990) □ Atkinson (1987) recommendation for abstraction of α-H atom ○ Atkinson (1987) recommendation for abstraction of β-H atom.

(1987). The overall kinetics of this reaction was studied by Kasner et al. (1990) in a static reactor using laser photolysis and laser-induced fluorescence spectroscopy from 294 to 789 K. Their measurements agree with the room-temperature study of Howard and Evenson (1976). The reaction was also analyzed by Cohen and Benson (1987) using transition state theory, and re-examined by Cohen and Westberg (1991). As seen in Figure 4.17, the rate coefficients for the overall reaction (k_{a+b}) recommended by previous investigators are in good agreement with the sum of the individual rate coefficients k_a and k_b determined individually by the SAR analysis of Atkinson (1987). Therefore, we believe that the rate parameters derived from the SAR analysis are reliable enough to warrant their use in combustion modeling.

$$\begin{aligned} OH + CH_3CHCl_2 &\longrightarrow CH_3CCl_2 + H_2O \text{ (a)} \\ &\longrightarrow CH_2CHCl_2 + H_2O \text{ (b)} \end{aligned}$$

The overall kinetics were studied by Howard and Evenson (1976) at room temperature. Based on their measurements Cohen and Westberg (1991) proposed a three-parameter rate expression for use in combustion modeling without any distinction between isomeric pathways. As seen from the BDE information in Table 4.1, reaction path (a) is about 40 kJ/mol more exothermic than reaction path (b). Consequently, one expects the abstraction rates of α hydrogens to be faster than those of β hydrogens. SAR analysis supports this (Figure 4.18). The rate coefficients derived from SAR analysis, and their sum k_{a+b}, are consistent with

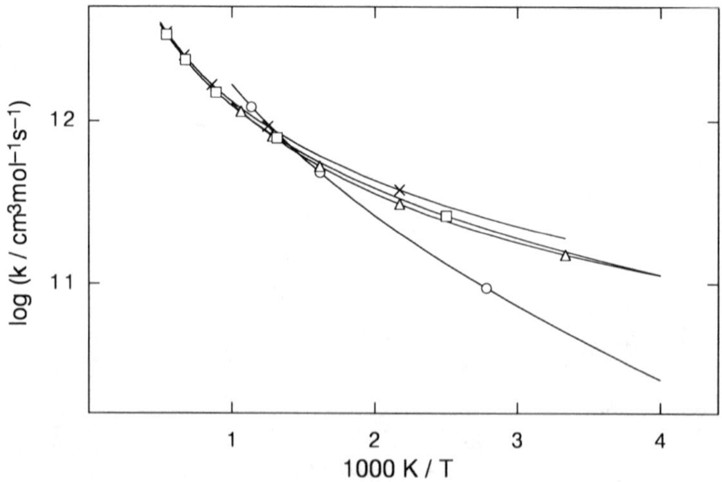

FIGURE 4.18. Rate coefficient expressions for OH reaction with CH_3CHCl_2. □ Cohen and Benson (1987), total rate coefficient × Cohen and Westberg (1991) and recommendation for total rate coefficient △ Atkinson (1987) and recommendation for abstraction of α-H atom ○ Atkinson (1987) and recommendation for abstraction of β-H atom.

the experimental data and the recommendation of Cohen and Westberg for k_{a+b}. It can be seen from Figure 4.18 that abstraction of α hydrogens—reaction (a)—dominates the overall process and $k_{a+b} \approx k_a$ at low temperatures, that is, in the room temperature experiments of Howard and Evenson (1976). At higher temperatures, however, the rate of β hydrogen abstraction competes with and even exceeds that of α hydrogen abstraction. This is not surprising, because the ratio of β hydrogens to α hydrogens is 3. We recommend the SAR-derived rate expressions for combustion modeling.

$OH + CH_2ClCH_2Cl \longrightarrow CH_2ClCHCl + H_2O$

The kinetics of this reaction was studied by Taylor et al. (1991) using laser photolysis and laser-induced fluorescence and by Xing et al. (1992) using a discharge flow reactor and resonance fluorescence spectroscopy. The results of these investigations agree with each other and with the earlier room temperature measurements of Howard and Evenson (1976) (Figure 4.19). This reaction provides an interesting case to test the accuracy of the SAR method, as there is only one type of hydrogen in the molecule. As seen in Figure 4.19, the SAR method slightly overpredicts the experimental results of Xing et al. and Taylor et al. However, considering the potential broad applicability of the SAR method, this is still a remarkably good agreement. Taylor et al. also provide a rate coefficient expression for this reaction with $m = 1$ which adequately represents the data over the temperature range studied. We recommend the use of Taylor et al.'s expression in combustion modeling.

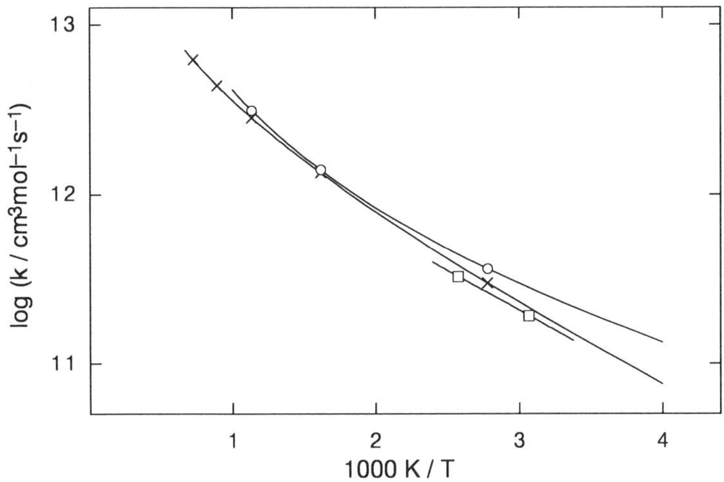

FIGURE 4.19. Rate coefficient expressions for reaction of OH with 1,2-dichloroethane. × Taylor et al. (1991) and recommendation O Atkinson (1987) □ Xing et al. (1992).

$$OH + CH_2ClCHCl_2 \longrightarrow CH_2ClCCl_2 + H_2O \text{ (a)}$$
$$\longrightarrow CHClCHCl_2 + H_2O \text{ (b)}$$

This reaction was studied by Jeong et al. (1984) in a discharge flow reactor from 250 to 470 K. Since the product distribution was not monitored, their measurements correspond to k_{a+b}. Cohen and Westberg (1991) subsequently analyzed their experimental data and proposed a three-parameter rate expression for the overall reaction. In this case reaction (a) is only about 17 kJ/mol more exothermic than reaction (b), and hence the difference in rates of abstraction of α and β hydrogens would be expected to be less than observed in CH_3CHCl_2. We also applied SAR analysis to this reaction to establish relative rates. As seen in Figure 4.20, the rate of abstraction of α hydrogens indeed is predicted to be higher than that for the abstraction of β hydrogens, consistent with Evans–Polanyi considerations. In addition, as expected, the relative rates of reactions (a) and (b) were closer to one another than those observed in the case of CH_3CHCl_2. Again the sum of k_a and k_b derived from SAR analysis agrees with the experimental k_{a+b}.

$$OH + CHCl_2CCl_2H \longrightarrow CHCl_2CCl_2 + H_2O$$

This reaction, which involves abstraction of only one type of hydrogen, was studied by Xing et al. (1992) in a discharge flow reactor from 293 to 481 K using resonance fluorescence spectroscopy. As seen in Figure 4.21, their results agree with SAR predictions. We adjusted the A–factor of the SAR formulation to make it coincide with the experimental data to arrive at our recommended rate expression.

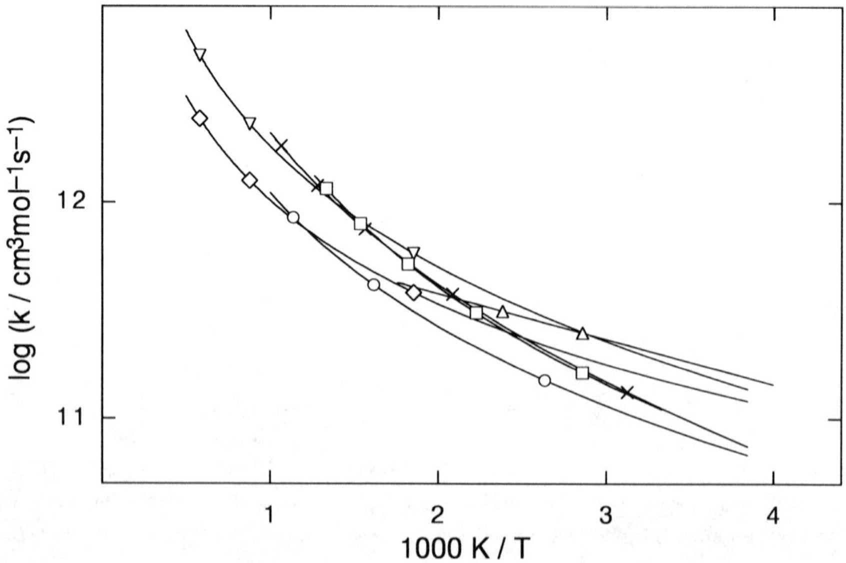

FIGURE 4.20. Rate coefficient expressions for reaction of OH with 1,1,2-trichloroethane. × Cohen and Benson (1987) ◇ Cohen and Westberg (1991) △ Jeong et al. (1984) □ Taylor et al. (1992) ○ and × Atkinson (1987) and our recommendations for abstraction of α-H and β-H atoms, respectively.

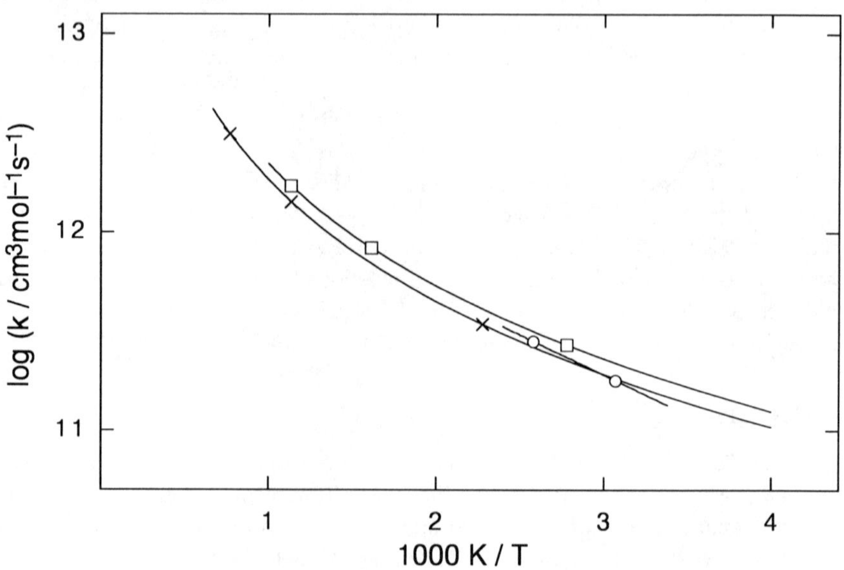

FIGURE 4.21. Rate coefficient expressions for reaction of OH with 1,1,2,2-tetrachloroethane. × Recommendation □ Atkinson (1987) ○ Xing et al. (1992).

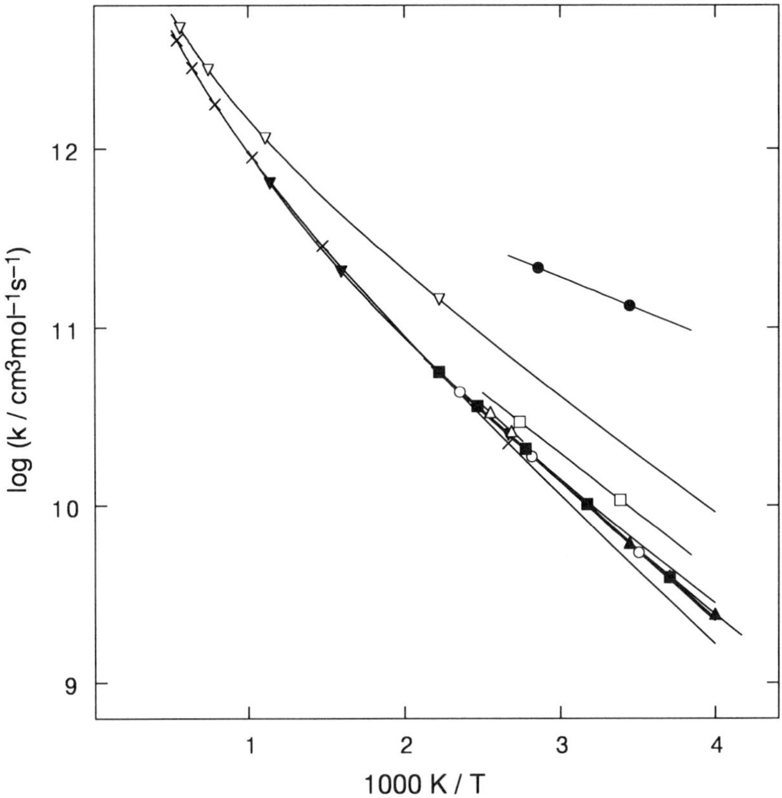

FIGURE 4.22. × Cohen and Westberg (1991) and recommendation ● Watson et al. (1977) ■ Jeong et al. (1984) ▽ Cohen and Benson (1987) ○ DeMore et al. (1992) ▼ Atkinson (1987) △ Nelson et al. (1990) □ Baulch et al. (1991) ▲ Atkinson et al. (1992).

$OH + CH_3CCl_3 \longrightarrow CCl_3CH_2 + H_2O$

This reaction was studied by Talukdar et al. (1992) using pulsed photolysis and laser-induced fluorescence. They found rates 15–20% below those of earlier studies. The results of experimental investigations by Nelson et al. (1990b) using pulse radiolysis and absorption spectroscopy and by Jeong et al. (1984) using discharge flow reactor and resonance fluorescence spectroscopy agree with one another. These and earlier experiments were evaluated by Atkinson et al. (1992) for use in atmospheric modeling and by Cohen and Westberg (1991) for higher-temperature applications. The latter review proposed a three-parameter rate expression, which we accept for our recommendation. The SAR result agrees with the experimental data and the Cohen and Westberg expression, as shown in Figure 4.22.

OH + unsaturated CHCs $\longrightarrow$ products

As shown in Table 4.5, studies of OH reactions with CHCs containing unsaturated bonds have been reported. As noted before, these reactions are complicated to evaluate because they involve chemically activated adducts, rates that may be pressure dependent, and endothermic reaction channels may be accessible to an energy-rich adduct. Theoretical estimation of reaction rate parameters requires energy transfer calculations for the prevailing composition and pressure. Such energy transfer calculations require knowledge of the high-pressure limiting reaction rate coefficients and the molecular parameters of the adduct for estimating rate coefficients. These issues are discussed in Section 4.7. Atkinson (1987) provides SAR parameters for estimating reaction rate parameters for addition of OH to unsaturated compounds at ambient conditions.

4.5.4 Reactions of Cl atoms

As is the case for O and OH, reactions of Cl atoms with hydrocarbons and CHCs have received sustained attention primarily because of their importance in industrial and atmospheric chemistry (Chiltz et al. 1963, Fettis and Knox 1964, Atkinson et al. 1992, DeMore et al. 1992, Hooshiyar and Niki 1995). Again as for OH and O, Cl attack on CHCs leads primarily to H rather than Cl abstraction, the latter generally having endothermicities greater than 20 kJ/mol. Because most Cl reactions have been studied at temperatures substantially below those encountered in combustion, the results have limited utility in combustion modeling. Accordingly, SAR methods have been developed to estimate Cl reaction rate parameters for combustion applications from bond dissociation energy and group considerations (Senkan and Quam 1992). The reactions of Cl with chlorinated alkanes were correlated in a manner similar to that described in Section 4.3 for OH reactions. That is, the total rate of H-atom abstraction by Cl is expressed as a sum of contributions from primary, secondary, and tertiary hydrogens using the form

$$\begin{aligned} k_{\text{total}}/\text{cm}^3\text{mol}^{-1}\text{s}^{-1} &= 24.9 \times 10^6 T^2 \text{e}^{(T_x-380)/T} \\ &+ 19.2 \times 10^6 T^2 \text{e}^{(T_x+T_y-35)/T} \\ &+ 6.4 \times 10^6 T^2 \text{e}^{(T_x+T_y+T_z+200)/T}, \end{aligned}$$

where the appropriate values for the fitting parameters T_x, T_y, and T_z are given in Table 4.6. For reaction with nonsubstituted alkanes, which have very small activation energies, a separate set of SAR parameters was developed, the equations then taking the form

$$\begin{aligned} k_{\text{total}}/\text{cm}^3\text{mol}^{-1}\text{s}^{-1} &= 2.3 \times 10^{13} \text{e}^{(T_x-76)/T} \\ &+ 6.8 \times 10^{13} \text{e}^{(T_x+T_y-210)/T} \\ &+ 1.3 \times 10^{13} \text{e}^{(T_x+T_y+T_z+415)/T}, \end{aligned}$$

TABLE 4.5. Reactions of OH radical.[a]

Reaction	$\Delta_r H$	A	m	E_a	ΔT	Method	Reference
$OH + H_2 \longrightarrow H_2O + H$	–62.8	1.0×10^{8}	1.6	13.8	1000–2500	Review	ABCHKT 1992
		1.0×10^{8}	1.6	13.8	1000–2500	Review	War 1984
$OH + HCl \longrightarrow H_2O + Cl$	–67.4	1.5×10^{12}		2.8	200–300	Review	ABCHKT 1989
		1.6×10^{12}		2.9	300	Review	DSHRGKHKM 1992
		2.3×10^{12}		4.2	210–500	Review	BDGM 1981
		2.8×10^{12}		4.1	240–295	FR, LP, RFS	MMS 1984
		2.7×10^{7}	1.65	–0.9	240–105	FR, LP, RFS	RWWT 1985
		1.3×10^{12}		2.3	258–334	DFR, RF	Key 1984
Recommendation		2.7×10^{7}	1.65	–0.9	240–1500		
$OH + HOCl \longrightarrow H_2O + ClO$	–105.0	1.8×10^{12}		4.1	300	Review	DSHRGKHKM 1992
		$1.0–5.7\times10^{13}$			298	DFR,RFS,MS	EB 1988
$OH + Cl_2 \longrightarrow HOCl + Cl$	7.7	8.4×10^{11}		7.5	300	Review	DSHRGKHKM 1992
		1.0×10^{12}		7.6	253–333	DFR, RF	BHW 1987
		4.0×10^{10}			298	DFR, RF	LA 1984
$OH + ClO \longrightarrow HO_2 + Cl$ (a)	–8.4						
$OH + ClO \longrightarrow HCl + O_2$ (b)	–232.6						
		6.6×10^{12} *b*		–1.0	200–373	Review	ABCHKT 1992
$OH+OClO \longrightarrow HOCl+O_2$ (a)	–236.0	2.7×10^{11} *b*		–6.7	290-480	Review	ABCHKT 1992
$OH+OClO \longrightarrow ClO+HO_2$ (b)	–21.8	2.7×10^{11} *b*		–6.7	293–473	DFR, MS	PZLeB 1986

[a] See footnote on page 402 for units and comments.

[b] Sum of channels (a) and (b).

TABLE 4.5. Reactions of OH radical.[a] (Continued)

Reaction	$\Delta_r H$	A	m	E_a	ΔT	Method	Reference
$OH + Cl_2O \longrightarrow$ Products		1.0×10^{12}		–3.5	230–400	LMR, DFR	SA 1992
		2.4×10^{11}			298	DFR, MS	EB 1988
$OH + CH_4 \longrightarrow CH_3 + H_2O$	–60.2	1.6×10^{6}	1.83	11.6	1000–2500	Review	ABCHKT 1992
		1.6×10^{7}	2.1	10.3	1000–2500	Review	War 1984
		2.4×10^{12}		15.7	240–300	Review	ABCHKT 1992
$OH+CH_3Cl \longrightarrow CH_2Cl+H_2O$	–84.1	1.1×10^{12}		9.3	240–300	Review	ABCHKT 1992
		1.6×10^{12}		2.4	250–450	Review	BDGM 1981
		5.0×10^{8}	1.4	2.4	250–2000	TST Analysis	CB 1987
		1.3×10^{12}		9.5	295–800	Review	DSHRGKHKM 1992
		5.1×10^{8}	1.38	10.0	292–800	LP, LIF	TD'AMKD 1989
		2.7×10^{6}	2	4.9	250–1000	SAR Analysis	Atk 1987
		9.3×10^{7}	1.6	8.7	250–2000	Review	CW 1991
Recommendation		9.3×10^{7}	1.6	8.7	250–2000		
$OH+CH_2Cl_2 \longrightarrow CHCl_2+H_2O$	–100.8	2.7×10^{12}		8.6	240–300	Review	ABCHKT 1992
		2.9×10^{12}		9.1	240–400	Review	BDGM 1981
		6.0×10^{7}	1.6	4.3	250–2000	TST Analysis	CB 1987
		3.5×10^{12}		9.2	298	Review	DSHRGKHKM 1992
		9.5×10^{7}	1.58	5.2	292–800	LP, LIF	TD'AMKD 1989
		2.6×10^{6}	2	2.8	250–1000	SAR Analysis	Atk 1987
		6.0×10^{7}	1.6	4.3	250–2000	Review	CW 1991
Recommendation		6.0×10^{7}	1.6	4.3	250–2000		

[a] See footnote on page 402 for units and comments.

TABLE 4.5. Reactions of OH radical.[a] (Continued)

Reaction	$\Delta_r H$	A	m	E_a	ΔT	Method	Reference
$OH + CHCl_3 \longrightarrow CCl_3 + H_2O$	−113.0	2.0×10^{12}		8.6	240–300	Review	ABCHKT 1992
		2.8×10^{12}		9.4	240–380	Review	BDGM 1981
		6.6×10^{6}	1.7	1.4	250–2000	TST Analysis	CB 1987
		2.6×10^{12}		9.2	295	Review	DSHRGKHKM 1992
		1.2×10^{4}	2.78	0.8	292–800	LP, LIF	TD'AMKD 1989
		1.1×10^{6}	2	1.3	250–1000	SAR Analysis	Atk 1987
		1.0×10^{4}	2.8	0.8	250–2000	Review	CW 1991
Recommendation		1.0×10^{4}	2.8	0.8	250–2000		
$OH + CCl_4 \longrightarrow CCl_3 + HOCl$	46.3	6.0×10^{11}		18.8	250–300	Review	ABCHKT 1992
		6.0×10^{11}		19.3	290–380	Review	ABCHKT 1981
$OH+CH_3CHO \longrightarrow CH_3CO+H_2O$	−138.1	1.0×10^{13}			298	SR, LP, RFS	B-GLeBMacL 1992
$OH+CH_2ClCHO \longrightarrow CH_2ClCO+H_2O$							
		1.8×10^{12}			298	SR, LP, RFS	B-GLeBMacL 1992
		1.9×10^{12}			298	SR, IS, GC	SZEN 1990
		1.9×10^{12}			298	SR, IS, G	SCS 1991
$OH+CHCl_2CHO \longrightarrow CHCl_2CO+H_2O$							
		1.5×10^{12}			298	SR, LP, RFS	B-GLeBMacL 1992
		1.7×10^{12}			298	SR, IS, GC	SZEN 1990
		1.4×10^{12}			298	SR, IS, GC	SCS 1991

[a] See footnote on page 402 for units and comments.

TABLE 4.5. Reactions of OH radical.[a] (Continued)

Reaction	$\Delta_r H$	A	m	E_a	ΔT	Method	Reference
$OH + CCl_3CHO \longrightarrow CCl_3CO + H_2$		5.2×10^{11}			298	SR, LP, RFS	B-GLML 1992
		9.6×10^{11}			298	SR, IS, GC	SCS 1991
		7.1×10^{12}		5.0	298–520	DFR, RFS	DKB 1989
		5.5×10^{9}			298	SR, PH, GC	NSSTN 1990
$OH+CH_3COCl \longrightarrow CH_2COCl+H_2O$		4.3×10^{10}			298	SR, PH, IR	NTS 1984
		5.5×10^{9}			298	SR, PH, GC	NSSTN 1990
$OH + C_2H_6–C_2H_5 + H_2O$	–79.7	7.2×10^{6}	2	37.1	1000–2500	Review	ABCHKT 1992
		6.3×10^{6}	2	2.7	1000–2500	Review	War 1984
		4.7×10^{12}		8.5	240–30	Review	ABCHKT 1992
$OH+C_2H_5Cl \longrightarrow CH_3CHCl+H_2O$ (a)	–115.9	2.6×10^{6}	2	0.5	250–1000	SAR Analysis	Atk 1987
Recommendation		2.6×10^{6}	2	0.5	250–1000		
$OH+C_2H_5Cl \longrightarrow CH_2ClCH_2+H_2O$ (b)	–70.7						
		2.7×10^{6}	2	3.9	250–1000	SAR Analysis	Atk 1987
Recommendation		2.7×10^{6}	2	3.9	250–1000		
		1.5×10^{13} [b]		9.0	295–360	PR, UVS	MN 1992
		6.5×10^{4}	2.6	–0.9	250–2000	Review	CW 1991
		6.8×10^{4}	2.59	–1.0	294–789	SR, LP, LIF	KTD 1990
		5.8×10^{7}	1.6	2.1	250–2000	TST Analysis	CB 1987
		2.4×10^{11}			293	DFR, RFS	HE 1976
Recommendation		5.8×10^{7}	1.6	2.1	250–2000		

[a] See footnote on page 402 for units and comments.

[b] Sum of channels (a) and (b).

TABLE 4.5. Reactions of OH radical.[a] (Continued)

Reaction	$\Delta_r H$	A	m	E_a	ΔT	Method	Reference
$OH+CH_3CHCl_2 \longrightarrow CH_3CCl_2+H_2O$ (a)	−103.3						
		1.1×10^6	2	−1.0	250–1000	SAR Analysis	Atk 1987
Recommendation		1.1×10^6	2	−1.0	250–1000		
$OH+CH_3CHCl_2 \longrightarrow CHCl_2CH_2+H_2O$ (b)	−61.9						
		2.7×10^6	2	3.9	250–1000	SAR Analysis	Atk 1987
Recommendation		2.7×10^6	2	3.9	250–1000		
		1.2×10^4	2.67	−3.1	294-800	LP, LIF	JTD 1992b
		2.1×10^7	1.6	0.0	298–2000	Review	CW 1991
		2.1×10^7	1.6	0.5	250–2000	TST Analysis	CB 1987
		1.6×10^{11}			293	DFR, RFS	HE 1976
Recommendation		2.1×10^7	1.6	0.0	298–2000		
$OH + CH_2ClCH_2Cl \longrightarrow CH_2ClCHCl + H_2O$	−90.5						
		5.4×10^{12}		9.0	293–418	DFR, RFS	XSQ 1992
		8.2×10^9	1	6.9	292–775	LP, LIF	TMcCD 1991
		5.2×10^6	2	1.8	250–1000	SAR Analysis	Atk 1987
		1.4×10^{11}			296	DFR, RFS	HE 1976
Recommendation		8.2×10^9	1	6.9	250–1500		

[a] See footnote on page 402 for units and comments.

TABLE 4.5. Reactions of OH radical.[a] (Continued)

Reaction	$\Delta_r H$	A	m	E_a	ΔT	Method	Reference
$OH+CH_2ClCHCl_2 \longrightarrow CH_2ClCCl_2+H_2O$ (a)							
	−108.4	1.1×10^{6}	2	0.3	250–1000	SAR Analysis	Atk 1987
Recommendation		1.1×10^{6}	2	0.3	250–1000		
$OH+CH_2ClCHCl_2 \longrightarrow CHCl_2CHCl+H_2O$ (b)							
	−90.4	2.6×10^{6}	2	1.8	250–1000	SAR Analysis	Atk 1987
Recommendation		2.6×10^{6}	2	1.8	250–1000		
		2.8×10^{4} [b]	2.64	−0.3	295–790	LP, LIF	TJD 1992
		1.6×10^{7}	1.6	−0.1	250–2000	Review	CW 1991
		1.6×10^{7}	1.7	0.8	250–2000	TST Analysis	CB 1987
		9.9×10^{11}		4.0	250–470	DFR, RF	JHJK 1984
$OH+CHCl_2CCl_2H \longrightarrow CHCl_2CCl_2+H_2O$							
	−94.6	2.3×10^{6}	2	0.3	250–1000	SAR Analysis	Atk 1987
		3.0×10^{12}		7.7	293–418	DFR, RF	XSQ 1992
Recommendation		1.9×10^{6}	2	0.3	250–1500		

[a] See footnote on page 402 for units and comments.
[b] Sum of channels (a) and (b).

TABLE 4.5. Reactions of OH radical.[a] (Continued)

Reaction	$\Delta_r H$	A	m	E_a	ΔT	Method	Reference
$OH + CH_3CCl_3 \longrightarrow CCl_3CH_2 + H_2O$							
	–61.5	1.1×10^{12}		12.9	243–379	PH, LIF	TMSWMR 1992
		1.7×10^{6}	2.08	8.9	298–761	LP, LIF	JTD 1992a
		1.9×10^{12}		13.8	240–300	Review	ABCHKT 1992
		2.1×10^{12}		13.0	260–400	Review	BDGM 1981
		3.0×10^{12}		15.0	250–460	Review	DSHRGKHKM 1992
		2.4×10^{9}	1.1	13.4	250–200	Review	CW 1991
		3.3×10^{12}		14.9	359–402	PR, AS	NSSTN 1990a
		2.7×10^{6}	2	8.5	250–1000	SAR Analysis	Atk 1987
		2.4×10^{9}	1.1	9.8	250–2000	TST Analysis	CB 1987
		3.0×10^{12}		14.9	250–470	DFR, RFS	JHJK 1984
		2.2×10^{12}		6.8	260–375	FR, LP, RFS	WMCWD 1977
		9.0×10^{9}			296	DFR, LMR	HE 1976
Recommendation		2.4×10^{9}	1.1	13.4	250–2000		
$OH + C_2HCl_5 \longrightarrow C_2Cl_5 + H_2O$	–104.2	1.4×10^{11}			292	DFR, RF	Qiu et al 1992
$OH{+}CH_2ClCH_2OH \longrightarrow$Products		7.7×10^{11}			298	SR, PH, RFS	WDK1988
$OH{+}CCl_3CH_2OH \longrightarrow$Products		1.5×10^{11}			298	SR, PH, RFS	WDK1988
$OH + C_2H_4 \longrightarrow C_2H_3 + H_2O$	–47.0	2.1×10^{13}		24.9	1000–2500	Review	ABCHKT 1992
		3.0×10^{13}		12.5	1000–2500	Review	War 1984
$OH{+}C_2H_3Cl \longrightarrow$Products		1.3×10^{12}		–2.9	313–588	FR, PR, RFS	LMJ 1989
		6.9×10^{11}		–4.4	299–426	FR, LP, RFS	PAP 1977
		3.6×10^{12}				296 DFR, LMR	How 1976

[a] See footnote on page 402 for units and comments.

TABLE 4.5. Reactions of OH radical.[a] (Continued)

Reaction	$\Delta_r H$	A	m	E_a	ΔT	Method	Reference
$OH + C_2H_3Cl \longrightarrow CH_2CCl + H_2O$ (b)	–43.5						
$OH + C_2H_3Cl \longrightarrow CHClCH + H_2O$ (c)	–38.0						
		8.4×10^6 [b]	2	10.0	723–1173	FR, PR, RFS	LMJ 1989
$OH + t\text{-}C_2H_4Cl_2 \longrightarrow$ Products		5.7×10^{11}		–2.3	240–400	SR, LP, RFS	ZLHK 1991
		1.3×10^{12}		0.2	297–367	DFR, UVS	AA 1991
$OH + cC_2H_4Cl_2 \longrightarrow$ Products		1.3×10^{12}		–0.5	240–400	SR, LP, RFS	ZLHK 1991
		1.2×10^{12}		–0.6	297–367	DFR, UVS	AA 1991
$OH + CH_2CCl_2 \longrightarrow$ Products		1.4×10^{12}		–3.9	240–400	SR, LP, RFS	ZLHK 1991
		9.6×10^{11}		–4.6	297–367	DFR, UVS	AA 1991
		1.4×10^{11}		–9.5	295–350	DFR, MS	KHOV 1990
$OH + C_2HCl_3 \longrightarrow$ Products		3.0×10^{13}		–3.7	234–420	Review	DSHRGKHKM 1992
		4.7×10^{11}		–2.0	298–459	DFR, MS	KHOV 1990
$OH + C_2Cl_4 \longrightarrow$ Products		5.7×10^{12}		10.0	297–420	Review	DSHRGKHKM 1992
		3.3×10^{12}		8.6	298–459	DFR, MS	KHOV 1990
$OH + CH_2ClCHCH_2 \longrightarrow$ Products		1.0×10^{13}			298	SR, PH, IRS	TAA 1990
		1.2×10^{13}			296	SR, PH, GC	EKC 1986
$OH + c\text{-}CH_2ClCHCHCl \longrightarrow$ Products		5.1×10^{12}			298	SR, PH, FTIR	TAAGW 1988
$OH + t\text{-}CH_2ClCHCHCl \longrightarrow$ Products		8.7×10^{12}			298	SR, PH, FTIR	TAAGW 1988

[a] See footnote on page 402 for units and comments.
[b] Sum of channels (b) and (c).

TABLE 4.5. Reactions of OH radical.[a] (Concluded)

Reaction	A	E_a	ΔT	Method	Reference
$OH + CH_3CH_2CH_2Cl \longrightarrow$ Products	6.0×10^{13}	12.0	295–360	PR, UVS	MN 1992
$OH + CH_3CHClCH_3 \longrightarrow$ Products	3.0×10^{13}	11.0	295–360	PR, UVS	MN 1992
$OH + CH_3(CH_2)_2CH_2Cl \longrightarrow$ Products	1.7×10^{13}	7.0	295–360	PR, UVS	MN 1992
$OH + CH_3(CH_2)_3CH_2Cl \longrightarrow$ Products	1.0×10^{13}	4.0	295–360	PR, UVS	MN 1992
$OH + CH_3(CH_2)_4CH_2Cl \longrightarrow$ Products	9.6×10^{12}	3.0	295–360	PR, UVS	MN 1992
$OH + (CH_2Cl)_2CCH_2 \longrightarrow$ Products	2.0×10^{13}		298	SR, PH, FTIR	TAAGW 1988
$OH + C_6H_5Cl \longrightarrow$ Products	3.3×10^{11}		297	SR, PH, GC	EKC 1986
OH + 1,2-$C_2H_4Cl_2 \longrightarrow$ Products	2.5×10^{11}		298	SR, PH, RFS	WZ 1983
OH + 1,3-$C_2H_4Cl_2 \longrightarrow$ Products	4.3×10^{11}		298	SR, PH, RFS	WZ 1983
OH + 1,4-$C_2H_4Cl_2 \longrightarrow$ Products	2.0×10^{11}		298	SR, PH, RFS	WZ 1982
OH + 1,2,3-$C_6H_3Cl_3 \longrightarrow$ Products	1.4×10^{12}	3.6	273–368	SR, PH, RFS	RZ 1984
$OH + C_6H_5CH_2Cl \longrightarrow$ Products	1.7×10^{12}		298	SR, PH, IRS	TAA 1990
	1.8×10^{12}		298	SR, PH, GC	EKC 1990
OH + 2,3,7,8–Tetrachlorodibenzodioxin $\longrightarrow$ Products	5.4×10^{12}		298	Estimate	Atk 1987
OH + Chlorinated biphenyls $\longrightarrow$ Products			298	Estimate	Atk 1987

[a] See footnote on page 402 for units and comments.

TABLE 4.6. Structure activity relationship parameters for H abstractions from substituted hydrocarbons by Cl attack.[a]

Group	T_x, T_y, T_z
CH_3	480
Cl	–280
CH_2Cl	–80
$CHCl_2$	–740
CCl_3	–1030
–OH	1260
$–OCH_3$	1620
$–OOCH_3$	1490
CH_2	1020
>CH–	1100
>C<	980

[a] Adapted from Senkan and Quam (1992).

and the T_x, T_y, T_z values are 0 ($–CH_3$), –7 ($>CH_2$), –86 (>CH–) and +28 (>C<) K. The SAR parameters in this case have preexponential temperature exponent $m = 0$, consistent with the experimental data. Again we shall make use of the SAR methodology when appropriate in discussing the reactions of Cl with hydrocarbons and chlorinated hydrocarbons.

$Cl + H_2 \longrightarrow HCl + H$

The kinetics of this reaction appear to be well understood. The reaction was studied by Miller and Gordon (1981) from 200 to 500 K and evaluated by Atkinson et al. (1992), DeMore et al. (1992), and by Baulch et al. (1981), all of whom recommended two-parameter rate expressions. The agreement between experiments is generally good, as is that between the evaluations. We recommend a three-parameter rate expression in order to account for the expected Arrhenius plot curvature at higher temperatures, as shown in Figure 4.23.

$Cl + HOCl \longrightarrow OH + Cl_2$ (a)
$\longrightarrow HCl + ClO$ (b)

This reaction was studied in a discharge flow reactor by Cook et al. (1981), who concluded that channel (a) is the primary reaction path. Ennis and Birks (1985) concluded that the channel is (b). We make no recommendation.

$Cl + HO_2 \longrightarrow HCl + O_2$ (a)
$\longrightarrow OH + ClO$ (b)

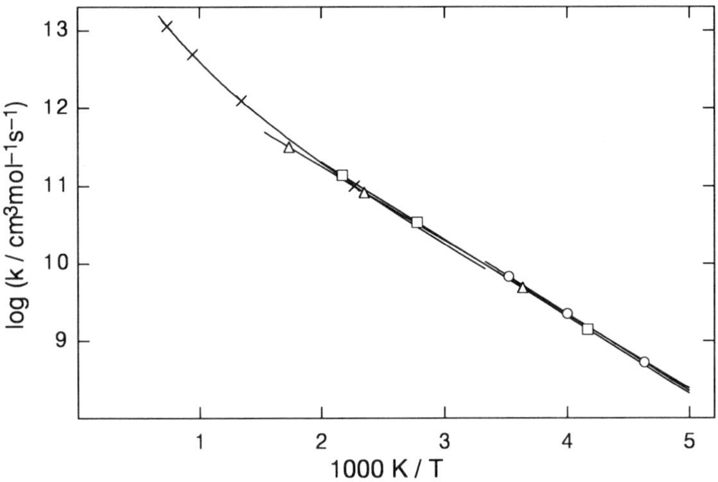

FIGURE 4.23. Rate coefficient expressions for reaction of Cl with H_2.
× Recommendation ○ Atkinson et al. (1992) △ Baulch et al. (1981)
□ Miller and Gordon (1981).

This reaction is important in early stages of ignition, while HO_2 radicals still play a significant role. As noted above, this reaction can be either a termination process with inhibitory effects—channel (a)—or a propagation step—channel (b). It was studied by Cox (1980) and by Lee and Howard (1982) in discharge flow reactors using molecular modulation spectroscopy and laser magnetic resonance spectroscopy, respectively. Lee and Howard measured the branching ratio and found that channel (a) dominates at low temperatures and has negative activation energy. The importance of channel (b) increased with temperature, becoming competitive with channel (a) above 600 K. Lee and Howard (1982) found that the overall reaction rate coefficient k_{a+b} was independent of temperature. Atkinson et al. (1992), however, preferred channel (a). Additional studies to establish the branching ratio would be useful for combustion modeling.

$Cl + H_2O_2 \longrightarrow HO_2 + HCl$

The reaction of Cl with H_2O_2 was studied by Michael et al. (1977) from 265 to 400 K using laser photolysis and resonance fluorescence spectroscopy, by Poulet et al. (1978) at 298 K in a discharge flow reactor coupled to a mass spectrometer, and by Keyser (1980), also in a discharge flow reactor, from 298 to 420 K using resonance fluorescence spectroscopy for Cl atom detection. Heneghan and Benson (1983) investigated it in a very low pressure pyrolysis reactor coupled to a mass spectrometer and found the reaction channels leading to production of HOCl + OH and ClO + H_2O to be unimportant. Evaluations by Atkinson et al. (1992) and by DeMore et al. (1992) agree with the measurements of Keyser (1980). As seen in Figure 4.24, although the measurements are in reasonable accord at room temperature, they deviate at higher temperatures. We make no recommendation.

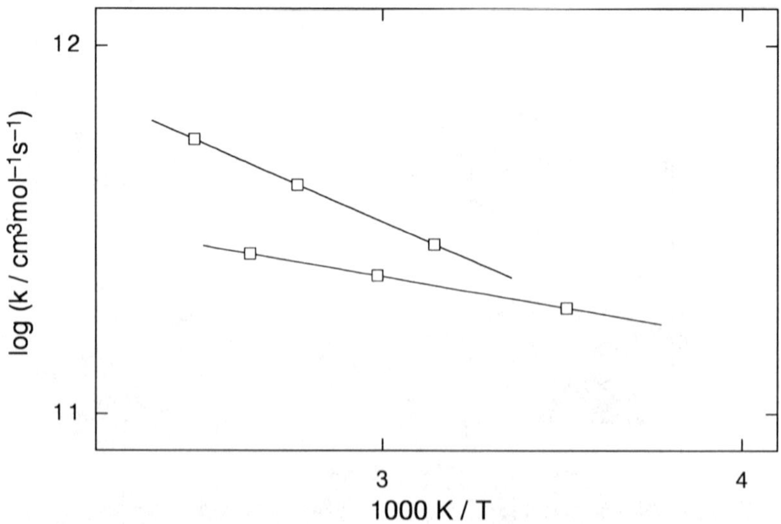

FIGURE 4.24. Rate coefficient expressions for $Cl + H_2O_2 \longrightarrow HCl + HO_2$. □ Keyser (1980) ○ Michael et al. (1977).

$Cl + COCl_2 \longrightarrow COCl + Cl_2$

This is an important reaction in the combustion of CHCs because of the formation of $COCl_2$ as an intermediate in CHC oxidation (Chang and Senkan 1989). The kinetics of this reaction has not been re-examined since early studies by Bodenstein et al. (1937) and Burns and Dainton (1952), which are in accord with one another and the known thermochemistry.

$Cl + CH_4 \longrightarrow CH_3 + HCl$

The reaction of Cl with CH_4 is unique among the alkanes in being endothermic, by 1.7 kJ/mol, and having a significant activation energy. The analogous reactions of other alkanes are exothermic and have no activation barriers (see below). This reaction is one of the most widely studied reactions in chemical kinetics. As seen in Figure 4.25, the results of the various investigators, both early and recent, are in agreement. Although the recommendations of Atkinson et al. (1992) and DeMore et al. (1992) provide reliable representations of the kinetics near ambient temperature, we prefer the three-parameter expression suggested by Baulch et al. (1981) for combustion modeling.

$Cl + CH_3Cl \longrightarrow CH_2Cl + HCl$

As discussed previously, the presence of a Cl atom weakens the C–H bond (see Table 4.1b), making the reaction exothermic. The experimental studies of this reaction, with the exception of that of Clyne and Walker (1973), which is believed to be in error (Manning and Kurylo 1977), generally agree with one another

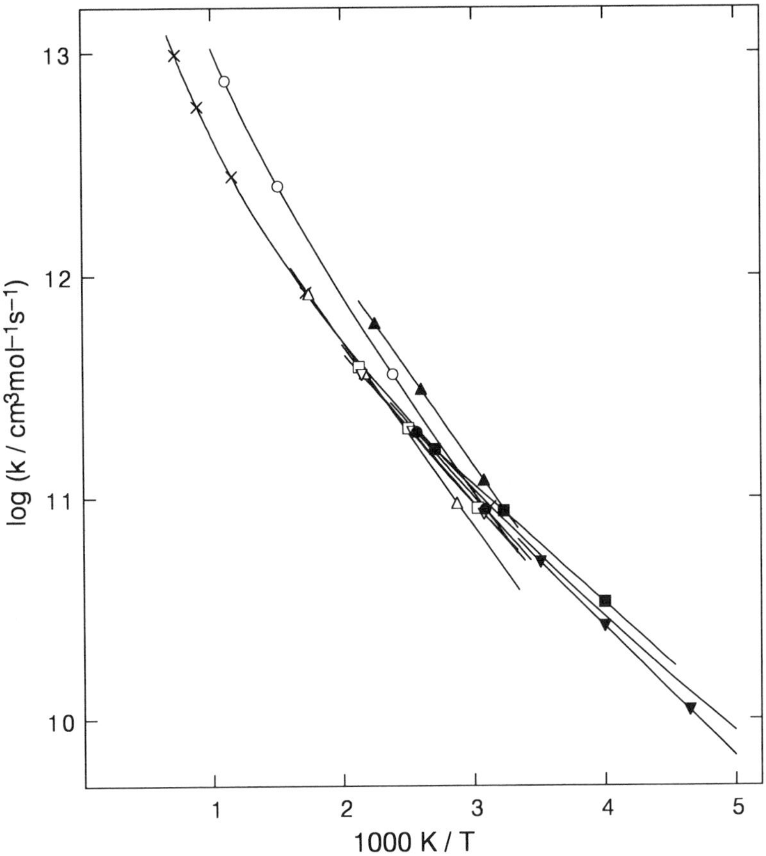

FIGURE 4.25. Rate coefficient expressions for reaction of Cl with CH_4. × Baulch et al. (1981) and recommendation ○ Heneghan et al. (1981) △ Fettis and Knox (1964) □ Zahniser et al. (1978) ▽ Russell et al. (1988) ● Keyser (1978) ▲ Chiltz et al. (1963) ■ Watson et al. (1976) ▼ Atkinson et al. (1992).

and with SAR predictions (Senkan and Quam 1992) and ambient temperature recommendations (Atkinson et al. 1992), as shown in Figure 4.26. We elected to use the three-parameter rate coefficient expression with $m = 2$ listed in Table 4.7.

Although the Cl + CH_3Cl reaction is exothermic by about 4 kJ/mol, it has a significant activation energy, which can be attributed to Cl···$ClCH_3$ repulsion in the transition state. One would also expect similar behavior with other chlorinated methanes. Increased chlorine substitution, however, also further decreases the C–H bond dissociation energy (Table 4.1b), and therefore the heat of reaction; this, in turn, would be expected to further decrease the activation energy in the Evans–Polanyi sense (Senkan and Quam 1992). As a result of the offsetting effect of Cl···Cl repulsion, however, the activation energy changes little from CH_3Cl to CH_2Cl_2 to $CHCl_3$.

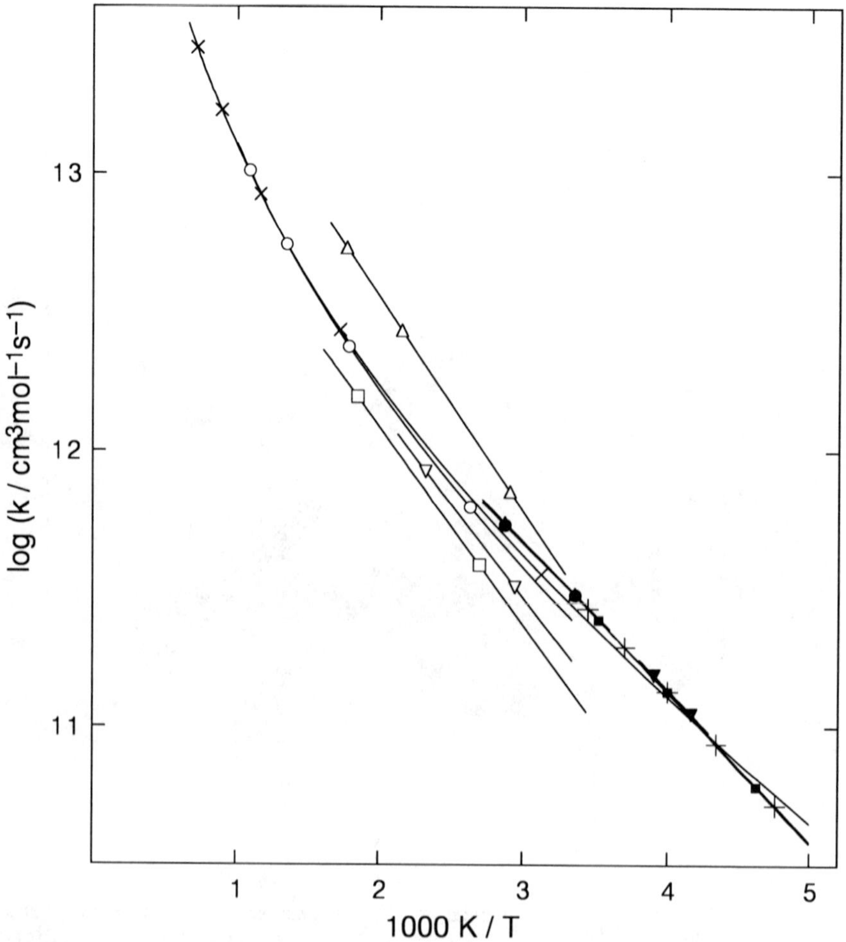

FIGURE 4.26. Rate coefficient expressions for reaction of Cl with CH_3Cl.
× Recommendation ○ Senkan and Quam (1992) △ Clyne and Walker (1973) □ Fettis and Knox (1984) ▽ Chiltz et al. (1963) ● Tschuikow-Roux et al. (1985) ▲ Tschuikow-Roux et al. (1985) ■ Atkinson et al. (1989) + Atkinson et al. (1992) ▼ Manning and Kurylo (1977).

$Cl + CH_2Cl_2 \longrightarrow CHCl_2 + HCl$

The experimental data for this reaction, with the exception of that of Clyne and Walker (1973), believed to contain errors (Manning and Kurylo 1977), agree with one another and with SAR predictions (Senkan and Quam 1992) and ambient temperature recommendations (Atkinson et al. 1992), as seen in Figure 4.27. We chose to represent the experimental data using the three-parameter rate expression given in Table 4.7.

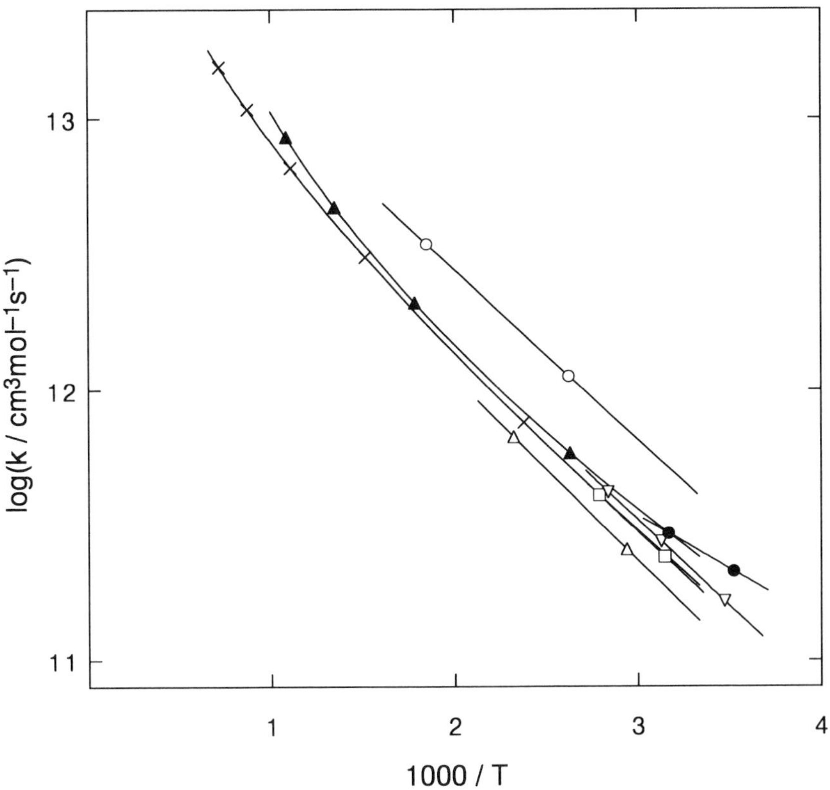

FIGURE 4.27. Rate coefficient expressions for reaction of Cl with CH_2Cl_2.
× Recommendation ○ Clyne and Walker (1973) □ Fettis and Knox (1964) △ Chiltz et al. (1973) ▽ Tschuikow-Roux et al. (1988) ● Atkinson et al. (1992) ▲ Senkan and Quam (1992).

$Cl + CHCl_3 \longrightarrow CCl_3 + HCl$

The experimental data available for this reaction are more limited than for the other chlorinated methanes. Discounting the results of Clyne and Walker (1973) owing to the apparent presence of systematic error, we are left with the relative rate measurements of Fettis and Knox (1964), who studied this reaction in a static reactor and determined product distributions using gas chromatography against CH_4 as an internal standard, and those of Chiltz et al. (1963). The original rate coefficient determinations of Fettis and Knox (1964) were based on a rate coefficient for Cl + CH_4 that is now known to be incorrect. Using the Table 4.7 expression for the Cl + CH_4 rate coefficient, we recalculated the rate of hydrogen abstraction from $CHCl_3$ as the basis for the recommended rate expression in Table 4.7. As seen in Figure 4.28, this recommendation agrees with SAR predictions (Senkan and Quam 1992) and ambient temperature recommendations. (Atkinson et al. 1992)

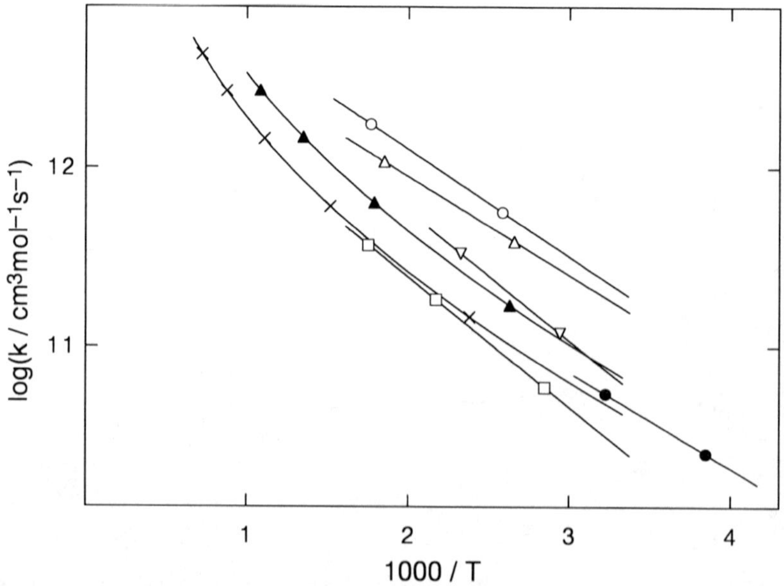

FIGURE 4.28. Rate coefficient expressions for reaction of Cl with $CHCl_3$. × Recommendation ○ Clyne and Walker (1973) △ Recalculated from Fettis and Knox (1964) □ Fettis and Knox (1964) ▽ Chiltz et al. (1963) ● Atkinson et al. (1992) ▲ Senkan and Quam (1992).

Cl + small oxygenated hydrocarbons (such as CH_3OH, CH_2O, C_2H_5OH, CH_3CHO and CH_3OCH_3)

Several investigators have studied this class of reactions, which are expected to lead to a variety of products because of the presence of chemically different H atoms. The experimental techniques used generally did not distinguish between different products, that is, only total reaction rate coefficients were reported. Most of these reactions exhibited no dependence on temperature over the ranges investigated. We discuss here only the rate information for CH_3OH and CH_2O.

$Cl + CH_3OH \longrightarrow$ products + HCl

The kinetics of the reaction with CH_3OH was studied by Michael et al. (1979b) using laser photolysis and resonance fluorescence spectroscopy from 200 to 500 K. Their results agree with room temperature studies by Nelson et al. (1990b), Payne et al. (1988), and Wallington et al. (1988b). Michael et al. observed no temperature dependence for this reaction and speculated that the primary reaction path is abstraction of H from CH_3. Since no product identifications have been made, the branching ratio to CH_2OH and CH_3O remains unknown. Evaluations by Atkinson et al. (1992) and DeMore et al. (1992) also suggest CH_2O + HCl formation as the main channel.

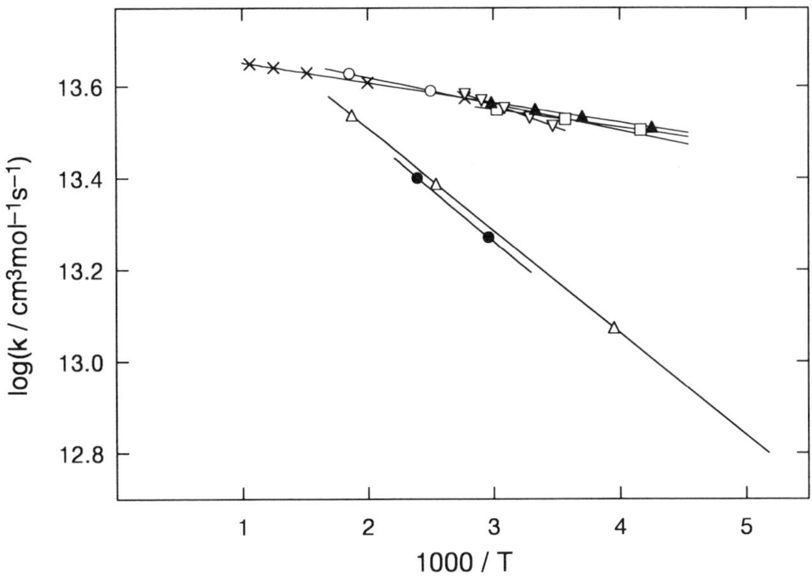

FIGURE 4.29. Rate coefficient expressions for reaction of Cl with C_2H_6.
× Recommendation ○ Lewis et al. (1980) △ Fettis and Knox (1964) □ Demore et al. (1990) ▽ Tschuikow-Roux et al. (1985a) ● Cillien et al. (1967) ▲ Atkinson et al. (1992).

$Cl + CH_2O \longrightarrow CHO + HCl$

This reaction was studied by Michael et al. (1979a) using laser photolysis and resonance fluorescence spectroscopy from 200 to 500 K. Their results agree with those of by Fasano and Nogar (1981) at 295 K and ambient temperature evaluations by Atkinson et al. (1992) and DeMore et al. (1992). The rate of this reaction also appears to be independent of temperature.

$Cl + C_2H_6 \longrightarrow C_2H_5 + HCl$

This reaction, exothermic by about 3 kJ/mol, has been used experimentally by several investigators as a reference for relative rate measurements. As seen in Figure 4.29, with the exception of earlier measurements by Fettis and Knox (1964) and Cillien et al. (1967), now thought to be incorrect, the experimental studies and ambient temperature evaluations are in general agreement. (DeMore 1990; Atkinson et al. 1992) The reaction was studied by Tschuikow-Roux et al. (1985a) in a static reactor from 280 to 368 K. Their results agree with room temperature measurements of Kaiser et al. (1992) and the rate parameters determined by SAR (Senkan and Quam 1992) . The results with C_2H_6 are consistent with the generalization that reactions of Cl with alkanes, with the exception of CH_4, are exothermic and have virtually no energy barrier.

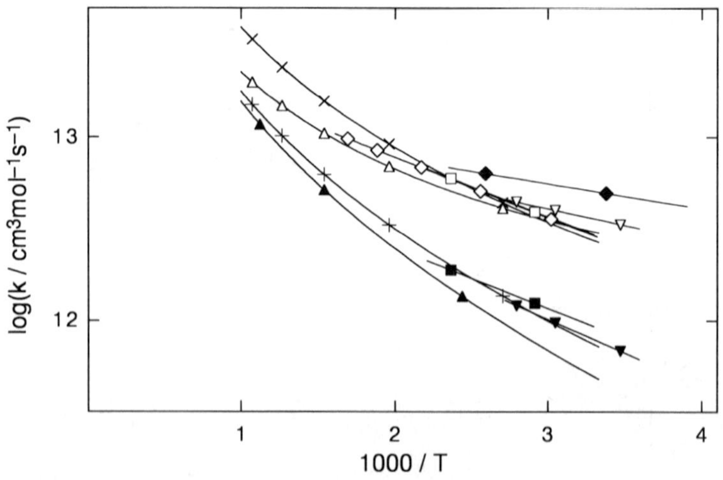

FIGURE 4.30. Rate coefficient expressions for $Cl + C_2H_5Cl \longrightarrow HCl + CH_3CHCl$ (a) and $Cl + C_2H_5Cl \longrightarrow HCl + CH_2CH_2Cl$ (b). × Recommendation (a) + recommendation (b) ◇ Fettis and Knox (1964, a+b) □ Cillien et al. (1967, a) ■ Cillien et al. (1967, b) ◆ Wine and Semmes (1983, a+b) ▽ Tschuikow-Roux et al. (1984b, a) ▼ Tschuikow-Roux et al. (1984b, b) △ Senkan and Quam (1992, a) ▲ Senkan and Quam (1992, b).

$Cl + C_2H_5Cl \longrightarrow CH_3CHCl + HCl$ (a)
$\longrightarrow CH_2ClCH_2 + HCl$ (b)

Reaction of Cl with C_2H_5Cl can lead to two different products, abstraction of α hydrogens being favored by bond dissociation energy considerations (cf. Table 4.1b). Cillien et al. (1967), using a static reactor and gas chromatography, established that CH_3CHCl formation is the more important route. They attributed this, however, to steric factors and not to reaction energetics, i.e., their reported rate coefficients for reactions (a) and (b) have the same activation energy but different pre-exponential factors. Tschuikow-Roux et al. (1984b) reinvestigated the reaction and established the contrary to be the case, consistent with SAR expectations (Senkan and Quam 1992) and other reactions. We adopted the results of Tschuikow-Roux et al. (1984b) to derive the three-parameter expression compared to experimental measurements and SAR predictions in Figure 4.30.

$Cl + CH_2ClCH_2Cl \longrightarrow CH_2ClCHCl + HCl$

Several investigators have studied this one-product reaction. As seen in Figure 4.31, the results of Wine and Semmes (1983), Cillien et al. (1967), and Tschuikow-Roux et al. (1986) agree with one another and with SAR predictions (Senkan and Quam 1992). We used the data of Tschuikow-Roux et al. (1986) from 280 to 368 K to derive a three-parameter rate expression with $m = 2$. It is expected to be valid at combustion temperatures.

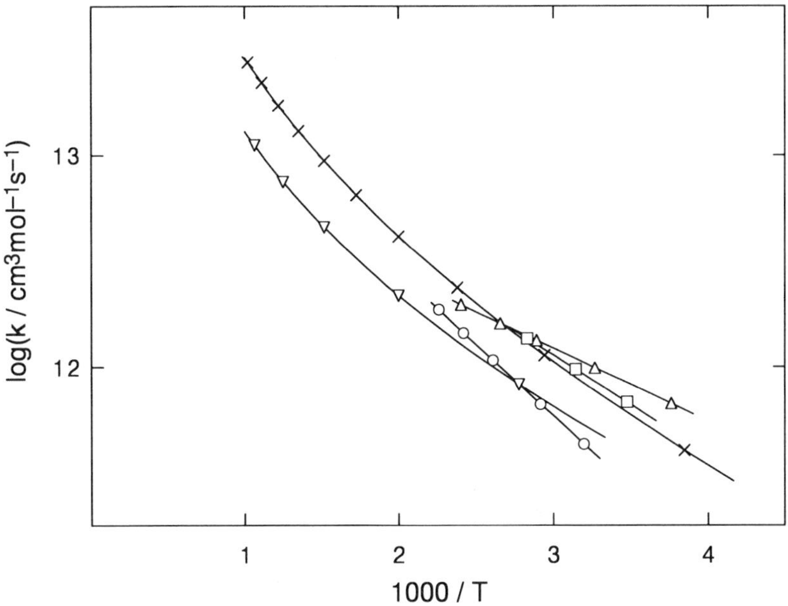

FIGURE 4.31. Rate coefficient expressions for Cl + $CH_2ClCH_2Cl \longrightarrow$ HCl + CH_2Cl CHCl. × Recommendation ○ Cillien et al. (1967) △ Wine and Semmes (1983) □ Tschuikow-Roux et al. (1986) ▽ Senkan and Quam (1992).

$$\begin{aligned} Cl + CH_3CHCl_2 &\longrightarrow CH_3CCl_2 + HCl\ (a) \\ &\longrightarrow CH_2CHCl_2 + HCl\ (b) \end{aligned}$$

In CH_3CHCl_2 the BDE difference between α and β hydrogens is almost 10 kJ/mol, suggesting greater selectivity in favor of α hydrogen abstraction. The measurements of Cillien et al. (1967) and Tschuikow-Roux et al. (1985a), as well as the SAR predictions of Senkan and Quam (1992), confirm that the difference in activation energy is the primary reason for the different reaction rates (Figure 4.32). We base our recommendation on the data of Tschuikow-Roux et al. (1985a), modified to a three-parameter form with $m = 2$.

$$\begin{aligned} Cl + CH_2ClCHCl_2 &\longrightarrow CH_2ClCCl_2 + HCl\ (a) \\ &\longrightarrow CHCl_2CHCl + HCl\ (b) \end{aligned}$$

The BDE difference between α and β hydrogens in $CH_2ClCHCl_2$ is about 4 kJ/mol, again suggesting preferential abstraction of α hydrogens, but less pronounced than seen in CH_3CHCl_2. The measurements of Tschuikow-Roux et al. (1986), which agree with those of Cillien et al. (1967), support this. As seen in Figure 4.33, SAR predictions (Senkan and Quam 1992), while slightly overpredicting the data, are in agreement with the experimental trend. The recommended expression is based on the measurements of Tschuikow-Roux et al. (1986). The results of Wine and Semmes (1983), who reported k_{a+b}, are consistent with the others at lower temperatures.

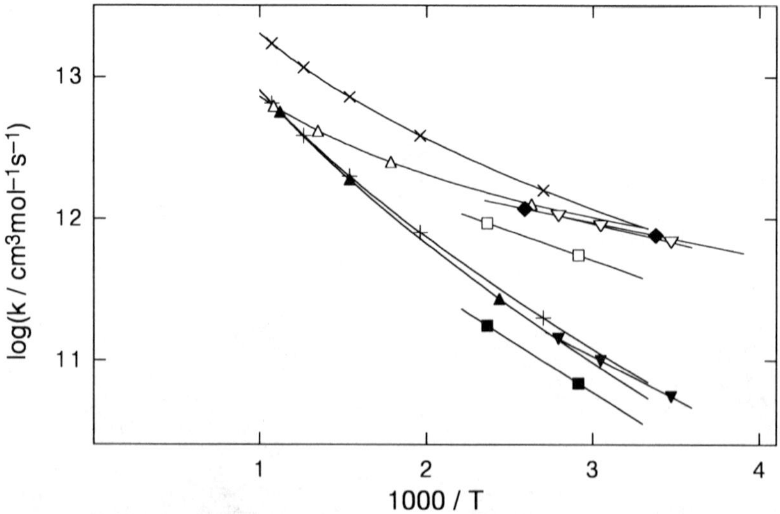

FIGURE 4.32. Rate coefficient expressions for Cl + CH_3CHCl_2 ⟶ HCl + CH_3CCl_2 (a) and Cl + CH_3CHCl_2 ⟶ HCl + $CHCl_2CH_2$ (b). × Recommendation (a) + Recommendation (b) □ Cillien et al. (1967, a) ■ Cillien et al. (1967, b) ◆ Wine and Semmes (1983, a+b) ▽ Tschuikow-Roux et al. (1985a, a) ▼ Tschuikow-Roux et al. (1985a, b) △ Senkan and Quam (1992, a) ▲ Senkan and Quam (1992, b).

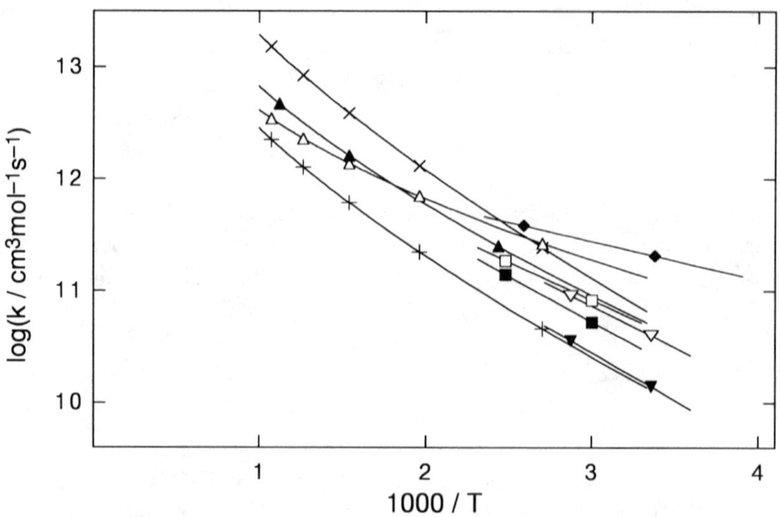

FIGURE 4.33. Rate coefficient expressions for Cl + $CH_2ClCHCl_2$ ⟶ HCl + CH_2Cl CCl_2 (a) and Cl + CH_2Cl $CHCl_2$ ⟶ HCl + $CHCl_2CHCl$ (b). × Recommendation (a) + Recommendation (b) □ Cillien et al. (1967, a) ■ Cillien et al. (1967, b) ◆ Wine and Semmes (1983, a+b) ▽ Tschuikow-Roux et al. (1986, a) ▼ Tschuikow-Roux et al. (1986, b) △ Senkan and Quam (1992, a) ▲ Senkan and Quam (1992, b).

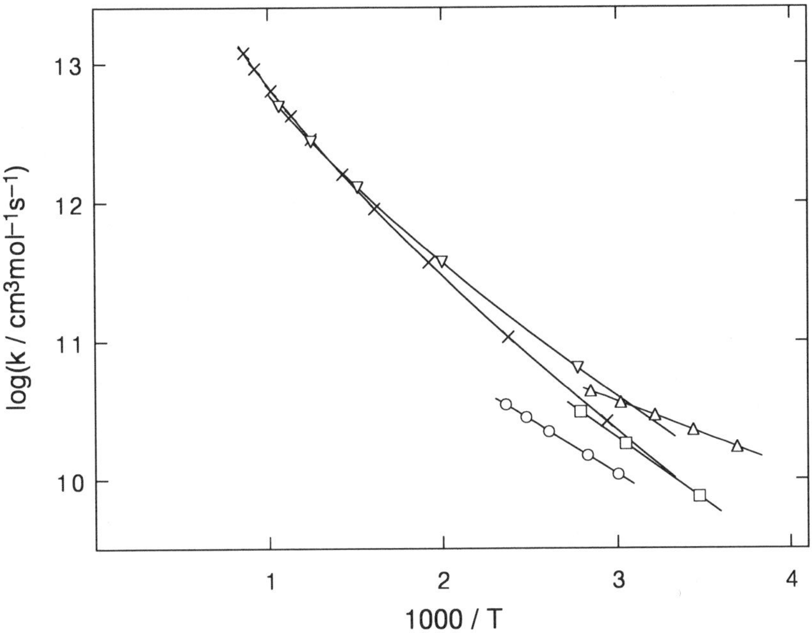

FIGURE 4.34. Rate coefficient expressions for Cl + CH_3CCl_3 ⟶ HCl + CH_2CCl_3. × Recommendation ○ Cillien et al. (1967) △ Wine et al. (1982) □ Tschuikow-Roux et al. (1985a) ▽ Senkan and Quam (1992).

Cl + CH_3CCl_3 ⟶ CH_2CCl_3 + HCl

This reaction is expected to have the highest activation energy of this series of reactions, because the C–H bond dissociation energy in CH_3CCl_2 is the largest of all the chlorinated ethanes (cf. Table 4.1b). The experimental results of Cillien et al. (1967) and Tschuikow-Roux et al. (1985a) are consistent with this. The 2.2 kJ/mol activation energy reported by Wine et al. (1982) seems low based on BDE considerations. Consequently, we have relied on the results of Tschuikow-Roux et al. (1985a) to establish a three-parameter rate coefficient expression with $m = 2$. As seen in Figure 4.34, the recommended expression agrees quite well both with the experimental data and with the SAR prediction reported by Senkan and Quam (1992).

Cl + $CHCl_2CHCl_2$ ⟶ CCl_2CHCl_2 + HCl

The only experimental study of this reaction is a relative rate measurement of Cillien et al. (1967) against $CHCl_3$. Since the rate of H abstraction from $CHCl_3$ is uncertain, the results of Cillien et al. (1967) should be accepted with caution. We used the Table 4.7 rate coefficient expression for the $CHCl_3$ reaction to derive a rate coefficient expression for this reaction; it compares favorably with SAR predictions (Senkan and Quam 1992) as shown in Figure 4.35.

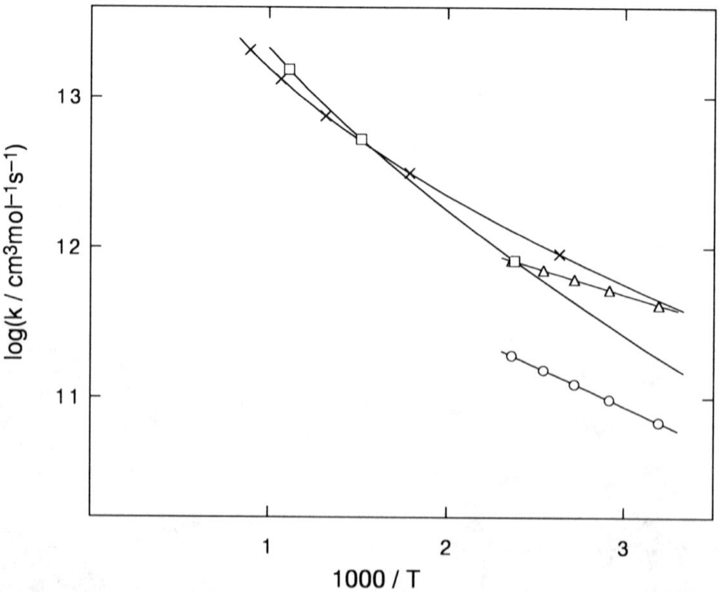

FIGURE 4.35. Rate coefficient expressions for $Cl + CHCl_2CHCl_2 \longrightarrow HCl + CCl_2CHCl_2$. × Recommendation ○ Cillien et al. (1967) △ Recalculated from Cilien et al. (1967) □ Senkan and Quam (1992).

$Cl + CH_2ClCCl_3 \longrightarrow CHClCCl_3 + HCl$

The only experimental study of this reaction is again a relative rate measurement of Cillien et al. (1967) against $CHCl_3$, and again the results of Cillien et al. (1967) should be accepted with caution. We used the Table 4.7 rate coefficient expression for the $CHCl_3$ reaction to derive a rate coefficient expression for this reaction; it is compared to the SAR prediction of Senkan and Quam (1992) in Figure 4.36. There is considerable discrepancy between the SAR predictions and the recommendation based on experimental data. This, however, is not altogether surprising, because for this reaction especially significant uncertainties exist in both the data and the SAR predictions. As discussed by Senkan and Quam (1992), the SAR parameters for CCl_3 are uncertain for lack of reliable rate data. Additional studies are needed to resolve this discrepancy.

$Cl + C_2HCl_5 \longrightarrow C_2Cl_5 + HCl$

The results of Cillien et al. (1967) agree with those of Fettis and Knox (1964). Both were determined from the same rate coefficient expression for the reference reaction. We recalculated the rate coefficient using the relative rates of Cillien et al. (1967) to derive the expression listed in Table 4.7. As seen in Figure 4.37, our recommendation again deviates significantly from the SAR prediction of Senkan and Quam (1992), because of the same problem discussed above: the SAR group parameters for CCl_3 are uncertain.

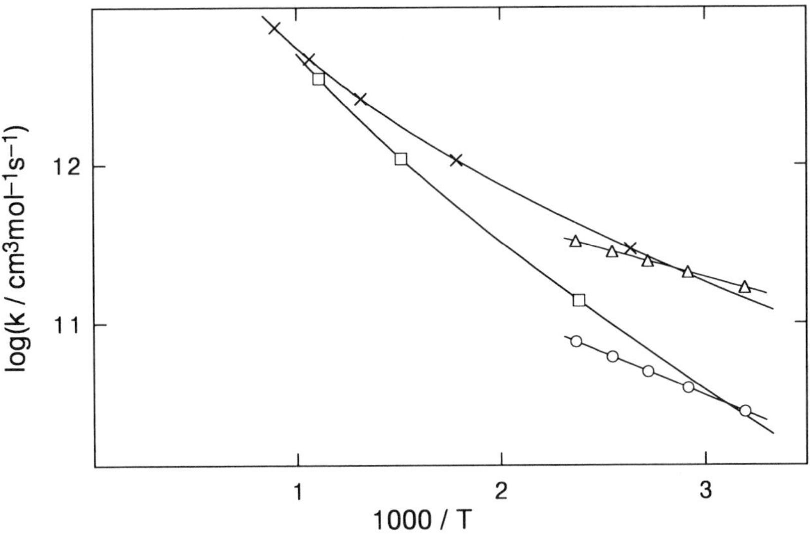

FIGURE 4.36. Rate coefficient expressions for Cl + CH_2ClCCl_3 ⟶ HCl + $CHClCCl_3$. × Recommendation ○ Cillien et al. (1967) △ Recalculated from Cilien et al. (1967) □ Senkan and Quam (1992).

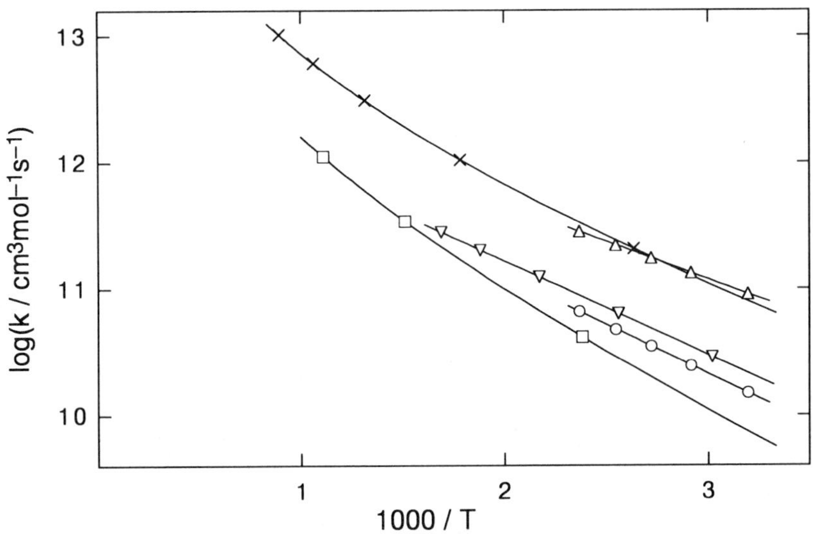

FIGURE 4.37. Rate coefficient expressions for Cl + $CHCl_2CCl_3$ ⟶ HCl + CCl_2CCl_3. × Recommendation ○ Cillien et al. (1967) △ Recalculated from Cilien et al. (1967) ▽ Fettis and Knox (1985) □ Senkan and Quam (1992).

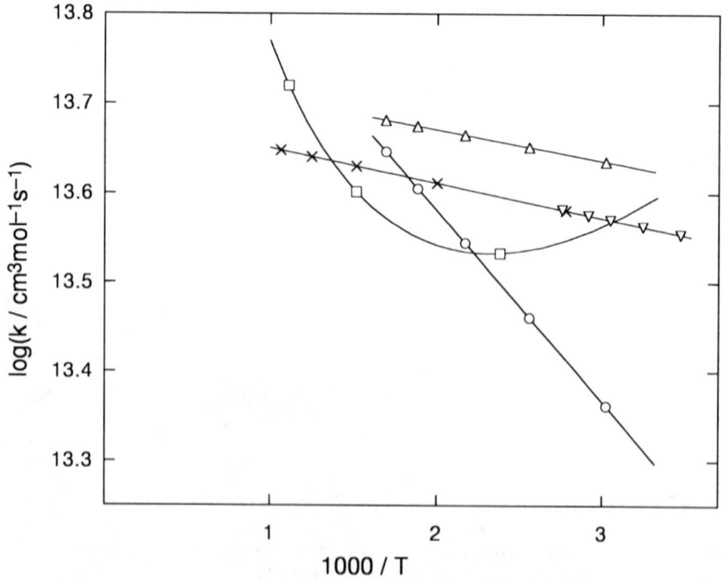

FIGURE 4.38. Rate coefficient expressions for Cl + $C_3H_8 \longrightarrow$ HCl + n-C_3H_7. × Recommendation ○ Fettis and Knox (1964) △ Recalculated from Fettis and Knox (1964) ▽ Tschuikow-Roux et al. (1985b) □ Senkan and Quam (1992).

$$\begin{aligned} Cl + C_3H_8 &\longrightarrow n\text{-}C_3H_7 + HCl \text{ (a)} \\ &\longrightarrow i\text{-}C_3H_7 + HCl \text{ (b)} \end{aligned}$$

These reactions were studied in static reactors over temperature ranges by Fettis and Knox (1964) and Tschuikow-Roux et al. (1985b) using relative rate methods against Cl + $CH_4 \longrightarrow CH_3$ + HCl. Their measurements are in accord with one another after recalculating the results of Fettis and Knox (1964) using the Table 4.7 rate coefficient expression for the reference reaction. We used the later measurements as the basis of the recommended expressions given in Table 4.7 and shown in Figures 4.38 and 4.39, which agree well with the SAR results reported by Senkan and Quam (1992).

Reactions of Cl with higher alkanes

Because the reactions of Cl with alkanes, with the exception of CH_4, have small or no activation energy, rate measurements at ambient conditions can be useful for combustion modeling. Accordingly, ambient temperature kinetic data for the reactions of Cl with higher molecular weight alkanes are reported in Table 4.7. As discussed above in connection with using the SAR formulation for C_2 hydrocarbons, Cl reactions with higher alkanes are expected to proceed with preferential abstraction of tertiary hydrogens, followed by secondary and finally primary hydrogens, from the parent molecule. In absence of experimental data the SAR method of Senkan and Quam (1992) can be used to estimate rate coefficients for individual reaction channels.

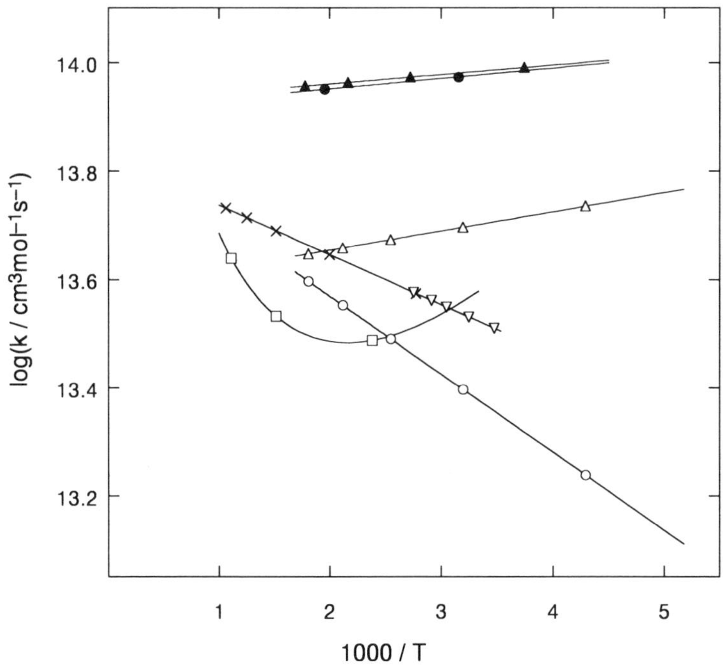

FIGURE 4.39. Rate coefficient expressions for $Cl + C_3H_8 \longrightarrow HCl + i\text{-}C_3H_7$. × Recommendation ● Lewis et al. (1980) for both isomers ▲ DeMore et al. (1990) for both isomers. ○ Fettis and Knox (1964) △ Recalculated from Fettis and Knox (1964) □ Senkan and Quam (1992) ▽ Tschuikow-Roux et al. (1985b).

4.5.5 Reactions of Cl_2

Reactions of Cl_2 with H, O and OH were discussed in earlier sections. As indicated in Table 4.8, there are few sources of information on reactions of Cl_2 with CHC radicals. These reactions were systematically studied first by Chiltz et al. (1963) in a static reactor and later by Gutman and coworkers in a flow reactor using laser photolysis and photoionization mass spectrometric detection (Timonen and Gutman 1986a, Timonen et al. 1986, 1987, Seetula et al. 1991). The data suggest several interesting trends. First, the activation energies for the general reaction $Cl_2 + R \longrightarrow RCl + Cl$ increase with increasing chlorine content of the radical R; for example, the activation energies for the chlorination of C_1 radicals increase from 0.2 kJ/mol for CH_3 to 5.0 kJ/mol for CCl_3(Figure 4.40). Second, these reactions conform to Evans–Polanyi relationships, the activation energies decreasing with increasing exothermicity. We adopted the experimental data of Gutman and coworkers in deriving recommended rate coefficient parameters for these reactions.

TABLE 4.7. Reactions of Cl atoms.[a]

Reaction	$\Delta_r H$	A	m	E_a	ΔT	Method	Reference
$Cl + H_2 \longrightarrow HCl + H$	4.2	2.2×10^{13}		19.1	200–300	Review	ABCHKT 1992
		3.0×10^{13}		21.3	298–1192	HTP, RFS	AF 1993
		1.5×10^{13}		18.3	200–650	Review	BBDGM 1981
		1.7×10^{13}		19.1	300	Review	DSHRGKHKM 1992
		2.2×10^{13}		19.1	200–500	LP, RFS	MG 1983
Recommendation		2.0×10^{7}	2	13.4	298–1500		
$Cl + O_2 + M \longrightarrow ClO_2 + M$	–20.1	1.2×10^{5}	–3.9	0.0	160–260	FR, LP, UVS	BHRSST 1991
$Cl + O3 \longrightarrow ClO + O_2$	–162.8	1.8×10^{13}		2.2	205–298	Review	ABCHKT 1992
$Cl_2 + M \longrightarrow Cl + Cl + M$	242.3	2.3×10^{13}		196.4	1550–2800	Review	BBDGM 1981
$Cl + Cl + M \longrightarrow Cl_2 + M$	–242.3	2.2×10^{14}		–7.5	195–520	Review	BBDGM 1981
$HCl + M \longrightarrow H + Cl + M$	431.0	4.4×10^{13}		342.0	2900–7000	Review	BBDGM 1981
$Cl + HO_2 \longrightarrow HCl + O_2$ (a)	–225.9	1.1×10^{13}		–1.4	250–420	DFR, LMR	LH 1982 and ABCHKT 1992
		4.1×10^{13}			298	DFR, MS	PLC 1978
$Cl + HO_2 \longrightarrow OH + ClO$ (b)	8.4	2.5×10^{13}		3.7	250–420	DFR, LMR	LH 1982
		4.2×10^{13} *b*		0.0	274–338	FR, PH, MMS	Cox 1980
$Cl + H_2O_2 \longrightarrow HCl + HO_2$ (a)	–74.5	6.6×10^{12}		8.2	300	Review	ABCHKT 1992
		6.6×10^{12}		8.2	300	Review	DSHRGKHKM 1992
		6.3×10^{12}		8.2	298–424	DFR, RFS	Key 1980
		2.4×10^{11}			298	DFR, MS	PLC 1978
		3.5×10^{11}			300	FR, LP, RFS	WMFD 1976
		7.5×10^{11}		3.2	265–400	FR, PH, RFS	MWLPS 1977

[a] See footnote on page 402 for units and comments.

b Sum of channels (a) and (b).

TABLE 4.7. Reactions of Cl atoms.[a] (Continued)

Reaction	$\Delta_r H$	A	m	E_a	ΔT	Method	Reference
$Cl + H_2O_2 \longrightarrow HOCl + OH$ (b)	−20.9	*b*				VLPP, MS	HB 1983
$Cl + H_2O_2 \longrightarrow ClO + H_2O$ (c)	−125.9	*b*				VLPP, MS	HB 1983
$Cl + HOCl \longrightarrow OH + Cl_2$ (a)	−7.9	2.1×10^{12}		1.1	243–365	DFR, MS	CELB 1981
$Cl + HOCl \longrightarrow HCl + ClO$ (b)	−37.7	*c*				DFR, MS	EB 1985
$Cl + Cl_2O \longrightarrow Cl_2 + ClO$	−107.9	5.9×10^{13}			300	Review	DSHRGKHKM 1992 and ABCHKT 1992
		3.6×10^{13}		−1.0	230–400	LMR, DFR	SA 1992
		5.9×10^{13}			298	DFR, MS, RFS	RKW 1980
$Cl + OClO \longrightarrow ClO + ClO$	−23.4	2.1×10^{13}		−1.3	300	Review	DSHRGKHKM 1992 and ABCHKT 1992
$Cl + ClOO \longrightarrow Cl_2 + O_2$ (a)	−225.9	8.4×10^{13}			300	Review	DSHRGKHKM 1992
$Cl + ClOO \longrightarrow ClO + ClO$ (b)	−23.4	4.8×10^{12}			300	Review	DSHRGKHKM 1992
$Cl + Cl_2O_2 \longrightarrow Cl_2 + ClOO$		6.0×10^{13}			230–298	Review	ABCHKT 1992
$Cl + CO + M \longrightarrow COCl + M$	−25.4	3.8×10^{13}		−6.7	185–260	FR, LP, RFS	NKW 1990
$ClO + CO \longrightarrow CO_2 + Cl$		1.0×10^{13}		31.4	298–400	Review	DSHRGKHKM 1992
$Cl + COCl \longrightarrow CO + Cl_2$	−215.1	1.3×10^{15}		13.9	300–600	Review	BBDGM 1981
		4.0×10^{14}		12.6	298–328	PH, DGA	BD 1952
$Cl + COCl_2 \longrightarrow COCl + Cl_2$	82.0	8.7×10^{12}	0.5	97.9	283–313	SR, DGA	BBS 1937
		3.2×10^{13}	0.5	85.8	298–328	PH, DGA	BD 1952

[a] See footnote on page 402 for units and comments.
b Unimportant compared to channel (a).
c Minor channel.

TABLE 4.7. Reactions of Cl atoms.[a] (Continued)

Reaction	$\Delta_r H$	A	m	E_a	ΔT	Method	Reference
$Cl + CH_2O \longrightarrow CHO + HCl$	–54.0	4.9×10^{13}			200–500	Review	ABCHKT 1992
		4.9×10^{13}		0.3	298	Review	DSHRGKHKM 1992
		4.5×10^{13}			295	FR, LP, IRC	FN 1981
		4.5×10^{13}		0.0	200–500	SR, PH, RFS	MNPS 1979a
Recommendation		4.5×10^{13}		0.0	200–1500		
$Cl + CHClO \longrightarrow COCl + HCl$	–62.3	7.2×10^{12}		6.8	265–325	Review	ABCHKT 1992
		8.4×10^{13}		7.2	266–322	SR, UVS	LZFB 1990
$Cl + CH_4 \longrightarrow CH_3 + HCl$	7.0	5.8×10^{12}		11.2	200–300	Review	ABCHKT 1992
		1.1×10^{7}	1.97	6.2	200–500	Review	BBDGM 1981
		6.6×10^{12}		11.6	300	Review	DSHRGKHKM 1992
		5.1×10^{6}	2.1	6.6	298–500	DFR, RFS	ZBK 1978
		2.1×10^{8}	1.76	11.1	300–1000	VLLP, MS	HKB 1981
		1.0×10^{13}		13.0	296–495	FR, LP, MS	RSSG 1988
		9.9×10^{12}		12.7	291–423	DFR, RFS	Key 1978
		4.5×10^{12}		10.2	218–401	FR, LP, RFS	WMFD 1976
		2.4×10^{13}		16.0	298–621	SR, PH, GC, IS	FK 1964
		5.0×10^{13}		16.3	300–470	Review	CGHMV 1963
Recommendation		1.1×10^{7}	1.97	6.2	200–1500		
$CH_3 + HCl \longrightarrow CH_4 + Cl$	–7.0	3.0×10^{11}		5.9	296–495	FR, LP, MS	RSSG 1988
		3.2×10^{11}		3.9	872–1085	Estimate	ML 1988
		2.3×10^{11}		9.6	300–500	Review	BBDGM 1981
$CH_2 + HCl \longrightarrow CH_3 + Cl$	–27.6	1.7×10^{12}		3.6	296–670	DFR, LMR	GTW 1990

[a] See footnote on page 402 for units and comments.

TABLE 4.7. Reactions of Cl atoms.[a] (Continued)

Reaction	$\Delta_r H$	A	m	E_a	ΔT	Method	Reference
$Cl + CH_3Cl \longrightarrow CH_2Cl + HCl$	–16.3	2.0×10^{13}		10.4	200–300	Review	ABCHKT 1992
		2.5×10^{7}	2	5.5	300–1000	SAR Analysis	SQ 1992
		2.1×10^{13}		10.5	200–300	Review	ABCHKT 1989
		1.8×10^{13}		10.1	281–368	SR, PH, GC, IS	T-R 1985
		1.9×10^{13}		10.2	281–368	SR, PH, GC, IS	T-R 1985
		2.0×10^{13}		10.4	233–263	LP, RFS	MK 1977
		1.3×10^{14}		14.9	300–604	MD, MS	CW 1973
		3.3×10^{13}		13.7	298–621	SR, GC, PH, IS	FK 1964
		3.2×10^{13}		13.0	300–470	Review	CGHMV 1963
Recommendation		2.3×10^{7}	2	5.0	200–1500		
$Cl + CH_2Cl_2 \longrightarrow CHCl_2 + HCl$	–32.6	5.2×10^{12}		7.6	270–330	Review	ABCHKT 1992
		1.9×10^{7}	2	4.9	300–1000	SAR Analysis	SQ 1992
		2.8×10^{13}		12.3	273–368	SR, PH, GC, IS	T-RFPNM 1988
		5.0×10^{13}		12.0	298–621	MD, MS	CW 1973
		2.6×10^{13}		12.4	298–378	SR, PH, GC, IS	FK 1964
		2.5×10^{13}		13.0	300–470	Review	CGHMV 1963
Recommendation		2.5×10^{10}	1	9.2	300–1500		
$Cl + CHCl_3 \longrightarrow CCl_3 + HCl$	–46.9	3.0×10^{12}		10.3	240–330	Review	ABCHKT 1992
		6.4×10^{6}	2	5.4	300–1000	SAR Analysis	SQ 1992
		2.0×10^{13}		11.5	297–652	MD, MS	CW 1973
		6.9×10^{12}		13.9	298–621	SR, PH, GC, IS	FK 1964
		1.1×10^{13}		10.5	298–621	Recalculated	FK 1964
		1.6×10^{13}		13.8	300–470	Review	CGHMV 1963
Recommendation		3.5×10^{6}	2	5.0	300–1500		
$Cl + CCl_4 \longrightarrow CCl_3 + Cl_2$	41.0	1.0×10^{14}		83.7	303–425	Review	Kon 1972
		1.0×10^{14}		83.3	358–563	Estimate	GJM 1958

[a] See footnote on page 402 for units and comments.

TABLE 4.7. Reactions of Cl atoms.[a] (Continued)

Reaction	$\Delta_r H$	A	m	E_a	ΔT	Method	Reference
$Cl + CH_3OH \longrightarrow CH_2OH + HCl$	–29.7	3.2×10^{13}		0.0	200–500	Review	ABCHKT 1992
		2.5×10^{7}	2	–7.3	300–1000	SAR Analysis	SQ 1992
		3.4×10^{13}		0.0	200–500	Review	DSHRGKHKM 1992
		2.9×10^{13}			298	SR, PH,UVS,IS	NSSTN 1990
		3.1×10^{13}			298		PBMS 1988
		2.8×10^{13}			298	SR, PH, GC, IS	WSSWJ 1988
		3.8×10^{13}		0.0	200–500	FR, PH, RFS	MNPS 1979b
Recommendation		3.2×10^{13}		0.0	300–1000		
$Cl + CH_2CO \longrightarrow CH_2Cl + CO$ (a)	–66.5						
$Cl + CH_2CO \longrightarrow CHO + HCl$ (b)	–61.1						
		4.1×10^{13}[b]			294	DFR, MS	EHW 1990
$Cl + C_2H_5OH \longrightarrow C_2H_5O + HCl$		5.1×10^{13}			298	SR, PH, GC, IS	WSSWJ 1988
$Cl + n\text{-}C_3H_9OH \longrightarrow n\text{-}C_3H_9O + HCl$		8.7×10^{13}			298	SR, PH, GC, IS	WSSWJ 1988
$Cl + t\text{-}C_4H_9OH \longrightarrow t\text{-}C_4H_9O + HCl$		2.0×10^{13}			298	SR, PH, GC, IS	WSSWJ 1988
		5.1×10^{13}			298	SR, PH, GC, IS	WSSWJ 1988
$Cl + CH_3OCH_3 \longrightarrow CH_3OCH_2 + HCl$	–38.1						
		2.5×10^{7}	2	–10.3	300–1000	SAR Analysis	SQ 1992
		1.2×10^{14}			298	SR, PH, GC, IS	WSSWJ 1988
		1.1×10^{14}		0.0	200–500	FR, PH, RFS	MNPS 1979b
$Cl+C_2H_5OC_2H_5 \longrightarrow C_2H_5OC_2H_4+HCl$		2.1×10^{14}			298	SR, PH, GC, IS	WSSWJ 1988
$Cl + MBTE \longrightarrow Products$		1.0×10^{14}			298	SR, PH, GC, IS	WSSWJ 1988

[a] See footnote on page 402 for units and comments.
[b] Sum of channels (a) and (b).

TABLE 4.7. Reactions of Cl atoms.[a] (Continued)

Reaction	$\Delta_r H$	A	m	E_a	ΔT	Method	Reference
$Cl+CH_3OOCH_3 \longrightarrow CH_3OOCH_2+HCl$		2.5×10^{7}	2	−9.2	300–1000	SAR Analysis	SQ 1992
		7.2×10^{13}		0.0	200–330	FR, PH, RFS	MNPS 1981
$Cl + CH_3CHO \longrightarrow CH_3CO + HCl$	−70.7	4.3×10^{13}		0.0	210–340	Review	ABCHKT 1992
		4.0×10^{13}		0.0	210–343	FR, PH, RFS	PNNS 1990
$Cl + C_2H_2 \longrightarrow C_2H_2Cl$	−88.7						
k_∞		5.0×10^{16}	−1	0.0	252–370	SR, PH, GC, IS	Kai 1992
		3.1×10^{13}			210–361	FR, PH, FRS	BS 1985
		1.4×10^{14}			295	SR, PH, GC, IS	WALKM 1990
$k_\circ/[M]$		0.49	−2.1		230	SR, PH, GC, IS	Kai 1992
		1.9×10^{18}			295	SR, PH, GC, IS	WALKM 1990
		1.9×10^{4}	−3.5	0.0	210–361	FR, PH, FRS	BS 1985
$Cl + C_2H_2 \longrightarrow C_2H + HCl$		1.6×10^{14}		70.7	1300	Estimate	WB 1984
$Cl + C_2H_4 \longrightarrow C_2H_3 + HCl$	20.1	1.9×10^{13}		23.3	263–338	VLPP, MS	PB 1988
		1.0×10^{14}		29.3	1300	Estimate	WB 1984
$Cl + C_2H_4 \longrightarrow C_2H_4Cl$	−84.5	1.1×10^{14}[b]			29	RC, IS, GC	IRR 1983
k_∞		1.8×10^{14}			295	SR, PH, GC, IS	WALKM 1990
$k_\circ/[M]$		5.8×10^{18}			295	SR, PH, GC, IS	WALKM 1990
$C_2H_3 + HCl \longrightarrow C_2H_4 + Cl$	−20.1	5.2×10^{11}		8.4	203–343	Estimate	DB 1991
		4.0×10^{11}		−2.8	298–495	FR, LP, MS	RS 1989a,b
		5.6×10^{11}		−3.5	278–370	LP, DLA	KW 1988
$Cl + C_2H_3Cl \longrightarrow CHClCH + HCl$		1.2×10^{14}		55.6	620–750	FR, PH, GC	SW 1986
$CH_2CCl + HCl \longrightarrow C_2H_3Cl + Cl$	−23.8	1.3×10^{11}		0.0	298–495	FR, LP, MS	RS 1989a,b

[a] See footnote on page 402 for units and comments.
[b] At 4000 Torr.

TABLE 4.7. Reactions of Cl atoms.[a] (Continued)

Reaction	$\Delta_r H$	A	m	E_a	ΔT	Method	Reference
$Cl + C_2H_6 \longrightarrow C_2H_5 + HCl$	−12.1	4.9×10^{13}		0.8	220–350	Review	ABCHKT 1992
		4.6×10^{13}		0.6	300–1000	SAR Analysis	SQ 1992
		4.6×10^{13}		0.8	220–350	Review	DSHRGKHKM 1992
		7.3×10^{13}		1.9	280–368	SR, PH, GC, IS	T-R 1985
		5.4×10^{13}		1.1	220–604	DFR, RFS	LSWW 1980
		4.3×10^{13}			298	FR, LP, TRISP	KRSL 1992
		8.9×10^{13}		4.4	303–453	SR, PH, GC, IS	CGHM 1967
		9.0×10^{13}		4.3	193–593	SR, PH, GC, IS	FK 1964
Recommendation		4.9×10^{13}		0.8	300–1000		
$Cl+C_2H_5Cl \longrightarrow CH_3CHCl+HCl$ (a)	−23.6						
		1.9×10^{7}	2	−1.4	300–1000	SAR Analysis	SQ 1992
		1.4×10^{13}		3.5	280–368	SR, PH, GC, IS	T-R 1984b
		3.5×10^{13}		6.3	303–453	SR, PH, GC, IS	CGHM 1967
Recommendation		4.5×10^{7}	2	1.0	300–1000		
$Cl+C_2H_5Cl \longrightarrow CH_2ClCH_2+HCl$ (b)	−3.6						
		2.5×10^{7}	2	3.8	300–1000	SAR Analysis	SQ 1992
		1.3×10^{13}		7.0	280–368	SR, PH, GC, IS	T-R 1984b
		1.1×10^{13}		6.3	303–453	SR, PH, GC, IS	CGHM 1967
		1.4×10^{13}[b]		2.6	257–426	FR, LP, RFS	WS 1983
		4.1×10^{12}			298	FR, LP, TRISP	KRSL 1992
		3.5×10^{13}		6.3	298–621	SR, PH, GC, IS	FK 1964
Recommendation		2.5×10^{7}	2	2.8	280–1000		

[a] See footnote on page 402 for units and comments.
[b] Sum of channels (a) and (b).

TABLE 4.7. Reactions of Cl atoms.[a] (Continued)

Reaction	$\Delta_r H$	A	m	E_a	ΔT	Method	Reference
$Cl + CH_2ClCH_2Cl \longrightarrow CH_2ClCHCl + HCl$	–23.0						
		1.9×10^{7}	2	3.3	300–1000	SAR Analysis	SQ 1992
		1.3×10^{13}		6.6	257–426	FR,LP,RFS	WS 1983
		6.3×10^{13}		13.0	303–453	SR,PH,GC,IS	CGHM 1967
		3.0×10^{13}		9.0	273–368	SR,PH,GC,IS	T-RFN 1986
Recommendation		5.0×10^{7}	2	4.6			
$Cl + CH_3CHCl_2 \longrightarrow CH_3CCl_2 + HCl$ (a)	–36.2						
		6.4×10^{6}	2	–1.0	300–1000	SAR Analysis	SQ 1992
		6.1×10^{12}		5.3	280–368	SR,PH,GC,IS	T-R 1985
		8.9×10^{12}		7.9	323–423	SR,PH,GC,IS	CGHM 1967
Recommendation		2.8×10^{7}	2	2.7	280–1000		
$Cl + CH_3CHCl_2 \longrightarrow CHCl_2CH_2 + HCl$ (b)	5.2						
		2.5×10^{7}	2	9.3	300–1000	SAR Analysis	SQ 1992
		7.1×10^{12}		11.6	280–368	SR,PH,GC,IS	T-R 1985
		1.0×10^{13}		14.2	323–423	SR,PH,GC,IS	CGHM 1967
Recommendation		2.2×10^{7}	2	8.4	280–1000		
		4.9×10^{12}[b]		4.6	257–426	FR,LP,RFS	WS 1983
		4.1×10^{12}			298	FR,LP,TRISP	KRSL 1992
$Cl+CH_2ClCHCl_2 \longrightarrow CH_2ClCCl_2+HCl$ (a)	–41.0						
		6.4×10^{6}	2	3.7	300–1000	SAR Analysis	SQ 1992
		1.2×10^{13}		14.2	273–368	SR,PH,GC,IS	T-RFN 1986
		8.9×10^{12}		13.0	303–433	SR,PH,GC,IS	CGHM 1967
Recommendation		8.0×10^{7}	2	11.7	300–1200		

[a] See footnote on page 402 for units and comments.
[b] Sum of channels (a) and (b).

TABLE 4.7. Reactions of Cl atoms.[a] (Continued)

Reaction	$\Delta_r H$	A	m	E_a	ΔT	Method	Reference
$Cl+CH_2ClCHCl_2 \longrightarrow CHCl_2CHCl+HCl$ (b)	–23.0						
		1.9×10^7	2	8.7	300–1000	SAR Analysis	SQ 1992
		1.0×10^{13}		16.3	273–368	SR,PH,GC,IS	T-RFN 1986
		1.4×10^{13}		15.5	303–433	SR,PH,GC,IS	CGHM 1967
Recommendation		1.0×10^7	2	10.5	300–1000		
		2.9×10^{12} [b]		1.56	257–426	FR,LP,RFS	WS 1983
$Cl + CH_3CCl_3 \longrightarrow CCl_3CH_2 + HCl$	5.6						
		2.5×10^7	2	11.8	300–1000	SAR Analysis	SQ 1992
		1.0×10^{13}		17.2	280–368	SR,PH,GC,IS	T-R 1985
		1.0×10^{12}		9.2	259–356	SR,PH,RFS	WSR 1982
		2.5×10^{12}		15.1	323–433	SR,PH,GC,IS	CGHM 1967
Recommendation		4.0×10^7	2	14.6	300–1200		
$Cl + CHCl_2CHCl_2 \longrightarrow CHCl_2CCl_2 + HCl$	–27.6						
		6.4×10^7	2	9.2	300–1000	SAR Analysis	SQ 1992
		3.6×10^{12}		10.3	303–433	SR,PH,GC,IS	CGHM 1967
		5.7×10^{12}		6.8		Recalculated from CGHM 1967	
Recommendation		2.7×10^7	2	4.6	300–1200		
$Cl + CH_2ClCCl_3 \longrightarrow CCl_3CHCl + HCl$	–19.5						
		1.9×10^7	2	11.2	300–1000	SAR Analysis	SQ 1992
		1.4×10^{12}		10.3	303–433	SR,PH,GC,IS	CGHM 1967
		2.3×10^{12}		6.8		Recalculated from CGHM 1967	
Recommendation		1.0×10^7	2	5.0	300–1200		

[a] See footnote on page 402 for units and comments.

[b] Sum of channels (a) and (b).

TABLE 4.7. Reactions of Cl atoms.[a] (Continued)

Reaction	$\Delta_r H$	A	m	E_a	ΔT	Method	Reference
$Cl + C_2HCl_5 \longrightarrow C_2Cl_5 + HCl$	–37.2	6.43×10^{6}	2	11.6	300–1000	SAR Analysis	SQ 1992
		4.5×10^{12}		14.9	303–433	SR,PH,GC,IS	CGHM 1967
		5.0×10^{12}		14.2	298–621	SR,PH,GC,IS	FK 1964
		7.1×10^{12}		11.4		Recalculated	from CGHM 1967
Recommendation		2.0×10^{7}	2	8.4	300–1200		
$Cl + C_3H_8 \longrightarrow n\text{-}C_3H_7 + HCl$ (a)	–33.5	4.6×10^{13}		0.6	300–1000	SAR Analysis	SQ 1992
		1.0×10^{14}		4.1	193–593	SR,PH,GC,IS	FK 1964
		5.5×10^{13}		0.7		Recalculated	from FK 1964
		4.9×10^{13}		0.8	281–368	SR,PH,GC,IS	T-R 1985
Recommendation		4.9×10^{13}		0.8	300–1000		
$Cl + C_3H_8 \longrightarrow i\text{-}C_3H_7 + HCl$ (b)	–37.7	6.8×10^{13}		1.8	300–1000	SAR Analysis	SQ 1992
		7.2×10^{13}		2.8	193–593	SR,PH,GC,IS	FK 1964
		3.8×10^{13}		–0.7		Recalculated	from FK 1964
		6.8×10^{13}		1.8	281–368	SR,PH,GC,IS	T-R 1985
Recommendation		6.8×10^{13}		1.8	300–1000		
		8.2×10^{13}[b]		–0.4	220–607	DFR, RFS	LSWW 1980
		8.1×10^{13}			296	SR, LP, GC	AA 1985
		8.4×10^{13}		–0.3	220–607	Review	DSHRGKHKM 1992
$Cl + c\text{-}C_3H_6 \longrightarrow c\text{-}C_3H_5 + HCl$		5.4×10^{13}		17.2	193–593	SR,PH,GC,IS	FK 1964

[a] See footnote on page 402 for units and comments.
[b] Sum of channels (a) and (b).

TABLE 4.7. Reactions of Cl atoms.[a] (Continued)

Reaction	A	E_a	ΔT	Method	Reference
$Cl + n\text{-}C_4H_{10} \longrightarrow n\text{-}C_4H_9 + HCl$ (a)	4.6×10^{13}	0.6	300–1000	SAR Analysis	SQ 1992
	8.4×10^{13}	3.2	193–593	SR,PH,GC,IS	FK 1964
	4.4×10^{13}	–0.2		Recalculated from	FK 1964
$Cl + n\text{-}C_4H_{10} \longrightarrow CH_3CH_2CHCH_3 + HCl$ (b)	6.8×10^{13}	1.7	300–1000	SAR Analysis	SQ 1992
	9.6×10^{13}	1.3	193–593	SR,PH,GC,IS	FK 1964
	4.6×10^{13}	–2.4		Recalculated from	FK 1964
	1.3×10^{14} *b*	0.1	298–598	LSWW 1980	
	1.2×10^{14}		296	SR, PH, GC	AA 1985
$Cl + i\text{-}C_4H_{10} \longrightarrow (CH_3)_2CH_2CH + HCl$ (a)	1.2×10^{14}	3.3	193–593	SR,PH,GC,IS	FK 1964
	6.9×10^{13}	–0.1		Recalculated from	FK 1964
$Cl + i\text{-}C_4H_{10} \longrightarrow (CH_3)_3C + HCl$ (b)	2.1×10^{13}	0.4	193–593	SR,PH,GC,IS	FK 1964
	1.3×10^{13}	–3.4		Recalculated from	FK 1964
	8.3×10^{13} *b*		296	SR, PH, GC	AA 1985
$Cl + c\text{-}C_4H_8 \longrightarrow c\text{-}C_4H_7 + HCl$	2.6×10^{14}	3.3	193–593	SR,PH,GC,IS	FK 1964
$Cl + C_4H_2 \longrightarrow C_4H_2Cl$					
k_∞	2.9×10^{13} *c*		298	DFR, MS	MBPS 1988
$Cl + i\text{-}C_5H_{12} \longrightarrow$ Products	1.2×10^{14}		296	SR, PH, GC	AA 1985
$Cl + neo\text{-}C_5H_{10} \longrightarrow neo\text{-}C_5H_9 + HCl$	1.7×10^{14}	3.8	193–593	SR,PH,GC,IS	FK 1964
	8.1×10^{13}	0.5		Recalculated from	FK 1964
	6.6×10^{13}		296	SR, PH, GC	AA 1985

[a] See footnote on page 402 for units and comments.
[b] Sum of channels (a) and (b).
[c] At 1 Torr.

TABLE 4.7. Reactions of Cl atoms.[a] (Concluded)

Reaction	A	E_a	ΔT	Method	Reference
$Cl + c\text{-}C_5H_{10} \longrightarrow c\text{-}C_5H_9 + HCl$	2.3×10^{14}	2.5	193–593	SR,PH,GC,IS	KN 1959
	1.2×10^{14}	–0.9		Recalculated	from FK 1964
$Cl + n\text{-}C_6H_{14} \longrightarrow$ Products	1.8×10^{14}		296	SR, PH, GC	AA 1985
$Cl + c\text{-}C_6H_{12} \longrightarrow c\text{-}C_6H_{11} + HCl$	1.9×10^{14}		296	SR, PH, GC	AA 1985
	1.1×10^{14}		298	SR, PH, GC	DBB 1970
$Cl + n\text{-}C_7H_{16} \longrightarrow$ Products	2.1×10^{14}		296	SR, PH, GC	AA 1985

[a] See footnote on page 402 for units and comments.

TABLE 4.8. Reactions of Cl_2.[a]

Reaction	$\Delta_r H$	A	m	E_a	ΔT	Method	Reference
$Cl_2 + CHO \longrightarrow CHClO + Cl$	–84.9	4.2×10^{12}		0.0	296–582	FR, LP, MS	TRG 1988
$Cl_2 + COCl \longrightarrow COCl_2 + Cl$	–79.7	1.7×10^{9}	0.5	33.9	283–313	Review	Kon 1972
		2.5×10^{12}		12.4	298–328	SR,PH,DGA	BD 1952
					300	SR, PH, IR	Ohta 1983
$Cl_2 + CH_3 \longrightarrow CH_3Cl + Cl$	–107.9	3.0×10^{12}		1.0	293–712	FR, LP, MS	TG 1986
		1.1×10^{13}		9.0	298–423	SR, PH, GC	TKK 1986
		2.9×10^{12}		2.0	296–712	FR, PH, MS	Tim 1988
		7.9×10^{12}		9.6		SR, PH, GC	CGHMV 1963
Recommendation		3.0×10^{12}		1.0			
$Cl_2 + CH_2Cl \longrightarrow CH_2Cl_2 + Cl$	–92.9	9.0×10^{11}		4.2	295–719	FR, LP, MS	SGLRS 1991
		4.0×10^{12}		12.6		SR, PH, GC	CGHMV 1963
Recommendation		9.0×10^{11}		4.2	295–1000		
$Cl_2 + CHCl_2 \longrightarrow CHCl_3 + Cl$	–75.6	5.1×10^{11}		10.3	357–719	FR, LP, MS	SGLRS 1991
		1.0×10^{12}		16.7		SR, PH, GC	CGHMV 1963
Recommendation		5.1×10^{11}		10.3	357–1000		
$Cl_2 + CCl_3 \longrightarrow CCl_4 + Cl$	–41.0	4.1×10^{11}		21.0	357–693	FR, LP, MS	TKK 1986
		2.5×10^{12}		25.1	300–670	Estimate	WB 1980
		5.8×10^{11}		25.1		SR, PH, GC	CGHMV 1963
Recommendation		4.2×10^{11}		21.0	357–1000		
$Cl_2 + C_2H_3 \longrightarrow C_2H_3Cl + Cl$	–147.3	5.2×10^{12}		–2.0	298–435	FR, LP, MS	TRSG 1987
		1.1×10^{13}		–2.0	298–435	FR, LP, MS	Tim 1988
Recommendation		5.2×10^{12}		–2.0			

[a] See footnote on page 402 for units and comments.

TABLE 4.8. Reactions of Cl_2.[a] (Concluded)

Reaction	$\Delta_r H$	A	E_a	ΔT	Method	Reference
$Cl_2 + C_2H_5 \longrightarrow C_2H_5Cl + Cl$	–110.0	7.6×10^{12}	–1.0	295–498	FR, LP, MS	TG 1986
		4.6×10^{12}		298	SR, LP, GC	KRW 1989
		1.5×10^{13}	4.0	298–423	SR, PH, GC	TKK 1986
Recommendation		7.6×10^{12}	–1.0			
$Cl_2 + C_2Cl_5 \longrightarrow C_2Cl_6 + Cl$	–45.2	2.0×10^{11}	23.0	298–423	SR, PH, GA	ACDHW 1961
		2.0×10^{11}	10.0	360–520	SR, PH, MS	DGMM 1961
$Cl_2 + CHCl_2CH_2 \longrightarrow CHCl_2CH_2Cl + Cl$	–110.5	6.3×10^{11}	3.7	298–423	SR, PH, GA	ACDHW 1961
$Cl_2 + CHCl_2CHCl \longrightarrow CHCl_2CHCl_2 + Cl$	–72.4	5.0×10^{11}	11.4	303–338	SR, PH, GA	ACDH 1962
$Cl_2 + (CH_3)_2CH \longrightarrow (CH_3)_2CHCl + Cl$	–109.1	1.5×10^{13}	–2.0	298–498	FR, LP, MS	TG 1986
Recommendation		1.5×10^{13}	–2.0			
$Cl_2 + C_3H_3 \longrightarrow C_3H_3Cl + Cl$	–61.9	1.7×10^{13}	28.0	298–498	FR, LP, MS	TG 1986
Recommendation		1.7×10^{13}	28.0			
$Cl_2 + C_3H_5 \longrightarrow C_3H_5Cl + Cl$	–49.0	9.3×10^{12}	18.0	487–693	FR, LP, MS	TRSG 1987
Recommendation		9.3×10^{12}	18.0			
$Cl_2 + CH_2CHCH_2 \longrightarrow CH_2CHCH_2Cl + Cl$		9.3×10^{12}	18.0	487–693	FR, LP, MS	TRSG 1987
$Cl_2 + HCCCH_2 \longrightarrow HCCCH_2Cl + Cl$		1.7×10^{13}	28.0	525–693	FR, LP, MS	TRSG 1987
Cl_2 + n,s-$C_4H_9 \longrightarrow$ n,s-$C_4H_9Cl + Cl$		4.1×10^{12}		298	SR, PH, IR	NL 1982
$Cl_2 + (CH_3)_3C \longrightarrow (CH_3)_3CCl + Cl$	–107.9	2.4×10^{13}	0.0	298–498	FR, LP, MS	TG 1986
Recommendation		2.4×10^{13}	0.0			
$Cl_2 + C_6H_5CH_2 \longrightarrow C_6H_5CH_2Cl + Cl$		3.4×10^{13}	3.7	295–372	FR, LP, LIF	NMcD 1982
Recommendation		3.4×10^{13}	3.7			

[a] See footnote on page 402 for units and comments.

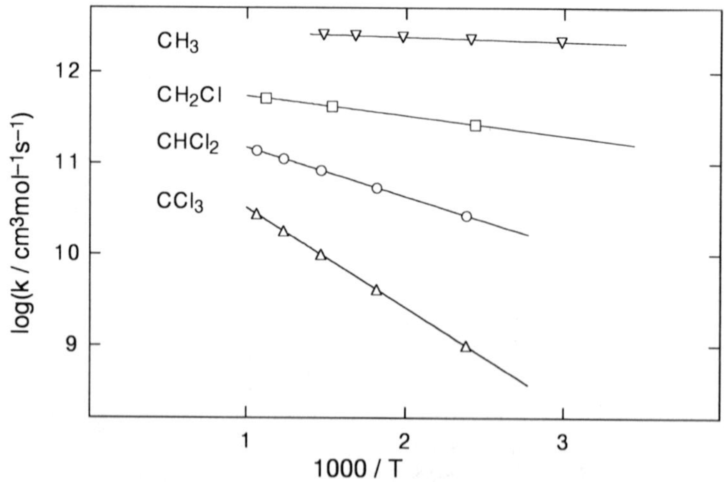

FIGURE 4.40. Rate coefficient expressions for reaction of Cl_2 with methyl and chlorine-substituted methyl radicals. ∇ , □ , O Seetula et al. (1991) Δ Timonen et al. (1986).

4.5.6 Unimolecular and pressure-dependent bimolecular reactions

The compilations of Benson and O'Neal (1970) and Benson (1976) are the main source for high-pressure limit rate parameters for the unimolecular decompositions of CHCs. CHCs generally decompose by 1,2 and 1,1 HCl eliminations understood to proceed via formation of tight cyclic transition states. These reactions generally exhibit activation energies about 25–35 kJ/mol greater than the endothermicity, the excess being attributed to the strain energy needed for formation of the transition state (Benson 1976). Under conditions encountered in combustion, unimolecular decomposition of C_1 to C_3 CHCs is pressure-dependent. Consequently, the high-pressure rate parameters in Table 4.9 generally have to be corrected for the appropriate extent of fall-off. Several methods have been used to obtain rate parameters for combustion conditions (Robinson and Holbrook 1972). If the high-pressure and low-pressure rate parameters are both available, one can estimate the applicable falloff parameters by analogy to those of known reactions and interpolate between the limits to obtain rate coefficients for combustion conditions. (Dean and Bozzelli, Chap. 2; Troe 1977a,b; Gardiner and Troe 1984) As can be seen in Table 4.9, experimental data for both high- and low-pressure limits of unimolecular reactions are available for only very few reactions of chlorine-containing species. Reliable measurements do exist for the high-pressure rate coefficients of numerous unimolecular decomposition reactions of similar molecules, however, and one can also make reasonable estimates for high-pressure limit rate parameters using transition state theory (Benson and O'Neil 1970; Benson 1976). Consequently, methods based on experimental knowledge of the high-pressure rate coefficient alone, taking the low-pressure rate coefficient from analogies, find considerable use in combustion modeling. A method that possesses both a reasonable level of rigor and simplicity of application is the quantum form of RRK theory (QRRK) (Dean and Bozzelli, Chap. 2; Robinson and Holbrook 1972). An extended version of QRRK theory has been developed and applied to a large variety of hydrocarbon and chlorinated hydrocarbon reactions by Dean (1985), Westmoreland et al. (1986), Dean and Westmoreland (1987), and Karra and Senkan (1988). Because the RRKM and QRRK methods presuppose the composition of the reaction medium to be known, different mixtures having different energy transfer rates, rate coefficient parameters are listed only for k_∞ and k_o (when available) with the understanding that the user of these recommendations must undertake the required RRKM or QRRK analysis to generate rate expressions appropriate for use in a specific application.

4.5.7 Reactions of large molecules and radicals

There is a substantial literature on reactions of partially chlorinated methyl and ethyl radicals and a smaller one on reactions of various radicals with partially or wholly chlorinated molecules. Many of the elementary reactions in this literature are reasonably well characterized, and the reported rate parameters, summarized in Table 4.10, can be used for study of models of toxic waste incineration.

4.6 DATA GAPS AND SUGGESTIONS FOR FUTURE WORK

As evident from the above discussion, although considerable information exists on the kinetics of reactions involving chlorine and chlorinated compounds, major areas exist in which our understanding needs further improvement. First, as noted earlier, relatively few studies exist in which both the rates and mechanisms of the reactions of chlorinated compounds at combustion temperatures were directly determined. Most studies have been undertaken at or near ambient temperature. Because reaction mechanisms at high temperatures can differ significantly from those derived from experiments done at lower temperatures, the study of reaction mechanisms at combustion temperatures should be attempted whenever possible. Such determinations will be useful in establishing the suitability of the use of the vast kinetic literature on the atmospheric reactions of chlorinated species in combustion modeling. Experimental studies at high temperatures will also be useful to test the applicabilities and limitations of theoretical methods used to extrapolate low-temperature data to combustion temperatures.

Getting high-temperature rate data for reactions of chlorine-containing species is difficult because of the rapid thermal decomposition of reactants and/or products at combustion conditions. In addition, commercial sources for many of the desired reactants, for example chlorinated acetylenes, are not available, further impeding developments in the field. New experimental techniques must be devised to

TABLE 4.9. Unimolecular reactions.[a]

Reaction	$\Delta_r H$	A	m	E_a	ΔT	Method	Reference
$CH_3Cl \longrightarrow CH_3 + Cl$ (a)	350.6	1.1×10^{15}		357.7	1000–1173	FR, GC	SS 1959
		4.9×10^{12}		298.0	1223–1373	SR, GC	ADS 1990
k_∞		1.2×10^{10}		231.4	1663–2059	ST, ARAS	LM 1993
$k_\circ/[M]$		3.6×10^{15} [b]		247.3	1600–2400	ST, UVS	KSM 1980
$CH_3Cl \longrightarrow CH_2 + HCl$ (b)	378.2						
$CHCl_3 \longrightarrow CCl_2 + HCl$ (a)	248.1						
k_∞		1.8×10^{14}		228.0	1050–1380	ST, OS	SWZ 1979
$k_\circ/[M]$		9.3×10^{8}			1200	ST, OS	SWZ 1979
		2.6×10^{11}		196.6	783–857	FR, GC	SS 1960
				234.3	873–1343	VLPP, MS	BS 1967
		5.8×10^{9}		171.5	725–800	FR, GC	SB 1957
$CHCl_3 \longrightarrow CHCl_2 + Cl$ (b)	304.2				1000–1200	ST, GC	Yan 1977
$CCl_4 \longrightarrow CCl_3 + Cl$	283.7						
k_∞		3.2×10^{2}	−2.29	288.6	1100–2200	RRKM	MLKK 1993
		2.2×10^{12}		230.5	773–973	FR, GC	SS 1959
$k_\circ/[M]$		1.2×10^{16}		183.2	1470–2200	ST, LS	MLKK 1993
		2.6×10^{16}		195.6	1100–1700	ST, ARAS	MLKK 1993
$CCl_3 + M \longrightarrow CCl_2 + Cl + M$	287.1						MLKK 1993
k_∞ [c]				234.3		FR, IR–LP, IRS	CW 1983
		6.3×10^{13}		236.8	713–733	SR, GC	KC 1983

[a] See footnote on page 402 for units and comments.
[b] For M = argon.
[c] $k_\circ/[M]$ 0.1 of CCl_4 decomposition rate.

TABLE 4.9. Unimolecular reactions.[a] (Continued)

Reaction	$\Delta_r H$	A	m	E_a	ΔT	Method	Reference
$C_2H_3Cl \longrightarrow C_2H_2 + HCl$	115.1						
k_∞		1.0×10^{14}		290.4	1400–1800	ST, UVS	Zab 1977
$k_\circ$		5.3×10^{16}		204.2	1400–1800	ST, UVS	Zab 1977
$C_2HCl_3 \longrightarrow CHClCCl + Cl$ (a)	372.4						
$C_2HCl_3 \longrightarrow CCl_2CH + Cl$ (b)	398.7						
$C_2HCl_3 \longrightarrow C_2Cl_2 + HCl$ (c)	159.4						
k_∞		7.0×10^{15}		351.5	1450–1900	ST, UVS	Zab 1974
$k_\circ$		1.4×10^{17} b		242.9	1450–1900	ST, UVS	Zab 1974
$C_2Cl_4 \longrightarrow C_2Cl_3 + Cl$ (a)	372.4						
$C_2Cl_4 \longrightarrow C_2Cl_2 + Cl_2$ (b)	259.4						
k_∞		4.5×10^{14} c		334.7	1450–1900	ST, UVS	Zab 1974
$k_\circ$		9.5×10^{16} c		212.5	1450–1900	ST, UVS	Zab 1974
$C_2H_5Cl \longrightarrow C_2H_4 + HCl$	74.1						
k_∞		6.3×10^{13}		240.2	1100–1400	SR, GC	DSBM 1982
		4.0×10^{13}		243.1	975–1213	VLPP, MS	FSKG 1981
		5.6×10^{13}		241.4	960–1100	ST, GC	EIT-R 1978
		2.1×10^{13}		235.6	714–767	SR, PM	HDJ 1975
		1.5×10^{13}		236.0	820–1000	ST, IS	T 1964
		1.6×10^{14}		248.9	671–766	SR, GC	BH 1949
		2.5×10^{13}		238.9	600–1000	Analysis	BB 1963
		1.6×10^{13}		236.4	600–1000	Review	BO'N 1970
$CH_2ClCH_2Cl \longrightarrow C_2H_3Cl + HCl$	63.2	1.2×10^{7}		115.1	743–823	FR, GA	KO 1962
		6.6×10^{10}		196.2	635–758	SR, GA	How 1952
$CHCl_2CH_2Cl \longrightarrow$ cis-CHClCHCl+HCl (a)							
k_∞		1.3×10^{14}		244.8	1100–1400	CO_2, GC, IS	DSBM 1982
$CHCl_2CH_2Cl \longrightarrow$ trans-CHClCHCl+HCl (b)							
k_∞		7.9×10^{13}		248.9	1100–1400	CO_2, GC, IS	DSBM 1982

[a] See footnote on page 402 for units and comments.
[b] For M = argon.
[c] Sum of channels (a) and (b).

TABLE 4.9. Unimolecular reactions.[a] (Continued)

Reaction	$\Delta_r H$	A	E_a	ΔT	Method	Reference
$CH_3CHCl_2 \longrightarrow C_2H_3Cl + HCl$	62.8					
k_∞		2.8×10^{13}	223.8	641–800	Review	BO'N 1970
$CH_3CCl_3 \longrightarrow CH_2CCl_2 + HCl$	51.7					
k_∞		7.08×10^{13}	225.5	587–658	SR, GC	HHV M 1989
		2.0×10^{14}	229.7	1773–2900	SR, GC	KP 1985
		1.3×10^{13}	207.1	1100–1400	CO_2, GC, IS	DSBM 1982
		1.5×10^{14}	216.3	890–1250	VLPP, MS	BS 1967
		1.3×10^{14}	226.8	683–1250	Review	BO'N 1970
$C_2HCl_5 \longrightarrow C_2Cl_4 + HCl$ (a)	33.1	1.3×10^{14}	249.4	300–700	Estimate	BW 1982
		4.0×10^{11}	201.7	820–865	FR, GC	HC 1975
$C_2HCl_5 \longrightarrow CHCl_2 + CCl_3$ (b)	300.4	4.0×10^{17}	320.9	300–700	Estimate	BW 1982
$C_2HCl_5 \longrightarrow Cl$ + Products		1.6×10^{16}	282.0	300–700	Estimate	BW 1982
$C_2HCl_4 \longrightarrow C_2HCl_3 + Cl$	84.1					
k_∞		5.0×10^{13}	85.4	385–490	Review	Kon 1972
$C_2Cl_5 \longrightarrow C_2Cl_4 + Cl$	66.5					
k_∞		3.2×10^{16}	92.9	360–320	Review	Kon 1972
		6.3×10^{15}	70.3	360–475	Review	Kon 1972
$C_2Cl_6 \longrightarrow C_2Cl_4 + Cl_2$ (a)	111.7	5.0×10^{13}	225.9	300–760	Estimate	WB 1980
$C_2Cl_6 \longrightarrow C_2Cl_5 + Cl$ (b)	287.9	1.0×10^{16}	289.1	300–760	Estimate	WB 1980
$C_2Cl_6 \longrightarrow CCl_3 + CCl_3$ (c)	274.5	6.3×10^{17}	285.8	300–760	Estimate	WB 1980
cis-CHClCHCl $\longrightarrow$ trans-CHClCHCl		2.5×10^{13}	220.5	1022–1198	ST, GC	Jef 1974
cis-CHClCHCl $\longrightarrow$ CHCCl + HCl		3.6×10^{12}	220.5	633–748	SR, GA	GH 1956
trans-CHClCHCl $\longrightarrow$ cis-CHClCHCl		2.2×10^{13}	236.0	828–1050	Estimate	ML 1989
trans-CHClCHCl $\longrightarrow$ CHCCl + HCl		3.6×10^{12}	220.5	633–748	SR, GA	GH 1956
$CH_3CH_2CH_2Cl \longrightarrow CH_3CHCH_2 + HCl$	58.6					
k_∞		2.8×10^{13}	230.1	672–733	Review	BO'N 1970

[a] See footnote on page 402 for units and comments.

TABLE 4.9. Unimolecular reactions.[a,b] (Continued)

Reaction	$\Delta_r H$	A	E_a	ΔT	Method	Reference
$CH_3CHClCH_3 \longrightarrow CH_3CHCH_2 + HCl$		3.6×10^{13}	213.0	637–690	SR, PM	HDJ 1975
$CH_3CH_2CHCl_2 \longrightarrow CH_3CHCHCl + HCl$	45.6	3.2×10^{13}	223.8	653–713	Review	BO'N 1970
$CH_3CCl_2CH_3 \longrightarrow CH_3CClCH_2 + HCl$	46.4	1.0×10^{14}	209.2	603–683	Review	BO'N 1970
$CH_3CHClCH_2Cl \longrightarrow CH_2CHCH_2Cl + HCl$ (a)		1.3×10^{13}	226.8	689–725	Review	BO'N 1970
		3.0×10^{13}	225.9	940–1060	ST, IS	TWB 1982
$CH_3CHClCH_2Cl \longrightarrow CH_3CHCHCl + HCl$ (b)		2.0×10^{13}	222.6	689–725	Review	BO'N 1970
trans		1.6×10^{13}	231.8	940–1060	ST, IS	TWB 1982
cis		9.6×10^{12}	223.4	940–1060	ST, IS	TWB 1982
$CH_3CHClCH_2Cl \longrightarrow CH_3CClCH_2 + HCl$ (c)		6.3×10^{12}	222.6	689–725	Review	BO'N 1970
		1.1×10^{13}	247.3	940–1060	ST, IS	TWB 1982
		5.0×10^{13} [c]	226.8	940–1060	ST, IS	TWB 1982
$CH_3(CH_2)_2CH_2Cl \longrightarrow CH_3CH_2CHCH_2 + HCl$	58.6	4.3×10^{13}	231.0	663–744	Review	BO'N 1970
$CH_3(CH_2)_3CH_2Cl \longrightarrow CH_3(CH_2)_2CHCH_2 + HCl$	59.4	6.5×10^{13}	231.4	662–729	Review	BO'N 1970
$(CH_3)_2CHCH_2Cl \longrightarrow (CH_3)_2CCH_2 + HCl$	47.7	7.9×10^{12}	222.6	600–700	Review	BO'N 1970
$(CH_3)_2CHCl \longrightarrow CH_3CHCH_2 + HCl$	65.7	4.0×10^{13}	213.8	600–940	Review	BO'N 1970
$(CH_3)_3CCl \longrightarrow (CH_3)_2CCH_2 + HCl$	71.1	6.3×10^{13}	188.3	547–950	Review	BO'N 1970
		8.5×10^{13}	191.2	576–622	SR, PM	HDJ 1975

[a] See footnote on page 402 for units and comments.

[b] The rate coefficients listed in this table are k_∞ except for those measured by the ST and SR methods.

[c] k_∞ for sum of channels (a), (b) and (c).

TABLE 4.9. Unimolecular reactions.[a,b] (Concluded)

Reaction	$\Delta_r H$	A	E_a	ΔT	Method	Reference
$CH_3CH_2CHClCH_3 \longrightarrow CCH_3CHCHCH_3 + HCl$ (a)	56.1	1.9×10^{13}	220.5	589–666	SR, GC	HR 1961
$CH_3CH_2CHClCH_3 \longrightarrow$ t-$CH_3CHCHCH_3 + HCl$ (b)	51.9	3.7×10^{13}	213.7	589–666	SR, GC	HR 1961
$CH_3CH_2CHClCH_3 \longrightarrow CH_3CH_2CHCH_2 + HCl$ (c)	63.2	4.4×10^{13}	216.3	589–666	SR, GC	HR 1961
$CH_3CH_2C(CH_3)_2CHCl \longrightarrow CH_3CH_2C(CH_3)CH_2 + HCl$ (a)						
$CH_3CH_2C(CH_3)_2CHCl \longrightarrow CH_3CHCC(CH_3)_2 + HCl$ (b)						
		5.9×10^{13}	184.1	533–593	SR, GC	CM 1980
$(CH_3)_2C(CH_3)_2Cl \longrightarrow (CH_3)_2CHC(CH_3)CH_2 + HCl$ (a)						
$(CH_3)_2C(CH_3)_2Cl \longrightarrow (CH_3)_2CC(CH_3)_2 + HCl$(b)						
		2.1×10^{13}[c]	175.3	533–593	SR, GC	CM 1980

[a] See footnote on page 402 for units and comments.

[b] Rate coefficients in this table are k_∞.

[c] Sum of channels (a) and (b).

TABLE 4.10. Reactions of large molecules and radicals.[a]

Reaction	$\Delta_r H$	A	m	E_a	ΔT	Method	Reference
$CH_3 + CH_2Cl \longrightarrow C_2H_5Cl$ (a)		3.0×10^{35}	–6.9	31.8	500–1700	Estimate *b*	KS 1987
$CH_3 + CH_2Cl \longrightarrow C_2H_4 + HCl$ (b)		1.5×10^{21}	–2.2	21.8	500–1700	Estimate *b*	KS 1987
$CH_3 + CH_2Cl \longrightarrow C_2H_5 + Cl$ (c)		3.2×10^{10}		19.7	500–1700	Estimate *b*	KS 1987
$CH_2Cl + CH_2Cl \longrightarrow CH_2ClCH_2Cl$ (a)		3.6×10^{37}	–7.6	38.5	500–1700	Estimate *b*	KS 1987
$CH_2Cl+CH_2Cl \longrightarrow C_2H_3Cl+HCl$ (b)		1.3×10^{24}	–3.3	34.3	500–1700	Estimate *b*	KS 1987
$CH_2Cl + CH_2Cl \longrightarrow CH_2ClCH_2Cl$ (a)		1.9×10^{17}	–1	40.6	500–1700	Estimate *b*	KS 1987
$CH_3 + CH_3Cl \longrightarrow CH_4 + CH_2Cl$	–23.4	1.3×10^{11}		48.5	396–469	SR, GC	MS 1979
		3.3×10^{11}		39.3	400–480	SR, PH, GC	RS 1952
$CH_3 + CH_2Cl_2 \longrightarrow CH_4 + CHCl_2$	–39.7	6.3×10^{11}		37.6	396–469	SR, GC	MS 1979
		6.6×10^{10}		30.1	402–482	SR, PH, GC	RS 1952
$CH_3 + CHCl_3 \longrightarrow CH_4 + CCl_3$	–53.6	3.2×10^{11}		28.0	396–469	SR, GC	MS 1979
		1.9×10^{10}		24.3	405–476	SR, PH, GC	RS 1952
$CH_3 + CCl_4 \longrightarrow CH_3Cl + CCl_3$	–66.9	1.6×10^{12}		41.4	330–523	SR, PH, GC	MTS 1982
		6.3×10^{13}		42.3	395–449	SR, GC	MS 1979
		1.6×10^{13}		56.1	363–413	SR, PH, GA	TGP 1964
$CH_3 + C_2H_3Cl \longrightarrow CH_3CH_2CHCl$		2.5×10^{11}		41.0	335–424	SR, GC	TWW 1972
$CH_3+C_6H_5CCl_3 \longrightarrow CH_3Cl+C_6H_5CCl_2$		1.9×10^{10}		31.8	363–418	Review	Kon 1972
$CH_2Cl + HCl \longrightarrow CH_3Cl + Cl$	16.6	2.0×10^{12}		32.6	360–420	Review	CGHMV 1963
$CHCl_2 + HCl \longrightarrow CH_2Cl_2 + Cl$	32.8	1.6×10^{12}		46.0	360–420	Review	CGHMV 1963
$CH_2Cl + C_2H_2 \longrightarrow CH_2ClCHCH$		1.0×10^{12}		33.5	1300	Estimate	WB 1989
$CCl_3 + HCl \longrightarrow CHCl_3 + Cl$ (a)	46.9	4.5×10^{11}		47.3	303–425	SR, GA	DH 1967
$CCl_3 + HCl \longrightarrow CCl_4 + H$ (b)	147.7	1.3×10^{12}		55.6	303–425	SR, GA	EGHMMS 1960
$CCl_3 + CCl_3 \longrightarrow C_2Cl_6$	–272.8	6.7×10^{9}	–1	0.0	253–623	SR, LP, UVS	DCVLL 1989

[a] See footnote on page 402 for units and comments.

[b] For 1 atm pressure.

TABLE 4.10. Reactions of large molecules and radicals.[a] (Concluded)

Reaction	$\Delta_r H$	A	E_a	ΔT	Method	Reference
$CCl_3 + C_2Cl_4 \longrightarrow CCl_3C_2Cl_4$		3.8×10^{10}	78.7	363–448	SR, RC, GC	HB 1979
$CCl_3 + C_2H_3Cl \longrightarrow CCl_3CH_2CHCl$		4.0×10^{11}	31.8	361–439	Review	KP 1972
$CCl_3 + c\text{-}C_5H_{10} \longrightarrow CHCl_3 + c\text{-}C_5H_9$		2.5×10^{11}	41.8	330–482	SR, PH, GC	MTS 1983
$CCl_3 + CH_4 \longrightarrow CHCl_3 + CH_3$	53.6	5.0×10^{12}	74.9	330–523	SR, PH, GC	MTS 1982
$CCl_3 + C_3H_8 \longrightarrow CHCl_3 + CH_3CH_2CH_2$	7.5	1.0×10^{12}	59.4	330–523	SR, PH, GC	MTS 1982
$CCl_3 + C_3H_8 \longrightarrow CHCl_3 + CH_3CHCH_3$	7.5	1.6×10^{11}	44.4	330–523	SR, PH, GC	MTS 1982
$CCl_3 + C_2H_6 \longrightarrow CHCl_3 + C_2H_5$	28.9	1.0×10^{12}	59.4	330–523	SR, PH, GC	MTS 1982
$CCl_3 + n\text{-}C_4H_{10} \longrightarrow CHCl_3 + CH_3(CH_2)_3$		1.6×10^{12}	59.8	330–523	SR, PH, GC	MTS 1982
$CCl_3 + n\text{-}C_4H_{10} \longrightarrow CHCl_3 + CH_3CHCH_2CH_3$		1.0×10^{12}	44.8	330–523	SR, PH, GC	MTS 1982
$CCl_3 + t\text{-}C_4H_{10} \longrightarrow CHCl_3 + (CH_3)_3C$		5.0×10^{10}	32.2	330–523	SR, PH, GC	MTS 1982
$C_2H_3 + O_2 \longrightarrow CH_2O + CHO$	–358.6	1.0×10^{12}	0.0	298–523	FR, LP, MS	SPHG 1984
$CH_2CCl + O_2 \longrightarrow CH_2O + COCl$	–389.5	3.0×10^{12}	1.4	298–648	FR, LP, MS	RS 1989a,b
$C_2Cl_3 + O_2 \longrightarrow COCl_2 + COCl$	–464.4	1.2×10^{12}	1.0	298–648	FR, LP, MS	RS 1989a,b
$C_2H_5 + CCl_4 \longrightarrow C_2H_5Cl + CCl_3$	–68.6	2.5×10^{12}	42.7	653–744	SR, GC	SWG 1989
		2.5×10^{12}	41.4	330–523	SR, PH, GC	MTS 1982
$C_2H_5Cl + HO_2 \longrightarrow CH_3CHCl + H_2O_2$ (a)	43.5	6.0×10^{12}	75.3	300–1000	Estimate	SYG 1992
$C_2H_5Cl + HO_2 \longrightarrow CH_2ClCH_2 + H_2O_2$ (b)	63.6	6.0×10^{12}	87.9	300–1000	Estimate	SYG 1992
$CH_3CHCl + O_2 \longrightarrow C_2H_3Cl + HO_2$	–46.0	2.0×10^{12}	25.1	300–1000	Estimate	SYG 1992
$CH_2ClCH_2 + O_2 \longrightarrow C_2H_3Cl + HO_2$	–66.1	2.0×10^{12}	20.9	300–1000	Estimate	SYG 1992
$C_2Cl_5 + HCl \longrightarrow C_2HCl_5 + Cl$	36.8	1.3×10^{11}	45.2	385–490	Review	Kon 1972
$i\text{-}C_3H_7 + CCl_4 \longrightarrow i\text{-}C_3H_7Cl + CCl_3$		1.0×10^{12}	42.7	330–523	SR, PH, GC	MTS 1982
$c\text{-}C_5H_9 + CCl_4 \longrightarrow c\text{-}C_5H_9Cl + CCl_3$		1.0×10^{12}	43.1	330–482	SR, PH, GC	MTS 1983
$C_4H_5 + HCl \longrightarrow C_4H_6 + Cl$		2.0×10^{12}	20.9	1300	Estimate	WB 1989

[a] See footnote on page 402 for units and comments.

overcome these problems and to acquire reliable data on the kinetics of reactions of interest in CHC combustion modeling. One successful approach has been to generate the stable reactants from suitable precursors under the actual reaction conditions, in a manner similar to those used for producing radicals (Russell et al. 1988).

Second, isomer-specific rate data are needed for developing detailed chemical kinetic mechanisms. The limited knowledge we have about reactions of specific CHC isomers, both stable species and radicals, impedes progress in the field. For example, consider the reaction of H atoms with C_2H_5Cl. Even in this simple case one must consider the formation of three distinct products:

$$\begin{aligned} H + C_2H_5Cl &\longrightarrow C_2H_5 + HCl \qquad (\Delta_r H = -18.9\,\text{kJ/mol}) \\ &\longrightarrow CH_3CHCl + H_2 \qquad (\Delta_r H = -6.7\,\text{kJ/mol}) \\ &\longrightarrow CH_2ClCH_2 + H_2 \qquad (\Delta_r H = -1.9\,\text{kJ/mol}). \end{aligned}$$

All channels are exothermic and therefore energetically favored. The subsequent reactions of the C_2 radicals formed, however, will be very different. Consequently, for the development of reaction mechanisms the rates of the individual, that is, isomer-specific, reactions of C_2H_5Cl must be determined. Similar remarks can be made for virtually all other CHCs. The reactions of radicals with chlorinated alkenes and alkynes also deserve special attention because these reactions involve chemically activated adducts. In addition to the possible presence of multiple addition sites, caused by the asymmetry of the groups at the double or triple bond, multiple reaction pathways may also be accessible to the energy-rich adduct. Although experimental determination of all isomer-specific reaction rate coefficients would be desirable, this clearly will be unreasonably expensive and in many instances impossible using current experimental techniques. Consequently, we must also consider alternative approaches to establish these rate coefficients. One approach is SAR, applied successfully to H atom abstractions from hydrocarbons and substituted hydrocarbons by OH (Atkinson 1986) and Cl (Senkan and Quam 1992). Since SAR methods can be extended to other types of reactions, their broader utilization in detailed chemical kinetic modeling is warranted.

Another productive approach to determine isomer-specific reaction rate information is computational quantum chemistry. It is now feasible to estimate thermochemical properties, most importantly relative energies, by electronic structure calculations using ab initio and semi-empirical (see for example Stewart 1989a,b) methods. Together with transition state theory, quantum chemistry can be used to estimate rate coefficients of elementary chemical reactions, including isomer-specific reaction channels, in a systematic manner. We have applied semi-empirical quantum chemistry to study the isomer-specific addition reactions of chlorinated C_1 and C_2 (vinylic) radicals to chlorinated ethylenes and acetylenes. (Shi and Senkan 1991) These studies indicated the presence of several interesting features and trends, with important implications for detailed mechanism development. For example, the activation energies of C_1 radical additions to chlorinated ethylenes and acetylenes increase with increasing chlorine substitution of the methyl radicals. For the additions to C_2H_4 the results were as follows:

Reaction			$\Delta_r H$ kJ/mol	E_a kJ/mol
$C_2H_4 + CH_3$	$\longrightarrow$	$CH_3CH_2CH_2$	–35.5	4.0
$+ CH_2Cl$	$\longrightarrow$	$CH_2ClCH_2CH_2$	–29.0	6.8
$+ CHCl_2$	$\longrightarrow$	$CHCl_2CH_2CH_2$	–22.9	9.6
$+ CCl_3$	$\longrightarrow$	$CCl_3CH_2CH_2$	–18.5	12.1.

These results suggest that with increased chlorine substitution of the C_1 radicals, molecular weight growth is increasingly dominated by radical recombination, which occurs with little or no activation energy, as opposed to radical additions to C_2 and higher molecular weight species. Experimental testing of this idea should be pursued.

Another result of the theoretical investigations was that the activation energies of C_1 radical additions to β sites were lower than those for additions to α sites, as illustrated by the addition for CH_2Cl to C_2H_3Cl:

Reaction			Site	$\Delta_r H$ (kJ/mol)	E_a (kJ/mol)
$CH_2Cl + C_2H_3Cl$	$\longrightarrow$	$CH_2ClCHClCH_2$	α	–25.4	8.6
	$\longrightarrow$	CH_2ClCH_2CHCl	β	–31.6	6.0

The activation energy calculations were consistent with the greater bond strengths of C–C bonds formed at β sites, relative to α sites, owing to the presence of highly electronegative Cl. Similar observations also were made for the addition of vinylic C_2 radicals, in which the barrier heights were found to be sensitive to the nature of the vicinal atom at the radical site. For example, consider the addition of C_2H_2Cl radicals to C_2H_4. For C_2H_2Cl, three distinct structures exist, that is, CH_2CCl, cis-CHClCH, and trans-CHClCH radicals. The activation energy barriers for the addition of these radicals to C_2H_4 were found by Shi and Senkan (1991) to be:

Reaction			$\Delta_r H$ (kJ/mol)	E_a (kJ/mol)
$C_2H_4 + CH_2CCl$	$\longrightarrow$	$CH_2CClCH_2CH_2$	–35.4	4.4
+ cis-CHClCH	$\longrightarrow$	$CHClCHCH_2CH_2$	–42.2	2.6
+ trans-CHClCH	$\longrightarrow$	$CHClCHCH_2CH_2$	–41.7	2.5

As indicated above, the activation energy for addition of the CH_2CCl radical, in which the Cl atom is in the vicinity of the radical site, is higher than the others, while its exothermicity is lower. The differences in activation energies for addition to chlorinated ethylenes are even more pronounced. (Shi and Senkan 1991) Again, experimental testing of these theoretical predictions is needed for improvement of computational methods and thus for the development of detailed kinetic models. Least-squares fits to semi- empirical quantum chemical predictions have also been developed. They resulted in the determination of the Evans–Polanyi relationships (Shi and Senkan 1991):

$$E_a = 0.39\Delta_r H + 19.6 \text{ kJ/mol (R = 0.96)}$$

for the addition of chlorinated C_1 and C_2 radicals to ethylenes, and

$$E_a = 0.34\Delta_r H + 20.8 \text{ kJ/mol} \ (R = 0.91)$$

for additions to acetylenes. Because Evans–Polanyi relationships allow the prediction of reaction rate parameters rapidly and systematically, their further exploration in detailed chemical kinetic modeling should be pursued.

Research needs and opportunities exist in both the experimental and the theoretical aspects of reactions of chlorine-containing species at combustion conditions. Determination of isomer-specific thermochemical properties as well as reaction rate parameters using a combination of experimental and theoretical approaches should be vigorously pursued, as this information is crucial for the development of detailed chemical kinetic mechanisms describing the combustion of CHCs in incinerators.

4.7 ACKNOWLEDGMENTS

Preparation of this review was supported in part by the National Science Foundation and the U.S. Environmental Protection Agency. We thank J.-W. Huang for help in the preparation of the figures.

4.8 REFERENCES

Abadzhev, S.S., Dzikh, I.P., & Shevchuk, V.U. (1990). Kinet. Catal., **30**, 893.

Abbatt, J.P.D., & Anderson, J.G. (1991). J. Phys. Chem., **95**, 2382.

Aders, W.K., Pangritz, D., & Wagner, H.Gg., (1975). Ber. Bunsenges. Phys. Chem., **79**, 90.

Adusei, G.Y., & Fontijn, A. (1993). J. Phys. Chem., **97**, 1409.

Ahmed, M., G., & Jones, W.E. (1985). Can. J. Chem., **63**, 2127.

Ambidge, P.F., Bradley, J.N., & Whytock, D.A. (1976). J. Chem. Soc. Faraday Trans., **72**, 1157.

Atkinson, R. (1985). Chem. Rev., **85**, 183.

Atkinson, R. (1986). Int. J. Chem. Kinet., **18**, 555.

Atkinson, R. (1987). Environ. Sci. & Technol., **21**, 305.

Atkinson, R. (1987). Int. J. Chem. Kinet., **19**, 799.

Atkinson, R., & Aschmann, S.M. (1985). Int. J. Chem. Kinet., **17**, 33.

Atkinson, R., Baulch, D.L., Cox, R.A., Hampson, R.F., Kerr, J.A., & Troe, J. (1992). Atmos. Environ., **26A**, 1187.

Atkinson, R., Baulch, D.L., Cox, R.A., Hampson, R.F., Kerr, J.A., & Troe, J. (1989). J. Phys. Chem. Ref. Data, **18**, 881.

Ayscough, P.B., Cocker, A.J., Dainton, F.S., & Hirst, S. (1962). Trans. Faraday Soc., **58,** 295.

Ayscough, P.B., Cocker, A.J., Dainton, F.S., Hirst, S., & Weston, M. (1961). Proc. Chem. Soc. (London), 244.

Baer, S., Hippler, H., Rahn, R., Siefke, N., Seitzinger, N., and Troe, J. (1991). J. Chem. Phys., **95,** 6463.

Balestra-Garcia, C., Le Bras, G., & Mac Leod, H. (1992). J. Phys. Chem., **96,** 3312.

Barasssin, J., & Combourieu, J. (1974). Bull. Soc. Chim. France, 1.

Barasssin, J., & Combourieu, J. (1973). Bull. Soc. Chim. France, 2173.

Barassin, J., Richoux, M., & Combourieu, J. (1977). Bull. Soc. Chim. France, 69.

Barton, D.H.R., & Howlett, K.E. (1949). J. Chem. Soc., 165.

Baulch, D.L., Cobos, C.J., Cox, R.A., Esser, C., Frank, P., Just, Th., Kerr, J.A., Pilling, M.J., Troe, J., Walker, R.W., and Warnatz, J. (1992). J. Phys. Chem. Ref. Data, **21,** No. 3, 411.

Baulch, D.L., Duxbury, J., Grant, S.J., & Montague, D.C. (1981). J. Phys. Chem. Ref. Data, **10,** Supp.1.

Benson, S.W. (1976). Thermochemical Kinetics, 2nd Edition, John Wiley & Sons, New York.

Benson, S.W., & Bose, A.N. (1963). J. Chem. Phys., **39,** 3463.

Benson, S.W., & O'Neal, H.E. (1970). Kinetic Data on Gas Phase Unimolecular Reactions, NSRDS-NBS 21, Feb. 1970.

Benson, S.W., & Spokes, G.N. (1967). 11th Symposium (International) on Combustion, p. 95.

Benson, S.W., & Weissman, M. (1982). Int. J. Chem. Kinet., **14,** 1287.

Bodenstein, M., Brendschede, W., & Schumacher, H.J. (1937). Z. Phys. Chem., **B35,** 382.

Boodaghians, R.B., Hall, I.W., & Wayne, R.P. (1987). J. Chem. Soc. Faraday Trans., **83,** 529.

Bradley, J.N., Whytock, D.A., & Zaleski, T.A. (1973). J. Chem. Soc. Faraday Trans., **69,** 1251.

Breitbarth, F.W., & Rottmeyer, S. (1986). Plasma Chem. Plasma Process., **6,** 381.

Brunning, J., & Stief, L.J. (1985). J. Chem. Phys., **83,** 1005.

Burns, W.G., & Dainton, F.S. (1952). Trans. Faraday Soc., **48,** 39.

Caballero, J.F., & Wittig, C. (1983). J. Chem. Phys., **78,** 7169.

Chang, W.D., & Senkan, S.M. (1985). Combust. Sci. Tech., **43,** 49.

Chang, W.D., & Senkan, S.M. (1989). Environ. Sci. Tech., **23,** 442.

Chase, M.W., Davies, C.A., Downey, J.R., Frurip, D.J., McDonald, R.A., & Syverud, A.N. (1985). J. Phys. Chem. Ref. Data, **14,** Supplement No.1.

Chiltz, G., Goldfinger, P., Huybrechts, G., Martens, G., & Verbeke, G. (1963). Chem. Rev., **63,** 355.

Chuchani, G., & Martin, I. (1980). J. Phys. Chem., **84,** 3188.

Cicek, B., & Senkan, S.M. (1993). Combust. Sci. Tech., **91,** 53.

Cillien, C., Goldfinger, P., Huybrechts, G., & Martens, G. (1967). Trans. Faraday Soc., **63,** 1631.

Clyne, M.A.A., & Walker, R.F. (1973). J. Chem. Soc. Faraday Trans., **69,** 1547.

Cohen, N., & Benson, S.W. (1987). J. Phys. Chem., **91,** 162 and 171.

Cohen, N., & Westberg, K.R. (1991). J. Phys. Chem. Ref. Data, **20,** 1211.

Combourieu, J., Le Bras, G., & Paty, C. (1973). 14th Symposium (International) on Combustion, p. 485.

Cook, J.L., Ennis, C.A., Leck, T.J., & Birks, J.W. (1981). J. Chem. Phys., **74,** 545.

Cox, R.A. (1980). Int. J. Chem. Kinet., **12,** 649.

Cui, J.P., He., Y.Z., & Tsang, W. (1989). J. Phys. Chem., **93,** 724.

Cvetanovic, R.J. (1987). J. Phys. Chem. Ref. Data, **16,** 261.

Dai, H-L., Specht, E., Berman, M.R., & Moore, C.B. (1982). J. Chem. Phys., **77,** 4494.

Danis, F., Caralp, F., Veyret, B., Loirat, H., & Lesclaux, R. (1989). Int. J. Chem. Kinet., **21,** 715.

Davis, D.D., Machado, G., Conaway, B., Oh, Y., & Watson, R. (1976). J. Chem. Phys., **65,** 1268.

Davis, D.D., Braun, W., & Bass, A.M. (1970). Int. J. Chem. Kinet., **2,** 101.

Dean, A.M. (1985). J. Phys. Chem., **89,** 4600.

Dean, A.M., & Westmoreland, P.R. (1987). Int. J. Chem. Kin., **19,** 207.

DeMare, G.R., & Huybrechts, G. (1968). Trans. Faraday Soc., **64,** 1311.

DeMore, W.B., Sander, S.P., Howard, C.J., Ravishankara, A.R., Golden, D.M., Kolb, C.E., Hampson, R.F., Kurylo, M.J., and Molina, M.J. (1992). JPL Publication 92-20, August 15, 1992.

Dixon-Lewis, G., & Williams, A. (1955) Presented at the 5th Symposium on Free Radicals, Uppsala. Dobe, S., Khachatryan, L.A., & Berces, L.A. (1989). Ber. Bunsenges. Phys. Chem., **93,** 847.

Dobis, O., & Benson, S.W. (1991). J. Amer. Chem. Soc., **113,** 6377.

Dusoleil, S., Goldfinger, P., Mahien-Van Der Auwera, A.M., & Martens, G. (1961). Trans. Faraday Soc., **57,** 2197.

Ebrecht, J., Hack, W., & Wagner, H.Gg. (1990). Ber. Bunsenges. Phys. Chem., **94,** 587.

Eckling, R., Goldfinger, P., Huybrechts, G., Martens, G., Meyers, L., & Smoes, S. (1960). Chem. Ber., **93**, 3014.

Ednay, E.O., Kleindrienst, T.E., & Corse, E.W. (1986). Int. J. Chem. Kinet., **18**, 1355.

Ennis, C.A., & Birks, J.W. (1985). J. Phys. Chem., **89**, 186.

Ennis, C.A., & Birks, J.W. (1988). J. Phys. Chem., **92**, 1119.

Evans, P.J., Ichimura, T., & Tschuikow-Roux, E. (1978). Int. J. Chem. Kinet., **10**, 855.

Fabry, L., Potzinger, P., Reimann, B., Ritter, A., & Steenbergen, H.P. (1986). Organometallics, **5**, 1231.

Fasano, D.M., & Nogar, N.S. (1981). Int. J. Chem. Kinet., **13**, 325.

Fettis, G.C., & Knox, J.H. (1964). in Prog. Reaction Kinetics 2, G. Porter, Ed., Pergamon, New York.

Fisher, E.M., Koshland, C.P., Hall, D., Sawyer, R., & Lucas, D. (1990). 23rd Symposium (International) on Combustion, p. 1273.

Force, A.P., & Wiesenfeld, J.R. (1981). J. Chem. Phys., **85**, 782.

Francisco, J.S., Steinfeld, J.I., King, K.D., & Gilbert, R.G. (1981). J. Phys. Chem., **85**, 4106.

Frerichs, H., Tappe, M., & Wagner, H.Gg. (1989). Z. Phys. Chem., **162**, 147.

Fujimoto, G.T., Umstead, M.E., & Lin, M.C. (1980). Chem. Phys., **51**, 399.

Gardiner, W.C. Jr., & Troe, J. (1984). in Combustion Chemistry, W.C. Gardiner, Jr., Ed., Springer–Verlag, New York.

Goldbach, A., Temps, F., & Wagner, H., Gg. (1990). Ber. Bunsenges. Phys. Chem., **94**, 1367.

Goldfinger, P., Jeunehomme, M., & Martens, G. (1958). J. Chem. Phys., **29**, 456.

Goodall, A.M., & Howlett, K.E. (1956). J. Chem. Soc., 2640.

Granada, A., Karra, S.B., & Senkan, S.M. (1987). Ind. Eng. Chem. Res., **26**, 1901.

Hart, L.W., Grunfelder, C., & Fristrom, R.M. (1974). Combust. Flame, **23**, 109.

Heneghan, S.P., & Benson, S.W. (1983). Int. J. Chem. Kinet., **15**, 1311.

Heneghan, S.P., Knoot, P.A., & Benson, S.W. (1981). Int. J. Chem. Kinet., **13**, 677.

Herron, J.T. (1988). J. Phys. Chem. Ref. Data, **17**, 967.

Herron, J.T., & Huie, R.E. (1969). J. Phys. Chem., **73**, 1326.

Herron, J.T., & Huie, R. E. (1973). J. Phys. Chem. Ref. Data, **2**, 467.

Heydtman, H., Dill, B., & Jonas, R. (1975). Int. J. Chem. Kinet., **7**, 973.

Heydtman, H., & Rinck, G. (1963). Z. Phys. Chem., **36**, 75.

Howlett,K.E. (1952). Trans. Faraday Soc., **48**, 25.

Ho, W., Yu, Q-R., & Bozzelli, J.W. (1992). Combust. Sci. Tech., **85**, 23.

Hooshiyar, P.A., & Niki, H. (1995). Int. J. Chem. Kin., **27**, 1197.

Horowitz, A., & Baruch, G. (1979). Int. J. Chem. Kinet., **11**, 1263.

Houser, T.J., & Cuzcano, T. (1975). Int. J. Chem. Kinet., **7**, 331.

Howard, C.J., & Evenson, K.M. (1976). J. Chem. Phys., **64**, 4303.

Howard, C.J. (1976). J. Chem. Phys., **65**, 4771.

Hranisavljevic, J., Adusei, G.Y., Xun, Y., & Fontijn, A. (1993). Combust. Sci. Tech., **101**, 231.

Huybrechts, G., Hubin, Y., & Van Mele, B. (1989). Int. J. Chem. Kinet., **21**, 575.

Iyer, R.S., Rogers, P.J., & Rowland, F.S. (1983). J. Phys. Chem., **87**, 3799.

Jeffers, P. M. (1974). J. Phys. Chem., **78**, 1469.

Jeong, K-M., Hsu, K-J., Jeffries, J.B., & Kaufman, F. (1984). J. Phys. Chem., **88**, 1222.

Jeong, S.C., Choo, K.Y., & Benson, S.W. (1991). J. Phys. Chem., **95**, 7282.

Jiang, Z., Taylor, P.H., & Dellinger, B. (1992a). J. Phys. Chem., **96**, 8961.

Jiang, Z., Taylor, P.H., & Dellinger, B. (1992b). J. Phys. Chem., **96**, 8964.

Jones, W.E., MacKnight, S.D., & Teng, L. (1973). Chem. Rev., **73**, 407.

Jones, W.E., & Ma, J.L. (1986). Can. J. Chem., **64**, 2192.

Kaiser, E.W. (1992). Int. J. Chem. Kinet., **24**, 179.

Kaiser, E.W., Rimai, L., Schwab, E., & Lim, E.C. (1992). J. Phys. Chem., **96**, 303.

Kaiser, E.W., Rimai, L., & Wallington, T.J. (1989). J. Phys. Chem., **93**, 4094.

Karra, S.B., & Senkan, S.M. (1988). Ind. Eng. Chem. Res., **27**, 447.

Karra, S.B., Gutman, D., & Senkan, S.M. (1988). Combust. Sci. Tech., **60**, 45.

Kasner, J.H., Taylor, P.H., & Dellinger, B. (1990). J. Phys. Chem., **94**, 3250.

Kerr, J.AA., & Parsonage, M.J. (1972). Evaluated Data on Gas Phase Addition Reactions, Butterworths, London.

Keyser, L.F. (1978). J. Chem. Phys., **69**, 214.

Keyser, L.F. (1980). J. Phys. Chem., **84**, 11.

Keyser, L.F. (1984). J. Phys. Chem., **88**, 4750.

Kim, H.J., & Choo, K.Y. (1983). Bull. Korean Chem. Soc., **4**, 203.

Kirchnner, K., Helf, DD., Ott, P., & Vogt, S. (1990). Ber. Bunsenges. Phys. Chem., **94**, 77.

Kita, D., & Stedman, D.H. (1982). J. Chem. Soc. Faraday Trans., **2**, 1249.

Kitabatake, M., & Onouchi, T. (1962). Kogyo Kagaku Zasshi, **65**, 931.

Knox, J.H., & Nelson, R.L. (1959). Trans. Faraday Soc., **55**, 937.

Ko, T., Fontijn, A., Lim, K.P., & Michael, J.V. (1992). 24th Symposium (International) on Combustion, p. 735,.

Kondo, O., Saito, K., & Mukarami, I. (1980). Bull. Chem. Soc. Japan, **53,** 2133.

Kondratiev, V.N. (1970). Konstanty Skorosti Gazofaznyx Reaktsii, Spravotchnik, Nauka, Moscow; translation as Rate Constants of Gas Phase Reactions, U.S. Department of Commerce National Bureau of Standards NSRDS Publication COM-72-10014, 1972.

Krueger, H., & Weitz, E. (1988). J. Chem. Phys., **88,** 1608.

Kubat, P., & Pola, J. (1985). Collect. Czech. Chem. Commun., **50,** 1537.

Lee, W.J., Cicke, B., & Senkan, S.M (1993). Environ. Sci. Tech., **27,** 949.

Lee, Y-P., & Howard, C.J. (1982). J. Chem. Phys., **77,** 756.

Leu, M-T. (1984). J. Phys. Chem., **88,** 1394.

Lewis, R.S., Sander, S.P., Wagner, S., & Watson, R.T. (1980). J. Phys. Chem., **84,** 2009.

Libuda, H.G., Zabel, F., Fink, E.H., & Becker, K.H. (1990). J. Phys. Chem., **94,** 5860.

Lim, K.P., & Michael, J.V. (1993). J. Chem. Phys., **98,** 3919.

Liu, A., Mulac, W.A., & Jonah, C.D. (1989). J. Phys. Chem., **93,** 4092.

Loewenstein, L.M., & Anderson, J.G. (1984). J. Phys. Chem., **88,** 6277.

Macken, K.V., & Sidebottom, H.W. (1979). Int. J. Chem. Kinet., **11,** 511.

Mahmud, K., Kim, J-S., & Fontijn, A. (1990). J. Phys. Chem., **94,** 2994.

Manion, J.A., & Louw, R. (1988). J. Chem. Soc. Perkin Trans., **2,** 1547.

Manion, J.A., & Louw, R. (1989). Rec. Trav. Chim. Pays-Bas, **108,** 235.

Mannig, R.G., & Kurylo, M.J (1977). J. Phys. Chem., **18,** 291.

Markert, F., & Nielsen, O.J. (1992). Chem. Phys. Lett., **194,** 123.

Martigan, J.J. (1984). J. Phys. Chem., **88,** 3638.

Matheson, I., Tedder, J., & Sidebottom, H. (1982). Int. J. Chem. Kinet., **14,** 1033.

Matheson, I., Tedder, J., & Sidebottom, H. (1983). Int. J. Chem. Kinet., **15,** 905.

Michael, J.V., Lim, K.P., Kumaran, S.S., & Kiefer, J.H. (1993). J. Phys. Chem., **97,** 1914.

Michael, J.V., Nava, D.F., Payne, W.A., & Stief, L.J. (1979a). J. Chem. Phys., **70,** 1147.

Michael, J.V., Nava, D.F., Payne, W.A., & Stief, L.J. (1979b). J. Chem. Phys., **70,** 3652.

Michael, J.V., Nava, D.F., Payne, W.A., & Stief, L.J. (1981). Chem. Phys. Lett., **77,** 110.

Michael, J.V., Whytock, D.A., Lee, J.H., Payne, W.A., & Stief, L.J. (1977). J. Chem. Phys., **67,** 3533.

Miller, J.A., & Bowman, C.T. (1989). Prog. Energy Combust. Sci., **15,** 287.

Miller, J.C., & Gordon, R.J. (1981). J. Chem. Phys., **75**, 5305.

Mitchell, M.B., Brunning, J., Payne, W.A., & Stief, L.J. (1988). J. Phys. Chem., **92**, 1502.

Moin, F.B., Yurkevich, Ya.P., & Drogobytskii, V.M. (1979). Kinet. Catal., **20**, 456.

Molina, M.J., Molina, L.T., & Smith, C.A. (1984). Int. J. Chem. Kinet., **16**, 1151.

Nelson, H.H., & McDonald, J.R. (1982). J. Phys. Chem., **86**, 1242.

Nelson, L., Rattigan, O., Neavyn, R., Treacy, J., & Nielsen, O.J. (1990a). Int. J. Chem. Kinet., **22**, 1111.

Nelson, L., Shanahan, I., Sidebottom, H.W., Treacy, J., & Nielsen, O.J. (1990b). Int. J. Chem. Kinet., **22**, 577.

Nelson, L., Treacy, J.J., & Sidebottom, H.W. (1984). in Physico-Chemical Behavior of Atmospheric Pollutants, Versino, B., & Angeletti, G., Eds., Kluwer Academic Publishers, Dordrecht, p. 258.

Nesbitt, D.J., & Leone, S.R. (1982). J. Phys. Chem., **86**, 4962.

Nicovich, J.M., Kreutter, K.D., & Wine, P.H. (1990). J. Chem. Phys., **92**, 3539.

Niedzielski, J., Tschuikow-Roux, E., & Yano, T. (1984). Int. J. Chem. Kinet., **16**, 621.

Ohta,T. (1983). Bull. Chem. Soc. Jpn., **5**, 869.

Oppelt, E.T. (1987). J. Air Pollut. Control Assoc., **37**, 558.

Parmar, S.S., & Benson, S.W. (1988). J. Phys. Chem., **92**, 2652.

Pauling, L. (1960). The Nature of the Chemical Bond, 3rd Ed., Cornell University Press.

Payne, W.A., Brunning, J., Mitchell, M.B., & Stief, L.J. (1988). Int. J. Chem. Kinet., **20**, 63.

Payne, W.A., Nava, D.F., Nesbitt, F.L., & Stief, L.J. (1990). J. Phys. Chem., **94**, 7190.

Perry, R.A., Atkinson, R.A., & Pitts, J.N. (1977). J. Chem. Phys., **67**, 458.

Pitz, W.J., Westbrook, C.K., Lutz, A.E., Kee, R.J., Seebold, J.G., & Senkan, S.M. (1994). Comb. Sci. Tech., **101**, 381.

Poulet, G., Le Bras, G., & Combourieu, J. (1978). J. Chem. Phys., **69**, 767.

Poulet, G., Zagogianni, H., & Le Bras, G. (1986). Int. J. Chem. Kinet., **18**, 847.

Qui, L-X., Shi, S.H., Xing, S.B., & Chen, X.G. (1992). J. Phys. Chem., **96**, 685.

Qun, M., & Senkan, S.M. (1994). Combust. Sci. Tech., **101**, 101.

Raal, F.A., & Steacie, E.W.R. (1952). J. Chem. Phys., **20**, 578.

Ravishankara, A.R., Wine, P.H., Wells, J.R., & Thompson, R.L. (1985). Int. J. Chem. Kinet., **17**, 1281.

Ray, G.W., Keyser, L.F., & Watson, R.T. (1980). J. Phys. Chem., **84**, 1674.

Rinke, M., & Zetzsch, C. (1984). Ber. Bunsenges. Phys. Chem., **88,** 55.

Robinson, P.J., & Holbrook, K.A. (1972). Unimolecular Reactions, John Wiley & Sons, New York.

Russell, J.J., Seetula, J.A., Gutman, D., & Senkan, S.M. (1989a). J. Phys. Chem., **93,** 1934.

Russell, J.J., Seetula, J.A., Gutman, D., Danis, F., Caralp, F., Lightfoot, P.D., Lesclaux, R., Melius, C.F., & Senkan, S.M. (1990b). J. Phys. Chem., **94,** 3277.

Russell, J.J., Seetula, J.A., Gutman, D., Melius, C.F., & Senkan, S.M. (1990a). 23rd Symposium (International) on Combustion, p. 163,.

Russell, J.J., Seetula, J.A., Senkan, S.M., & Gutman, D. (1988). Int. J. Chem. Kinet., **20,** 759.

Russell, J.J., Senkan, S.M., Seetula, J.A., & Gutman, D. (1989b). J. Phys. Chem., **93,** 5184.

Russell, J.J., Senkan, S.M., Seetula, J.A., & Gutman, D. (1989c). 22nd Symposium (International) on Combustion, p. 1007.

Schneider, M., & Wolfrum, J. (1986). Ber. Bunsenges. Phys. Chem., **90,** 1058.

Schug, K.P., Wagner, H.Gg., & Zabel, F. (1979). Ber. Bunsenges. Phys. Chem., **83,** 167.

Scollard, D., Corrigan, M., & Sidebottom, H.W. (1991). STEP/AFEAS Workshop, Dublin (Ireland), May 14–16.

Seetula, J.A., Gutman, D., Lightfoot, P.D., Rayes, M.T., & Senkan, S.M. (1991). J. Phys. Chem., **95,** 10688.

Seidel, W., Marin, H., & Meitzner, F.G. (1965). Z. Phys. Chem. NF, **47,** 348.

Semeluk, G.P., & Bernstein, R.B. (1957). J. Amer. Chem. Soc., **79,** 46.

Senkan, S.M. (1988). Environ. Sci. Tech., **22,** 368.

Senkan, S.M., & Quam, D. (1992). J. Phys. Chem., **96,** 10837.

Senkan, S.M., Robinson, J.M., & Gupta, A.K. (1983). Combust. Flame, **49,** 305.

Senkan, S.M., Yildirim, R., & Gutman, D. (1992). 24th Symposium (International) on Combustion, p. 749.

Shi, Y., & Senkan, S.M. (1991). J. Phys. Chem., **95,** 5181.

Shilov, A.E., & Sabirova, R.D. (1959). J. Phys. Chem. Soc. Russ., **33,** 1365.

Shilov, A.E., & Sabirova, R.D. (1960). Russ. J. Phys. Chem., **34,** 408.

Singleton, D.L., & Cvetanovic, R.J. (1981). Int. J. Chem. Kinet., **13,** 945.

Slagle, I., & Gutman, D. (1985). J. Am. Chem. Soc., **107,** 5342.

Slagle, I., Gutman, D., & Gilbert, J.R. (1975). 15th Symposium (International) on Combustion, p. 785.

Slagle, I., Park, J.Y., Heaven, M., & Gutman, D. (1984). J. Am. Chem. Soc., **106,** 4356.

Starcke, J., Zabel, F., Elsen , L., Nilsen, W., Barnes, I., & Becker, K.H. (1990). in Physico-Chemical Behavior of Atmospheric Pollutants, Restelli, G., & Angeletti, G., Eds., Kluwer Academic Publishers, Dordrecht, p. 172.

Stevens, P.S., & Anderson, J.G. (1992). J. Phys. Chem., **96**, 1708.

Su, M.C., Lim, K.P., Michael, J.V., Hranisavljevic, J., & Fontijn, A. (1994). J. Phys. Chem., **98**, 8411.

Szirovicza, L., Woog, J., & Bardi, I. (1989). Acta Chim. Acad. Sci. Hungary, **126**, 117.

Talukdar, R.K., Mellouki, A., Schmoltner, A., Watson, T., Montzka, S., & Ravishankara, A.R. (1992). Science, **257**, 227.

Taylor, P.H., D'Angelo, J.A., Martin, M.C., Kasner, J.H., & Dellinger, B. (1989). Int. J. Chem. Kin., **21**, 829.

Taylor, P.H., Jiang, Z., & Dellinger, B. (1992). J. Phys. Chem., **96**, 1293.

Taylor, P.H., McCarron,, S., & Dellinger, B. (1991). Chem. Phys. Lett., **177**, 27.

Tedder, J.M., Walton, J.C., & Winton, K.D.R. (1972). J. Chem. Soc. Faraday Trans., **68**, 1866.

Timonen, R. (1988). Ann. Acad. Sci. Finn. Ser. A2, **218**, 5.

Timonen, R.S., & Gutman, D. (1986). J. Phys. Chem., **90**, 2987.

Timonen, R., Kalliorinne,K., & Koskikallio,J. (1986a). Acta Chem. Scand., **40**, 459.

Timonen, R.S., Ratajczak, E., & Gutman, D. (1988). J. Phys. Chem., **92**, 651.

Timonen, R.S., Russell, J.J., & Gutman, D. (1986b). Int. J. Chem. Kinet., **18**, 1193.

Timonen, R.S., Russell, J.J., Sarzynski, D., & Gutman, D. (1987). J. Phys. Chem., **91**, 1873.

Tomkinson, D.M., Galvin, J.P., & Pritchard, H.O. (1964). J. Phys. Chem., **68**, 541.

Troe, J. (1977a). J. Chem. Phys., **66**, 4745.

Troe, J. (1977b). J. Chem. Phys., **66**, 4758.

Tsang, W. (1967). J. Chem. Phys., **41**, 2487.

Tsang, W., & Walker, J.A. (1991). 23rd Symposium (International) on Combustion, p. 139.

Tsang, W., Cui, J.P., & Walker, J.A. (1990). Current Topics in Shock Waves: Proc. 17th Symp. Shock Tubes and Waves, Kim, Y.W., Ed., American Institute of Physics, New York, p. 205.

Tsang, W., He, Y.Z., Mallard, W.G., & Cui, J.P. (1988). Proc. 16th Symp. Shock Tubes and Waves, Gronig, H., Ed., VCH Press, Weinheim, p. 467.

Tsang, W., Walter, J.A., & Braun, W. (1982). J. Phys. Chem., **86**, 719.

Tschuikow-Roux, E., & Niedzielski, J. (1984a). J. Photochem., **27**, 141.

Tschuikow-Roux, E., Faraji, F., & Niedzielski, J. (1986). Int. J. Chem. Kinet., **18,** 513.

Tschuikow-Roux, E., Faraji, F., Paddison, S., Niedzielski, J., & Miyokawa, K. (1988). J. Phys. Chem., **92,** 1488.

Tschuikow-Roux, E., Niedzielski, J., & Faraji, F. (1985a). Can. J. Chem., **63,** 1093.

Tschuikow-Roux, E., Yano, T., & Niedzielski, J. (1984b). J. Phys. Chem., **88,** 1408.

Tschuikow-Roux, E., Yano, T., & Niedzielski, J. (1985b). J. Chem. Phys., **82,** 65.

Tuazon, E.C., Atkinson, R.A., & Aschmann, S.M. (1990). Int. J. Chem. Kinet., **22,** 981.

Tuazon, E.C., Atkinson, R., Aschmann, S.M., Goodman, M.A., & Winer, A.M. (1988). Int. J. Chem. Kinet., **20,** 241.

Tuazon, E.C., Atkinson, R., & Corchnoy, S.B. (1992). Int. J. Chem. Kinet., **24,** 639.

Ung, A.Y.M., & Schiff, H.I. (1962). Can. J. Chem., **40,** 486.

Valeiras, H., Gupta, A.K., & Senkan, S.M. (1984). Combust. Sci. Tech., **36,** 123.

Vance, J.E., & Bauman, W.C. (1938). J. Chem. Phys., **6,** 811.

Wagner, H. Gg., Welzbacher, U., & Zellner, R. (1976). Ber. Bunsenges. Phys. Chem., **80,** 902.

Wahner, A., & Zetzsch, C. (1982). in Physico-Chemical Behavior of Atmospheric Pollutants, Versino, B., & Ott, N., Eds., Kluwer Academic Publishers, Dordrecht, p. 138.

Wahner, A., & Zetzsch, C. (1983). J. Phys. Chem., **87,** 4945.

Wallington, T.J., Andino, J.M., Lorkovic, I.M., Kaiser, E.W., & Marston, G. (1990). J. Phys. Chem., **94,** 3644.

Wallington, T.J., Dagaut, P., & Kurylo, M.J. (1988a). J. Phys. Chem., **92,** 5024.

Wallington, T.J., Skewes, L.M., Siegl, W.O., Wu, C-H, & Japar, S.M. (1988b). Int. J. Chem. Kinet., **20,** 867.

Warnatz, J. (1984). in Combustion Chemistry, W.C. Gardiner, Jr., Ed., Springer-Verlag, New York.

Washida, N., & Hatakeyama, S. (1987). Bull. Chem. Soc. Jpn., **60,** 3751.

Wategaonkar, S.J., & Setser, D.W. (1989). J. Chem. Phys., **90,** 251.

Watson, R.T., Machado, G., Conaway, B., Wagner, S., & Davis, D.D. (1977). J. Phys. Chem., **81,** 256.

Watson, R., Machado, G., Fischer, S., & Davis, D.D. (1976). J. Chem. Phys., **65,** 2126.

Wecker, D., Johanssen, R., & Schindler, R.N. (1982). Ber. Bunsenges. Phys. Chem., **86,** 532.

Weissman, M., & Benson, S.W. (1980). Int. J. Chem. Kinet., **12**, 403.

Weissman, M., & Benson, S.W. (1984). Int. J. Chem. Kinet., **16**, 307.

Weissman, M., & Benson, S.W. (1989). Prog. Energy Combust. Sci., **15**, 273.

Westbrook, C.K. (1982) 19th Symposium (International) on Combustion, p. 127.

Westenberg, A.A., & De Haas, N. (1975). J. Chem. Phys., **62**, 3321.

Westenberg, A. A., & De Haas, N. (1977). J. Chem. Phys., **66**, 4900.

Westmoreland, P.R., Howard, J.B., Longwell, J.P., & Dean, A.M. (1986). Amer. Inst. Chem. Eng. J., **32**, 1971.

Wilson, W.E., & O'Donovan, J.T. (1967). J. Chem. Phys., **47**, 5455.

Wine, P.H., & Semmes, D.H. (1983). J. Phys. Chem., **87**, 3572.

Wine, P.H., Nicovich, J.M., & Ravishankara, A.R. (1985). J. Phys. Chem., **89**, 3914.

Wine, P.H., Semmes, D.H., & Ravishankara, A.R. (1982). Chem. Phys. Lett., **90**, 128.

Xieqi, M., Cicek, B., & Senkan, S.M. (1993). Combust. Flame, **94**, 131.

Xing, S-B., Shi, S-H, & Qiu, L-X. (1992). Int. J. Chem. Kinet., **24**, 1.

Yano, T. (1977). Bull. Soc. Chem. Japan, **50**, 1272.

Zabel, F. (1974). Ber. Bunsenges. Phys. Chem., **78**, 232.

Zabel, F. (1977). Int. J. Chem. Kinet., **9**, 651.

Zahniser, M.S., Berquist, B.M., & Kaufman, F. (1978). Int. J. Chem. Kinet., **10**, 15.

Zellner, R.E. (1984). in Combustion Chemistry, W.C. Gardiner, Jr., Ed., Springer-Verlag New York.

Zhang, Z., Liu, R., Huie, R.E., & Kurylo, M.J. (1991). J. Phys. Chem., **95**, 194.

5

Ideal Gas Thermochemical Data for Combustion and Air Pollution Use

Alexander Burcat [1]

William C. Gardiner, Jr. [2]

5.1 INTRODUCTION

There are two widespread misconceptions about thermochemical data:

The first is the wrong belief that thermochemical data are always available and that there are no outstanding problems in this domain. In fact, thermochemical data of any kind are available for at most 10000 to 15000 chemical species out of the several million ones known, that is, not even 1%.

The second fallacy is that whatever thermochemical data may not be available from experiments reported in the literature can be estimated readily with accuracy sufficient for combustion modeling by group additivity methods (Benson 1976; Pedley et al., 1986; Ritter 1990; Ritter and Bozzelli 1990, 1991; Ritter 1991; Stein et al. 1991; Cohen and Benson 1992; Stein 1994; Lay et al. 1995; Muller et al. 1995; Cohen 1996) or by molecular quantum mechanics. The truth about group additivity is that it cannot be used for many important species, or can be used only with high uncertainty, for lack of trustworthy group and bond data. The truth about molecular quantum mechanics is that it is successful only for finding molecular geometries and vibrational frequencies—bond energies, and hence standard enthalpies of formation, can only be computed accurately enough by quantum mechanics for quite small molecules and radicals, and then only at great computer expense and by experts.

Three fundamental thermochemical properties are used in combustion. Values

[1] Faculty of Aerospace Engineering, Technion—Israel Institute of Technology, Haifa 32000 Israel

[2] Department of Chemistry and Biochemistry, The University of Texas at Austin, Austin, Texas 78712

for the standard molar enthalpy of formation $\Delta_f H_T^\circ$ of fuel molecules and other chemical species of interest are required to compute enthalpy changes of reaction $\Delta_r H^\circ$, commonly called heats of reaction, from which the heat evolution in a system can be computed. To compute equilibrium compositions one must also know the standard free energy change of reaction, which requires one to have values for the standard molar entropy S_T° of all species thought to be present. Together with the molar heat capacity of the system, which can be computed for any composition from the molar heat capacities C_{Pi} and mole fractions X_i of the species present from $\sum X_i C_{Pi}$, adiabatic flame temperatures and other flame properties can be calculated.[1] For nearly all applications one assumes that the thermodynamic state is governed by ideal gas behavior; in occasional circumstances related to propellants or explosives one has to correct ideal gas thermochemical properties for real-gas compressibility effects (Ficket and Cowan 1955).

Since calculations are done automatically with digital computers nowadays, the needed thermochemical data have to be provided in a form suitable for computer access. A polynomial form of representing thermochemical data has been found to be suitable (Burcat 1984). An alternate procedure adopted by some authors is to store room-temperature values for standard molar enthalpies of formation $\Delta_f H_{298}^\circ$ (or $\Delta_f H_{300}^\circ$) and standard molar entropies S_{298}° (or S_{300}°) together with a list of molar $C_p(T_i)$ values at higher temperatures. One then computes the needed thermochemical values at nonstored temperatures by interpolation and use of standard thermodynamic equations. If extrapolation to temperatures beyond those for which data has been stored is needed, it must be done in a physically correct way.[2] This method has both merits and shortcomings. The principal merit is availability of $\Delta_f H_{300}^\circ$, S_{300}°, and $C_p(T_i)$ values without having to evaluate polynomials. The main shortcoming is that it requires users of the data, who may not have experience with the behavior of thermochemical functions, to write thermochemical interpolations or risky extrapolations into their computer codes.

---

[1] For optimizing polynomial representations of thermochemical data, the heat capacity and enthalpy are treated independently (McBride and Gordon 1992). Physically, the relation $C_P \equiv (\partial H/\partial T)_P$ always connects them to one another (cf. Section 5.9). While the fundamental temperature-dependent properties enthalpy and entropy completely define the thermochemistry of ideal gases, they are often arranged into alternative thermochemical functions for convenience in carrying out calculations or preparing tabulations.

[2] The problems inherent in extrapolating thermochemical values to the high temperatures encountered in combustion were discussed by Burcat (1984). The Wilhoit (1975) method for extrapolating thermochemical properties has become a standard procedure (McBride and Gordon 1992). It is suitable for all cases where thermochemical data extending to 1000 K or beyond are available. For situations where the data extends only to 600 or 700 K the Wilhoit method is likely to fail. An alternate extrapolation method that can be used safely for such cases has been developed by Ritter and Bozzelli (1990).

5.2 THERMOCHEMICAL DATABASE

An enlarged and critically updated version of our earlier database of thermochemical polynomial coefficients (Burcat 1984) has been developed. An overview of the species included in it, together with the sources of the standard enthalpy of formation of these species, is provided in the table at the end of this chapter. The size of the database—now over 700 species—precludes listing the coefficients themselves. Procedures for obtaining the complete database over the Internet are described in Section 5.11.

Data for many additional species important in combustion have been added since 1984. These include organic free radicals, large organic molecules commonly used as standard reference fuels or assumed to play important roles in PAH and soot formation, and halogenated compounds important in the chemistry of waste incineration, fire extinguishing and air pollution. New sulphur compounds were added. The database includes substantial contributions from many authors: a number of C_4 and C_8 molecules and radicals reported by Hall and Colket (1992); several free radicals such as the C_4H_9 family calculated by W. Tsang; liquid fuels such as benzene, toluene, octane, and Jet-A calculated by B. McBride, and radicals (such as vinylidene and cyclo-C_5H_5) and molecules (ranging from vinyl alcohol to buckminsterfullerene) calculated by A. Burcat. Other authors cited in the table also contributed calculations for various species. Revised coefficients for a number of species were computed to take into account new thermochemical data or improve fitting accuracy. High-temperature values were used directly whenever possible; extrapolations, where neccessary, were done in an expert way using the methods of Wilhoit (1975) or Ritter and Bozzelli (1990, 1991).

The date and place where the current polynomials were generated are indicated by the bracketed notations in the table at the end of this chapter. Thus [NASA 1993] means that the polynomial coefficients in the database were reported at NASA-Lewis Research Center in 1993.

5.3 SOURCES OF THERMOCHEMICAL DATA

The sources of ideal gas thermochemical data that the combustion research community has generally felt to be most reliable are those that are first of all based on carefully analyzed experimental data and secondly have been published as joint products of long-term collaborations of groups of thermochemistry experts. In the United States such collaborations have been carried out over the years at the National Bureau of Standards in Washington—now the National Institute of Standards and Technology in Gaitherburg—at the Bureau of Mines in Pittsburgh, at the Dow Chemical Company in Midland, Michigan, and at Texas A & M University in College Station. In the former Soviet Union similar large-scale efforts were undertaken by a group led by the late L.V. Gurvich. From the acronyms of the sponsors and publications of these groups one speaks respectfully of API, JANAF, NBS, TRC, and TSIV thermochemical data.

Unfortunately for contemporary combustion science, the best efforts of measurers, compilers, and analyzers of thermochemical data have not yet succeeded in determining accurate property values for all of the important species in combustion. Such species either do not appear in the standard compilations at all or their property values are assigned such large error bounds that one has to be concerned about whether computations which have to use them can be trusted. In this situation it is necessary to use to theoretical methods for estimating thermochemical properties of species in the ideal gas state.

The estimation problem comprises two essentially different parts. The first is a primarily physical one, computing the standard molar entropy and heat capacity. The underlying statistical thermodynamic theory for doing this is well understood and computationally trivial, so what has to be estimated are the molecular constants of the species of interest—its product of inertia, vibrational frequencies, and where applicable the constants describing internal rotations or low-energy electronic states. A summary of the applicable equations can be found in Chase et al. (1985) together with references to sources of further information. In nearly every case one can compute the molecular geometry and force constants with more than just acceptable accuracy by comparisons to other molecules (the "parent molecule method") or by quantum-mechanical structure and vibrational frequency calculations.

The second and far more difficult part of the estimation procedure is the inherently chemical one of estimating the strengths of chemical bonds, which expresses itself in thermochemical property tables as the standard enthalpy of formation of each species at one reference temperature, usually 0 or 298.15 K. Three routes to this information have been developed: The parent molecule method, in which the standard enthalpy of formation for a species is computed from that of a related molecule and estimates of chemical bond dissociation energies; the group additivity method, in which the standard enthalpy of formation of a species is computed from consensus values of group and bond energies;[1] and quantum-mechanical calculations. Quantum-mechanical calculations provide accurate moments of inertia and vibrational frequencies that can be made accurate enough for thermochemical purposes by means of empirical corrections (Foresman and Frisch 1996); computed values for standard enthalpies of formation can be adjusted to be accurate enough to be useful for combustion modeling if one compensates for systematic computational errors by using corrections derived from calculations done for molecules with known thermochemical properties.

All of the above resources for establishing thermochemical data are represented in the table. Where more than one data source was available for a species, the

[1] The use of molecular structures to extend thermochemical knowledge beyond that provided by experiments has a long history, described critically by Cox and Pilcher (1970) and Pedley et al. (1986). Recent formulations of particular interest to combustion are described by Benson 1976; Stein 1985; Ritter 1990, 1991; Ritter and Bozzelli 1990, 1991; Stein et al. 1991; Cohen and Benson 1992; Stein 1994; Lay et al. 1995; Muller et al. 1995; and Cohen 1996.

one chosen was the one thought to be most reliable at the time the polynomial coefficients were determined. If no data were available in known tabulations or in the archival literature, the literature was searched for experimental or quantum mechanical calculated vibrational frequencies and moments of inertia, and independent calculations were performed. If no data at all could be found, estimation methods were used to determine reasonable values for the needed molecular constants and standard enthalpy of formation.

5.4 THERMOCHEMICAL POLYNOMIALS

The accuracy of thermochemical property values found in compilations, as in the scientific literature generally, varies widely.

Some properties, such as for atoms, N_2 or HD, have been calculated to very high accuracy—far beyond the level needed for combustion research—while others have only been estimated with approximate methods. The rigid rotor harmonic oscillator (RRHO) approximation has been accepted, with very occasional exceptions, as the standard for polyatomic species. It was used for the JANAF (1985), TSIV, Thermodynamic Research Center, and Stull, Westrum, and Sinke (1969) compilations and many others. In cases where the necessary spectroscopic information has been reported, this method can be and occasionally has been supplemented by taking anharmonic corrections and vibration-rotation interactions into account. In some cases the RRHO approximation has to be extended to account for internal rotation(s) or low-energy electronic states. Such calculations can be considered for combustion purposes to give "accurate" thermochemical values.[1] The extrapolation methods used to extend the temperature range of thermochemical data beyond the experimentally accessible one, those of both Wilhoit (1975) and Ritter and Bozzelli (1990), have been shown by comparisons with detailed calculations to give generally excellent results; typically, the maximum deviations from RRHO calculations of $C_p(T)$ are less than 0.5%.

Thermochemical property errors are expected to be greater for species whose thermochemistry has to be estimated. It is safe to assume that the parent molecule method used for estimating the properties of free radicals (Burcat and Kudchadker 1979, Burcat 1982, Burcat et al. 1983, 1985) has maximum errors of ± 3 % for $C_p(T)$, while group additivity methods, as implemented by Stein (1994), Muller et al. (1995), and Ritter and Bozzelli (1991), give error ranges for unknown species in the ± 10 kJ/mol range for $\Delta_f H_{298}$. These are felt to be maximum values—for molecules whose structure is similar to the structures of species used in computing group values, the errors are assumed to be smaller (cf. Section 5.7 and Reid et al. 1988, p. 196).

[1] Spectroscopic levels of accuracy are only rarely extensible to determinations of standard enthalpies of formation; see Section 5.7.

The fitting accuracy given by the polynomial forms varies from one species to the next. For most applications of combustion modeling, however, errors introduced by uncertainty of thermochemical properties, including those introduced by using polynomial fitting forms, are smaller than uncertainties of the modeling introduced by model limitations or systematic errors in experimental data.[1]

The database includes estimates of the accuracy to which the polynomials represent the original calculated tabular values for a number of species.[2] It is expressed in terms of the maximum fitting error among the three properties heat capacity, entropy, and enthalpy. This information is given, together with the temperature at which it is observed, on header records preceding the polynomial coefficients. In most cases the heat capacity shows the maximum fitting error. Unusually large fitting errors indicate that the underlying tabulated values are not smooth functions of temperature.

Because of different machine roundoffs, word lengths, and values of constants used, completely identical values are seldom calculated from the same polynomial coefficients on different computers. Accuracy can be improved if double precision computation is consistently used (on other than 64 bit word machines) and if the polynomial is coded to be evaluated in a form like

$$\mathtt{CPR} = (((\mathtt{A5} * \mathtt{T} + \mathtt{A4}) * \mathtt{T} + \mathtt{A3}) * \mathtt{T} + \mathtt{A2}) * \mathtt{T} + \mathtt{A1}$$

to deter the compiler from generating code that lets intermediate values escape from the floating point unit data stack.

5.5 CALCULATION PROCEDURES

The polynomial coefficients were calculated with four successive versions of the McBride and Gordon PAC program as it was modified and extended over the years. Documentation for the most recent version used for preparing the table is given by McBride and Gordon (1992). Entries prepared before 1984 were produced using the PAC3 version, in which contributions from internal rotations had to be added separately. The PAC4 version written in 1984 allowed the contributions of internal rotations to be computed automatically.

The wide temperature range needed for combustion thermochemistry requires that different sets of polynomial coefficients be used at low and high temperatures. The two polynomials could in principle meet at different temperatures for different species (Burcat 1984). The THERM program (Ritter 1990, 1991; Ritter and

---

[1] See, for example, the data on methane ignition and flame properties surveyed at the URL `http://euler.me.berkeley.edu/ gri_mech.`

[2] The fitting error, the errors in molecular constants, and the error in standard enthalpy of formation contribute independently to the overall error of the polynomials. The fitting error dominates only for a few species whose properties have been defined spectroscopically or whose standard enthalpy of formation has been measured with unusual precision.

Bozzelli 1990, 1991) optimizes this temperature and fixes the two polynomials there. Most programs that use thermochemical polynomials, however, adopt a constant joining temperature for all species. In the PAC programs it was arbitrarily set to 1000 K. In PAC4 the polynomials are fixed at the 1000 K source values, and therefore reproduce them exactly, while the values at 298.15 K are not forced to be the source values but are permitted in the optimization procedure to be slightly different, depending on the local errors of the polynomial coefficients. In the latest versions, PAC90 and PAC97, the low-temperature polynomial is fixed at 298.15 K to the source values and at 1000 K the two polynomials are pinned to each other, not at the source values. These versions of the PAC program also include Wilhoit-type extrapolations, as discussed by Burcat (1984), and optional use of estimated additivity group properties as described by Stein (1985). The values for some species in the table were estimated using the THERM program (Ritter 1990, 1991; Ritter and Bozzelli 1990, 1991), while others were estimated using Stein's (1985) method as implemented in the NIST database program (Stein et al., 1991; Stein 1994). The two methods differ computationally as well as in the selection of group values. The computational difference is that in the THERM program property values for each group are computed at all temperatures to be fit, the group values are added, and the polynomial coefficients are determined by a choice of three methods, Wilhoit's among them; while in the NIST program polynomial coefficients for each of the fragment groups are added together to obtain coefficients for the species of interest.

5.6 ACCURACY OF THE DATABASE

The reliability of the database entries has been classified according to the methods used to generate the underlying thermochemical values.[1] The category assigned to each species is coded by a letter appearing in column 68 of its first data record:

- **A** denotes the most accurate calculations and is reserved for results obtained with the "direct summation" method, for atoms and diatomic molecules, and for species for which anharmonicity corrections were included.
- **B** denotes calculations done using the RRHO approximation, including internal rotations where present, and taking low-lying electronic excitations into account. Results obtained with the "parent molecule" method for computing the properties of radicals are included in this category.
- **C** includes species whose properties were calculated by the RRHO method using shortcuts. Results of RRHO calculations with estimated vibrational frequencies were assigned to this category, as were cases where internal rotor contributions had been estimated as

[1] Aside from the standard enthalpy of formation $\Delta_f H^\circ_{298}$ value, which has to be evaluated separately as explained in Section 5.7.

those of a a free rotor. Some of the entries based on the TRC/API tables are included in this category because it was not possible to determine how they had been calculated.

D and **E** categories include data estimated from group additivity methods. Such species were normally assigned to the **E** group, but if reliable supplementary information could be used, the **D** category was assigned.

An **F** category was reserved for rough approximations using group additivity or other types of estimation with large error limits for species where the applicability of the methods is uncertain.

Users of the database are cautioned that their computational results may be sensitive to the necessarily limited accuracy of the thermochemical data. If species assigned to the lower accuracy categories are thought to play important roles in an application, it is wise to determine the sensitivity of the results to this thermochemistry by direct tests.

5.7 ACCURACY OF STANDARD ENTHALPIES OF FORMATION

The standard enthalpies of formation of the species in the database fall into four categories:

(a) those that were measured directly with reliable experimental methods;
(b) those derived from a difference between their standard enthalpy of formation value and that of another compound;
(c) those estimated through comparisons with measured values of related compounds;
(d) those estimated on the basis of other estimated compounds or structural groups.

Standard enthalpies of formation are quoted and requoted by successive authors, sometimes making it difficult to discover which category applies. When values change with time owing to improved or disproved experiments or to correction of errors in previous measurements or calculations, the enthalpy of formation values of compounds in categories (b–d) whose estimation was based on those values have to be changed. There are no means to effect these changes other than ongoing and painstaking examination of each value. There are disputes between groups of researchers supporting different standard enthalpies of formation for important species. In this compilation, many decisions about which value to adopt had to be done arbitrarily for lack of established criteria.

Very few species of interest in combustion can be assigned standard enthalpies of formation with error limits so narrow that for combustion modeling purposes they may be taken to be exact (Cox et al. 1989; Cox and Pilcher 1970; Cohen 1996). The most accurately known of all (aside from the elements in their reference states, for which the value 0 is defined to be exact) are those based on carefully recorded molecular electronic spectra supplemented by quantum-mechanical analysis. Among these the hydrogen atom stands out, being known to the accuracy with which Planck's constant is known; a few diatomic and triatomic species whose electronic spectra have been successfully analysed to establish the dissociation limit also belong in the exact category. An overview of the uncertainties of the standard enthalpies of formation of the key combustion-relevant atomic and triatomic species that have been exhaustively studied by calorimetric and spectroscopic methods is given in Table 5.1. One sees that the accuracy benchmark set by this group of species, in effect setting a standard for what can be achieved in measuring or computing standard enthalpies of reaction, is in the vicinity of 0.1 to 0.2 kJ/mol.

TABLE 5.1. Standard enthalpies of formation in kJ/mol at 298.15 K for small gas-phase species of interest in combustion (from Cox et al. 1989 and [NO and NO_2] Chase et al. 1985).

Species	$\Delta_f H^\circ_{298}$	Species	$\Delta_f H^\circ_{298}$
C(g)	716.68 ±0.45	NO(g)	90.29 ±0.17
H(g)	217.998 ±0.006	CO(g)	−110.53 ±0.17
O(g)	249.18 ±0.10	H_2O(g)	-241.826 ±0.040
N(g)	472.68 ±0.40	CO_2(g)	-393.51 ±0.13
S(g)	276.98 ±0.25	SO_2(g)	-296.81 ±0.20
Cl(g)	121.190 ±0.008	NO_2(g)	33.10 ±0.8

The number of species important in combustion for which "directly measured" experimental values of standard enthalpies of formation can be assigned is small. All are based one way or another on chemical reactions to which enthalpy changes of reaction can be assigned either calorimetrically or from the temperature dependence of equilibrium constants. As far as stable molecules of the elements carbon, hydrogen, oxygen, and nitrogen are concerned, it is fortunate that combustion reactions themselves serve for this purpose, as the standard enthalpies of formation of the combustion products carbon dioxide and water have been painstakingly evaluated and reaction can usually be arranged to occur with accurately measured stoichiometry (Cox and Pilcher 1970).

Even for the most favorable cases, however, the error bounds that have to be accepted are larger than one would wish. This is illustrated in Table 5.2, adapted from Cohen and Benson (1992), who give references to the archival literature. Here one sees that the "best available" standard enthalpy of formation values for the small hydrocarbons come with error ranges that imply significant uncertainty in equilibrium constants. (A 1 kJ uncertainty in the enthalpy change of a reaction at 1000 K implies an uncertainty of 11% in its equilibrium constant.) Not only are the uncertainty ranges asserted by the evaluators larger than one would wish, the differences between the values obtained with the two most trustworthy calorimetric techniques are seen on close inspection to differ from one another by more than the sum of the stated uncertainty ranges for three of the five cases. Aside from these discrepancies, the asserted uncertainty ranges are about twice as large as for the values listed in Table 5.1.

TABLE 5.2. Standard enthalpies of formation in kJ/mol at 298.15 K for small hydrocarbons (after Cohen and Benson 1992).

Species	Bomb Calorimeter	Flame Calorimeter
$CH_4(g)$	-74.85 ± 0.29	-74.48 ± 0.42
$C_2H_6(g)$	-84.68 ± 0.50	-83.85 ± 0.29
$C_3H_8(g)$	-103.89 ± 0.59	-104.68 ± 0.50
n-$C_4H_{10}(g)$	-127.03 ± 0.67	-125.65 ± 0.67
i-$C_4H_{10}(g)$	-135.60 ± 0.54	-134.18 ± 0.63

The values are less well known for most of the other stable species of interest in combustion, and still less well known for unstable ones. Among the unstable species, the thermochemistry of free radicals has attracted particular interest in combustion modeling because of their roles as chain centers. An overview of current knowledge of the standard enthalpies of formation of some of the common ones is given in Table 5.3. In contrast to the stable hydrocarbons, where the standard enthalpy of formation is based on one or another of the direct calorimetries, values for radicals come from all sorts of measurements ranging from photoionization mass spectroscopy to reaction rates. It is no surprise that the results are more contentious and less accurate. An overview of the present status for some of the common free radicals is given in Table 5.3, where the uncertainty ranges can be seen to be typically an order of magnitude greater than one finds for stable hydrocarbons.

The error limits noted in the table for 239 species are displayed in their table order in Figure 5.1. While most of the high values correspond to exotic species of peripheral interest to combustion, even the typical values shown underscore that there is serious thermochemical work to do. The main implication of these uncertainties is illustrated in Figure 5.2, where thermochemical errors are translated, using the van't Hoff isochore $\ln K = -\Delta_r H^\circ/RT + \Delta_r S^\circ/R$, into their implications for equilibrium constants and therefore for reverse reaction rate coefficients computed from $k_r = k_f/K_{eq}$. Comparing the two figures leaves no doubt that the thermochemical problems discussed in Section 5.1 are severe and deserve far more attention than they get.

TABLE 5.3. Standard enthalpies of formation in kJ/mol at 298.15 K for common radicals. Except for CH and CH_2, the hydrocarbon-based radical data is from Tsang 1996; the other data are from Chase et al. 1985.

Species	$\Delta_f H^\circ_{298}$	Species	$\Delta_f H^\circ_{298}$
OH(g)	38.4 ±1.2	CH(g)	594.128 ±17.5
CN(g)	435.1 ±10	NH(g)	376.56 ±16.7
SH(g)	139.33 ±5	CH_2OH(g)	-9 ±4
HO_2(g)	2.1 ±8.4	CH_3O(g)	17 ±4
CHO(g)	42 ±4	C_3H_3(g)	339 ±4
CH_2(g)	386.39 ±4.9	C_2O(g)	286.6 ±63
CH_3(g)	147 ±1	C_3H_5(g)	171 ±3
C_2H(g)	556 ±8	n-C_3H_7(g)	100 ±2
C_2H_3(g)	299 ±5	i-C_3H_7(g)	90 ±2
C_2H_5(g)	119 ±2	C_6H_5(g)	339 ±8

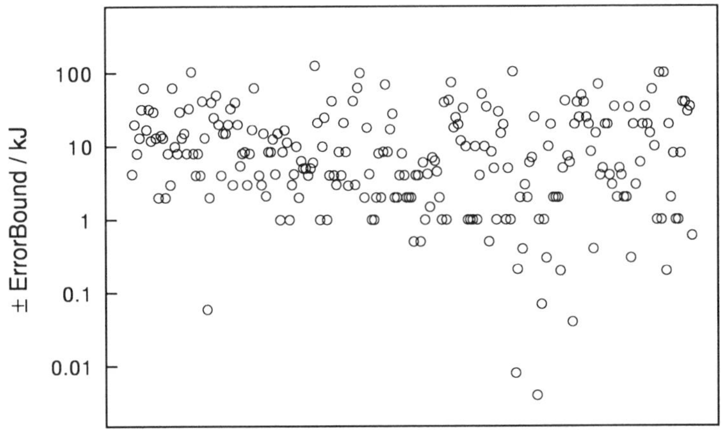

FIGURE 5.1. Standard enthalpy of formation error ranges, in the database order, for the 239 species for which error ranges are tabulated. While many of the values shown were assigned on the basis of experimenter's or theoretician's feelings rather than detailed assessment of uncertainty, the situation depicted still represents the accuracy of current knowledge of the thermochemistry of combustion-relevant species.

For hydrocarbons and their various derivatives containing oxygen and nitrogen atoms, a long history of thermochemical investigation has left a legacy of experimental standard enthalpy of formation values. (Some 3000 have been compiled by Pedley et al. 1986.) The uncertainty level of this legacy varies considerably because of the fluctuating care given to the (mostly) calorimetric measurements and problems of reagent purity and reaction stoichiometry. From early on there have been successful efforts to systematize the database in terms of molecular structure. (Reviewed in detail by Cox and Pilcher 1970.) As a result, one can compute a standard enthalpy of formation value for "ordinary" compounds that have not

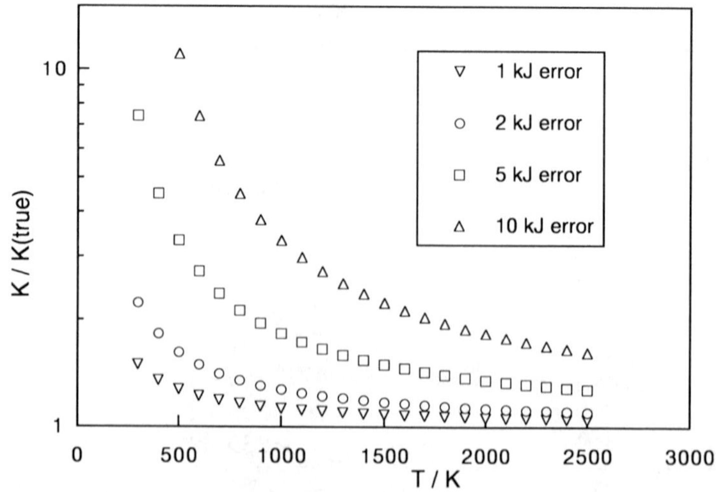

FIGURE 5.2. Ratio of computed to true equilibrium constant that would result from using a standard enthalpy change of reaction less than the true value by the amounts indicated in the legend. For the case of overestimating the standard enthalpy change of reaction, the y-axis would represent the ratio of true to computed equilibrium constant.

been studied experimentally with almost the same confidence one can place in the experimental values themselves. A large number of entries in the present tabulation have been derived by the NIST or THERM group additivity programs that offer current embodiments of this idea. The capabilities and limitations of group additivity methods for stable organic molecules have been reviewed by Pedley et al. (1986) and Cohen (1996); discussions of the issues involved in making group additivity estimates for radicals are given by Muller et al. (1995) and Lay et al. (1995).

Unfortunately, many of the most interesting molecules and radicals used in combustion modeling are not ordinary at all, but have highly strained rings or electronic structures that are not well represented in the experimental data base used for setting group additivity parameters. For such molecules and radicals we recommend cautious use of the group additivity methods anyway, or, if that appears infeasible, even more cautious use of semi-empirical or semi-theoretical molecular electronic structure calculations. An extensive survey of the results of such calculations has been given by Dewar et al. (1993).

5.8 OTHER SOURCES OF THERMOCHEMICAL DATA

While the database provides sufficient data for characterizing a wide range of combustion systems, it does not by any means represent all of the ideal gas thermochemical data in the literature.

Data for hydrocarbons with up to 20 carbon atoms can be found in the TRC (Thermodynamics Research Center) compilation, in Stull et al. (1969), and in the NIST computerized database (Stein 1994), which has data for about 2800 species and is on the World Wide Web (Afeefy et al. 1997; Mallard and Linstrom 1997). Data for aromatic hydrocarbons are reported by Emdee et al. (1992). Extensive standard enthalpy of formation data at 298 K is given by Pedley et al. (1986). Articles reporting data for additional species appear from time to time in the archival literature such as the *Journal of Physical and Chemical Reference Data,* so that there is some prospect of success in undertaking a systematic search of the chemical literature. A database of molecular properties computed by quantum mechanical electronic structure methods for about 3500 species, some of them transition states, is available (Melius 1993). Unfortunately, for most of them only limited thermochemical information ($\Delta_f H^\circ_{298}$ and $\Delta_f G^\circ_{298}$) is provided.

Thermochemical information for additional nitrogen-, sulfur-, and chlorine-containing compounds is provided in Chapters 2, 3 and 4 of this book.

Extensive thermochemical data have been estimated, mostly by group additivity methods, for use in modeling ignition and flame properties of liquid fuels. This information can often be obtained by contacting the authors of such articles, who are usually willing to provide it in immediately useful forms, that is, as ascii computer files. (An example of such an article is that of Marinov et al. 1996.)

Data for pure solids and liquids can be found in the JANAF (Chase et al. 1985; Chase 1998) compilation and in polynomial form in the report of McBride et al. (1993a,b).

Ideal gas thermochemical data for silicon-containing species can be found in the CHEMKIN database (Kee et al. 1992).

Ideal gas thermochemical data for diatomic and triatomic ions can also be found in the JANAF compilation and the NIST computerized database. The information for the latter was taken from Lias et al. (1988).

Methods for computing thermochemical values theoretically are discussed in a symposium report edited by Irikura and Frurip (1998).

5.9 FORMAT OF THE DATABASE

The thermodynamic and thermochemical data are represented as two sets of polynomial coefficients for each species, the first one valid above 1000 K and the second valid from room temperature to 1000 K. Both polynomials give the same value at 1000 K for all properties.

The following standard molar thermodynamic and thermochemical functions can be obtained from the polynomial coefficients:

$$\frac{C_p^\circ}{\mathrm{R}} = a_1 + a_2T + a_3T^2 + a_4T^3 + a_5T^4$$

$$\frac{H_T^\circ}{\mathrm{R}T} = a_1 + \frac{a_2T}{2} + \frac{a_3T^2}{3} + \frac{a_4T^3}{4} + \frac{a_5T^4}{5} + \frac{a_6}{T}$$

$$\frac{S_T^\circ}{\mathrm{R}} = a_1 \ln T + a_2T + \frac{a_3T^2}{2} + \frac{a_4T^3}{3} + \frac{a_5T^4}{4} + a_7$$

$$\frac{G_T^\circ}{\mathrm{R}T} = \frac{H_T^\circ}{\mathrm{R}T} - \frac{S_T^\circ}{\mathrm{R}} = a_1(1 - \ln T) - \frac{a_2}{2}T - \frac{a_3}{6}T^2 - \frac{a_4}{12}T^3 - \frac{a_5}{20}T^4 + \frac{a_6}{T} - a_7 .$$

The H_T° function is not the $H_T^\circ - H_{Tref}^\circ$ function often tabulated but rather

$$H_T^\circ = \Delta_f H_{298}^\circ + \int_{298}^{T} \mathrm{d}T\ C_p^\circ(T),$$

so that $H_T^\circ/\mathrm{R}T$ values calculated from the polynomials can be used directly to compute standard enthalpies of reaction. In other words, the a_6 and a_7 constants have been adjusted such that the standard enthalpy of formation and the standard entropy at 298.15 are reproduced when the low-temperature polynomial is evaluated at $T = 298.15$ K. Similarly, the $G^\circ/\mathrm{R}T$ function can be used directly to compute equilibrium constants in terms of concentration through

$$K_c = (\mathrm{R}'T)^{-\Delta\nu} \exp\left(\Delta a_1(\ln T - 1) + \frac{\Delta a_2}{2}T + \frac{\Delta a_3}{6}T^2 + \frac{\Delta a_4}{12}T^3 + \frac{\Delta a_5}{20}T^4 - \frac{\Delta a_6}{T} + \Delta a_7\right).$$

The mole number change and the coefficient changes are defined by

$$\Delta\nu = \sum \nu_j \qquad \text{and} \qquad \Delta a_i = \sum \nu_j a_{ij} ,$$

where the summations are over all reactant and product species j with the stoichiometric coefficients ν_j taken to be positive for products and negative for reactants, and R' denotes the gas constant in units appropriate to pressure in bar and volume as used to measure concentrations, for example, 83.1451 when concentrations are expressed as mol/cm^3.

As explained in Section 5.5, polynomials generated after 1990 were fixed to reproduce the source 298.15 K values exactly; polynomials generated with earlier versions of the PAC program give 298.15 K values that differ only slightly from the source values, that is, by amounts much less than the experimental uncertainties in standard entropies and enthalpies of formation.

The reference state is the ideal gas at one bar pressure. The gradual acceptance of using 1 bar rather than 1 atm as the thermochemical reference state, and defining

1 atm to be exactly 101 325 N/m^2 = 1.01325 bar, has simplified accounting for thermal and mechanical energies in calculations by enabling all quantities to be expressed consistently in SI units. This modification to traditional practice affects only the standard molar thermochemical values for gases, and for them it affects only thermochemical functions that contain the entropy. The standard molar entropy when pressure is expressed in bar is higher than when pressure is expressed in atm by Rln(1.01325) or 0.013163R. This correction—0.10944 J/K or 0.02616 cal/K—is primarily a matter of form, the difference being so small that for practical applications in combustion no harm would result from combining data based on two different reference states.

Data for a few condensed phase species, denoted for liquids by an (l) and for solids by an (s) on the first data record, are also included in the database. Computing equilibrium or dynamic properties for systems containing condensed phases must be done with different methods than pertain to systems containing only ideal gases (Van Zeggeren and Storey 1970; Gordon and McBride 1992).

The information given prior to the coefficient cards indicates sources and in many cases values of molecular constants used to generate input thermochemical values for the fitting program (McBride and Gordon 1992). The notations preceding some of the polynomial coefficient sets provide molecular constants in standard spectroscopic notation as follows:

`SIGMA` Symmetry number.

`STATWT` Statistical weight. The normally encountered values are 1 for saturated molecules and 2 for doublet radicals.

`IA, IB, IC` or `IAIBIC` Principal moments of inertia in g cm^2 $\times 10^{39}$ units or product of inertia in g^3 cm^6 $\times 10^{117}$ units.

`AO, BO, CO` Rotational constants in cm^{-1} units.

`BE` Rotational constant of a diatomic species in cm^{-1} units.

`NU( )` Vibration frequencies, with degeneracies in parenthesis. In the case of diatomic species with anharmonic data the notations are `WE, WEXE, ... ALFAE, ALFAB1...`, etc.

`X11, X12,` and `ALFA1, ALFA2, ...` Anharmonicity constants.

`TO` Energy of an excited state in cm^{-1} units, with statistical weight in parentheses.

`ROSYM` Symmetry of an internal rotor.

`IR` Internal rotation moment of inertia in gcm^2 $\times 10^{39}$ units.

`BROT` Internal rotational constants in cm^{-1}.

`V()` Potential barrier for an internal rotation in cm^{-1} or cal.

`HF`T Standard enthalpy of formation at temperature T in kJ or kcal.

`REF` Literature reference. Separate references are given in cases where the source of the $\Delta_f H_T^\circ$ value is different from that for the molecular constants.

`Max Lst Sq Error` The maximum difference between tabular thermochemical values and values generated by the polynomial.

The arrangement of the records in the database is as follows:

The first record contains the name of the species, coded information about the source of the data (see below), the date of calculation of the polynomial coefficients, formula composition information required by equilibrium-solving programs, the temperature range over which the polynomials were fitted, the evaluation assignment (Section 5.6), and the molecular weight.

On the second record the high-temperature polynomial coefficients a_1 through a_5 are listed, followed on the third record by the corresponding a_6 and a_7 values and then a_1 through a_3 for the low-temperature range. The fourth record continues with a_4 through a_7 for the low-temperature range and finally $\Delta_f H^\circ_{298}/R$.

The coding of data sources is A = Hall and Colket 1992; C = Sandia National Laboratory CHEMKIN database; CODA = Cox et al. 1989; J = JANAF tables, Chase et al. 1985; L = NASA TM-4513 (1993), where references to the archival literature may be found; R= TSIV Tables, P = Thermodynamic Research Center publications (American Petroleum Institute Project 44); T = Technion TAE Reports 411, 476, 504 and 804.

5.10 CONVERSION FACTORS

The following physical constants and conversion factors were used in creating the database:

1 cal = 4.184 J

R = 8.31451 J/mol K = 1.987216 cal/mol K

1 cal/mol = 2.8591434 cm^{-1}

1 Bohr radius (B) = $5.2917706 \times 10^{-11}$ m

amu-B^2 = $21.505458 \times 10^{-39}$ g cm^2

amu-Å2 = $6.0221367 \times 10^{-39}$ g cm^2

1 Hz = $(1/2.99792458 \times 10^{10})$ cm^{-1}.

5.11 INTERNET TRANSFER OF THE DATABASE

The current version of the database is available for downloading in ASCII form as the file BURCAT.THR at

`ftp://ftp.technion.ac.il/pub/supported/aetdd/thermodynamics.`

The file is approximately 450 kB long.

A short UPDATE.INF file lists the recently added species.

A table generator program in FORTRAN written by B. McBride is also available at the same URL. It enables generation of thermochemical property tables from the polynomial coefficients. These tables provide values of C_p, S°_T, $H^\circ_T - H^\circ_{298}$, and $-(G^\circ_T - H^\circ_{298})/T$, and optionally of $\Delta_f H^\circ$ and $\log K_f$ as well, at user-requested temperatures within the polynomial's temperature limits.

5.12 REFERENCES

Afeefy, H.Y., Liebman, J.F., & Stein, S.E. (1997). "Neutral Thermochemical Data", in *NIST Chemistry WebBook, NIST Standard Reference Database Number 69, February 1997,* W.G. Mallard and P.J. Linstrom, Eds., at the URL `http://webbook.nist.gov/chemistry`.

Atkinson, B., & Stedman, M. (1962). J. Chem. Soc., 512.

Bauer, S.H., Javanovic, S., Yu, C.-L., and Cheng, H.-Z. (1997). Energy and Fuels, **11,** 1204.

Baulch, D.L., Cox, R.A., Crutzen, P.L., Hampson, R.F., Jr., Kerr, J.A., Troe, J., & Watson, R.T. (1982). J. Phys. Chem. Ref. Data, **11,** 327.

Benson, S.W. (1976). *Thermochemical Kinetics, Second Edition,* Wiley, New York.

Benson, S.W. (1978). Chem. Rev., **78,** 23.

Benson, S.W., & O'Neal, H.E. (1970). *Kinetic Data on Gas Phase Unimolecular Reactions,* NSRDS–NBS 21.

Burcat, A. (1980). *Ideal Gas Thermodynamic Functions of Hydrides and Deuterides. Part I,* TAE Report No. 411, Technion, Haifa.

Burcat, A. (1982). *Ideal Gas Thermodynamic Properties of C_3 Cyclic Compounds, Part II,* TAE Report No. 476, Technion, Haifa.

Burcat, A. (1984). "Thermochemical Data for Combustion Calculations", Chap. 8 of *Combustion Chemistry,* W.C. Gardiner, Ed., Springer-Verlag, New York.

Burcat, A. (1992). *Ideal Gas Thermodynamic Data for Buckminsterfullerene C_{60} and C_{70}.* TAE Report No 680, Technion, Haifa.

Burcat, A., & Kudchadker, S.A. (1979). Acta Chimica Hung., **101,** 249.

Burcat, A., & McBride, B. (1997) *Ideal Gas Thermodynamic Data for Combustion and Air-Pollution Use,* TAE Report No. 804, Technion, Haifa.

Burcat, A., Miller, D., & Gardiner, W.C. (1983). *Ideal Gas Thermodynamic Properties of C_2H_nO Radicals, Part III,* TAE Report No. 504, Technion, Haifa.

Burcat, A., Zeleznik, F.J., & McBride, B. (1985). *Ideal Gas Thermodynamic Properties of Phenyl, Deuterated Phenyl, Phenoxy and o-Biphenyl Radicals,* NASA TM-83800.

Chao, J., & Hall, K.R. (1980). Personal communication.

Chao, J., Hall, K.R., Marsh, K.N., & Wilhoit, R.C. (1986). J. Phys. Chem. Ref. Data, **15,** 1369.

Chao, J., Rodgers, A.S., Wilhoit, R.C., & Zwolinski, B.J. (1974). J. Phys. Chem. Ref. Data, **3,** 141.

Chao, J., Wilhoit, R.C., & Hall, K.R. (1980). Thermochimica Acta, **41,** 41.

Chao, J., Wilhoit, R.C., & Zwolinski, B.J. (1973). J. Phys. Chem. Ref. Data, **2,** 427.

Chao, J., & Zwolinski, B.J. (1975). J. Phys. Chem. Ref. Data, **4,** 251.

Chao, J., & Zwolinski, B.J. (1978). J. Phys. Chem. Ref. Data, **7,** 363.

Chase et al. (1985). See JANAF Thermochemical Tables (1985).

Chase, M.W. (1996a). J. Chem. Ref. Data, **25,** 551.

Chase, M.W. (1996b). J. Chem. Ref. Data, **25,** 1069.

Chase, M.W. (1996c). J. Chem. Ref. Data, **25,** 1297.

Chase, M.W., Jr. (1998). *NIST-JANAF Thermochemical Tables, Fourth Edition,* J. Phys. Chem. Ref. Data Monograph No. 9, Parts I and II, American Chemical Society and American Physical Society, Washington.

Chen, S.S., Kudchadker, S.A., & Wilhoit, R.C., (1979). J. Phys. Chem. Ref. Data, **8,** 527.

Chen, S.S., Rodgers, A.S., Chao, J., Wilhoit, R.C., & Zwolinski B.J. (1976). J. Phys. Chem. Ref. Data, **4,** 441.

Chen, S.S., Wilhoit, R.C., & Zwolinski B.J. (1975). J. Phys. Chem. Ref. Data, **4,** 859.

Chen, S.S., Wilhoit, R.C., & Zwolinski B.J., (1977). J. Phys. Chem. Ref. Data, **6,** 105.

Chen, Y., Rauk, A., & Tschuikow-Roux, E. (1990). J. Chem. Phys., **93,** 1187.

Chen, Y., Rauk, A., & Tschuikow-Roux, E. (1991). J. Chem. Phys., **95,** 2774.

CODATA. See Cox et al. (1989).

Cohen, N. (1992). J. Phys. Chem. **96**, 9052.

Cohen, N. (1996). J. Phys. Chem. Ref. Data **25,** 1411.

Cohen, N., & Benson, S.W. (1992). Chap. 6 of *The Chemistry of Alkanes and Cycloalkanes,* S. Patai and Z. Rappoport, Eds., Wiley, New York.

Cox, J.D., & Pilcher, G. (1970). *Thermochemistry of Organic and Organometallic Compounds,* Academic Press, London.

Cox, J.D., Wagman, D.D., & Medvedev, V.A. (1989). *CODATA Key Values for Thermodynamics,* Hemisphere, New York. Also available at www.cisti.nrc.ca/programs/codata/databases/key1.html.

Das, A., Frenkel, M., Gadalla, N.A.M., Kudchadker, S., Marsh, K.N., Rodgers, A.S.. & Wilhoit, R.C. (1993). J. Phys. Chem. Ref. Data, **22,** 659.

Dewar, M.J.S., Jie, C., & Yu, J. (1993). Tetrahedron, **49,** 5003.

Domalski E.S., & Hearing E.D. (1993). J. Phys. Chem. Ref. Data, **22,** 805.

Dorofeeva, O.V., Gurvich, L.V., & Jorish, V.S. (1986). J. Phys. Chem. Ref. Data, **15,** 437.

Dorofeeva, O.V., & Gurvich, L.V. (1995). J. Phys. Chem. Ref. Data, **24,** 1351.

Draeger, J.A., Harrison, R.H., & Good, W.D. (1983). J. Chem. Thermodynam., **15,** 367.

Draeger, J.A., & Scott, D.W. (1981). J. Chem. Phys., **74,** 4748.

Duff, R.E., & Bauer, S.H. (1962). J. Chem. Phys., **36,** 1754.

Emdee, J., Brezinsky, Ka,. & Glassman, I. (1992). J. Phys. Chem. **96,** 2151.

Ficket, W., & Cowan, R.D. (1955) J. Chem. Phys., **23,** 1349.

Foresman, J.B., & Frisch, Æ, (1996) *Exploring Chemistry with Electronic Structure Methods, Second Edition,* Gaussian, Pittsburgh.

Gordon, S., and McBride, B.J. (1993). *Computer Program for Calculation of Complex Chemical Equilibrium Compositions and Applications. Part I. Analysis,* NASA RP-1311.

Gurvich, L.V., Veyts, I.V., & Alcock, C.B. (1989). *Thermodynamic Properties of Individual Substances, Fourth Edition, Vol. 1,* Hemisphere, New York.

Gurvich, L.V., Veyts, I.V., & Alcock, C.B. (1991). *Thermodynamic Properties of Individual Substances, Fourth Edition, Vol. 2,* Hemisphere, New York.

Hall, R.J., & Colket, M.B. (1992). J. Aerosol. Sci., **23,** S129–S132.

Holtzclaw, J.R., Harris, W.C., & Bush, S.F. (1980). J. Raman Spectr., **9,** 257.

Irikura, K.K., & Frurip, D.J. (1998). *Computational Thermochemistry, Prediction and Estimation of Molecular Thermodynamics,* American Chemical Society, Washington.

JANAF Thermochemical Tables, Third Edition. (1985) Edited by M.W. Chase, Jr., C.A. Davies, J.R. Downey, Jr., D.J. Frurip, R.A. McDonald and A.N. Syverud. J. Phys Chem. Ref. Data, **14,** Supplement No. 1, Vols. 1 and 2. This reference combines earlier (1971, 1975, 1978 and 1982) supplements to the Second Edition (NSRDS-NBS 37, 1971, edited by D.R. Stull and H. Prophet) and presents all data in S.I. units referred to the 1 bar ideal gas standard state. It is superseded by Chase (1998).

Karni, M., Oref, I., & Burcat, A. (1991). J. Phys. Chem. Ref. Data, **20**, 665.

Kee, R.J. Rupley, F.M., & Miller, J.A. (1992). *The Chemkin Thermodynamic Data Base*, Sandia National Laboratory Report SAND87-8215B UC-4.

Kiefer, J.H., Sidhu, S.S., Kern, R.D., Xie, K., Chen, H., & Harding, L.B. (1992). Comb. Sci. Technol., **82,** 101.

Kolesov, V.P., & Papina, T.S. (1983). Russ. Chem. Rev. **52,** 754.

Kudchadker, S.A., Kudchadker, A.P., Wilhoit, R.C., & Zwolinski, F. (1978). J. Phys. Chem. Ref. Data, **7,** 417.

Kudchadker, S.A., Kudchadker, A.P., & Zwolinski, F. (1979). J. Chem. Thermodynam., **11,** 1051.

Lay, T.H., Bozzelli, J.W., Dean, A.M., & Ritter, E.R. (1995). J. Phys. Chem., **99,** 14514.

Lias, S.G., Bartmess, J.E., Holmes, J.L., Levin, R.D., Liebman, J.F., and Mallard, W.G. (1988) *Gas-Phase Ion and Neutral Thermochemistry,* J. Phys. Chem. Ref. Data, **17,** Supplement 1.

Liebman, J.F. (1988). J. Phys. Chem. Ref. Data **17,** Supplement 1.

Mallard, W.G., & Linstrom, P.J. (1997). *NIST Chemistry WebBook,* NIST Standard Reference Database Number 69, February 1997, on the World Wide Web at the URL `http://webbook.nist.gov/chemistry`.

Marinov, M.N., Pitz, W.J., Westbrook, C.K., Castaldi, M.J., and Senkan, S.M. (1996). Comb. Sci. Technol. **116-117,** 211.

McBride, B.J., & Gordon, S. (1992). *Computer Program for Calculating and Fitting Thermodynamic Functions,* NASA RP 1271.

McBride, B.J., Gordon S,, & Reno, M.A., (1993a). *Thermodynamic Data for Fifty Reference Elements,* NASA TP-3287.

McBride, B.J., Gordon, S., & Reno, M.A. (1993b). *Coefficients for Calculating Thermodynamic and Transport Properties of Individual Species,* NASA TM-4513.

McMillen, D.F., & Golden, D.M. (1982). Annu. Rev. Phys. Chem. **33,** 493.

Melius, C. (1993). Database of thermochemical properties for 3500 species calculated by the BAC/MP4 method at Sandia National Laboratories. Available at the URL `http://herzberg.ca.sandia.gov/~melius/index.html`.

Moore, C.E., *Atomic Energy Levels as Derived from the Analyses of Optical Spectra, Vols. I–III,* NSRDS-NBS 35 (1971).

Moore, C.E., *Selected Tables of Atomic Spectra,* NSRDS-NBS 34, Sec. 3 (1970); Sec. 6 (1972); Sec. 5 (1975); Sec. 7 (1976).

Muller, C., Michel, V., Scacchi, G., & Côme, G.M. (1995). J. Chim. Phys., **92,** 1154.

NASA (1993). See McBride et al. (1993b).

NIST (1991, 1994). See Stein et al. (1991), Stein (1994), Afeefy et al. (1997), and Mallard and Linstrom (1997).

Pamidimukkala, K.M., Rogers, D., & Skinner, G.B. (1982). J. Phys. Chem. Ref. Data, **11,** 83.

Pedley, J.B. Naylor, R.D., & Kirby S.P. (1986). *Thermochemical Data of Organic Compounds, Second Edition,* Chapman and Hall, London.

Pedley, J.B., & Rylance, J. (1977). *Computer Analysed Thermochemical Data for Organic and Organo-metallic Compounds,* University of Sussex, Brighton.

Reid, R.C., Prausnitz, J.M., & Poling, B.E. (1988). *The Properties of Gases and Liquids, Second Edition.* McGraw-Hill, New York.

Ritter, E.R. (1990). *THERM User's Manual.* Department of Chemical Engineering, New Jersey Institute of Technology, Newark, NJ.

Ritter, E.R. (1991). J. Chem. Info. Comp. Sci. **31,** 400.

Ritter, E.R., & Bozzelli, J.W (1990). *THERM: Thermodynamic Property Estimation for Gas Phase Radicals and Molecules.* 12th International CODATA Conference, July 1990.

Ritter, E.R., & Bozzelli, J.W. (1991). Int. J. Chem. Kinet., **23,** 767.

SANDIA. See Kee et al. (1992).

Sidhu, S.S., Kiefer, J.H., Lifshitz, A., Tamburu, C., Walker, J.A., & Tsang, W. (1991). Int. J. Chem. Kinet. **23,** 215.

Stein, S.E. (1985). J. Phys. Chem., **89,** 3714.

Stein, S.E. (1994). *NIST Structures, & Properties Version 2.0 Computerized Database 25,* National Institute of Standards and Technology, Gaithersburg.

Stein, S.E, Rukkers, J.M., & Brown, R.L. (1991). *NIST Structures & Properties Database and Estimation Program,* National Institute of Standards and Technology, Gaithersburg.

Stull, D.R., Westrum, E.F., Jr., & Sinke, G.C. (1969). *The Chemical Thermodynamics of Organic Compounds, Second Edition,* Wiley, New York; Reprinted by Krieger, Malabar, 1987.

TRC/API. See TRC–Thermodynamic Research Center.

Tsang, W. (1985). J. Am. Chem. Soc., **107**, 2872.

Tsang, W. (1996). "Heats of Formation of Organic Free Radicals by Kinetic Methods", in *Energetics of Organic Free Radicals,* J.A. Martinho Simoes, A. Greenberg & J.F. Liebman, Eds., Blackie, London, pp. 22–58.

TSIV, *Thermodynamic Properties of Individual Substances, Vols. 1, 2, 3 and 4* (1978, 1978, 1979, 1982). Edited by L.V. Gurvich et al., Nauka, Moscow (in Russian).

TRC–Thermodynamic Research Center, *Thermodynamic Tables of Hydrocarbons;* formerly American Petroleum Institute Project 44. Published irregularly as loose leaf sheets by Texas A & M University, College Station, TX. The URL is `http://www-chen.tamu.edu`.

Van Zeggeren, F., & Storey, S.H. (1970). *The Calculation of Chemical Equilibriums,* Cambridge University Press, New York.

Weissman, M., & Benson, S.W. (1989). Prog. Energy Combust. Sci., **15**, 273.

Wilhoit, R.C, (1975). "Ideal Gas Thermodynamic Functions", TRC Current Data News **3,** # 2.

5.13 TABLE OF STANDARD ENTHALPIES OF FORMATION

Species	$\Delta_f H^\circ_{298}$	Comments
Al(s, l)	0	The liquid is the reference state above 933.61 K. Cox et al. 1989. [NASA 1993]
Al(g)	329	329.7±4.2 kJ JANAF 6/83 [NASA 1993]
AlH	259	259.4±20 kJ JANAF 6/63 [NASA 1993]
AlO	67	66.9±8 kJ JANAF 12/79 [NASA 1993]
AlOH	−180	179.9±13 kJ JANAF 12/67 [NASA 1993]
AlO_2	−86	–86.2±32 kJ JANAF 12/79 [NASA 1993]
AlO_2H	−460	–460±63 kJ JANAF 12/68 [NASA 1993]
Al_2O	−145	–145.2±17 k JANAF 12/79 [NASA 1993]
Al_2O_2	−395	–394.6±32 kJ JANAF 12/79 [NASA 1993]
Al_2O_3(s)	−1676	JANAF 12/79 [NASA 1993]
Al_2O_3(l)	−1700	JANAF 12/79 [NASA 1993]
Ar	0	Reference state Moore 1971 [NASA 6/88]
Ar^+	1527	Moore 1971 gives 1526.8 kJ [NASA 10/92]
B	560	560±12 kJ JANAF 6/83 [NASA 1993]
B(s, l)	0	The solid is the reference state below 2350 K. JANAF 3/79 [NASA 1993]
BCl	141	JANAF 12/64 [NASA 1993]
BClF	−314	–314±30 kJ JANAF 12/64 [NASA 1993]
BCl_2	−79	–79.5±13 kJ JANAF 6/72 [NASA 1993]
BCl_3	−403	Trichloroborane –403.0±2 kJ JANAF 12/64 [NASA 1993]
BF	−116	–115.9±14 kJ JANAF 12/64 [NASA 1993]
BF_2	−590	–589.9±13 kJ JANAF 6/72 [NASA 1993]
BF_3	−1136	Trifluoroborane –1135.6±2 kJ JANAF 6/69 [NASA 1993]
BH	443	442.7±8 kJ JANAF 12/64 [NASA 1993]
BHF_2	−734	–733.9±3 kJ JANAF 12/65 [NASA 1993]
BH_2	201	201±63 kJ JANAF 12/64 [NASA 1993]
BH_3	107	Borane 106.7±10 kJ JANAF 12/64 [NASA 1993]
BO	0	0±8 kJ JANAF 6/68 [NASA 1993]
BOCl	−316	–316.3±30 kJ JANAF 12/65 [NASA 1993]
BOF	−602	–602±13 kJ JANAF 12/65 [NASA 1993]
BOF_2	−837	–837±15 kJ JANAF 12/66 [NASA 1993]
BO_2	−285	–285±8 kJ JANAF 6/68 [NASA 1993]
B_2	830	829.7±33 kJ JANAF 3/79 [NASA 1993]
B_2O	96	96±105 kJ JANAF 6/66 [NASA 1993]
B_2O_2	−456	$(BO)_2$ –456.1±8 kJ JANAF 12/64 [NASA 1993]
B_2O_3(l)	−1253	JANAF 6/71 [NASA 1993]
B_2O_3	−836	–836.0±4 kJ JANAF 6/71 [NASA 1993]
$B_3O_3Cl_3$	−1632	$(BOCl)_3$ –1632±8 kJ JANAF 3/65 [NASA 1993]

Species	$\Delta_f H^\circ_{298}$	Comments
$B_3O_3F_3$	−2365	$(BOF)_3$ −2365.2±4 kJ JANAF 3/65 [NASA 1993]
$B_3O_3H_3$	−1218	Boroxin −1218±42 kJ JANAF 3/65 [NASA 1993]
$H_3B_3O_6$	−2272	Boric acid $(HBO_2)_3$ −2272±13 kJ JANAF 12/64 [NASA 1993]
Br	112	111.86±0.06 kJ JANAF 6/82 [NASA 1993]
BrO	126	125.8±2 kJ Chase 1996b [Technion 02/97]
BrO_2	108	Br–O–O 108.0±40 kJ Chase 1996b [Technion 02/97]
BrO_2	152	O–Br–O 152.0±25 kJ Chase 1996b [Technion 02/97]
BrO_3	221	221.0±50 kJ Chase 1996b [Technion 02/97]
Br_2	0	Reference state JANAF 6/82 [NASA 1993]
Br_2O	168	BrBr–O 168.0±20 kJ Chase 1996b [Technion 02/97]
Br_2O	108	Br–O–Br 107.6±4 kJ Chase 1996b [Technion 02/97]
C(gr)	0	Reference state TRC/API 4/83 [NASA 1993]
C	717	Moore 1970 gives 716.68±0.45 kJ [NASA 7/88]
C^+	1809	Moore 1970 gives 1809.4 kJ [NASA 7/88]
$CBrClF_2$	−435	Bromochlorodifluoromethane Halon 1211 CFC-12b1 −435±15 kJ TSIV 1979 [NASA 1993]
$CBrF_3$	−649	Bromotrifluoromethane CFC 1301 TSIV 1979 [NASA 1993]
CBr_2F_2	−380	Dibromodifluoromethane Halon 1202 CFC-12b2 −380±15 kJ TSIV 1979 [NASA 1993]
CCl	502	Chloromethylidene 502.1±20 kJ JANAF 12/69 [NASA 1993]
CClF	26	Chlorofluoromethylene TSIV 1978 [Technion 7/95]
COClF	−427	Carbonic chloride fluoride −426.8±33 kJ JANAF 6/61 [NASA 1993]
$CClF_2$	−275	Chlorodifluoromethyl radical TSIV 1979 [NASA 1993]
$CClF_3$	−708	Chlorotrifluoromethane CFC-13 −707.93±3 kJ JANAF 6/69 [NASA 1993]
ClCN	138	Cyanogen chloride JANAF 6/66 [NASA 1993]
COCl	−63	Carbonyl chloride −62.8±40 kJ JANAF 12/65 [NASA 1993]
CCl_2	239	Dichloromethylene 238.5±20 kJ JANAF 12/68 [NASA 1993]
CCl_2F	−105	TSIV 1979 [NASA 1993]
CCl_2F_2	−493	Dichlorodifluoromethane CFC-12 TRC/API 12/75 (−477.4±5.4 kJ Pedley and Rylance 1977) [NASA 1993]
$COCl_2$	−220	Phosgene Gurvich 1991 (Baulch et al. 1982) [NASA 1993]

Species	$\Delta_f H^\circ_{298}$	Comments
CCl_3	80	79.5=±8 kJ JANAF 6/70 [NASA 1993]
CCl_3F	−285	Trichlorofluoromethane CFC-11 TRC/API 12/75 (–267.8±8.4 kJ Pedley and Rylance 1977) [NASA 1993]
CCl_4	−96	Carbon tetrachloride Gurvich 1991 (–97.1±3 kJ Pedley and Rylance 1977) [NASA 12/81]
CD	593	Burcat 1980 [Technion 2/80]
CDH_3	−78	Monodeuteromethane Burcat 1980 [Technion 05/79]
CDO	41	Deuteroformyl radical $\Delta_f H^\circ_0$ = 40.3 kJ TSIV 1978 [Technion 5/92]
CD_2	383	Methylene-D2 Burcat 1980 [Technion 05/80]
CD_2O	−115	Methanal-D2 (formaldehyde–D2) Chao et al. 1980 [Technion 8/81]
CD_3	138	Methyl-D3 radical Burcat 1980 138.69 kJ [Technion 11/79]
CD_4	−89	Deuteromethane Burcat (1980) [Technion 05/79]
CF	255	255.2±8 kJ JANAF 6/70 [NASA 1993]
FCN	36	Cyanogen fluoride 36.0±17 kJ JANAF 6/69 [NASA 1993]
COF	−172	Carbonylfluoride –171.5±63 kJ JANAF 12/65 [NASA 1993]
CF_2	−182	–182.0± 6 kJ JANAF 6/70 [NASA 1993]
COF_2	−640	Carbonic difluoride Gurvich 1991 [NASA 1993]
CF_3	−470	Trifluoromethyl radical –470.3±4 kJ JANAF 6/69 [NASA 1993]
CF_3I	−589	Trifluoroiodomethane –589.1±3 kJ JANAF 6/69 [NASA 1993]
CF_4	−933	Tetrafluoromethane CFC-14 TRC/API 6/83 (–934.7 kJ Pedley and Rylance 1977) [NASA 1993]
CH	597	Methylidene TSIV 1979 (594.128±17.5 kJ JANAF 1985) [NASA 1993]
$CHBrF_2$	−422	Bromodifluoromethane HBCFC-22b1 TSIV 1979 [NASA 1993]
CHCl	308	Chloromethylene radical TSIV 1979 [NASA 3/84]
CHClF	−84	Chlorofluoromethyl radical TSIV 1979 [NASA 4/86]
$CHClF_2$	−475	Chlorodifluoromethane HCFC-22 –475.0±15 kJ TSIV 79 [NASA 1993]
$CHCl_2$	74	Dichloromethyl radical TSIV 1979 [NASA 3/84]
$CHCl_2F$	−285	Dichlorofluoromethane CFC-21 TRC/API 12/75 [NASA 1993]
$CHCl_3$	−103	Chloroform (Trichloromethane) TRC/API 6/81 (–104.6±2.1 kJ Pedley and Rylance 1977) [NASA 1993]
CHD_3	−85	Methane-D3 Burcat [Technion 05/97]

Species	$\Delta_f H^\circ_{298}$	Comments
CHF_2	−254	Difluoromethyl radical TSIV 1979 [NASA 1993]
CHF_3	−693	Fluoroform (Trifluoromethane, HCFC-23) TRC/API 6/81 (−694.5±8.4 kJ Kudchadker et al. 1978) [NASA 1993]
HCN	135	Hydrocyanic acid TSIV 1979 (132±4 kJ Gurvich 1991) [NASA 7/88]
HNC	194	Isocyanic acid $\Delta_f H^\circ_0 = 194 \pm 9$ kJ Gurvich 1991 (200.8±8.4 kJ Pau and Hehre, J. Phys. Chem. 86 (1982) 321) [NASA 11/92]
HNCO	−117	Hydrogen isocyanate East and Allen, J. Chem. Phys. 99 (1993) 4638. (−104.6±12.6 kJ proposed by Spiglanin, Perry and Chandler, J. Phys. Chem. 90 (1986) 6184) [Technion 6/94]
CHN_2	320	Cyanamide radical HN*–CN 319.8±23 kJ Melius [Technion 3/93]
HCO	42	Formyl radical 42±4 kJ TSIV 1979 (44.8 kJ from Becker, Lippmann and Schurath, Ber. Bunsenges. Phys. Chem. 81 (1977) 567.) [NASA 12/89]
CHO^+	833	Formyl ion JANAF 12/70 [NASA 1993]
COOH	−213	Carboxyl radical TSIV 1979 (−209.2±4.2 kJ Benson 1976) [NASA 1993]
HCOO	−151	Formic acid radical Benson 1976 [Technion 04/97]
HCS	300	$\Delta_f H^\circ_0$ = 300 kJ Bauer et al. 1997 [Technion 05/97]
CH_2	390	Triplet methylene TSIV 1979 (386.39±4.9 kJ JANAF 1985) [NASA 1993]
CH_2	425	Singlet methylene Melius [Sandia 12/87]
CH_2BrCl	−45	Bromochloromethane Halon 1011 −45±15 kJ TSIV 1979 [NASA 1993]
CH_2Cl	117	Chloromethyl radical TSIV 1979 [NASA 8/84]
CH_2ClF	−264	Chlorofluoromethane CFC-31 TRC/API 12/75 (−95.8±1 kJ Pedley and Rylance 1977) [NASA 1993]
CH_2Cl_2	−95	Dichloromethane TRC/API 12/81 [NASA 1993]
CH_2D_2	−82	Methane-D2 Burcat [Technion 05/79]
CH_2F	−32	Fluoromethyl radical TSIV 1979 [NASA 1993]
CH_2F_2	−453	Difluoromethane CFC-32 TRC/API 6/81 (−451.9 ±8.4 kJ Kudchadker et al 1978) [NASA 1993]
H_2CN	240	H2C=N* radical Bauer et al. 1997 [Technion 05/97]
HCNH	299	Trans H*C=NH radical Bauer et al. 1997 [Technion 05/97]
HCNH	320	Cis H*C=NH radical Bauer et al. 1997 [Technion 05/97]
H_2NCO	−23	H2N–C(*)=O radical −23.3 ±10 kJ Melius [Technion 09/96]

Species	$\Delta_f H^\circ_{298}$	Comments
CH_2NO	173	CH2=N–O* radical 173.4 ±21 kJ Melius [Technion 9/96]
CH_2NO	224	H2C*N=O radical 223.9 ±8 kJ Melius [Technion 9/96]
CH_2N_2	136	Cyanamide H_2N–CN 135.9 ±20 kJ Melius [Technion 3/93]
CH_2N_2	149	Carbodiimide HN=C=NH 149.0 ±15 kJ Melius [Technion 3/93]
CH_2N_2	286	Diazomethane H_2C=N=N 286.4 ±25 kJ Melius (230.1±16.7 kJ Vogt, Williamson and Beauchamp J. Am. Chem. Soc. 100 (1978) 3478) [Technion 3/93]
CH_2N_2	320	Diazirene H_2(CNN) 320.1 ±20 kJ Melius (264.8±11.3 kJ Laufer and Okabe, J. Phys. Chem. 76 (1972) 3504.) [Technion 3/93]
CH_2O	−109	Formaldehyde TSIV 1979 [NASA 8/88]
HCOOH	−379	Methanoic (Formic) acid monomer CH_2O_2 Chao and Zwolinski 1978 [NASA 8/88]
H_2CS	115	$\Delta_f H^\circ_0$ = 118.4 kJ Bauer et al. 1997 [Technion 05/97]
CH_3	147	Methyl radical Pamidimukala et al. 1982 (146.8±1 kJ Cohen and Benson 1992) [NASA 8/88]
CH_3Br	−38	Methyl bromide Stull et al. 1969 [Technion 2/94]
CH_3Cl	−82	Methyl chloride TRC/API 12/81 [NASA 1993]
CH_3F	−238	Methyl fluoride TRC/API 6/81 [NASA 1993]
CH_3I	211	Methyl iodide Stull et al. 1969 [Technion 2/94]
CH_3N	96	Methanimine (CH_2NH) Sidhu et al. 1991 (94.4 kJ NIST 1994; 133.9 kJ DeFrees and Hehre, J. Phys. Chem. 82 (1978) 391.) [Technion 2/92]
CH_3NO	79	Nitrosomethyl 79.0 ±7 kJ Melius (71.1±3 kJ Batt and Milne, Int. J. Chem. Kinet. 5 (1973) 1067.) [Technion 12/92]
$OCHNH_2$	−195	Formamide O=CH–NH_2 –195.3 ±10 kJ Melius (–186.1 kJ NIST 1994) [Technion 12/92]
CH_2NOH	29	Formaldehyde–oxime CH_2=N–OH NIST 1991. [Technion 12/92]
NCH_3O	−148	Formimidic acid HN–CH–OH –148.4 ±11 kJ Melius [Technion 12/92]
H_3CNO	59	CH_2–NH=O 59.0 ±12 kJ Melius [Technion 12/92]
CH_3NO_2	−75	Nitromethane Stull et al. 1969 [Technion 1/81]
CH_3NO_3	−120	Methyl nitrate Stull et al. 1969 (–123.8±4.2 kJ NIST 1994) [Technion 09/81]
CH_3N_2	248	CH_3N=N* Methyl diazine radical 247.7±10 kJ Melius [Technion 9/96]
CH_3O	16	Methoxy radical 16.3±4 kJ Burcat and Kudchadker 1979; 14.6 kJ Benson 1976. [NASA 8/88]

Species	$\Delta_f H^\circ_{298}$	Comments
CH_2OH	−9	Hydroxymethylene radical (CH_2OH) −9±4 kJ Setula and Gutman, J. Chem. Phys. 96 (1992) 5401. [Technion 9/92]
CH_3O_2	28	Methylperoxide radical (CH_3OO) Benson 1976 [NASA 1/84]
CH_3S	125	Methyl sulfide radical Nicovich et al., J. Chem. Phys. 96 (1992) 2518. (138.9±6.3 kJ McMillen and Golden 1982) [Technion 5/93]
CH_4	−75	Methane −74.85±0.42 kJ JANAF 5/61 [NASA 1993]
CH_4N	188	CH3NH* Methyl amino radical 187.6±5 kJ Melius [Technion 9/96]
CH_3OH	−201	Methanol (CH_3OH) Chen et al. 1977 [NASA 8/88]
CH_4O_2	−140	Peroxymethane (CH_3OOH) −139.8±5 kJ Lay et al., J. Phys. Chem. 100 (1996) 8240. (−124.6 kJ Melius) [Technion 11/96]
CH_4S	−23	Methanethiol, (methyl mercaptan) −22.5 ±4 kJ Melius (−23.3 kJ NIST 1994) [Technion 4/93]
CH_5N	−23	Methylamine (CH_3NH_2) Stull et al. 1969 [Technion 09/81]
CH_5N_2	215	Methyl hydrazine radical $CH_3N^*NH_2$ 215.1±5 kJ Melius [Technion 9/96]
CH_6N_2	95	Methylhydrazine CH_3–NH–NH_2 Stein et al. NIST 1991. [Technion 2/92]
CN	445	Cyanide radical 437.1±6 kJ Huang, Barts and Halpern, J. Phys. Chem. 96 (1992) 425. [Technion 6/94]
CNO	132	East and Allen, J.Phys. Chem. 99 (1993) 4638. [Technion 6/94]
CNN	401	TSIV 1979 (585±126 kJ JANAF) [NASA 1993]
NCN	501	$\Delta_f H^\circ_0$ = 500 kJ in Gurvich 1991 and 472±21 kJ in JANAF [NASA 12/89]
CO	−111	Carbon monoxide −110.53±0.17 kJ TSIV 1979 [NASA 1993]
COS	−138	−138.4±1 kJ JANAF 3/61 [NASA 1993]
CO_2	−394	Carbon dioxide −393.51±0.13 kJ TSIV 1979 [NASA 7/88]
CP	72.5	Gurvich 1991 72.5 kJ The earlier JANAF value was 449.9±10 kJ. [NASA 1993]
CS	280	Carbon sulfide 280.3±25 kJ JANAF 12/76 [NASA 1993]
CS_2	117	Carbon disulfide 116.9±1 kJ JANAF 12/76 [NASA 1993]
C_2	831	TSIV 1979 [NASA 1993]
$C_2Br_2F_4$	−791	1,2-Dibromotetrafluoro ethane, Halon 2402 −790.8 ±4 kJ Kolesov and Papina 1983 [Technion 8/95]
C_2Cl	494	Chloro ethynyl radical TSIV 1979 [NASA 1993]

Species	$\Delta_f H^\circ_{298}$	Comments
C_2Cl_2	210	Dichloroacetylene 209.6=±42 kJ JANAF 12/68 [NASA 1993]
$C_2Cl_2F_4$	−900	Dichlorotetrafluoroethane CFC-114 TRC/API 6/89 (–924.7±4 kJ Kolesov and Papina 1983; –889.4 kJ NIST 1994) [NASA 1993]
C_2Cl_3	190	Trichlorovinyl radical TSIV 1979 [NASA 1993]
$C_2Cl_3F_3$	−706	Trichlorotrifluoroethane CFC-113 TRC/API 6/89 (–728±3 kJ Kolesov and Papina 1983) [NASA 1993]
$C_2Cl_3F_3$	−741	1,1,1-Trichloro-2,2,2-trifluoro ethane CF_3–CCl_3 (CFC-113a) TRC/API 6/89 (–723.8±8.4 kJ Kolesov and Papina 1983; –765.2 kJ NIST 1994) [NASA 1993]
C_2Cl_4	−12	Tetrachloroethylene TRC–12/82 (–17.5 NIST 1994) [NASA 10/87]
C_2Cl_5	39	Pentachloroethyl radical TSIV 1979 [NASA 1993]
C_2Cl_6	−139	Hexachloroethane Chao et al. 1974 (–150.6±4 kJ Kolesov and Papina 1983; –183.2 kJ NIST 1994) [NASA 10/87]
C_2D_2	222	Acetylene-D2 $\Delta_f H^\circ_0$ = 222.7 kJ derived from $\Delta_f H^\circ_0$ of C_2H_2 in JANAF 1971. [Technion 8/80]
C_2D_2O	40	Ketene-D2 Derived from Benson's value for C_2H_2O. [Technion 10/82]
C_2D_4	30	Ethylene-D4 Burcat 1980 [Technion 12/79]
C_2OD_4	−181	Ethanal-D4 (acetaldehyde–D4) Chao et al. 1980 [Technion 8/81]
C_2D_6	−111	Ethane-D6 Burcat 1980 [Technion 05/80]
$C_2D_6N_2$	119	Azomethane-D6 (CD_3NNCD_3) Pamidimukkala et al. 1982 119.2 kJ [NASA 8/84]
C_2D_6O	−209	Dimethyl ether-D6 $\Delta_f H^\circ_0$ = –192.1 kJ from the $\Delta_f H^\circ_0$ of C_2H_6O of Stull et al. 1969 [Technion 12/82]
C_2F	354	Fluoro ethynyl TSIV 1979 [NASA 1993]
C_2F_2	21	Difluoroacetylene 20.9±21 kJ JANAF 12/67 [NASA 1993]
C_2F_3	−228	Trifluorovinyl radical TSIV 1979 (–192±8.4 kJ Spyrov, Sauers and Christophorou, J. Chem. Phys. 78 (1983) 7200.) [NASA 1993]
C_2F_4	−659	Tetrafluoroethylene CFC-1114 –658.56±2.9 kJ JANAF 6/69 (–648.4 NIST 1994) [NASA 1993]
C_2F_5	−891	Pentafluoroethyl radical Chen et al. 1990 [Technion 01/92]
C_2F_6	−1344	Hexafluoroethane CFC-116 –1343.9± 5 kJ JANAF 6/69 (-1322.1 kJ NIST 1994) [NASA 1993]
C_2H	547	Ethynyl radical Greene, Kinsey and Field, J. Chem. Phys. 98 (1989) 5160. (543.6±8 kJ NIST 1994) [Technion 04/93]

Species	$\Delta_f H^\circ_{298}$	Comments
C_2HCl	214	Chloroacetylene 213.8±42 kJ JANAF 12/68 (187.2 kJ NIST 1994) [NASA 1993]
$CHClCF_2$	−334	Chlorodifluoro ethylene CFC-1122 TSIV 1979 [NASA 1993]
C_2HClF_2	−324	Cis–chlorodifluoro ethylene cis-CFCl=CHF CFC-1131 TSIV 1979 [NASA 1993]
C_2HClF_2	−323	Trans–CFCl=CHF TSIV 1979 [NASA 1993]
C_2HClF_4	−903	1-Chloro-1,1,2,2-tetrafluoro ethane (HCFC-124a) TRC/API 89 [NASA 1993]
C_2HClF_4	−925	2-Chloro-1,1,1,2-tetrafluoro ethane (HCFC-124) TRC/API 89 [NASA 1993]
$C_2HCl_2F_3$	−744	4,2-Dichloro-1,1,1-trifluoro ethane (HCFC-123) TRC/API 89 [NASA 1993]
$C_2HCl_2F_3$	−710	1,2-Dichloro-1,1,2-trifluoro ethane (HCFC-123a) TRC/API 89 [NASA 1993]
$C_2HCl_2F_3$	−702	1,1-Dichloro-1,2,2-trifluoro ethane TRC/API 89 [NASA 1993]
C_2HCl_3	−4	Trichloroethylene TSIV 1979 (-13.7 kJ NIST 1994; –18.8±3 kJ Papina and Kolesov, Zh. Fiz. Khim. 59 (1985) 2169.) [NASA 1993]
C_2HCl_4	45	Tetrachloroethyl radical ($CHCl_2CCl_2$*) Calculated with THERM program [Technion 07/93]
C_2HF	126	Fluoroacetylene 125.5±63 kJ JANAF 12/67 [NASA 1993]

Species	$\Delta_f H^\circ_{298}$	Comments
C_2HF_5	−1105	Pentafluoroethane (HCFC-125) Chen et al. 1975 [Technion 9/92]
HCCN	610	$\Delta_f H^\circ_0$ = 609.2±100 kJ Gurvich 1989 [NASA 1993]
C_2HO	177	Ketyl radical 177.4 ± 9 kJ Oakes, Jones, Bierbaum and Ellison, J. Phys. Chem. 87 (1983) 4810. [Technion 6/94]
H_2C_2	415	Vinylidene CH_2C $\Delta_f H^\circ_0$ = 403.91 kJ Chen, Jonas, Kinsey and Field, J. Chem. Phys. 91 (1989) 3976. [NASA 12/89]
C_2H_2	227	Acetylene TSIV 1978 [NASA 8/88]
$C_2H_2F_2$	336	1,1-Difluoroethylene 1,1-$C_2H_2F_2$ (CFC-1132a) TSIV 1979 [NASA 1993]
$C_2F_2H_2$	313	Cis-difluoroethylene cis-$C_2H_2F_2$ TSIV 1979 [NASA 1993]
$C_2F_2H_2$	310	Trans-difluoroethylene trans-$C_2H_2F_2$ CFC-1132 TSIV 1979 [NASA 1993]
$C_2F_3H_2$	−517	β–Trifluoroethyl radical Chen et al. 1990 [Technion 1/92]
C_2H_2O	−48	Ketene –47.7±2 kJ Vogt, Williamson and Beauchamp, J. Am. Chem. Soc. 100 (1978) 3478. [Technion 6/94]
HCCOH	93	Ethynol (hydroxyacetylene) HCC–OH 93.2±18 kJ Melius [Technion 4/93]

Species	$\Delta_f H^\circ_{298}$	Comments
t-$C_2H_2O_2$	−212	Trans-glyoxal CHO–CHO Pedley et al. 1986 [Technion 9/92]
c-$C_2H_2O_2$	−212	Cis-glyoxal CHO–CHO Pedley et al. 1986. [Technion 9/92]
C_2H_3	300	Vinyl radical 299±5 kJ Ervin, J. Am. Chem. Soc. 112 (1990) (294.0 kJ NIST 1994; 265.3±4.2 kJ Kiefer, Wei, Kern and Wu, Int. J. Chem. Kinet. 17 (1985) 225.) [Technion 06/93]
C_2H_3Cl	23	Chloroethylene TSIV 1979 (21.2 kJ NIST 1994) [NASA 1993]
$C_2H_3F_2$	−303	α-Difluoroethyl radical (CH_3CF_2) Chen et al. 1990 [Technion 12/91]
$C_2H_3F_3$	−746	1,1,1–Trifluoroethane 1,1,1–$C_2H_3F_3$ (CFC-143a) Stull et al. 1969 [Technion 1/92]
C_2H_3N	88	Methylcyanide (acetonitrile) (CH_3CN) Stull et al. 1969 (75.3±1 kJ An and Mansson, J. Chem. Thermodyn. 15 (1983) 287) [Technion 04/80]
C_2H_3O	−10	Acetyl radical CH_3CO Niiaranen, Gutman and Krasnoperov, J. Phys. Chem. 96 (1992) 5881. [Technion 9/92]
C_2H_3O	25	CH_2CHO radical 26.0 kJ derived from Benson and O'Neal, NSRDS–NBS 1970 [Technion 4/83]
C_2H_3O	140	Oxirane (ethylene oxide) radical calculated with THERM program using (JANAF) ethylene oxide value of 9/1965 minus a secondary proton [Technion 9/92]
C_2H_4	52	Ethylene Chao and Zwolinsky 1978; adjusted by Burcat [NASA 4/85]
C_2H_4F	−72	α-Fluoroethyl radical (CH_3CHF) Chen et al. 1990 [Technion 12/91]
C_2H_4Cl	90	β-Chloroethyl radical (CH_2ClCH_2) THERM program [Technion 7/93]
C_2H_4O	−125	Vinyl alcohol (ethenol) Holm and Losing, J. Am. Chem. Soc. 104 (1982) 2648. [NASA 8/89]
C_2H_4O	−53	Oxirane (ethylene oxide) –52.6±1 kJ JANAF 1985 [NASA 8/88]
CH_3CHO	−166	Acetaldehyde Chao et al. 1986 (–164.3 kJ NIST 1994) [NASA 8/88]
$C_2H_4O_2$	−432	Acetic acid Chao and Zwolinski [NASA 4/85]
$C_2H_4O_2Cl_2$	−231	α-Dichloroperoxyethane CH_3CCl_2O–OH Lay et al., J. Phys. Chem. 100 (1996) 8240. [Technion 01/97]
$(HCOOH)_2$	−821	Formic acid dimer $(HCOOH)_2$ Chao and Zwolinski 1978 [NASA 4/85]
C_2H_5	119	Ethyl radical Chen et al. 1990 (118.8±2 kJ Cohen and Benson 1992) [Technion 12/91]

Species	$\Delta_f H^\circ_{298}$	Comments
C_2H_5Br	−64	Ethyl bromide Stull et al. 1969 [Technion 2/94]
C_2H_5Cl	−112	Chloroethane Stull et al. 1969 [Technion 1/92]
C_2H_5F	−262	Ethyl fluoride Stull et al. 1969 [Technion 2/94]
C_2H_5I	−8	Ethyl iodide Stull et al. 1969 [Technion 2/94]
C_2H_5O	−17	Ethyl oxide radical (CH_3CH_2O) Benson 1976 [Technion 11/82]
C_2H_5O	−21	CH_2CH_2OH radical Benson 1976 [Technion 4/83]
C_2H_5O	−21	CH_3CHOH radical Benson 1976 [Technion 4/83]
C_2H_5O	−8	CH_2OCH_3 radical Benson 1976 [Technion 11/82]
$C_2H_5O_2Cl$	−213	α–Chloroperoxyethane CH_3CHClO–OH Lay et al., J. Phys. Chem. 100 (1996) 8240. [Technion 01/97]
C_2H_6	−84	Ethane –83.85±0.5 kJ Chao et al. 1973 [NASA 5/84]
C_2H_6N	159	Dimethylazide radical $(CH_3)_2N$ NIST 1994 [Technion 7/95]
$C_2H_6N_2$	149	Azomethane (CH_3NNCH_3) Pamidimukkala et al. 1982 [Technion 8/81]
C_2H_5OH	−235	Ethanol Chao et al. 1986 Equilibrium mixture of one trans and two gauche isomers [NASA 8/88]
CH_3OCH_3	−184	Dimethylether Chao et al. 1986 [NASA 9/88]
$C_2H_6O_2$	−174	Peroxyethane C_2H_5O–OH Lay et al., J. Phys. Chem. 100 (1996) 8240. [Technion 10/96]
C_2H_6S	−46	Ethanethiol C_2H_5SH Stull et al. 1969 [Technion 4/93]
C_2H_6S	−38	Methyl sulfide CH_3–S–CH_3 Stull et al. 1969 [Technion 4/93]
$C_2H_6S_2$	−24	Methyl disulfide CH_3–S–S–CH_3 Stull et al. 1969 [Technion 4/93]
$C_2H_7N_2$	168	$(CH_3)_2N$–NH* Unsymmetrical dimethyl hydrazine radical; estimate based on THERM and Stein's method 1985 [Technion 3/92]
$C_2H_8N_2$	94	Symmetrical dimethyl hydrazine CH_3NH–$NHCH_3$ Melius 94.5±8 kJ (89.6 kJ NIST 1994; 92.0 kJ Pedley and Rylance 1977) [Technion 7/93]
$C_2H_8N_2$	53	Unsymmetrical dimethyl hydrazine $(CH_3)_2N$–NH_2 Durig and Harris, J. Chem. Phys. 51 (1969) 4457. (84.6 kJ NIST 1994; 83.7±2 kJ Pedley and Rylance 1977) [Technion 09/91]
CCN	605	TSIV 1979 [NASA 1993]
CNC	655	TSIV 1979 [NASA 1993]
C_2N_2	309	Ethanedinitrile TSIV 1979 [NASA 1993]
C_2O	291	TSIV 1979 (286.6±63 kJ JANAF 1985) [NASA 1993]
C_3	840	TSIV 1979 [NASA 1993]
C_3D_4	263	Cyclopropene-D4 Burcat(1982) 263.6 kJ [Technion 2/82]

Species	$\Delta_f H^\circ_{298}$	Comments
C_3D_6	33	Cyclopropane-D6 from $\Delta_f H^\circ_0$ of C_3H_6 [Technion 12/81]
C_3F_6	−1125	Hexafluoropropene Nielsen, Claassen and Smith, J. Chem. Phys. 20 (1952) 1916. [Technion 3/95]
C_3F_7	−1335	$CF_3CF^*CF_3$ Approximation using THERM program to 800 K CF_3CHFCF_3 minus a secondary proton. [Technion 11/94]
C_3F_8	−1760	Octafluoropropane (CFC-218) Domalski and Hearing, J. Phys. Chem. Ref. Data 22 (1993) 1065. (−1727.9 kJ NIST 1994; 1782.4±8.4 kJ Pedley and Rylance 1977) [Technion 1/94]
C_3H	684	C_3H radical estimated from C_2H, C_4H and C_6H Kiefer et al. 1992 [Technion 07/93]
C_3HF_7	−1527	1,1,1-2-3,3,3-Heptafluoro propane CF_3–CHF–CF_3 (CFC-227ea) NIST 1994 [Technion 12/94]
C_3HN	354	Cyanoacetylene HCC–CN Hall and Colket 1992 [Technion 6/92]
C_3H_2	846	HCC–CH** radical using Stein's additivity coefficients. McBride and Gordon 1992 [NASA 8/89]
C_3H_2	679	*HC=C=CH* Stein's method used; allene less 2 vinyl H. McBride and Gordon 1992 678.6±70 kJ NIST 1994. [Technion 7/94]
C_3H_2N	406	Cyanoethylene radical HC*=CH–CN Hall and Colket 1992 [Technion 6/92]
C_3H_3	346	Propargyl radical Tsang, Int. J. Chem. Kinet. 10 (1978) 687. (340.6±8.4 kJ McMillen and Golden 1982; these authors also quote 439.7±17.2 kJ for the cyclic isomer.) [NASA 8/89]
C_3H_3Cl	147	1-Chloro-1-propyne Cl–CC–CH_3 Stull et al. 1969 [Technion 1/96]
C_3H_3Cl	163	3–Chloro–1–propyne H–CC–CH_2Cl Stull et al. 1969 [Technion 1/96]
C_3H_3Cl	218.3	3–Chlorocyclopropene Kumaran et al., Israel J. Chem. 36 (1996) 223. [Technion 5/97]
C_3H_3Cl	160.9	Chloroallene ClHC=C=CH_2 Kumaran et al., Israel J. Chem. 36 (1996) 223. [Technion 5/97]
C_3H_3N	184	Cyanoethylene (acrylonitrile) H_2C=CH–CN Hall and Colket 1992 (184.1 kJ Chu, Nguyen and King, J. Phys. Chem. 86 (1982) 443.) [Technion 3/93]
C_3H_3O	302	CH_2=C*–CH=O radical of acrolein THERM [CD–(CO)(H)];[CO–(CD)(H)];[CD–(H)(H)] minus vinyl H [Technion 6/92]
C_3H_4	185	Propyne (methyl acetylene) CH_3CCH TRC–API [Technion 2/90]
C_3H_4	191	Allene H_2C=C=CH_2 TRC(1988). [NASA 8/89]
C_3H_4	277	Cyclopropene Dorofeeva et al. 1986 [Technion 12/81]
C_3H_4Cl	236	3-Chloropropenyl-1 (*CH=CH–CH_2Cl) Weissman and Benson (1989) [Technion 2/92]

Species	$\Delta_f H^\circ_{298}$	Comments
ClC_3H_4	107	1-Chloroallyl ($CHCl$–C^*H–CH_2) Weissman and Benson (1989) [Technion 2/92]
C_3H_4N	195	2-Propionitrile radical CH_3–CH^*CN NIST 1994 [C–(H)3]; [C.–(C)(CN)(H)] [Technion 10/94]
$C_3H_4N_2$	141	1,3-Diazole (imidazole, glyoxaline) CAS 288-32-4 141±28 kJ Melius (146.4±2 kJ Sabbah, Thermochimica Acta 41 (1980) 33.) [Technion 9/96]
C_3H_4O	−74	Propenal (acrolein, acrylaldehyde) [CD–(CO)(H)]; [CO–(CD)(H)]; [CD–(H)(H)]. –74.5±4 kJ NIST 91–94 [Technion 6/92]
C_3H_5	164	Allyl radical, symmetric, resonance–stabilized CH_2–C^*H–CH_2 Wu and Kern, J. Phys. Chem. 91 (1987) 6291. (171±3 kJ Tsang 1996) [Technion 9/96]
t-C_3H_5	238	Tertiary nonsymmetric allyl radical (CH_2=C^*CH_3) Wu and Kern, J. Phys. Chem. 91 (1987) 6291. [Technion 6/96]
s-C_3H_5	263	Secondary allyl radical (CH_3–CH=CH*) Wu and Kern J. Phys. Chem. 91 (1992) 6291. [Technion 6/96]
C_3H_5Cl	−27	3–Chloropropene–1 Weissman and Benson 1989 [Technion 2/92]
C_3H_5N	51	Ethyl cyanide (propionitrile) C_2H_5CN Stull et al. 1969 (51.5 kJ NIST 1994) [Technion 3/93]
$C_3H_5NO_2$	27	Nitrocyclopropane Holtzclaw et al. 1980 [Technion 11/81]
C_2H_5CO	−36	C_2H_5CO radical THERM program based on propionaldehyde Chao et al. 1986 minus aldehyde H (–46.86 kJ NIST 1994) [Technion 9/92]
CH_2COCH_3	−14	Acetone radical *CH_2COCH_5 THERM [Technion 5/92]
C_3H_5O	110	*CH_2–$CHOCH_2$ propylene oxide radical THERM program propylene oxide (Stull et al.1969) minus a proton. [Technion 9/92]
C_3H_6	20	Propylene TRC/API [NASA 4/85]
C_3H_6	53	Cyclopropane Dorofeeva et al. 1986 [Technion 12/81]
C_2H_5CHO	−192	Propionaldehyde Stull et al. 1969 [Technion 9/92]
CH_3COCH_3	−217	Acetone (CH_3–CO–CH_3) Chao and Zwolinski J. Phys. Chem. Ref. Data 5 (1976) 319. [Technion 5/92]
C_3H_6O	−93	Propyleneoxide Stull et al. 1969 [NASA 9/85]
C_3H_6O	−81	Trimethylene oxide (cyclo) Stull et al. 1969 [Technion 4/93]
C_3H_6O	−108	Vinyl methyl ether C_2H_3–O–CH_3 (–108.0±8 kJ NIST 1993) [Technion 07/95]
C_3H_6O	−114	Cyclopropanol NIST 1994 [Technion 07/95]

Species	$\Delta_f H^\circ_{298}$	Comments
C_3H_6S	61	Thiethane cylic C_3H_6S Pedley et al. 1986 and Bauer et al. 1997 obtain the identical value by different methods. [Technion 05/97]
n-C_3H_7	101	n–Propyl radical Tsang 1985 (98±4 kJ Cohen and Benson 1992) [NASA 9/84]
i-C_3H_7	93	Secondary propyl radical Tsang 1985 (83.7±2 kJ Cohen and Benson 1992) [NASA 9/84]
1-C_3H_7I	−34	1-Iodopropane $\Delta_f H^\circ_0$ = –10.2 ±2 kJ Brand et al., Chem. Phys. 76 (1983) 114. (–29.7 kJ NIST 1994) [Technion 9/96]
2-C_3H_7I	−45	2–Iodopropane $\Delta_f H^\circ_0$ = –20.1±2 kJ Brand et al., Chem. Phys. 76 (1983) 114. (–41.3±2 kJ NIST 1994) [Technion 9/96]
$C_3H_5NH_2$	77	Cyclopropylamine ($C_3H_5NH_2$) Draeger et al. 1983 (72.7 kJ NIST 1994)(77.0±0.5 kJ Pedley and Rylance 1977) [NASA 2/84]
C_3H_7N	98	Azetidine cyclic C_3H_6N–H Kamo et al., Nippon Kagakai Shi 8 (1987) 1560. [Technion 05/97]
C_3H_7O	−38	n–Propoxy radical NIST 1994 [Technion 3/96]
C_3H_8	−104	Propane –103.89±0.59 kJ Chao et al. 1973. (–104.7 kJ Pitam and Pilcher J. Chem. Soc. Faraday Trans. I, 68 (1972) 2224.) [NASA 4/85]
C_3H_8O	−255	1–Propanol TRC/API [NASA 9/88]
C_3H_8O	−273	2-Propanol $(CH_3)_2CHOH$ Chao et al. 1986 [NASA 9/88]
C_3O_2	−94	Carbon suboxide TRC/API 1984 [NASA 7/88]
C_4	1034	TSIV 1979 [NASA 1993]
C_4F_6	−1004	Perfluoro–1,3–butadiene Atkinson and Stedman, J. Chem. Soc. (1962) 512. [Technion 12/82]
C_4F_6	−1211	Perfluorocyclobutene Atkinson and Stedman, J. Chem. Soc. (1962) 512. [Technion 12/82]
C_4F_8	−1514	Perfluorocyclobutane NIST 1994 [Technion 11/94]
C_4F_{10}	−2137	Perfluorobutane (CFC-3–1–10) NIST 1994 [Technion 12/94]
C_4H	803	Kiefer et al. 1992 [Technion 12/91]
C_4H_2	440	Butadiyne Stein et al. NIST 1991. [NASA 9/89]
C_4H_3	484	Removing a stabilized vinyl H from 1-buten-3-yn using Stein's method. McBride and Gordon 1993 [NASA 9/89]
C_4H_4	308	1-Buten-3-yn $\Delta_f H^\circ_0$ = 315.1 kJ Torneng et al., Spectrochim. Acta 36a (1989) 975. (289.4 NIST 1994) [NASA 9/89]
C_4H_4	385	Cyclobutadiene Dorofeeva et al. 1986 [Technion 2/90]

Species	$\Delta_f H^\circ_{298}$	Comments
$C_4H_4N_2$	196	Pyrazine (six membered ring with N atoms in para position) CAS 290–37–9 195.8±4 kJ Pedley et al. 1986 [Technion 9/96]
$C_4H_4N_2$	197	Pyrimidine (1,3 diazine) (six membered ring with N atoms in ortho position) CAS 289–95–2 196.7±4 kJ Pedley et al. 1986 [Technion 9/96]
C_4H_4O	−35	Furan Stull et al. 1969 [Technion 03/97]
C_4H_4S	115	Thiophene CAS 110–02–1 Dorofeeva and Gurvich 1995 (115.1±0.5 kJ Kudchadker et al., API monograph 717–81 1981) [Technion 03/97]
n-C_4H_5	345	CH_2=CHCH=CH radical Colket (unpublished) 345.1±6 kJ [Colket 8/83]
i-C_4H_5	331	CH_2=CHC=CH_2 radical Colket (unpublished) [Colket 8/83]
t-C_4H_5	302	CH_2CH=C=CH_2 radical Colket (unpublished) [Colket 8/83]
C_4H_5N	108	Pyrrole (azole, imidole) CAS 109–97–7 Das et al., J. Phys. Chem. Ref. Data 22 (1993) 659. [Technion 8/95]
C_4H_5N	184	Cyclopropane carbonitrile C_3H_5–CN 184.1±1 kJ Fuchs, Hallman and Perlman, Can. J. Chem. 60 (1982) 1832. [Technion 5/97]
C_4H_6	146	2–Butayn (dimethylacetylene) Stull et al. 1969 (145.3 kJ NIST 1994) [Technion 12/82]
1, 3-C_4H_6	109	1,3-Butadiene Colket (unpublished). 110.9 kJ Weissman and Benson 1989 [Colket 8/83]
1, 2-C_4H_6	165	1,2–Butadiene Stull et al. 1969 [Colket 8/83]
C_4H_6	157	Cyclobutene Dorofeeva et al. 1986 [Technion 1/90]
$C_4H_6Cl_2$	−58	1,4-Dichlorobutene-1 Weissman and Benson 1989 [Technion 2/92]
$C_4H_6Cl_2$	−69	3,4-Dichlorobutene-1 Weissman and Benson 1989 [Technion 2/92]
C_4H_6O	−109	2,5-Dihydrofuran (1-oxolene) CAS 1708-29-8 Kudchadker et al. 1978 Key Chemicals Data Book–Furan, Dihydrofuran, Tetrahydrofuran (–66.9±4.2 kJ Allinger et al., J. Org. Chem. 46 (1981) 658.) [Technion 3/97]
2,5-C_4H_6S	87	2,5-Dihydrothiophen Dorofeeva and Gurvich 1995 (87.0±1.5 kJ Kudchadker et al., API monograph 717–81 1981) [Technion 3/97]
n-C_4H_7	192	CH_2=$CHCH_2CH_2$ radical NIST 1994 [Technion 2/96]
i-C_4H_7	127	CH_2=$CHCHCH_3$ radical Colket (unpublished) [Colket 8/83]
t-C_4H_7	128	CH_2CH=$CHCH_3$ radical Colket (unpublished) [Colket 8/83]
C_4H_7	121	*$CH_2C(CH_3)$=CH_2 radical NIST 1994 [Technion 2/96]

Species	$\Delta_f H^\circ_{298}$	Comments
C_4H_7	214	Cyclobutyl radical NIST 1994 [Technion 02/96]
C_4H_7O	11	2–Butanone radical CH_3–CH*CO–CH_3 THERM program 2–butanone minus alkoxy bond [Technion 6/92]
C_4H_7O	52	H_2C=C(CH_3)CH_2O* 2–Methyl allyl alcohol radical THERM program for 2 methyl allyl alcohol using [C/CD/H3]; [CD/H2]; [CD/C2]; [O/C/H]; [C/CD/H2/O] minus alkoxy bond. [Technion 1/96]
C_4H_8		1-Butene CH_2=CH–CH_2–CH_3 Chao and Hall 1980 [Technion 6/83]
C_4H_8	−17	Isobutene CH_2=C(CH_3)$_2$ Chao and Hall 1980 [Technion 6/83]
C_4H_8	−11	Trans–2–Butene CH_3–CH=CH–CH_3 Chao and Hall 1980 (–13.4 NIST 1994) [Technion 6/83]
C_4H_8	−7	Cis–2–butene CH_3–CH=CH–CH_3 Chao and Hall 1980 [Technion 6/83]
C_4H_8	28	Cyclobutane Dorofeeva et al. 1986 (27.3 kJ NIST 1994) [Technion 1/90]
C_4H_8O	−166	2–Methyl allyl alcohol H_2C=C(CH_3)CH_2OH NIST 1994 [C/CD/H2/O]; [CD/C2]; [C/CD/H3]; [CD/H2] [Technion 1/96]
C_4H_8O	−238	2–Butanone C_2H_5–CO–CH_3 Stull et al. 1969 [Technion 5/92]
C_4H_8O	−130	Trans–2,3–dimethyl–oxyrane (trans–dimethyl–ethylene–oxide) NIST 1991 [Technion 8/93]
C_4H_8O	−117	Ethyl oxirane NIST 1991 [Technion 9/93]
C_4H_8O	−184	Tetrahydrofuran Kudchadker et al. 1978 TRC Key Chemicals Data Book– Furan, Dihydrofuran, Tetrahydrofuran [Technion 3/97]
$C_4H_8O_2$	−314	1,4-Dioxane –314.4±7 kJ Melius [Technion 03/97]
$C_4H_8O_4$	−929	Acetic acid dimer (CH_3COOH)$_2$ Chao and Zwolinski 1978 [NASA 4/85]
C_4H_8S	−34	Tetrahydrothiophen Dorofeeva and Gurvich 1995 [Technion 3/97]
1, 4-$C_4H_8S_2$	0	1,4-Dithiane Dorofeeva and Gurvich 1995 [Technion 03/97]
1, 3-$C_4H_8S_2$	−10	1,3-Dithiane Dorofeeva and Gurvich 1995 [Technion 03/97]
n-C_4H_9	67	n–Butyl radical TRC/API 10/84 (76.9 kJ Cohen 1992) [NASA 1993]
i-C_4H_9	57	Isobutyl radical $CH_3C(CH_3)(H)CH_2$* TRC/API 10/84 (67±6.3 kJ Cohen 1992) [NASA 1993]
s-C_4H_9	70	2-Butyl radical Tsang, J. Phys. Chem. Ref. Data 19 (1990) 86. (67±6.3 kJ Cohen 1992) (62.8±4.5 kJ Cohen and Benson 1992) [Technion 07/95]

Species	$\Delta_f H^\circ_{298}$	Comments
t-C_4H_9	52	t-Butyl radical Tsang, J. Am. Chem. Soc. 107 (1985) 2872. (46.14±6.3 kJ Cohen 1992) (39.3±2 kJ Cohen and Benson 1992) [NASA 1/93]
C_4H_9N	−4	Pyrrolidine (tetrahydropyrrole, tetramethyleneimine) –3.6±1 kJ Das et al J. Phys. Chem. Ref. Data 22 (1993), 659. [Technion 3/95]
n-C_4H_9O	−58	n-Butoxy radical NIST 1994 [Technion 1/96]
i-C_4H_9O	−68	i-Butoxy radical (2–methyl propoxy radical) $(CH_3)_2CHCH_2O$* NIST 1994 [Technion 1/96]
s-C_4H_9O	−73	s-Butoxy–2 radical $CH_3CH(O^*)CH_2CH_3$ NIST 1994 [Technion 3/96]
t-C_4H_9O	−92	t–Butoxy radical $(CH_3)_3CO$* NIST 1994 [Technion 1/96]
n-C_4H_{10}	−126	n–Butane –125.65±0.67 kJ Chen et al. 1975 [NASA 4/85]
i-C_4H_{10}	−135	Isobutane (2–methylpropane) –134.18±0.63 kJ Chen et al. 1975 [NASA 4/85]
$C_4H_{10}FO_2P$	−963	Sarin $CH(CH_3)_2OP(O)FCH_3$ –963.2±40 kJ Melius [Technion 9/96]
n-$C_4H_{10}O$	−275	1-Butanol Chao et al. 1986 [Technion 07/96]
s-$C_4H_{10}O$	−293	2-Butanol (d,l) Chao et al. 1986 [Technion 07/96]
t-$C_4H_{10}O$	−313	2-Methyl-2-propanol Chao et al. 1986 [Technion 07/96]

Species	$\Delta_f H^\circ_{298}$	Comments
C_4N_2	533	Carbon subnitride (2–butynedinitrile) 533.5±1 kJ JANAF 3/61 [NASA 1993]
C_5	1051	TSIV 1979 [NASA 1993]
C_5F_{12}	−2543	Perfluoropentane (CFC-4112) NIST 1994 [Technion 12/94]
C_5H	778	$\Delta_f H^\circ_0$ = 775.7 kJ Duff and Bauer 1962 [Technion 12/81]
C_5H_2	691	$\Delta_f H^\circ_0$ = 690.4 kJ Duff and Bauer 1962 [Technion 12/81]
C_5H_3	603	1,3-Pentadiyne-5-yl radical 602.6 kJ THERM program using 1,3 pentadiyne minus H [Technion 2/92]
C_5H_3	565	1,4-Pentadiyne-3-yl radical (HCCCH*CCH) 564.6±43 kJ Melius [Technion 3/94]
C_5H_3	698	Cyclopentatriene-yl (radical of nonsymmetrical 1,2,4–cyclopentatriene) 697.8±75 kJ Melius [Technion 3/94]
C_5H_3N	416	Cyano vinyl acetylene HCC–CH=CH–CN Hall and Colket 1992 [Technion 6/92]
C_5H_4	400	1,3-Pentadiyne HCC–CC–CH_3 Stein et al., NIST 1991 (435.3 kJ NIST 1994) [Technion 2/92]
C_5H_4	465	1,4–Pentadiyne HCC–CH_2–CCH 464.8±18 kJ Melius [Technion 3/94]
C_5H_4	482	Pentatetraene CH_2=C=C=C=CH_2 Stein et al., NIST 1991 [Technion 2/92]

Species	$\Delta_f H^\circ_{298}$	Comments
C_5H_4	444	1,2–Pentadiene–4–yne CH_2=C=CHCCH 444.0±25 kJ Melius [Technion 3/94]
C_5H_4	551	1,2,4-Cyclopentatriene, nonsymmetrical Melius 551.5±20 kJ [Technion 3/94]
C_5H_4N	477	*CH=CH–CH=CH–CN radical Hall and Colket 1992 477.0±12 kJ [Technion 7/92]
C_5H_4N	383	Meta-pyridyl radical Hall and Colket 1992 383.25 kJ For ortho and para radicals add –6 kJ. [Technion 6/92]
C_5H_4O	−7	Cyclopentadiene-1-one THERM program using [CO/CD2]; 2×[CD/CD/H]; 2×[CD/C/H]; [cy/c5/de] (–33.5±33.5 kJ Liebman 1988) [Technion 2/94]
C_5H_5	406	3-Pentene-1-yne-5-yl radical 406.3±10 kJ NIST 1994 [Technion 4/94]
C_5H_5	266	Cyclopentadienyl radical Karni et al. 1991 [Technion 12/89]
C_5H_5N	140	Pyridine (azine) 140.4±1 kJ Das et al. 1993 (138.1±1 kJ Kudchadker and Kudchadker API monograph 710–79 1979) [Technion 3/95]
C_5H_4OH	87	1-Hydroxy-2,4-cyclopentadiene-1-yl radical Karni et al. 1991 [Technion 12/89]
1, 3-C_5H_5O	60	1-Oxyl-1,3-cyclopentadiene radical Karni et al. 1991 [Technion 4/91]
1, 4-C_5H_5O	103	1-Oxyl-1,4-cyclopentadiene radical Karni et al. 1991 [Technion 4/91]
C_5H_6	249	1-Penten-3-yne NIST 1991 [Technion 4/94]
C_5H_6	256	3–Penten–1–yne NIST 1991 The method does not differentiate between cis and trans isomers. [Technion 4/94]
C_5H_6	134	Cyclopentadiene Pedley et al. 1986 [Technion 1/90]
$C_5H_6N_2$	119	2-Amino-pyridine Melius. (117.5±1 kJ Binkerton, Pilcher and Al–Takhin, J. Chem. Thermodyn. 16 (1984) 373.) [Technion 9/96]
2, 4-C_5H_5OH	8	2,4-Cyclopentadiene-1-ol Karni et al. 1991 [Technion 4/91]
1, 3-C_5H_5OH	−24	1,3-Cyclopentadiene-1-ol Karni et al. 1991 [Technion 4/91]
1, 4-C_5H_5OH	−27	1,4-Cyclopentadiene-1-ol Karni et al. 1991 [Technion 4/91]
C_5H_7	222	1,3-Pentadienyl-5 radical H_2C=CH–HC=CH–CH_2* Weissman and Benson 1989 [Technion 2/92]
C_5H_7	167	Cyclopentenyl radical THERM program using cyclopentene from Dorofeeva et al. 1986 minus cypene bond. [Technion 4/94]
C_5H_7Cl	71	5-Chloropentadiene-1,3 (CH_2=CHCH=CHCH_2Cl) Weissman and Benson 1989. [Technion 2/92]

Species	$\Delta_f H^\circ_{298}$	Comments
$C_5H_7Cl_2$	62	1,5-Dichloropentene-1-yl-3 *$ClCHCH_2CH$=$CHCH_2Cl$ Weissman and Benson 1989. [Technion 7/95]
C_5H_7NO	−109	2-Methyl-3-oxo-butyro-nitryl CH_3–C=O–CH(CH_3)–CN NIST 1994 using [C–(C)(CN)(CO)(H)]=[C–(C)2(CN)(H)]; [CO–(C)2]; 2X[C–(H)3] [Technion 10/94]
C_5H_8	76	1,3-Pentadiene H_2C=CH–HC=CH–CH_3 Weissman and Benson 1989 [Technion 2/92]
C_5H_8	76	Isoprene (2-methyl-1,3-butadiene) Stull et al. 1969 [Technion 5/96]
C_5H_8	33	Cyclopentene Dorofeeva et al. 1986 [Technion 2/90]
C_5H_8Cl	79	5-Chloropentene-1yl-3 (*CH_3CH_2CH=$CHCH_2Cl$) Weissman and Benson 1989 [Technion 2/92]
$C_5H_8N_4O_{12}$	−376	Pentaerithritoltetranitrate PETN C(CH_2ONO_2)$_4$ Melius [Technion 10/96]
C_5H_9	115	Cyclopentyl radical Stein et al., NIST program 1991. [Technion 2/92]
C_5H_9	159	Penten-2-yl-5 CH_3CH=CHC_2CH_2* Weissman and Benson 1989. [Technion 2/92]
C_5H_9	113	Penten-2-yl-1 *CH_2CH=$CHCH_2CH_3$ Weissman and Benson 1989. [Technion 2/92]
C_5H_9	81	3-Methyl-1-buten-3-yl H_2C=CH–C*(CH_3)–CH_3 NIST 1994 [Technion 5/96]
C_5H_9	212	3-Methyl-1-buten-1-yl *HC=CH–CH(CH_3)–CH_3 NIST 1994 [Technion 5/96]
C_5H_9	177	3-Methyl-1-buten-4-yl H_2C=CH–CH(CH_3)–CH_2* NIST 1994 [Technion 5/96]
C_5H_9N	75	1,2,3,6-Tetrahydropyridine 75.3±10 kJ THERM program [Technion 2/92]
C_5H_{10}	−21	1-Pentene TRC/API 12/52 [NASA 1993]
C_5H_{10}	−32	2-Pentene-trans TRC/API 12/52 [NASA 1993]
C_5H_{10}	−36	2-Methyl-1-butene Stull et al. 1969 [Technion 11/95]
C_5H_{10}	−43	2-Methyl-2-butene Stull et al. 1969 [Technion 11/95]
C_5H_{10}	−29	2-Methyl-3-butene Stull et al. 1969 [Technion 5/96]
C_5H_{10}	−78	Cyclopentane Dorofeeva et al. 1986 (–76.6 kJ NIST 1994) [Technion 2/10]
$C_5H_{10}O$	−224	Tetrahydropyran –223.8±1 kJ Stull et al. 1969 [Technion 4/93]
n-C_5H_{11}	56	n-Pentyl radical Cohen 1992 [Technion 03/97]
s-C_5H_{11}	46	s-Pentyl radical Cohen 1992 [Technion 03/97]
t-C_5H_{11}	33	t-Pentyl radical Tsang, J. Am. Chem. Soc. (1985) 2872. (26.6 kJ Cohen 1992) [NASA 5/87]

Species	$\Delta_f H^\circ_{298}$	Comments
C_5H_{11}	34	2,2,m,m-Propyl (neopentyl) radical Cohen 1992 (37.2±4 kJ Cohen and Benson 1992) [Technion 03/97]
C_5H_{12}	−147	n-Pentane TRC/API 10/85 [NASA 1993]
i-C_5H_{12}	−154	Isopentane (2-Methyl butane) $(CH_3)_2CH–CH_2–CH_3$ TRC/API 10/85 [NASA 1993]
C_5H_{12}	−167	Neopentane $CH_3C(CH_3)_2CH_3$ TRC/API 10/85 [NASA 1993]
C_6Cl_6	−34	Hexachlorobenzene Stull et al. 1969 [Technion 1/92]
C_6D_5	307	Phenyl-D5 radical Burcat et al. 1985 [NASA 12/84]
C_6D_6	58	Benzene-D6 Burcat et al. 1985 [NASA 12/84]
C_6F_6	−957	Hexafluorobenzene Stull et al. 1969 (–948.7 kJ NIST 1994) [Technion 03/97]
C_6F_{14}	−2949	Perfluorohexane (FC–5114) NIST 1994 and THERM program [Technion 12/94]
C_6H	1038	Kiefer et al. 1992 [Technion 3/92]
C_6H_2	701	Hexatriyne Kiefer et al. 1992 (653.8 kJ NIST 1994) [Technion 3/92]
C_6H_3	682	Duff and Bauer 1962 [Technion 2/90]
C_6H_4	481	Benzyne Pollack and Hehre, Tetrahedron Lett. 21 (1980) 2483. [Technion 2/92]
C_6H_4	526	Trans-1,5-Hexadiyne-3-ene HCC–HC=CH–CCH Stein et al., NIST 1991 [Technion 2/92]
C_6H_4	515	Cis-1,5-hexadiyne-3-ene NIST 1991 [Technion 2/92]
C_6H_4	129	Hexane pentaene $H_2C=C=C=C=C=CH_2$ THERM program [Technion 2/92]
C_6H_4	328	1,2,3-Hexatrieneyne-5 $H_2C=C=C=CH–CCH$ THERM program [Technion 2/92]
C_6H_5	531	Hexa–1,3–dien–5–yn–yl radical (C_6H_5 chain) Dewar, Gardiner, Frenklach and Oref, J. Am. Chem. Soc. 109 (1987) 4456. [Technion 09/90]
C_6H_5	328	Phenyl radical Burcat et al. 1985 (333.0 kJ NIST 1994; 339±8 kJ Tsang 1996) [NASA 12/84]
C_6H_5	862	Fulvenyl radical THERM program using fulvene (Melius) (5-methylene-1,3 pentadiene) minus vinyl H. Radical is on the methylene group. [Technion 05/97]
C_6H_5	490.3	Fulvenyl Radical (Methylene cyclopenta-2,4-diene-2-yl) $–C(=CH_2)C^*=CHCH=CH–$ 490.3±52 kJ Melius. Radical on the ring on carbon 2. [Technion 05/97]
C_6H_5Br	105	Bromobenzene Stull et al. 1969 [Technion 1/92]
C_6H_5Cl	52	Chlorobenzene Stull et al. 1969 [Technion 1/92]
C_6H_5F	−117	Fluorobenzene Stull et al. 1969 [Technion 1/92]

Species	$\Delta_f H^\circ_{298}$	Comments
C_6H_5NO	201	Nitrosobenzene Choo et al. Int. J. Chem. Kinet 7 (1975) 713. [Technion 7/95]
C_6H_5O	48	Phenoxy radical Burcat et al. 1985 [NASA 12/84]
C_6H_5OO	166	Peroxyphenyl radical Melius [Technion 03/97]
C_6H_6(l)	49	Benzene liquid TRC/API 4/83 [NASA 1993]
C_6H_6	83	Benzene Burcat et al. 1985 [NASA 12/84]
C_6H_6	237	Fulvene (5-methylene-1,3-cyclopentadiene) 236.8±10 kJ Melius [Technion 03/97]
C_6H_5OH	−96	Phenol Burcat et al. 1985 [NASA 4/84]
C_6H_7	201	1,4-cyclo-C_6H_7 diene radical CAS 15819–51–9 200.6±35 kJ Melius (217.1 kJ NIST 1994; 209.2 kJ Benson 1976) [Technion 6/93]
C_6H_7	389	Chain radical THERM using 1,3,5-hexatriene as parent molecule [Technion 12/91]
C_6H_7N	87	Aniline Stull et al. 1969 [Technion 2/92]
C_6H_8	102	3-Methyl cyclopentadiene NIST 1994 using 2× [CD–(C)(H)]; [CD–(C)(CD)]; [CD–(CD)(H)]; [C–$(CD)_2$(H)2]; [C–$(H)_3$] ring correction [Technion 10/94]
C_6H_8	152	1,3,5-hexatriene; equilibrium mixture of three isomers J.D. Vaughan private communication [NASA 8/89]
C_6H_8	106	1,3-Cyclohexadiene Dorofeeva et al. 1986 (106.3±0.5 kJ Pedley and Rylance 1977) [Technion 2/90]
C_6H_8	109	1,4-Cyclohexadiene Dorofeeva et al. 1986 (110 kJ NIST 1994) [Technion 2/90]
C_6H_9	184	1,3-Hexadiene-yl-5 CH2=CHCH=CHCH*CH3 Weissman and Benson 1989 [Technion 2/92]
C_6H_9	251	1,3-Hexadiene-yl-6 CH_2=CHCH=$CHCH_2CH_2$* Weissman and Benson 1989 [Technion 2/92]
C_6H_9	126	Cyclohexenyl-3 Weissman and Benson 1989 [Technion 2/92]
C_6H_{10}	56	1,3-Hexadiene Weissman and Benson 1989 [Technion 2/92]
C_6H_{10}	−5	Cyclohexene Dorofeeva et al. 1986 [Technion 2/90]
C_6H_{11}	142	Trans-Hexene-3-yl-6 CH_3CH_2CH=$CHCH_2CH_2$* Weissman and Benson 1989 [Technion 2/92]
C_6H_{11}	93	2-Methylene-1-pentene radical CH_2=C(CH_2*)-C_3H_7 NIST 1994 [Technion 11/95]
C_6H_{11}	150	2-Methyl-1-pentene-5-yl radical CH_2=C(CH_3)C_3H_6* NIST 1994 [Technion 11/95]
C_6H_{11}	81	2-Methylene-2-pentene radical CH_3C(CH_2*)=CHC_2H_5 NIST 1994 [Technion 11/95]

Species	$\Delta_f H^\circ_{298}$	Comments
C_6H_{11}	142	2-Methyl-2-pentene-5-yl radical $(CH_3)_2C{=}CHC_2H_4$* NIST 1994 [Technion 11/95]
C_6H_{12}	−53	Trans-hexene-3 $C_2H_5CH{=}CHC_2H_5$ Weissman and Benson 1989. [Technion 2/92]
C_6H_{12}	−42	1-Hexene TRS/API [Technion 5/83]
C_6H_{12}	−59	2-Methyl-1-pentene Stull et al. 1969, 1987 correction [Technion 11/95]
C_6H_{12}	−67	2-Methyl-2-pentene Stull et al. 1969, 1987 correction [Technion 11/95]
C_6H_{12}	−57	4-Methyl-2-pentene (cis) Stull et al. 1969, 1987 correction [Technion 5/96]
C_6H_{12}	−61	4-Methyl-2-penten trans Stull et al. 1969, 1987 correction [Technion 11/95]
C_6H_{12}	−123	Cyclohexane Dorofeeva et al. 1986 [Technion 2/90]
n-C_6H_{13}	25	n-Hexyl radical TRC/API 10/83 (36.1 kJ Cohen 1992) [NASA 1993]
2-C_6H_{13}	23	2-Hexyl radical adjusted to n-C_6H_{13} NIST 1994 based on 2[C-(H)(H)(H)], 2[C-(C)(C)(H)(H)], 1[C-(C)(C.)(H)(H)], 1[C.-(C)(C)(H)] (20.4 kJ Cohen 1992) [Technion 12/95]
C_6H_{13}	29	2-Methyl-pentane-1-yl radical $CH_2CH(CH_3)C_3H_7$ NIST 1994 (25.3 kJ Cohen 1992) [Technion 11/95]
C_6H_{13}	33	2-Methyl-pentane-5-yl radical $CH_3CH(CH_3)C_2H_4CH_2$ NIST 1994 (27.7 kJ Cohen 1992) [Technion 11/95]
s-C_6H_{13}	15	2–Methyl–4–pentyl (secondary) radical NIST 1994 [Technion 2/96]
t-C_6H_{13}	9	2–Methyl–2–pentyl (tertiary) radical NIST 1994 (8.24 kJ Cohen 1992) [Technion 1/96]
C_6H_{14}	−167	n-Hexane TRC 1985 [Technion 12/91]
C_6H_{14}	−175	2-Methylpentane TRC 1985 [Technion 12/91]
C_6H_{14}	−172	3-Methylpentane TRC 1985 [Technion 12/91]
C_6H_{14}	−185	2,2-Dimethylbutane TRC 1985 [Technion 12/91]
C_6H_{14}	−177	2,3-Dimethylbutane TRC 1985 [Technion 12/91]
C_7F_{16}	−3384	Perfluoroheptane Domalski and Hearing, J. Phys. Chem. Ref. Data 22 (1993) 1059. [Technion 12/94]
C_7H_5N	219	Phenyl-CN (benzonitrile) Stull et al. 1969 [Technion 3/93]
C_7H_6O	−37	Benzaldehyde Phenyl-CHO Ambrose, Connett, Green, Hales, Head and Martin, J. Chem. Thermo. 7 (1975) 1143. [NASA 3/86]
C_7H_7	211	Benzyl radical Phenyl-CH_2 Hippler and Troe, J. Phys. Chem. 94 (1990) 3803. (205.1 kJ NIST 1994) [Technion 08/90]

Species	$\Delta_f H^\circ_{298}$	Comments
C_7H_8	12	Liquid toluene TRC/API 10/86 [NASA 1993]
C_7H_8	50	Toluene Stull et al. 1969 [NASA 6/87]
C_7H_8	248	Norbornadiene 2,5-bicycloheptadiene Steele, J. Chem. Thermo. 10 (1978) 919. [Technion 2/95]
C_7H_8	183	1,3,5-Cycloheptatriene Dorofeeva et al. 198 (187.1 kJ NIST 1994) [Technion 2/95]
C_7H_8	396	1,6-Heptadiyne HCC–$(CH_2)_3$–CCH NIST 1994 [Technion 2/95]
C_7H_8O	−132	Cresol (Methyl phenol) Kudchadker et al. 1978 Because the differences between cis and trans m-cresol are slight, they were combined. Other isomers are o-cresol (cis and trans) and p-cresol. [NASA 6/87]
C_7H_8O	−100	Benzyl alcohol (Phenyl-OH) NIST 1991 [NASA 7/87]
C_7H_{10}	67	3,5-Dimethyl-cyclo-pentadiene NIST 1994 from 2×[C–(H)3]; [CD–(C)2]; [CD–(C)(H)]; [CD–(C)(CD)]; [CD–(CD)(H)]; [C–(CD)2(H)2], cyclo correction in enthalpy of formation 25.1 kJ. [Technion 10/94]
C_7H_{12}	−54	Norbornane (1,4-bicycloheptane) –53.7±5 kJ Rogers et al. J. Phys. Chem. 84 (1980) 1810. [Technion 2/95]
C_7H_{12}	−9	Cycloheptene Dorofeeva et al. 1986 [Technion 2/95]
C_7H_{14}	−62	1-Heptene TRC/API 12/52 [NASA 1993]
C_7H_{14}	−118	Cycloheptane Dorofeeva et al. 1986. Polynomial results do not agree well with the original data. [Technion 2/95]
C_7H_{15}	4	n-Heptyl radical TRC/API 10/83 [NASA 1993]
C_7H_{16}	−188	n-Heptane TRC/API 4/81 [NASA 1993]
C_8H	1162	THERM program from C_8H_2 [Technion 2/92]
C_8H_2	934	Octatetrayne Kiefer et al. 1992. [Technion 2/92]
C_8H_6	328	Phenylacetylene C_6H_5CCH Melius [Technion 9/96]
C_8H_6O	27	Benzofurane Stein and Barton Thermochim. Acta 44 (1981) 265. [Technion 1/94]
C_8H_6S	166	Benzothiophene Pedley et al. 1986 [Technion 12/93]
C_8H_7N	157	Indole (1-benzazole, 2,3-benzapyrrole) 156.5±1 kJ Das et al., J. Phys. Chem. Ref. Data 22 (1993) 658. [Technion 03/95]
C_8H_8	652	Cubane (pentacyclo[4.2.0.0.0.0]octane) 651.7±30 kJ Melius [Technion 12/94]
C_8H_8	147	Styrene $C_6H_5CH{=}CH_2$ NIST 1994 and Pedley et al. 1986 obtain identical the same value by different methods. [Technion 12/94]
C_8H_9	252	$C_6H_5CH_2CH_2$* radical 252.4±15 kJ Melius [Technion 9/96]

Species	$\Delta_f H^\circ_{298}$	Comments
C_8H_{10}	29	Ethyl benzene NIST 1994 [Technion 9/96]
C_8H_{10}	18	1,4-Dimethylbenzene Draeger and Scott 1981 [NASA 2/84]
C_8H_{16}	−83	1-Octene TRC/API 12/52 [NASA 1993]
n-C_8H_{17}	−16	n-Octyl radical TRC/API 10/83 [NASA 1993]
C_8H_{18}(l)	−250	n-Octane liquid TRC/API 10/80 [NASA 1993]
n-C_8H_{18}	−209	n-Octane Stull et al. 1969 TRC/API 4/85 [NASA 1993]
i-C_8H_{18}	−224	2,2,4-Trimethylpentane (iso-octane) TRC/API 4/85 [NASA 1993]
C_9H_7	286	Indenyl radical 285.6±20 kJ Melius [Technion 9/96]
C_9H_8	164	Indene Melius (163.2±1 kJ Kudchadker and Kudchadker API monograph 714–80 1980) [Technion 9/96]
C_9H_{10}	113	α-Methylstyrene (1-methyl-1-phenyl ethylene) Stull et al. 1969 and NIST 1994 [Technion 1/96]
n-C_9H_{19}	−37	n-Nonyl radical TRC/API 10/83 [NASA 1993]
$C_{10}D_8$	118	Naphthalene-D8 Chen et al. 1979 [Technion 9/82]
$C_{10}H_8$	280	Azulene Stull et al. 1969 [Technion 9/82]
$C_{10}H_8$	151	Naphthalene Chen et al. 1979 (150.6±0.6 kJ NIST 1997) [Technion 9/82]

Species	$\Delta_f H^\circ_{298}$	Comments
$C_{10}H_{16}$	−60	JP-10 (tetrahydro-di-cyclopentadiene) –60.3±5 kJ Boyd, Sanwal, Shary-Tehrany and McNally, J. Phys. Chem. 75 (1971) 1264. [Technion 12/91]
n-$C_{10}H_{21}$	−58	n-Decyl radical TRC/API 10/83 [NASA 1993]
o-$C_{12}D_9$	387	o-Biphenyl radical-D9 Burcat et al. 1985 [NASA 12/84]
$C_{12}D_{10}$	138	Biphenyl-D10 Burcat et al. 1985 [NASA 12/84]
$C_{12}H_9$	428	Ortho-biphenyl radical Burcat et al. 1985 427.73 kJ [NASA 12/84]
$C_{12}H_9Cl$	149	Chlorobiphenyl (position of Cl undefined) THERM program [Technion 2/92]
$C_{12}H_{10}$	182	Biphenyl Burcat et al. 1985 [NASA 12/84]
$C_{14}H_{10}$	230	Anthracene Kudchadker et al. 1979 (218.4 kJ NIST 1994) [Technion 1/94]
$C_{14}H_{10}$	207	Phenanthrene Kudchadker et al. 1979 (209.2 kJ NIST 1994) [Technion 1/94]
$C_{14}H_{14}$	143	Bibenzyl $C_6H_5C_2H_4C_6H_5$ Colket (unpublished) [Colket 6/83]
$C_{16}H_{10}$	226	Pyrene Smith et al., J. Chem. Thermo. 12 (1980) 919. (225.7±1.5 kJ NIST 1997; 217.6±1 kJ Kudchadker and Kudchadker, API monograph 709-79 1979) [Technion 10/96]

Species	$\Delta_f H^\circ_{298}$	Comments
$C_{24}H_{17}$	623	Triphenylbenzene radical (outer phenyl radical in para position) Parent molecule $C_{24}H_{18}$ approximated by NIST 1994 was used with THERM program [Technion 5/94]
$C_{24}H_{18}$	373	Triphenylbenzene NIST 1994 [Technion 2/92]
C_{60}	2586	Buckminsterfullerene 2585.7±105 kJ Beckhaus et al., Angew. Chem. Int. Ed. 31 (1992) 63. [Technion 6/93]
C_{70}	2895	(elipsoid) Burcat TAE No. 680 (1992) [Technion 6/93]
Jet-A(l)	−303	Liquid Jet-A fuel Faith, Ackerman and Henderson, NASA CR-72951 July 1971. [NASA 6/88]
Jet-A(g)	−250	Garcia-Salcedo et al. NASA TM 101475 1988. [NASA 6/88]
Cl	121	121.3±0.008 kJ JANAF 6/82 [NASA 1993]
DCl	−93	Deuterochloric acid –93.345±0.21 kJ JANAF 6/77 [NASA 1993]
DOCl	−78	–78.2±2 kJ JANAF 3/79 [NASA 1993]
ClF	−50	–50.3±0.4 kJ JANAF 6/77 [NASA 1993]
ClF_3	−159	–158.67±3 kJ JANAF 9/65 [NASA 1993]
ClO	101	101.2±2 kJ JANAF 6/61 [NASA 1993]
ClO_2	105	O–Cl–O 104.6±6 kJ JANAF 3/61 [NASA 1993]
ClOO	96	ClOO radical Burcat estimate [NASA 4/84]
Cl_2	0	Reference state [TRC/API 89]

Species	$\Delta_f H^\circ_{298}$	Comments
Cl_2O	88	Chlorine oxide 87.9±7 kJ JANAF 12/65 [NASA 1993]
Cl_2O_2	139	Cl–O–O–Cl $\Delta_f H^\circ_0$ =143.1 kJ Lee et al. J. Chem. Phys. 97 (1992) 6593. (117.2 ±25.1 kJ Benson 1976) [Technion 2/94]
D	222	Deuterium atom 221.720±0.004 kJ JANAF 3/82 [NASA 1993]
DF	−276	Deuterofluoric acid –275.5±1 kJ JANAF 6/77 [NASA 1993]
HD	0	Protodeuterium Gurvich 1989 TSIV 1989 [NASA 1993]
HDO	−245	Gurvich 1989 [NASA 5/95]
HDO_2	−140	Gurvich 1989 $\Delta_f H^\circ_0$ = –134.4 kJ [NASA 5/95]
OD	37	Deuteryl radical $\Delta_f H^\circ_0$ = 36.9 kJ Gurvich 1989 [NASA 1993]
OD^-	−145	Deuteryl ion $\Delta_f H^\circ_0$ = –139.2 kJ Gurvich 1989 [NASA 1993]
DO_2	13	$\Delta_f H^\circ_{298}$ modified to 12.6 kJ to agree with 15.9 kJ for HO_2 from Fisher and Armentrout, J. Chem. Phys. 94 (1990) 4396. [Technion 8/90]
D_2	0	Reference state TRC/API 1989 [NASA 1993]
D_2O	−249	Deuterated water –249.2±0.07 kJ JANAF 6/77 [NASA 1993]
D_2O_2	−144	Deuterium peroxide TSIV 1978 [NASA 1993]

Species	$\Delta_f H^\circ_{298}$	Comments	Species	$\Delta_f H^\circ_{298}$	Comments
D_2S	−24	Deuterium sulfide −23.9±1 kJ JANAF 6/77 [NASA 1993]	HOF	−97	$\Delta_f H^\circ_0$ = −94 kJ Gurvich 1989 TSIV 1989 [NASA 1993]
e	0	Electron Reference state [NASA 6/88]	HI	26	Hydroiodic acid 26.4±0.2 kJ JANAF 9/61 [NASA 1993]
F	79	79.4±0.3 kJ JANAF 6/82 [NASA 1993]			
FO	109	109±10 kJ Chase 1996a [Technion 02/97]	HNO	102	$\Delta_f H^\circ_0$ = 105 kJ Gurvich 1989 [NASA 12/89]
FO_2	379	O–F–O 378.6±20 kJ Chase 1996a [Technion 02/97]	HNO_2	−78	Nitrous acid $\Delta_f H^\circ_0$ = −72.8 kJ Gurvich 1989 [NASA 1993]
FO_2	25	F–O–O 25.4±2 kJ Chase 1996a [Technion 02/97]	HNO_3	−134	Nitric acid Gurvich 1989 $\Delta_f H^\circ_0$ = −124.2 kJ [NASA 4/90]
F_2	0	Reference state TRC/API 89 [NASA 1993]			
F_2O	25	F–O–F 24.5±2 kJ Chase 1996a [Technion 02/97]	OH	39	Hydroxyl radical 38.4±1.2 kJ TSIV 1978 [NASA 1993]
F_2O_2	19	F–O–O–F 19.2±2 kJ Chase 1996a [Technion 02/97]	OH^+	1290	Hydroxyl ion TSIV 1978 [NASA 1993]
H	218	$\Delta_f H^\circ_0$ = 211.801±0.006 kJ Moore 1972 [NASA 6/94]	OH^-	−143	Hydroxyl ion $\Delta_f H^\circ_0$ = −150.805 kJ [NASA 3/93]
H^+	1536	$\Delta_f H^\circ_0$ = 1530.048 kJ Moore 1972 [NASA 7/88]	HPO	−57	Phosphorus oxyhydride $\Delta_f H^\circ_0$ = −53 kJ Gurvich 1989 [NASA 1993]
H^-	139	$\Delta_f H^\circ_0$ = 132.834 kJ JANAF 1982 [NASA /7/88]	SH	139	139.3±5 kJ JANAF 6/77 [NASA 1993]
HBr	−36	Hydrobromic acid Gurvich 1989 [NASA 1993]	SOH	−21	S–OH radical −20.9±42 kJ Melius [Technion 4/93]
HCl	−92	Hydrochloric acid Gurvich 1989 [NASA 1993]			
HOCl	−76	$\Delta_f H^\circ_0$ = −72.8 kJ Gurvich 1989 TSIV 1989 [NASA 1993]	HSO	−5	HS=O radical 4.8±7.3 kJ Melius [Technion 4/93]
HF	−273	Hydrofluoric acid Gurvich 1989 TSIV 1989 [NASA 1993]	HO_2	13	Hills and Howard, J. Chem. Phys. 81 (1984) 4458. (2.1±8.4 kJ Tsang 1996) [NASA 5/89]

Species	$\Delta_f H^\circ_{298}$	Comments
HO_2S	−255.9	HOSO radical −255.9±6 kJ Melius [Technion 11/96]
HSO_3	−385	$HOSO_2$ radical Margitan, J. Phys. Chem. 88 (1984) 3314. (Melius −282.9 kJ) [Technion 3/93]
HS_2	27	Luke and McLean, J. Phys. Chem. 89 (1985) 4592. [Technion 3/94]
H_2	0	Reference state TSIV 1978 [NASA 1993]
H_2F_2	−570	$\Delta_f H^\circ_0$ = −566.5 kJ Gurvich 1989 [NASA 1993]
$H_2O(l)$	−286	Cox et al. 1989 [NASA 8/89]
H_2O	−242	Wooley, J. Res. NBS 92 (1987) 35. −241.826 ±0.040 kJ based on $\Delta_f H^\circ_{298}$(l) from Cox et al. 1989, p. 21, and heat of vaporization from Haar, Gallagher and Kell, NBS/NRC Tables, Hemisphere 1984. [NASA 5/89]
H_2O_2	−136	Hydrogen peroxide TSIV 1978 [NASA 2/93]
H_2S	−21	Hydrogen sulfide Gurvich 1989 [NASA 1993]
$H_2SO_4(l)$	−814	Sulfuric acid liquid JANAF 9/77 [NASA 1993]
H_2SO_4	−733	Sulfuric acid $\Delta_f H^\circ_0$ = −719. kJ Gurvich 1989 [NASA 1993]
H_2S_2	17	HS–SH Burcat rough estimate. 15.9 kJ Benson Chem. Rev. 78 (1978) 23 [Technion 3/94]
H_3F_3	−884	$\Delta_f H^\circ_0$ = −873 kJ Gurvich 1989 [NASA 1993]
H_3O^+	598	Hydronium ion Gurvich 1989 [NASA 1993]
H_4F_4	−1187	$\Delta_f H^\circ_0$ = −1174 kJ Gurvich 1989 [NASA 1993]
H_5F_5	−1490	$\Delta_f H^\circ_0$ = −1475 kJ Gurvich 1989 [NASA 1993]
H_6F_6	−1806	$\Delta_f H^\circ_0$ = −1788 kJ Gurvich 1989 [NASA 1993]
H_7F_7	−2100	Gurvich 1989 $\Delta_f H^\circ_0$ = −2080 kJ [NASA 1993]
He	0	Reference state [NASA 10/90]
He^+	2378	Moore 1970 [NASA 10/92]
I	107	106.76±0.04 kJ JANAF 6/82 [NASA 1993]
IO	126	126±20 kJ Chase 1996c [Technion 02/97]
IO_2	117	I–O–O 116.5±40 kJ Chase 1996c [Technion 02/97]
IO_2	159	O–I–O 159.3±25 kJ Chase 1996c [Technion 02/97]
IO_3	242	241.9±50 kJ Chase 1996c [Technion 02/97]
I_2	62	Not reference state below 457.7 K. JANAF 9/61 [NASA 1993]
I_2O	107	I–I–O 106.7±40 kJ Chase 1996c [Technion 02/97]
I_2O	120	I–O–I 119.5±25 kJ Chase 1996c [Technion 02/97]
Kr	0	Reference state [NASA 10/90]
Kr^+	1357	[NASA 10/92]

Species	$\Delta_f H^\circ_{298}$	Comments
N	473	472.68±0.40 kJ Cox et al. 1989, p. 22. [NASA 6/88]
ND	375	375.3±20 kJ JANAF 6/77 [NASA 1993]
NHD	188	187.9±2 kJ Burcat from $(NH_2 + ND_2)/2$ [Technion 04/97]
ND_2	185	185.4±8.4 kJ JANAF 6/77 [NASA 1993]
ND_2H	−51	Burcat $ND_2 + NH_3 - NH_2$ [Technion 04/97]
ND_3	−59	–58.6±0.4 kJ JANAF 6/77 [NASA 1993]
NF	233	Gurvich 1989 $\Delta_f H^\circ_0$ = 233 kJ [NASA 1993]
NF_2	34	$\Delta_f H^\circ_0$ = 37 kJ Gurvich 1989 [NASA 5/95]
NF_3	−132	Gurvich 1989 [NASA 5/95]
NH	357	Anderson, J. Phys. Chem. 93 (1989) 530. (376.56±16.7 kJ JANAF 1985) [NASA 11/89]
NH^+	1666	$\Delta_f H^\circ_0$ = 1656.3 kJ Gurvich 1989 [NASA 2/89]
NHF	112	Gurvich 1989 [NASA 1993]
NHF_2	−103	Gurvich 1989 [NASA 1993]
NH_2	189	Amidogen radical $\Delta_f H^\circ_0$ =192.0 kJ Anderson, J. Phys. Chem. 93 (1989) 530. [NASA 12/89]
NH_2D	−44	Burcat $NH_2 + ND_3 - ND_2$ [Technion 04/97]
NH_2F	−75	Gurvich 1989 [NASA 1993]
NH_3	−46	Ammonia Gurvich 1989 [NASA 1993]
NH_2OH	−50	Hydroxylamine TSIV 1978 and Gurvich 1989 quote the same value. [NASA 1993]
NH_4^+	645	Ammonium ion TSIV 1978 [NASA 1993]
NO	91	$\Delta_f H^\circ_0$ = 82.1±0.17 kJ Gurvich 1989 [NASA 1993]
NO^+	991	$\Delta_f H^\circ_0$ = 982.1 kJ Gurvich 1989 [NASA 1993]
NOCl	64	Gurvich 1989 $\Delta_f H^\circ_0$ = 54.6 kJ [NASA 5/95]
NOF	−65	Gurvich 1989 [NASA 5/95]
NOF_3	−187	Gurvich 1989 [NASA 1993]
NO_2	34	$\Delta_f H^\circ_0$ = 37±0.8 kJ Gurvich 1989 [NASA 7/88]
NO_2^-	−200	Gurvich 1989 [NASA 1993]
NO_2Cl	13	Gurvich 1989 [NASA 5/95]
NO_2F	−109	Gurvich 1989 [NASA 5/95]
NO_3	71	JANAF 12/64 [NASA 1993]
NO_3^-	−311	$\Delta_f H^\circ_0$ = –298 kJ Gurvich 1989 [NASA 1993]
NO_3F	15	Gurvich 1989 [NASA 1993]
N_2	0	Reference state TSIV 1978 [NASA 1993]
N_2D_2	207	Cis–N_2D_2 JANAF 6/77 [NASA 1993]
N_2F_2	62	Cis and trans isomers mixed; included as excited states. $\Delta_f H^\circ_0$ = 67 kJ Gurvich 1989 [NASA 1993]

Species	$\Delta_f H^\circ_{298}$	Comments
N_2F_4	−22	Equilibrium mixture of trans and gauche isomers. Gurvich 1989 [NASA 1993]
N_2H	250	NNH radical 249.5±15 kJ Melius [Technion 07/93]
N_2H_2	212	Gurvich 1989 [NASA 5/90]
NH_2NO_2	−26	Gurvich 1989 [NASA 1993]
N_2H_3	225	Hydrazine radical 225.1±70 kJ Melius [Technion 7/93]
N_2H_4	95	Hydrazine 95.2±4 kJ Gurvich 1989 [NASA 5/90]
N_2O	82	Nitrous oxide Gurvich 1989 [NASA 7/88]
N_2O^+	1333	JANAF 12/70 [NASA 1993]
N_2O_3	87	$\Delta_f H^\circ_0$ = 91.2 kJ Gurvich 1989 [NASA 4/90]
N_2O_4	11	$\Delta_f H^\circ_0$ = 20.4 kJ Gurvich 1989 [NASA 4/90]
N_2O_5	13	Gurvich 1989 [NASA 4/90]
N_3	436	Azide Gurvich 1989 [NASA 7/88]
N_3H	294	Gurvich 1989 [NASA 7/88]
Ne	0	Reference state [NASA 10/90]
Ne^+	2087	Moore 1970 [NASA 10/92]
O	249	249.18±0.10 kJ Moore 1976 [NASA 1/90]
O^-	102	$\Delta_f H^\circ_0$ = 95.09 kJ Gurvich 1989 [NASA 1/90]
O_2	0	Reference state Gurvich 1989 [NASA 4/89]
O_2^+	1172	$\Delta_f H^\circ_0$ = 1165 kJ Gurvich 1989 [NASA 1993]

Species	$\Delta_f H^\circ_{298}$	Comments
O_2-	−48	$\Delta_f H^\circ_0$ = −42.5 kJ Gurvich 1989 [NASA 4/89]
O_3	142	Ozone Gurvich 1989 [NASA 5/90]
P	316	JANAF 12/82 [NASA 1993]
PCl_3	−289	−288.7±5 kJ JANAF 6/70 [NASA 1993]
PF	−52	−52.3±20 kJ JANAF 6/77 [NASA 1993]
PF_2	−488	−488.3±20 kJ JANAF 6/77 [NASA 1993]
PF_3	−958	−958.4±4 kJ JANAF 12/69 [NASA 1993]
PF_5	−1594	−1594.4±3 kJ JANAF 12/69 [NASA 1993]
PH	236	McBride (The 253.6±35 kJ value from JANAF 6/62 is in error.) [NASA 1993]
PH_3	5	Phosphine McBride (The 22.9±2 kJ value from JANAF 6/62 is in error.) [NASA 1993]
PN	105	104.8±5 kJ JANAF 9/62 [NASA 1993]
PO	−30	McBride (The −23.6±4 kJ value from JANAF 6/71 is in error.) [NASA 1993]
PO_2	−315	JANAF 9/62 [NASA 1993]
P_2	144	143.7±2 kJ JANAF 6/61 [NASA 1993]
P_4	59	58.9±2 kJ JANAF 6/61 [NASA 1993]
P_4O_6	−2145	$(P_2O_3)_2$ McBride (The −2214.3±34 kJ value from JANAF 12/62 is in error.) [NASA 1993]
P_4O_{10}	−2904	$(P_2O_5)_2$ −2904.1+/9 kJ JANAF 12/65 [NASA 1993]
$P_4O_{10}(s)$	−3010	JANAF 12/65 [NASA 1993]

Species	$\Delta_f H^\circ_{298}$	Comments	Species	$\Delta_f H^\circ_{298}$	Comments
S(s)	0	Reference state below 388.36 K. JANAF 9/77 [NASA 1993]	SO_2ClF	−556	Sulfuryl chloride fluoride –556.5±20 kJ JANAF 6/71 [NASA 1993]
S	277	277.0±0.3 kJ JANAF 9/82 [NASA 1993]	SO_2Cl_2	−355	Sulfuryl chloride –354.8±2 kJ JANAF 6/71 [NASA 1993]
SCl	156	156.5±20 kJ JANAF 6/78 [NASA 1993]	SO_2F_2	−759	Sulfuryl fluoride –758.6±8 kJ JANAF 6/71 [NASA 1993]
SCl_2	−18	–17.6±3 kJ JANAF 6/78 [NASA 1993]	SO_3	−396	–395.8±1 kJ JANAF 9/65 [NASA 1993]
SD	138	138.5± 5 kJ JANAF 6/77 [NASA 1993]	S_2	128	128.6±1 kJ JANAF 9/77 [NASA 1993]
SF	13	13.0±6 kJ JANAF 6/76 [NASA 1993]	S_2Cl	79	S–S–Cl 78.6±8 kJ JANAF 6/78 [NASA 1993]
SF_2	−297	–296.6±20 kJ JANAF 6/76 [NASA 1993]	S_2Cl_2	−17	Cl–S–S–Cl [NASA 4/93]
SF_3	−503	–503.0±35 kJ JANAF 6/77 [NASA 1993]	S_2F_2	−401	Thiothionyl fluoride S–S–F_2 –401.4±40 kJ JANAF 6/76 [NASA 1993]
SF_4	−763	—-763.2±20 kJ JANAF 6/76 [NASA 1993]	FS_2F	−336	Fluorodisulfane FS–SF 336.4±40 kJ JANAF 6/76 [NASA 1993]
SF_5	−908	Pentafluorosulfur –908.5±15 kJ JANAF 12/77 [NASA 1993]	S_2F_{10}	−2064	–2064.4±30 kJ JANAF 12/77 [NASA 1993]
SF_5Br	−973	–972.8±60 kJ JANAF 12/77 [NASA 1993]	S_2O	−56	S–S–O –56.5±35 kJ JANAF 9/65 [NASA 1993]
SF_5Cl	−1039	–1038.9±10 kJ JANAF 12/77 [NASA 1993]	S_8	100	100.4±0.6 kJ JANAF 9/77 [NASA 1993]
SF_6	−1221	Sulfur hexafluoride –1220.5±1 kJ JANAF 6/76 [NASA 1993]	Xe	0	Reference state Moore 1970 [NASA 12/91]
SN	264	263.6±100 kJ JANAF 6/61 [NASA 1993]	Xe^+	1177	Xenon ion [NASA 10/92]
SO	5	5.0±1 kJ JANAF 6/77 [NASA 1993]			
SOF_2	−544	Thyonyl fluoride –543.9±100 kJ JANAF 6/72 [NASA 1993]			
SO_2	−297	O–S–O –296.8±0.2 kJ JANAF 6/61 [NASA 1993]			

Index